World Soils Book Series

Series Editor

Alfred E. Hartemink
Department of Soil Science, FD Hole Soils Laboratory
University of Wisconsin–Madison
Madison, WI
USA

The World Soils Book Series publishes peer-reviewed books on the soils of a particular country. They include sections on soil research history, climate, geology, geomorphology, major soil types, soil maps, soil properties, soil classification, soil fertility, land use and vegetation, soil management, soils and humans, soils and industry, future soil issues. The books summarize what is known about the soils in a particular country in a concise and highly reader-friendly way. The series contains both single and multi-authored books as well as edited volumes. There is additional scope for regional studies within the series, particularly when covering large land masses (for example, The Soils of Texas, The Soils of California), however, these will be assessed on an individual basis.

More information about this series at http://www.springer.com/series/8915

Ryusuke Hatano • Hitoshi Shinjo •
Yusuke Takata
Editors

The Soils of Japan

Springer

Editors
Ryusuke Hatano
Research Faculty of Agriculture
Hokkaido University
Sapporo, Hokkaido, Japan

Hitoshi Shinjo
Graduate School of Global Environmental Studies
Kyoto University
Kyoto, Japan

Yusuke Takata
National Agriculture and Food Research Organization
Tsukuba, Ibaraki, Japan

ISSN 2211-1255 ISSN 2211-1263 (electronic)
World Soils Book Series
ISBN 978-981-15-8228-8 ISBN 978-981-15-8229-5 (eBook)
https://doi.org/10.1007/978-981-15-8229-5

This Springer imprint is published by the registered company Springer Nature Singapore Pte Ltd.
The registered company address is: 152 Beach Road, #21-01/04 Gateway East, Singapore 189721, Singapore

Foreword by Takashi Kosaki

Who knows soils were studied and classified into nine grades based on productivity or fertility and those data were compiled in a series of books named "Fudoki" for all territories controlled by the Emperor of Japan in the eighth century? The description of such soil evaluation in the Far East dates back to the sixth century BC when "Yu Gong (Tribute of Yu the Great in Xia Dynasty in ca. 2000 BC)" was written as a chapter of the Book of Xia section of the Book of Documents, whose compilation was attributed to Confucius in China. The "Fudokis" were, of course, complied with the ideas originally developed in such textbooks in ancient China and used as a base for governing the ancient nation of Japan.

Who knows the first modern soil map, Agronomic Map of the Kai Province, was published in 1885, only two years after V. V. Dokuchaev's classic "Russian Chernozem" was published? Who knows one Japanese soil scientist started serving for the International Society of Soil Science (ISSS) as an executive officer when it took off in 1924 at Rome, Italy? Who knows the Japanese Society of Soil Science and Plant Nutrition was born in 1927, only three years younger than ISSS? In the modern soil science history as well, Japan has been running together with and leading other members in the world.

It is, however, a thousand pities that not many people knows about natural as well as human history of Japan for about twenty centuries. It is particularly true about the soils thereof. One of the reasons must be that Japan has been located very remotely from the major political, economic or cultural centers of the world. The other should be the uniqueness in soil forming processes and agricultural production systems attributed to its humid climate and young and changing topography with steep mountains and valleys, i.e., strong acidification, preference for practicing rice-based production system under submerged condition and wide coverage of volcanic materials. The combination of the above is not common in the areas where the doctrine of modern soil science was originally developed and widely tested in the world.

I do not blame any soil scientist for being unfamiliar with soils of Japan. Because Japan is not only a mysterious country with unreadable smile but also there have been not many good books telling about the soils of Japan published in English before. They, however, made it today. This book "Soils of Japan" is now available to you all. It is the very products of research, experience, technology and even philosophy about soil and soil sciences that the Japanese has fostered for the last twenty centuries in their history. It is the Bible which provides you with all necessary and reliable information on soils and their management which has been and will be a clue for the Japanese to be the steward of soil and natural resources and may also be the one for the readers to be so. From this day forth, you never excuse for being unfamiliar with soils of Japan due to the lack of information.

Nagoya, Japan
January 2019

Takashi Kosaki
President, International Union of Soil Sciences

Foreword by Kazuyuki Inubushi

Future Soil Research: A Perspective for Soil Science and Plant Nutrition

As described in this book, the soils of Japan are quite diverse due to the islands' active geological activity, unique island climate and predominantly mountainous landscape with limited but cascade- or integratory-used flat plains. However, intensive agriculture and land use change have transformed many natural soils to anthropogenic soils over both the short and long term. Fertilization and irrigation also change the soil properties and influence not only the agroecosystem but also the global environment, including aspects such as climate change and ozone layer destruction. Modern industrialization has caused serious soil contamination that is not easily rehabilitated.

To solve such problems, the Japanese Society of Soil Science and Plant Nutrition was established in 1927, under the original name of the Society of the Science of Soil and Manure, after preliminary activities as a part of the Japanese Society of Agronomy. At that time, there were 1441 members, and there are now more than 2000 including student members. The first volume of the *Journal of the Science of Soil and Manure* was published in October 1927 and contained 10 papers, with titles such as "Distribution and formation of Andosols," "Composition and function of humic substances," "Soya bean residue as alternative fertilizer," "Fertilizer policy in Japan," "Effect of phosphorous fertilizer" and "Effect of silicate on plant disease." Such topics are still important today.

The English-language *Journal of Soil Science and Plant Nutrition* started in 1955 as *Soil and Plant Food*, which was also published by the Society of the Science of Soil and Manure, Japan. In its very first issue, a total of 29 papers were published covering soil chemistry and biology, as well as the application of fertilizers, lime, slag and manure, with titles such as "Glassy volcanic-ash soils in Japan," "Some Characteristics of the humus in soil types," "A proposed method to determine adequate amount of fertilizers to be applied to crops" and "Radioactive contamination of plants in Japan." It may be surprising that there were so many active studies presented in English, even from the beginning of this journal.

Over nearly 100 years, the circumstances of agriculture and the natural environment have changed drastically. Additionally, technologies in soil science and related areas have developed tremendously, in association with interdisciplinary fields of science and technology. However, in terms of human nutrition and welfare, our direction in soil science has not changed substantially, even with some shift from food security to environmental friendly agriculture and sustainable development.

The future of soil science and plant nutrition is not easy to predict, but I hope it will move forward in directions that will use cutting-edge modern technology and also in an integrated and more systematic direction with a philosophy harmonized with social awareness regarding soil science and plant nutrition. In these perspectives, this book is really the useful chart for progress and development of soil science and plant nutrition in the new era.

Learn from yesterday, live for today, hope for tomorrow. (*Albert Einstein*)

Chiba, Japan
May 2019

Kazuyuki Inubushi
President of the Japanese Society of Soil Science and Plant Nutrition

Foreword by Katsutoshi Sakurai

Future Soil Research: A Pedologist's Perspective

Even today in the twenty-first century, we still live on a planet made up of sky, ocean and land. Among these three, most human beings live the surface of consolidated land—soils—for our entire lives. Soils have been the indispensable basis of the prosperity of humans since the times of our ancestors, and this will not change greatly for our offspring in the future.

Most of our food has been produced on Earth through the blessing of sunshine. Energy supplied by the sun is the strongest and long-lasting resources on Earth, now and in the future. Energy for sustaining our lives on Earth is supplied from outside our planet. We can utilize this energy without any modifications. On the other hand, the soil resources for producing our food are not perpetual, since the soil properties are changed, modified and often degraded by various activities of humans. For most people around the world, this is common sense. We must, at the very least, maintain the basic quality of soils.

Soils are one of the most important features of Earth: They sustain life, including urban life, through the support of agriculture; maintain a healthy natural environment, physically support buildings and infrastructure; and provide us with places for recreation. Furthermore, soils are closely related to global climate and culture. In sum, soils will always be with us as long as we survive on Earth.

Environmental pollution is easy to recognize because its dangerous toxicity harms human health. However, soil degradation is difficult to recognize without any outside symptoms, such as an eroded soil surface (soil erosion) or salt crystallization on the soil surface (salinization). In other words, we have to utilize soil resources through careful management practices. For this reason, soil research should be continued not only for appropriate utilization but also for conservation through proper management. Pedology is the most important cutting-edge field for this purpose. Without knowledge of pedology, any practical management of soil-related activities will fail.

Considering the future life of human beings on this planet, pedologists should make clear the importance of basic soil research. To secure the world's continued and sustainable prosperity, pedologists must draw on all aspects of the natural sciences. Every dataset or procedure, such as soil databases, soil maps, Landsat data and GIS systems, should be utilized properly in order to (1) establish an integrated soil classification system; (2) conduct appropriate land use planning based on the evaluation of the pedological properties of soil resources; (3) evaluate ecosystems in terms of soil–vegetation–water systems; and (4) establish an overview of landscape ecology based on the relationship between soils, the natural environment and local communities.

Soil research may appear to be a basic and slowly progressing area of science. However, it will play a critical role in allowing human beings to survive and prosper. Future soil research should follow each path as far as possible for the good of humankind and the planet.

This book clearly reveals how Japanese pedologists have understood the soils of Japan and, I strongly believe, will provide readers with directions of future soil research.

Kochi, Japan
May 2019

Katsutoshi Sakurai
President of the Japanese Society of Pedology

Preface

Japan has numerous islands, including the four major islands of Hokkaido, Honshu, Shikoku and Kyushu. The major part of Japanese agriculture has been rice farming due to the presence of a monsoon climate in all regions except part of Hokkaido. Although most Japanese islands are affected by a monsoon climate, they still have various climate zones, which range from subarctic to subtropical climates. The islands also have various landscapes, from mountain terrain to alluvial plains within a single region. The soils of Japan are affected not only by these various climates and landscapes, but also by fresh volcanic tephra, which has a significant impact. According to these variations in climate, landscapes and soils, Japan's cultural regions are divided into six regions: (1) Hokkaido; (2) Tohoku; (3) Kanto and Koushinetsu; (4) Tokai and Hokuriku; (5) Kinki, Chugoku, and Shikoku; and (6) Kyushu and Okinawa. Each of these regions has developed its own agricultural management practices.

This book will explain the pedogenesis and properties of land use in Japan following these variations in climate, landscapes and geology, as well as human activity and other factors. This book was written by researchers from the Japanese Society of Soil Science and Plant Nutrition (JSSSPN) and the Japanese Society of Pedology (JSP) with the aim of presenting the results of their work as simply and accessibly as possible. Chapter 1 overviews natural environments including soils and land use of Japan. Chapter 2 explains factors in soil formation in Japan, and Chap. 3 describes soil classification system of Japan and distribution of soils in Japan. In Chap. 4, the general physical and chemical properties of the major soil types in Japan are explained. From Chaps. 5–10, each chapter describes the characteristics of soil management in each of the aforementioned six cultural regions of Japan.

We wish the readers could enjoy full of knowledge and achievements accomplished on soils of Japan by JSSPN and JSP. On behalf of all the authors contributing to this book, we sincerely acknowledge the support by JSSPN and JSP.

Sapporo, Japan — Ryusuke Hatano
Kyoto, Japan — Hitoshi Shinjo
Tsukuba, Japan — Yusuke Takata

Contents

Contributors

Tomonori Abe Furukawa Agricultural Experiment Station, Miyagi Prefectural Government, Furukawa City, Miyagi Prefecture, Japan (Subsection 6.6.2)

Isao Akagi Faculty of Agriculture, Kagoshima University, Kagoshima City, Kagoshima Prefecture, Japan (Subsection 10.4.4)

Kazuto Ando Yanmar HeliAgri CO., LTD, Osaka City, Osaka Prefecture, Japan (Subsection 6.7.1)

Tadashi Ando Yamagata Integrated Agricultural Research Center, Yamagata Prefectural Government, Yamagata City, Yamagata Prefecture, Japan (Subsection 6.5.3)

Takayuki Ando Horticultural Experiment Station, Yamagata Integrated Reserch Center, Yamagata Prefectural Government, Yamagata City, Yamagata Prefecture, Japan (Subsection 6.3.3)

Masakazu Aoyama Faculty of Agriculture and Life Science, Hirosaki University, Hirosaki City, Aomori Prefecture, Japan (Subsection 6.3.1)

Yusuke Arakawa Division of Upland Farming Research, Kyushu Okinawa Agricultural Research Center, NARO, Miyakonojo City, Miyazaki Prefecture, Japan (Subsection 10.3.8, Subsection 10.4.1)

Daiji Asaka Agricultural Research Department, Hokkaido Research Organization, Sapporo City, Hokkaido, Japan (Subsection 5.8.1)

Susumu Asakawa Graduate School of Bioagricultural Sciences, Nagoya University, Nagoya City, Aichi Prefecture, Japan (Chapter 8 (Edited))

Kumiko Baba Yamanashi Prefectural Agricultural Technology Center, Yamanashi Prefectural Government, Kai City, Yamanashi Prefecture, Japan (Subsection 7.1.2_2, Subsection 7.1.3_2, Subsection 7.3.2_2)

Weiguo Cheng Faculty of Agriculture, Yamagata University, Tsuruoka City, Yamagata Prefecture, Japan (Subsection 6.5.2)

Tsuneyoshi Endo Faculty of Agriculture, Tottori University, Tottori City, Tottori Prefecture, Japan (Subsection 9.2.4)

Nobuhiko Fueki Agricultural Research Department, Hokkaido Research Organization, Sapporo City, Hokkaido, Japan (Subsection 5.3.3)

Hiroshi Fujii Faculty of Agriculture, Yamagata University, Tsuruoka City, Yamagata Prefecture, Japan (Subsection 6.7.1)

Kazumichi Fujii Division of Forest Soils, Forestry and Forest Products Research Institute, Tsukuba City, Ibaraki Prefecture, Japan (Section 4.4)

Shigeto Fujimura Agricultural Radiation Research Center, Tohoku Agriculture Research Center, NARO, Morioka City, Iwate Prefecture, Japan (Subsection 6.6.4, Subsection 6.6.5)

Haruki Fujisawa Aomori Prefectural Industrial Technology Research Center, Kuroishi City, Aomori Prefecture, Japan (Subsection 6.2.4)

Syunrokurou Fujiwara Former Kurokawa Field Science Center, Meiji University, Kawasaki City, Kanagawa Prefecture, Japan (Subsection 7.5.4)

Eiji Gotou Agricultural Research Department, Hokkaido Research Organization, Sapporo City, Hokkaido, Japan (Subsection 5.2.2)

Masayuki Hara Mie Prefectural Agricultural Research Institute, Mie Prefectural Government, Matsusaka City, Mie Prefecture, Japan (Subsection 8.2.3_(3))

Yoshitaka Hara Kyushu Okinawa Agricultural Research Center, NARO, Koshi City, Kumamoto Prefecture, Japan (Subsection 10.3.2)

Hiroyuki Hasukawa Shiga Prefecture Agricultural Technology Promotion Center, Ohmihachiman City, Shiga Prefecture, Japan (Subsection 9.3.1)

Ryusuke Hatano Research Faculty of Agriculture, Hokkaido University, Sapporo City, Hokkaido, Japan (Editor, Chapter 1, Subsectoin 5.7.1)

Hiroaki Hayashi The Kunisaki Peninsula Usa GIAHS Promotion Association, Kunisaki City, Oita Prefecture, Japan (Subsection 10.3.8_(2))

Tetsuo Hayashi Agricultural Research Department, Hokkaido Research Organization, Sapporo City, Hokkaido, Japan (Subsection 5.4.2_(1))

Yuji Hikasa HOKUREN Federation of Agricultural Cooperatives, Sapporo City, Hokkaido, Japan (Subsection 5.4.1)

Kiyoshi Hiraoka Western Region Agricultural Research Center, NARO, Zentsuji City, Kagawa Prefecture, Japan (Subsection 9.2.6)

Takashi Hirayama Fukushima Prefecture Agricultural Technology Centre, Fukushima Prefectural Government, Koriyama City, Fukushima Prefecture, Japan (Subsection 6.6.5)

Yuhei Hirono Institute of Fruit Tree and Tea Sciences, NARO, Shimada City, Shizuoka Prefecture, Japan (Subsection 8.2.3_(2))

Tomoyoshi Hirota Hokkaido Agricultural Research Center, NARO, Sapporo City, Hokkaido, Japan (Subsection 5.9.3);
Faculty of Agriculture, Kyushu University, Fukuoka City, Fukuoka Prefecture, Japan

Kosuke Homma Faculty of Agriculture, Field Center for Sustainable Agriculture and Forestry, Niigata University, Niigata City, Niigata Prefecture, Japan (Subsection 7.4.1)

Yukio Hosobuchi Agricultural Research Department, Hokkaido Research Organization, Sapporo City, Hokkaido, Japan (Subsection 5.7.2_(3))

Kouichi Hosokawa Regional Agriculture Division, Fukui Prefectural Government, Fukui City, Fukui Prefecture, Japan (Subsection 8.2.1_(1))

Hirotaka Ihara Division of Agro-Environment Research, Kyushu Okinawa Agricultural Research Center, NARO, Koshi City, Kumamoto Prefecture, Japan (Subsection 10.3.5)

Masaharu Ikeba Ibaraki Agricultural Center, Ibaraki Prefectural Government, Mito City, Ibaraki Prefecture, Japan (Subsection 7.3.1_(1))

Akihiro Imaya Forestry Division, Japan International Research Center for Agricultural Sciences, Tsukuba City, Ibaraki Prefecture, Japan (Section 4.9)

Katsuhiro Inoue Nagasaki Agricultural & Forestry Technical Development Center, Nagasaki Prefectural Government, Isahaya City, Nagasaki Prefecture, Japan (Subsection 10.3.4)

Chiharu Ito Akita Prefectural Agricultural Experiment Station, Akita Prefectural Government, Akita City, Akita Prefecture, Japan (Subsection 6.1.2)

Masashi Ito Akita Prefectural Agricultural Experiment Station, Akita Prefectural Government, Akita City, Akita Prefecture, Japan (Subsection 6.5.1)

Toyoaki Ito Faculty of Agriculture, Niigata Agro-Food University, Tainai City, Niigata Prefecture, Japan (Section 4.3, Subsection 6.6.1)

Kenji Kanazawa Central Region Agricultural Research Center, NARO, Tsukuba City, Ibaraki Prefecture, Japan (Subsection 7.2.1)

Minako Kanda Reinan Regional Bureau, Fukui Prefectural Government, Fukui City, Fukui Prefecture, Japan (Subsection 8.2.1_(1))

Shinji Kaneko Kansai Research Center, Forestry and Forest Products Research Center, Kyoto City, Kyoto Prefecture, Japan (Subsection 2.3.1, Section 4.8)

Yoshihiro Kaneta Faculty of Bioresource Sciences, Akita Prefectural University, Akita City, Akita Prefecture, Japan (Subsection 6.2.2)

Nobuyuki Kanuma Gunma Agricultural Technology Center, Gunma Prefectural Government, Isesaki City, Gunma Prefecture, Japan (Subsection 7.3.1_(3))

Masahiro Kasuya Aichi Agricultural Research Center, Aichi Prefectural Government, Nagakute City, Aichi Prefecture, Japan (Subsection 8.2.3_(1))

Tetsuya Katagiri KONICA MINOLTA, INC., Tokyo, Japan (Subsection 6.7.1)

Taku Kato Department of Agricultural Chemistry, Faculty of Applied Bioscience, Tokyo University of Agriculture, Tokyo, Japan (Subsection 4.10.1)

Masayuki Kawahigashi Department of Geography, Tokyo Metropolitan University, Hachioji City, Tokyo, Japan (Section 4.1, Subsection 7.2.2)

Kimihiro Kida Department of Geography, Tokyo Metropolitan University, Hachioji City, Tokyo, Japan (Section 4.1)

Kazutoshi Kinjo Faculty of Agriculture, University of Ryukyus, Nishihara Town, Okinawa Prefecture, Japan (Subsection 10.4.5)

Rintaro Kinoshita Research Center for Global Agromedicine, Obihiro University of Agriculture and Veterinary Medicine, Obihiro City, Hokkaido, Japan (Section 4.2, Subsection 5.9.5)

Makoto Kobayashi Field Science Center for Northern Biosphere, Hokkaido University, Sapporo City, Hokkaido, Japan (Subsection 5.6.4)

Takayuki Kobayashi College of Bioresource Sciences, Nihon University, Fujisawa City, Kanagawa Prefecture, Japan (Subsection 7.5.3)

Nobuhisa Koga Headquarters, NARO, Tsukuba City, Ibaraki Prefecture, Japan (Subsection 10.3.1, Subsection 10.5.1)

Kazunori Kohyama Institute for Agro-Environmental Sciences, NARO, Tsukuba City, Ibaraki Prefecture, Japan (Section 4.7)

Jun Koike Toyama Prefectural Agricultural, Forestry and Fisheries Research Center, Toyama Prefectural Government, Toyama City, Toyama Prefecture, Japan (Subsection 8.2.1_(3))

Masataka Komatsu Nagano Prefecture Farmers Academy, Nagano Prefectural Government, Nagano City, Nagano Prefecture, Japan (Subsection 7.1.2_(2), Subsection 7.1.3_(2), Subsection 7.3.2_(1))

Tomohiro Kondo Faculty of Regional Innovation, University of Miyazaki, Miyazaki City, Miyazaki Prefecture, Japan (Subsection 10.3.8_(3));
Graduate School of Agriculture, Kyoto University, Kyoto City, Kyoto Prefecture, Japan

Katashi Kubo Agricultural Radiation Research Center, Tohoku Agriculture Research Center, NARO, Morioka City, Iwate Prefecture, Japan (Subsection 6.6.5)

Hideo Kubotera Institute for Agro-Environmental Sciences, NARO, Tsukuba City, Ibaraki Prefecture, Japan (Subsection 2.3.2, Subsection 10.1.3)

Kanta Kuramochi Research Faculty of Agriculture, Hokkaido University, Sapporo City, Hokkaido, Japan (Subsection 5.7.2_(2))

Takashi Kusaba Division of Agro-Environment Research, Kyushu Okinawa Agricultural Research Center, NARO (Retired), Koshi City, Kumamoto Prefecture, Japan (Subsection 10.1.1, Subsection 10.1.2)

Yuji Maejima Institute for Agro-Environmental Sciences, NARO, Tsukuba City, Ibaraki Prefecture, Japan (Section 4.6, Section 4.8)

Hiroyuki Maki Hyogo Prefectural Technology Center for Agriculture, Forestry and Fisheries, Hyogo Prefectural Government, Kasai City, Hyogo Prefecture, Japan (Subsection 9.3.2_(1))

Kunio Maruoka Saitama Prefectural Honjo Agriculture and Forestry Promotion Center, Saitama Prefectural Government, Kumagaya City, Saitama Prefecture, Japan (Subsection 7.3.1_(4))

Takehiko Matsumoto Agricultural Research Department, Hokkaido Research Organization, Sapporo City, Hokkaido, Japan (Subsection 5.7.3_(3))

Yumi Matsumoto Faculty of Agriculture, Yamagata University, Tsuruoka City, Yamagata Prefecture, Japan (Subsection 6.7.1)

Hisaya Matsunami Agricultural Radiation Research Center, Tohoku Agriculture Research Center, NARO, Morioka City, Iwate Prefecture, Japan (Subsection 6.6.5)

Minoru Matsuyama Hyogo Prefectural Technology Center for Agriculture, Forestry and Fisheries, Hyogo Prefectural Government, Kasai City, Hyogo Prefecture, Japan (Subsection 9.3.2_(2))

Nobuhiko Matsuyama Faculty of Agriculture and Life Science, Hirosaki University, Hirosaki City, Aomori Prefecture, Japan (Subsection 6.1.3)

Kojiro Mitsugi Kumamoto Prefectural Agricultural Research Center, Kumamoto Prefectural Government, Koshi City, Kumamoto Prefecture, Japan (Subsection 10.5.2)

Hideki Miura National Institute of Polar Research, Tachikawa City, Tokyo, Japan (Subsection 2.1.3, Section 2.2)

Naoko Miyamaru Sugarcane and Ordinary Crops Section, Okinawa Prefectural Agricultural Research Center, Itoman City, Okinawa Prefecture, Japan (Subsection 10.2.2, Subsection 10.4.5)

Naruo Miyazaki Tochigi Prefectural Agricultural Experiment Station, Tochigi Prefectural Government, Utsunomiya City, Tochigi Prefecture, Japan (Subsection 7.3.1_(2))

Hidetoshi Mochizuki Division of Agro-Enviroment Research, Western Region Agricultural Research Center, NARO, Fukuyama City, Hiroshima Prefecture, Japan (Subsection 9.2.1)

Shizuka Mori Faculty of Agriculture, Yamagata University, Tsuruoka City, Yamagata Prefecture, Japan (Subsection 6.2.3, Subsection 6.7.1)

Mikio Morioka Yamagata Integrated Agricultural Research Center, Yamagata Prefectural Government, Yamagata City, Yamagata Prefecture, Japan (Subsection 6.3.2)

Naoki Moritsuka Graduate School of Integrated Arts and Sciences, Kochi University, Nankoku City, Kochi Prefecture, Japan (Subsection 9.2.2)

Hirotatsu Murano Faculty of Agriculture, Meijo University, Nagoya City, Aichi Prefecture, Japan (Section 8.1)

Jun Murase Graduate School of Bioagricultural Sciences, Nagoya University, Nagoya City, Aichi Prefecture, Japan (Chapter 8 (Edited))

Osamu Nagata Hokkaido Agricultural Research Center, NARO, Sapporo City, Hokkaido, Japan (Subsection 5.7.3_(1))

Makoto Nagatomo Kagoshima Prefectural Institute for Agricultural Development, Minamisatsuma City, Kagoshima Prefecture, Japan (Subsection 10.3.6)

Shinpei Nakagawa Akita Prefectural Agricultural Experiment Station, Akita Prefectural Government, Akita City, Akita Prefecture, Japan (Subsection 6.1.2)

Miyuki Nakajima Tohoku Agricultural Research Center, NARO, Morioka City, Iwate Prefecture, Japan (Subsection 6.5.2)

Hiroshi Nakamoto Agricultural Research Department, Hokkaido Research Organization, Sapporo City, Hokkaido, Japan (Subsection 5.8.2)

Keiko Nakano Division of Lowland Farming Research, Kyushu Okinawa Agricultural Research Center, NARO, Chikugo City, Fukuoka Prefecture, Japan (Subsection 10.2.5)

Toshiro Nakatsuji Agricultural Research Department, Hokkaido Research Organization, Sapporo City, Hokkaido, Japan (Subsection 5.7.3_(2))

Hisashi Nasukawa Graduate School of Agricultural Science, Tohoku University, Sendai City, Miyagi Prefecture, Japan (Subsection 6.6.1)

Hiroshi Niimi Division of Upland Farming Research, Kyushu Okinawa Agricultural Research Center, NARO, Miyakonojo City, Miyazaki Prefecture, Japan (Subsection 10.2.4, Subsection 10.4.2)

Mizuhiko Nishida Graduate School of Agricultural Science, Tohoku University, Sendai City, Miyagi Prefecture, Japan (Subsection 6.2.1, Subsection 6.4.1)

Katsuhisa Niwa Zukosha Co., Ltd, Obihiro City, Hokkaido, Japan (Subsection 5.9.1)

Akira Noguchi College of Bioresource Sciences, Nihon University, Fujisawa City, Kanagawa Prefecture, Japan (Subsection 7.5.1)

Hiroshi Obara Institute for Agro-Environmental Sciences, NARO, Tsukuba City, Ibaraki Prefecture, Japan (Section 4.5, Subsection 4.10.2, Subsection 4.10.3, Subsection 4.10.4)

Naoto Ogawa Graduate School of Integrated Science and Technology, Shizuoka University, Shizuoka City, Shizuoka Prefecture, Japan (Chapter 8 (Edited))

Tetsushi Ohshio Hyogo Prefectural Technology Center for Agriculture, Forestry and Fisheries, Hyogo Prefectural Government, Kasai City, Hyogo Prefecture, Japan (Subsection 9.3.2_(3))

Ryo Ohtomo Hokkaido Agricultural Research Center, NARO, Sapporo City, Hokkaido, Japan (Subsection 5.9.4)

Norikuni Oka Hokkaido Agricultural Research Center, NARO, Sapporo City, Hokkaido, Japan (Subsection 5.9.4)

Masayuki Onodera Agricultural Research Department, Hokkaido Research Organization, Sapporo City, Hokkaido, Japan (Subsection 5.8.2)

Takeshi Ota Agricultural Radiation Research Center, Tohoku Agriculture Research Center, NARO, Morioka City, Iwate Prefecture, Japan (Subsection 6.6.4, Subsection 6.6.5); Bio-Oriented Technology Research Advancement Institution, NARO, Kawasaki City, Kanagawa Prefecture, Japan

Masaya Oya Okayama Prefectural Technology Center for Agriculture, Forestry and Fisheries, Akaiwa City, Okayama Prefecture, Japan (Subsection 9.3.3)

Jiwan Palta College of Agricultural and Life Sciences, University of Wisconsin-Madison, Madison, USA (Subsection 5.9.5)

Yuichi Saeki Faculty of Agriculture, University of Miyazaki, Miyazaki City, Miyazaki Prefecture, Japan (Subsection 10.2.3)

Toshiya Saigusa College of Agriculture, Food and Environment Sciences, Rakuno Gakuen University, Ebetsu City, Hokkaido, Japan (Subsection 5.5.1)

Masanori Saito Graduate School of Agricultural Science, Tohoku University, Sendai City, Miyagi Prefecture, Japan (Subsection 6.6.1, Subsection 6.6.6)

Takashi Saito Hama Agricultural Regeneration Research Centre, Fukushima Prefecture Agricultural Technology Centre, Soma City, Fukushima Prefecture, Japan (Subsection 6.6.4, Subsection 6.6.5)

Masashi Saitou Fukui Agricultural Experiment Station, Fukui Prefectural Government, Fukui City, Fukui Prefecture, Japan (Subsection 8.2.1_(1))

Masami Sakaguchi Agricultural Research Department, Hokkaido Research Organization, Sapporo City, Hokkaido, Japan (Subsection 5.4.2_(2))

Osamu Sakai Agricultural Research Department, Hokkaido Research Organization, Sapporo City, Hokkaido, Japan (Subsection 5.5.3)

Michihiko Sakurai Agricultural Research Department, Hokkaido Research Organization, Sapporo City, Hokkaido, Japan (Subsection 5.3.2)

Tetsuya Sano Faculty of Engineering, Tohoku Institute of Technology, Sendai City, Miyagi Prefecture, Japan (Subsection 2.3.1)

Hidetaka Sasaki Fukui Public Works Office, Fukui Prefectural Government, Fukui City, Fukui Prefecture, Japan (Subsection 8.2.1_(1))

Yuka Sasaki Faculty of Agriculture, Yamagata University, Yamagata City, Yamagata Prefecture, Japan (Subsection 6.1.1)

Takashi Sato Faculty of Bioresource Sciences, Akita Prefectural University, Akita City, Akita Prefecture, Japan (Subsection 6.4.3)

Takumi Sato RIKEN BioResource Research Center, Tsukuba City, Ibaraki Prefecture, Japan (Subsection 6.3.4)

Fuyuki Satoh Field Science Center for Northern Biosphere, Hokkaido University, Sapporo City, Hokkaido, Japan (Subsection 5.6.1)

Takuji Sawamoto College of Agriculture, Food and Environment Sciences, Rakuno Gakuen University, Ebetsu City, Hokkaido, Japan (Subsection 5.7.2_(1))

Hideaki Shibata Field Science Center for Northern Biosphere, Hokkaido University, Sapporo City, Hokkaido, Japan (Subsection 5.6.2)

Makoto Shibata Graduate School of Global Environmental Studies, Kyoto City, Kyoto Prefecture, Japan (Section 4.3)

Hiroyuki Shiga HOKUREN Federation of Agricultural Cooperatives, Sapporo City, Hokkaido, Japan (Subsection 5.1.1)

Takeo Shima Division of Upland Farming Research, Kyushu Okinawa Agricultural Research Center, NARO, Miyakonojo City, Miyazaki Prefecture, Japan (Subsection 10.4.6)

Teruo Shima Iwate Agricultural Reserch Center, Kitakami City, Iwate Prefecture, Japan (Subsection 6.2.6, Subsection 6.6.3)

Mariko Shimizu Civil Engineering Research Institute for Cold Region, Public Works Research Institute, Sapporo City, Hokkaido, Japan (Subsection 5.9.2)

Takuro Shinano Agricultural Radiation Research Center, Tohoku Agriculture Research Center, NARO, Morioka City, Iwate Prefecture, Japan (Subsection 6.6.4, Subsection 6.6.5); Graduate School of Agriculture, Hokkaido University, Spporo City, Hokkaido, Japan

Hitoshi Shinjo Graduate School of Global Environmental Studies, Kyoto University, Kyoto City, Kyoto Prefecture, Japan (Editor, Chapter 1, Sectoin 3.1)

Fumie Shinmachi College of Bioresource Sciences, Nihon University, Fujisawa City, Kanagawa Prefecture, Japan (Subsection 7.5.1)

Hiroyuki Shiono Yamagata Prefecture Murayama Area General Branch Administration Office, Yamagata City, Yamagata Prefecture, Japan (Subsection 6.5.3)

Yutaka Shiratori Niigata Agricultural Research Institute, Niigata Prefectural Government, Nagaoka City, Niigata Prefecture, Japan (Subsection 7.1.2_(3), Subsection 7.1.3_(3), Subsection 7.3.2_(3))

Kazuko Someya Gunma Agricultural Technology Center, Gunma Prefectural Government, Isesaki City, Gunma Prefecture, Japan (Subsection 7.3.1_(3))

Hiroaki Sumida College of Bioresource Sciences, Nihon University, Fujisawa City, Kanagawa Prefecture, Japan (Subsection 7.1.4)

Ryousuke Tajima Graduate School of Agricultural Science, Tohoku University, Sendai City, Miyagi Prefecture, Japan (Subsection 6.6.1)

Kentaro Takagi Field Science Center for Northern Biosphere, Hokkaido University, Sapporo City, Hokkaido, Japan (Subsection 5.6.3)

Reiji Takahashi College of Bioresource Sciences, Nihon University, Fujisawa City, Kanagawa Prefecture, Japan (Subsection 7.5.2)

Tadashi Takahashi Department of Biological Environment, Faculty of Bioresource Sciences, Akita Prefectural University, Akita City, Akita Prefecture, Japan (Section 4.3, Subsection 6.1.3)

Tomoki Takahashi Tohoku Agricultural Research Center, NARO, Morioka City, Iwate Prefecture, Japan (Subsection 6.4.4)

Fumiaki Takakai Faculty of Bioresource Sciences, Akita Prefectural University, Akita City, Akita Prefecture, Japan (Subsection 6.4.2)

Yusuke Takata Institute for Agro-Environmental Sciences, NARO, Tsukuba City, Ibaraki Prefecture, Japan (Editor, Chapter 1, Section 3.2, Section 4.6, Subsection 7.1.1, Subsection 7.1.2, Subsection 7.1.4)

Kunihiko Takehisa Shiga Prefecture Agricultural Technology Promotion Center, Ohmi-hachiman City, Shiga Prefecture, Japan (Subsection 9.3.1)

Hiroshi Takesako Former Kurokawa Field Science Center, Meiji University, Kawasaki City, Kanagawa Prefecture, Japan (Subsection 7.1.4)

Harunobu Takeuchi Agricultural Research Department, Hokkaido Research Organization, Sapporo City, Hokkaido, Japan (Subsection 5.1.2)

Kenji Tamura Faculty of Life and Environmental Sciences, University of Tsukuba, Tsukuba City, Ibaraki Prefecture, Japan (Subsection 2.1.1, Subsectoin 2.1.2)

Masayuki Tani Research Center for Global Agromedicine, Obihiro University of Agriculture and Veterinary Medicine, Obihiro City, Hokkaido, Japan (Section 4.2, Subsection 5.9.5)

Norimasa Tanikawa Aomori Prefectural Industrial Technology Research Center, Kuroishi City, Aomori Prefecture, Japan (Subsection 6.2.5)

Keitaro Tawaraya Faculty of Agriculture, Yamagata University, Yamagata City, Yamagata Prefecture, Japan (Subsection 6.3.4)

Naoko Tokuchi Field Science Education and Research Center, Kyoto University, Kyoto City, Kyoto Prefecture, Japan (Subsection 9.2.7)

Mitsuru Toma Graduate School of Sciences and Technology for Innovation, Yamaguchi University, Yamaguchi City, Yamaguchi Prefecture, Japan (Section 9.1)

Yo Toma Graduate School of Agriculture, Ehime University, Matsuyama City, Ehime Prefecture, Japan (Subsection 9.2.5)

Takashi Tsukamoto Chiba Prefectural Agriculture and Forestry Research Center, Chiba Prefectural Government, Chiba City, Chiba Prefecture, Japan (Subsection 7.3.1_(5))

Yasutaka Tsukamoto Agricultural Research Department, Hokkaido Research Organization, Sapporo City, Hokkaido, Japan (Subsection 5.2.1)

Tomoji Uchiyama DGCbase, Toyonaka City, Osaka Prefecture, Japan (Subsection 9.2.4)

Ichiro Uezono Kagoshima Prefectural Institute for Agricultural Development Osumi Branch, Kanoya City, Kagoshima Prefecture, Japan (Subsection 10.2.1)

Fumio Uno Ishikawa Agriculture and Forestry Research Center, Ishikawa Prefectural Government, Kanazawa City, Ishikawa Prefecture, Japan (Subsection 8.2.1_(2))

Toru Uno Graduate School of Agricultural Science, Tohoku University, Sendai City, Miyagi Prefecture, Japan (Subsection 6.6.1)

Kentaro Wada Tokushima Agriculture, Forestry and Fisheries Technology Support Centre, Ishii Town, Tokushima Prefecture, Japan (Subsection 9.2.3)

Shokichi Wakabayashi Tohoku Agriculture Research Center, NARO, Morioka City, Iwate Prefecture, Japan (Subsection 6.6.4)

Akira Watanabe Graduate School of Bioagricultural Sciences, Nagoya University, Nagoya City, Aichi Prefecture, Japan (Chapter 8 (Edited))

Hitoshi Watanabe Gifu Prefectural Research Institute for Forests, Gifu Prefectural Government, Mino City, Gifu Prefecture, Japan (Subsection 8.2.2)

Yuji Watanabe Agricultural Research Department, Hokkaido Research Organization, Sapporo City, Hokkaido, Japan (Subsection 5.3.1)

Tetsuo Yagi Agricultural Research Department, Hokkaido Research Organization, Sapporo City, Hokkaido, Japan (Subsection 5.5.2)

Yasunao Yamada Nagasaki Agricultural & Forestry Technical Development Center, Nagasaki Prefectural Government, Isahaya City, Nagasaki Prefecture, Japan (Subsection 10.3.3)

Noriko Yamaguchi Division of Soil Science and Plant Nutrition, Central Region Agricultural Research Center, NARO, Tsukuba City, Ibaraki Prefecture, Japan (Subsection 10.3.8_(1))

Tsuyoshi Yamane Division of Agro-Environmental Research, Hokkaido Agricultural Research Center, NARO, Sapporo City, Hokkaido, Japan (Subsection 10.4.3)

Hiroko Yamaya-Ito College of Bioresource Sciences, Nihon University, Fujisawa City, Kanagawa Prefecture, Japan (Subsection 7.5.1)

Junta Yanai Graduate School of Life and Environmental Sciences, Kyoto Prefectural University, Kyoto City, Kyoto Prefecture, Japan (Section 9.1)

Koichi Yoshida Okinawa Agricultural Technology & Development Co., Ltd., Itoman City, Okinawa Prefecture, Japan (Subsection 10.3.7)

Part I
Introduction

Overview

1

Ryusuke Hatano, Hitoshi Shinjo, and Yusuke Takata

Abstract

This chapter provides the overview of soil types, land uses, and land managements of Japan. The soils of Japan are immature soils developed from the flesh parent materials such as volcanic tephra or ejecta or eroded or deposited materials in various landscapes from mountain terrain to alluvial plains within about 10,000 years. Japanese soils are generally acidic, since the basic cations of soils are leached by the high precipitation of the monsoon climate. According to the soil classification system of Japan newly developed in 2017, Brown Forest soils, Andosols, Fluvic soils, Red-Yellow soils, and Regosols cover 33.2, 30.3, 13.7, 7.6, and 6.9% of the country, respectively. In land uses of Japan in 2016, forests occupy 63.5% of the total land area of Japan (37.8 million ha). Cultivated land area was 4.47 million ha, which accounts for 12.0%. However, the cultivated land area has continuously decreased from 6.08 million ha in 1961. In Japan, from 1959 to 1978, the Fundamental Soil Survey for Soil Fertility Conservation was carried out and classified the crop production potential of cultivated land soils into four grades. Based on the results of the soil survey, field improvements have been performed to ameliorate the limiting factors for crop production. This has made it possible to set the fertilization standards for each crop in each region. The fertilization standards are defined as the amount of fertilizer that achieves the target yield without causing environmental issues and are the bases of the environmentally friendly agriculture. Based on the efforts in each region, in 1999, Japanese government enacted the Act on Promotion of Introduction of Sustainable Agricultural Production Practices.

Keywords

Environmentally friendly agriculture • Field improvement • Land uses • Land managements • Soil survey • Fertilization standard

1.1 Climate, Landscapes, Geology, and Soils of Japan

Japan is located in East Asia. According to the variations in climate, landscapes, geology, and soils, Japanese cultural regions are divided into six regions: (1) Hokkaido; (2) Tohoku; (3) Kanto and Koushinetsu; (4) Tokai and Hokuriku; (5) Kinki, Chugoku, and Shikoku; and (6) Kyushu and Okinawa. Each of these regions has developed its own agricultural management practices (Fig. 1.1).

1.1.1 Climate

Japan is generally rainy with a monsoon climate, but it also has a wide vertical and horizontal distribution of land. Japan is located at 20°25′31″–45°33′26″ N and elevations range from about 0 m to over 3000 m above sea level. These variations give rise to six major climate regions in Japan: (1) Hokkaido; (2) the Nansei (or "Southwest") Islands; (3) the Sea of Japan side; the (4) the Pacific Ocean side; (5) the Central Highlands; and (6) the Seto Inland Sea. The climate of Hokkaido basically features low temperature and low precipitation; the mean annual temperature and mean annual precipitation are 6.3 °C and 1013 mm, respectively.

R. Hatano (✉)
Research Faculty of Agriculture, Hokkaido University, Sapporo City, Hokkaido, Japan
e-mail: hatano@chem.agr.hokudai.ac.jp

H. Shinjo
Graduate School of Global Environmental Studies, Kyoto University, Kyoto City, Kyoto Prefecture, Japan

Y. Takata
Institute for Agro-Environmental Sciences, NARO, Tsukuba City, Ibaraki Prefecture, Japan

R. Hatano et al. (eds.), *The Soils of Japan*, World Soils Book Series,
https://doi.org/10.1007/978-981-15-8229-5_1

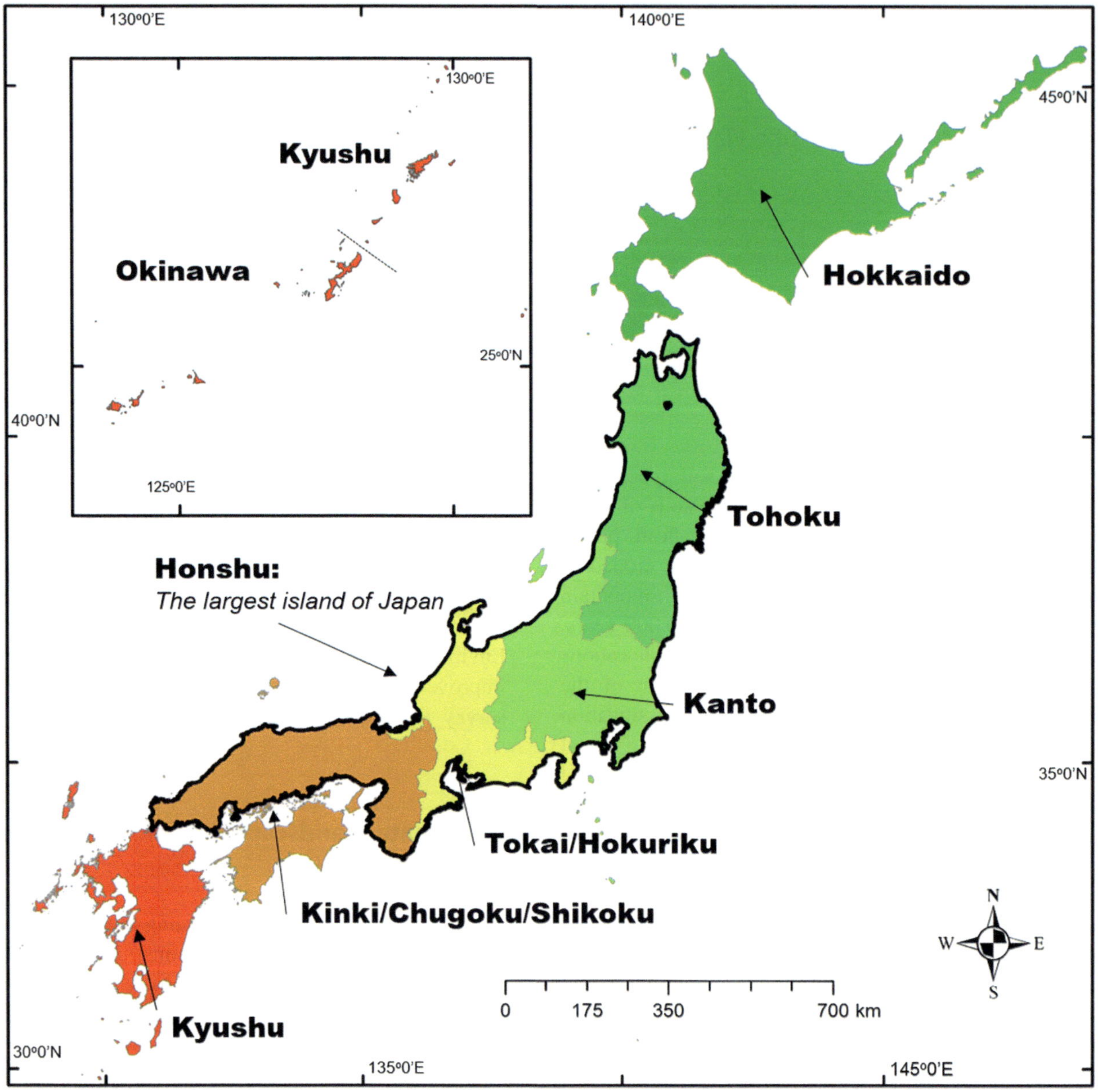

Fig. 1.1 Japan's cultural regions (Figure supplied by Yusuke Takata)

The western part of Hokkaido receives plentiful snowfall, whereas the eastern part is colder and drier. The climate of the Nansei Islands, which include the islands of Okinawa and Ogasawara, is rainy subtropical; the mean annual temperature and mean annual precipitation are 23.1 °C and 2040 mm, respectively. The climate of the Sea of Japan side, which does not include the Hokkaido and Kyushu regions, is snowy temperate. The mean annual temperature and mean annual precipitation are 13.6 °C and 2760 mm, respectively. The climate of the Pacific Ocean side, which covers the Tohoku, Kanto, Tokai, and Kinki regions, the Pacific side of the Shikoku region, and part of the Kyushu region, is temperate with hot and rainy summers. The mean annual temperature and mean annual precipitation are 15.4 °C and 1580 mm, respectively. The climate of the Central Highlands, which is affected by the Japanese Alps of the Koushinetsu region, is characterized by relatively low temperatures and low precipitation. The mean annual temperature and mean annual precipitation are 11.8 °C and 1030 mm, respectively. The climate of the Seto Inland Sea, which is located between the Chugoku Mountains and the Shikoku Mountains, is temperate with relatively low precipitation. The mean annual temperature and mean annual precipitation are 16.3 °C and 1080 mm, respectively. The natural climax vegetation under these climatic conditions is basically forest, except in high mountains and wet lowlands, where the vegetation is grassland or sphagnum.

1.1.2 Landscapes and Geology

The Japanese islands are covered with volcanic ejecta along its volcanic belt, which is part of the orogenic belt, and have a lot of earthquakes and volcanic activity. The Japanese islands are part of the circum-Pacific volcanic belt and contain two distinct volcanic belts. One is the East Japan volcanic belt, which runs from east to west through Hokkaido, the west side of Tohoku, the northern part of Koushinetsu to the Izu Islands and Iou Islands, and to the Mariana Trench. The other is the West Japan volcanic belt, which runs from the Sea of Japan side of the Chugoku region (San'in area), through Kyushu and to the Nansei Islands. Volcanic rocks are generally composed of silica-rich andesite and rhyolite, although alkaline basalts are seen in Kyushu. Terrain changes, such as mountain erosion and plain formation due to sedimentation in flat areas, are fast and intense under the humid monsoon climate.

1.1.3 Soils

The soils of Japan are immature soils of about 10,000 years in age. Their parent materials are volcanic tephra or ejecta or eroded or deposited materials. In 2017, Japanese Society of Pedology developed a new soil classification system, the "Soil Classification System of Japan," to consider these factors (Fifth Committee for Soil Classification and Nomenclature 2017). According to that classification system, Brown Forest soils cover 33.2% of the country, and Andosols, Fluvic soils, Red-Yellow soils, and Regosols cover 30.3%, 13.7%, 7.6%, and 6.9%, respectively. Peat soils are also distributed around the rivers and coasts of Hokkaido and cover 3% of that island's area. Brown Forest soils do not have andic properties, and although those rock structures are lost, clay formation and soil structure development are observed. Brown Forest soils mainly correspond to Cambisols in the World Reference Base for Soil Resources (WRB). Andosols are derived from volcanic ash, have andic properties, and correspond to Andosols in the WRB. Fluvisols correspond to Fluvisols in the WRB; their parent materials are sediment deposits in rivers, seas, and lakes. Paddy field soils are also classified as Fluvisols in the Japanese taxonomic system, but those soils are classified as Anthrosols in the WRB. Red-Yellow soils are well weathered and relatively developed soils in Japan. Some Red-Yellow soils have argic horizons and also include soils that have cambic horizons. The criterion of cation exchange capacity per 1 kg of clay is not defined in our system, and so, there is no soil group corresponding to Acrisols, Lixisols, Alisols, or Luvisols in the WRB. Regosols correspond to Regosols and Lithosols in the WRB, and Peat soils to Histosols. Japanese soils are generally acidic, since the basic cations of soils are leached by the high precipitation of the monsoon climate.

The crystalline clay mineral composition of Fluvisols tends to be 2:1 in Northern Japan and 1:1 in Southern Japan. Minerals with a size of 1.4 nm are dominant in thc Hokkaido, Tohoku, and Hokuriku regions, and minerals with a size of 0.7 nm are dominant in the Chugoku, Shikoku, and Kyushu regions, and the composition of those minerals are balanced in the Kanto, Tokai, and Kinki regions (Soil Genesis and Classification Laboratory 1996). Andosols have a high phosphate fixation capacity and are divided into allophanic Andosols and non-allophanic Andosols, the latter of which have low pH. allophanic Andosols are distributed in the Hokkaido, Tohoku, Kanto, and Kyushu regions, where Holocene volcanic ejecta are deposited, whereas non-allophanic Andosols are distributed in the Tokai, Hokuriku, Kinki, Chugoku, and Shikoku regions, where Holocene volcanic ejecta is absent.

1.2 Material Cycling and Environmental Problems in Agriculture

(1) **Material cycling through soil**

In ecosystems, energy, water, and materials constantly enter and leave, and produce soil (shown schematically in Fig. 1.2). In plants, solar energy is consumed by photosynthesis and transpiration, atmospheric carbon dioxide (CO_2) is absorbed via the leaves, water and nutrients are absorbed via the roots and plant growth occurs. The nutrients absorbed by plants include ammonia (NH_3) synthesized from nitrogen gas (N_2) in the atmosphere by nitrogen-fixing bacteria and minerals supplied by the hydrolysis of rock minerals (CO_2 dissolved in rainwater acts as a weathering agent by supplying protons from carbonic acid). When a plant dies and is decomposed by microorganisms in the soil, CO_2 is discharged, and weathering is accelerated in the soil. At the same time, NH_3 and minerals are released into the soil. The NH_3 dissolves in the soil solution to form ammonium ions (NH_4^+), and a part of it is converted into nitrate ions (NO_3^-) by nitrifying bacteria and is reabsorbed by plants. Inorganic nitrogen, minerals that are not absorbed by plants, and bicarbonate ions produced by the dissolution of CO_2 are subsequently leached out due to the penetration of water in the soil, before moving into groundwater and flowing out to rivers.

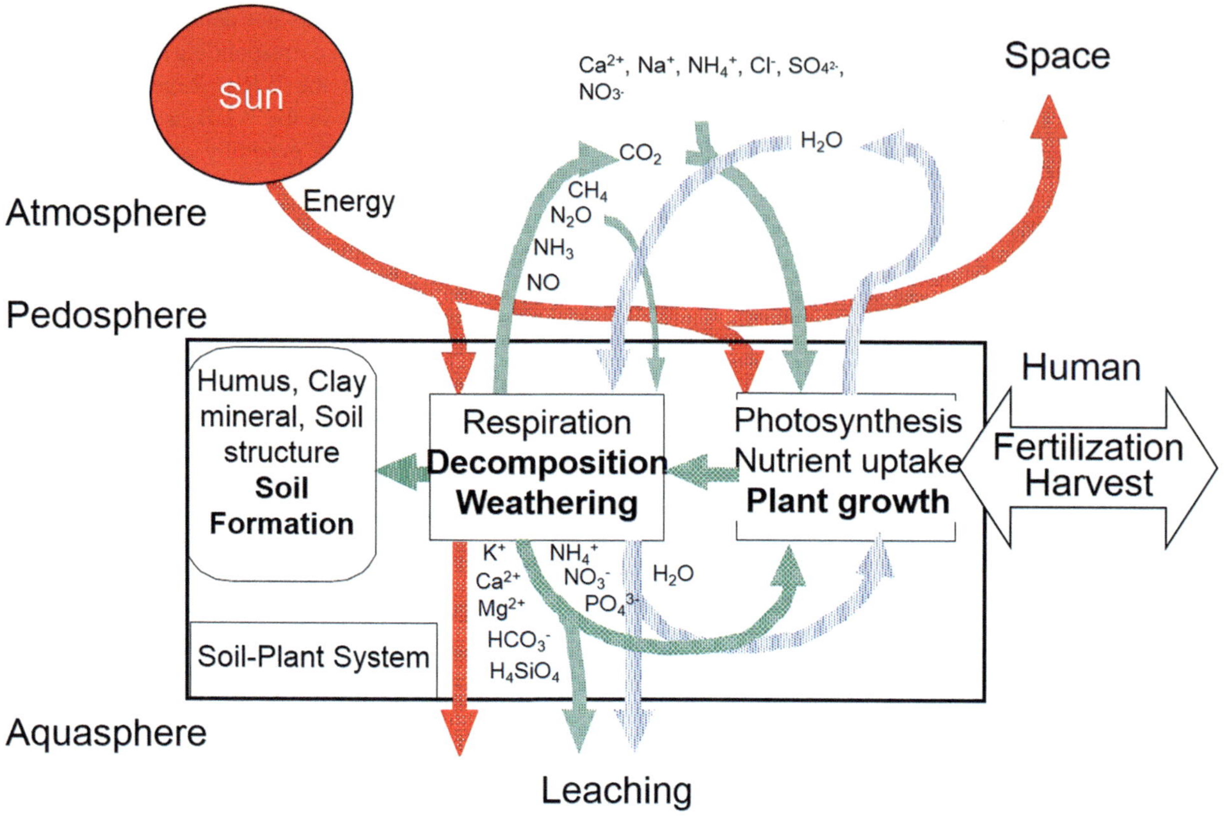

Fig. 1.2 Cycles of energy, water, and materials cross through soil (Figure supplied by Ryusuke Hatano)

Under anaerobic conditions such as those in wetlands, methane (CH_4) is emitted, which has a global warming potential 25 times that of CO_2. Meanwhile, methane-oxidizing bacteria in the forest soil turn CH_4 to CO_2. In soil with a low oxygen content, denitrifying bacteria convert NO_3^- into N_2 gas. During nitrification and denitrification, nitrous oxide (N_2O), which has a global warming potential 298 times that of CO_2, and nitric oxide (NO), which produces ozone, are also emitted.

During such material cycling, clay minerals are formed from rock minerals, humus is generated in soil from the decomposition of dead plant bodies, and levels of soil organic carbon and nitrogen increase. Clay and humus are negatively charged and buffer acids, adsorb basic cations such as potassium, calcium, and magnesium, and also inhibit the leaching of these cations. Clay and humus form organo-mineral complexes and cross-link large particles such as sand to form aggregates. As a result, a soil structure is formed with coarse pores between the aggregates, which improves drainage and aeration, and capillary pore spaces within the aggregates, which improves water retention, both of which improve the growth of plants. In agriculture, this natural material cycling system is utilized to cultivate crops and feed livestock.

(2) **Influence of agriculture on the material cycle**

Agriculture has a strong influence on material cycling via the input and output of substances through fertilization and harvesting. If an amount of organic matter corresponding to the amount of organic matter removed by harvesting is not input to the soil by compost, and so on, then soil humus is decomposed by microorganisms, and the organic carbon stored in the soil is thereby converted to CO_2 and released to the atmosphere, contributing to global warming. Additionally, the soil structure deteriorates and drainage and aeration are reduced, causing the growth of crops to stagnate. However, if a large amount of compost is applied along with fertilization, an increase in the volatilization of NH_3 occurs, which in turn increases the nitrogen fallout to the forest and acidifies the soil (because nitrification releases protons). Furthermore, leaching of NO_3^- increases in agricultural land, causing groundwater contamination and the eutrophication of coastal areas.

1.3 Perspective of Regional Soil Management

There are various regional soil management practices in Japan, since each region has various climates, landscapes, and soils.

The island of Hokkaido, in Northern Japan, is the most recently developed (150 years ago) region in the country (1869). The highly snowy central and southern areas are used for rice cultivation, while the seasonally soil freezing eastern and northern areas are grassland used for dairy farming. The southern half of the island is covered by volcanic ash and has recently become a popular area for upland crop cultivation and vegetable production.

The Tohoku region, located in the north of Honshu Island, is the main rice production area in the snowy cold part of Japan. The Pacific Ocean side of Tohoku was severely affected by the Great East Japan Earthquake of March 2011, through the impact of a tsunami and the accident at the Fukushima Daiichi Nuclear Power Station operated by the Tokyo Electronic Power Company. Fruit production and livestock production are thriving in cold areas in the north and hilly areas covered with volcanic ash.

The Kanto–Koushinetsu region is located in the center of Japan. The Kanto region, which includes Mt. Fuji, is covered with Kanto loam (a volcanic ash layer) and mainly produces vegetables. In the Koushinetsu region, the Central Highlands produces fruit and livestock, and the Sea of Japan side mainly produces rice on fluvial plains (the Echigo Plain).

The Tokai–Hokuriku region is located in the center of Honshu Island. The Tokai region, located on the Pacific Ocean side, is warm and has Red-Yellow soils. Tokai produces green tea and oranges. The Hokuriku region, located on the Sea of Japan side, has warm summers and snowy winters, and produces rice. The people in Hokuriku experienced "*itai-itai*" disease, which was caused by cadmium-contaminated rice, and worked to eradicate it.

The Kinki–Chugoku–Shikoku region consists of the Kansai region, located in the western part of Honshu Island, and the Chugoku and Shikoku regions neighboring the Seto Inland Sea. The major agricultural management practices on the Pacific Ocean side are a two-cropping system under a warm and rainy climate condition, while fruit and warm-land type livestock farming is practiced in the warm and relatively dry area around the Seto Inland Sea, and rice production and forestry are practiced on the Sea of Japan side, which experiences warm summers and cold winters.

The Kyushu–Okinawa region is located in the southwestern part of Japan. The Okinawa region is in a subtropical climate. Rice and vegetable production is common in the alluvial plains and reclaimed land of Kyushu, while sweet potato and forage crops are dominant in its southern part, which is covered by volcanic ash. Additionally, sugarcane and pineapple production is thriving in Okinawa.

1.3.1 Land Use

The following is an overview of the properties of Japanese land use referring to statistical data from the Japanese government (Ministry of Agriculture, Forestry and Fisheries 2018a, b; Forestry Agency 2018).

(1) **Forests**

Forests occupy 63.5% of the total land area of Japan (37.8 million ha), and 41.0% of forests are planted forests. Larch trees are planted in Hokkaido, and *Cryptomeria* and Hinoki cypress (*Chamaecyparis obtusa*) are planted in other areas. The ratio of forest in urban areas of the Kanto and Osaka regions is low (40%). This ratio also tends to be low in the Kyushu–Okinawa region. The ratio of planted forest tends to be more than 50% in many prefectures of the Tokai, Kinki, and Kyushu regions.

(2) **Cultivated land**

In 2016, the area of cultivated land in Japan was 4.47 million ha, which accounts for 12.0% of the country's total land area. A total of 54.4% (2.43 million ha) of the cultivated land is paddy field, and the rest is upland cropping field (2.03 million ha). Upland cropping field is divided into ordinary upland cropping field (56.4%), orchard (14.1%), and grazing field (29.6%). Of the total cultivated land area, 25.5% is located in Hokkaido (1.14 million ha), 18.9% in Tohoku (0.84 million ha), 16.2% in Kanto–Koushinetsu (0.72 million ha), and each 13.0% in the other three regions (0.60 million ha).

The total area of cultivated land has decreased by an average of 29,000 ha per year, from 6.08 million ha in 1961. The area of abandoned farmland has increased by an average of 8570 ha per year since 1980, reaching 0.42 million ha in 2015 (9.4% of the total cultivated land area).

(3) **Paddy field**

Paddy fields are present in all the regions of Japan except for the northeastern part of Hokkaido. In 2016, the area of rice planting was 1.47 million ha, accounting for 33.1% of the total cultivated land area. Of the total area of rice planting, 25.3% is located in Tohoku, 20.2% in Tokai–Hokuriku, 18.3% in Kanto–Koushinetsu, 17.7% in Kinki–Chugoku–Shikoku, 11.3% in Kyushu–Okinawa, and 7.1% in Hokkaido.

It should also be noted that 60.8% of the paddy field area is now used solely for rice cropping. The rest is used under rotation of paddy rice and upland crops. This rotation derives from the "*gentan*" policy that was started in 1971. This policy resulted in the conversion of paddy fields from rice production to the production of other upland crops in order to counteract the overproduction of rice. The improvement of drainage was promoted to cultivate upland crops in wet and semi-wet paddy field. In 2016, the proportion of paddy fields in upland crop farming reached 52.8% in Hokkaido, 47.2% in Kyushu–Okinawa, 41.9% in Kinki–Chugoku–Shikoku, 37.9% in Tohoku, 33% in Kanto–Koushinestu, and 31% in Tokai–Hokuriku.

The area of rice cropping field in Japan decreased from 3.13 million ha in 1961 to 1.48 million ha in 2016, a 52.8% decrease. However, the harvest of brown rice decreased by only 32.5%, from 11.9 to 8.04 million tons, since the yield increased by 43%, from 3.80 to 5.44 tons/ha. The drying of wet paddy field allows for appropriate field management and could contribute to an increase in rice yield. Japan could potentially produce 13.2 million tons of rice if all paddy fields were used for rice cropping. Small paddy fields tend to be merged into larger paddy fields (i.e., $\geq$4.4 ha). The yield of brown rice is as low as 5.06 tons/ha in Kyushu, compared with 5.67 tons/ha in Tohoku. Thus, Southern Japan tends to have a lower brown rice yield, and a further reduction in yield may occur due to global warming. Additionally, soil compaction and a decrease in soil nitrogen levels are concerns, since paddy rice-upland rotational management decreases levels of soil organic matter (Nira 2013).

(4) **Upland crop fields**

The total area of ordinary upland crop field is 1.15 million ha, of which 36.2% is located in Hokkaido, 22.8% in Kanto–Koushinetsu, 15.9% in Kyushu–Okinawa, 11.3% in Tohoku, 7.5% in Tokai–Hokuriku, and 6.2% in Kinki–Chugoku–Shikoku. Of a total orchard area (287,000 ha), 27.6% is distributed in Kinki–Chugoku–Shikoku, 20.9% in Kyushu–Okinawa, 17.3% in Kanto–Koushinetsu, 16.7% in Tokai–Hokuriku, 16.5% in Tohoku, and 1% in Hokkaido. On the other hand, the total area of grazing field is 603,000 ha, of which 83.7% is located in Hokkaido, 10% in Tohoku, 3.4% in Kyushu–Okinawa, 1.5% in Kanto–Koushinetsu, and less than 1% in Tokai–Hokuriku and Kinki–Chugoku–Shikoku.

The cropping area of wheat, which is a major upland crop, fell from 0.65 million ha in 1961 to 0.21 million ha in 2016, a decrease of 67.8%. However, the yield increased by 49% in the same period, from 2.75 to 4.10 tons/ha. The cropping area of soybean also fell, decreasing by 47.7% from 0.29 to 0.15 million ha between 1961 and 2016, but the yield of soybean increased by 17.7% in the same period, from 1.35 to 1.59 tons/ha. The cropping area of vegetables decreased by 15.7% between 1980 and 2016, falling to 0.56 million ha. The cropping area has also tended to decrease for all upland crops.

On the other hand, the area of grazing field and upland crop field for livestock increased from 0.15 million ha in 1996 to 0.74 ha in 2016. This was due to an increase of livestock production; between 1961 and 2016, the numbers of dairy cattle increased by 52.0%, beef cattle increased by 7.2%, pigs increased by 258%, and egg-laying hens increased by 144%.

1.3.2 Potential of Crop Production

From 1959 to 1978, the Fundamental Soil Survey for Soil Fertility Conservation was carried out by the prefectural agricultural experiment stations under the Japanese Ministry of Agriculture, Forestry and Fisheries, and cultivated soil maps at a scale of 1:50,000 and evaluation maps of the potential of crop production were developed. Soil surveys were performed at 220,000 measurement points (one point in every 25 ha), covering 2.89 million has of paddy field and 1.80 million ha of upland crop field. This survey identified the physicochemical properties of soils that limited crop production. Out of these survey points, 3323 points of representative soil profile data, including descriptions of soil profile and physicochemical properties, were digitized (Oda et al. 1987). The Fundamental Soil Survey for Soil Fertility Conservation classified the crop production potential into four grades:

Grade I: Soils have neither limitations nor hazards and have high potential for crop production without any improvement;
Grade II: Soils have some limitations or hazards for crop production. They require some improvement to achieve good production;
Grade III: Soils have many limitations or hazards for crop production. They require intensive improvement;
Grade IV: Soils have such great limitations or hazards that they can hardly be used for agriculture. They require very intensive improvement.

The factors used to define the soil grade were thickness of topsoil (t), effective rooting depth of soil (d), gravel content in topsoil (g), difficulty of plowing (p), saturated water permeability (l), redox properties (r), wetness of land (w), dryness of land ((w)), inherent fertility (f), available nutrients (n), risk of growth disorders (i), risk of disaster (a), slope (s), and erodibility (e).

Field improvement was performed to ameliorate the limiting factors for crop production. This improvement included large-scale field improvement such as rezoning, flattening, installation of underdrains, improvement of efficiency of machine work and productivity, mineral soil dressing in peatland, removal of gravel, drainage improvement, the application of phosphate fertilizer to Andosols, and the application of lime to adjust the pH of acidic soils. However, the beneficial effects of these improvements were not always sustained for long. Soils that have strong limiting factors easily return to being problematic soils. For example, drained peat can shrink and decompose, and then, land subsidence can occur; after subsidence, the groundwater table can rise again, which returns the field to a state of poor drainage. Another problem is that iron compounds stick and deposit to drainage pipes in fields with poor drainage. Plowing gradually decreases the soil organic matter content and degrades soil structures, and large machines destroy macropores in soil and soil compactness increases in clay-rich soils. Additionally, in recent years, the merging of small fields into large ones has caused concerns about the problem of field heterogeneity.

1.3.3 Fertilization Standards

The fertilization standards were developed following the Fundamental Soil Survey for Soil Fertility for each prefecture in Japan. The fertilization standards are defined as the amount of fertilizer that achieves the target yield without causing environmental issues. The fertilization standards indicate the appropriate amount of fertilizer based on soil diagnosis referring to soil chemical properties, soil type, crop management history, and target crop. If a soil is diagnosed to be non-usable, it should be conditioned before farming starts.

The Organisation for Economic Co-operation and Development (OECD 2008) reported that the amount of excess fertilizer nitrogen in Japan (calculated as nitrogen fertilizer input minus nitrogen output by harvest) was 174 kg ha^{-1} between 2000 and 2004, the second highest value globally behind the Republic of Korea. However, only 5% of measurement points showed higher levels of nitrate-nitrogen than the drinking water standard (NO_3^--N < 10 mg N L^{-1}) in Japan, while that value was 20% in Italy, Korea, Denmark, Belgium, and the Netherlands. Although Japan uses excessive amounts of nitrogen fertilizer, the groundwater tends to be less polluted. The possible reasons for this are that cultivated land occupies a relatively small portion of land (12%), precipitation in the growing period is high, and 33.1% of cultivated land is paddy field. The pollution of groundwater by NO_3^--N is a serious problem around the world. The proportion of points exceeding drinking water standards is more than 10% in Europe and USA. Consequently, organic farming—that is, farming that does not use chemical fertilizers—is promoted. However, the share of organic farming has remained low in Japan, at only 0.2% of cultivated land area.

Nevertheless, the excessive application of organic materials can cause excessive nitrogen accumulation, as with chemical fertilization. Efforts in environmentally friendly agriculture in Japan started from negotiations of the "agriculture and environment" relationship agreed in the Uruguay Round of GATT in 1993, after undergoing "sustainable development" proposed by the World Commission on Environment and Development (Brundtland Commission) in 1987.

The Japanese government defined environmentally friendly agriculture as "agricultural production practice that has a low impact on the environment with reduced chemical fertilizers and pesticides, by taking measures such as soil conditioning ('*Tsuchi-Zukuri*' in Japanese) and emphasizing harmony of the environment with agriculture and fully understanding the material cycling function of agriculture."

In 1999, the Act on Promotion of Introduction of Sustainable Agricultural Production Practices and the Act on the Appropriate Treatment and Promotion of Utilization of Livestock Manure were enacted. The fertilization standards were defined following these acts. For example, the Hokkaido Fertilizer Recommendations 2015 is now used in Hokkaido (see Chapter 5).

References

Fifth Committee for Soil Classification and Nomenclature of Japanese Society of Pedology (2017) Soil classification system of Japan. (in Japanese)

Forestry Agency in Ministry of Agriculture, Forestry and Fisheries (2018). Forestry coverage and planted forest coverage data at 2012/3/31. (in Japanese) http://www.rinya.maff.go.jp/j/keikaku/genkyou/h24/1.html

Ministry of Agriculture, Forestry and Fisheries, (2018a). Statistics of crops. (in Japanese) http://www.maff.go.jp/j/tokei/kouhyou/sakumotu/

Ministry of Agriculture, Forestry and Fisheries. (2018b). Livestock statistics. (in Japanese) https://www.e-stat.go.jp/stat-search/files?page=1&layout=datalist&toukei=00500222&tstat=000001015614&cycle=7&year=20160&month=0&tclass1=000001020206&tclass2=000001088855

Nira R (2013) A new frame work for study of irrigated paddy rice and upland crops rotation farming and its relation to soil and plant nutrition science. 4 The declines of soil fertility of drained paddy fields under crops rotation and its fertility management. Japanese J Soil Sci Plant Nutr 84:487–492 (in Japanese)

Oda K, Miwa E, Iwamoto A (1987) Database of representative soil profiles surveyed in fundamental soil survey for soil fertility. Japanese J Soil Sci Plant Nutr 58:112–131 (in Japanese)

OECD (2008) Environmental performance of agriculture in OECD countries since 1990. p 575. OECD Publication, Paris

Soil genesis and classification laboratory, National Institute for Agro-Environmental Sciences (1996) typology and regional characteristics of clay mineral composition in lowland soil in Japan. http://www.naro.affrc.go.jp/archive/niaes/sinfo/result/result13/result13_02.html. National Institute for Agro-Environmental Sciences, Tsukuba. (in Japanese)

Part II

General Features

2 Soil-Forming Factors

Kenji Tamura, Hideki Miura, Shinji Kaneko, Tetsuya Sano, and Hideo Kubotera

Abstract

Because the current climate zone of Japan ranges from subtropical to subarctic, soil temperature regime also varies from hyperthermic to frigid. The annual precipitation ranges from less than 1000 to over 2500 mm. The Pleistocene climate also affected the development of soils of Japan through formation and deposition of soil parent materials. The outline of the landforms has mainly been formed by the subduction of the tectonic plates in the vicinity, leading to a long and complicated history of the geology. The Japanese Islands constitute the most prominent volcanic area in the world with more than 111 active volcanoes. The plains are divided into the coastal plains and the inland mountain basin. A wide range of wind-blown dust, including *Kosa*, has been added to various sediments. The vegetation is characterized by a large number of species and a high percentage of endemic species due to a large variation in landscape, geology and climate. The majority of mountainous, hilly and volcanic areas are covered with forests. Most agricultural lands in cities situated in flat areas and suburban alluvial lowlands are mainly used for paddy fields, and higher terraces are used for upland cultivation.

K. Tamura (✉)
Faculty of Life and Environmental Sciences, University of Tsukuba, Tsukuba City, Ibaraki Prefecture, Japan
e-mail: tamura.kenji.gn@u.tsukuba.ac.jp

H. Miura
National Institute of Polar Research, Tachikawa City, Tokyo, Japan

S. Kaneko
Kansai Research Center, Forestry and Forest Products Research Center, Kyoto City, Kyoto Prefecture, Japan

T. Sano
Faculty of Engineering, Tohoku Institute of Technology, Sendai City, Miyagi Prefecture, Japan

H. Kubotera
Institute for Agro-Environmental Sciences, NARO, Tsukuba City, Ibaraki Prefecture, Japan

R. Hatano et al. (eds.), *The Soils of Japan*, World Soils Book Series,
https://doi.org/10.1007/978-981-15-8229-5_2

Keywords

Agriculture • Climate • Geology • Geomorphology • Landscape • Vegetation

2.1 Climate

2.1.1 Current Climate

Climate influences soil development through differences in mean annual and seasonal temperature and moisture and through extremes in these parameters (Bockheim and Hartemink 2017). Matsui (1988) emphasized that the climate formed various temperature and moisture regimes in soils which affect their physical, chemical and biological processes. Soils distributed according to climate zone are called zonal soils (Matsui 2001). The climate zone of Japan ranges from subtropical to subarctic. Therefore, soil is distributed from podzols to red–yellow soils. Podzols are distributed in northern Hokkaido (Fig. 2.1) and in the subalpine zone of Central Japan, whereas red–yellow soils are distributed mainly in the Nansei Islands (Fig. 2.2).

The mean annual air temperature in Japan varies considerably, as shown in Fig. 2.3. In northern Hokkaido, the average annual temperature is less than 5 °C, while in the Southwest Islands it is over 20 °C. In the Köppen climate classification, the Hokkaido region belongs to the subarctic zone, while the Honshu, Shikoku and Kyushu regions belong to the temperate zone.

Figure 2.4 shows the annual precipitation in Japan. In the eastern part of Hokkaido, the annual precipitation is less than 1000 mm, while in the Hokuriku, Shikoku and Kyushu regions it is over 2500 mm. Since Japan experiences a humid climate, soil nutrients are eluted with precipitation and subsequently transported downward. As a result, the base saturation of the soil is low and the soil is acidic. On the Sea of Japan side, large amounts of snow fall in winter,

Fig. 2.1 Podzol developed on sand dune in northern Hokkaido (Figure supplied by Kenji Tamura)

while on the Pacific side it is sunny in winter. Furthermore, rainfall patterns are different on both sides.

Kyuma (1985) calculated the soil temperature regimes of soils in various parts of Japan using the 50-cm soil temperature data of Japan's Ministry of Agriculture, Forestry and Fisheries (MAFF) and plotted the soil temperature regimes on the map of Japan. Additionally, Takata et al. (2011) drew a detailed soil temperature regime map based on the soil temperature data of two publications (Fig. 2.5) and divided the Japanese soil temperature regime (Soil Survey Staff 2014) into four classes, frigid, mesic, thermic and hyperthermic.

The frigid soil temperature regime is distributed in the eastern part of Hokkaido. This region is the coldest region in Japan and is affected by the Okhotsk Current. Podzols are distributed even in low-elevation areas. The mesic soil temperature regime area is mainly distributed in eastern Japan, while the thermic soil temperature regime area is distributed in western Japan. The soil temperature regime of the Nansei Islands is hyperthermic.

2.1.2 The Changing Climate

The Japan Meteorological Agency reported that the annual average temperature in Japan is predicted to increase by 4.5 °C at the end of twenty-first century and that it will rise significantly across the country (Fig. 2.6) (Japan Meteorological Agency 2017). The global annual average temperature rise is about 3.7 °C by the end of twenty-first century; however, the temperature rise near Japan is expected to be larger than the world average because temperature is projected to rise more at relatively high latitudes. Temperature is projected to rise by 4.9 °C on the Pacific Ocean side of Northern Japan and by 3.3 °C in the Nansei Islands. The average temperature for each season shows the same trend as the annual average temperature. However, the temperature rise is greater in summer than in the winter, when melting sea ice and snowfall mitigate the temperature rise.

A rise of 4 to 5 °C in temperature leads to a change in the climate zone; a subarctic climate turns into a temperate climate, and a temperate climate turns into a subtropical climate. The soil temperature rises with increasing air temperature, and it is expected that the soil temperature will also greatly affect soil formation in the future. For example, the distribution of soils may change in the future. By examining the relationship between past climate change and pedogenesis, the influence of global warming on soil distribution may be clarified.

2.1.3 Past Climate

The soils of Japan have been affected not only by the current Holocene climate but also by the Pleistocene climate during and after the formational or depositional era of soil parent materials. Therefore, it is very important to know the past climate situation (paleoclimate) to understand soil-forming processes. Given that the oldest plains of the Japanese Islands were formed in the Quaternary/Pleistocene around 630,000 years ago, the Japanese soils on the plain were formed sometime after this. In this section, we first describe the outline of past global climate change, mainly after the Middle Pleistocene (about 770,000 years ago), and then summarize the paleoclimate of the Japanese Islands from the last interglacial age (about 125,000 years ago) to the present (Holocene/postglacial period) corresponding to the last glacial–interglacial cycle.

(1) Outline of global climate change since the Middle Pleistocene

Research on past global climate change and sea-level change originally commenced using the chronology and stratigraphy of the regional terrestrial geology/geomorphology of glacial

Fig. 2.2 Red–yellow soil distributed in Nansei Islands (Figure supplied by Kenji Tamura)

and marine landforms around the world. However, since the mid-1950s, the oxygen isotope ratios of microfossils in marine sediments and of glacial ice have proved to be effective as indicators of global climate change in the past (Emiliani 1955; Dansgarrd 1964). In particular, the variation of the oxygen isotope ratio ($\delta^{18}O$) of benthic foraminiferal shells in marine sediment is related to changes in the volume of the continental ice sheet, corresponding to increases and decreases in the amount of seawater. Since benthic foraminiferal $\delta^{18}O$ values show almost the same variation pattern anywhere on Earth, it became clear that the climate change that caused the change of the volume of the continental ice sheet occurred on a global scale (Schackleton 1969; Lisiecki and Raymo 2005; Railsback et al. 2015; Fig. 2.7).

As described in Sect. 2.2, in the plains of the Japanese Islands the oldest marine terraces were formed in marine isotope stage (MIS) 13 (about 500,000 years ago) and the oldest fluvial terraces were probably formed in MIS 16–17 (about 690,000–630,000 years ago) (Machida and Chinzei 2001). Therefore, the landforms and soils of the Japanese plains are considered to have formed sometime after this time.

(2) **Climate change after the last interglacial period in the Japanese Islands**

The regional paleoclimate has been reconstructed primarily from plant fossils (pollen, seeds, fruit, wood, phytolith, diatoms, etc.), fossil periglacial phenomena and minerals in the various terrestrial and marine sediments and landforms. These records have been compared with the timing and magnitude of global climate change (Fig. 2.8). The period from MIS 5e (the last interglacial period) to MIS 4–2 (the last glacial period) is the era for which detailed past global climatic records have been obtained, even in the Pleistocene. Therefore, paleoenvironmental reconstruction after the last interglacial period is also important in order to estimate the paleoenvironment of older glacial and interglacial periods. However, it is not easy to quantitatively reconstruct the specific temperature and precipitation for this period.

Fig. 2.3 Mean annual air temperature pattern (°C) in Japan (mesh average value estimated from normal year value from 1981 to 2010. *Source* Japan Meteorological Agency Web site (http://www.data.jma.go.jp/obd/stats/etrn/view/atlas.html)

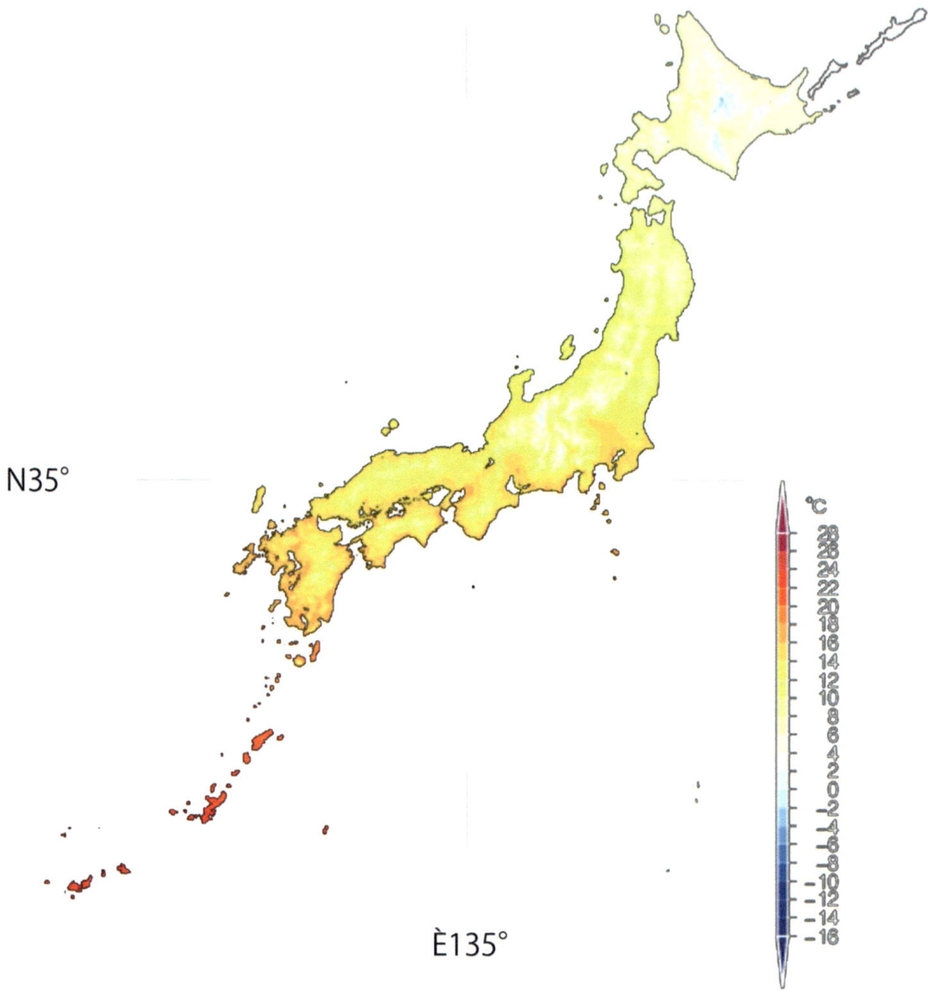

It is important to know whether the climate of the last interglacial period (MIS 5e) was milder than the current interglacial period (MIS 1) in order to consider the history of soil formation, such as that of relic red soils, and the shift of past climate zones around the Japanese Islands. In general, the distribution of plants is not determined by climatic conditions alone, but the comparative study of two interglacial periods—the last interglacial period (MIS 5e) and the current interglacial period (MIS (1)—using vegetation reconstruction obtained from plant fossils would provide some insights into climate conditions. In a study of pollen composition obtained from marine sediment cores off Kashima in the Pacific Ocean (MD 01-2421: 36° 01.40′ N, 141° 46.80′ E; water depth 2224 m; about 100 km off the east coast of the Honshu Island, Central Japan), the temperature index value (Tp: pollen temperature index) of pollen from the MIS 5e level was found to be lower than that of pollen from the MIS 1 level (Igarashi and Oba 2006). On the other hand, Okuda et al. (2010) applied the modern analog method (calculation based on a data set of seasonal changes in temperature and precipitation based on a pollen fossil database: Nakagawa et al. 2002) to the analysis of pollen from the Lake Biwa core (35° 13.28′ N, 136°00.63′ E; 1422 m deep from the lake bottom to the basement; central part of Lake Biwa, the largest and oldest freshwater lake in Japan, located in Western Honshu) and reported that the average temperature of the MIS 5e level was not much different from that of the MIS 1 and MIS 11 levels, but that the yearly difference in temperature was large. Additionally, Tsuji et al. (1984) studied sediments from the last interglacial period (MIS 5e) collected in the southern part of the Kanto region, Central Japan, and reported that the fossils of seeds and fruits of deciduous broad-leaved trees, which are now distributed in the coastal region of southwestern Japan and the central and southern parts of China, were present. They also reported that

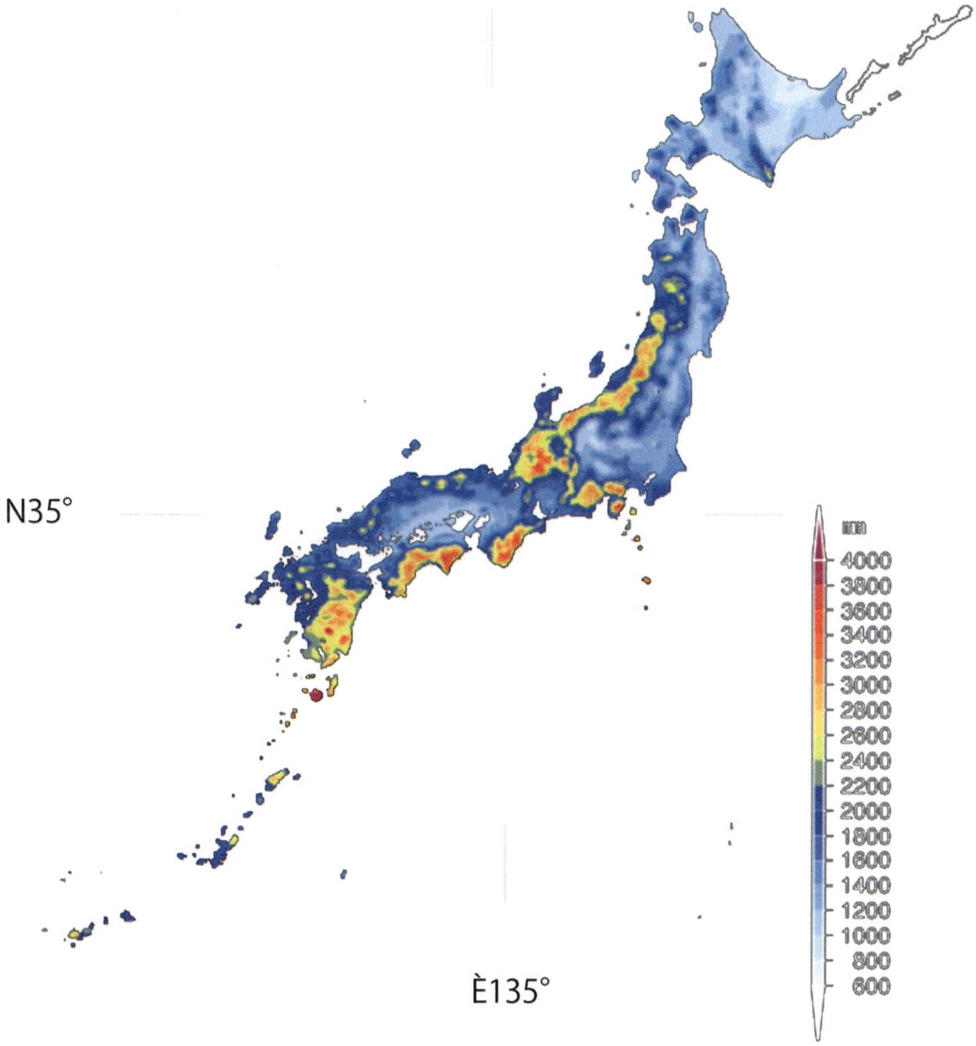

Fig. 2.4 Annual precipitation (mm) pattern in Japan (mesh average value estimated from normal year value from 1981 to 2010. *Source* Japan Meteorological Agency Web site (http://www.data.jma.go.jp/obd/stats/etrn/view/atlas.html)

the pollen of the evergreen broad-leaved trees that dominated in the Holocene pollen group was rare and that the pollen of deciduous broad-leaved trees was predominant. Moreover, Sakaguchi and Okumura (1986) showed that the climate during the last interglacial period (MIS 5e) was slightly colder than the present climate (MIS1) in the eastern part of Hokkaido, Northern Japan, based on an analysis of pollen from peat sediments from the last interglacial period.

In the last glacial period (MIS 4 to 2), the ice sheet expanded not only in polar regions but also in the high- and midlatitude regions of North America and Northern Europe, and the sea level decreased by about 120–130 m in the last glacial maximum (MIS 2). As the sea level declined, the East China Sea almost dried up and the Kuroshio Current flowed to the east on the south side of the Ryukyu Islands. For this reason, the Sea of Japan was in a state close to that of a lake, and the Japanese Islands formed the edge of the east continent, a situation similar to continental vegetation. In this period, coniferous trees of the Pinaceae (pine family) dominated; forests with mixed deciduous broad-leaved trees were established in Honshu, and taiga (boreal forest) of *Larix gmelinii* and *Pinus pumila* with grasslands spread in Hokkaido (Fig. 2.9). The altitude of the snow line (the average position of the boundary between snow- or ice-covered areas and areas not covered by snow or ice throughout the year) declined, and mountain glaciers formed in the central region of the Japanese Islands and the Hidaka Mountains of Hokkaido (Figs. 2, 3 and 4: Iwata and Koaze, 2001). In this period, tree lines also greatly declined—in the Chubu region to around 1000 to 1200 meters above sea level and in Hokkaido to about 200 m or less above sea level (Figs. 2.9 and 2.10)—and permafrost formed in the northern and eastern part of the plains of Hokkaido (as determined on the basis of fossil periglacial phenomena (Miura and Hirakawa 1995)). In areas above the tree line, grasslands and bare ground spread, and in such places the movement of surface material progressed via periglacial processes.

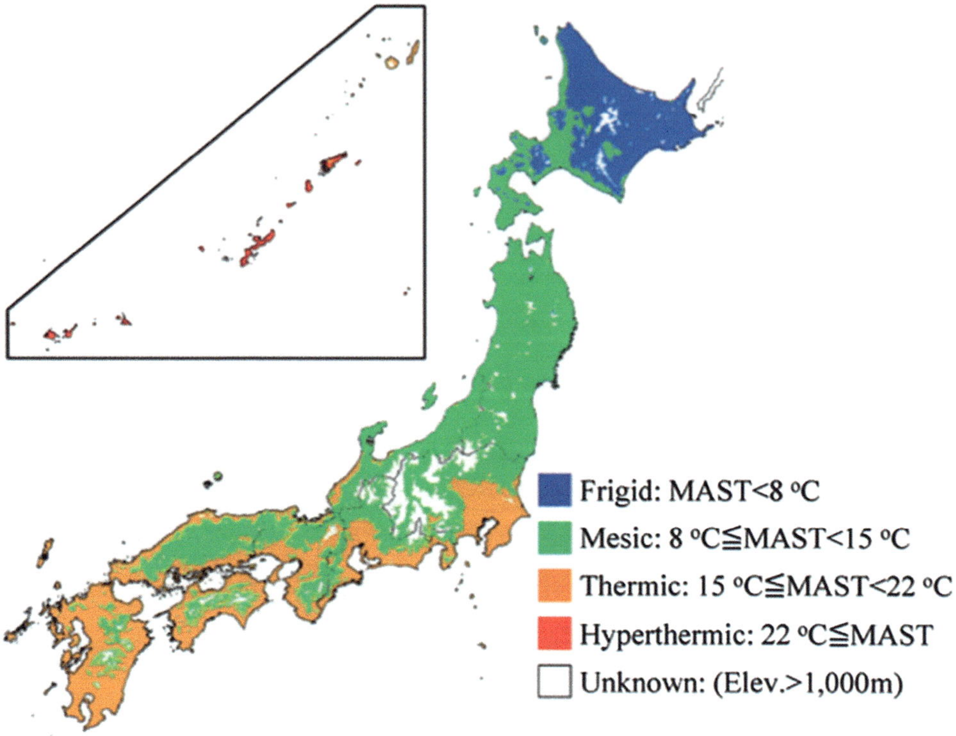

Fig. 2.5 Soil temperature regime map of Japan (Takata et al. 2011), Copyright 2011, with permission from Taylor and Francis Group

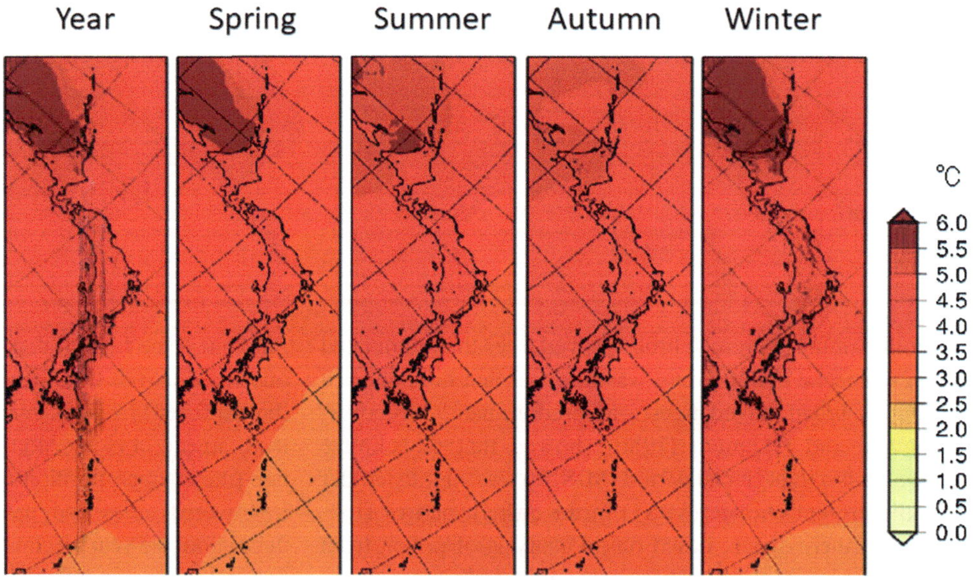

Fig. 2.6 Forecast of the temperature in Japan. It is indicated as the difference (°C) between the average value from 1980 to 2010 and the end of the twenty-first century. *Source* Japan Meteorological Agency Web site (http://www.data.jma.go.jp/cpdinfo/GWP/Vol9/pdf/02.pdf)

After the end of the last glacial period, sea level rose due to the melting of the continental ice sheet and outside seawater again flowed into the Sea of Japan. The change in the $\delta^{18}O$ values of planktonic foraminifera measured in marine sediment cores from the Sea of Japan revealed that the Oyashio Current also known as Okhotsk Current flowed into the Sea of Japan since 15,500 years ago (Oba et al. 1991; Okumura et al. 1996; Matsui et al. 1998). The Tsushima Current began to flow into the Sea of Japan at full strength around 10,000 years ago or later (Oba and Akasaka 1990). After the last glacial maximum (MIS 2), the lowland vegetation of the Japanese Islands from Honshu to Hokkaido changed, largely around 14,000 years ago and between 12,000 and 11,000 years ago in conformity with the arrival of the Kuroshio Current and the Tsushima Current (Tsuji 1997). After about 9000 years ago, the influence of the

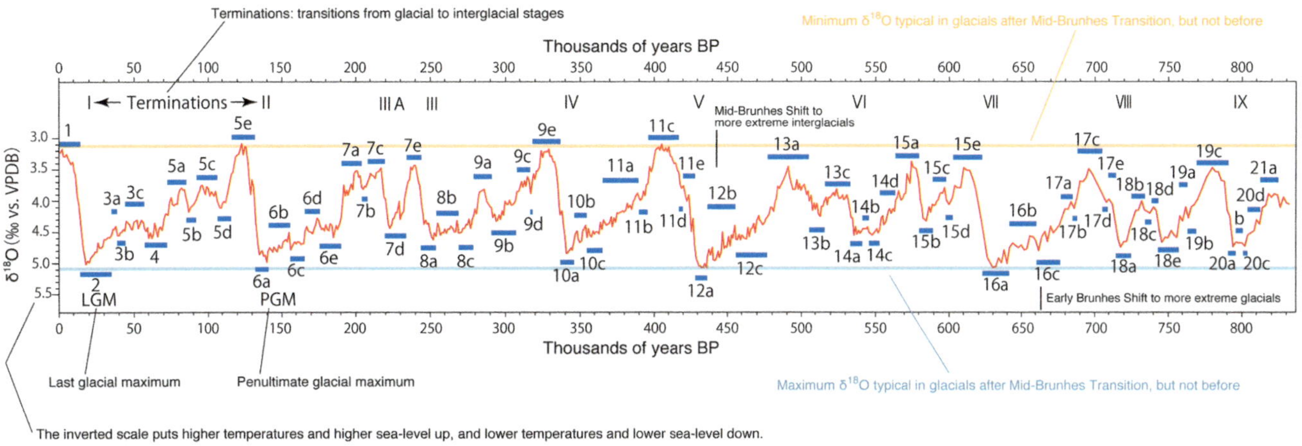

Fig. 2.7 Marine isotope stages and substages after ca. 800 ka (modified from Railsback (2015), Copyright 2015 Railsback, with permission from Railsback). Each glacial period and interglacial period are numbered as the marine oxygen isotope stage (MIS). The odd numbered stages other than MIS 3 indicate the interglacial period, and the even numbered stages indicate the glacial period. The current interglacial period (Holocene/postglacial period) is represented by MIS1, and the interglacial period including the period from 80,000 years ago to 125,000 years ago is displayed at MIS 5, respectively. Each stage is divided into substages which are further alphabetized. The levels of isotopic ratios at the peak period of the interglacial and glacial periods differ slightly depending on the timing, and the details of the variation are also different. In particular, the temperate interglacial period comparable to the Holocene (MIS 1) occurred repeatedly 4 times (MIS 11, 9e, 7 and 5e) approximately every 100,000 years from around 430,000 years ago

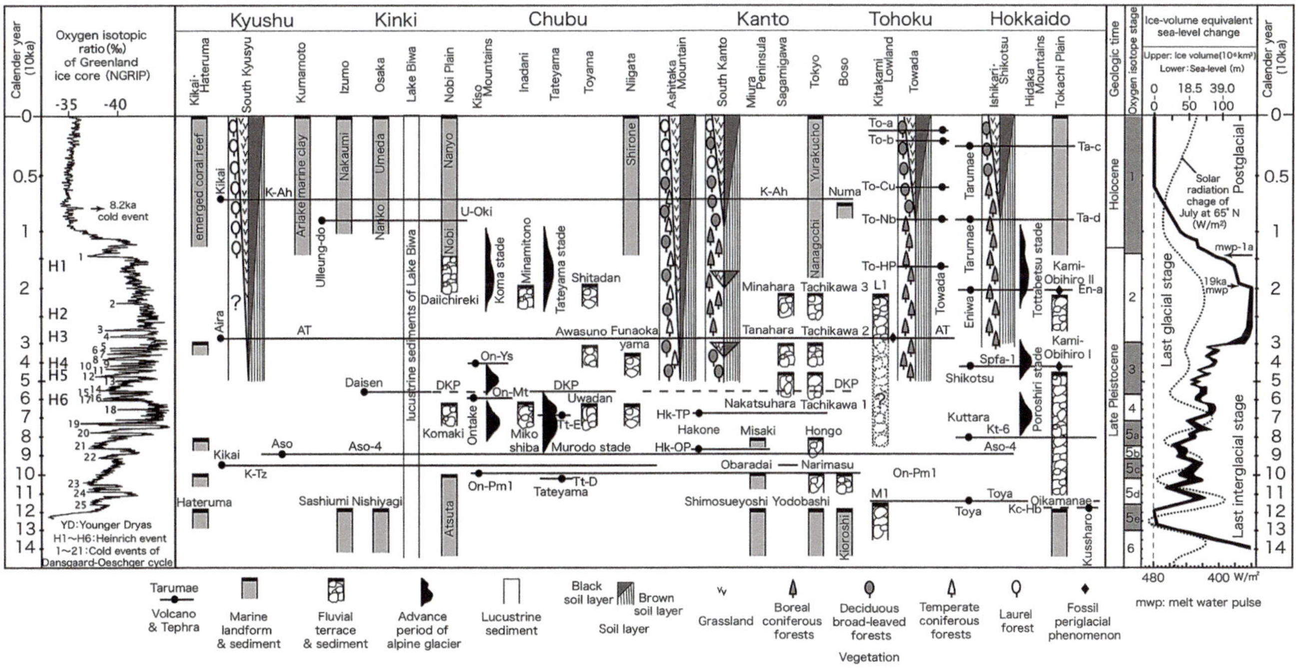

Fig. 2.8 Chronology of various paleoenvironmental phenomena of the Japanese Islands since last interglacial with the oxygen isotope of Greenland ice core and ice volume changes (modified from Miura (2017), Copyright 2017 Asakura Publishing Co., Ltd., with permission from Asakura Publishing Co., Ltd.)

ocean current further strengthened, and the evergreen forests in the western part of the Chubu region, and the beech and *Quercus* (oak) forests in Northern Japan, became prominent.

Through the oxygen isotope analysis of fine quartz in the soil, the past wind system and the transition of monsoon activity are debated. During the last glacial maximum (MIS 2), many fine particles originating from Precambrian rocks in the northern part of the Asian continent were transported to the area north of the Seto Inland Sea in Japan; however, in regions to the south of the Seto Inland Sea, particles derived

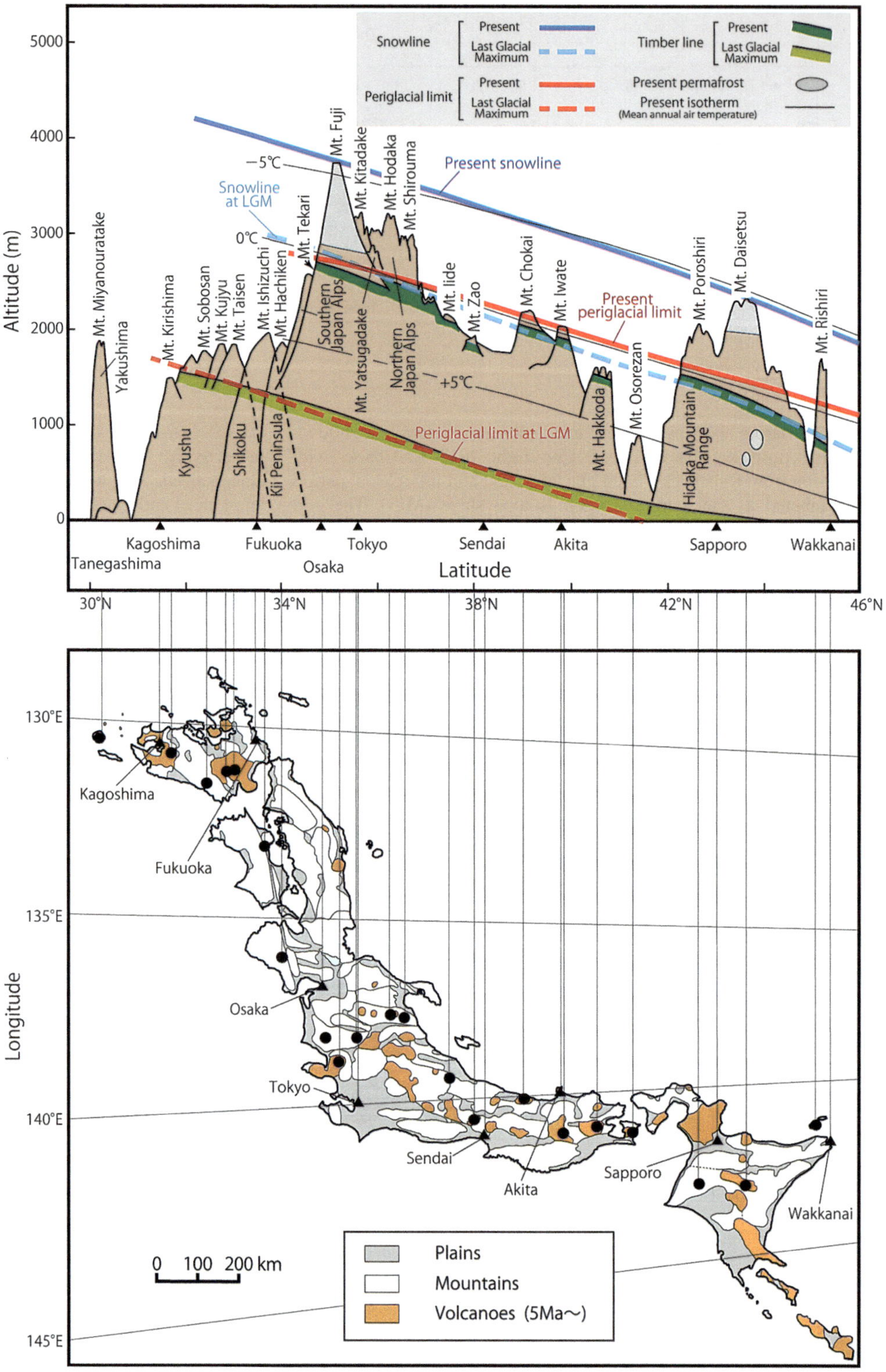

Fig. 2.9 Vertical distribution of snow line, periglacial environment, timber line at last glacial (MIS4-2) and present (MIS1) (upper figure: modified from Iwata and Koaze (2001) which partially revised Kaizuka and Chinzei (1995), lower figure: modified from National Astronomical Observatory of Japan (2018))

from arid parts of China were transported. In the postglacial period (MIS 1), the total amount of particles decreased in every region, and wind dust particles that were blown from dry areas of China made up the main proportion. This is attributed to the East Asian monsoon wind system moving between the glacial period and the postglacial period, and the

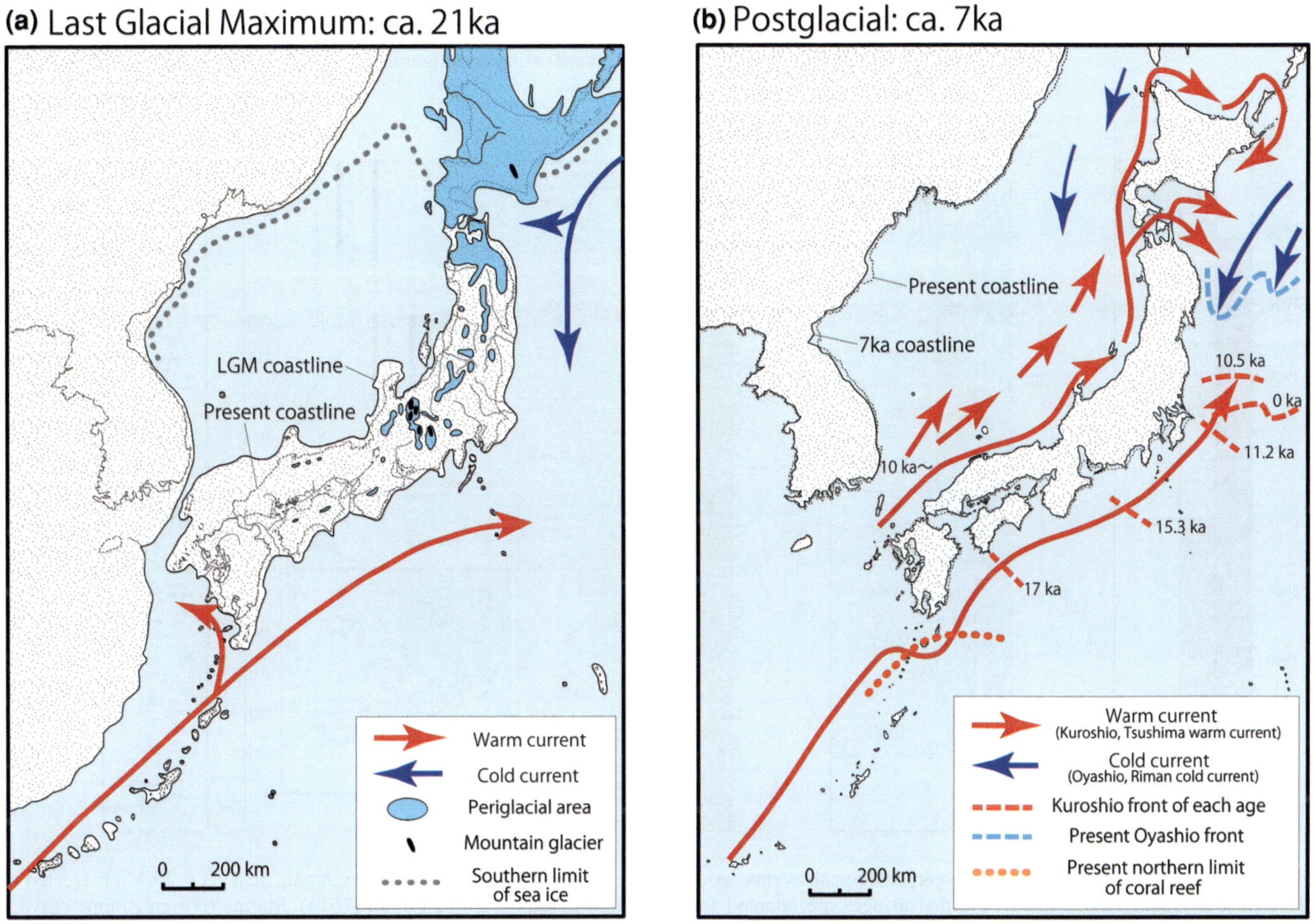

Fig. 2.10 Paleogeographical maps of the Japanese Islands at last glacial maximum (ca. 21 ka) and postglacial (ca. 7 ka) (modified from Machida and Chinzei (2001) which partially revised Kaizuka (1977) and Japan Association for Quaternary Research (1987)). The location of the ocean current is based on Oba et al. (1991), Okumura et al. (1996) and Matsui et al. (1998)

position of the polar front moving to the north and south (Toyoda and Naruse 2002).

2.2 Geology and Geomorphology

In order to clearly understand the origin of the parent rocks and the parent materials of soils, and their formation time and environments in the Japanese Islands, it is necessary to know the contents and the ages of the geology and the landforms.

The Japanese Islands consist of the four main islands of Hokkaido, Honshu, Shikoku and Kyushu, as well as the adjacent Izu-Ogasawara Islands, which extend southward from central Honshu, and the Nansei Islands, which extend from Kyushu to Taiwan. The geographical distribution of these major islands ranges from 24°27′ to 45°33′ N latitude and from 148°46’ to 122°56’ E longitude. The territory of Japan extends to Minami-Tori-shima (24°17′N, 153°59′E) in the far east from Honshu and Okinotorishima (20°25′N, 136° 05′E) far off the southern coast of Honshu. The Japanese Islands developed a high-relief mountain range, which belongs to the mobile belt where seismic activity, volcanic activity and crustal deformation are very active. A temperate humid monsoon climate dominates around the Japanese Islands, and the rainfall intensity at the time of typhoon passage and the rainy season is high; therefore, the topography change due to erosion and deposition is fast and intense.

In this section, an outline is given of the formation of the Japanese Islands as a macro-landform (on a horizontal scale of about 100 to 1000 km) associated with basement rocks/strata. Then, an explanation is given of the formation of volcanoes, mountains and plains as meso- to mini-landforms (on scales of about 1 to 100 km) with their sediments. Finally, the characteristics of the landform-covered sediments in the Japanese Islands are described, such as air-fall sediments (pyroclastics/ tephras and eolian dust) and sediments of terrestrial plant origin. The reader is referred to Fig. 2.11 and also Fig. 2.7 in Sect. 2.1.3 for the names and timescales of the geological ages in this section.

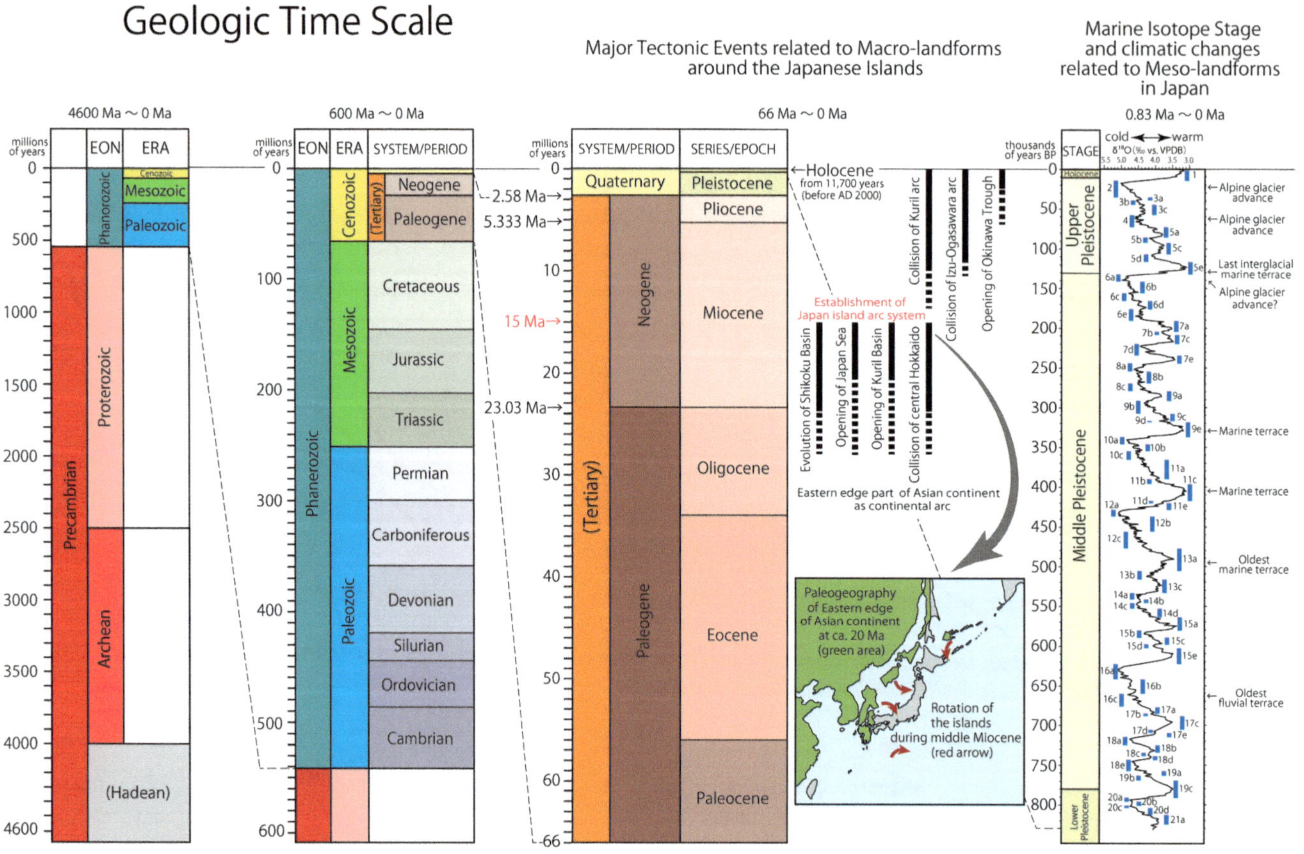

Fig. 2.11 Geological timescale and major geological events around the Japanese Islands. Data source: Numerical ages are adapted from International Chronostratigraphic Chart by International Commission on Stratigraphy (2018). Paleogeographic map at ca. 20 Ma is created by author based on Kimura et al. (2014). Marine oxygen isotope curve is based on Lisiecki and Raymo (2005)

2.2.1 Macro-landforms and Basement Rocks/Strata Around the Japanese Islands

(1) Macro-landforms formed by plate subduction

At least four tectonic plates are in the vicinity of the Japanese Islands: Pacific Plate, Philippine Sea Plate, North American Plate (Okhotsk Plate) and Eurasian Plate (Amurian Plate) (Fig. 2.12). The outline of the landforms of the Japanese Islands has mainly been formed by the subduction of these plates (Ueda and Sugimura 1970).

The planar shape of the Japanese Islands protrudes in an arc from the Asian continent toward the Pacific Ocean and is termed an "island arc" based on its shape. From the viewpoint of the land continuity, the island arc extending from the Japanese Islands is divided into the Kuril arc from the Kamchatka Peninsula toward Hokkaido, the Honshu arc from Hokkaido toward Honshu, the Ryukyu arc from Kyushu toward Taiwan, the Izu-Ogasawara arc from Central Honshu toward the Ogasawara Islands, and the Mariana arc further south. Based on the oceanic trenches, volcanic distribution and seismic activity, the Honshu arc is further subdivided into the Northeast Japan arc and Southwest Japan arc (Kaizuka 1988; Matsuda and Yoshikawa 2001; Fig. 2.13).

In the offshore areas to the east and south of the Japanese Islands, the Kuril–Kamchatka Trench, the Japan Trench, the Mariana Trench and the Ryukyu Trench exceed depths of 6000 m; offshore of Southwest Japan from Honshu to Kyushu, three troughs similar to the trenches are distributed—the Sagami Trough, the Suruga Trough and the Nankai Trough—although their depths are slightly shallower than the trenches. These trenches and troughs are distributed along the Pacific Ocean side of the island arc, and together the island arc and trenches are referred to as an "island arc–trench system". This island arc–trench system, a typical example of a recent mobile belt, features not only a large topographical rise caused by crustal deformation but also active seismic and volcanic activity (Ueda and Sugimura 1970; Fig. 2.12). The arrangement of the macro-landforms and the basement rocks/strata along the island arc–trench system of the Japanese Islands can be understood as unified phenomena in the subduction zone of the plate.

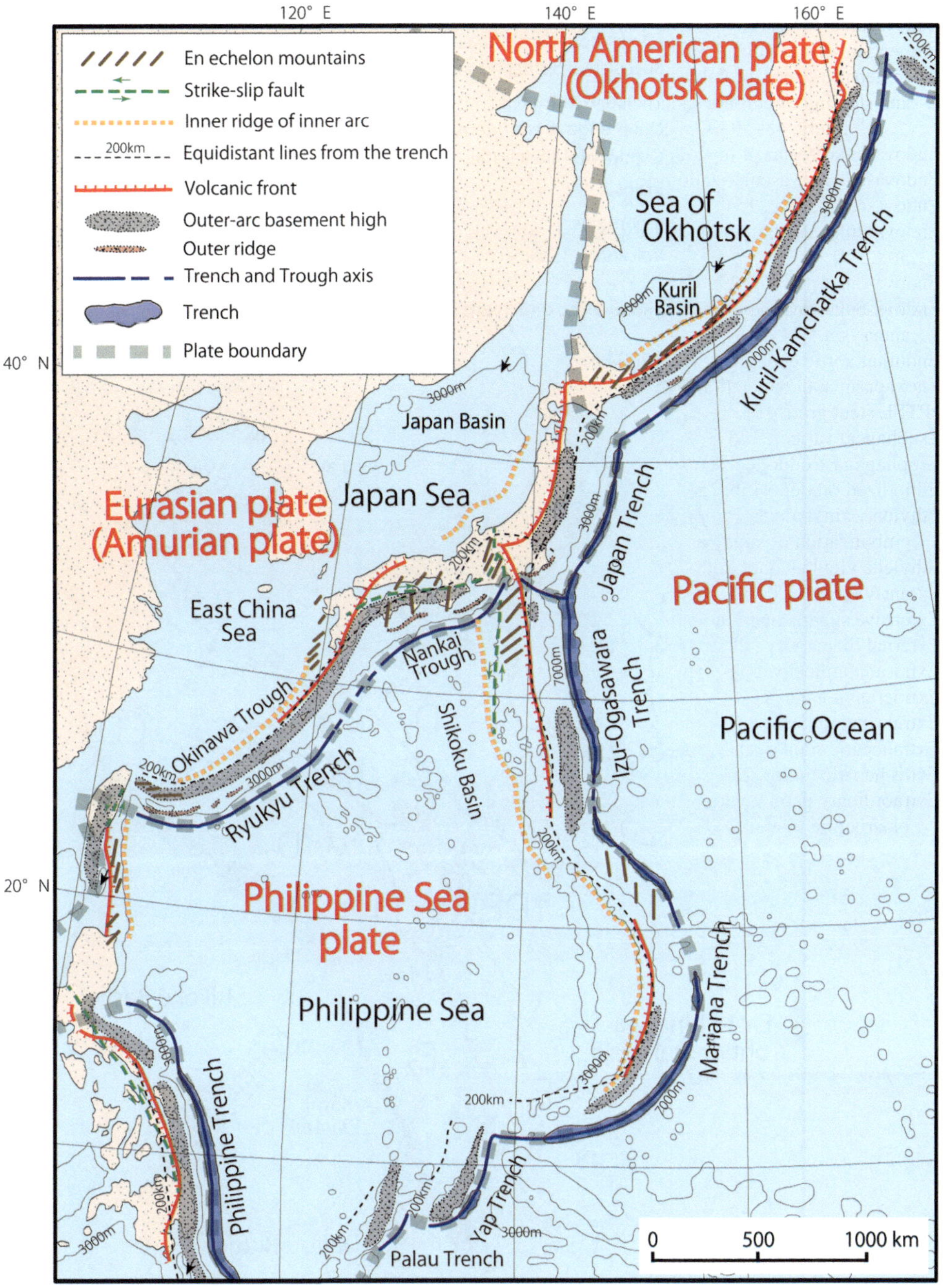

Fig. 2.12 Macro-landforms and of island arc and plate placement around the Japanese Islands (modified from Kaizuka (1972), Copyright 1972 Iwanami Shoten, Publishers., with permission from Iwanami Shoten, Publishers.)

Between the Japanese Islands and the Asian continent are marine areas, consisting of the Sea of Okhotsk, the Sea of Japan, the East China Sea and other bodies, with submarine basins and continental shelves with water depths shallower than 4000 m (Fig. 2.12). Since these marine areas are located on the margin of the Pacific Ocean, they are referred to as "marginal seas". Their submarine topography is called "back-arc basin", in the sense of the basin behind the island arc. There is no deep submarine topography such as trenches between the Japanese Islands and these marginal seas.

The terrestrial landforms of the island arc are characterized by active volcanoes that have erupted within the past 10,000 years or have active fumarolic activity at present, and/or Quaternary volcanoes that have erupted within the past 2.58 million years. Many Quaternary volcanoes are distributed in strips along the direction of extension of the island arc, forming a volcanic zone. Volcanoes are distributed from the central part of the island arc to the Sea of Japan side (inside). The distribution limit of the volcanoes to the Pacific Ocean side is clear, and there is no active volcano on the Pacific side from the limit line. The limit line of the sea side is called the "volcanic front" (Fig. 2.12).

The island arc is divided by the volcanic front into the "non-volcanic outer arc" ("fore-arc" or "outer zone": from

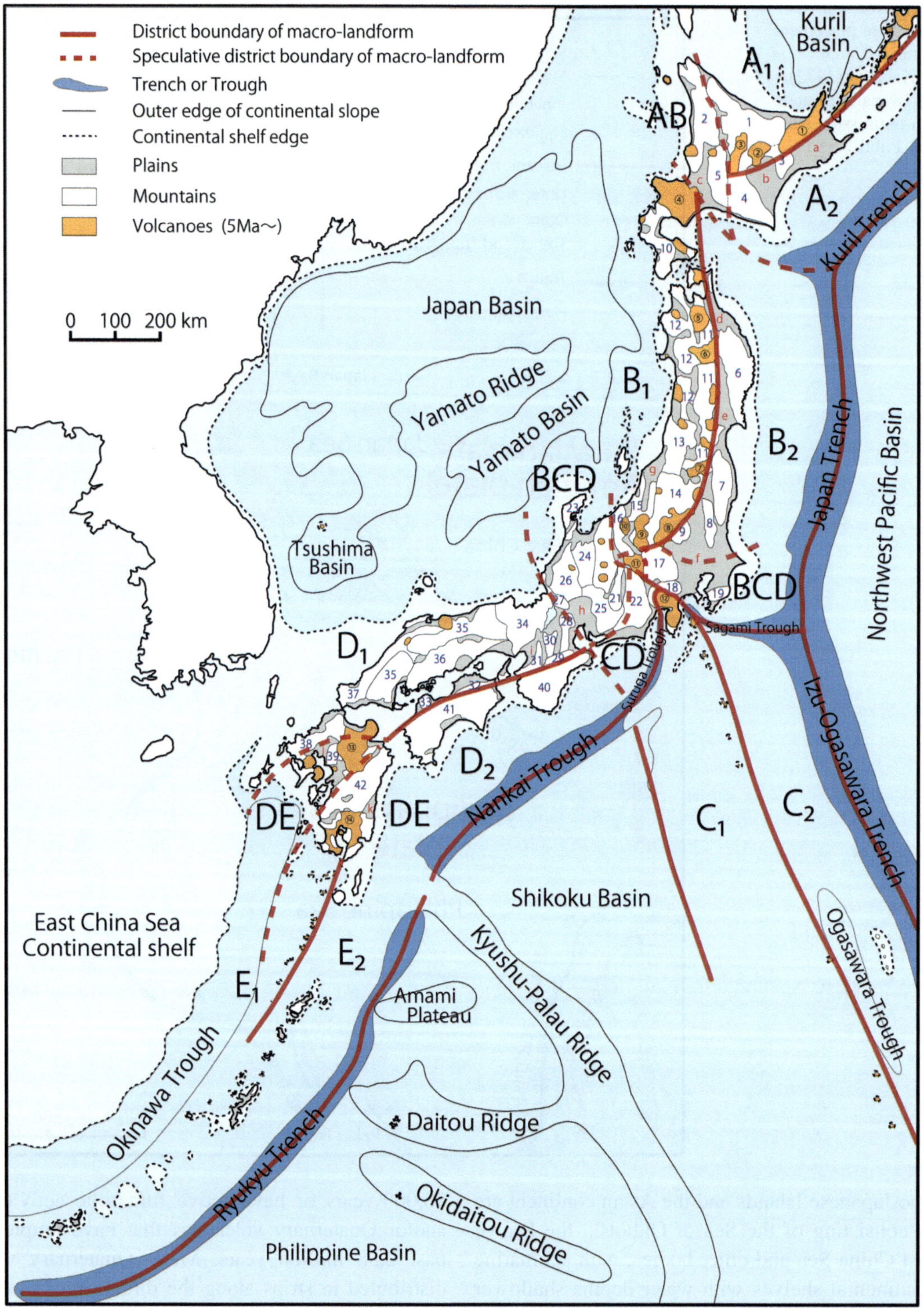
District boundary of macro-landform
Speculative district boundary of macro-landform
Trench or Trough
Outer edge of continental slope
Continental shelf edge
Plains
Mountains
Volcanoes (5Ma~)
0 100 200 km
Kuril Basin
Japan Basin
Yamato Ridge
Yamato Basin
Tsushima Basin
Kuril Trench
Japan Trench
Northwest Pacific Basin
Sagami Trough
Suruga Trough
Nankai Trough
Izu-Ogasawara Trench
Ogasawara Trough
Shikoku Basin
Kyushu-Palau Ridge
Amami Plateau
Daitou Ridge
Okidaitou Ridge
Philippine Basin
Ryukyu Trench
Okinawa Trough
East China Sea Continental shelf
AB
BCD
CD
DE

◀ **Fig. 2.13** Japanese Island arc and landform classification of the Japanese Islands (modified from Kaizuka (1988) and National Astronomical Observatory of Japan (2018)). Distinct boundary of macro-landform—A1: Kuril inner arc, A2: Kuril outer arc, AB: overlapping area between Kuril arc and Northeast Japan arc, B1: Northeast Japan inner arc, B2: Northeast Japan outer arc, C1: Izu-Ogasawara inner arc, C2: Izu-Ogasawara outer arc, CD: overlapping area between Northeast Japan arc and Izu-Ogasawara arc, BCD: overlapping area between Northeast Japan arc, Izu-Ogasawara arc and Southwest Japan arc, D1: Southwest Japan inner zone, D2: Southwest Japan outer zone, DE: overlapping area between Southwest Japan arc and Ryukyu arc, E1: Ryukyu inner arc, E2: Ryukyu outer arc. Major volcanoes ① Shirctoko–Akan, ② Shikaribetsu, ③ Taisetsu–Tokachi, ④ Shikotsu-Toya, ⑤ Hakkoda–Towada, ⑥ Hachimantai–Iwate, ⑦ Azuma-Bandai, ⑧ Nikko–Akagi–Haruna, ⑨ Kustasushirane–Asama, ⑩ Myoko, ⑪ Yatsugatake, ⑫ Fuji–Izu, ⑬ Central Kyushu, ⑭ Southern Kyushu. Major plains—a: Konsen Plain and Kushiro Plain, b: Tokachi Plain, c: Ishikari Plain, d: Kamikita Plain, e: Kitakami Basin and Sendai Plain, f: Kanto Plain, g: Echigo Plain, h: Nobi Plain, i: Osaka Plain, j: Chikushi Plain, k: Miyazaki Plain. Major mountains—1: Kitami Mountains, 2: Teshio Mountains, 3: Shiranuka Hills, 4: Hidaka Mountain Range, 5: Yubari Mountains, 6: Kitakami Mountains, 7: Abukuma Mountains, 8: Yamizo Mountains, 9: Ashio Mountains, 10: Oshima Mountains, 11: Ou Mountain Range, 12: Dewa Mountains, 13: Asahi–Iide Mountains, 14: Echigo Mountain Range, 15: Echigo Hills, 16: Nishikubiki–Chikuma Mountains, 17: Kanto Mountains, 18: Tanzawa–Misaka–Tenshu Mountains, 19: Boso–Miura Hills, 20: Hida Mountain Range, 21: Kiso Mountain Range, 22: Akaishi Mountain Range, 23: Noto Hills, 24: Hida Highland, 25: Mino–Mikawa Highlands, 26: Kamo–Mino Mountains, 27: Ibuki Mountains, 28: Suzuka–Nunobiki Mountains, 29: Takami Mountains, 30: Kasagi Mountains, 31: Ikoma–Kongo–Izumi Mountains, 32: Sanuki Mountain Range, 33: Ishizuchi Range, 34: Tanba Highland, 35: Chugoku Mountains, 36: Kibi Highland, 37: Iwami Highland, 38: Chikushi Mountains, 39: Minou–Chikushi Mountains, 40: Kii Mountains, 41: Shikoku Mountains, 42: Kyushu Mountains

the top of the land-side slope of the trench to the volcanic front) and the "volcanic inner arc" ("back-arc" or "inner zone": from the volcanic front to the edge down to the edge of the marginal sea) (Figs. 2.12 and 2.13). However, in the Southwest Japan arc, the volcanic front that distinguishes the fore-arc/outer zone and the back-arc/inner zone is not clear, and the "median tectonic line" is conventionally used to separate the fore-arc/outer zone from the back-arc/inner zone.

Crustal movement, as an endogenetic process, plays an important role in the formation of macro-landforms, such as island arcs with lengths of 500 to 1000 km and widths of 300 to 500 km, and the arrangement of the mountains, basins and plains constituting the island arcs. On the other hand, in the case of meso-landforms and mini-landforms, the speed of the geomorphic change due to exogenetic process such as rivers, waves, glaciers and mass movements often exceeds the speed of crustal movement. For this reason, meso-landforms and mini-landforms are basically formed by exogenetic processes.

(2) **Basement rocks/strata**

The geology of the Japanese Islands, which are located in a plate subduction zone at the boundary between the Eurasian continent and the Pacific Ocean, has a long and complicated history (Fig. 2.11). This is largely divided into two categories: older rocks that formed when Japan was still a part of the continent (Fig. 2.14) and younger rocks/strata that formed after Japan separated as an island arc, bordered by marginal seas such as the Sea of Japan, after the Miocene epoch (Fig. 2.13).

As shown in Fig. 2.14, basement rocks and strata which formed before the birth of the island arc are basically arranged in strips almost parallel to the extending direction of the Japanese Islands. The basement rocks include intrusive granitic rocks.

Sedimentary rocks in the basement of the Japanese Islands are mainly composed of cherts, mudstones, sandstones and the like, coexisting with basaltic volcanic rocks and reef limestones. Based on analysis of Radiolarians of floating microfossils contained in this stratum, its age is mainly from the Late Paleozoic to the Jurassic (Mesozoic), with the latter being especially widely distributed (Fig. 2.14). From Kyushu to the Chubu region, this band structure is distributed almost parallel to the direction of the current island arc; however, in the southern part of the Kanto region and the Tohoku region, through the collision zone with the Izu-Ogasawara arc in central Honshu, this band structure becomes oblique to the direction of the island arc. In Hokkaido, it is believed that the continuing part of the Northeast Japan arc collides with the Kuril arc at almost a right angle.

The Japanese Islands, as an island arc, were established in the Miocene (Neogene), around 15 million years ago, when Japan broke away from the continent and marginal seas, such as the Sea of Japan and the East China Sea, appeared. The rocks/strata which formed after the birth of the island arc consist of the following materials: marine sediments and volcanic rocks associated with the appearance of new marginal seas, strata deposited around the fore-arc basin (the sea basin distributed parallel to the trench; Fig. 2.13) facing the Pacific Ocean, rocks and strata of the Izu-Ogasawara arc and Kuril arc which are colliding with the Japanese Islands, and non-marine sediments and volcanic rocks deposited on land and in shallow water on the Japanese Islands (Fig. 2.13).

These rocks and strata are often covered by weathering materials and Quaternary deposits. In the field of soil science and Quaternary geology, they are collectively referred to as "basement rocks" or "basement geology." Generally, in the plains, there are not many places where the basement geology is directly exposed at the ground surface, since the basement geology is eroded and transported from

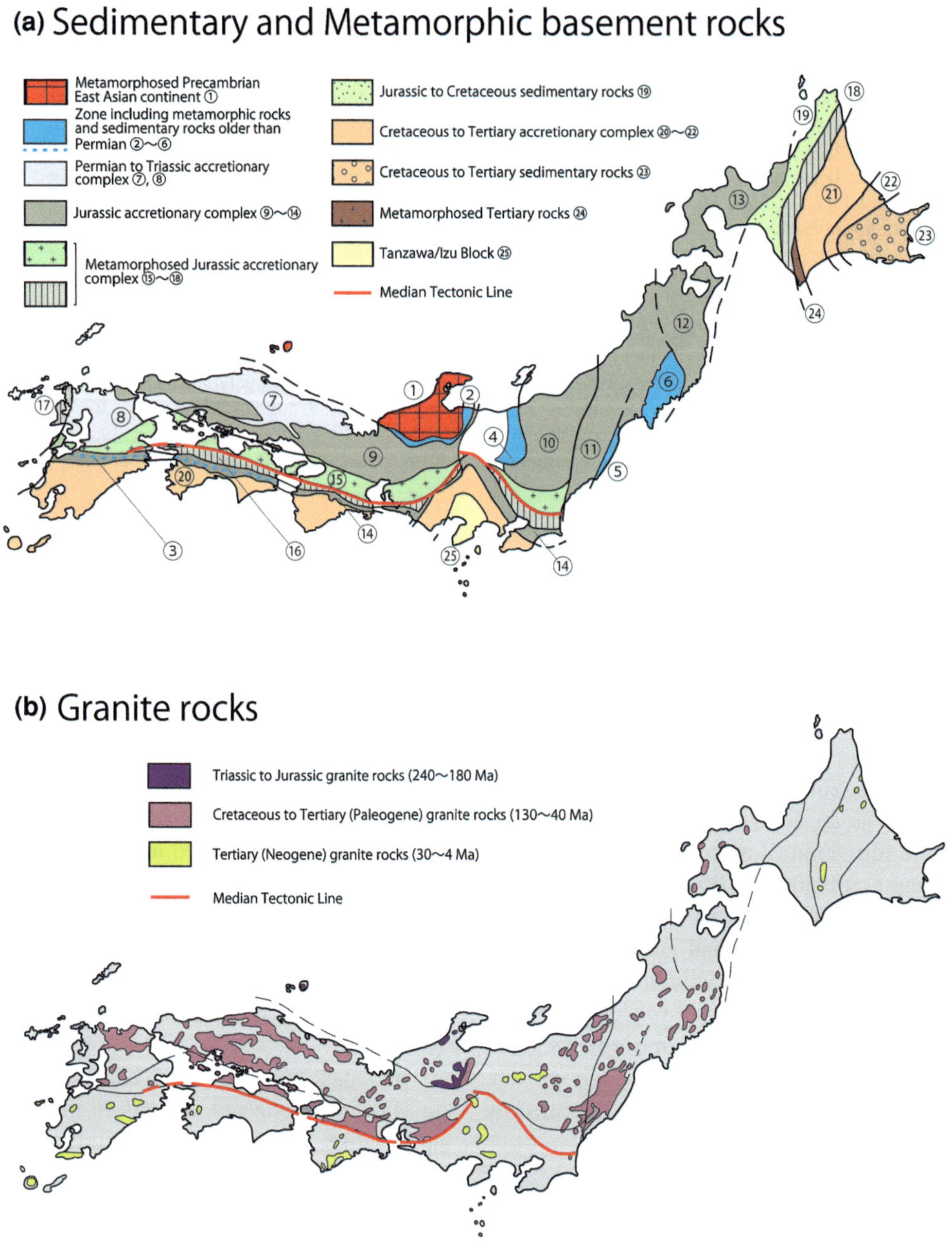

Fig. 2.14 Large-scale basement geological structure of the Japanese Islands (modified from Saito (1992), Copyright 1992 Iwanami Shoten, Publishers., with permission from Iwanami Shoten, Publishers.). **a** Sedimentary and metamorphic basement rocks. This figure removes sediments of the Quaternary and Neogene series and intruded granite rocks from the Japanese Islands. The geology of the Japanese Islands formed before the island arc is subdivided into as follows: Part of the metamorphosed Precambrian East Asian continent (① Hida Belt, Oki Belt), zone including metamorphic rocks and sedimentary rocks older than the Permian (② Hida Marginal Belt, ③ Kurosegawa Belt, ④ Joetsu Belt, ⑤ Eastern Abukuma Marginal Belt, ⑥ Southern Kitakami Belt), Permian to Triassic accretionary complex (⑦ Sangun Belt, Maizuru Belt, ⑧ Yamaguchi Belt), Jurassic accretionary complex (⑨ Mino–Tamba Belt, ⑩ Ashio Belt/Yamizo Belt, ⑪ Abukuma Belt, ⑫ Northern Kitakami Belt, ⑬ Oshima Belt, ⑭ Chichibu Belt), the metamorphosed Jurassic accretionary complex (⑮ Ryoke Belt, ⑯ Sanbagawa Belt, ⑰ Nagasaki Belt, ⑱ Kamuikotan Belt), Jurassic to Cretaceous sedimentary rocks (⑲ Sorachi/Ezo Belt), Cretaceous to Tertiary accretionary complex (⑳ Shimanto Belt, ㉑ Takayuki Belt, ㉒ Tokoro Belt), Cretaceous to Tertiary sedimentary rocks (㉓ Nemuro Belt), the metamorphosed Tertiary rocks (㉔ Hidaka Metamorphic Belt), Tanzawa/Izu Block (㉕). Among them, the accretionary complex of the Triassic and Jurassic layers in the latter half of the Jurassic period (from the north ⑨ Mino/Tamba Belt, ⑮ Ryoke Belt, ⑯ Sanbagawa Belt, ⑭ Chichibu Belt) is the most widely distributed and occupies in the central part of the Japanese Islands. In the middle part, there is ⑮ Ryoke Belt as metamorphic rock which is heated by the intrusion of granite, ⑯ Sanbagawa Belt which was buried deep under the ground and metamorphosed by high pressure, which is same sediments as ⑨ Mino/Tamba Belt and ⑭ Chichibu Belt in the southern part. **b** Granite rocks. This figure shows the distribution of granite rocks among plutonic rocks which are currently exposed on the ground. Granite rocks older than Paleogene intruded when the Japanese Islands were located on the eastern edge of Asian continent. The magma activity of this period forms the main skeleton of the Japanese Islands together with the accretionary complex. On the other hand, Tertiary (Neogene) granite rocks intruded after the establishment of the Japan Island arc

mountainous areas by the action of running water, waves, wind, mass movement and so forth, and the eroded material is deposited in the plain. Therefore, the sediments constituting the plains include clastics and weathered materials derived from the basement geology upstream. As these sediments are exposed near the surface of lowland plains or redeposited as wind dust, materials derived from the basement geology also affect soil parent materials.

2.2.2 Meso- to Mini-landforms in the Mobile Zone: Arrangement of Mountains, Basins, Plains and Volcanoes

(1) **Arrangement of mountains, basins and plains as features of the island arc**

The Japanese Islands comprise the Eastern Japan island arc system (Kuril arc, Northeast Japan arc and Izu-Ogasawara arc), which is active in connection with the subduction of Pacific Plate, and the Western Japan island arc system (Southwest Japan arc and Ryukyu arc), which is active in connection with the subduction of the Philippine Sea Plate. The ongoing tectonic movement manifests as earthquakes and geodetic crustal movements. Earthquakes repeatedly occur over a long period, crustal deformation is ongoing, uplift or subsidence accumulates at each place, and landforms and geological structures such as basins, mountain ranges, island arcs and trenches have been formed. Mountains are areas of relative uplift, and plains are areas of relative subsidence.

In areas where arcs meet other arcs, unique geological structures and topographical arrangements arise due to mutual collision and the duplication of properties (Figs. 2.12, 2.13 and 2.14). The terrestrial area of the Japanese Islands consists mainly of an outer arc (fore-arc) and an inner arc (back-arc). The lowland between the outer and inner arcs is an area of relative subsidence and in places consists of marine areas, such as the Seto Inland Sea.

The outer arc is located outside the volcanic front and is a broad zone of slow uplift with large, long-wavelength undulations. The terrestrial area of this zone is a mainly mountainous area composed of basement rocks of pre-Paleogene age; the Neogene formation is barely distributed inside the mountains. This area was generally subjected to land erosion through the Late Cenozoic. The terrestrial areas of the outer arc are as follows: Shikotan; Habomai Island; the Nemuro Peninsula; Kushiro and the Kitakami Mountains; the Abukuma Mountains; the Ogasawara Islands included in the Eastern Japan island arc system; the Akaishi Mountains; the Kii Mountains; the Shikoku Mountains and Amami Islands; the Okinawa Islands; and the Sakishima Islands included in the Western Japan island arc system (Fig. 2.13).

The inner arc is located inside the volcanic front and is a region with marked crustal activity characterized by volcanic activity. Uplift and subsidence with short wavelengths are dominant, and faults and folds are prominent. Long north–south uplift parts include the Ou Mountains and Dewa Mountains, and alternating areas of subsidence occur between them, which is visible in the topography (Fig. 2.15). In the subsidence basin, thick Neogene and Quaternary materials are deposited. A part of the formation afterward became a constituent of the uplifted mountains. From the northeastern part of Hokkaido, the main part of the Kuril Islands, including Kunashiri Island and Etorofu Island, and the Tohoku region west of the Ou Mountains, constitutes the inner arc of the Eastern Japan arc system. In the Southwest Japan arc, like in the Eastern Japan arc system, volcanic activity and the topographical and geological features are not distinct. In the Ryukyu arc, volcanic islands such as Iōjima, Kuchinoerabu-jima and the Tokara Islands are inner arc islands.

At the meeting point of the island arcs, the band structure of the island arcs is disturbed by plate collision and lateral movement of the land-side plate, resulting in topography and structure characterized by compression and extension toward different directions (Fig. 2.13). In the central part of Hokkaido, the collision of the Okhotsk Plate with the Northeast Japan arc formed a folding mountain range in the central axis of Hokkaido in the mid-Cenozoic Era, stretching from the Teshio Mountains to the Yubari Mountains. Later, the westward movement of the outer arc of the Kuril Island arc formed the Hidaka Mountains by fault movement. In the central part of Honshu, the collision of the Southwest Japan arc with the Philippine Sea Plate formed the flexible arrangement of Akaishi Mountains and the Kanto Mountains in the Late Cenozoic Era. In Kyushu, the Ryukyu arc and the Southwest Japan arc are in contact with each other, and the characteristics of the inner arc of the Ryukyu arc (Beppu–Shimabara rift belt crossing the central part of Kyushu) overlap with the old belt structure of the Southwest Japan arc (Fig. 2.13).

(2) **Geographical distribution of volcanoes**

In the Japanese Islands, the occurrence and rise of magma that manifests as volcanic activity are caused by the subduction of the oceanic plate at the convergent boundary, and the volcanic line is connected to the shaft of the island arc. In the process of the massive penetration of felsic magmas to the upper crust, which creates formations of acidic rocks, a wide area of uplift in the rock mass distribution area sometimes forms mountains. The wide area of uplift in the

Fig. 2.15 Fuji volcano which is a typical stratovolcano in Japan (Photograph credit: Copyright 2019 by Hideki Miura). The location corresponds to major volcano ⑫ Fuji–Izu in Fig. 2.13

Hida Mountains (Fig. 2.13) without a geological structure that uplifts the mountains is an example of this magmatic activity.

The Japanese Islands constitute the most prominent volcanic area in the world, with more than 111 active volcanoes, including stratovolcanoes and caldera volcanoes, particular to the subduction zone of the plate, and over 200 Quaternary volcanoes with original surfaces (Figs. 2.15 and 2.16). The volcanoes in the Japanese Islands are most densely distributed on the volcanic front at a distance of 200 to 300 km inside the trench (Fig. 2.13). Some island arcs may have second and third rows of volcanoes; however, the distribution density of volcanoes decreases away from the volcano front. The content of sodium and potassium with respect to silica in the volcanic products tends to increase toward the inner arc side due to the increased depth of magma formation (Kuno 1966). This is important in considering the chemical properties of volcanic ash soils in and around the Japanese Islands.

2.2.3 Meso- to Mini-landforms Created by Quaternary Climatic and Sea-Level Changes

The meso- to mini-landforms of the Japanese Islands are classified into a mountainous region and a plain, depending on the difference in altitude and undulation. In the mountainous region, the altitude is high and undulations and slopes are large, and the region is basically composed of ridges and valleys (slopes), whereas the plain has low altitudes and small undulations, exhibiting a flat ground surface.

As mentioned in Sect. 2.1.3, since the 1950s, a powerful way to reconstruct climate change from marine sediment cores and ice cores has emerged, allowing Quaternary climate change to be clarified in detail. As a result, it has been determined that climate change on the scale of tens of thousands of years surely occurred at almost the same time on a global scale. The landform change and the formation

Fig. 2.16 Major source volcanos of Quaternary tephras in and around the Japanese Islands (reprinted from Machida, (2002), Copyright 2019 Japan Academic Association for Copyright Clearance, with permission from Japan Academic Association for Copyright Clearance). 1, Large caldera with diameter of more than 10 km; 2, caldera volcanoes with diameters of less than 10 km often associated with composite stratovolcanoes; 3, stratovolcanoes; 4, central cones within calderas; 5, buried calderas; 6, trench and trough axis (plate boundary)

history of the Japanese Islands can be characterized from the viewpoint of such climate change.

(1) **Denudation of mountains and volcanoes**

A total of 62% of Japanese terrestrial land is composed of mountainous and volcanic regions (Fig. 2.13). In the mountains of the Japanese Islands, mass movement in which slope material moves downward only under the influence of gravity (i.e., without forces such as running water, waves, wind, glaciers, etc.) plays an important role in changing the slope topography. This mass movement includes the following phenomena: creep (slope material slowly moves downward), rockfalls (one or a number of rock fragments instantaneously fall down the slope separately from a cliff or steep slope), landslip (material suddenly undergoes brittle fracture, is peeled off from the slope and rapidly falls down the slope in a disturbed state), landslide (the top lump slips from a continuous slip surface inside the slope) and debris flow (one or a number of rock fragments instantaneously fall down the slope, falling apart from a cliff or steep slope).

When rain falls on a slope, the soil particles are first transported by sheet surface flow caused by raindrops. As the velocity and the depth of the surface flow of the depression increase, a rill is formed by the grooved erosion, which eventually turns into a gully. On mountain slopes in Japan, valleys and gullies are densely distributed, so mass movements in the valley heads and side walls are the most important factors in erosion and sediment production in mountainous areas. The slopes of large undulating mountains are an aggregate of previously collapsed terrain because mass movement occurs with high density during heavy rainfall. Due to landslips and landslides, the slopes of mountainous areas gradually retreat, causing a decrease in the altitude of mountain ridges. The debris produced by the collapse of a slope temporarily accumulates at the bottom of the slope and the valley bottom, but it is transported to downstream areas via debris flows due to flooding. Even in stratovolcanoes, the central part of the volcanic body, which consists of steep slopes, is engraved in a deep radiant valley, and the produced rock debris flows downward, mainly as a debris flow, to form an "alluvial fan" at the foot of the volcano. In this way, the material on the slopes of the mountains and volcanoes of the Japanese Islands is always in an unstable state, and generally the soil on the slopes is easily eroded or moved.

Additionally, past glacial landforms are recognized in the uppermost part of the mountains and river valleys in the central mountains of the Japanese Islands (the Hida, Kiso and Akaishi Mountains) and the Hidaka Mountains of Hokkaido (Fig. 2.13) (see Figs. 2.10, 2.12 and 2.13 in Sect. 2.1.3). The age determined by tephra stratigraphic analysis indicates that the mountain glaciers in Japan advanced during MIS 4, 3 and 2 in the last glacial period. However, in the central mountains (the Hida, Kiso and Akaishi Mountains) and the Hidaka Mountains, the glacier advance in MIS 2 was smaller than that in MIS 4 and 3. In both mountainous areas, glacial landforms formed in MIS 6 have been reported, but glacial landforms formed before MIS 8 have not yet been certified. Most of the glacial landforms are small; cirques (horseshoe-shaped depressions formed by glacial erosion) are shallow with steep floors, and typical U-shaped valleys are scarce; the moraine (debris that accumulated at the edge and surface of a glacier) at the end of the glacier is small in scale, and the development of downstream outwash terraces is rare.

Above the current tree lines in Japan, sand and gravel slopes with sparse vegetation are under periglacial environments, where mechanical weathering and creep due to freeze–thaw action on the ground surface, and snow, affect soil conditions. On these slopes, various formations are produced on the surface. During the last glacial period (MIS 4 to 2), the altitude of the tree lines declined, so these periglacial regions also declined to lower altitudes, and permafrost was formed in the plains in the northern and eastern parts of Hokkaido (see Figs. 2.10, 2.12 and 2.13 in Sect. 2.1.3). In the postglacial period (MIS 1), due to warming and increasing rainfall, mountain slopes were again covered by forest vegetation and surface erosion was extremely slow. Moreover, in the postglacial period, as the frequency of heavy rainfall increased, water streams with increasing tractive force began to strongly undercut valley bottoms. The transitional areas from the gentle upper slope to the steep lower slope are gravitationally unstable and collapse during earthquakes or heavy rainfall. In this convex break of the slope, the collapse erodes the debris made in the last glacial period or before; this is called the "postglacial dissection front" (Hatano 1979).

In rivers flowing on mountains with a glaciated valley in the upper stream, moraine and terraces have not been observed to be directly connected. However, river terraces and alluvial fans developing from a valley to the downstream basin/plain are known to have formed by erosion and deposition almost simultaneously everywhere (see Fig. 2.8 in Sect. 2.1.3) and were therefore probably formed under the influence of global climate change. Thus, as described in the next section, it is possible to discuss changes in the supply and transport of debris on the basis of fluvial terraces during the glacial to interglacial period. This knowledge will also provide useful information about the formation age and the environments of fluvial terraces and associated soils.

(2) **Formation and components of plains**

The plains of the Japanese Islands are roughly divided into the coastal plains facing the coast and the inland mountain basin. Both plains are dominated by sedimentary plains where sediments bury structural depressions (basins and inner bays formed by relative subsidence). Mountains and plains in the Japanese Islands are small in size, and the plains are distributed between mountains in a mosaic pattern (Fig. 2.13). In the mobile belt, under the temperate humid climate of the Japanese Islands, and due to the large relief and high denudation rate of the mountainous areas, the massive amounts of sand and gravel produced by erosion are transported by steep rivers and accumulate in upstream and midstream rivers or mountain basins; otherwise, they are transported downstream and accumulate on the coast facing the inner bay and the open ocean to form alluvial plains. Along the coast, "coastal plains", formed mainly by erosion and marine sedimentation processes such as waves and tidal currents, are also distributed. Aquatic areas, such as lakes and lagoons, are left as part of the plain in the basins and the inland bays where the sedimentation of sand and gravel is insufficient. The landforms of the plain are divided into lowlands, terraces (uplands) and hills according to the altitude, relief, flatness and formation age. The boundaries of these geomorphic units are generally clearly classified by the break of slope angle and orientation (the concave break of the slope and the piedmont line) (Fig. 2.17).

Lowlands are low-relief landforms, distributed along rivers and coasts, at a low relative height above river and sea levels. Lowlands are young landforms formed during the Holocene (MIS 1) and are divided into fluvial lowland (alluvial plain) and marine lowland (coastal plain) by major

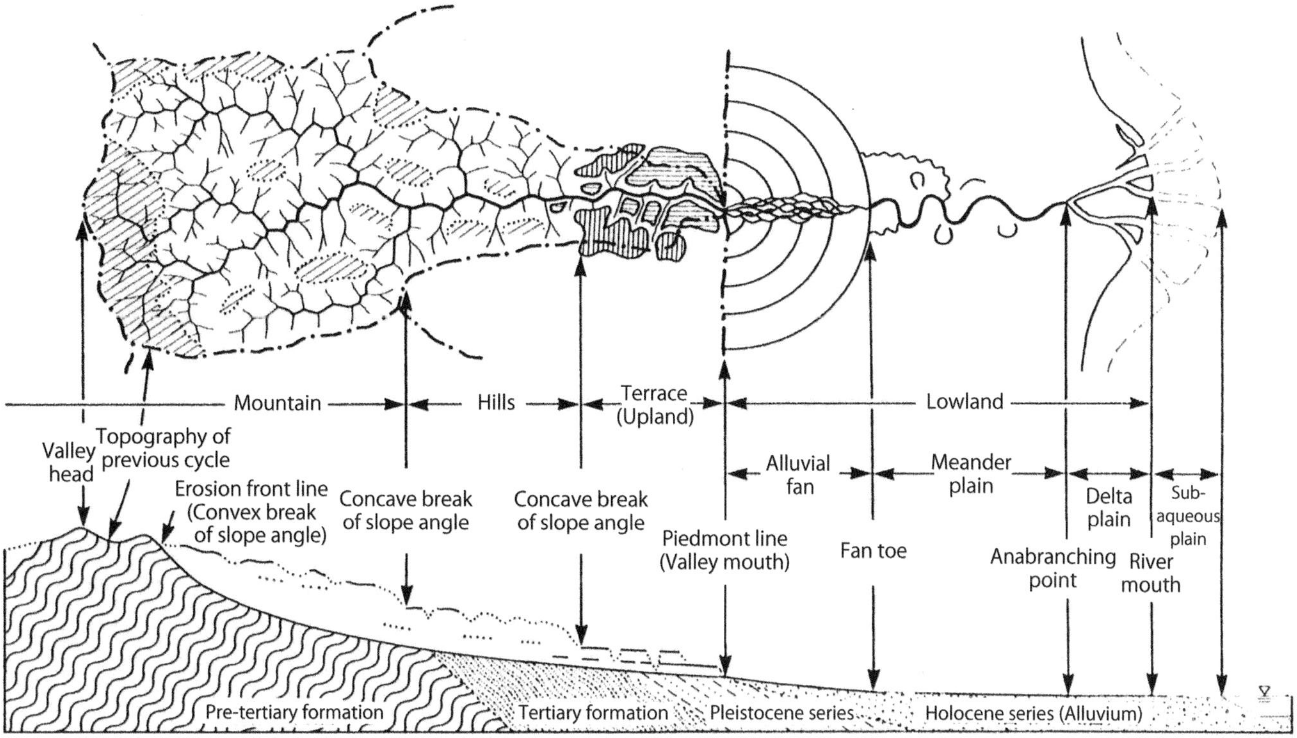

Fig. 2.17 Schematic diagram showing the general arrangements of landforms in a drainage basin in the Japanese main islands located in the tectonically active and climatically humid-temperate zone (modified from Suzuki (1997), Copyright 1997 Takasuke Suzuki, with permission from Takasuke Suzuki). Upper: Plan view and lower: cross-sectional view

geomorphological processes. Among the plains facing the coast, in areas where wind action is strong, sand is supplied from the coast, forming sand dunes. In lowland areas, sediments from rivers and beaches directly become soil parent material on the surface.

Terraces (uplands) occur above river and sea levels; their upper surface is flat, and some or all of them are surrounded by cliffs and steep slopes. The flat surface of a terrace is called the terrace surface, and the surrounding cliffs and steep slopes are called terrace cliffs. The terrace of the Japanese Islands is a landform which formed during the Middle Pleistocene to the Late Pleistocene (about 770,000 years ago to 10,000 years ago) and is divided into fluvial terrace and maritime terrace by major geomorphological processes.

Following the formation of a terrace, its surface is dissected by various erosional processes, the area of the terrace surface decreases, and it turns into a hill mainly composed of sloped features such as ridges and valleys. In Japan's terraces and hills, exposed sediments in rivers and beaches are usually covered with eolian sediments such as air-fall pyroclastics (tephras) and various eolian dusts, which rarely become soil parent materials directly on the land surface.

Additionally, the alluvial fans formed by the sand and gravel that are eroded and transported from volcanoes usually contain various sediments from debris flows, pyroclastic flows, lava flows and so forth, and so are called "alluvial fans at the foot of volcanoes" to distinguish them from the alluvial fans formed by river streams.

(3) **Formation of fluvial terraces**

The lowlands formed by river sedimentation (alluviation) are called "fluvial lowlands" or "alluvial lowlands". In the Japanese Islands, where the mountains and plains are adjacent, the erosion process is dominant in mountainous areas, so river valleys are narrow and rivers form erosional valleys. On the other hand, when the volume of the transported sediment becomes large, sedimentary valleys with wide valley bottoms are formed. River channels become easy to move in mountain basins and at the foot of mountains, and river beds become wider (or water depths become shallower). Therefore, the transporting power of rivers decreases, and the transported sand and gravel are deposited to form alluvial fans (Fig. 2.17). Alluvial fans are the typical fluvial landforms in the Japanese Islands, where the processes of production, transportation and sedimentation of

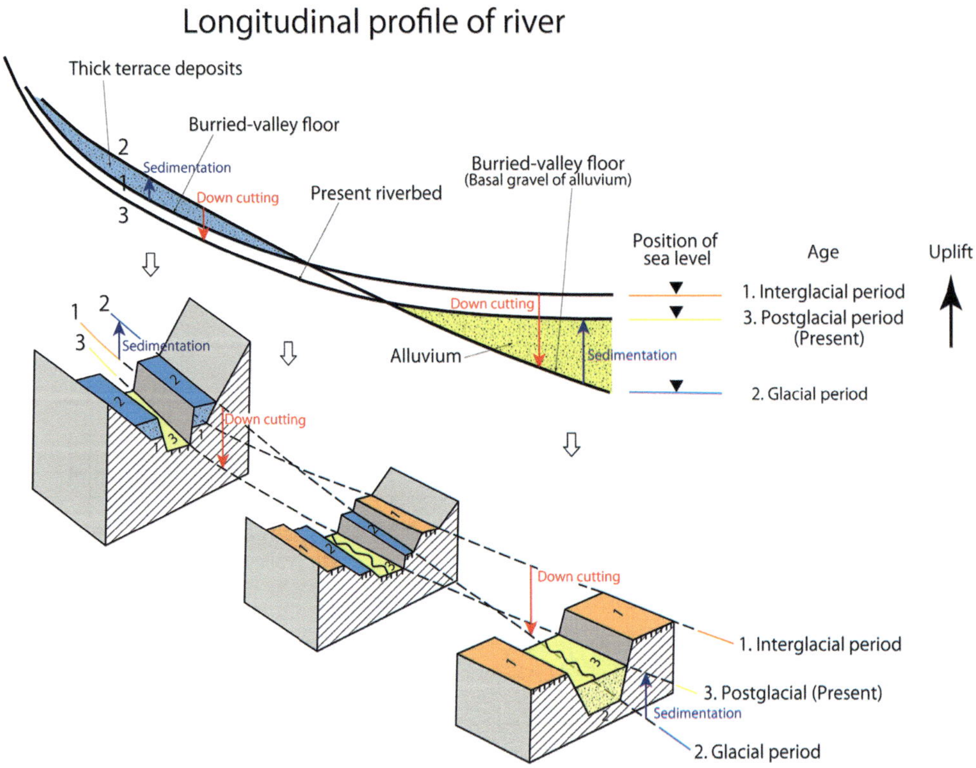

Fig. 2.18 Schematic model of fluvial terrace formation through the glacial–interglacial cycle (modified from Kaizuka (1977), Copyright 1977 Iwanami Shoten, Publishers., with permission from Iwanami Shoten, Publishers.)

sand and gravel are very active. Worldwide, alluvial fans are mainly distributed in subduction zones, such as the Japanese Islands, or in collision zones.

Downstream of an alluvial fan, the river creates "meanders" (Fig. 2.17); moreover, in the farthest downstream regions, "river deltas" (Fig. 2.17) are formed where the river flows into still water areas such as the sea or a lake. However, in rapidly flowing rivers such as the Fuji River and the Kurobe River in Central Japan, where the mountains and the ocean are in close proximity, alluvial fans face the sea directly and the rivers have no deltas.

In areas facing the ocean, under the influence of the sea-level change that occurred in the glacial to interglacial period (see Fig. 2.8 in Sect. 2.1.3), inner bays and the drowned valleys expanded during the interglacial period; inner bays and the continental shelf formed land due to the sea level lowering during the glacial period (MIS 4 to 2) and an extended river is formed (Fig. 2.18). During the postglacial period (MIS 1), the area that became the marine area due to the transgression will be converted to land by the sedimentation process of the river and the beach forming the fluvial lowland (alluvial plain) and marine lowland (coastal plain).

On the other hand, in the middle upstream part of a river, the water flow generally cuts deeply into the basement rock, and a valley landscape with a narrow valley bottom and steep valley wall is formed. In this part of the river, flat river terraces often developed in a higher position than the current riverbed, meaning that a wider flood plain had spread before river incision (Fig. 2.18). Many terraces consist of thick sediments widely filling valleys that had been deeply incised by a river. Such a relationship between the changes in river processes and climate change has been identified in the northeastern part of the Japanese Islands using tephrochronology (Machida et al. 2006). Generally, during the period of MIS 5e, the rivers in the middle upstream part displayed a remarkable trend of cutting downward, as in the MIS 1 period. After the MIS 5d period, the valley was gradually buried at the maximum rate of the MIS 4 period; it was slightly incised during the MIS 3 period and then did not change much until the first half of MIS 2. After the MIS 2 period, the river quickly incised. Such a change of river processes implies that the balance between the supplies of debris was strongly influenced by climate change and that the transporting ability of running water (frequency and intensity of flood) was involved.

(4) Formation of marine terraces

If the sea level is stable for a certain period, landforms corresponding to the sea level are formed. As shown in Fig. 2.19, on the rocky coast, notches and sea caves are formed at the shoreline by wave erosion, and in the slopes whose base was eroded a steep cliff is formed by collapse. In the intertidal zone, weathering effects due to the repetition of wave erosion and dry and wet conditions may form a "shore

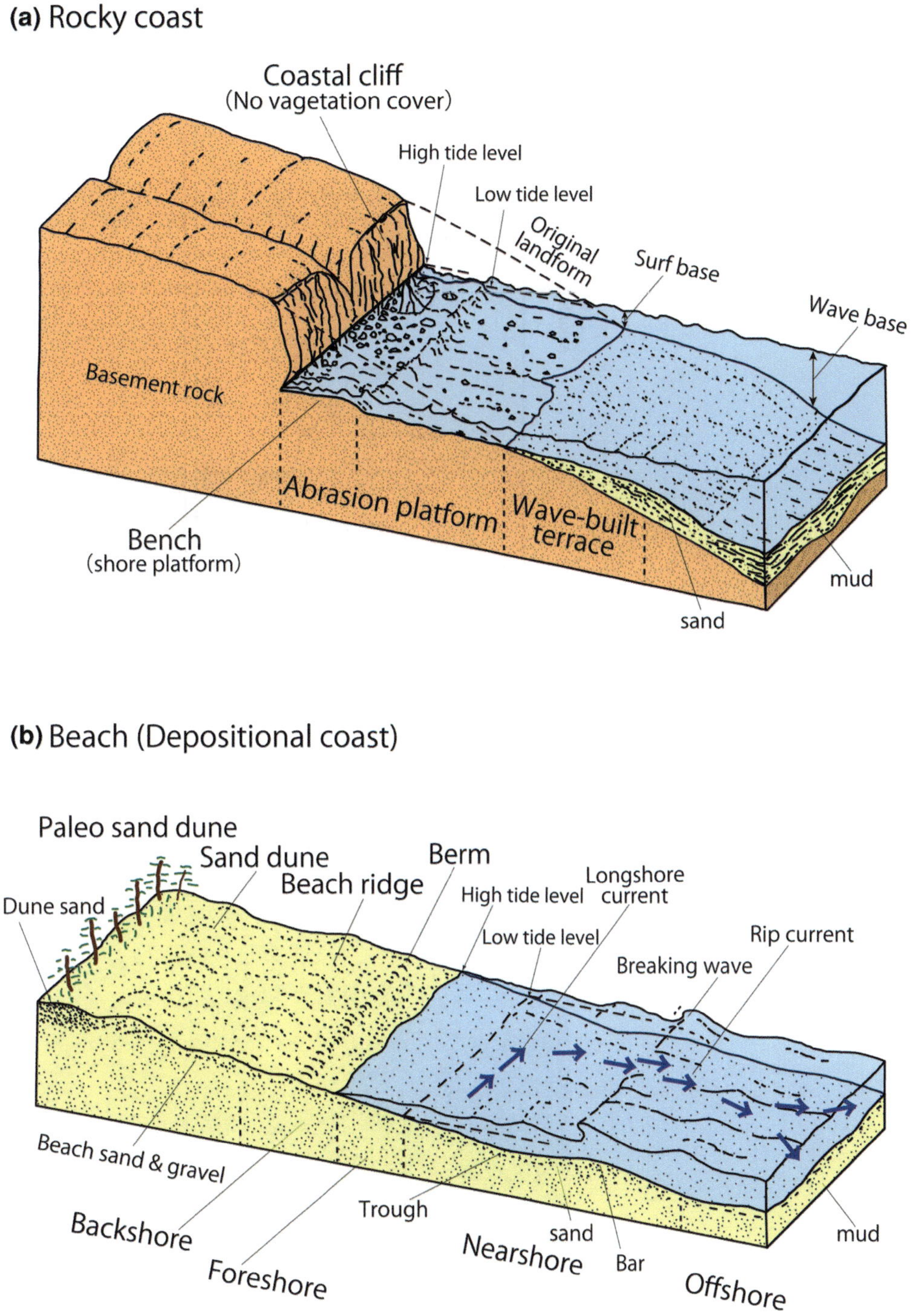

Fig. 2.19 Schematic diagram of coastal landforms (modified with permission from Kaizuka (1992), Copyright 1992 Iwanami Shoten, Publishers., with permission from Iwanami Shoten, Publishers.). Upper: rocky coast and lower: beach (depositional coast)

platform." On the ocean side, "abrasion platforms" that have gentle slopes toward the sea side due to wave erosion under the average sea level are formed down to water depths of 10 to 15 m, the lower limit of wave erosion.

As these landforms emerge, stepped landforms with flat terrace surfaces and steep cliffs are formed. In areas of prominent uplift such as the Japanese Islands, the records of sea-level change repeated during the Late Quaternary (see Fig. 2.8 in Sect. 2.1.3) appear on land as different marine terraces. The earlier the emergence, the higher the position of the marine terraces will be, due to crustal uplift (Fig. 2.20). On the coast of the Japanese Islands, a number of marine terraces have developed in various areas (Fig. 2.21); in particular, the marine terraces formed during the high sea level of MIS 5e (the last interglacial period) are widely developed, and the terrace surface is usually well preserved. This terrace surface is accompanied by transgression sediments and is clearly separated clearly from the

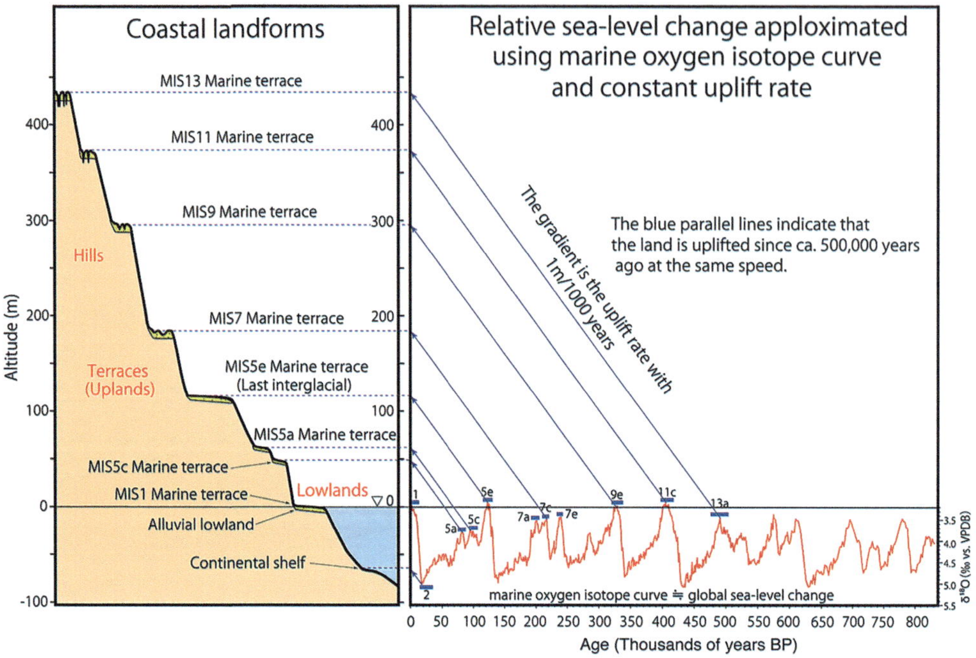

Fig. 2.20 Schematic diagram of the formation of marine terraces through the sea-level changes in the uplift area (modified with permission from Kaizuka (1992), Copyright 1992 Iwanami Shoten, Publishers., with permission from Iwanami Shoten, Publishers.). The parallel blue line is shown as the uplift of the land continued from ca. 500,000 years ago at the same speed

Fig. 2.21 Marine terraces developed in Cape Muroto, Kochi Prefecture, Shikoku (Photograph credit: Copyright 2019 Hideaki Maemoku). The location corresponds to the cape at the southwest end of major mountain no. 41 (Shikoku Mountains) included in D_2 (Southwest Japan outer zone) in Fig. 2.13

upper and lower terrace by terraced cliffs. Above MIS 5e terrace, there are MIS 7, MIS 9 and rarely older marine terraces on the higher level, which also accompany transgression marine sediments. In places where the uplift rate was high (generally 0.5 m/ky or more), young terraces corresponding to MIS 5c and MIS 5a were formed below the terrace of MIS 5e age. In such places, Holocene marine terraces formed during postglacial transgression (MIS 1) are also distributed at the lowest level.

Coral reefs and coral reef terraces are distributed on the Nansei Islands south of 29° N (Tanegashima, south of Yakushima) and the south coast of the Ogasawara Islands (see Fig. 2.10b in Sect. 2.1.3). Coral reefs are formed by reef corals that grow in shallow waters where the average water temperature in the coldest month is 18 °C or higher. Coral reef terraces (uplifted coral reefs) are often found on the Nansei Islands in the uplift zone. The age of emergence of coral reef terraces has been estimated from the measurement of the

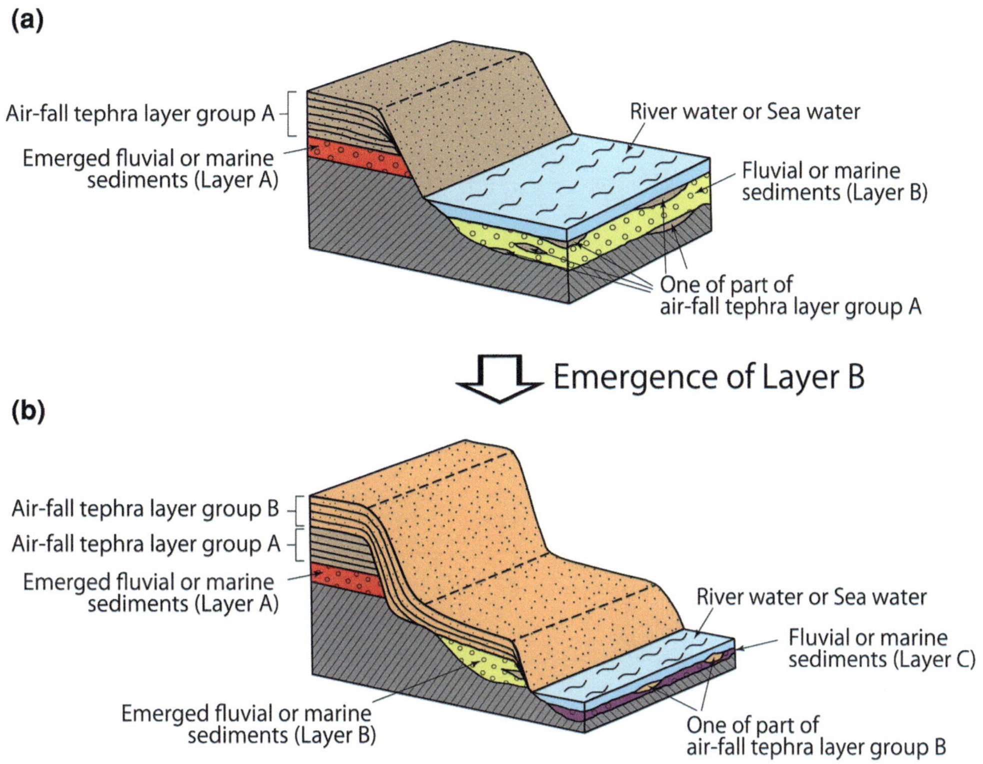

Fig. 2.22 Schematic diagram showing the relationship between emerged terraces and air-fall deposits (modified from Machida (1977), Copyright 1977 Hiroshi Machida, with permission from Hiroshi Machida). (A) When the eolian deposits such as tephra layers were fallen in the active formation stage of the Layer B (fluvial or marine sediments), the eolian sediments were deposited on the Layer A which was terraced before the Layer B formation. However, in layer B, eolian sediments are hardly remained by the action of running water or exist only on the patch. (B) When the sea level or the riverbed lowered and the deposition of the Layer B stopped, the eolian deposit such as tephra layers begins to deposit on the Layer B as soon as the geomorphic surface of Layer B emerged. Furthermore, the slope between the A and B terraces begins to be covered by new eolian sediments. Therefore, at the cliff edge of the A terrace, the eolian sediments such as tephra layers, which had fallen until just before the geomorphic surface B became a terrace, may cross the new eolian sediments such as tephra layers

uranium series ages of coral fossils, and the relationship between the terrace formation period and interglacial transgressions has been directly obtained (Konishi et al. 1974; Ota and Omura 1992). However, the ages of terraces older than MIS 5e and MIS 7 are still not well confirmed.

2.2.4 Air-Fall Pyroclastics and Eolian Dust as Landform-Covered Materials

In regions where air-fall deposits are distributed, geomorphic surfaces are generally covered with air-fall deposits immediately after the emergence of marine and/or fluvial landforms (Fig. 2.22). In the Japanese Islands, where the volcanic activity is remarkable and westerlies including the jet stream are dominant, pyroclastic products from volcanoes and eolian dust transported from neighboring areas and the Eurasian continental landmass often cover geomorphic surfaces. The integrated consideration of the formation age of landforms, the landform materials and the landform-covering materials provide the key to understanding the soil formation age and the origin of the soil parent materials.

(1) **Air-fall pyroclastics: tephra**

Generally, in volcanic eruptions, gas and solid debris escaping from magma, as well as magma itself passing through the crater, are ejected. Purmice, scoria and volcanic ash that are ejected by explosive eruption as pieces of loose solids together with volcanic gas are called "tephras" or "pyroclastics," as distinguished from lava that is ejected as a continuous liquid.

In the Japanese Islands, where many volcanoes are distributed, air-fall tephras widely cover both terrestrial and oceanic areas in a very short time due to explosive volcanic eruptions; most of these can be identified, and so can be used as time indicators of the formation and landforms (Fig. 2.23). The oldest marine terraces of the Japanese Islands whose age can be confirmed with tephras are the Byobugaura terrace (MIS 13, about 500,000 years ago) and the Konan terrace (MIS 11, about 400,000 years ago) in the southern part of the Kanto region. The oldest fluvial terraces are identified in the Sayama Hills, the Azuyama Hills and the Kitsuregawa Hills (probably MIS 16-17, approximately 690,000-630,000 years ago) in the southern and northern part of the Kanto region (Kaizuka et al. 2000).

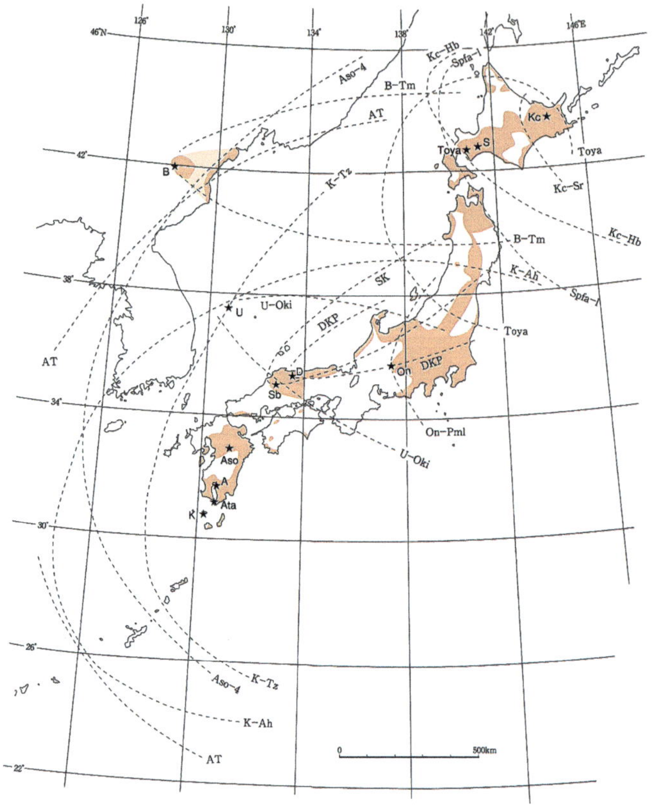

Fig. 2.23 General distribution of the representative marker-tephras of the Late Pleistocene and volcanic ash soil ("loam layer") with total thickness of more than 2 m in Japan and adjacent areas (stipples) (reprinted from Machida(2002), Copyright 2019 Japan Academic Association for Copyright Clearance, with permission from Japan Academic Association for Copyright Clearance). Volcano symbol: name (location): A: Aira (Southern Kyushu), Aso: Aso (Central Kyushu), Ata: Ata (Southern Kyushu), B: Baekdusan (North Korea/China), D: Daisen (Western Honshu), U: Ulreung (Korea), K: Kikai (Southern Kyushu), Kc: Kutcharo (Eastern Hokkaido), On: Ontake (Central Honshu), S: Shikotsu (Central Hokkaido), Sb: Sambe (Western Honshu), Toya: Toya (Central Hokkaido). Tephra symbol: name (age) Aso-4: Aso-4 (80–90 ka), AT: Aira–Tanzawa (26-29 ka), B-Tm: Baekdusan–Tomakomai (1 ka), DKP: Daisen–Kurayoshi (> 55 ka), K-Ah: Kikai–Akahoya (7.3 ka), K-Tz: Kikai–Tozurahara (95 ka), Kc-Hb: Kutcharo–Haboro (115-120 ka), Kc-Sr: Kutcharo–Shoro (35-40 ka), On-Pm1: Ontake-1 (100 ka), SK: Sambe-Kisuki (110-115 ka), Spfa-1: Shikotsu-1 (40-45 ka), Toya: Toya (112-115 ka), U-Oki: Ulreung–Oki (10.7 ka)

Because air-fall tephra covers the landforms of the mountains and plains and constitutes the present surface, it is also important as a soil parent material. Many active volcanoes of the Japanese Islands produce andesitic and basaltic ejecta and erupt about once every several decades or several centuries. As a result, the tephra accumulating from the foot of a volcano to the leeward plain forms a thick stratum. A large number of these tephra layers blown by wind and deposited on the surface are weathered as time elapses after deposition, lose their original sedimentary structure and become seemingly non-stratified. The Kanto Plain is a typical example for such deposits, and it is difficult to distinguish individual layers of many thin scoria ejected from Mt. Fuji (Machida et al. 1971; Fig. 2.24). With increasing distance from a volcano, the apparent sedimentation rate decreases and the chance of contamination by eolian dust and other particles moving from the adjacent slope becomes greater, resulting in a stratum with massive facies. Such strata are universally found around volcanic bodies in the Japanese Islands and have been called "*loam layer*" in the field of Quaternary science in Japan. However, "loam" was originally defined as a term referring to a soil texture with an equal mixture of sand and silt, and normally the particle size of the "*loam layer*" decreases with the distance from the volcano. Thus, it is important to note that "*loam layer*" as a stratum name is not meaningful.

(2) Eolian dust

Strong wind induces dust to rise from the ground. It also transports fine grains of landform materials formed by running water, waves, mass movement, etc., and redeposits them to cover another geomorphic surface. Fine-grained substances transported by wind are generally called eolian dust. Among eolian dust, sand mainly comes from continental sandstorms, and that which continues to be transported to a distant place such as Japan is called "tropospheric dust." In order for eolian dust to be fixed on a ground

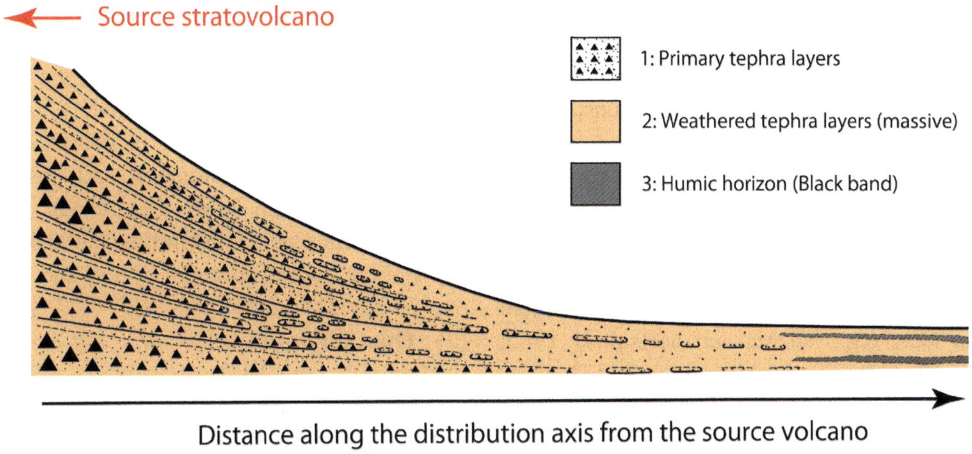

Fig. 2.24 Schematic diagram of facies change of tephras corresponding to distance from volcanic vent (modified from Machida et al. (1971), Copyright 1971 Japan Association for Quaternary Research, with permission from Machida Hiroshi and Japan Association for Quaternary Research). The humic horizons (black band) were not formed downward only by humus accumulation during the long rest period of the eruption; however, they were formed upward by the fine grain tephra sedimentation and the humus accumulation in parallel

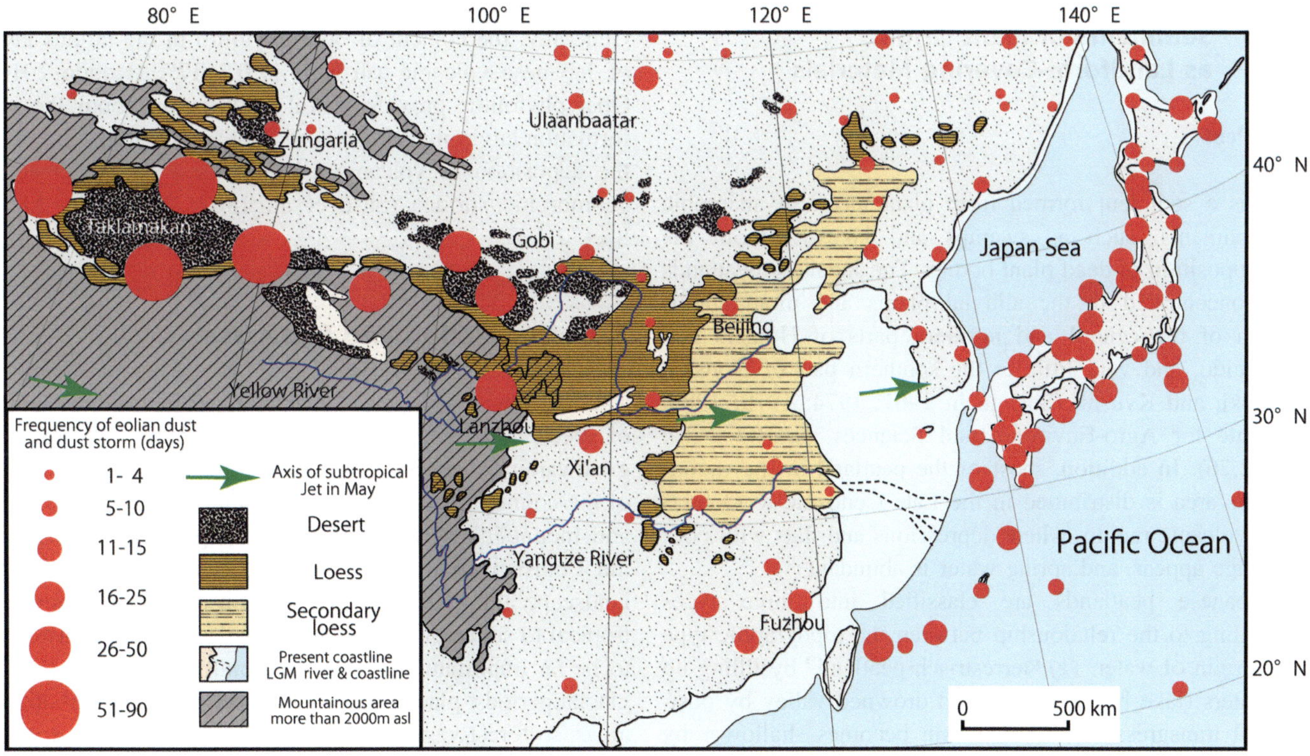

Fig. 2.25 Distribution of loess, frequency of the days of dust storm and eolian dust in 1975 in the East Asia (modified from Naruse and Inoue (1990), Copyright 1990 Naruse Toshiro, with permission from Naruse Toshiro). Yellow sand and sandstorms occur intensively in deserts around 40 degrees north. Since the yellow sand that soared rides the westerly wind and spreads over time, yellow sand is confirmed in a wide area in Japan. During the last glacial period, the coastline extends to 200 km west of the Nansei Islands, and the Yellow River and the Yangtze River carried and transported secondary loess to the estuary. As a result, the continental shelf together with the inland desert area, became the main source of wind dust during the last glaciation

surface, a plant covering that reduces the wind speed on the surface is necessary.

In Japan, from the latter half of the 1960s, the mineralogy of "*Kosa*" ("yellow sand" in Japanese) which is transported from March to May, in the early spring, has been examined. Inoue (1981) found that *Kosa* is mainly composed of quartz, feldspar and illite and had a similar mineral composition to that of "loess" distributed in China. Additionally, from meteorological observations at the time of the arrival of the *Kosa*, it was revealed that it originated from the Loess Plateau and the desert in mainland China (Fig. 2.25: Ishizaka et al. 1981). Furthermore, oxygen isotope analysis of quartz in the 0.01–0.001 mm grain size fraction of volcanic ash showed that the quartz originated not only from volcanic rocks, but mainly from a wide range of wind-blown dust from China (Mizota and Matsuhisa 1985; Mizota and Inoue 1988). It has also been pointed out that a wide range of wind-blown dust, including *Kosa*, is added to various sediments distributed throughout Japan (see, e.g., Mizota and Matsuhisa 1985; Inoue and Naruse 1987; Inoue and Mizota 1988.). The sedimentation rate of tropospheric dust on the Japanese Islands after the last interglacial period has been approximately 10^0–10^{-1} mg cm^{-2} yr^{-1}, and in cold periods (glacial period and stadial) the rate has tended to increase by three to five times compared with mild temperate periods (interglacial period, interstadial, postglacial period) (Yoshinaga 1999).

2.2.5 Sediments of Plant Origin as Landform-Covering Materials

(1) **Peat**

Peat is a sediment formed from plant tissue in anaerobic environments, such as swamps and lakes, without the decomposition of dead plant bodies. The Japanese peatlands are concentrated in the alluvial plains and mountainous region of the central and northern parts of Honshu and Hokkaido, and very little in the southern part of Honshu, Shikoku and Kyushu (Sakaguchi 1961; 1974; 1979; 1989; Institute for Agro-Environmental Sciences, NARO 2019: Fig. 2.26). In addition, most of the peatland in the mountainous area is distributed in the Late Cenozoic to Quaternary volcanic region, where depressions and flats with poor drainage appear, and spring water is abundant.

Japanese peatlands are classified into three types according to the relationship between the topographic type and origin of water: (1) "terrestrisch peatland" by filling up of waters (lake basin, lagoon and drowned valley by postglacial transgression), where basin becomes shallower by sedimentation, (2) "telmisch peatland" by paludification of dry land (back marsh of flood plain), which is formed by periodical flood, and (3) "telmisch peatland" by paludification of dry land (valley head, slope and watershed), which is formed by heavy snow, spring water, clouds and fog (Sakaguchi 1961, 1974, 1979, 1989: Fig. 2.26).

Peat deposits are also subdivided into "autochtonous peat“, during the formation of which marsh plants die and are deposited in situ, and "allochthonous peat," where peat separated by erosion is carried in running water and accumulates in another place. In Japan, the depositional rate of autochtonous peat is approximately 0.5 to 1.5 mm/year (Sakaguchi 1961, 1974). It is possible to reconstruct the climate, hydrological situation and past vegetation at the time of deposition from the change and decomposition degree of the plants constituting the peat substance, as well as from the compositions of pollen and diatoms in the peat.

(2) **Opal phytolith**

Opal phytoliths are structures composed of fine, amorphous particles of silica, and are formed by hydrous silicic acid being deposited in plant cells and tissue pores of gramineous grasses and trees (Kondo 2010). The particle size is very wide, from 5 to 200 μm. Although the particle size is small, even after the plant body decomposes in the ground, it remains in the soil accumulated by the deposition of air-fall tephra and eolian dust, even under dry oxidizing conditions. Generally, the content of opal phytolith in the humus layer of Japanese volcanic ash soils is 0.3–17% (Kondo 2010). Opal phytoliths provide useful information about the soil formation environment, because they are left in the soil with the deposition of volcanic ash and loess on the emerged geomorphic surface (Sase et al. 1987; 2008: Fig. 2.8 in Sect. 2.1.3).

2.3 Vegetation and Land Use

The vegetation of Japan is characterized by a large number of species and a high percentage of endemic species. The great variety of species is related to the geographical position of the country. The Japanese archipelago is a group of hilly and mountainous islands, located along the eastern end of the Eurasian continent, which shows a large variation in landscape and geology. The variation of climate from subtropical to subarctic along the archipelago is also considered to be an important factor in the floral biodiversity. During glacial periods, glacial cover developed in only high-altitude areas, so forest vegetation remained in the glacial age in the Pleistocene. It is thought that humans started living in the Japanese archipelago in the Paleolithic age, 120,000 years ago, and that rice cultivation was initiated in the Yayoi period (800 BC–250 AD). Despite human influence, forest vegetation persisted in the mountainous areas which cover two-thirds of Japan's territory; however, the area of the original natural forests has gradually decreased and has been replaced by early successional deciduous broad-leaved trees

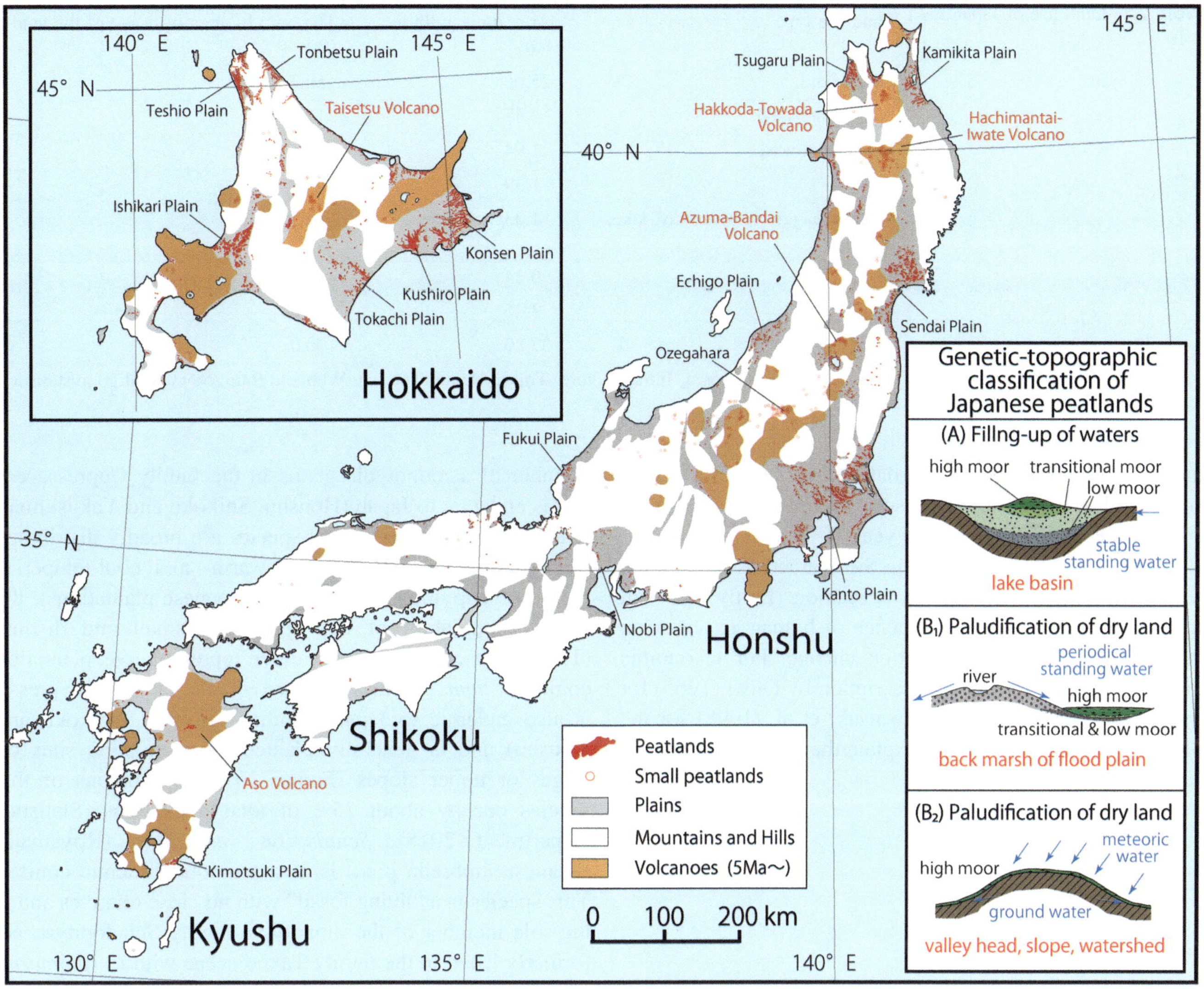

Fig. 2.26 Outline of distribution of peatlands and schematic diagram of genetic–topographic classification of peatland in the Japanese Island. Solid red areas show peatlands, and red small circles show small peatlands which are mostly distributed in the mountain regions (mostly Late Neogene volcanic regions). This distribution map includes areas where peat has disappeared because the area has already been farming or urbanizing. Data source: Distribution of peatlands is created by author based on Sakaguchi (1961, 1974, 1979, 1989) and Institute for Agro-Environmental Sciences, NARO (2019). Landform classification is created by author based on Geospatial Information Authority of Japan (2019) and National Astronomical Observatory of Japan (2018). Genetic–topographic classification of Japanese peatlands is created by author based on Sakaguchi (1974; 1979; 1989) (Figure supplied by Hideki Miura)

and pine. The planting of coniferous species was initiated in the Edo period (1603–1868 AD), and planted coniferous forests account for 40% of total forests in recent years.

Reflecting the complex topography of Japan, the distribution patterns of land use are also very complicated. The majority of mountainous, hilly and volcanic areas are covered with forests, while some of these areas are used for farmland, such as pasture and orchard. Most agricultural lands in cities situated in flat areas and suburban alluvial lowlands are mainly used for paddy fields, and higher terraces are used for upland cultivation, fruit orchards and tea fields. In the northern and the eastern parts of Hokkaido Island, flat areas are used for farmland, forest and pasture, since they are not suitable for rice farming due to the low temperature. The distribution of fruit orchards and tea fields is quite different from region to region, due to the influence of social and economic conditions as well as natural conditions.

According to the White Paper on Land, 2018 (Ministry of Land, Transport and Tourism 2018), 66.3% of Japan is forest, 11.8% is agricultural land and 5.11% is residential land (Table 2.1). Between 1975 and 2016, the proportion of agricultural land decreased from 14.8% to 11.8%. On the other hand, the proportion of residential land increased from 4 to 5.1% and the area of roads increased from 2.4 to 3.7%

Table 2.1 Land use of Japan in 2016*

Land use type	Area (unit: million ha)	The ratio by area of usage of the land (%)
Forest land	25.06	66.3
Agriculture field	4.47	11.8
Residential area	1.94	5.1
Road	1.39	3.7
Water area (river, waterway, lake, etc.)	1.33	3.5
Wilderness	0.34	0.9
Others	3.25	8.6
Total	37.80	100.0

**Source* Ministry of Land, Infrastructure, Transport and Tourism Web site (http://www.mlit.go.jp/statistics/file000006.html)

during the same period. These data suggest that the expansion of urban areas might cause a decrease in the area of agricultural land in the coming years.

In Sect. 2.3.1, we will explain the characteristics of flora, vegetation history and natural vegetation. Lastly, we will present an example of the influence of human activities on soil formation through vegetation change. For taxonomic nomenclature, we follow the description by Ohwi (1965) for graminoid plants and that by Iwatsuki et al. (1998) for the others. In Sect. 2.3.2, we will explain the history and present state of agriculture in Japan.

2.3.1 Vegetation

(1) **Flora**

The Japanese archipelago is phytogeographically classified into the Eastern Asiatic floristic region, which belongs to the Holarctic Kingdom (Takhtajan 1986). As with other surrounding countries which are included in the Eastern Asiatic floristic region, the flora of Japan is extremely rich and distinctive, and is characterized by a large number of endemic species. In Japan, wild vascular plants number about 5700 species, 30% of which are endemic to the country (Kato and Ebihara 2011). Because of the lack of extensive glaciations during the Quaternary, Japanese flora contains survivors of the Arcto-Tertiary Geoflora. The number of plant species in this region is much larger than in the glaciated regions of North America and Europe.

The forest flora of Japan is characterized by the presence of many tertiary relict conifers. About 46% of gymnosperms are endemic (Kato and Ebihara 2011), and several of them are important in silviculture and widely planted in Japan. For example, the principal conifer used in Japanese forestry, *Cryptomeria japonica* (*Sugi*, Japanese cedar), is the sole member of a monotypic genus in the family Cupressaceae and is endemic to Japan (Honshu, Shikoku and Yakushima). Although plantations of this species are broadly distributed throughout mesic sites in the warm- and cool-temperate region (occupying ca. 44% of all Japanese plantations), the natural population of this species is small and distinct (Thomas et al. 2013). The other representative plantation conifer, *Chamaecyparis obtusa* (*Hinoki,* Japanese cypress), is also endemic to Japan (southern Honshu, Shikoku and Kyushu) and is generally planted more in xeric sites on ridges or upper slopes (Farjon 2013). Plantations of this species occupy about 25% of total plantations (Statistics Department 2018). *Sciadopitys verticillata* (Koyamaki, Japanese umbrella pine) is also a unique endemic conifer. This species is a "living fossil" with no close relatives and is the sole member of the monotypic family Sciadopityaceae, formerly listed in the family Taxodiaceae with *Cryptomeria*. The natural distribution of this species is restricted to southern Honshu, Kyushu and Shikoku, and is most commonly found in mixed broad-leaved and coniferous forest on rocky, cool, moist ravines and valleys together with other endemic conifers such as *Chamaecyparis pisifera*, *Tsuga sieboldii* and *Abies firma* (Katsuki et al. 2013). Despite the value of wood, the planted area of these species is small compared with that of *C. japonica* and *C. obtusa*.

The distribution of these tertiary relict conifers ranges from the warm- to cool-temperate zone and particularity occurs in the intermediate area between the warm-temperate evergreen forest zone and the cool-temperate deciduous broad-leaved forest zone. Most of these species grow as small patchy populations in harsh environments or disturbed sites, such as scree slopes in deep ravines, rocky ridges and cliffs, where the invasion of modern trees (broad-leaved species) is scarce. The complex mountain topography may have provided diverse habitats as refugia for these tertiary relict conifers in the glacial and interglacial periods (Ohsawa 1987; 2006).

As with special topographical features, the flora associated with particular substrates, such as calcareous and serpentine soil (or rocks), is rich in endemic species. Limestone is rich in calcium, and serpentine is rich in magnesium, while both tend to be deficient in essential macronutrients (N, P, K). Under edaphically stressful conditions, flora in these habitats tends to contain recently formed taxa, that is, other types of endemics different from the relict species.

Serpentine is an ultramafic rock derived from oceans on the margins of former tectonic plates and is particularly distributed along tectonic lines in Japan. Horie (2002) defined 65 taxa from 22 families as "serpentine plants" in Japan; a total of 46 of these taxa occur in Hokkaido, where the largest serpentine area in Japan is located along the Kamuikotan and Hidaka metamorphic belt (Horie 2002, Mizuno et al. 2009). Most of these members are herbs or shrubs, such as Asteraceae (Compositae), Primulaceae, Ranunculaceae and Rosaceae. *Betula apoiensis* and *Picea glehnii* are characteristic woody species in serpentine areas.

Masses of limestone are also patchy distributed and are somewhat near to serpentine. Shimizu (1962, 1963) studied 200 taxa linked to limestone areas in Japan and Taiwan, and reported 75 taxa which were endemic to limestone but not found on other substrates, such as *Crepidiastrum yoshinoi, Senecio furusei* and *Adenophora maximowicziana*. *Betula chichibuensis* and *Carpinus turczaninovii* are characteristic woody species of limestone areas.

(2) **Historical aspect of vegetation**

Vegetation in the last glacial periods has been estimated by analysis of fossil pollen in soils (Yasuda and Miyoshi 1998). *Cryptomeria japonica* was a dominant species until 100,000 years ago. After that, the species dramatically decreased, while birch and coniferous species such as fir, hemlock and pine increased. Even in the coldest times in the last glacial periods, glaciers were only distributed in higher mountains, while forest persisted in most of the Japanese territory throughout whole glacial periods. After the glacial periods ended, the distribution of beech expanded (Yasuda and Miyoshi 1998); it occupied most of northeast Honshu and the southern parts of Hokkaido. The distribution of evergreen broad-leaved trees expanded in Western Japan, where the population density was high enough that forests might be influenced by human activities. In the Jōmon period (13,000–800 BC), shifting cultivation must have taken place in forest areas. Rice cultivation was introduced in the Yayoi period (800 BC–250 AD), and the area of original natural forests gradually decreased and was replaced by early successional deciduous broad-leaved trees (Matsugi 2007). In the Yayoi period, the density of red pine (*Pinus densiflora*) increased.

The impact of humans on forests increased along with population growth and the development of civilization. In western Honshu, bare land was formed in granite hills where severe erosion occurred due to the excessive utilization of forests in the Edo period (Chiba 1991). In the Meiji period (1868–1912 AD), the new government decided to undertake afforestation on bare land in order to prevent soil erosion. The rehabilitation effort continued into the Showa period (1925–1988 AD), and the bare land changed to forest; however, at present biomass is poor and soil is still immature in forests (Kaneko et al. 2007). The Meiji government also promoted coniferous plantations to increase timber supply, and after World War II coniferous plantation became more popular since large amounts of timber were required for house construction to support economic development. Because of this, a great number of trees were cut down in natural forests, with the trees being replaced with coniferous trees such as *C. japonica* and *Chamaecyparis obtusa*. Larch (*Larix kaempferi*) was also planted in high-altitude areas and in Hokkaido and the northern Honshu. Spruce (*Picea jezoensis*) and fir (*Abies sachalinensis*) were planted in Hokkaido, which it is not suitable for the growth of Japanese cedar and Japanese cypress. As of 2012, 41% of Japan's total forest area of 250,810 km^2 was planted coniferous forest, with the remaining fraction consisting of natural forest (Statistics Department 2018). Most of the natural forests are secondary, while the primary forests are distributed in only low-accessibility locations such as high-altitude areas and religious sanctuaries.

(3) **Forest vegetation zones**

Forest ecosystems in the Japanese archipelago are divided into four forest zones: subtropical evergreen broad-leaved, warm-temperate evergreen broad-leaved, cool-temperate deciduous broad-leaved and cold temperate/subalpine coniferous forest zones (Fig. 2.27). Since most of Japan receives adequate precipitation for the development of forest, indices using accumulated temperature are often used to explain the distribution of forest zones. For example, the warmth index (WI) and coldness index (CI) proposed by Kira (1949; 1977) are useful in defining the boundaries of forest zones in Japan and throughout the monsoon area of East Asia. The indices are calculated by totaling monthly mean temperatures (t) as follows:

$$\mathrm{WI} = \sum (t-5), \text{ for months in which } t > 5^{\circ}\mathrm{C}$$
$$\mathrm{CI} = -\sum (5-\mathrm{t}), \text{ for months in which } \mathrm{t} < 5^{\circ}\mathrm{C}$$

The composition and structure of each forest zone are described in the following sections on the basis of Numata (1974) and Shimizu (2014).

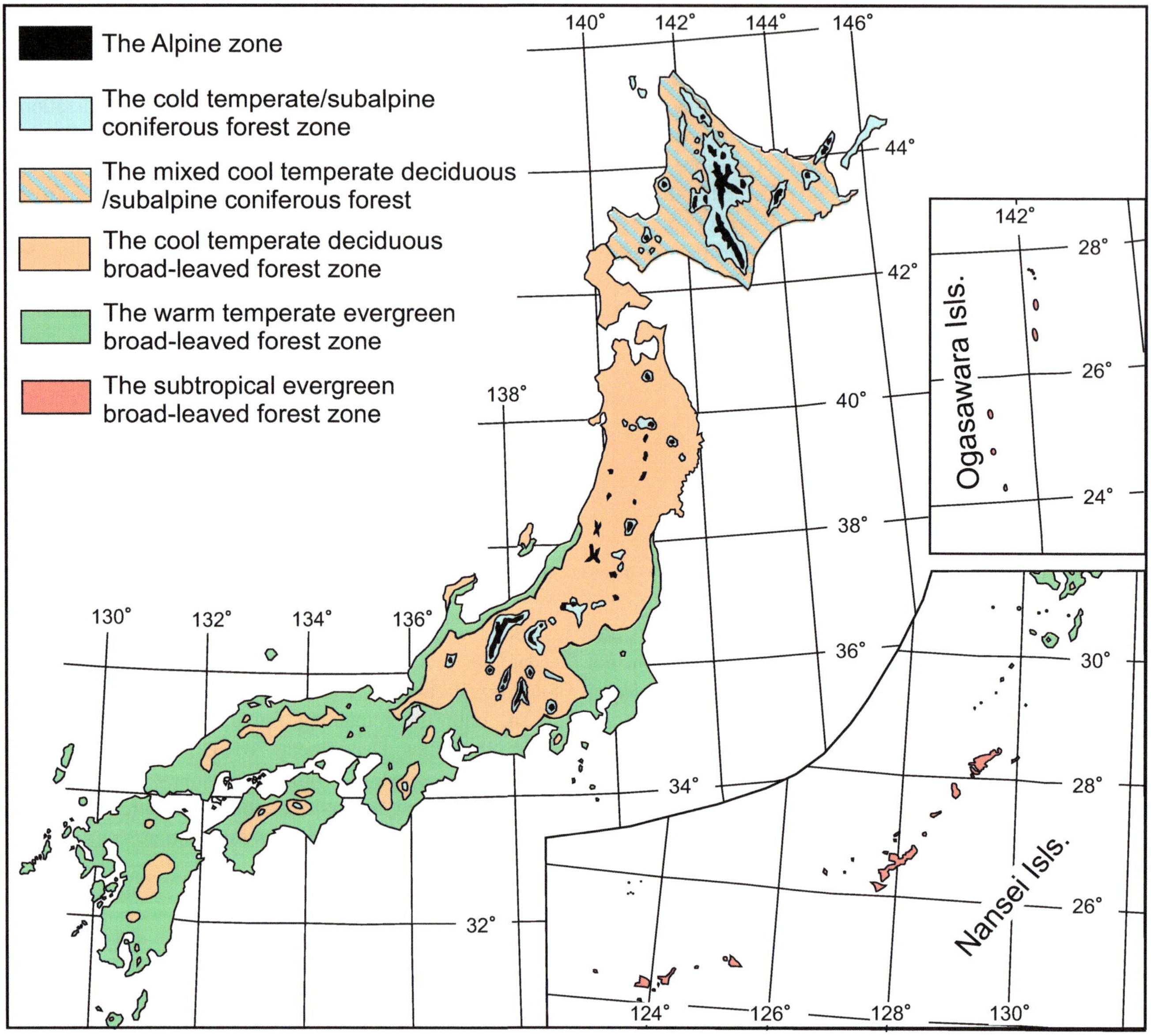

Fig. 2.27 Vegetation map of Japan. Redrawing of the vegetation map of Japan (Yoshioka 1973). We added the map in subtropical area and partially changed the forest distribution of the original map (Figure supplied by Tetsuya Sano)

1. **Subtropical evergreen broad-leaved forest zone**

The subtropical evergreen broad-leaved forest zone corresponds to WI values of 180 to 240, and in Japan includes the two chains of small islands between 20 and 30° N: the Nansei and Ogasawara Islands. While the Nansei Islands have been intermittently connected to the Asian continent as a result of sea-level changes during the geological past, the Ogasawara Islands are isolated oceanic volcanoes. The flora and ecology of the two island chains are therefore different. The coastal vegetation of the Nansei Islands is characterized by the presence of mangrove species, such as *Kandelia candel*, *Sonneratia alba*, *Bruguiera gymnorrhiza*, *Rhizophora mucronata* and *Heritiera littoralis*, although the number of mangrove species decreases drastically as the latitude increases. In contrast, the Ogasawara Islands lack mangrove species. Both groups of islands are characterized by evergreen broad-leaved forest including fig (*Ficus* spp.) and tree fern (*Cyathea* spp.). The mountain forests of the Nansei Islands are dominated by lauro-fagaceous trees (*Machilus*, *Litsea*, *Castanopsis*, *Lithocarpus* and *Quercus*) associated with understory trees such as Myrsinaceae, Symplocaceae and Theaceae. However, the natural forests of the Ogasawara Islands lack fagaceous species such as those in the genus *Quercus* that have difficulty dispersing seeds over long distances.

2. **Warm-temperate evergreen broad-leaved forest zone**

The warm-temperate evergreen broad-leaved forest zone corresponds to WI values of 85 to 180, and in Japan stretches from 30° N to nearly 38° N. Here, forests with species strongly similar to those of subtropical forest and lauro-fagaceous trees are dominant. The diversity of evergreen broad-leaved species becomes less rich with increasing latitude. Near the northern or upper limit of evergreen forests, there are particular types of forest which lack both tall evergreen broad-leaved trees and a representative tree in the cool-temperate zone, *Fagus crenata*, but are instead characterized by warm-temperate deciduous trees such as *Quercus serrata*, *Carpinus* spp. and *Fagus japonica*, sometimes mixed with temperate conifers such as *Abies firma*. This forest is called "warm temperate deciduous forest" (Kira 1949, 1977) and is explained as an area where the warmth of the growing season is still within the range of the warm-temperate zone (WI > 85) but cannot support the growth of broad-leaved evergreen tree species due to increased winter coldness (CI < −10).

3. **Cool-temperate deciduous broad-leaved forest zone**

The cool-temperate deciduous broad-leaved forest zone corresponds to WI values of 45 to 85, and in Japan extends northward to the southern part of Hokkaido. This zone is dominated by deciduous f*agaceous* species such as *F. crenata* and *Quercus crispula* mixed with *Acer* and *Carpinus*. On the Sea of Japan side, where the snow cover is heavy in winter, the dominancy of *F. crenata* is high. Other associated tree species are *Viburnum furcatum, Lindera umbellata* var. *membranacea*, etc., in the shrub layer. The shrub layer is also characterized by the dense growth of a dwarf bamboo, *Sasa kurilensis*, and the occurrence of evergreen broad-leaved shrubs such as *Camellia rusticana* and *Ilex leucoclada* that can endure winter coldness under snow cover. On the Pacific Ocean side, the canopy layer is dominated by species such as *F. crenata*, *F. japonica*, *Q. crispula*, *Carpinus tschonoskii* and *Betula grossa*. The shrub layer in most areas is dominated by dwarf bamboos such as *Sasa nipponica* or *Sasamorpha borealis*. In the region northward from the northern limit of *F. crenata*, the lowlands of Hokkaido, mixed cool-temperate deciduous/subalpine coniferous forest occurs, dominated by *Q. crispula*, *Acer pictum* subsp. *mono*, *Betula* spp., *Abies sachalinensis* and *Picea jezoensis*.

4. **Cold temperate/subalpine coniferous forest zone**

The cold temperate/subalpine coniferous forest zone corresponds to WI values of 15 to 45 and is dominated by evergreen conifers such as *Abies* and *Picea*. This forest zone is not found in Kyushu. The dominant species of the forest zone in Hokkaido is *A. sachalinensis* and *Picea jezoensis,* whereas *P. glehnii* also dominates in wetlands and rocky areas with shallow soil. *Abies mariesii* is the dominant species in the north of Honshu. In central Honshu, both *A. mariesii* and *A. veitchii* are commonly dominant. *A. mariesii* is not found on Shikoku or the Kii Peninsula of Honshu. In the forest zones, except in Hokkaido and northern Honshu, other conifers, such as *Chamaecyparis pisifera, Thuja standishii, Tsuga diversifolia* and *P. jezoensis* var. *hondoensis* (not in Shikoku) are also common species present.

5. **Alpine zone**

As an altitudinal vegetation zone, the regions above the tree line of the cold temperate/subalpine coniferous forest zone have been conventionally called the "alpine zone" in Japan. Many of the components of this zone came from subarctic and arctic zones in the glacial period and were left as relics on the summits of its high mountains. There are various types of vegetation in the alpine zone, and their distribution is governed by wind and snow as well as edaphic and topographic factors.

Most representative vegetation in the habitats where snow melts early is alpine scrubs dominated by *Pinus pumila*, a dwarf pine that occurs widely in northeastern Asia including Hokkaido and Honshu. Mt. Tekari (35° 20′ N) of Akaishi Mountains is the southernmost point at which *P. pumila* is found throughout the world, although this species is not found on Mt. Fuji (35° 21′ N). This species forms carpet-like monodominant stands of about 1 m in height, sometimes mixed with evergreen *Rhododendron*, such as *R. brachycarpum* and *R. aureum*, and dwarf tree components of the subalpine zone such as *Alnus viridis* subsp. *maximowiczii*, *Betula ermanii*, *Acer tschonoskii* and *Sorbus commixta* in the canopy layer.

(4) **Human influence on soil formation through vegetation change**

In Japan, forest has been converted to grassland by burning to obtain materials such as manure for agricultural production, feedstuff for livestock and material for making thatched roofs in the Edo period (Ogura 2009). A recent study revealed that savanna-like grasslands with small trees and bushes were widely distributed in areas surrounding human settlements throughout the country until the early Meiji period (the late nineteenth century) (Ogura 2009). Here, we will introduce a hypothesis that human activities for maintaining grassland vegetation affect the formation of "*black soils.*"

It is well known that soil derived from volcanic ash has a very high organic carbon content and the color of the surface horizon is often deep black. In Japan, such soils have traditionally been called "*kuroboku soils*" by agricultural soil scientists and "black soils" by forest soil scientists (Fig. 2.28). Forest scientists have considered that grassland vegetation plays an important role in the formation of "black soils," because these soils are found under grassland vegetation, while "brown forest soils" are distributed in areas under forest vegetation where two types of soils existed side by side (Forest Experimental Station 1952). Kawamuro et al. (1986) studied the composition of phytolith and pollen in "*black soils*" and a "*brown forest soil,*" both of which were situated in a forest on a slope of Mt. Kurohime, a volcano in Central Japan. They estimated that the former vegetation of the black soils was a grassland of *susuki* (*Miscanthus sinensis*) or *tigaya* (*Imperata cylindrica*) (Fig. 2.29), while that of the "*brown forest soil*" was forest; therefore, they surmised that the difference in vegetation had affected the soil types at the two sites. Both *susuki* and *tigaya* are species

Fig. 2.28 Susuki grass on andosols in Ushiku City, Ibaraki Prefecture (Figure supplied by Shinji Kaneko)

Fig. 2.29 "*Black soils*" in Towada City, Aomori Prefecture (Figure supplied by Shinji Kaneko)

of the grass family Poaceae and are C4 plants. Ishizuka et al. (2014) analyzed stable carbon isotope ratios ($\delta^{13}C$) of "*black soils*" collected throughout Japan and obtained the following results. The contribution of C4 grass to soil organic carbon (SOC) in black soils was generally high (> 44.6%) according to mass balance calculations from the $\delta^{13}C$ of SOC. The melanic index, an index of the black color of humus, was negatively correlated with $\delta^{13}C$ values. According to these results, Ishizuka et al. (2014) concluded that C4 grass played an important role in generating the dark-colored organic matter in "*black soils*."

The "*black soils*" were distributed throughout Japan, and the area of these soils accounts for 17% of the country's territory (Okamoto 2009). The color of humus in "*black soils*" has been attributed to "A type" humic acids (Kumada 1987), which are characterized by a deep black color. Due to this spectroscopic property, "*black soils*" are classified as Melanudands in the U.S. Soil Taxonomy (Soil Survey Staff 2010) and correspond to melanic Andosols in the World Reference Base for Soil Resources 2014 (WRB) (IUSS Working Group WRB, 2015). In the Soil Classification System of Japan (The Fifth Committee for Soil Classification and Nomenclature, 2017), "*black soils*" are mainly classified as Humic Non-Allophanic Andosols or Humic Allophanic Andosols.

In order to maintain grassland vegetation, grassland was burned in early spring every year. As a result, small grains of charcoal are often found in black soils (Okamoto 2009). According to Shindo (2012), the amount of charcoal grains comprises up to 33% of SOC, and a high correlation ($r = 0.777$ where r is correlation coefficient) was observed between the amount of charcoal grains and total SOC. Shindo et al. (1986) compared physicochemical and spectroscopic properties of humic acids obtained from the charred residue of *susuki* grass, and that from the "A horizon" of black soils, and found that both were quite similar in their elemental composition and spectroscopic properties. This evidence suggests that the black color of humus might be derived from charcoal particles which were formed by the burning of grass.

The "*black soils*" have been found to have a very high humus content: The concentration of SOC in the A horizon reaches 30% (Shindo et al. 1986), and the average carbon stock at a depth of 0–30 cm was measured as 13.8 Mg/m^2, which is second only to peaty soils among forest soils (Morisada et al. 2004). As for Andosols, the large accumulation of organic matter results from a combination of high detritus input associated with the generally high fertility and productivity of Andosols and from the effective stabilization of soil organic matter (SOM) against decomposition (Takahashi and Dahlgren 2016). The stabilization of SOM in Andosols has been attributed to the incorporation of SOM into organo-mineral and/or organometallic (Al/Fe–humus) complexes (Takahashi and Dahlgren 2016).

2.3.2 Agriculture

(1) **History and present state of agriculture in Japan**

The Köppen climate classification of Japan, except for the northernmost island of Hokkaido, is mostly Cfa or Dfa, characterized by warm temperatures in summer and a certain amount of rainfall throughout the year. This climatic condition is suitable for the production of paddy rice, an excellent staple food crop with a large yield and good nutritional quality and taste. In addition, paddy fields can maintain good productivity after continuous cultivation over a number of years and are free from growth injury due to continuous cropping.

Some archaeological studies have suggested that agriculture in Japan started in the Jōmon period, between around 15,000 years ago and the fourth century BC—the era preceding the Yayoi period. However, it is thought that hunting and gathering was the dominant means of obtaining food in the Jōmon period. The beginning of full-fledged agriculture in Japan was in the Yayoi period, between the fourth century BC and the third century AD, although recent studies using ^{14}C dating by the accelerator mass spectrometry method have pointed out that the beginning of the Yayoi period might be the tenth century BC (Fujio et al. 2005). The Yayoi period is defined by widespread paddy rice farming that was introduced from the Asian continent.

In the Yamato period, which started in the fourth century AD, governors constructed various political schemes concerning the development and possession of paddy fields in order to manage people's agricultural activities and collect taxes. In the sixth century AD (the Nara period), local governors surveyed their land fertility and compiled *Fudoki* in response to an imperial order. For example, "*Harima Fudoki*" classified land fertility into 9 grades in each *sato* (small village or hamlet) in Hyogo Prefecture of the Kinki region. Tax payments by rice in kind were the fundamental basis of the Japanese taxation system until the nineteenth century. As for upland farming, shifting cultivation such as slash-and-burn and semipermanent farming in uplands had been conducted from ancient times. However, the principal means of food production was paddy rice farming, due to its advantages as described above, and upland farming was the alternative in fields for which paddy rice production was difficult, until the late modern period. After the Meiji Restoration in 1868, upland farming was developed with the formation of a market for upland crops under modern capitalism as background (Cho 1986).

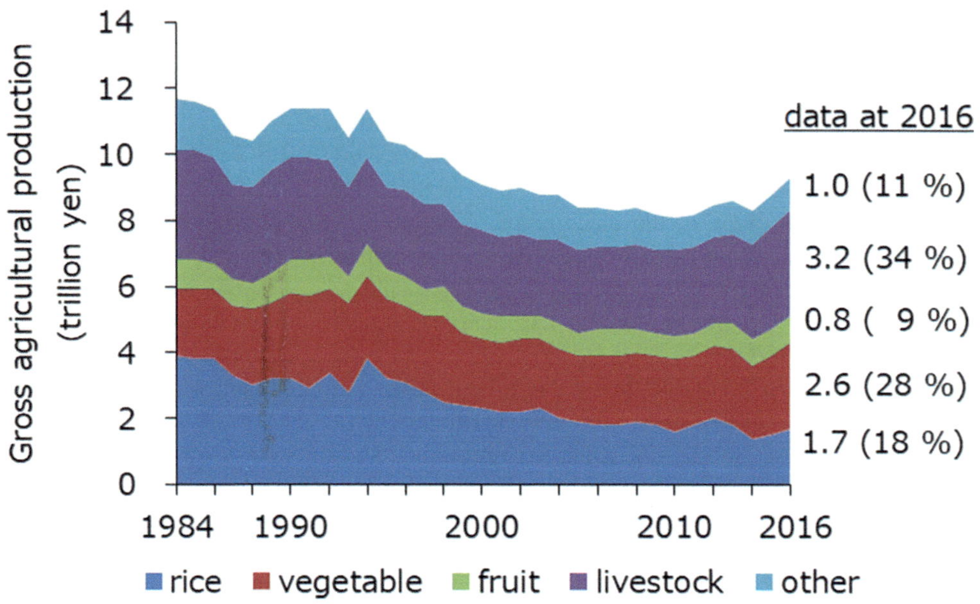

Fig. 2.30 Farming area of Japan. Data source. http://www.maff.go.jp/j/wpaper/w_maff/h28/h28_h/trend/part1/other/P044_h28_d0_2_21.xls

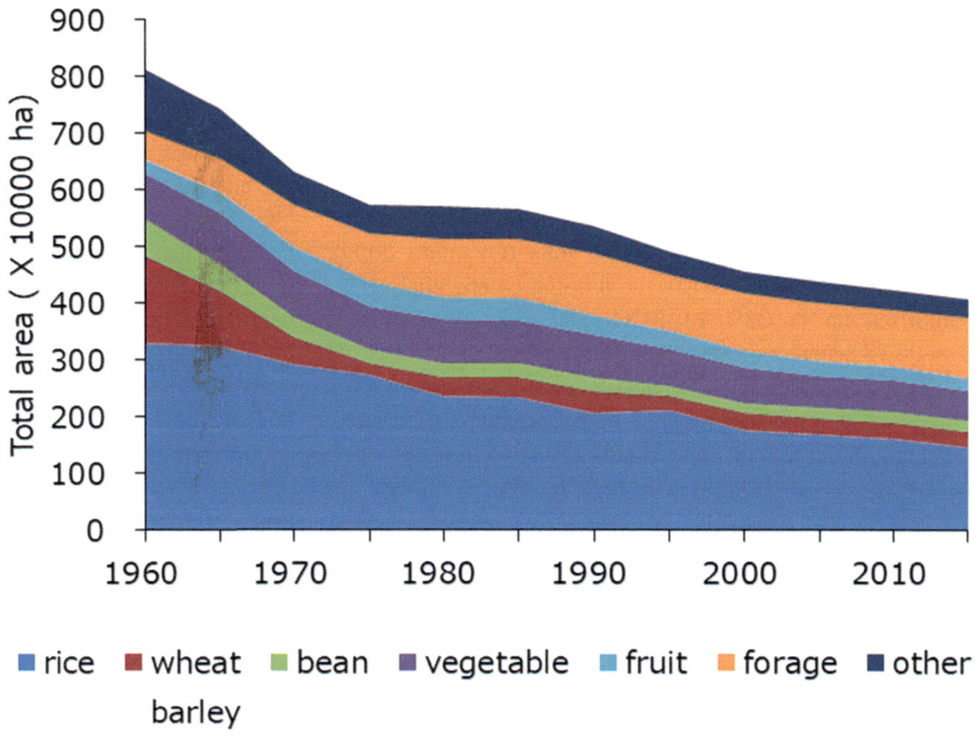

Fig. 2.31 Gross agricultural production of Japan. †100 yen = 0.94 US$ = 0.80 EUR (rate at 9/September/2020) Data source. https://www.e-stat.go.jp/stat-search/filedownload?statInfId=000031701883&fileKind=0

Hereinafter, data on agricultural production and farming areas are cited from statistics prepared by the MAFF. Recent gross agricultural production and farming areas of Japan are shown in Figs. 2.30 and 2.31, respectively. The area of paddy fields was 3.38 million ha in 1960 and decreased to 2.43 million ha by 2016. The paddy rice-growing area was 3.31 million ha in 1960 and was halved to 1.48 million ha by 2016. Paddy rice production was 3.9 trillion yen (equal to 34% of all agricultural production) in 1984 and had decreased to 1.7 trillion yen (18%) in 2016. The decrease of paddy rice farming was principally caused by a decrease in the consumption of rice in Japan with changes in Japanese dietary habits. In the 1970s, the domestic production of rice became sufficient to meet demand. In 1971, the Japanese government started to introduce the *gentan* policy, which involved subsidizing farmers who stopped growing rice for consumption as a staple and instead switched to producing other crops including non-staple rice in their paddy fields.

The *gentan* policy was continued for over 40 years, though it was abolished in 2018. Under the *gentan* policy, many paddy fields were converted to upland field, and some of them were used as paddy–upland rotational fields.

The area of upland fields was 2.69 million ha in 1960 and 2.04 million ha in 2016. The production of upland crops (vegetables and other in Fig. 2.31) was 3.6 trillion yen (equal to 39% of all agricultural production). The growing areas (in ha) for major upland crops in 2016 were as follows: 276,000 for wheat and barley, 150,000 for soybean, 77,000 for potato, 36,000 for sweet potato, 32,000 for radish, 18,000 for carrot, 26,000 for onion, 35,000 for cabbage and 24,000 for sweet corn. Some upland crops, especially wheat, barley and soybean, are also widely cultivated in converted paddy fields as alternatives to rice.

In livestock farming, headages for dairy cattle, beef cattle, swine, laying hen and broiler chicken in 2016 were 1.345 million, 2.479 million, 9.313 million, 173.3 million and 134.4 million, respectively. Related to livestock, and especially to cattle, the growing areas for pasture grass and maize in 2016 were 744,000 and 94,000 ha, respectively. In orchards, the growing area of *mikan* (*Citrus unshiu*; satsuma mandarin) and apple in 2016 was 44,000 and 38,000 ha, respectively. The growing areas of grape, chestnut and persimmon were each around 20,000 ha, while that of tea fields was 43,000 ha.

The food self-sufficiency ratio of Japan was 78% in 1961. This gradually decreased during the latter part of the twentieth century and has hovered around 40% in the twenty-first century. This ratio is very low among industrialized nations and similar to that of the Republic of Korea.

(2) **Cultivated fields and soils**

The area of each soil type for Japanese cultivated fields is shown in Table 2.2. As mentioned above, the area of paddy fields in 2016 was 2.43 million ha. Paddy fields are most widely distributed in alluvial areas of plains; however, the slopes of hills and mountains have also been used as terraced paddy fields. Almost all paddy fields in Japan are equipped with irrigation systems, and rice is cultivated with irrigation water rather than groundwater. Recently, farmers' demand for direct seeding, which can reduce labor, has been increasing, and its technological development is being advanced. However, in 2013, 98% of rice production was still performed with transplanting by farm machinery. Some paddy fields were converted to upland fields, or cultivated rotationally as paddy and upland, for the production adjustment of rice. The scale of paddy fields has been increased by land consolidation. The percentage of paddy fields with a standard scale (0.3 ha) or larger, which was 2.4% in 1964, had increased to 62.1% in 2010, by which time large-scale (1 ha or larger) fields accounted for 8.4% of the total. The major soil type of paddy fields is fluvic soils, including fluvic paddy soils, gley fluvic soils and gray fluvic soils.

Upland fields are widely distributed, from alluvial lowlands to terraces, hills and mountains. In upland fields also, terraces were formed on slopes in order to improve the workability of farm machinery. As for soil groups, nearly half of upland fields are Andosols, and approximately 30% are terrestrial soils such as brown forest soils.

(3) **Human impacts on soils by agricultural activities**

A typical human impact on soils by agricultural activities in Japan is the influence of water management in paddy fields. In Japan, upland rice is scarce and most rice is cultivated in paddy fields. Rice production usually takes place once a year, with planting in May to June and harvesting in autumn. In warm regions, wheat, barley or vegetables are cultivated in paddy fields as succeeding crops during winter, whereas the dual cropping of rice (two growing seasons in a year) is rare. In the rice-growing season from spring to autumn, paddy fields are saturated with irrigation water that moves from the surface downward and soils are in a reductive condition. In contrast, paddy fields between autumn and spring are in an oxidative condition similar to upland fields

Table 2.2 Distribution area of great soil groups in Japanese agricultural land (%) (Table supplied by Yusuke Takata)

	Paddy	Upland field	Meadow	Orchard	Agricultural land
Organic soils	5.0	1.5	6.4	0.1	3.9
Andosols	13	49	55	28	29
Podzols	0.0	0.0	0.1	0.0	0.0
Fluvic soils	71	22	16	15	47
Red–yellow soils	2.3	6.7	3.0	17	4.6
Stagnic soils	5.4	4.0	6.0	2.3	4.9
Eutrosols	0.0	2.6	0.2	0.5	0.7
Brown forest soils	2.2	11	11	33	8.0
Regosols	0.5	3.3	2.1	4.0	1.7

because irrigation is not conducted in this period. Additionally, the seasonal variation of groundwater level also influences the soil moisture conditions. As a result of these water actions, paddy soils have unique characteristics, such as separation into an oxidized layer and a reduced layer, the formation of mottles and concretion of iron and manganese, and gley and pseudo-gley horizons.

By contrast, the impact of water is small for upland fields. However, the topsoil of upland fields is frequently disturbed by plowing because untilled cropping is not dominant in Japan. Deep tillage for the enlargement of the root zone and the improvement of water permeability is also conducted, especially in terrestrial soils with problematic physical properties. Additionally, upland soils are subject to chemical impact due to the input of fertilizer and organic matter such as compost; this contrasts with the situation in paddy fields, because the demand for fertilizer in upland cropping, especially for vegetables, is generally larger than it is for paddy rice. The input of animal manure compost is especially large in areas where livestock farming is active and the utilization of animal waste is required, such as southern Kyushu.

The farming of some upland crops has particular effects on soils. Because acidic soils are suitable for tea cultivation, tea garden soils are usually acidified to pH (H_2O) 4.5 or lower by soil management, such as the heavy application of ammonium sulfate. Even for potato, for which acidic soils are not suitable, the soil reaction is still managed to be acidic, in order to counter the common scab, a serious soilborne disease that can be suppressed by soil acidity. In Andosol areas, farmers prefer growing potato or sweet potato in brown subsoil or transferred non-Andosol, rather than black topsoil, because it is assumed that the appearance of the potato skins is better in light-colored soils. The swapping of subsoil to topsoil or soil dressing is often conducted in potato-producing Andosol areas for this purpose. In burdock and Chinese yam fields, planting rows are tilled deeply by a trenching machine. Trenches about 20 cm in width and 1 m in depth, filled with a mixture of topsoil and subsoil, are observed in soil profile surveys.

References

Astronomical Observatory of Japan (ed) (2018) Chronological Scientific Tables 2019 (Rika Nenpyo 2019). Maruzen, Tokyo. (in Japanese)

Bockheim JG, Hartemink AE (2017) The Soils of Wisconsin, pp 393, Springer

Broecker WS (1984) Terminations. In: Berger A, Imbrie J, Hays J, Kukla G, Saltzman B (eds) Milankovitch and Climate, Reidel, Dordrecht, pp 687–698

Chiba T (1991) Hageyama no kenkyu (A study of bald mountains), revised edition. Sosiete, Tokyo (in Japanese)

Cho K (1986) Fundamental problems and conditions for expansion of upland farming. Jpn. J Farm Manage 23(3):4–17 (In Japanese)

Dansgaard W (1964) Stable isotopes in precipitation. Tellus 16:436–468

Division 3 of Soils (1977) Classification of cultivated soils. 2nd Approximation. National Institute for Agricultural Sciences. (in Japanese)

Emiliani C (1955) Pleistocene temperatures. Journal Geol 63:538–578

Farjon A (2013) Chamaecyparis obtusa. The IUCN Red List of Threatened Species 2013. e.T42212A2962056.http://dx.doi.org/10.2305/IUCN.UK.2013-1.RLTS.T42212A2962056.en. Accessed 28 Sep 2018

The Fifth Committee for Soil Classification and Nomenclature of Japanese Society of Pedology (2017) Soil classification system of Japan. (in Japanese)

Forest Experimental Station (1952) Forest Soil Profiles. Forest Experimental Station, Tokyo (in Japanese with English summery)

Forest Soil Division (1976) Classification of forest soils in Japan (1975). Bulletin of the government forest experiment station 280:1–28 (in Japanese with English summary)

Fujio S, Imamura M, Nishimoto T (2005) When did the wet-rice cultivation begin in Japanese Archipelago? Sokendai Review of Cultural and Social Studies 1:69–96 (in Japanese)

Geospatial Information Authority of Japan (2019) Geospatial information library. https://maps.gsi.go.jp/ (June 2019) (in Japanese)

Hatano S (1979) Landform classification map of postglacial dissection front and prediction of slope failure. Abstract of research presentation, Japan Society of Erosion Control Engineering, pp 16–17. (in Japanese)

Horie K (2002) Studies on the chemical characteristics of the ultramafic plants in Hokkaido. J Rakuno Gakuen University, Natural Science 26:155–264(in Japanese with English summary)

Igarashi Y, Oba T (2006) Fluctuations in the East Asian monsoon over the last 144 ka in the northwest Pacific based on a high-resolution pollen analysis of IMAGES core MD01-2421. Quaternary Science Reviews 25:1447–1459

Inoue K (1981) Implications of eolian dusts to 14Å minerals in the volcanic ash soils in Japan. Pedologist 25:97–118 (in Japanese with English summary)

Inoue K, Mizota C (1988) Eolian origin of 2:1 layer silicates and fine quartz in Andsols and red-yellow soils developed on limestones and basalts. Clay Science(Nendo Kagaku), 28:30–47. (in Japanese with English summary)

Inoue K, Naruse T (1987) Physical, chemical, and mineralogical characteristics of modern eolian dust in Japan and rate of dust deposition. Soil Science and Plant Nutrition 33:327–345

Institute for Agro-Environmental Sciences, NARO (2019) Soil inventory in Japan. http://soil-inventory.dc.affrc.go.jp/ (June 2019) (in Japanese)

International Commission on stratigraphy (2018) International Chronostratigraphic Chart. http://stratigraphy.org/ICSchart/ChronostratChart2018-07.pdf (July 2018)

Ishizaka T, Ono A, Kadowaki R (1981) Characteristics of the aeolian dust over Japan and its origin. Tenki 28:651–665 (in Japanese)

Ishizuka S, Kawamuro K, Imaya A, Torii A, Morisada K (2014) Latitudinal gradient of C4 grass contribution to Black Soil organic carbon and correlation between $\delta_{13}C$ and the melanic index in Japanese forest stands. Biogeochemistry 118:339–355

IUSS Working Group WRB (2014) World Reference Base for Soil Resources 2014. International soil classification system for naming

soils and creating legends for soil maps. World Soil Resources Reports No. 106, FAO, Rome
IUSS Working Group WRB (2015) World reference base for soil resources 2014. Update 2015. World Soil Resources Reports No. 103. FAO, Rome
Iwata S, Koaze T (2001) Mountain Landforms of the Japanese Islands. In: Yonekura N, Kaizuka S, Nogami M, Chinzei K (eds) Regional Geomorphology of the Japan Islands Vol.1 Introduction to Japanese Geomorphology, pp 131–177. University of Tokyo Press(in Japanese)
Iwatsuki K, Yamazaki T, Bufford DE, Ohba H (eds) (1998) Flora of Japan. Kodansha, Tokyo
Japan Association for Quaternary Research (ed) (1987) Quaternary Maps of Japan. University of Tokyo Press. Tokyo.(in Japanese)
Japan Meteorological Agency (2017) Global Warming Prediction Information Vol 9, http://www.data.jma.go.jp/cpdinfo/GWP/Vol9/pdf/02.pdf (May 2018)
Japan Meteorological Agency (2018) Mesh normal year value map, http://www.data. jma.go.jp/obd/stats/etrn/view/atlas.html (May, 2018)
Kaizuka S (1972) Macro-landforms of island arc and plate tectonics. Kagaku 42:573–581 (in Japanese)
Kaizuka S (1977) Landforms of Japan: characteristics and origin. Iwanami Shoten, Publishers, Tokyo (in Japanese)
Kaizuka S (1988) Landform classification of Japan and its surroundings. Monthly Earth (Gekkan Chikyu) 10:99–101 (in Japanese)
Kaizuka S (1992) Reading of coast and plain. Iwanami Shoten, Publishers, Tokyo (in Japanese)
Kaizuka S, Chinzei K (eds) (1995) Nature of the Japan Islands, vol 2. Iwanami Shoten, Publishers. Tokyo, Mountains of Japan (New edition) (in Japanese)
Kaizuka S, Koike K, Endo K, Yamazaki H, Suzuki T (eds) (2000) Regional Geomorphology of the Japan Islands, vol 4. Geomorphology of Kanto and Izu-Ogasawara. University of Tokyo Press, Tokyo (in Japanese)
Kaneko S, Akieda N, Naito F, Tamai K, HiranoY (2007) Nitrogen budget in a rehabilitation forest on degraded granitic hills. J For Res 12:38–44
Kato M, Ebihara J (2011) Endemic plants of Japan. Tokai University Press, Kanagawa Prefecture
Katsuki T, Luscombe D, Farjon A (2013) Sciadopitys verticillata. The IUCN Red List of Threatened Species 2013. e. T34111A2846623. http://dx.doi.org/10.2305/IUCN.UK.2013-1. RLTS.T34111A2846623.en. Accessed 28 Sep 2018
Kawamuro K, Torii A (1986) Difference in past vegetation between black soils and brown forest soils derived from volcanic ash at Mt. Kurohime, Nagano Pref. Japan. The Quaternary Research 25:81–98 (in Japanese)
Kimura G, Hashimoto Y, Kitamura Y, Yamaguchi A, Koge H (2014) Middle Miocene swift migration of the TTT triple junction and rapid crustal growth in southwest Japan: A review. Tectonics 33:1219–1238
Kira T (1949) Forest zones of Japan. Ringyo Gizyutu Kyokai, Tokyo (in Japanese)
Kira T (1977) A climatological interpretation of Japanese vegetation zones. In: Miyawaki A, Tüxen R (eds) Vegetation Science and Environmental Protection. Maruzen, Tokyo, pp 21–30
Kondo R (2010) Graphic chart of opal phytoliths: Introduction to opal phytolith science by scanning electron microscope photographs. Hokkaido University Press, Sapporo (in Japanese)
Konishi K, Omura A, Nakamichi O (1974) Radiometric coral ages and sea level records from the late Quaternary reef complexes of the Ryukyu Islands. Proceedings 2nd International Coral Reef Symposium. Australia, pp 595–613
Kumada K (1987) Chemistry of soil organic matter. Japan Scientific Press-Elsevier, Tokyo (in Japanese)
Kuno H (1966) Lateral variation of basalt magma type across continental margins and island arcs. Bull Volc 29:195–222
Kyuma K (1985) Soil temperature regime of Japanese soils. Soil Sci Plant Nutr 31:43–468
Lisiecki LE, Raymo ME (2005) A Pliocene-Pleistocene stack of 57 globally distributed benthic records. Paleoceanography, 20, PA1003, https://doi.org/10.1029/2004pa001071
Machida H (1977) Story of tephra: Natural history of volcano and plains. Sojyu-Syobo, Tokyo (in Japanese)
Machida H (2002) Volcanoes and tephra in the Japan area. Global Environmental Research 6:290–305
Machida H (2009) Perspectives from Paleoclimate Studies. In Japan Association for Quaternary Research (ed) Digital book: Progress in Quaternary Research, pp 3-3-3-49 (in Japanese)
Machida H, Chinzei K (2001) Historical geomorphology of the Japanese Islands. In Yonekura, N., Kaizuka, S., Nogami, M., Chinzei, K. (eds) Regional Geomorphology of the Japan Islands Vol 1 Introduction to Japanese Geomorphology, pp 297–322. University of Tokyo Press (in Japanese)
Machida H, Matsuda T, Umitsu M, Koizumi T (eds) (2006) Regional Geomorphology of the Japan Islands, vol 5. Geomorphology of Chubu Region. University of Tokyo Press, Tokyo (in Japanese)
Machida H, Suzuki M, Miyazaki A (1971) Chronology of the preceramic age in South Kanto with special reference to tephrochronology, radiocarbon dating and obsidian dating. The Quaternary Research (Daiyonki-Kenkyu), 10:1–14. (in Japanese with English summary)
Matsuda T, Kikkawa M (2001) Distribution of M 5 shallow earthquakes and active faults on the Japanese islands. Active Fault Research(Katsudanso-Kenkyu), 20:1–22.(in Japanese with English summary)
Matsui T (1988) Introduction to Soil Geography, p 316, Tsukiji Shokan, Tokyo. (in Japanese)
Matsui T (2001) Distribution of major soil type. In: Asami S (ed) Soil Geography, pp 24, Kokonshoin, Tokyo. (in Japanese)
Matsugi T (2007) A collection of Japanese history 1. A story of the creation of the archipelago, Shogakukan Inc, Tokyo (in Japanese)
Matsui H, Tada R, Oba T (1998) Low-salinity isolation event in the Japan Sea in resoponse to eustatic sea-level drop during LGM reconstruction based on salinity-balance model. The Quaternary Research (Daiyonki-Kenkyu), 37:221–233. (in Japanese with English summary)
Ministry of Agriculture, Forestry and Fishery (2017) Statistical Information. http://www.maff.go.jp/j/tokei/index.html (in Japanese)
Ministry of Land, Transport and Tourism (2018) White Paper on Land in Japan,2018. http://www.mlit.go.jp/statistics/file000006.html(in Japanese). Accessed 28 Sep 2018
Miura H (2017) Ice sheets and Landforms. In Koike, K. et al. (eds) Encyclopedia of Physical Geography, pp. 290–291.(in Japanese)
Miura H, Hirakawa K (1995) Origin of fossil periglacial wedges in northern and eastern Hokkaido, Japan. Journal of Geography (Chigaku Zasshi) 104:189–224 (in Japanese with English summary)
Mizota C, Inoue K (1988) Oxygen isotope composition of eolian quartz and sediments-its significance as a tracer of eolian components-. Clay Science(Nendo Kagaku), 28:38-54.(in Japanese with English summary)
Mizota C, Matsuhisa Y (1985) Eolian additions to soils and sediments of Japan. Soil Science and Plant Nutrition 31:369–382
Mizuno T, Horie K, Nosaka S, Obata H, Mizuno N (2009) Serpentine plants in Hokkaido and their chemical characteristics. Northeastern Naturalist 16(special volume 5):65–80

MOE (Ministry of the Environment), Japan (2014) Enforcement status of Agricultural Land–Soil Pollution Prevention Law in 2014 fiscal year. MOE, Japan, Available at

Morisada K, Ono K, Kanomata H (2004) Organic carbon stock in forest soils in Japan. Geoderma 119:21–32

Nakagawa T, Kitagawa H, Yasuda Y, Tarasov RE, Gotanda K, Suwai Y (2005) Pollen/event stratigraphy of the varved sediment of Lake Suigetsu, central Japan from 15,701 to 20,217 SG vyr BP (Suigetsu varve years before present): Description, interpretation and correlation with other regions. Quaternary Science Reviews 24:1691–1701

Nakagawa T, Tarasov RE, Nishida K, Gotanda K, Yasuda Y (2002) Quantitative pollen-based climate reconstruction in central Japan: Application to surface and late Quaternary spectra. Quaternary Science Reviews 21:2099–2113

Naruse Y, Inoue K (1990) Aolian deposits in the Ryukyus. In The Research Group on Coral Reef Regions in Japan (eds) Coral islands in Japan Vol 1 Tropics in Japan: Natural history of coral islands, pp 248–262. Kokon Shoin, Publishers. Tokyo. (in Japanese)

North Greenland Ice Core Project members (2004) High-resolution record of Northern Hemisphere climate extending into the last interglacial period. Nature 431:147–151

Numata M (ed) (1974) The flora and vegetation of Japan. Kodanshya, Tokyo, Elsevier, Amsterdam

Oba T, Akasaka N (1990) Paleoenvironmental change of the Japan Sea based on Organic Carbon Contents of Two Piston Cores. The Quaternary Research (Daiyonki-Kenkyu), 29:417–425. (in Japanese with English summary)

Oba T, Kato M, Kitazato H, Koizumi I, Omura A, Sakai T, Takayama T (1991) Paleoenvironmental changes in the Japan Sea during the last 85,000 years. Paleoceanography 6:499–518

Obara H, Ohkura T, Takata Y, Kohyama K, Maejima Y, Hamazaki T (2011) Comprehensive soil classification system of Japan. First Approximation. Bulletin of National Institute for Agro-Environmental Sciences 29:1–73 (in Japanese with English summary)

Ogura J (2009) Forest vegetation and fire in Edo and Meiji period. Forest Science 55:5–9 (in Japanese)

Ohsawa M (1987) Habitat differentiation and the ecological niche in vegetation. Physiol Eco Japan 24:15–27

Ohsawa M (2006) Climate change impacts on Vegetation in Humid Asian Mountains. Global Environmental Research 10(1):13–20

Ohwi J (1965) Flora of Japan. Smithsonian Institution, Washington, D. C

Okamoto T (2009) Charcoal in forest soils as an indicator of past fires. Shinrin Kagaku 55:18–23 (in Japanese)

Okuda M, Nakagawa T, Takemura K (2010) Present status of pollen-based paleoclimate reconstruction from Lake Biwa sediment. The Quaternary Research (Daiyonki-Kenkyu), 49:133–146. (in Japanese with English summary)

Okumura S, Minagawa M, Oba T, Ikehara K (1996) Paleoenvironmental analysis of two sediment cores off Akita City in the Japan Sea based on oxygen, carbon and nitrogen isotopes. The Quaternary Research (Daiyonki-Kenkyu), 35:349–358. (in Japanese with English summary)

Ota Y, Omura A (1992) Contrasting style and rates of tectonic uplift of coral reef terraces in the Ryukyu and Daito Island, southwestern Japan. Quaternary International 15(16):17–29

Railsback LB (2015) Marine isotope stages and substages. http://railsback.org/Fundamentals/SFMGSubstages01.pdf (Febrary 2015)

Railsback LB, Gibbard PL, Head MJ, Voarintsoa NRG, Toucanne S (2015) An optimized scheme of lettered marine isotope substages for the last 1.0 million years, and the climatostratigraphic nature of isotope stages and substages. Quaternary Science Reviews 111:94–106

Saigusa M, Matsuyama N, Abe A (1992) Distribution of Allophanic Andosols and Nonallophanic Andosols in Japan based on the data of soil survey reports on reclaimed land. Jpa J Soil Sci Plant Nutr 63:646–651 (In Japanese with English summary)

Saito Y (1992) Reading of the history of the Japanese Islands. Iwanami Shoten, Publishers, Tokyo (in Japanese)

Sakaguchi Y (1961) Paleogeographical studies of peat bogs in northern Japan. Journal of Faculty of Science, University of Tokyo, Section II 12:421–513

Sakaguchi Y (1974) Earth science of peatlands. University of Tokyo Press, Tokyo (in Japanese)

Sakaguchi Y (1979) Distribution and genesis of Japanese peatlands. Bulletin of the Department of Geography, University of Tokyo, 11:17–42

Sakaguchi Y (1989) Natural history of Ozegahara. Chuokoron-sha, Inc., Tokyo

Sakaguchi Y, Okumura K (1986) Interglacial climate and relic red soils in northern Japan based on pollen records of interglacial deposits in eastern Hokkaido. Bulletin of the Department of Geography, University of Tokyo, 18 29–48

Sase T, Hosono M, Utsugawa T, Kato S, Komahara M (1987) Opal phytolith analysis of the Kanto Loam formation in Musashino Upland, Kanto Plain, Japan. The Quaternary Research (Daiyonki-Kenkyu), 26:1–11.(in Japanese with English summary)

Sase T, Machida H, Hosono M (2008) Fluctuations of opal phytolith assemblage in the Tachikawa and Musashino Loam formations in southwest Kanto, central Japan: Changes in vegetation, climate, terrace, and soil-facies since marine isotope stage 5.1. The Quaternary Research (Daiyonki-Kenkyu), 47:1–14.(in Japanese with English summary)

Shackleton NJ (1969) The last interglacial in the marine and terrestrial record. Proceedings of the Royal Society of London, B. 174:135–154

Shimizu T (1962) Studies on the limestone flora of Japan and Taiwan, Part I. J Fac Textile Sci Technol Shinshu Univ Ser A11:1–105

Shimizu T (1963) Studies on the limestone flora of Japan and Taiwan, Part II. J Fac Textile Sci Technol Shinshu Univ Ser A12:1–88

Shimizu Y (2014) Process of the formation of Japanese forest and typification of vegetation zone-From an East Asian viewpoint. Regional Views 27:19–75 (in Japanese with English summary)

Shindo H (2012) Role of polyphenols and charred plant materials in the formation and accumulation of black humic acids (JSSPN Awards). Jpn J Soil Sci Plant Nutr 83:515–518 (in Japanese)

Shindo H, Matsui Y, Higashi T (1986) A possible source of humic acids in volcanic ash soils in Japan – charred residue of Miscanthus sinensis. Soil Sci 141:84–87

Soil Survey Staff (2010) Key to soil taxonomy, 11th edn. Natural Resources Conservation Service, USDA, Washington, DC

Soil Survey Staff (2014) Keys to Soil Taxonomy, 12th edn. USDA-Natural Resources Conservation Service, Washington, DC

Statistics department (2018) The 91th statistical yearbook of ministry of agriculture, forestry and fisheries Japan (MAFF). MAFF, Tokyo

Suzuki T (1997) Introduction to map reading for civil engineers: Volume 1 Geomorphological basis for mapping reading. Kokon Shoin, Publishers. Tokyo. (in Japanese)

Takahashi T, Dahlgren RD (2016) Nature, properties and function of aluminum–humus complexes in volcanic soils. Geoderma 263:110–121

Takata Y (2010) Estimation of soil carbon stock changes in Japanese agricultural soils using national resource inventory. Proceedings of an international workshop on evaluation and sustainable management of soil carbon sequestration in Asian countries. Indonesian Soil Research Institute (ISRI), Food & Fertilizer Technology Center for the Asian and Pacific Region (FFTC) and National Institute for Agro-environmental Sciences (NIAES), pp 15–28

Takata Y, Kuwagata T, Kohyama K, Obara H (2011) Delineation of Japanese soil temperature regime map. Soil Sci. Plant Nutr. 57:294–302

Takhtajan A (1986) Floristic Regions of the World. University of California Press, Berkeley

Thomas P, Katsuki T Farjon A(2013) Cryptomeria japonica. The IUCN Red List of ThreatenedSpecies2013. e.T39149A2886821.http://dx.doi.org/10.2305/IUCN.UK.2013-1.RLTS.T39149A2886821.en. Accessed 28 Sep 2018

Toyoda S, Naruse T (2002) Eolian dust from the Asian deserts to the Japanese islands since the Last Glacial Maximum: the basis for the ESR method. Transactions Japanese Geomorphological Union (Chikei) 23:811–820

Tsuiki M, Harada Y (1997) Present condition and future issues on the generation of animal wastes in Japan. In: Nishio M (ed) Environment Conservation and New Animal Production, pp. 15–29. Agriculture, Forestry and Fisheries Technical Information Society, Tokyo (in Japanese)

Tsuji S (1997) A land ecosystem in the transition to the Jomon Age. The Quaternary Research (Daiyonki-Kenkyu), 36, 309–318.(in Japanese with English summary)

Tsuji S, Minaki M, Osawa S (1984) Paleobotany and Paleoenvironment of the Late Pleistocene in the Sagami Region, Central Japan. The Quaternary Research (Daiyonki-Kenkyu), 22:279–296

Ueda S, Sugimura A (1970) Island Arc. Iwanami Shoten, Publishers, Tokyo (in Japanese)

Ugawa S, Takahashi M, Morisada K et al (2012) Carbon stocks of dead wood, litter, and soil in the forest sector of Japan: general description of the National Forest Soil Carbon Inventory. Bulletin of FFPRI 11:207–221

Yasuda Y, Miyoshi N (eds) (1998) Zusetsu Nihonretto Syokuseishi (illustrated vegetation history of Japanese archipelago). Asakura Publishing, Tokyo (in Japanese)

Yonekura N, Kaizuka S, Nogami M, Chinzei K (2001) Regional Geomorphology of the Japanese Islands: Volume 1 Introduction to Japanese Geomorphology. University of Tokyo Press. Tokyo. (in Japanese)

Yoshinaga S (1995) Origin of parent matetial of volcanic ash soils in Japan. Volcano (Kazan) 40:153–166 (in Japanese with English summary)

Yoshinaga S (1998) Accumulation rate of tropospheric dust in and around the Japan Islands during the latest Quaternary. The Quaternary Research (Daiyonki-Kenkyu), 37, 205–210.(in Japanese with English summary)

Yoshioka K (1973) Plant geography. Kyoritsu Shuppan Co.Ltd, Tokyo (in Japanese)

3 Soil Classification and Distribution

Hitoshi Shinjo and Yusuke Takata

Abstract

The Japanese Society of Pedology released the "Soil Classification System of Japan," the latest version of the comprehensive soil classification system in Japan. According to this classification system, the area of each soil group is calculated and soil map is produced. Andosols are found along volcanic areas, Brown Forest soils and Cambic Red-Yellow soils (refer to Cambisols) are distributed in steep mountainous zones far from active volcanoes, and Fluvic soils cover narrow lowlands. Cultivated land accounts for 12% of the total land surface and is mainly distributed in the plains. Thus, Fluvic soils show the largest distribution area (47%), followed by Andosols (29%) in the cultivated land. Nutrient imbalance, soil organic matter decline, heavy metal contamination, and soil sealing are major threats to Japanese agricultural soils.

Keywords

Soil classification • Soil map • Soil resources

The original version of this chapter was revised. Text that was inadvertently published in Table 3.1 has been deleted. The correction to this chapter can be found at https://doi.org/10.1007/978-981-15-8229-5_11

H. Shinjo (✉)
Graduate School of Global Environmental Studies, Kyoto University, Kyoto City, Kyoto Prefecture, Japan
e-mail: shinjo.hitoshi.4x@kyoto-u.ac.jp

Y. Takata
Institute for Agro-Environmental Sciences, NARO, Tsukuba City, Ibaraki Prefecture, Japan

3.1 Soil Classification

3.1.1 History of Soil Classification in Japan[1]

The development of a natural soil classification system in Japan was initiated in 1882 by the provincial soil survey conducted under the leadership of Dr. Max Fesca, a German scholar. This survey produced soil maps for all prefectures except Hokkaido and Okinawa by 1948. However, its classification used the criteria of surface geology, soil texture, and humus content following the agro-geological view of soil, rather than the view of soil formation factors that had recently emerged in those days. After World War II, soil classification in Japan, closely associated with various soil investigation projects, has developed uniquely for each land use, namely forest, upland field, and paddy field. Some of these achievements can be represented by the Classification of Forest Soils in Japan (1975) (Forest Soil Division 1976) and the Classification of Cultivated Soils in Japan (2nd Approximation) (Division 3 of Soils 1977). Although the basic land classification survey commenced in 1954 in order to produce a soil map covering all the above land uses, their classification criteria were not uniform across forest and cultivated land.

Against this background, the Japanese Society of Pedology noted the significance of having a comprehensive and internationally referable classification system and began working toward its realization in 1963. As a result, the first and second approximations were released in 1986 and 2003, respectively. Based partly on the second approximation, Obara et al. (2011) proposed to replace the classification of cultivated soils with a more comprehensive classification system to classify the soils of Japan regardless of land use. By revising this proposal, the Japanese Society of Pedology

[1]This section is mainly indebted to the preface to the Unified Soil Classification System of Japan, 2nd Approximation (2002), written by the late Prof. Dr. T. Hamazaki, the then President of the Japanese Society of Pedology.

R. Hatano et al. (eds.), *The Soils of Japan*, World Soils Book Series,
https://doi.org/10.1007/978-981-15-8229-5_3

released the "Soil Classification System of Japan" (The Fifth Committee for Soil Classification and Nomenclature 2017), which is the latest version of the comprehensive soil classification system in Japan. In the following section, some features of this classification system are described.

3.1.2 Latest Soil Classification System in Japan

This system defines the following categories: soil great group, soil group, and soil subgroup, using the key-out rule. The keys employed are defined as diagnostic horizons, diagnostic properties, and diagnostic materials that are determined objectively and quantifiably. Their main characteristics can be summarized as follows.

(1) **Diagnostic Horizon**

Albic horizon: An eluvial horizon with a light color from which clay and/or free Fe oxides have been removed.

Albic paddy epipedon: A plowed epipedon in paddy fields with bleaching.

Anthraquic-gray horizon: A subsurface horizon with epiaquic properties that has formed under irrigation in paddy fields. Commonly found in clayey paddy fields with shallow groundwater.

Anthraquic-iron horizon: A subsurface horizon with Fe illuviation typically developed in paddy fields with good percolation of water. Usually accompanied by Mn illuviation.

Argic horizon: A subsurface horizon with a significantly higher percentage of clay than the overlying soil material.

Cambic horizon: A subsurface horizon resulting from physical alterations, chemical transformations, or removals, or from a combination of two or more of these processes.

Dark Eutric epipedon: A dark-colored epipedon enriched in bases and organic matter.

Gley horizon: A horizon which is saturated with water for the majority of the time in normal years and showing a blue-gray color with the rapid positive dipyridyl reaction.

Humic epipedon: A dark-colored epipedon with high organic matter content.

Hyperhumic epipedon: A humic epipedon that has particularly high organic matter content.

Paddy epipedon: A plowed epipedon in paddy fields without bleaching.

Pseudogley horizon: A horizon naturally saturated with water for some time in a year and under reduced conditions with epiaquic properties.

Red-Yellow cambic horizon: A cambic horizon with red-color or yellow-color weathering.

Spodic horizon: A subsurface horizon that has formed as a result of podsolization. It shows a darker and/or more reddish color than the underlying horizon due to the presence of organic matter and/or Fe and Al oxides leached from the overlying horizon.

Thapto-humic horizon: A dark subsurface horizon with organic matter that was buried by newly deposited materials such as Fluvic and eolian sediments.

(2) **Diagnostic Properties**

Andic soil properties: Properties that show low bulk density, large phosphorus fixation and accumulation of organic matter due to the presence of active Al typically formed during weathering of volcanic glass.

Endoaquic properties: Properties that appear when soil is saturated with groundwater for some time in a year. They are exclusive from gley properties and show mottles of the oxidized form of Fe on the surfaces of peds and voids.

Epiaquic properties: Properties that show reduced conditions from saturation with surface water for part of the year.

Gley properties: Properties that show a blue-gray color due to the presence of the reduced form of Fe under saturated conditions.

Non-allophanic andic soil properties: Properties characterized by the predominance of 2:1 type clay minerals in addition to andic soil properties. Active Al is derived from Al–humus complexes. Soil with non-allophanic andic soil properties exhibits strong acidity due to the presence of large amounts of exchangeable Al.

Physical immaturity: Properties that rarely undergo dehydration process after sediments under water become land.

Red-Yellow soil properties: Properties that show reddish or yellowish soil color and low organic matter content due to the large extent of weathering processes.

Regosolic andic soil properties: Properties that show weathering of volcanogenous materials and accumulation of organic matter to some extent.

Thionic soil properties: Properties that exhibit strong acidity with pale yellow mottles of jarosite. The strong acidity is a result of the oxidation of lacustrine and marine sediments containing sulfides.

(3) **Diagnostic Materials**

Artefacts: Human-made materials that are not naturally present on the Earth's surface.

Calcaric sediments: Unconsolidated sediments containing carbonates equivalent to or more than 40% $CaCO_3$.

Fluvic materials: Fluvic, marine, and lacustrine unconsolidated sediments of Holocene age.

Hetero-soil materials: Soil materials classified as a soil great group other than that of the bulk soil materials.

Organic materials: Materials mainly composed of decomposed or undecomposed plant litter on the land surface. The materials should contain 20% or more of organic carbon and should not show andic soil properties.

Peat materials: Materials mainly composed of decomposed or undecomposed plant litter accumulated under water. The materials should contain 12% or more of organic carbon and should not show andic soil properties.

Pro-thionic materials: Materials that are under saturated conditions if not drained and that show strong acidity once oxidized.

Volcanogenous materials: Pyroclastic materials including volcanic ash, lapilli, pumice, scoria, and sediments deposited by pyroclastic flow.

3.1.3 Classification of a Soil

First, a soil is classified as one of the 10 soil great groups according to the keys. The simplified diagram is shown in Fig. 3.1. The soil in a soil great group is further classified as one of the 1–6 soil group(s) and the soil in the soil group is finally classified as either of 2–9 soil subgroups. In total, the system comprises 10 soil great groups, 26 soil groups, and 120 subgroups. Table 3.1 shows the names of the soil great groups, soil groups, and soil subgroups with reference to the WRB (IUSS Working Group WRB 2015) and Soil Taxonomy (Soil Survey Staff 2014). Furthermore, the three types of soil phases are available in this system: "soil temperature," "reclamation," and "forest soil." These phases can be attached to the classified name of soil at each level of soil great group, group, and subgroup. For example, the "reclamation" and "forest soil" phases can be optionally used to classify the soils in reclaimed lands and forests, respectively.

3.2 Soil Resources in Japan

3.2.1 Spatial Distribution Pattern of Soil Groups in Japan

A variety of climatic, vegetation, topographical, and geological conditions have influenced the diverse soil cover of Japan. From the global geological viewpoint, the Japanese

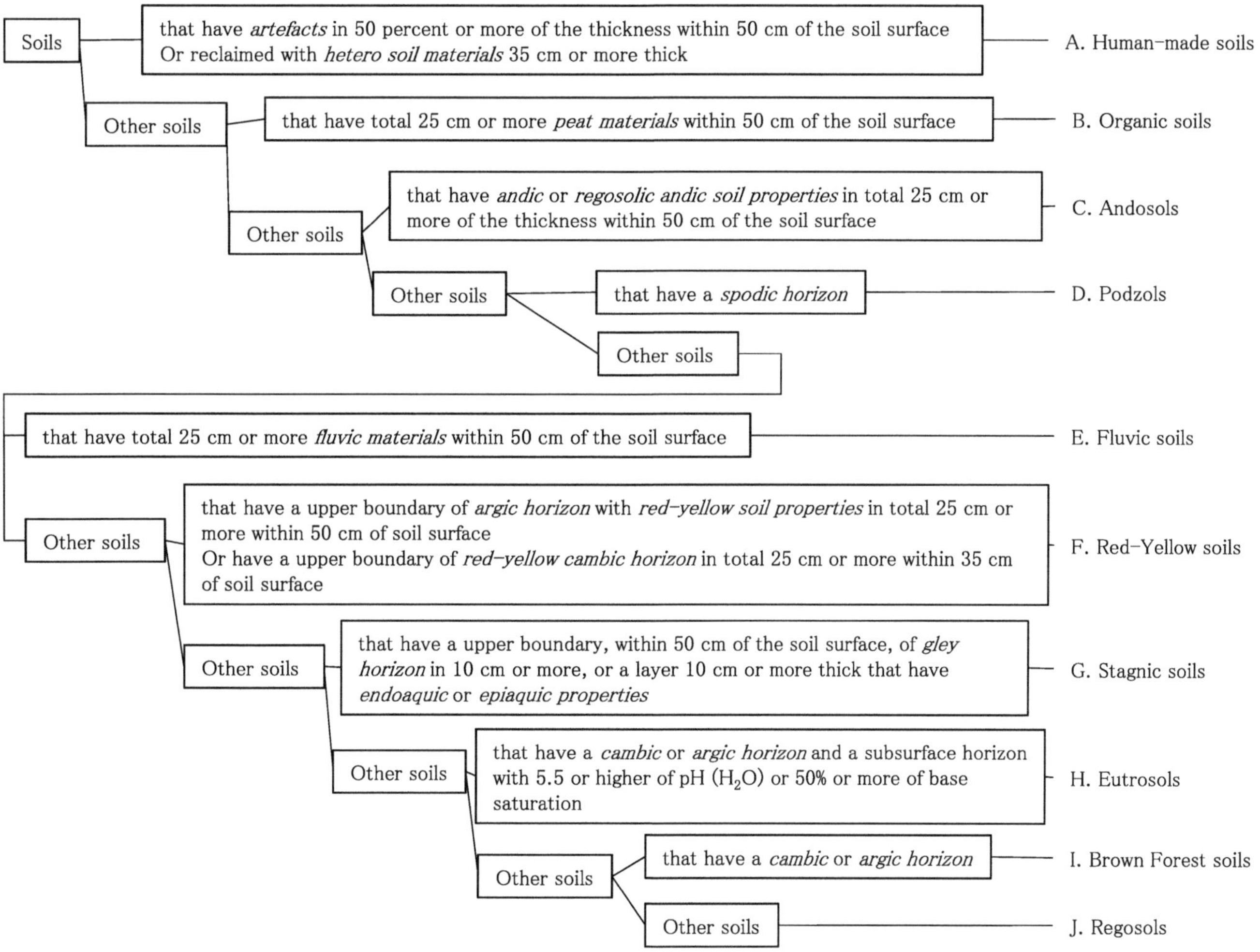

Fig. 3.1 Simplified flow diagram for classifying a soil at the level of Soil Great Group. Terms in italic are diagnostic horizons, diagnostic properties or diagnostic materials defined by the system. (Figure supplied by Hitoshi Shinjo)

Table 3.1 Corrleation of soil classification system of Japan (2017) with WRB (2015) and Soil Taxonomy (2014)

Soil classification system of Japan (2017)			World reference base for soil resources (2015)	USDA Soil Taxonomy (2014)
Great group				
	Group			
		Subgroup		
A. 【Human-made soils】			Technosols, Regosols	(Entisols)
	Artifactural soils		Technosols	(Udorthents)
		Organic	Gabric Technosols	
		Ekrantic	Linic Technosols Ekranic Technosols	
		Mineral	Urbic Technosols Spolic Technosols	
	Reformed soils		Regosols (Transportic)	(Udorthents)
		Upland	Regosols (Transportic)	
		Lowland	Regosols (Transportic)	
B. 【Organic soils】			Histosols	Histosols
	Peat soils		Histosols	Histosols
		Sapric	Saplic Histosols	Haplosaprists
		High-moor	Fibric Histosols Hemic Histosols	Shagnofibrists Haplofibrists Haplohemists
		Transitional-moor	Fibric Histosols Hemic Histosols	Haplofibrists Haplohemists
		Low-moor	Fibric Histosols Hemic Histosols	Haplofibrists Haplohemists
C. 【Andosols】			Andosols, Podzols	Andisols, Spodosols
	Podzolic Andosols		Andic Podzols	Aquods, Humods, Orthods, Aquands, Vitrands, Udands
		Epi-peaty	Andic Histic Podzols	Histic Endoaquods, Histic Endoaquands
		Epi-pseudogleyic	Andic Stagnic Podzols	Lithic Epiauods, Andic Epiaquods, Typic Epiaquods, Vitraquands, Melanaqunds, Epiaquands
		Aquic	Andic Gleyic Podzols	Lithic Epiauods, Andic Epiaquods, Typic Epiaquods, Aquic Haplorthods, Vitraquands, Melanaqunds, Epiaquands, Aquic subgroups in Andiosols
		Haplic	Andic Podzols	Andic Haplohumods, Andic Haplorthods, Andisols except Aquands and Aquic subgroups
	Regosolic Andosols		Vitric Andosols	Vitrands, Aquands
		Aquic	Gleyic Vitric Andosols	Vitraquands
		Humic	Melanic Vitric Andosols Umbric Vitric Andosols	Humic Udivitrands
		Thapto-humic	Vitric Andosols	Thaptic Udivitrands
		Haplic	Vitric Andosols	Typic Udivitrands
	Gleyed Andosols		Gleysols (Andic)	Aquands
		Peaty	Histic Gleysols (Andic)	Endoaquands
		Cumulic	Dystric Gleysols (Andic)	Melanaquands Endoaquands
		Haplic	Dystric Gleysols (Andic)	Typic Endoaquands

(continued)

Table 3.1 (continued)

Soil classification system of Japan (2017)			World reference base for soil resources (2015)	USDA Soil Taxonomy (2014)
Great group				
	Group			
		Subgroup		
	Wet Andosols		Gleyic Silandic Andosols Gleyic Aluandic Andosols	Aquands
		Peaty	Histic Gleyic Silandic Andosols Histic Gleyic Aluandic Andosols Thapto-histic Gleyic Silandic Andosols Thapto-histic Gleyic Aluandic Andosols	Histic Endoaquands
		Thapto-upland	Gleyic Silandic Andosols Gleyic Aluandic Andosols	Typic Endoaquands
		Endofluvic	Gleyic Silandic Andosols Gleyic Aluandic Andosols	Typic Endoaquands
		Cumulic	Histic Gleyic Silandic Andosols Histic Gleyic Histic Aluandic Andosols Gleyic Silandic Andosols (Melanic) Gleyic Aluandic Andosols (Melanic)	Melanaquands Endoaquands
		Haplic	Gleyic Silandic Andosols Gleyic Aluandic Andosols	Typic Endoaquands
	Non-allophanic Andosols		Aluandic Andosols	Alic Hapludands Melanudands
		Anthraquic	Aluandic Andosols	Anthraquic Melanudands Anthraquic Hapludands
		Cumulic	Aluandic Andosols (Melanic) Aluandic Andosols	Pachic Melanudands Alic Hapludands
		Thapto-humic	Aluandic Andosols	Alic Hapludands
		Humic	Aluandic Andosols	Alic Hapludands
		Brown-humic	Aluandic Andosols (Fulvic)	Pachic Fulvudands Thaptic Fulvudands Typic Fulvudands Hydric Fulvudands
		Aquic	Gleyic Aluandic Andosols	Aquic Hapludands Oxyaquic Hapludands
		Haplic	Aluandic Andosols	Alic Hapludands
	Allophanic Andosols		Silandic Andosols	Udands
		Anthraquic	Silandic Andosols	Anthraquic Melanudands Anthraquic Hapludands
		Thapto-upland	Silandic Andosols	Typic Hapludands Ultic Hapludands
		Endofluvic	Silandic Andosols	Typic Hapludands
		Cumulic	Siluandic Andosols (Melanic) Siluandic Andosols	Pachic Melanudands Typic Hapludands
		Cumulic	Silandic Andosols	Thaptic Hapluands
		Humic	Silandic Andosols	Typic Hapludands

(continued)

Table 3.1 (continued)

Soil classification system of Japan (2017)			World reference base for soil resources (2015)	USDA Soil Taxonomy (2014)
Great group				
	Group			
		Subgroup		
		Brown-humic	Silandic Andosols (Fulvic)	Pachic Fulvudands Thaptic Fulvudands Typic Fulvudands Hydric Fulvudands Typic Hapludands Hydric Hapludands
		Aquic	Gleyic Silandic Andosols	Aquic Hapludands Oxyaquic Hapludands
		Haplic	Silandic Andosols	Typic Hapludands Hydric Hapludands
D. 【Podzols】			Podzols	Spodosols
	Podzols		Podzols	Spodosols
		Epi-peaty	Histic Podzols	Histic Epiaquods Histic Endoaquods
		Aquic	Gleyic Podzols	Endoaquods Aquic Haplorthods
		Epi-pseudogleyic	Stagnic Podzols	Epiaquods
		Pseudogleyic	Stagnic Podzols	Aquic Haplorthods
		Haplic	Haplic Podzols	Haplorthods Haplocryods
E. 【Fluvic soils】			Fluvisols, Anthrosols	Inceptisols, Entisols
	Fluvic Paddy soils		Hydragric Anthrosols (Fluvic)	Anthraquic Eutrudepts Aeric Epiaquepts
		Albic	Hydragric Anthrosols (Fluvic)	Anthraquic Eutrudepts Aeric Epiaquepts
		Epi-gleyed	Hydragric Anthrosols (Fluvic)	Typic Epiaquepts
		Endoaeric	Hydragric Anthrosols (Fluvic)	Anthraquic Eutrudepts Aeric Epiaquepts
		Aquic	Hydragric Anthrosols (Fluvic)	Typic Epiaquepts
		Haplic	Hydragric Anthrosols (Fluvic)	Anthraquic Eutrudepts Aeric Epiaquepts
	Gley Fluvic soils		Fluvic Gleysols	Aquents, Aquepts, Aquepts
		Thionic	Fluvic Thionic Gleysols	Sulfic Endoaquepts Sulfic Hydraquents Sulfic Fluvaquents
		Peaty	Fluvic Histic Gleysols	Thapto-Histic Hydraquents Thapto-Histic Fluvaquents
		Humic	Fluvic Umbric Gleysols	Typic Hydraquents Mollic Fluvaquents
		Epi-gray	Fluvic Gleysols	Typic Fluvaquents
		Strong	Fluvic Reductigleyic Gleysols	Typic Hydraquents
		Mottled	Fluvic Oxygleyic Gleysols	Typic Hydraquents Typic Fluvaquents
	Gray Fluvic soils		Gleyic Fluvisols	Aquepts, Aquents
		Thionic	Gleyic Fluvisols	Sulfaqepts Sulfic Endoaquepts Sulfic Fluvaquents

(continued)

Table 3.1 (continued)

Soil classification system of Japan (2017)			World reference base for soil resources (2015)	USDA Soil Taxonomy (2014)
Great group				
	Group			
		Subgroup		
		Peaty	Gleyic Histic Fluvisols	Fluvaquentic Endoaquepts Thapto-Histic Fluvaquents
		Humic	Gleyic Umbrisols	Humaquepts Mollic Fluvaquents
		Epi-gleyed	Gleyic Fluvisols	Fluvaquentic Endoaquepts Typic Endoaquepts Typic Fluvaquents
		Gleyic	Gleyic Fluvisols	Typic Endoaquepts Typic Fluvaquents
		Thapto-andic	Gleyic Fluvisols	Aquandic Endoaquepts Aquandic Fluvaquents
		Haplic	Gleyic Fluvisols	Fluvaquentic Endoaquepts Typic Fluvaquents Typic Psammaquents
	Brown Fluvic soils		Fluvisols	Udifluvents, Psamments
		Aquic	Fluvisols (Oxyaquic)	Oxyaquic Udifluvents Aquic Udipsamments
		Humic	Gleyic Umbrisols	Mollic Udifluvents
		Protoanthraquic	Fluvisols (Oxyaquic)	Oxyaquic Udifluvents Aquic Udipsamments
		Haplic	Fluvisols	Typic Udifluvents Typic Udipsamments
	Regosolic Fluvic soils		Fluvisols	Udifluvents, Psamments
		Aquic	Fluvisols (Oxaquic)	Oxyaquic Udifluvents Oxyaquic Udipsamments
		Haplic	Fluvisols	Typic Udifluvents Typic Udipsamments
F. 【Red-Yellow soils】			Alisols, Acrisols, Cambisols	Udults, Udepts
	Argic Red-Yellow soils		Alisols, Acrisols, Lixisols	Udults
		Anthraquic	Alic Stagnosols Acric Stagnosols	Anthraquic Paleudults Aquic Hapludults
		Albic	Stagnic Albic Alisols Albic Acrisols	Typic Paleudults Aquic Paleudults Typic Hapludults Aquic Hapludults
		Pseudogleyic	Stagnic Alisols Stagnic Acrisols	Aquic Paleudults Aquic Hapludults
		Aquic	Gleyic Alisols Gleyic Acrisols	Aquic Paleudults Aquic Hapludults
		Humic	Alic Umbrisols Acric Umbrisols	Typic Haplohumults Humic Hapludults
		Reddish	Chromic Alisols Chromic Acrisols	Typic Paleudults Typic Hapludults
		Haplic	Haplic Alisols Haplic Acrisols	Typic Paleudults Typic Hapludults

(continued)

Table 3.1 (continued)

Soil classification system of Japan (2017)			World reference base for soil resources (2015)	USDA Soil Taxonomy (2014)
Great group				
	Group			
		Subgroup		
	Cambic Red-Yellow soils		Cambisols	Udepts
		Anthraquic	Stagnosols	Anthraquic Eutrudepts
		Albic	Gleyic Cambisols	Aquic Dystrudepts Typic Dystrudepts
		Pseudogleyic	Stagnic Cambisols	Aquic Dystrudepts Oxyaquic Dystrudepts
		Aquic	Gleyic Cambisols	Aquic Dystrudepts Oxyaquic Dystrudepts
		Humic	Cambic Umbrisols	Humic Dystrudepts
		Reddish	Chromic Cambisols	Typic Dystrudepts
		Andic	Dystric Cambisols	Andic Dystrudepts
		Haplic	Dystric Cambisols	Oxyaquic Dystrudepts Typic Dystrudepts
G. 【Stagnic soils】			Gleysols, Stagnosols, Anthrosols	Aquepts, Aquults, Aquents
	Stagnogley soils		Gleysols, Anthrosols	Epiaquepts, Endoaquepts, Endoaquents
		Anthraquic	Anthraquic Gleysols Hydroargic Anthrosols	Typic Epiaquepts
		Epi-peaty	Histic Gleysols	Typic Epiaquepts Typic Endoaquents
		Humic	Umbric Gleysols	Typic Humaquepts Humaquptic Endoaquents
		Haplic	Haplic Gleysols	Typic Epiaquepts Typic Endoaquents
	Pseudogley soils		Stagnosols, Gleysols	Aquepts, Aquults, Aquents
		Anthraquic	Anthraquic Stagnosols	Typic Epiaquepts Typic Epiaquults
		Groundwater-aquic	Endogleyic Stagnosols Anthraquic Gleysols Haplic Gleysols	Typic Endoaquepts Typic Endoaquents
		Humic	Umbric Stagnosols	Typic Humaquepts Typic Umbraquults
		Aeric	Haplic Stagnosols	Aeric Epiaquepts Aeric Epiaquults
		Haplic	Haplic Stagnosols	Typic Epiaquepts Typic Epiaquults
H. 【Eutrosols】			Luvisols, Cambisols	Udalfs, Udepts
	Magnesian Eutrosols		Luvisols, Cambisols	Udalfs, Udepts
		Argic	Leptic Luvisols (Chromic) Leptic Luvisols (Rhodic) Haplic Luvisols (Chromic)	Typic Paleudalfs Typic Rhodualfs Typic Hapludalfs

(continued)

Table 3.1 (continued)

Soil classification system of Japan (2017)			World reference base for soil resources (2015)	USDA Soil Taxonomy (2014)
Great group				
	Group			
		Subgroup		
			Haplic Luvisols (Rhodic)	
		Haplic	Leptic Cambisols (Eutric) Haplic Cambisols (Eutric)	Lithic Eutrudepts Typic Eutrudepts
	Calcareous Eutrosols		Luvisols, Cambisols	Udalfs, Udepts
		Argic	Leptic Luvisols Haplic Luvisols	Typic Paleudalfs Typic Rhodualfs Typic Hapludalfs
		Haplic	Leptic Cambisols (Eutric) Haplic Cambisols (Eutric)	Lithic Eutrudepts Typic Eutrudepts
I. 【Brown Forest soils】			Camibisols, Stagnosols	Udepts
	Brown Forest soils		Camibisols, Stagnosols	Udepts
		Anthraquic	Anthraquic Stagnosols	Anthraquic Eutrudepts Aquic Dystrudepts
		Andic	Dystric Cambisols	Andic Dystrudepts Andic Eutrudepts Lithic Dystrudepts
		Podzolic	Dystric Cambisols	Typic Dystrudepts Lithic Dystrudepts
		Humic	Cambic Umbrisols Dystric Cambisols (Humic)	Humic Dystrudepts
		Thapto-red-yellow	Dystric Cambisols Haplic Alisols	Typic Dystrudepts Typic Paleudults Inceptic Hapludults Typic Hapludults
		Aquic	Gleyic Cambisols (Oxaquic)	Aquic Dystrudepts Oxyaquic Dystrudepts Lithic Dystrudepts
		Epi-gleyed	Gleyic Cambisols	Aquic Dystrudepts Oxyaquic Dystrudepts Lithic Dystrudepts
		Eutric	Eutric Cambisols	Lithic Eutrudepts Typic Eutrudepts
		Haplic	Dystric Cambisols	Typic Dystrudepts Lithic Dystrudepts
J. 【Regosols】			Regosols, Arenosols, Leptosols, Phaeozems	Entisols, Mollisols
	Volcanogeous Regosols		Tephric Regosols	Orthents
		Aquic	Gleyic Tephric Regosols	Aquic Udorthents
		Haplic	Tephric Regosols	Vitrandic Udorthents
	Sandy Regosols		Arenosols	Udipsamments
		Calcaric	Calcaric Arenosols	Typic Udipsamments
		Aquic	Gleyic Arenosols	Aquic Udipsamments Oxyaquic Udipsamments
		Haplic	Arenosols	Typic Udipsamments
	Lithosols		Leptosols	Udorthents, Rendolls

(continued)

Table 3.1 (continued)

Soil classification system of Japan (2017)			World reference base for soil resources (2015)	USDA Soil Taxonomy (2014)
Great group				
	Group			
		Subgroup		
		Calcaric	Calcaric Rendzic Leptosols Calcaric Leptosols	Lithic Haprendolls Lithic Udorthents
		Aquic	Leptosols (Gleyic)	Lithic Udorthents
		Haplic	Leptosols	Lithic Udorthents
	Terrestrial Regosols		Regosols	Udorthents
		Marlitic	Calcaric Regosols Calcaric Leptosols	Typic Udorthents Lithic Udorthents
		Calcaric	Calcaric Regosols Calcaric Phaeozems	Typic Haprendolls Typic Udorthents
		Para-lithic	Dystric Regosols Dystric Leptosols	Typic Udorthents Lithic Udorthents
		Granitic	Skeletic Regosols Skeletic Leptosols	Typic Udorthents Lithic Udorthents
		Haplic	Dystric Regosols	Typic Udorthents

Please note that it shows only representative classification names corresponding to each classification and does not cover all cases

islands are located in one of the most active parts of the circum-Pacific Ring of Fire (see Sect. 2.2) and lie at the junction of two oceanic plates and two continental plates. The Japanese islands experience seismic activity, tectonic uplift, and rapid downcutting by streams, which have resulted in a steep topography and narrow lowlands. About 10% of the world's active volcanic mountains (111) are found in Japan (Fig. 2.16). Volcanic tephra has been deposited in large quantities on hilly and mountainous terrain and on Pleistocene terraces and is a main parent material of Japanese soil (Andosols). Therefore, these topographical and geological characteristics are one of the most noteworthy contributing factors to the spatial distribution pattern of soil groups in Japan (Fig. 3.2). For example, Andosols are found along volcanic areas, Brown Forest soils and Cambic Red-Yellow soils (refer to Cambisols) are distributed in steep mountainous zones far from active volcanoes and Fluvic soils cover narrow lowlands.

(1) **Brown Forest Soils**

Table 3.2 shows the summarized extent of soil great groups and soil groups in Japan. The most widely spread soil great group is Brown Forest soils, which account for about 33% of Japan's total land area. Brown Forest soils are mainly distributed in Western Japan and cover 47% of the Kinki–Chugoku–Shikoku region, which is characterized by a temperate climate and hilly-mountainous areas with little volcanic ash deposition. Brown Forest soils are also the main soil great group in the northern part of Hokkaido, the western part of Tohoku region, and the southern part of the Tokai–Hokuriku region. These areas have a subarctic temperate climate and hilly-mountainous areas without fresh volcanic ash deposition, conditions which coincide well with the global distribution pattern of Cambisols (Food and Agriculture Organization 2001).

(2) **Andosols**

Andosols have the second largest distribution area among the soil great groups in Japan. The total distribution area of Andosols in Japan is estimated at some 0.1 million km^2, which is about 10% of the global Andosol distribution area. Andosols are mainly distributed in the southern part of Hokkaido, in the northeastern part of Tohoku, and in the Kanto and Kyushu regions. As shown in Fig. 3.3, Volcanogenous Regosols are distributed around volcanoes. Regosolic Andosols and then Allophanic Andosols are distributed with increasing distance from the source of volcanic ash. Allophanic Andosols are the most dominant soil groups in the Andosol great group, and about 60% of the Andosol distribution areas are covered by the Allophanic group. Allophanic Andosols are mainly distributed on the east side of volcanoes (Fig. 3.3). Because high-altitude

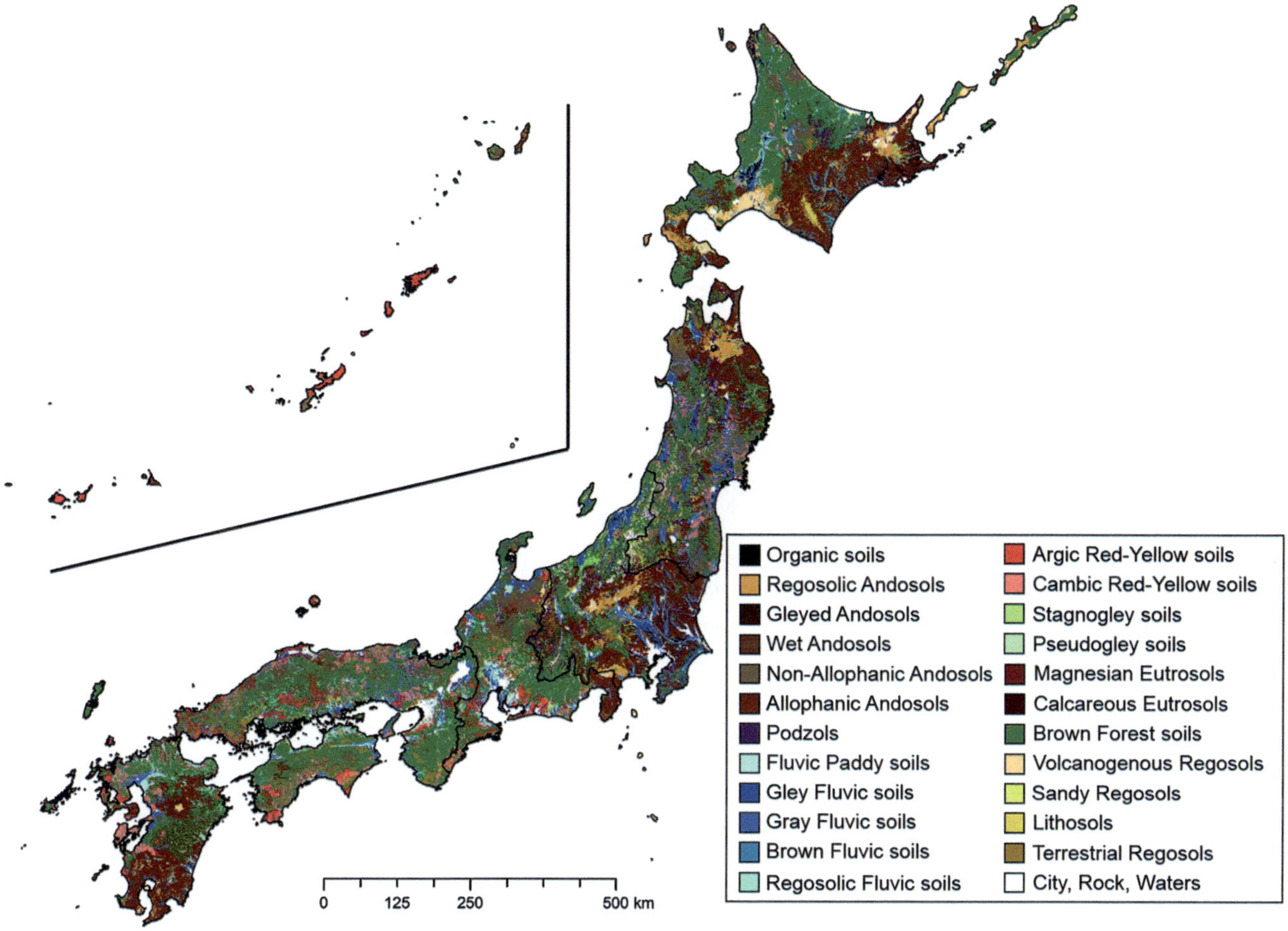

Fig. 3.2 Soil map of Japan. (Figure supplied by Yusuke Takata)

winds are commonly strong westerlies, volcanic materials blown into the air by eruptions fall predominantly to the east of the volcanoes. Therefore, Regosolic Andosols and Allophanic Andosols are mainly developed on the east side of the volcanic line. Non-allophanic Andosols are the second most dominant, and this soil group is distributed in the northern part of Hokkaido, the northern part of Tokai–Hokuriku, and the northern part of Kinki–Chugoku–Shikoku (Fig. 3.4). These areas partially coincide with the distribution area of Brown Forest soils where fresh volcanic ash was not deposited. It is thought that Allophanic Andosols contain short-range order clay minerals such as allophane and imogolite generated from fresh volcanic ash after the Holocene, while Non-allophanic Andosols contain crystalline minerals due to contamination with eolian dust and/or Non-allophanic parent materials and a decrease in the formation of short-range order clay minerals from fresh volcanic ash (Saigusa et al. 1992; Matsuyama et al. 1994; Fujita et al. 2007; Okuda et al. 2007).

Due to the limitations of soil survey data, the distribution pattern of Podzolic Andosols was not able to be delineated in the national soil map. However, Nishiue et al. (2014) reported that weak podzolization on Allopanic Andosols was recognized on the high-elevation plateau of the Tohoku region. In the northern part of the Tohoku region, the podzolization of ash-derived soils appeared in association with the *hiba* tree (*Thujopsis dolabrata* var. Hondai Makino), a strong podzolizer, at elevations of 150–300 m (Takahashi et al. 1989). Further soil surveys are required to delineate the spatial distribution pattern of Podzolic Andosols.

(3) **Fluvic Soils**

Fluvic soils are the third dominant soil great group in Japan and cover 13.7% of the total area of Japan. Fluvic soils are found along rivers and lakes and in areas of recent marine deposits. The distribution areas of Gley Fluvic soils and Gray Fluvic soils show a close spatial relationship. Gley Fluvic soils have a nearly permanently saturated gley horizon, and Gray Fluvic soils have a mottled horizon formed under seasonal saturation of groundwater. The difference between Gley and Gray soils is the groundwater level, and

Table 3.2 Distribution area of soil groups in Japan (%)

	Great group	Hokkaido	Tohoku	Kantou	Tokai–Hokuriku	Kinki–Chugoku–Shikoku	Kyushu–Okinawa	Whole
A	*Human-made soils*	*<0.1*	*<0.1*	*<0.1*	*<0.1*	*<0.1*	*0.2*	*<0.1*
A1	Artifactual soils	<0.1	<0.1	<0.1	<0.1	<0.1	<0.1	<0.1
A2	Reformed soils	<0.1	<0.1	<0.1	<0.1	<0.1	0.2	<0.1
B	*Organic soils*	*3.3*	*1.2*	*1.0*	*0.2*	*0.1*	*0.1*	*1.1*
B1	Peat soils	3.3	1.2	1.0	0.2	0.1	0.1	1.1
C	*Andosols*	*41.8*	*35.8*	*40.0*	*21.8*	*8.6*	*34.0*	*30.3*
C1	Podzolic Andsols	<0.1	<0.1	<0.1	<0.1	<0.1	<0.1	<0.1
C2	Regosolic Andosols	7	3	4	0	<0.1	2	3
C3	Gleyed Andosols	<0.1	0	1	<0.1	<0.1	<0.1	0
C4	Wet Andosols	0	1	1	0	1	1	1
C5	Non-allophanic Andosols	6	11	3	16	7	1	7
C6	Allophanic Andosols	28	20	30	5	1	31	19
D	*Podzols*	*1.2*	*5.5*	*4.0*	*1.7*	*<0.1*	*<0.1*	*2.1*
D1	Podzols	1.2	5.5	4.0	1.7	<0.1	<0.1	2.1
E	*Fluvic soils*	*10.3*	*13.8*	*16.6*	*15.7*	*14.2*	*12.6*	*13.7*
E1	Fluvic Paddy soils	0.2	0.1	0.3	4.0	2.1	4.1	1.5
E2	Gley Fluvic soils	1.4	6.6	8.0	7.7	3.3	2.7	4.7
E3	Gray Fluvic soils	2.6	6.0	6.0	2.7	6.3	4.0	4.7
E4	Brown Fluvic soils	5.3	0.8	1.9	0.6	1.3	0.5	2.0
E5	Regosolic Fluvic soils	0.8	0.3	0.4	0.8	1.1	1.4	0.8
F	*Red-Yellow soils*	*0.9*	*5.9*	*0.5*	*10.6*	*16.0*	*14.4*	*7.6*
F1	Argic Red-Yellow soils	<0.1	0.3	0.1	7.1	7.8	8.8	3.5
F2	Cambic Red-Yellow soils	0.9	5.6	0.5	3.6	8.2	5.6	4.1
G	*Stagnic soils*	*1.8*	*0.0*	*1.3*	*0.9*	*0.2*	*0.0*	*0.8*
G1	Stagnogley soils	<0.1	<0.1	1.2	<0.1	<0.1	<0.1	0.2
G2	Pseudogley soils	1.8	<0.1	0.2	0.9	0.2	<0.1	0.6
H	*Eutrosols*	*<0.1*	*<0.1*	*<0.1*	*0.2*	*0.0*	*1.1*	*0.1*
H1	Magnesian Eutrosols	<0.1	<0.1	<0.1	0.1	<0.1	<0.1	<0.1
H2	Calcareous Eutrosols	<0.1	<0.1	<0.1	0.1	<0.1	1.1	0.1
I	*Brown Forest soils*	*31.9*	*27.8*	*22.7*	*36.0*	*47.6*	*30.4*	*33.2*
I1	Brown Forest soils	31.9	27.8	22.7	36.0	47.6	30.4	33.2
J	*Regosols*	*8.1*	*6.1*	*6.0*	*7.5*	*7.9*	*4.4*	*6.9*
J1	Volcanogenous Regosols	4.9	0.5	0.6	0.4	<0.1	0.7	1.4
J2	Sandy Regosols	0.6	0.8	0.6	1.1	0.3	0.4	0.6
J3	Lithosols	0.9	2.7	1.7	0.7	0.7	0.3	1.2
J4	Terrestrial Regosols	1.8	2.1	3.1	5.3	6.9	3.0	3.7
Others		*0.6*	*3.9*	*7.7*	*5.3*	*5.2*	*2.8*	*4.1*
Z1	Rocks and stones	<0.1	2.1	2.7	0.8	0.5	0.4	1.1
Z3	City, Water	0.6	1.8	5.0	4.5	4.7	2.4	3.1

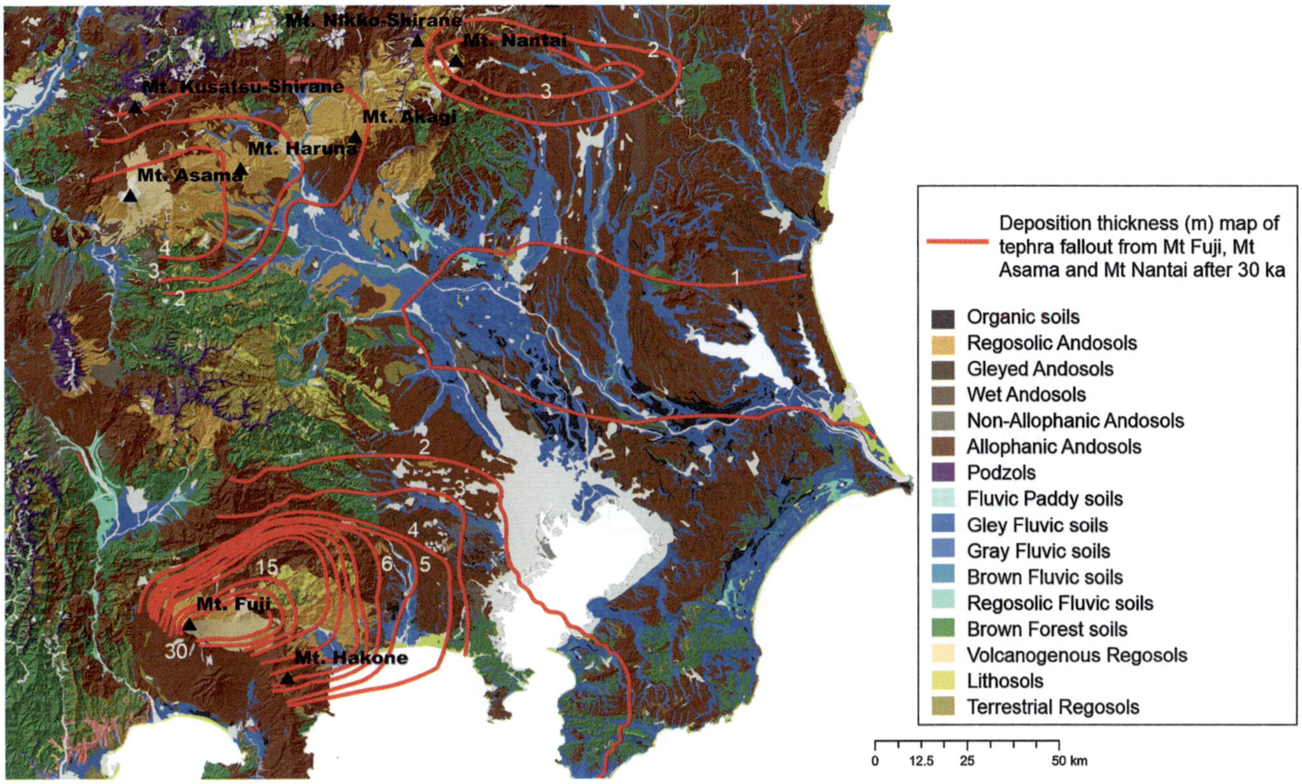

Fig. 3.3 Distribution of volcanic tephra and Soil Groups in Kanto district. Data source of active volcanic mountains: Japan Meteorological Agency website (https://www.data.jma.go.jp/svd/vois/data/tokyo/STOCK/kaisetsu/katsukazan_toha/katsukazan_toha.html). Data source of the depth volcanic ash: Uesugi et al. (1983). (Figure supplied by Yusuke Takata)

both soil groups are mainly used as paddy field. As mentioned below, Fluvic soils cover about 50% of Japanese cultivated land, meaning that they are one of the most important soil resources for Japanese food security.

(4) **Red-Yellow Soils**

Red-Yellow soils cover 7.6% of the total area of Japan and are divided into two soil groups, namely Argic Red-Yellow soils and Cambic Red-Yellow soils (refer to Cambisols). Argic Red-Yellow soils are mainly distributed in the southwestern part of Japan. Argic Red-Yellow soils correspond to several reference soil groups (Alisols, Acrisols, Luvisols, and Lixisols) of the WRB2014 because our classification system definitions do not refer to the activity of clay minerals and base saturation. Kanda et al. (2018) reported that soils with low-activity clay are dominant in western Kyushu (including Okinawa) and in Chubu, whereas soils with high-activity clay are dominant in the other regions. Thus, Argic Red-Yellow soils correspond to Acrisols in western Kyushu and Chubu and to Alisols in all the other regions (Kanda et al. 2018). In Japan, the distribution area of Cambic Red-Yellow soils is larger than that of the argic soil group. However, this spatial distribution pattern is opposite in the western Kyushu and Chubu regions; in these areas, argic horizons developed on stable landscape, while cambic horizons developed on mountains or hill ranges. Takata et al. (2010) reported that in Japan, the silt-to-clay ratio and cation exchange capacity (1 kg clay base) of argic horizons are significantly lower than those of cambic horizons in Red-Yellow soils. These results indicate that the development of argic horizons requires long pedogenetic time and landscape stability and could be an indicator of soil weathering in Japan.

(5) **Other Soil Groups**

Regosols cover 6.9% of Japanese territory. This soil great group is divided into four soil groups, namely Volcanogenous Regosols, Sandy Regosols, Lithosols, and Terrestrial Regosols. Volcanogenous Regosols are found at active volcanic mountains/islands, Sandy Regosols are distributed in coastal dunes, and Lithosols are mainly distributed in high mountainous areas. Terrestrial Regosols frequently occur in areas with landslides, especially those where weathered granite is distributed; in general, granitic

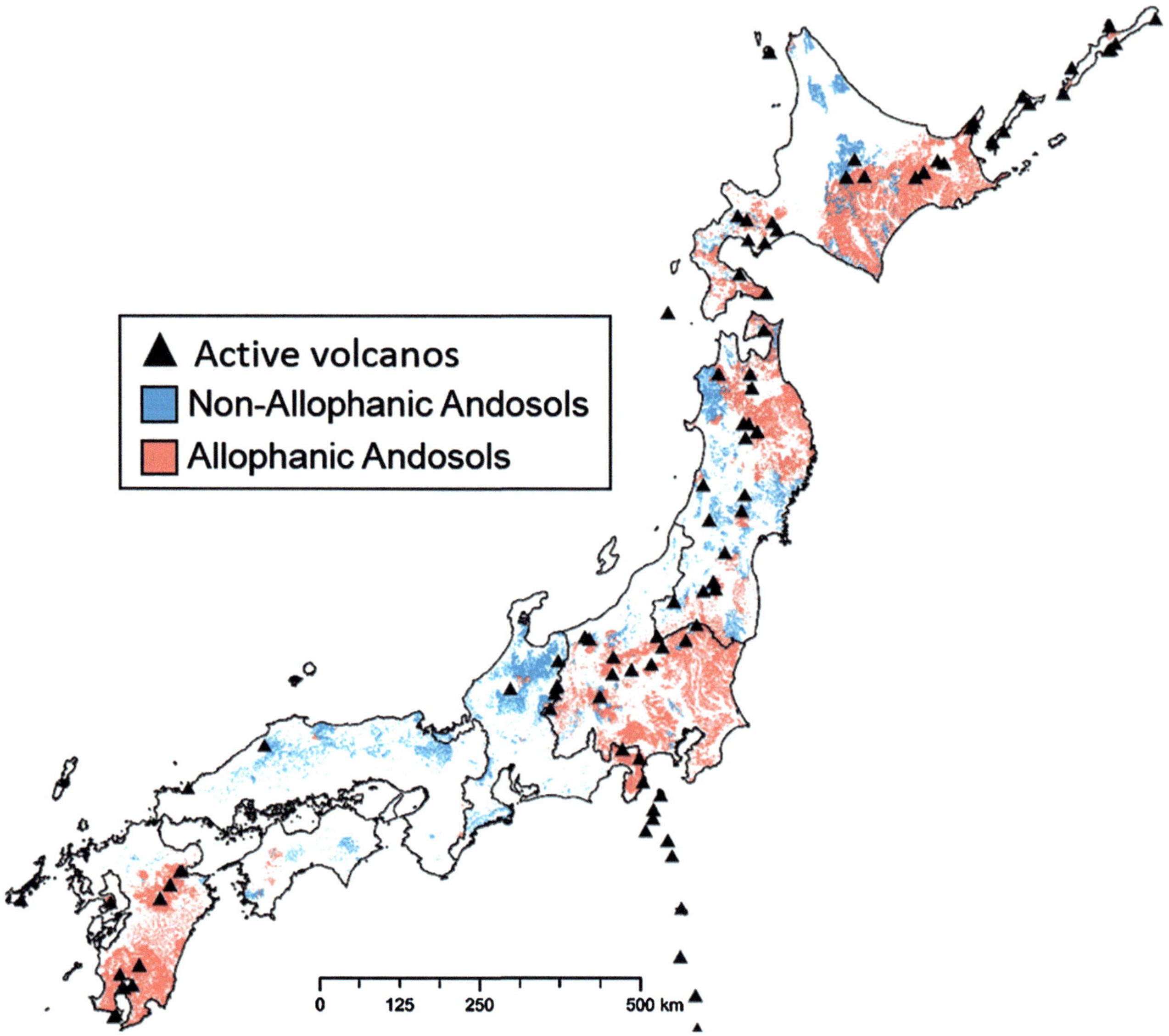

Fig. 3.4 Distribution area of Non-allophanic Andosols and Allophanic Andosols. Data source of active volcanic mountains: Japan Meteorological Agency website (https://www.data.jma.go.jp/svd/vois/data/tokyo/STOCK/kaisetsu/katsukazan_toha/katsukazan_toha.html). (Figure supplied by Yusuke Takata)

rocks are deeply weathered (10–30 m), and innumerable landslides in weathered granites have occurred in Japan. In these areas, *Masado* (Granitic Terrestrial Regosols) are recognized.

The distribution area of podzols is very small in Japan (2% of the Japanese territory), and this soil is recognized in only high mountainous areas, except in the northern part of Hokkaido. Organic soils cover only 1% of Japan's land area and are mainly distributed in the Hokkaido, Tohoku, and Kanto regions. Stagnic soils and Eutrosols account for less than 1%. Stagnic soils are mainly used as terraced paddy field.

3.2.2 Status of Cultivated Soil Resources

Japan's total land area is about 378,000 km^2. About 72% of this land area is mountainous, and rivers are characterized by their steep gradients and relatively short lengths. About two-thirds of the total land area consists of forest, while plains cover about 28% of the total land area. Most plains are located along the seacoast. Cultivated land accounts for 12.1% of the total land surface and is mainly distributed in the plains.

Table 2.2 in Chapter 2 shows the summarized extent of soil great groups in Japanese cultivated land. Fluvic soils show the largest distribution area (47%), followed by Andosols (29%). Fluvic soils are mainly used as paddy field (71%). About half of the uplands and meadow fields are covered by Andosols. Both Fluvic soils and Andosols are mainly developed on flat plains, and these areas are suitable for cultivation. Brown forest soils cover only 8% of the total cultivated area and are mainly utilized as orchard. Despite the small distribution area of organic soils and stagnic soils (Table 2.2 in Chapter 2) in Japan, both of these soil great groups are well utilized as paddy fields.

3.2.3 Soil Threats in Japanese Agricultural Land

Agricultural soil is not only an important resource that secures food supply, but also provides several ecological services. However, the functions of agricultural soils are threatened by a wide range of processes, and a number of soil threats have been identified in Japan. In Chaps. 5–10, there are several reports about the current status and trends in the condition of agricultural soil in each region of Japan. Nutrient imbalance, SOM decline, heavy metal contamination, and soil sealing are reported as major threats to Japanese agricultural soils.

(1) **Nutrient Imbalance**

Nutrient imbalance has been one of the most important issues for sustainable agriculture in Japan. The MAFF revised the Fundamental Guidelines for Fertility Improvement in 2008. These guidelines pointed out that the excessive application of nitrogen and phosphorus fertilizers was having a significant environmental impact. Today, the surface soil nitrogen (N) and phosphorus (P) balance has been improved in Japanese cultivated land (Mishima et al. 2009, 2010). The application of chemical fertilizer declined consistently from 1985 to 2005, while the rate of application of livestock manure peaked in 1990 and declined thereafter. Crop production remained constant during this period. Between 1985 and 2005, the surplus N and P (positive value of soil surface N balance) declined from 89.9 to 49.3 kg N ha^{-1} and from 153 to 105 kg P ha^{-1}, respectively (Mishima et al. 2009, 2010). However, this trend was not consistent at the regional level because organic amendment application was largely affected by the distribution of livestock excreta (Mishima et al. 2009) and soil types (Leon et al. 2012). High surplus P and low crop P-uptake compared with N and P input for crop production could be reduced, thereby reducing negative environmental effects such as the eutrophication of soil and water and conserving limited P resources. According to the Basic Soil-Environmental Monitoring Project, there was excess soil Ca in paddy fields, upland fields, and orchard fields during 1979–1998 (Ministry of Agriculture, Forestry and Fishery 2008); however, the soil Mg deficit gradually increased in this period. The balance of Ca and Mg has thus been getting worse in Japanese cultivated land. Moreover, the soil pH of paddy fields gradually decreased from 5.8 to 5.7 during 1979–1998 (Ministry of Agriculture, Forestry and Fishery 2008).

(2) **Soil Organic Matter Decline**

Soil organic matter (SOM) decline has been widely recognized as a major threat to world soil resources, because of the vital role played by the organic material in many soil functions, such as food production and the terrestrial reservoir of carbon (Intergovernmental Technical Panel on Soils 2015). The problem of SOM decline has thus become a hot topic in soil surveys in the last decade. Because SOC in Japanese cultivated land has been monitored since 1979, this data can provide very valuable information on this topic. A cultivated soil monitoring project named "Basic Soil-Environmental Monitoring" in Japan has repeated monitoring at five-year intervals for about 20,000 fixed points. The measured rate of changes in the total SOC stock evaluated by the monitoring approach was 2.3 Tg C yr^{-1} from 1979 to 1989, indicating that SOC increased in cultivated land. However, this approach also detected a loss of SOC, at a rate of −2.3 Tg C yr^{-1}, from 1989 to 1998. During this period, the cultivated land area decreased from 5.4 to 5.0 million ha and the soil carbon content of arable land gradually increased from 88 to 90 t C ha^{-1}. These results indicate that the decline of SOC stock in Japanese cultivated land was mainly caused by the decrease of cultivated land area.

(3) **Soil Contamination**

In Japan, rapid industrialization during the 1960s led to pollution of arable soil by heavy metals such as cadmium (Cd), copper (Cu), and arsenic (As). There were four main sources of heavy metal pollution for arable land, namely mining activity, factories and incinerators, fertilizers, and precipitation and irrigation water (Makino et al. 2010). In 1970, the Japanese government enacted the Agricultural Land Soil Pollution Prevention Law to regulate heavy metal pollution. The Agricultural Land Soil Pollution Prevention Law designated Cd, As, and Cu as hazardous substances to be regulated. The allowable limit of Cd was set as the Cd concentration in rice grains (1 mg kg^{-1}), and the allowable limits of As and Cu were set to 15 (1 M HCl soluble) and 125 (0.1 M HCl soluble) mg kg soil $^{-1}$, respectively. The amount of bioavailable Cd in soil is affected by many factors, so setting an allowable soil Cd content is impractical (Asami

1981). The area of polluted arable land reached 7592 ha (Cd: 7050 ha; Cu: 1405 ha; As: 391 ha). About 92% of the total polluted area has been remediated by uncontaminated soil dressing (Ministry of the Environment 2014).

(4) **Soil Sealing**

Soil sealing—the covering of the soil by construction—has been rapidly perceived as a threat to global food security. In Japan, urban sprawl and other changes in land use (including abandonment of agricultural land) have caused the loss of about 1 million hectares of cultivated area between 1973 and 2001. Urbanization (sealing) advanced in flat lowland areas (paddy fields) where Fluvic soils were distributed in this period. Additionally, declines in the area of Andosols with expanding urbanization were widely observed over the flat upland fields of the Kanto region. In contrast, many upland fields on steep slopes in Western Japan, with Brown Forest soils, were abandoned (Takata et al. 2011).

References

Asami T (1981) Maximum allowable limits of heavy metals in rice and soil. In: Kitagawa K, Yamane I (eds) Heavy metal pollution in soils of Japan. Japan Scientific Societies Press, Tokyo, pp 257–274

Division 3 of Soils (1977) Classification of cultivated soils. 2nd Approximation. National Institute for Agricultural Sciences. (in Japanese)

Food and Agriculture Organization (2001) Major soils of the world. World reference base for soil resources: Atlas. CD-ROM, Rome, Italy

Forest Soil Division (1976) Classification of forest soils in Japan (1975). Bull Gov Exp Station 280:1–28 (in Japanese with English summary)

Fujita T, Okuda T, Fujie K et al (2007) Influence of eolian dust brought from northern Asia continent on the parent materials in a fine-textured soil developed on the bedrock of the tertiary rock near cape Saruyama, in Noto peninsula, central Japan. Pedologist 51:97–103 (In Japanese with English summary)

Intergovernmental Technical Panel on Soils (2015) Status of the World's soil resources (Technical Summary). FAO, Rome, Italy

IUSS Working Group WRB (2015) World reference base for soil resources 2014. Update 2015. World Soil Resources Reports No. 103. FAO, Rome

Kanda T, Takata Y, Kohyama K, Ohkura T, Maejima Y, Wakabayashi S, Obara H (2018) New soil maps of Japan based on the comprehensive soil classification system of Japan—first approximation and its application to the world reference base for soil resources 2006. Jpn Agric Res Quart 52:285–292

Leon A, Kohyama K, Mishima S et al (2012) Factors controlling organic amendment application rate and long-term change in application rate in Japanese paddy field using longitudinal questionnaire survey dataset (the Basic Soil Environment Monitoring Project, Stationary Monitoring, 1979–1998). Soil Sci Plant Nutr 58:104–120

Makino T (2010) Heavy metal pollution of soil and risk alleviation methods based on soil chemistry. Pedologist 53:38–49

Ministry of Agriculture, Forestry and Fishery (2008) Trend and issue of cultivated soil in Japan. Available at: http://www.maff.go.jp/j/study/dozyo_kanri/01/pdf/ref_data1.pdf

Mishima S, Endo A, Kohyama K (2009) Recent trend in residual nitrogen on national and regional scales in Japan and its relation with groundwater quality. Nut Cycl Agroecosyst 83:1–11

Mishima S, Endo A, Kohyama K (2010) Recent trends in phosphate balance nationally and by region in Japan. Nutr Cycling Agroecosys 86:69–77

Nishiue A, Nanzyo M, Kanno H et al (2014) Properties and classification of volcanic ash soils around Lake Kawanuma on the eastern footslope of Mt. Funagata in Miyagi prefecture, northeastern Japan. Soil Sci Plant Nutr 60:848–862

Obara H, Ohkura T, Takata Y, Kohyama K, Maejima Y, Hamazaki T (2011) Comprehensive soil classification system of Japan. First Approximation. Bull Natl Inst Agro-Environ Sci 29:1–73 (in Japanese with English summary)

Okuda T, Fujita T, Fujie K et al (2007) Influence of eolian dust brought from the Precambrian area in northern Asia on the parent materials in a fine-textured soil developed on the tertiary rock in Mt. Horyu, Noto peninsula, central Japan. Pedologist 51:104–110 (In Japanese with English summary)

Saigusa M, Matsuyama N, Abe A (1992) Distribution of Allophanic Andosols and Nonallophanic Andosols in Japan based on the data of soil survey reports on reclaimed land. Jpn J Soil Sci Plant Nutr 63:646–651 (In Japanese with English summary)

Soil Survey Staff (2014) Keys to soil taxonomy, 12th edn. USDA-Natural Resources Conservation Service, Washington, DC

Takahashi T, Shoji S, Sato A (1989) Cleyey Spodosols and Andisols showing a biosequential relation from Simokita Peninsula, northeastern Japan. Soil Sci 148:204–218

Takata Y, Maejima Y, Ohkura T et al (2010) Locating the taxonomical position of argic and cambic horizons in Japanese Red-Yellow soils. Pedologist 54:11–20 (in Japanese with English summary)

Takata Y, Obara H, Nakai M et al (2011) Process of the decline in the cultivated soil area with land use changes in Japan. Jpn J Soil Sci Plant Nutr 82:15–24 (in Japanese with English summary)

The Fifth Committee for Soil Classification and Nomenclature of Japanese Society of Pedology (2017) Soil classification system of Japan. (in Japanese)

4 Major Soil Types

Yusuke Takata, Masayuki Kawahigashi, Kimihiro Kida, Masayuki Tani, Rintaro Kinoshita, Toyoaki Ito, Makoto Shibata, Tadashi Takahashi, Kazumichi Fujii, Akihiro Imaya, Hiroshi Obara, Yuji Maejima, Kazunori Kohyama, and Taku Kato

Abstract

This chapter presents the genesis, characteristics, classification, distribution, and land uses of the 10 soil grade groups (Human-made soils, Organic soils, Andosols, Podzols, Fluvic soils, Red-Yellow soils, Stagnic soils, Eutrosols, Brown Forest soils, and Regosols) in the Soil Classification System of Japan. Human-made soils are referred to soils formed through an anthropogenic process, especially in anthropized areas such as urban, industrial, traffic, mining, and waste disposal areas. Organic soils are formed under wetland conditions where the rate of organic matter production exceeds the rate of organic matter decomposition. Andosols are mainly derived from volcanic ejecta or tephra and form during a long period of pedogenetic soil formation process have unique and distinctive properties among soils, such as their fluffy and light surface properties, high humus content, and very high phosphorus-fixing capacity. Podzols are characterized by the eluviation of aluminum (Al) and iron (Fe) from the albic E horizon and their illuviation in the spodic B (Bs) horizon. Fluvic soils are young soils developed from recent alluvial deposits. Red-Yellow soils are a red or yellow color, low accumulation of organic matter, low base saturation, and strong weathering. Stagnogley soils are distributed in terraces, hills, and mountainous areas having a shallow groundwater table. Eutrosols are characterized by high base saturation, that is, with a eutric condition. Brown Forest soils are relatively young soils, are widely distributed in a rather wide range of temperate, and warm temperate zones in humid (high precipitation) climates without fresh volcanic ash deposition. Regosols are very weak weathering soils, which are divided into four groups: Volcanogenous Regosols, Sandy Regosols, Lithosols, and Terrestrial Regosols.

Keywords

Soil classification system of Japan • Soil great groups • Soil genesis • Soil characteristics • Soil-land use

Y. Takata (✉) · H. Obara · Y. Maejima · K. Kohyama
Institute for Agro-Environmental Sciences, NARO, Tsukuba City, Ibaraki Prefecture, Japan
e-mail: takatay@affrc.go.jp

M. Kawahigashi · K. Kida
Department of Geography, Tokyo Metropolitan University, Hachioji City, Tokyo, Japan

M. Tani · R. Kinoshita
Research Center for Global Agromedicine, Obihiro University of Agriculture and Veterinary Medicine, Obihiro City, Hokkaido, Japan

T. Ito
Faculty of Agriculture, Niigata Agro-Food University, Tainai City, Niigata Prefecture, Japan

M. Shibata
Graduate School of Global Environmental Studies, Kyoto University, Kyoto City, Kyoto Prefecture, Japan

T. Takahashi
Department of Biological Environment, Faculty of Bioresource Sciences, Akita Prefectural University, Akita City, Akita Prefecture, Japan

K. Fujii
Division of Forest Soils, Forestry and Forest Products Research Institute, Tsukuba City, Ibaraki Prefecture, Japan

A. Imaya
Forestry Division, Japan International Research Center for Agricultural Sciences, Tsukuba City, Ibaraki Prefecture, Japan

T. Kato
Department of Agricultural Chemistry, Faculty of Applied Bioscience, Tokyo University of Agriculture, Tokyo, Japan

R. Hatano et al. (eds.), *The Soils of Japan*, World Soils Book Series,
https://doi.org/10.1007/978-981-15-8229-5_4

4.1 Human-Made Soils

4.1.1 Pedogenesis and Characteristics of Human-Made Soils

(1) Artifactual soils and Reformed soils

The term Human-made soils refer to soils formed through an anthropogenic process, in contrast to other soils, which are usually named after their pedogenic processes or parent materials. The soils form through anthropogenic activities, especially in anthropized areas such as urban, industrial, traffic, mining, and waste disposal areas. Generally, soils under agricultural management, such as paddy fields, are excluded from the definition of Human-made soils, despite the anthropogenic land reformation on these agricultural fields. The central concept of Human-made soils is characterized by a soil profile that has been drastically changed with an unexpected soil pedon differing from any profiles of natural soils. Soils having features of pedogenic processes, such as mottle formation, clay accumulation, soil reduction, podzolization, should be classified into another soil great group characterized by specific soil formation processes. Human-made soils are further divided into the Artifactual soil group and the Reformed soil group. Artifactual soils are usually identified by the contents of artifacts and their position in depth, while Reformed soils are covered with intentionally transported soil materials. Both of these soil groups are intentionally or occasionally formed mainly due to land-forming processes and infrastructure constructions commonly using powerful heavy machinery.

(2) **Definition and properties of artifacts**

Artifactual soils are defined as soils which include a horizon containing "artifacts" over 20% of the area in the horizon with a cumulative thickness of over 25 cm or an impervious horizon consisting of asphalt or concrete materials within 30 cm of the surface (Fig. 4.1). Since the term "artifacts" is used in the definition and refers to the diagnostic materials, it is important to maintain a consensus on the definition of this term for translation between international classification systems. Artifacts are defined as human-manufactured materials that do not occur naturally on the Earth's surface. Examples include industrial wastes, such as mining wastes and construction debris consisting of asphalt and concrete released from industrial activities, and non-industrial wastes, which include home garbage and materials from synthetic polymers such as vinyl, plastics, metals, and porcelains. For example, Artifactual soils can be formed by burying burned residues at a waste disposal site or by burying construction debris in a reclamation area.

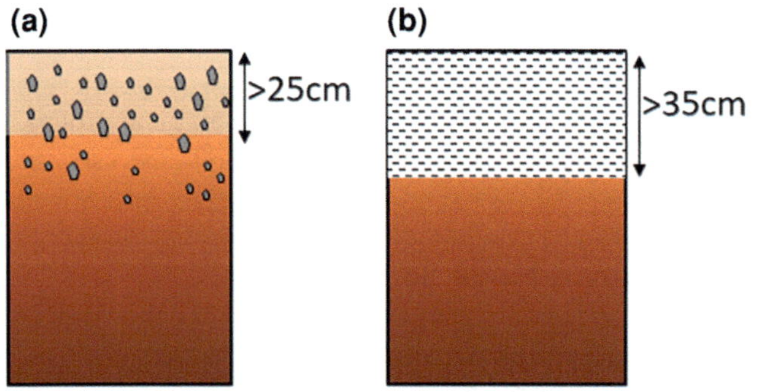

Fig. 4.1 Illustration of a soil profile of Human-made soils. **a** Artifactual soils containing a horizon exceeding 25 cm thickness comprised of artifacts over 20% of area in the horizon. **b** Reformed soils with a horizon piled over 35 cm thickness of different soil materials transported far from another place. The horizon containing over 20% of artifacts makes the soil as an Artifactual soils. (Figure supplied by Masayuki Kawahigashi and Kimihiro Kida)

(3) **Properties of Human-made soils**

Civil engineering for land reclamation leads to specific chemical and physical soil properties. Buried materials consisting of construction debris usually increase soil alkalinity due to mixing with cementing agents originating from lime. Under the climate conditions of Japan, which are characterized by excessive precipitation relative to evapotranspiration, the soil reaction gradually becomes acidic over time due to the downward movement of soil water (Kida and Kawahigashi 2015). This is a natural soil formation process occurring in Human-made soils. Some soil formation is allowed in Human-made soils, such as Technosols in the World Reference Base for Soil Resources (WRB; Schad 2018). The compaction of the soil after mixing solid waste into the soil is also a common physical property in reclamation areas and results in reductive soil conditions, especially in flat land fill areas. The reductive soil on the reclaimed land is not preferable for any land uses because of low ground stability, methane emission, and adverse plant growth conditions. Some ideas to solve this problem have been suggested, such as mixing wood chips into soils and installing a tubing system to release gases in reclamation areas. Although the establishment of greenery areas is a possible alternative land use in reclamation areas, the land is not favorable for plants due to its condition. Thus, concomitant land uses to achieve both waste disposal and plant growth are not easy to establish. The management of Human-made soils with physical and chemical properties that are unfavorable for plant growth is required in Japan, where little land area is available for waste disposal.

(4) **Special features of Human-made soils in greenery areas**

In greenery areas in cities, adding soil cover as planting soil makes the soil "Reformed soil" because of the banking of soil materials that have features different from those of the original base soils. Since greenery areas are constructed through civil engineering processes, very hard compacted subsoils are a common feature beneath sites of tree planting. Such subsoils will have a densic horizon that is too hard for roots to grow in. The distribution of roots will be concentrated in the surface horizons due to the compaction of subsoils. Densic horizons also impede the downward movement of water, resulting in high soil moisture with reductive soil conditions unfavorable for plant growth. When the artifact content is higher in subsoil horizons, the soil is referred to as Artifactual soil. Artifacts originating from construction debris promote soil alkalinization, adversely affecting plant growth.

If the covered soil has properties similar to those of base soils taken from near the reclamation area, it cannot be classified as Human-made soil, but rather as the soil great group of the base soils with supplementary description of a soil phase. For example, although urban forestry parks in Tokyo have undergone various changes in land use throughout their history, accompanying large-scale earthworks and constructions that have left records in the soil, the soil is classified not as Human-made soil, but rather as Andosol (Kawai et al. 2015).

The area of human-modified land extends along river and sea coasts because of protection against flood and tsunami and on mountain hills because of construction of recreation areas such as golf courses and ski slopes accompanied by land form modification. Human-modified soils associated with large-scale civil engineering works are also highly compacted, leading to a high bulk density and undeveloped soil structure (Ono et al. 2018). The soil condition leads to severe surface soil erosion under low vegetation density, resulting in a low stock of organic carbon and low-fertility soils (Uoi et al. 2013). After construction of the land, periodic assessment for monitoring of the physical and chemical properties of the soil is necessary for disaster prevention.

4.1.2 Significance in the Classification System

(1) **Change in the concept of Human-made soils**

In Japanese soil classification systems, as well as in the United States Department of Agriculture (USDA) and WRB soil classification systems, new revisions of the soil classification system always include major and/or minor changes regarding Human-made soils. This means that the definition, criteria, and concept of Human-made soils are still under debate in most soil classification systems. Since the impact of human activity on agricultural soils has long been recognized in Japan, Reformed soils were separated from other soil great groups in the revised second version of the Classification of Cultivated Soils in Japan (Cultivated Soil Classification Committee 1983). The agricultural reformation process was considered as a special activity to produce flat land for agriculture in Japan, which is dominated by mountainous terrain. On the other hand, in the same period, soils affected by anthropogenic activities have been classified as subgroups in the immature soil group [Regosols in WRB (FAO 1998) and Entisols in the USDA (Soil Survey staff 1999)] in the old version of the soil classification system released by the USDA and FAO. In the twenty-first century, the impact of humans on soil formation processes has been discussed in a broad sense, including non-agricultural areas in the soil classification systems. In the Unified Soil Classification System of Japan (second approximation) of 2002 (The Fourth Committee for Soil Classification and Nomenclature, Japanese Society of Pedology 2002), artifact contents and banking of transported soil materials were addressed as diagnostics to identify a surveyed soil as a Human-made soil at the first stage when the soil is classified. The Human-made soil great group is keyed out first in the Soil Classification System of Japan because human activity has been considered as a strong affecting factor to differentiate Human-made soils from the other soil great groups, which arise from pedogenic processes under natural conditions and/or agricultural management. This classification system is considered well about human impact on soil in categorizing Human-made soils, because human activity had not been sufficiently discussed to classify Human-made soils in the international soil classification systems.

(2) **Relationship between Human-made soils and other soil groups**

When diagnostic properties relating to pedogenic processes are recognized in soils under anthropogenic activities, the soils are not classified as Human-made soils, but rather as another soil type having specific diagnostic properties, indicating that no soil formation processes have ever been recognized for Human-made soil. Soil formation processes occurring in a soil have higher priorities in the classification and/or characterization of soils than any soil construction processes. Although human activity has a crucial influence on soils, it is not one of the soil-forming factors in the history of the Japanese soil classification system. Soils that have been artificially constructed by human activity are also naturally processed under climatic pedogenic factors. The concept of Human-made soil, which is considered as immature soil without any changes in soil, is the same as the concept of anthropogenic soils in the USDA Soil Taxonomy (Soil Survey Staff 2014) and WRB (IUSS Working Group WRB 2006).

4.1.3 Consideration of Human-Made Soils from the Viewpoint of International Classification Systems

(1) Categorization of Human-made soils

The concept of soils under intensive human activity is probably largely divergent between different national soil classification systems. Such difference can be found in the position of the soil category in each classification system. Human-made soils are placed as the highest category as a soil great group and are placed in the first position among soil great groups in the Soil Classification System of Japan. In the WRB classification system, the corresponding soils are also placed as the highest category of reference soil group, being given the name Technosols, which are mainly defined by the contents of artifacts in a soil horizon and are placed in the third position following the organic soils Histosols and the human-impacted agricultural soils Anthrosols (IUSS Working Group WRB 2015). Both the Japanese and WRB classification systems recognize the specificity of human impacts on soils. In particular, the WRB system considers human activity as a soil formation factor (IUSS Working Group WRB 2006), together with "Climate," "Parent materials," "Land form," "Organisms," and "Time."

On the other hand, in the USDA classification system (Soil Survey Staff 2014), human-impacted soils can consist of original soils that contain anthropogenic materials and/or possess characteristics arising from modification of the soils by human activities. Soils with these qualifying properties at lower levels can be called "human-altered and human-transported" (HAHT) soils in the USDA classification system. Qualifiers signifying Artifactual influences can be added at lower subgroup levels in the classification system.

(2) Translation of Human-made soils between international classification systems

Information regarding anthropogenic impacts on soils cannot be confirmed in the soil names of the soil great group under the USDA soil classification system, whereas the soil name of Technosols in the WRB classification system does not preserve information on the original soil name and properties unless supplementary qualifiers are added. Supplementary qualifiers are included in the WRB classification system to provide information about the properties of the original soils. As in the WRB classification system, soil names with subgroup categories of Human-made soils contain no information about the original soil under the Japanese classification system. Information regarding the original soil can be added to express soil phases in the Japanese classification system. The differences in the technical terms used to express anthropogenic properties in soil names among USDA, the WRB, and the Japanese classification systems make it difficult to translate soil names between the systems. To solve this translation issue, every classification system will be required to fit the definition of artifacts, the position of a soil surface, and soils that they classify. Upgrading the HAHT soils to the first category in the USDA classification system is now being considered to achieve compatibility between the various soil classification systems (Galbraith 2018).

(3) Soils beneath constructions or in pots

Since the Japanese classification system excludes soils beneath constructions such as roads and buildings, soils cannot be named in most urban areas. The WRB classification system enables the naming of soils covered with technic hard materials as "Ekranic Technosols" for the reference soil group with a principal qualifier (Fig. 4.2). Soil in a pot or in a tree pit is not included in the Soil Classification System of Japan, but it can be named by the WRB classification system using a principal qualifier of "linic" or "isolatic" accompanied by "relocatic" as a supplementary qualifier in the subcategory. Such soils can also be named by the USDA classification system, according to the origin of the soil beneath constructions or transferred soil materials filled into a pit. Here, it is also important to understand the difference between the concepts of soil classification systems. Artifacts and technic hard materials added or constructed are the main qualifier used to classify Human-made

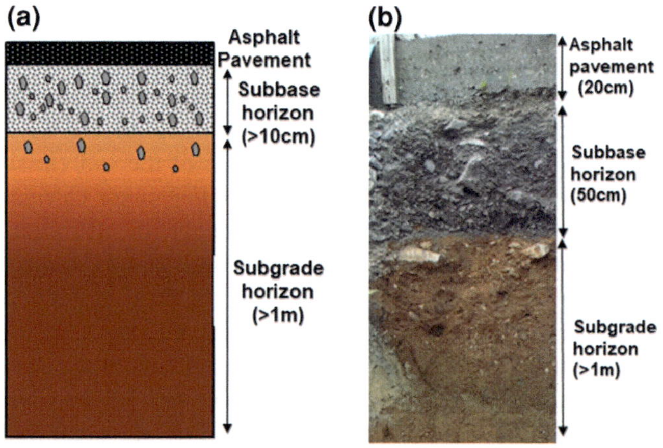

Fig. 4.2 Soils beneath the road construction. **a** Pavement is usually consisting of asphalt layer with thickness over 5 cm. The thickness of the pavement and the subbase layer change depending on traffic density. **b** The profile beneath the road construction with heavy traffic in Kanagawa prefecture (Ekranic Technosols). The profile consists of thick asphalt pavement reaching to 20 cm thickness underlain by thick subbase horizon mixing with crushed stones on the subgrade horizon originated from Andosol. (Figure supplied by Masayuki Kawahigashi and Kimihiro Kida)

soils in the WRB classification system, whereas the alteration of land and transferred soils are the main diagnostic characteristic for soils in anthropized environments in the USDA Soil Taxonomy. The central concept of the Japanese classification system does not translate well to the USDA classification system, but is relatively close to that of the WRB classification system with a narrower range of applications. In the Japanese classification, the concept of soils which have the potential to be responsible for primary productivity can be close to that in the USDA classification system (Kida and Kawahigashi 2017).

The definition of the top of a soil surface directly influences the stated vertical position of any diagnostic horizons for soil classification (Kida and Kawahigashi 2017). The definition of soil surface in the Japanese classification system is the mineral soil surface, as opposed to the ground surface which is the definition used for soil surface in the WRB classification system. The difference in the definitions of soil surface between classification systems differentiates the position of diagnostic horizons in a soil horizon, affecting the soil name and the translation of the soil name between the systems.

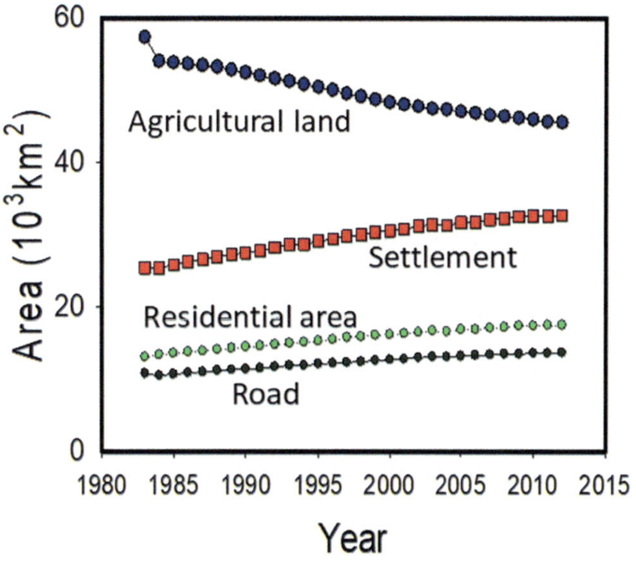

Fig. 4.3 Change in the area of agricultural land and settlement area in land use change in Japan from 1984 to 2012. The settlement area consisting of residential and road areas does not include recreation areas, greenery areas, and reclaimed lands. The figure was created by the authors based on data from white papers on land published by Ministry of Land, Infrastructure, Transport and Tourism of Japan (MLIT 2014)

4.1.4 Distribution of Human-Made Soils in Japan

(1) **Settlement areas in Japan**

Areas of human settlement, which occupy almost 10% of Japan's land area, have been artificially modified by large-scale civil engineering works. The total settlement area has increased during the past several decades, accompanied by a decrease in the area of agricultural fields (Fig. 4.3). In civil engineering, a large volume of soil, reaching approximately 300 million m^3 every year, is transferred from one site to another, mainly to make flat land (Katsumi 2015). The process of cutting and banking soils forms a new soil profile, leading to an increase in the area of Artifactual soils and Reformed soils. However, the processed area covered with constructions, which occupies approximately 80% of settlement areas (Ministry of Land, Infrastructure and Transport 2014; Fig. 4.4), is excluded from classification under the Soil Classification System of Japan. This means that most of the land area in urban areas, which is covered with artifacts, is not considered to be areas where Human-made soils are distributed.

The remaining 20% of settlement areas is occupied by greenery areas, which include recreation areas such as golf courses and ski slopes that required land-forming construction (Fig. 4.4). These artificially formed soils are frequently excluded from the definition of Human-made soils because the soils used on-site to flatten land for construction of road

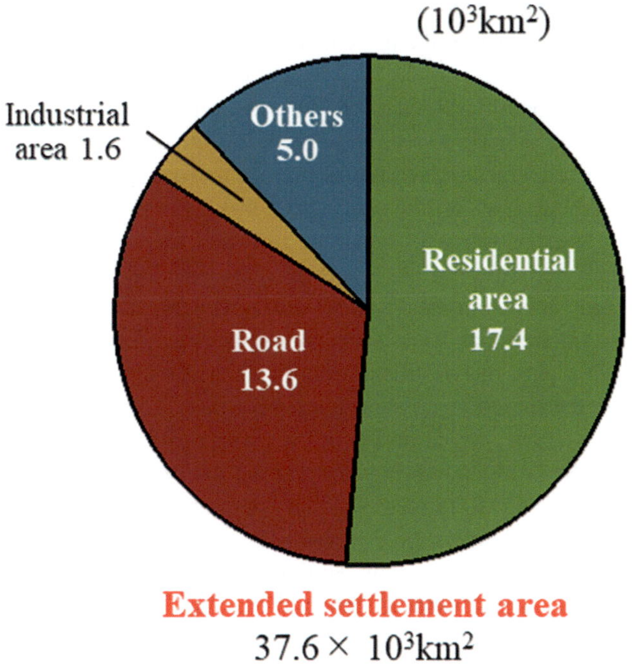

Fig. 4.4 Composition in land uses in the extended settlement of Japan. The figure was created by the authors based on data from "Greenhouse Gas Inventory Report of Japan 2012" published from Greenhouse Gas Inventory Office of Japan (GIO) and databases from Ministry of Land, Infrastructure, Transport and Tourism of Japan (MLIT 2014). Residential, Road, and Industry areas were the statistical data of white papers on land (MLIT 2014). These three areas were subtracted from total area of the extended settlement (GIO 2012) to calculate the area of "others" which include greenery areas, golf courses, reclaimed land areas and unknown land uses

and buildings or to create artificial topography for recreation areas consist of the same soil materials as at the excavated site. In such cases, the soil profile can be classified as the original soil with a description of the soil phase as a Human-made soil phase. Banking soil is not called Reformed soil when there is consistency between the dressed soil and the underlying soil; instead, it is usually named according to the properties of the filled and covered soils. Soil dressing using soil excavated from nearby hills, mountains, and terraces is a common practice when constructing greenery areas. The soils formed by these civil engineering works do not fit the definition of Reformed soils or Artifactual soils because of the vertical consistency of soil materials.

(2) **Distribution of Human-made soils in Japan**

Based on the Soil Classification System of Japan, the area of Human-made soils is estimated to be less than 1% of Japan's total land area, indicating that the Human-made soils are sporadically distributed in civil engineering areas. The Soil Classification System of Japan states that the distribution of Human-made soils is unknown because of the scarcity of information about soil descriptions and general soil physicochemical properties of such soils (The Fifth Committee for Soil Classification and Nomenclature 2017). Furthermore, there is no surveillance of the present status of Human-made soils in urban, industrial, traffic, mining, and waste disposal areas. The acquisition of further data and information about these soils is required in order to recognize them in spite of the difficulties involved with conducting soil surveys in these types of areas.

About half of the excavated soil released from civil engineering works in Japan are not used for banking at the excavated site, but rather are moved to landfill areas, commonly along the sea coast. Some greenery areas in reclamation areas can have Artifactual soils in the case of mixing with plenty of construction debris as artifacts within a proposed depth. Reclamation areas also have a role as a final disposal site, which results in the presence of artifacts in the soil. Soils mounded on ground using excavated soil can be classified as Reformed soils in greenery areas. However, irregular thickness of covered soil volume can cause the sporadic distribution of Human-made soils in artificial greenery areas used as urban parks and artificial forests. Although a land use map is one possible way to estimate the area and distribution of Human-made soils, the irregularity of artifacts or top-dressed soil materials brought by the process of civil engineering works make it difficult to estimate them through soil surveys. This is also one of the barriers to recognizing Human-made soils and is different from the case of other soils in Japan. Other soil types are usually expressed on a soil map containing topographical information because this can indicate the distribution of their soil parent materials.

4.1.5 Land Uses of Human-Made Soils

Unlike most soils, the properties of Human-made soils usually depend on the land use. Any land uses that require appropriate soil management create Human-made soils through civil engineering processes. However, in order to achieve stable ground for land-use purposes, civil engineering usually forms Artifactual soils with highly compacted subsoil and a soil profile consisting of unexpected combinations of soil horizons. In this case, the Human-made soil is not suitable for plant growth. After the construction of Human-made soils, the soil is usually no longer suitable for primary productivity. A suitable civil engineering process is necessary to construct a greenery area on landfill with waste disposal or on soil remaining after civil engineering.

Although land uses such as urban, industry, and traffic have replaced land uses for primary production in Japan's modern history, this trend will not continue in the future. It is necessary to consider land use change from the opposite direction. We are facing population decline accompanied by the downsizing of urban areas. A strategy for shifting land use may be required in the near future. Soil information for anthropized land should be collected for future land-use planning.

4.2 Organic Soils

4.2.1 Definition and Classification of Organic Soils in Japan

Organic soils form under wetland conditions where the rate of organic matter production exceeds the rate of organic matter decomposition, and when composed of partially decomposed hygrophytes, they are regarded as peats (Buol et al. 2011).

The Soil Classification System of Japan states that soils of the Organic soil great group must meet one of the following criteria (The Fifth Committee for Soil Classification and Nomenclature 2017):

(1) The thickness of Organic soil material totals 25 cm or more between the surface and a depth of 50 cm; *or*
(2) When non-organic soil material is dressed for less than 35 cm, the thickness of organic soil material total 25 cm or more consecutively below this non-organic soil material.

Organic soil material is mainly made up of undecomposed or decomposed plant materials that accumulated underwater and satisfy all of the following criteria:

(1) Contains more than 12% organic carbon; *and*
(2) Does not have andic soil properties, except when plant fibers comprise more than one sixth of the volume after rubbing in the hands.

Therefore, all of the soils under Organic soils are classified in the Peat soil group. Most of these soils are classified as Histosols in the USDA Soil Taxonomy. However, it requires the thickness of "organic soil material" to be more than 40 cm, and therefore, some of the Peat soils may not be classified as Histosols under Soil Taxonomy (Soil Survey Staff 2014). This is because Japanese Peat soils are strongly affected by inputs of inorganic material from the flooding of rivers and volcanic ash deposition, as will be discussed later (refer to Sect. 4.2.2).

Peat soils have received dressings of mineral soils for improving the chemical and physical characteristics of rice paddies and grasslands in Hokkaido, where large areas are occupied by Organic soils (refer to Sect. 4.2.4). When the thickness of dressed soil is less than 35 cm and organic soil material extends below the layer, a soil is classified as Organic soil because of the large influence of organic soil material. A soil is classified as Human-made soil when the thickness of the dressed soil layer is more than 35 cm.

In Japan, Peat soils had been classified as High-moor Peat soils, Transitional-moor Peat soils, and Low-moor Peat soils according to the formation process and the types of plant species included in their composition. This system was based on the peat soil classification of Germany, and the subsequent translations. Peat soils are additionally classified by their nutrient status into Eutrophic Peat soils and Oligotrophic Peat soils, as well as Mesotrophic Peat soils, which are intermediate between the previous two. Low-moor Peat soils correspond to Eutrophic Peat soils and High-moor Peat soils correspond to Oligotrophic Peat soils.

The Soil Classification System of Japan is based on the degree of decomposition of Peat soils and the type of component plant species and is divided into the following four soil subgroups:

- Sapric Peat soils: Peat soils that have soil layers of organic soil material with the highest fraction of sapric soil material being between the soil surface and a depth of 50 cm; *or*
- High-moor Peat soils: other Peat soils in which the uppermost 25 cm of soil layers is composed of High-moor organic soil material; *or*
- Transitional-moor Peat soils: other Peat soils in which the uppermost 25 cm of soil layers is composed of organic soil material with transitional-moor soil materials; *or*.
- Low-moor Peat soils: other Peat soils.

The organic soil materials mentioned above are defined based on the degree of decomposition of Peat soils as well as the type of component plant species.

- Fibric soil materials: those that contain three quarters or more by volume of fibers after rubbing.
- Sapric soil materials: those that contain less than one-sixth by volume of fibers after rubbing.
- Hemic soil materials: other organic soil materials that are not classified as fibric soil materials or sapric soil materials.
- High-moor soil materials: those that have the highest fraction (area) of the combination of *Sphagnum*, Horomui sedge (*Carex middendorffii*), small cranberry (*Vaccinium oxycoccos*), whitebeak sedge (*Rhynchospora alba*), and Rannoch-rush (*Scheuchzeria palustris*).
- Transitional-moor soil materials: those that have the highest fraction (area) of Japanese moor grass (*Molin-iopsis japonica*), hare's-tail cottongrass (*Eriophorum vaginatum*), Myrica gale (*Myrica gale* var. tomentosa), and Sakhalin spruce (*Picea glehnii*) combined in organic soil materials.
- Low-moor soil materials: other organic soil materials that are not classified as high-moor organic soil materials or transitional-moor organic soil materials.

Under the classical Japanese soil classification system, the Classification of Cultivated Soils in Japan (Cultivated Soil Classification Committee 1995), Sapric Peat soils were classified as Muck soils, which have highly decomposed organic soil materials and a high degree of humification, which makes it difficult to identify plant fibers, and the soil colors appear black to dark brown. These soils are classified as Saprists in the USDA Soil Taxonomy based on plant fiber content and sodium pyrophosphate index (PI; Soil Survey Staff 2014). In Japan, the input of inorganic soil materials from the flooding of rivers and from volcanic ash deposition is often observed, accelerating the decomposition of Peat soils above and below the incorporated inorganic soil materials.

High-moor, Transitional-moor, and Low-moor Peat soils are composed of fibric or hemic soil materials and classified into three soil subgroups according to the type of component plant species. In the USDA Soil Taxonomy, peat soils that are composed of fibric soil materials are classified as Fibrists and other peat soils that are composed of hemic soil materials are classified as Hemists (Soil Survey Staff 2014). In Japan, much of the peatlands are formed in the back swamp of an alluvial plain. Low-moor Peat soils dominate and are mainly composed of common reed (*Phragmites australis*) and Japanese alder (*Alnus japonica*). The areas occupied by High-moor or Transitional-moor Peat soils are relatively small.

4.2.2 Distribution and Formation of Organic Soils in Japan

In Japan, Organic soils cover approximately 4400 km^2 and account for roughly 1.2% of the land area. The majority of the Organic soils are found in the Hokkaido and Tohoku regions of Northern Japan, where the climate is relatively cool. These soils are rare in the Kyushu and Shikoku regions of Western Japan, where the climate is warm (Obara et al. 2016). More than 90% of Organic soils in Japan are distributed in the Hokkaido, Tohoku, and Kanto regions (Fig. 4.5). In all of these regions, Low-moor Peat soils are the most dominant at the soil subgroup level, and they comprise 95% of the Organic soils in the Tohoku region. Sapric Peat soils are widespread in the Kanto region, but are rarely seen in the Hokkaido and Tohoku regions (Fig. 4.6).

Peatlands that have Organic soils can be classified into paludization-type or terrestrialization-type depending on the formation process (Sakaguchi 1974). Most of the Japanese peatlands are paludization-type and are either fluvial peatland in the back swamp of an alluvial plain or seepage peatland in marshy ground. Such peatlands are widespread in Northern Japan, where the climate is cool. Conversely, small-scale terrestrialization-type peatlands are formed in highland limnic basins and mountainous basins or swales. They are scattered throughout young volcanic mountainous areas across Japan.

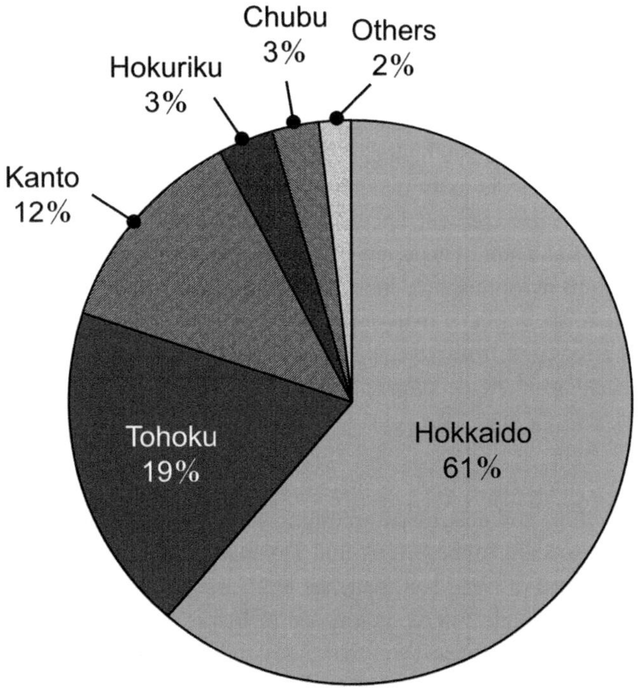

Fig. 4.5 Distribution of Organic soils in different regions of Japan. Based on the data in the Soil Classification System of Japan. *Data source* The Fifth Committee for Soil Classification and Nomenclature (2017). Others include Kinki, Chugoku, Shikoku, Kyusyu, and Okinawa regions. Large distribution of organic soils is found in Hokkaido and Tohoku regions where the climate is relative cool, and the two regions account for 80% of the total organic soil areas

Paludization-type peatlands occur when the soil surface environment becomes humid and suitable for hygrophytes to prevail, and Peat soils are formed. Unlike the typical evolutional process of Low-moor Peat soils → Transitional-moor Peat soils → High-moor Peat soils of the terrestrialization-type peatlands, the nutrient content of the supplied water permits the sole formation of Low-moor Peat soils or High-moor Peat soils, while inputs of inorganic materials may also reverse High-moor Peat soils back to Low-moor Peat soils (Kondo 1997). As later discussed in Sect. 4.2.3, in Japan, inputs of inorganic materials from the flooding of rivers and the deposition of volcanic ash supply nutrient salts to peats, allowing eutrophic plant species including the common reed and Japanese alder to prevail and making it suitable for Low-moor Peat soils to form (Fig. 4.7b). Most of these Japanese Organic soils are distributed in paludization-type peatlands.

Terrestrialization-type peatlands form in limnic basins or in lakes and marshes. First, emerged plants, including Manchurian wild rice (*Zizania latifolia*) and broadleaf cattail (*Typha latifolia*), and floating leaved plants, including Indian lotus (*Nelumbo nucifera*) and Japanese water chestnut (*Trapa japonica*), supply organic material, causing the water depth to become shallower, which allows common reeds and large sedges to grow. Their remains form Low-moor Peat soils. Next, a thick accumulation of Low-moor Peat soil causes the groundwater level to decrease and dries the soil, allowing hare's-tail cottongrass, Japanese moor grass, and Myrica gale to grow. The remains of these plants form Transitional-moor Peat soils. The further accumulation of Peat soil causes the groundwater level to become even deeper (i.e., further away from the soil surface); this in turn causes oligotrophic conditions, which allows *Sphagnum* and sedges that can survive with only precipitation to thrive, and their remains form High-moor Peat soils (Fig. 4.7a). In Japan, small-scale terrestrialization-type peatlands are present, but they cover a significantly smaller area compared with paludization-type peatlands.

Organic soils form under over-humid or low-temperature conditions where the rate of organic matter production by hygrophytes exceeds the rate of organic matter decomposition (Buol et al. 2011). Water is the most important factor controlling the formation and characteristics of Organic soils (Sakaguchi 1974). The formation and distribution of Organic soils in Japan are influenced by the interaction between climatic and topographic conditions.

Most of the land area of Japan is under a temperate climate, with four distinct seasons having significant climatic differences between summer and winter and with large annual precipitation showing a temperate humid climate.

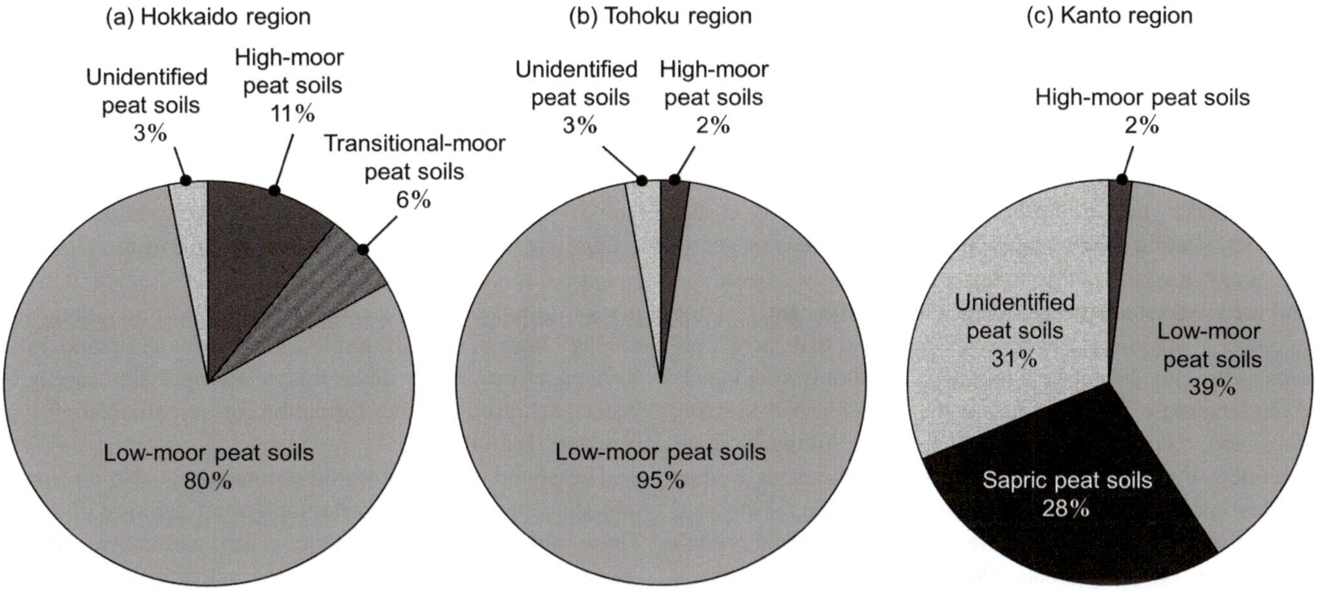

Fig. 4.6 Distribution of soil suborders of Organic soils in **a** Hokkaido, **b** Tohoku, and **c** Kanto regions of Japan. Based on the data in The Soil Classification System of Japan. *Data source* The Fifth Committee for Soil Classification and Nomenclature (2017). Unidentified peat soils were assigned when available data were not sufficient for classification. In all regions, low-moor peat soils have the largest distributions and account for up to 95% in the Tohoku region. Unlike Hokkaido and Tohoku regions, Sapric peat soils are widely distributed in Kanto region

Fig. 4.7 Peat soils of Hokkaido peatlands. **a** Typical High-moor Peat soils composed of *Sphagnum* in Bekanbeushi peatland (Akkeshi Town, Akkeshi County). Groundwater is found approximately 40 cm below the soil surface, and fibric peats of *Sphagnum* are accumulated at the surface. **b** Low-moor peat soils composed of common reed in Kiritappu peatland (Hamanaka Town, Akkeshi County). Incorporation of volcanic ash is observed approximately 15 cm below the soil surface. (Figures supplied by Masayuki Tani)

Additionally, seasonal winds from the southeast and the northwest in the summer and winter, respectively, cause heavy rainfall on Japan's Pacific side in the summer and heavy snowfall on the Sea of Japan coast in the winter. Moreover, there are rainy seasons in June to July and in September, when typhoons move up the Japanese archipelago causing strong winds and severe rain storms.

The Japanese archipelago experiences a varied climate, with the northern areas under a subarctic climatic zone and the southern areas in a subtropical climatic zone. Soil temperature regimes are frigid (<8 °C) in eastern and northern Hokkaido and mesic (8–15 °C) in western and southern Hokkaido and northern Honshu. In central and western Honshu, the soil temperature regime in highlands and mountainous areas is mesic, while that of plain and plateau areas are thermic (15–22 °C). In subtropical areas of the Nansei Islands, the soil temperature regime is hyperthermic ($\geq$22 °C).

As explained in Sect. 2.2, the topography of Japan has specific characteristics in contrast to other parts of the world. It is located in the circum-Pacific orogeny, and 70% of the land area is composed of steep, mountainous areas or gentle hilly areas; lowlands only occupy approximately 14% of the land area. Mountainous areas run through the center of the Japanese archipelago like a spine, meaning that large portions of the islands are steep mountains, and only small areas of flat land exist. This also means that Japan's river systems have short watercourses and steep slopes, resulting in small catchment areas. The rainy seasons and typhoons, as well as melting snow in early spring, cause large changes in river flow volume, which results in frequent natural disasters such as flooding. Flooding creates floodplains and natural levees in the alluvial plains of the lowlands. Behind these natural levees, back swamps form that contain fine particles such as clay and silt and are poorly drained.

In Northern Japan, including the Hokkaido and Tohoku regions, where climatic conditions are cool, hygrophytes, such as the common reed and Japanese alder, grow, and Low-moor Peat soils form due to low soil temperature limiting the decomposition of organic material (Fig. 4.8). In Hokkaido, frigid soil temperature regimes exist, especially in the eastern and northern parts of the island, but also in the Ishikari, Teshio, Sarobetsu, and Kushiro River basins, where large back swamps and marshy ground have formed. Large areas of Organic soils can be found here. High-moor Peat soils and Low-moor Peat soils are present in the Ishikari peatland, whereas low-moor peatland occupies much of the Sarobetsu and Kushiro peatlands (Sakaguchi 1974). The Kushiro peatland, one of the largest peatlands in Japan, is preserved under relatively natural conditions because of its poor geographical setting. At the base of the peat layer, a layer containing shell exoskeletons of marine origin has been found, showing that the area is a large shallow marshy ground created after the marine regression in the Holocene Epoch. Subsequently, the Kushiro River drained into the marshy ground, flooding it repeatedly. In addition, there was seepage from surrounding hilly areas (diluvial uplands) that were capes previously under marine transgression supplying water, resulting in the formation of a paludization-type peatland (Fig. 4.8).

The abundance of volcanic mountains is also an important characteristic of the topography and geology of Japan. These not only deposit volcanic ash and other volcanic ejecta during eruptions, but also occasionally form depressions and gently sloping topography in mountainous and highland areas as a result of lava and mud flows. Areas where young volcanoes of Quaternary age are present have topographical features formed by lava and mud flows that easily collect water, including dammed basins and/or shallow swale features. Subsequently, abundant seepage water is supplied from the slope and foot of a volcano, forming lakes and wetlands. In mountainous and highland areas with high elevation, the rate of decomposition of plant remains is slow due to low soil temperature, and terrestrialization-type peatlands form (Sakaguchi 1974).

The Oze Marshland, in Honshu, is located in the mountainous area of the Gunma–Fukushima–Niigata border at an elevation of 1400 m a.s.l. It is the largest typical high-moor

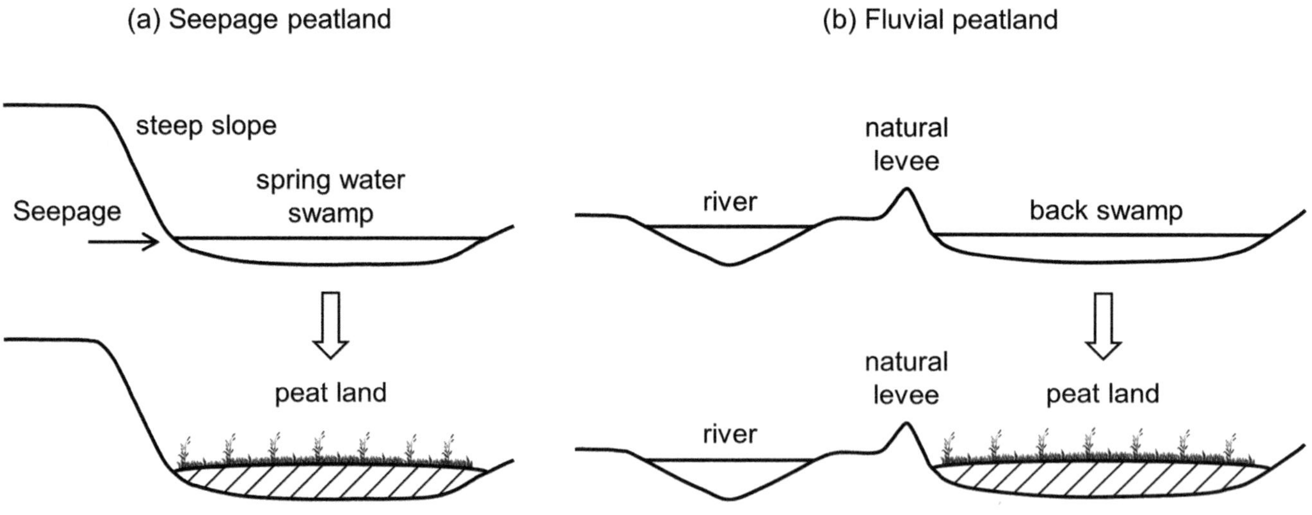

Fig. 4.8 Formation of typical paludaization-type peatland in Japan. **a** Seepage peatland formed by the growth of hygrophyte in swamp with seepage water from diluvial upland and terrace cliffs. **b** Fluvial peatland formed by the growth of hygrophyte such as common reed in back swamp of alluvial plain. These types of peatlands are common in northern Japan in the Hokkaido and Tohoku regions where the climate is cool, and they are often large scale. (Figures supplied by Masayuki Tani)

peatland in Honshu, Japan, and covers 8 km^2 in area. High-moor Peat soils composed of *Sphagnum* dominate, and have micro-topographical features that are characteristic of high-moor peatland. These High-moor Peat soils were formed by volcanic activity of a young active volcano, called Hiuchigatake, that produced lava and mudflows, forming flat plains where the climate is subarctic humid due to the high elevation (Sakaguchi 1974).

Organic soils in the west of Honshu, including the Kyushu and Shikoku regions, are terrestrialization-type peatlands formed in the mountains and highlands of volcanoes active in the Quaternary Period. These peatlands occupy a significantly smaller area compared to the paludization-type peatlands of Northern Japan and are also scattered throughout the area.

4.2.3 Characteristics of Organic Soils in Japan

The chemical and physical characteristics of Organic soils, including Peat soils, are controlled by the plant community (component plant species) of the peats, as well as the degree of decomposition and humification (Kondo 1997).

The plant communities of peatlands are dependent on the type of water and nutritional conditions available. The type of water that recharges peatlands is categorized into precipitation, surface water, and groundwater. In low-moor peatlands, the recharge water can belong to all of these water types, but the contribution of precipitation is low compared to that of surface water and groundwater. By contrast, in high-moor peatlands, the recharge water is almost exclusively derived from precipitation. Since surface water and groundwater supply nutrient salts to peatlands, eutrophic plant communities of herbaceous plants, including common reed and large sedges, as well as trees including Japanese alder and Manchurian ash (*Fraxinus mandshurica*), predominate. Since precipitation lacks nutrient salts, oligotrophic plant communities such as *Sphagnum* and Horomui sedge predominate (Sakaguchi 1974).

In peatlands, many plant species flourish and approximately 200 species are known to grow in the Kushiro peatland. However, the dominant plant species that comprise peats number around only 20 species, while the others are only accompanying plants, which grow in the environment but are not component plants species of peats (Kondo 1997). In the peatlands of Hokkaido, the main component plant species are common reed, Japanese alder, hare's-tail cottongrass, Japanese moor grass, Myrica gale, Rannoch-rush, cinnamon fern (*Osmundastrum cinnamomeum*), small cranberry, Horomui sedge, and *Sphagnum* (Sakaguchi 1974).

As discussed in Sect. 4.2.2, most of Japan's peatlands are paludization-type, which that can be fluvial peatlands forming in the back swamp of an alluvial plain or seepage peatlands forming in the marshy ground left after marine regression. Therefore, recharge waters are mainly surface and groundwater, which causes a eutrophic plant community to dominate, forming Low-moor Peat soils with common reed and Japanese alder as component plant species. Particularly in the paludization-type peatlands of Hokkaido, common reed is the predominant component plant species of peats. On the contrary, Transitional-moor Peat soils with Japanese moor grass and hare's-tail cottongrass or High-moor Peat soils with *Sphagnum* and Horomui sedge are relatively sparse, except in some parts of the Ishikari peatland and the Sarobetsu peatland in Hokkaido (Sakaguchi 1974; Kondo 1997).

Along with the component plant species, the degree of decomposition is another controlling factor determining the characteristics of peats. The acceleration of decomposition is caused by: (1) the reduction of plant biomass production by reaching the limit of peatland formation; (2) eutrophication by the input of inorganic materials such as volcanic ash and clays, as well as the neutralization of acidity; (3) eutrophication and improvements of aeration by the input of coarse particles such as pumice and sand; and (4) improvements of aeration by the lowering of groundwater (Sakaguchi 1974).

Paludization-type peatlands, the dominant type in Japan, receive frequent inputs of inorganic materials from flooded river sediments. Additionally, inputs of volcanic ash and pumice often occur near young volcanoes and create favorable environments for the decomposition of peats through eutrophication. Due to the presence of westerlies, peatlands on the eastern side of volcanoes receive volcanic ash, which is subsequently incorporated into the peat, thereby accelerating the decomposition of the peat above and below the incorporated inorganic materials (Fig. 4.7b).

As fully explained in the USDA Soil Taxonomy, the degree of decomposition of peat can be classified based on the plant fiber content after rubbing and on the PI (Soil Survey Staff 2014). Fiber content is assessed by washing a hand-rubbed peat sample over a 100-mesh sieve (0.15 mm aperture) and measuring the volume fraction that remains. This technique is also used for classifying soil subgroups in the Japanese soil classification system (refer to Sect. 4.2.1). The PI is judged against the Munsell soil color chart using the value (lightness) and chroma of a filter paper used for paper chromatography, after the paper is used to absorb an extract of a peat sample which was saturated in sodium pyrophosphate overnight. The peat sample is classified as fibric soil material when the difference between the value and chroma is large, and it is classified as sapric soil material when the difference is small (Soil Survey Staff 2014). PI indicates that more humification took place due to the greater decomposition of the peat. It is judged by the small difference between the value and chroma indicative of melanized humic substances. In Japan, the assessment of humification due to the advancement of peat decomposition is suggested to follow the modified method proposed by Kaila (1956), which is an index based on the extraction by the shaking of peat samples in 0.025 M sodium pyrophosphate solution overnight and measuring the absorbance of centrifuged and filtered extracts at 340 and 550 nm (Kondo and Endo 1993).

The degree of decomposition and humification of peat increases from High-moor Peat soil to Transitional Peat soil and to Low-moor Peat soil in Hokkaido, when their physicochemical characteristics and the degree of decomposition are compared (Kondo 1997). Following the same sequence, ash and total nitrogen (N) contents also increase, while total carbon (C) and C:N ratio decrease (Table 4.1). The formation of Low-moor Peat soil occurs under eutrophic conditions with the supply of surface water and groundwater recharge and is strongly influenced by the addition of incoming inorganic materials, leading to high N and ash contents, and also increased decomposition and humification. When peat samples from the Pacific coast of the Tokachi area of Hokkaido, with or without inorganic material inputs, are compared using the absorbance at 550 nm, samples with inorganic material input are clearly more humified, indicating that inorganic material functions as a catalyst that advances the humification of peats (Fig. 4.9).

As discussed, the Peat soils of Japan have high ash content due to the input of clay and sand from flooded river sediments and volcanic ash deposition. When the elemental composition of Peat soils from Hokkaido and Russia (then the USSR) was compared, the Peat soils of Hokkaido contained higher amounts of aluminum (Al) and phosphorus (P) (Sakaguchi 1974). Furthermore, when Peat soils of Hokkaido and the USA were compared, the former contained higher amounts of silicon (Si), Al, and iron (Fe), but had lower contents of calcium and P (Kondo 1997). In particular, while Eutrophic Peat soils of the USA contained on average 5 g kg^{-1} of Fe, Low-moor Peat soils of Hokkaido contained on average 15 g kg^{-1}, three times more. Most of the Fe in the Peat soils of Hokkaido can be extracted with sodium pyrophosphate or acid ammonium oxalate, which shows that a large quantity of Fe is present as Fe-humus complex and non-crystalline Fe hydrous oxide (Tani et al. 2001a). In highly humified peatlands, a portion of the Fe can dissolve into peatland water by binding to dissolved organic matter, some of which is transported to the ocean through rivers and is suggested to contribute to the growth of seaweed, as well as to the enrichment of fishing grounds (Tani et al. 2001b).

4.2.4 Uses of Organic Soils in Agriculture

In Japan, the area of agricultural land containing Organic soils is approximately 1800 km^2, accounting for roughly 40% of the total area of Organic soils in Japan (Kanda et al. 2017). As discussed, the majority of Organic soils in the

Table 4.1 Degree of decomposition, humification, and physicochemical characteristics of organic soils in Hokkaido

Soil suborder	Fiber content (%)	Degree of humification	Ash content (%)	Total carbon (%)	Total nitrogen (%)	C/N ratio
Low-moor peat soil	37 ± 14	55 ± 39	32 ± 16	40 ± 10	1.92 ± 0.49	22 ± 6
Transitional-moor Peat soil	42 ± 14	21 ± 9	20 ± 16	46 ± 10	1.91 ± 0.37	25 ± 8
High-moor Peat soil	65 ± 24	14 ± 12	9 ± 7	50 ± 3	1.16 ± 0.57	53 ± 22

Data source Modified from data in Kondo (1997)

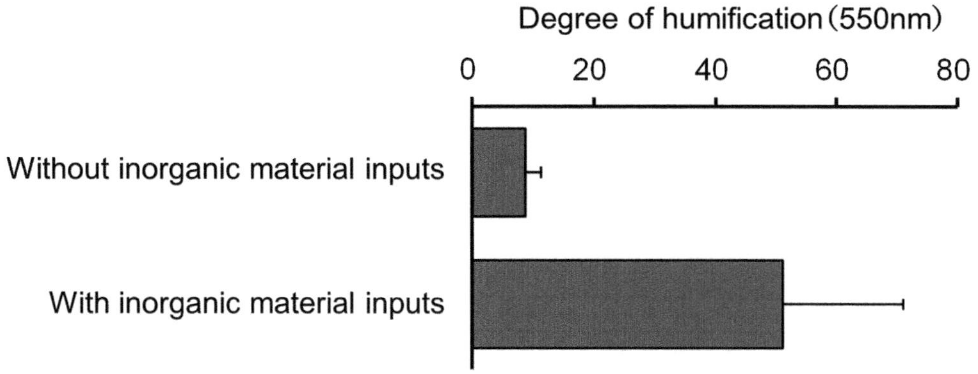

Fig. 4.9 Effects of inorganic material input on the degree of humification in low-moor peat soils of the Pacific coast of Tokachi district, Hokkaido. A low degree of humification was observed for a sample with little inorganic material inputs, showing an average value of 8.7 at the wavelength of 550 nm, while high degree of humification was observed for a sample with inorganic material inputs showing an average value of 51.1 at the wavelength of 550 nm. *Data source* Tani et al. (2001a)

Hokkaido and Tohoku regions, but also in the Kanto region, formed in the back swamps of alluvial plains, where the flat topography is suitable for irrigation; accordingly, much of this land has been used as rice paddies and arable fields. Most of the peatlands in the Kanto region are used for residential and urban areas.

In Hokkaido, the total area of Organic soil is approximately 2700 km^2, of which roughly 990 km^2 (37%) is used for agricultural land. Natural peatlands (including wetlands) occupy approximately 660 km^2 (Fujita et al. 1997). The total agricultural land area of Hokkaido is approximately 11,500 km^2, of which Organic soils account for roughly 9% (Kanda et al. 2017). Peat soils under natural conditions are often over-humid and have a high capacity to absorb and retain water. Moreover, many of their chemical and physical characteristics are not favorable for agricultural production, including excessive amounts of soil organic material, strong acidity, and low-bearing capacity. In order to ameliorate these disadvantages, large-scale drainage, soil dressing, and lime application have been carried out in Hokkaido (Hokkaido Regional Development Bureau 1979; Kondo 1997).

In Hokkaido, active surveys of peatlands used for agricultural purposes have been undertaken since the early twentieth century. In the 1950s, the Ishikari River Basin Development Program was proposed by the government and funded by the World Bank. Drainage and irrigation channels were developed, soil dressing was carried out, and much of the land were reclaimed for rice paddies (Sakaguchi 1974; Kondo 1997). In the late 1970s, the adjustment of rice production forced some of the rice paddies to be converted into upland fields, and land subsidence became problematic due to drainage and tillage (Fig. 4.10). In contrast, the Sarobetsu and Kushiro peatlands, located in northern and eastern Hokkaido, respectively, have unfavorable climatic conditions for rice cultivation, and thus parts of these peatlands have been converted to grassland (Kondo 1997). In particular, the Sarobetsu peatland was surveyed and received improvements between the 1950s and the 1970s; the channel correction and widening of the Sarobetsu River prevented flooding and overflow, while improvements of drainage channels lowered the groundwater level. As a result, the groundwater level of adjacent undeveloped peatlands also lowered, and an invasion of *Sasa* bamboo from the surrounding hilly areas has significantly accelerated the drying and decomposition of peats (Fig. 4.11).

In the Tohoku region, Organic soils occupy approximately 830 km^2, roughly 650 km^2 (78%) of which is used as agricultural land. In the Kanto region, Organic soils occupy 540 km^2, much of which is used for residential and urban areas and roughly 20% of which is used for agricultural land.

Fig. 4.10 Agricultural use of Ishikari peatland in Hokkaido. **a** Paddy rice cultivation on dressed soil to the depth of 25–30 cm, tile drained (Low-moor Peat soils with phase "reclamation by soil dressing") (drainage pipes can be seen on the right-hand side of the soil profile; Iwamizawa City). **b** Asparagus production without soil dressing, directly cultivating on surface peats (Low-moor Peat soils) (Bibai City) (Figures supplied by Masayuki Tani)

Fig. 4.11 Invasion of Sasa bamboo in Sarobetsu peatland in Hokkaido after drainage improvements. Drainage channels are installed in the middle of the picture. The left-hand side of the channel is established grassland after drainage improvements, while the invasion of Sasa bamboo can be observed on the right-hand side. Continuous subsiding of the peatland is underway due to acceleration of drying and decomposition due to Sasa bamboo invasions. (Figures supplied by Masayuki Tani)

There was even a time when Organic soils were present in central Tokyo, in valleys where seepage peatlands formed; in metropolitan Tokyo, certain place names carry the word "*ya*", such as Shibuya and Setagaya, which shows that such places were originally "*yachi,*" meaning peatlands in Japanese (Sakaguchi 1974).

4.2.5 Future of Organic Soil in Japan

Most of Japan's peatlands are Low-moor Peat soils that formed in the Holocene under cool climates in the northern regions of Hokkaido or Tohoku in back swamps of alluvial plains or marshy grounds after marine regression in the Holocene, with common reed and Japanese alder as the main component plant species, whereas high-moor peatland and transitional-moor peatland are relatively scarce. Frequent inputs of inorganic materials such as sand and clays are common from the flooding of rivers and volcanic ash deposition; these accelerate decomposition and humification and contain large quantities of Si and Fe. By contrast, regions adjacent to young volcanoes have terrestrialization-type peatlands in shallow lakes and marshes that were formed in catchment basins and swales created by lava and mud flows in mountainous and highland areas, and which are scattered across Japan at small scales.

Paludization-type peatlands in alluvial plains and lowlands have been used as agricultural land and residential areas; this ceases the formation and accumulation of peats, and these peatlands are therefore referred to as "disappearing peatlands." "Living peatlands," which remain in parts of the lowlands and mountainous areas of Hokkaido and the mountainous areas of Honshu and Kyushu, play important roles in the storage of groundwater, the prevention of flooding, the sequestration of soil carbon, and the maintenance of biodiversity. Therefore, understanding the formation and distribution of Organic soils in Japan is important for environmental conservation.

4.3 Andosols

4.3.1 Genesis and Characteristics

(1) **Terminology of Andosols**

Soils that are mainly derived from volcanic ejecta or tephra and form during a long period of pedogenetic soil formation process have unique and distinctive properties among soils, such as their fluffy and light surface properties, high humus content, and very high phosphorus-fixing capacity. Such soils are called "Kurobokudo" (i.e., Kuroboku soil), which means black (kuro) and fluffy (boku) soil in Japanese. These soils are widely distributed and are representative soils in the arable land of Japan. According to the World Reference Base for Soil Resources (FAO 2015) (WRB) and the USDA Soil Taxonomy (Soil Survey Staff 1999), these soils are classified as Andosols and Andisols, respectively; these names derive from "ando soils" which denotes dark (black) soil in Japanese (Shoji et al. 1993a).

(2) **Characteristics of morphology**

Soil organic matter accumulation is a distinctive property of Andosols (Wada 1985). The morphological characteristics of soil profiles in Japanese Andosols are as follows: (1) the formation of a humic horizon rich in organic matter and dark in color; (2) mainly under forest vegetation, the formation of a humic horizon rich in organic matter and brown in color; (3) the formation of a cumulic humic horizon; and (4) the formation of a buried humic horizon.

The accumulation of large amounts of humus and the development of a cumulic profile morphology (cumulic humus horizon, formation of buried humus horizon) are a pedogenetic process that is specific to Andosols, whose parent material is mainly volcanic ash. Due to intermittent ashfall accompanying with volcanic activity and the supply of large amounts of organic matter from vegetation, humic horizons develop upward, resulting in the development of a cumulic profile morphology. When volcanic deposits of a single short eruptive cycle are thick, the outermost humic horizon is buried by a newly formed volcanic ash layer (formation of buried humic horizon). In contrast, when volcanic deposits of a single short eruptive cycle are thin, humus accumulates in a whole newly formed volcanic ash layer, which connects with the underlying old humus horizon and forms a thick humus horizon.

Volcanic ash is prone to weathering (Shoji et al. 1993b) because volcanic glass, which is an abundant component of volcanic ash, has a very high specific surface area. Silica and aluminum (Al) dissolved from volcanic glass react rapidly, and form short-range order minerals (allophane and imogolite) which become main clay minerals. On the one hand, Al bonds with organic matter supplied from plant debris, resulting in the formation of a stable Al–humus complex (Wada and Higashi 1978). As shown in Fig. 4.12, in soils whose parent material is volcanic ash, organic matter becomes resistant to microbial attack due to the formation of an Al–humus complex and accumulates in the soil (Takahashi and Dahlgren 2016).

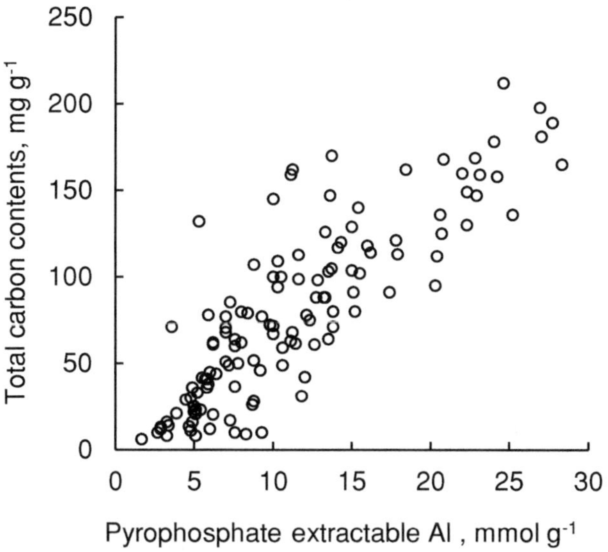

Fig. 4.12 Relationship between pyrophosphate-extractable Al (Al-humus) and total carbon contents in Japanese Andosols. *Data source* Shoji et al (1985), Ito et al. (1991a) and Wada (1986a)

(3) **Characteristics and genesis of humus**

The color of Andosol humic horizons is related to the content of humus and the degree of humification. The degree of humification of humic acid extractable with alkali can be presented as absorbance per carbon (Kumada 1987). This "A-type" humic acid, having a large value of absorbance per carbon, exhibits strong darkness per carbon, which is characteristic of Japanese Andosols (Kumada 1987). In contrast to foreign Andosols, Japanese Andosols have a dark humic horizon that contains mostly A-type humic acid (Shoji et al. 1987). Strongly dark-colored humic horizons are classified as melanic epipedon in the USDA Soil Taxonomy and the WRB based on the content of organic matter, soil color (moist soil color with value and chroma of 2 or less), and melanic index, by which the type of humic acid can be judged. A melanic index value (which shows the degree of darkness of humic acid extracted with alkali) of less than 1.70 indicates the dominance of A-type humic acid (Honna et al. 1988).

A biosequential study in Alaska (Shoji et al. 1988a), and the correlation between plant opals of Japanese pampas grass (*Miscanthus sinensis*) and organic matter content shown by Sase (1986), both suggest the formation of melanic epipedon is strongly related to herbaceous vegetation such as Japanese pampas grass. It has also been proved that not only C4 herbaceous plants but also C3 plants (Hiradate et al. 2004) and carbonized plant matter formed by forest fires (Shindo et al. 2005) contribute to the source of organic matter for the formation of melanic epipedon.

Moreover, studies of Andosols under forest vegetation in New Zealand (Shoji et al. 1987) and of Japanese volcanic ash soil under beech vegetation (Shoji et al. 1988b) revealed that the color of humic horizons is brown regardless of the content of organic matter. In these Andosols, the portion of humic acid in alkali-extracted humus is small, and humic acid is of B-type or P-type, both of which have a low degree of humification. Brown-colored Andosols, regardless of their organic matter content, are characterized as belonging to the Fulvic group, separate from the Melanic group, in the USDA Soil Taxonomy (Soil Survey Staff 1999).

(4) **Physical, chemical, and clay mineralogical characteristics**

Andosols are unique among soil types in terms of their physical and chemical properties (Shoji et al. 1993a), namely (1) their fluffy and light texture; (2) their high water-holding capacity; (3) their high reactivity with fluorine and high phosphate absorption; (4) their high cation exchange capacity (CEC) and predominated variable charge (the charged amount depends on pH); and (5) their low holding capacity of basic cation and acid under humid climate. These unique properties are closely related to the presence of short-range order minerals such as allophane, imogolite, and/or humus.

The five unique physical and chemical properties explained above are derived from the existence and nature of short-range order minerals and large amounts of humus (Wada 1985). Short-range order minerals (allophane, imogolite, and ferrihydrite) and humus largely contribute to the fluffiness and low bulk density, as well as the high water-holding capacity, of Andosols, through the formation of soil aggregate. As shown in Fig. 4.13, short-range order minerals and Al/Fe–humus complexes that can be selectively extracted with acid oxalate solution contain abundant Al/Fe (active Al/Fe) that can react with phosphate and fluoride ions (Wada and Gunjigake 1979). Such Al and Fe on colloidal surfaces have many hydroxyl groups, which are replaced by phosphate and fluoride ions through ligand exchange reactions, resulting in strong adsorption of those ions. A hydroxyl group on the surface of short-range order minerals and a carboxyl group at the edge of humus dissociate protons and express high negative charge when pH is high (Wada and Okamura 1980). However, absorbed cations are easy to desorb and are susceptible to leaching from soil horizons because pH-dependent negative charge becomes small when pH is low. Under acidic conditions, hydroxyl groups on the surface of short-range-order minerals express positive charge.

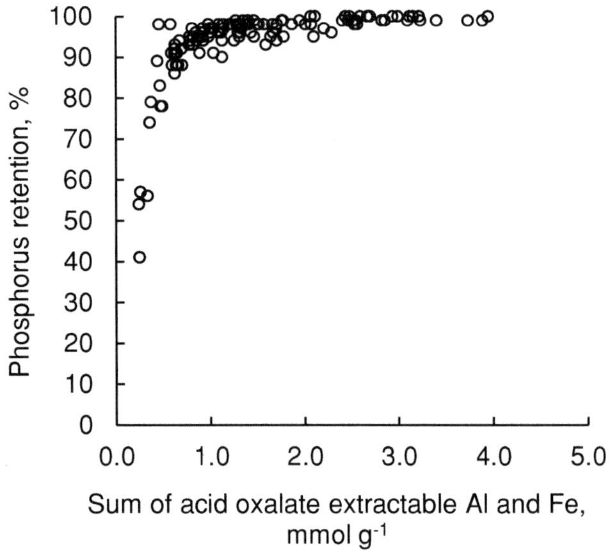

Fig. 4.13 Relationship between contents of acid oxalate extractable Al + Fe (active Al and Fe) and P retention in Japanese Andosols *Data source* Shoji et al (1985), Ito et al. (1991a) and Wada (1986a)

In this way, the distinctive physical and chemical properties of Andosols are strongly related to the nature of short-range order minerals and Al/Fe–humus complexes. Based on this, an up-to-date edition of the international soil classification system adopts the active Al and Fe present in large amounts as a central defining property of Andosols and Andisols (IUSS Working Group WRB 2015; Soil Survey Staff 2014). Active Al/Fe derived from short-range order minerals and Al/Fe–humus complexes can be quantitatively analyzed by the acid oxalate extraction method, in which active Al and Fe are selectively dissolved.

(5) **Formation of clay minerals in Andosols**

Volcanic glass, which is an abundant component of volcanic ejecta and has a very large specific surface area, enhances chemical weathering rates. Consequently, the concentration of released Al and Si in soil solution becomes high, so that Al and Si react rapidly, resulting in the formation of allophane and imogolite(Ugolini and Dahlgren 2002). Thus, in zones where volcanic ejecta are thickly deposited under a humid climate, soils rich in allophane and imogolite develop.

However, when large amounts of humus exist, Al forms complexes with humus. As a result, the hydrolysis and polymerization of Al and the bonding of Al and Si are prohibited, which in turn inhibits the formation of allophane and imogolite (Inoue and Huang 1984). When the soils contain large amounts of Al/Fe–humus complexes, they show specific properties commonly observed in Andosols rich in allophane, although allophanic clay is not present (Shoji et al. 1985). Japanese researchers have shown that non-allophanic Andosols tend to be distributed in areas which are less affected by Quaternary volcanic ash (such as the Sea of Japan side) (Saigusa and Matsuyama 1998), and only a small amount of allophane is produced if the Thornthwaite's index of potential evapotranspiration is large (intense leaching condition) in the area where identical volcanic ash is distributed (Takahashi and Shoji 1996). Because of these reasons, the formation of Al–humus complexes proceeds under the condition where the deposition of volcanic ash—the source of the Al—is relatively small and/or leaching is so intense that soil pH decreases. Non-allophanic Andosols have some similarities with Allophanic Andosols (Shoji et al. 1985), whereas they exhibit strong acidity (Saigusa et al. 1980) because

2:1 to 2:1:1 clay minerals are the main components of clay minerals, leading to the presence of exchangeable Al^{3+} under base-unsaturated conditions. Soil showing the unique properties of Andosols and containing almost no allophane was formerly reported by Uchiyama et al. (1960) and Kato (1962). The report by Shoji and Ono (1978) of non-allophanic Andosols in the Tohoku region in Japan became the driving force for modifying the central property of Andosols in the domestic and worldwide soil classification systems. The origin of the 2:1 to 2:1:1 clay minerals which are the main clay mineral components of non-allophanic Andosols is considered to be aeolian dust from inland China (Inoue 1981), the alteration of volcanic glass by weathering, and secondary deposition after hydrothermal alteration in a closed environment. Aeolian dust contains fine-grained quartz as well as 2:1 clay minerals. Based on the oxygen isotope ratio of fine quartz in Japanese Andosols, the 2:1 clay minerals in Andosols are assumed to be derived from aeolian dust from the drylands of inland China (Mizota et al. 1991).

4.3.2 Classification of Andosols and Related Soils

(1) Outline of Andosols in the Soil Classification System of Japan

In Japan, several soil classification systems are used, such as the Classification of Cultivated Soils in Japan (1995) for arable land soil and the Classification of Forest Soils in Japan (1976) for forest soil. This section will explain the position of Andosols in the up-to-date soil classification system (The Fifth Committee for Soil Classification and Nomenclature 2017), which does not depend on land use type. This classification system includes three categories—10 great groups, 26 groups, and 120 subgroups—and adopts a "key-out" system by using diagnostic horizons and characteristics, and places emphasis on compatibility with the WRB soil classification system.

For the great groups and groups, soils having distinctive characteristics and soils whose locations are easily specified due to the significant effect of specific soil-forming factor are keyed out earlier. Diagnostic horizons and characteristics are defined using distinctive characteristics and properties that can greatly affect crop productivity, and based on this, the unique soils are keyed out. Andosols are keyed out as the third great group, after Human-made soils, which are strongly influenced by humans and do not occur naturally, and Organic soils affected by water.

The central concept of an Andosol is a soil whose main parent material is volcanic ejecta and is weathered to some degree or a soil where large amounts of active Al and/or Fe, which are highly reactive with phosphate, are produced by in situ weathering. According to the Soil Classification System of Japan (The Fifth Committee for Soil Classification and Nomenclature 2017) or Comprehensive Soil Classification System (Obara et al. 2011), Andosols are soils that have a layer of 25 cm or more in thickness within 50 cm of the soil surface that has regosolic andic soil properties, representing weak pedogenetic horizontation mainly derived from volcanic ejecta, or that has andic soil properties, representing the presence of large amounts of active Al and/or Fe.

Regosolic andic soil properties meet all of the following requirements:

(1) consists of non-weathered volcanic gravel comprising less than 50% (by area); and
(2) more than 60% (by weight) of the soil consists of volcanic ejecta (volcanic ash, pumice, scoria, etc.); and
(3) has a phosphate absorption coefficient of between 3.00 and 15.0 mg P_2O_5/g (phosphate retention of at least 25% but less than 85%, or Al plus 1/2 Fe content (by acid ammonium oxalate) of at least 0.4% but less than 2.0%) or an organic carbon content of 3% or more if the phosphate absorption coefficient is less than 3.00 mg P_2O_5/g (phosphate retention of less than 25%, or Al plus 1/2 Fe content (by acid ammonium oxalate) of less than 0.4%).

Andic soil properties meet all of the following requirements:

(1) a phosphate absorption coefficient of 15.0 mg P_2O_5/g or more (phosphate retention of 85% or more, or Al plus 1/2 Fe content (by acid ammonium oxalate) of 2.0% or more); and
(2) a fiber content of less than one sixth of the original volume after rubbing.

In this classification system, the following soil groups are defined: "Podzolic Andosols," representing Podzols derived from volcanic ash (Fig. 4.14); "Regosolic Andosols," representing poorly weathered soils derived from volcanic ash (Fig. 4.15); "Gleyed Andosols," having gleyic properties; "Wet Andosols," having a groundwater–gley property (Fig. 4.16); "non-allophanic Andosols," representing acidic soils without allophanic minerals (Fig. 4.17); and "Allophanic Andosols," representing typical Andosols (Fig. 4.18).

(2) Classification of Podzols derived from volcanic ash

The pedogenetic soil formation process of Andosols is essentially different from that of Podzols. Andosols are formed by andosolization—the in situ formation of short-range-order minerals and/or Al(Fe)–humus complexes; Podzols are formed by podzolization—the formation of

Fig. 4.14 Shimokita soil in Mutsu City, Aomori Prefecture: Podzolized Andosol. *Photograph* H. Kanno (Tohoku University)

short-range order minerals and/or Al(Fe)–humus complexes derived from humus, with Al and Fe leached from the surface horizon. The WRB classification system and the USDA Soil Taxonomy give priority to podzolization, so that Podzols derived from volcanic ash are not included in Andosols or Andisols. On the other hand, the Soil Classification System of Japan gives priority to specific properties—a large amount of active Al and/or Fe and humus—different from other soil groups, so the Andosol great group is keyed out before the Podzolic great group and Podzols derived from volcanic ash are included as Andosols. The Podzolic group was defined as soil having "an eluvial horizon," "spodic horizon (morphological features)," or "spodic horizon (analytical results)."

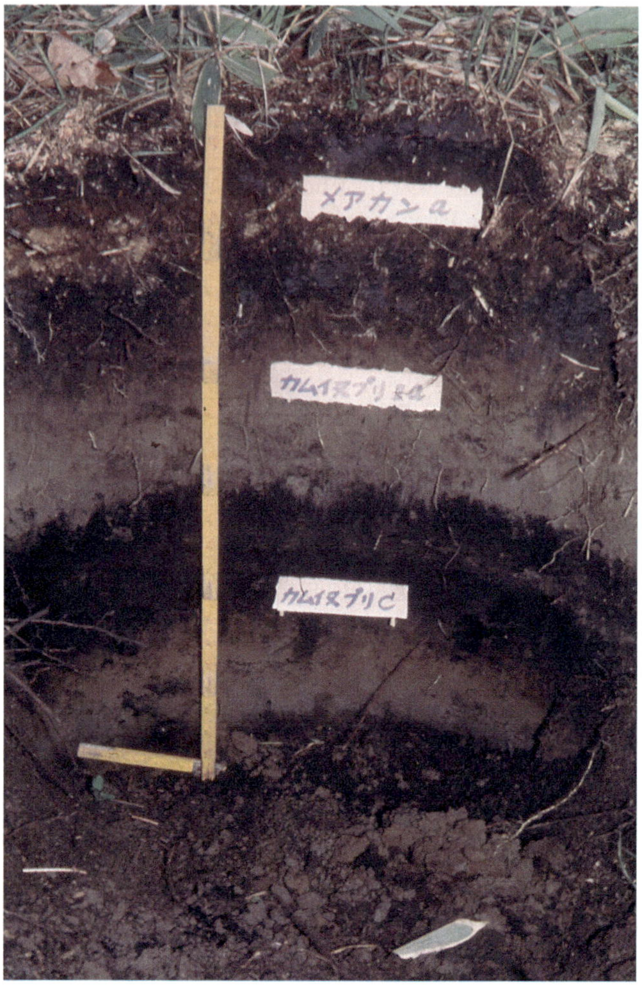

Fig. 4.15 Nijibetsu soil in Bekkai Town, Hokkaido: Regosolic Andosol. *Photograph* M. Saigusa (Formerly Tohoku University)

(3) **Classification of Brown soils derived from volcanic ash under forest vegetation**

For the classification of Japanese soils, the distinction between the Andosol great group and Brown Forest soils derived from volcanic ash was a critical issue. According to the Forest Soil Classification System of Japan (1976), Brown Forest soils—brown soils developed under forest vegetation—are the most abundant soils among forest soils, which make this soil type one of the most important. Because existing forest soil maps cannot separate Brown Forest soils derived from volcanic ash and those derived from non-volcanic ash, some researchers have considered it impractical to include Brown Forest soils that have been strongly affected by volcanic ash as Andosols. On the other hand, brown-colored highly humified soils are classified as Fulvic Andosols in the WRB classification and as Fulvudands in the USDA Soil Taxonomy, both of which

Fig. 4.16 Mihara soils in Bekkai Town, Hokkaido: Wet Andosol *Photograph* T. Ito

correspond to the Andosol group in the Soil Classification System of Japan.

In the Soil Classification System of Japan, brown-colored soils developed under forest vegetation, which contains large accumulations of humus, are classified as Andosols and included in the Brown-humic subgroup. This classification system enables the distinction between Andosols in Forest and Brown Forest soils derived from non-volcanic ash. This contributes to a more accurate evaluation of the carbon sequestration potential of soil, which has received considerable attention as a countermeasure to global warming. Imaya et al. (2010a) reported that the carbon sequestration potential at a depth of 1 m is 6.6 kg m^{-2} for Brown Forest soils with a small amount of active Al (acid oxalate extractable Al [Al_o]; non-volcanic ash soil), which contrasts with a value of 24 kg m^{-2} for Brown Forest soils with large amounts of active Al (Al_o of 2% or more; volcanic ash soil). It was also important to include the Brown-humic subgroups into the Andosol great group to facilitate comparison with international soil classification systems.

Fig. 4.17 Mukaiyama soil in Oosaki City, Miyagi prefecture: Non-allophanic Andosol. *Photograph* T. Ito

(4) **Classification of Non-allophanic Andosols**

Non-allophanic Andosols had become a driving force for a paradigm shift in the central concepts for volcanic ash soils, from "the existence of a lot of allophane" to the "existence of a lot of active Al and/or Fe" in the soil classification systems inside and outside of Japan. Non-allophanic Andosols are strongly acidic soils mainly consisting of 2:1 to 2:1:1 intermediate clay minerals and are defined as having non-allophanic andic soil properties.

Non-allophanic andic soil properties meet all of the following requirements:

(1) have andic soil properties; and
(2) have either:

Fig. 4.18 Oosawa soil in Imaichi City, Tochigi prefecture: Allophanic Andosol. *Photograph* T. Ito

(a) a y1 (exchangeable acidity) value of 5 or more; or
(b) a Si_o (acid oxalate extractable silicon) value of less than 0.6% or an Al_p/Al_o (ratio of pyrophosphate-extractable aluminum to acid oxalate extractable aluminum) value of 0.5 or more.

In Japan, about 30% of Andosols are non-allophanic Andosols, whose crop productivity is markedly different from that of Allophanic Andosols (Saigusa and Matsuyama 1998). In non-allophanic Andosols, it is difficult for crop roots to reach the subsoil; accordingly, crops cannot take up water and nutrients (such as inorganic nitrogen) from the subsoil, which results in low crop growth and yield. It is essential to correct not only surface soil acidity but also subsoil acidity to improve crop productivity in non-allophanic Andosols.

4.3.3 Correlation with International Classification Systems

As already noted, the Andosol great group in the Soil Classification System of Japan consists of six groups (Table 4.2). In this system, soils with both podzolic and andic properties are classified as Andosols (Fig. 4.14). These soils are classified as Podzols in the World Reference Base for Soil Resources (WRB) (IUSS Working Group WRB 2015). In the WRB classification system, Andosols are keyed out before Podzols, but soils with spodic horizons are excluded from Andosols. Similarly, these podzolized soils are classified as Spodosols in the USDA Soil Taxonomy (Soil Survey Staff 2014) when they have distinct bleached horizons (albic horizons). Without albic horizons, they are classified as Andisols.

Regosolic Andosols are soils that have undergone pedogenesis to some extent from volcanic ejecta and contain limited amounts of short-range-order minerals and organo-metallic complexes (Fig. 4.15). This group of Andosols corresponds to Vitric Andosols in the WRB and Vitrands, Vitraquands, or Udivitrands in Soil Taxonomy.

Gleyed Andosols are defined as Andosols in which a gley horizon affected by groundwater appears within a depth of

Table 4.2 Subgroups in each group of Andosols

Group	Podzolic Andosols	Regosolic Andosols	Gleyed Andosols	Wet Andosols	Non-allophanic Andosols	Allophanic Andosols
Sub group	Epi-peaty	Aquic	Peaty	Peaty		
	Epi-pseudogleyic					
					Anthraquic	Anthraquic
				Thapto-upland		Thapto-upland
				Endofluvic		Endofluvic
			Cumulic	Cumulic	Cumulic	Cumulic
		Thapto-humic			Thapto-humic	Thapto-humic
		Humic			Humic	Humic
					Brown-humic	Brown -humic
	Aquic				Aquic	Aquic
	Haplic	Haplic	Haplic	Haplic	Haplic	Haplic

50 cm from the soil surface. Wet Andosols are defined as Andosols in which a soil horizon showing endoaquic properties appears within a depth of 50 cm from the soil surface. These groups of Andosols are referred to as Aquands in Soil Taxonomy. However, the Gleyed Andosols in the Japanese classification system are classified as Gleysols in the WRB system, because Gleysols are keyed out before Andosols (IUSS Working Group WRB 2015).

In the Japanese classification system, the other Andosols are keyed out as non-allophanic Andosols (Fig. 4.16) and thereafter as Allophanic Andosols (Fig. 4.17). These correspond to Aluandic Andosols and Silandic Andosols, respectively, in the WRB. Both groups of Andosols are referred to as Udands in the USDA Soil Taxonomy.

The soil names "Andosols" and "Andisols" are derived from the Japanese characters *an* meaning dark and *do* meaning soil. Therefore, these soil names originally connote the typically dark soils derived from volcanic ejecta, which correspond to Humic Andosols in the Japanese classification system. Subdivision based on soil color is defined at the subgroup level because color is less important for andic soil properties. Among Humic Andosols in the Japanese classification system, surface horizons having a high organic carbon content (6% or more) are equivalent to "melanic" horizons in the WRB classification and USDA Soil Taxonomy.

4.3.4 Distribution of Andosols in Japan

Of the approximately 1500 active volcanoes in the world, 111 are located in Japan (Japan Meteorological Agency 2017). In the vicinity of the Japanese Archipelago, two plates, the Pacific Plate and the Philippine Sea Plate, are subducting. In parallel the subduction zone, the East Japan Volcanic Belt and the West Japan Volcanic Belt are distributed. The former includes volcanoes in the Kuril Islands, Hokkaido, Tohoku, the Chubu and Kanto regions, and the Izu-Ogasawara Islands, while the latter is composed of volcanoes in the Chugoku and Kyushu regions and the Nansei Islands (refer to Sect. 2.2).

Reflecting this fact, the spatial distribution of soils belonging to the Andosol great group is the second largest among all soil great groups in Japan, behind that of the Brown Forest soil great group (Table 2.4 in Sect. 2.5); Andosols are estimated to occupy about 30% of Japan's territory. Previously, the presumed distribution of Andosols was 17% of Japan's territory, whereas that of Brown Forest soils was 53% (Kanno et al. 2008). The change in the estimated distribution is due to a change in the Japanese soil classification system; now, Brown Forest soils that possess andic properties are classified as Andosols.

By region, the distribution of Andosols is particularly high in Hokkaido (42% by area), Kanto (40%), Tohoku (36%), and Kyushu (34%), while the distribution is also considerable in Tokai–Hokuriku (22%) and Kinki–Chugoku–Shikoku (15%). The high percentage of Andosol present in the Hokkaido, Tohoku, Kanto, and Kyushu regions is due to the high concentration of volcanoes in these areas, which provide the parent materials of Andosols.

Andosols are distributed nationwide in Japan, including areas with a near absence of volcanoes. This is partially due to the influence of widespread tephra. Machida (2002) identifies 31 types of widespread Quaternary tephra. Among them, the representative ones are the Kikai-Akahoya (K-Ah) tephra (Kikai volcano, southern Kyushu), the Aira-Tanzawa (AT) tephra (Aira volcano, southern Kyushu), the Shikotsu-1 tephra (Shikotsu caldera, central Hokkaido), the Aso-4 tephra (Aso caldera, central Kyushu), and the Toya tephra (Toya caldera, central Hokkaido) (Machida 2002). These tephras are distributed over areas covering from between one quarter and three quarters of the Japanese territory (Machida 2002).

As noted already, Andosols are often divided into two groups based on the major colloidal composition of surface horizons: Allophanic Andosols, dominated by allophanic clays (allophane and imogolite), and non-allophanic Andosols, dominated by Al–humus complexes and often containing 2:1 layer silicates. Allophanic Andosols preferentially form in thick Holocene tephra deposits in Hokkaido, northern Tohoku, Kanto, and Kyushu regions. On the other hand, non-allophanic Andosols form in areas with minimal Holocene tephra deposition (Saigusa and Matsuyama 1998) (Fig. 2.35 in Sect. 2.5).

4.3.5 Environmental Significance and Utility of Andosols

(1) **Organic carbon (OC) accumulation**

The accumulation of soil organic matter (SOM) is a characteristic property of Andosols. Andosols cover 0.84% of the Earth's land surface, but they contain approximately 1.8% of the global soil organic carbon (SOC). Even though the contribution of Andosols to the SOC stock is not so large at a global level, it is relatively important in Japan because Andosols cover 30% of the country's territory.

The large accumulation of SOC in Andosols results from a combination of high detritus input resulting from the generally high productivity of Andosols and from the effective stabilization of SOM against decomposition. Stabilization mechanisms of SOM were summarized to be attributed to: (1) the formation of SOM in organo-mineral or organo-metallic complexes; (2) the low activity of soil microorganisms due to low soil pH, aluminum toxicity, low base cation content, and/or phosphorus deficiency; (3) the physical protection of the SOM in stable microaggregates; (4) the sorption and deactivation of exoenzymes involved in the extracellular depolymerization process of SOM decomposition; (5) the burial of surface horizons with high SOM contents by additional tephra deposition; and (6) the presence of microbial-resistant charcoal (Takahashi and Dahlgren 2016).

According to a statistical analysis using data from 293 "A horizons," it is considered that, in the humic horizons of many Andosols, Al–humus complexation strongly contributes to SOM accumulation, and low soil pH and Al toxicity may be partially responsible for humus accumulation (Takahashi and Dahlgren 2016).

(2) **Water storage and water quality improvement**

Andosols have well-developed soil structures with various pore diameters and therefore can hold a large amount of plant-available water. These soils can supply enough water to grow most agricultural plants and natural vegetation. Andosols under forests can decrease flooding during the rainy season and can relieve water shortage during the dry season. Shoji and Takahashi (2002) indicated that water storage is remarkably enhanced when both Andosols and their substratum are very thick, by using the analytical data of several factors contributing to the water storage capacity of forest soils, such as type of soil, depositional mode, type of substrata, altitude, topography, and so on.

Andosols commonly contain a large amount of short-range-order minerals (allophane, imogolite, ferrihydrite) which have a large anion exchange capacity and contain active Al and Fe. These minerals can contribute to the maintenance and improvement of the quality of groundwater, river water, and other water bodies by removing and retaining nitrate and phosphate originating from fertilizers and other sources.

(3) **Amenity in Andosol areas**

Among Japan's 111 active volcanoes, 62 volcanoes are located in national parks (21 parks), while another 10 active volcanoes are located in quasi-national parks (9 parks). One of the purposes of national parks and quasi-national parks is to restrict development projects and other human activities to exceptional national landscapes, and another is to foster a joyful experience of nature, including an appreciation of landscape (Ministry of the Environment 2017). Thus, many volcanoes and their surrounding areas including Andosols are valuable resources for sightseeing and outdoor activities because of their unique topography and vegetation.

(4) **Ancient human activity and Andosols**

Intermittent volcanic deposition and repeated pedogenesis commonly create multi-storied soils in Japan. Multi-storied Andosols notably resist soil disturbance, truncation, and displacement by natural and human activities. These facts contribute to the preservation of heritage.

Many archaeological remains, and artifacts have been found in Holocene Andosols in Japan. There has been found to be a close relationship between the distribution of Jomon-age ruins (ca. 15,000–2300 YBP) and the distribution of dark-colored Andosols in Japan (Edamura and Kumagai 2009). Therefore, Melanic Andosols are thought to be created under ecosystems largely influenced by human beings (Hosono and Sase 2015). Thus, dark-colored Andosols can be regarded as a heritage.

(5) **Agricultural use of Andosols**

Andosols are among the most productive soils in the world (Shoji and Takahashi 2002). Andosols in Japan that formed under a humid-temperate climate were originally regarded as poorly productive soils because of their very low content of plant-available phosphorus, low concentrations of exchangeable bases, and strong acidity. However, these chemical constraints are easily overcome by the application of phosphorus fertilizer and liming. Some Andosols are especially productive for upland crops, and their notable properties are as follows: soil morphology is pachic (thick humus-rich horizon); physical properties show a medium-grained texture, friable consistency, free drainage, and high plant-available water content; and chemical properties show a high base saturation, a pH(H_2O) more than 6, and an absence of toxic Al (Shoji and Takahashi 2002).

Nowadays, Andosols are most extensively used for high-value horticultural crops, such as cabbage, Chinese cabbage, Chinese yam, edible burdock, Japanese radish, and ginger. These are described in detail in the "Andosols in Tohoku Region" chapter of this book.

4.4 Podzols

4.4.1 Introduction

Podzols, as defined in the Soil Classification System of Japan and the WRB (IUSS Working Group WRB 2015), or Spodosols, as defined in the USDA Soil Taxonomy (Soil Survey Staff 2014), are characterized by the eluviation of aluminum (Al) and iron (Fe) from the albic E horizon and

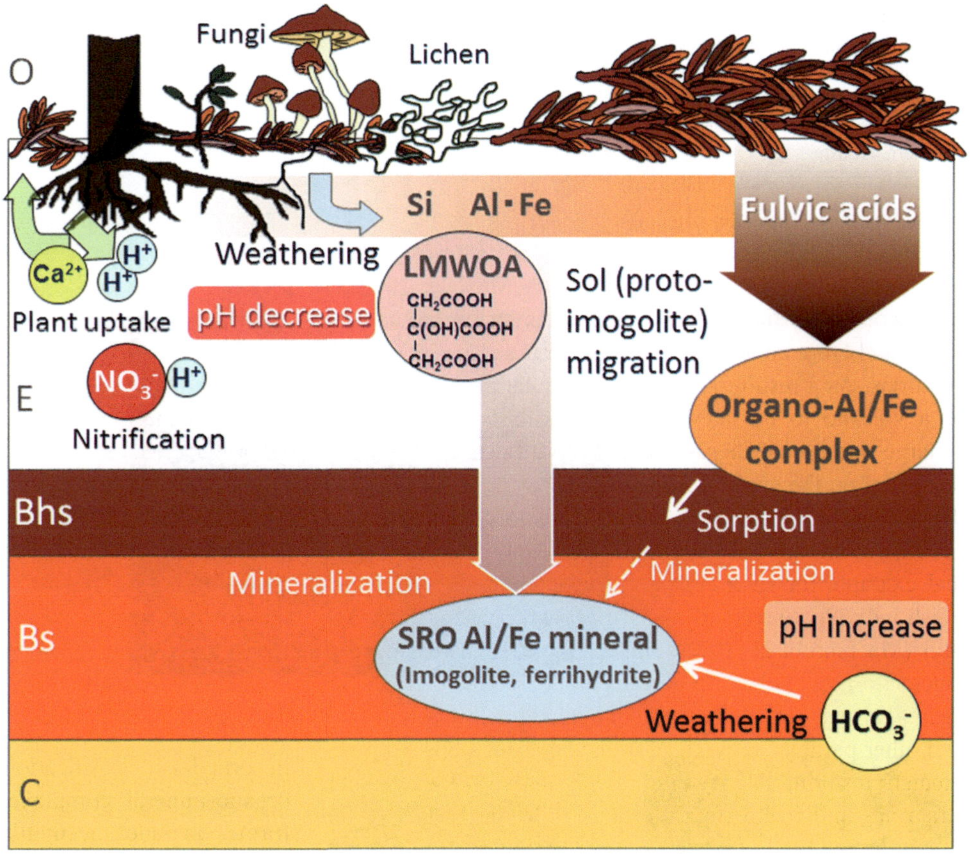

Fig. 4.19 Soil-forming processes potentially involved in podzolization. (Figure supplied by Kazumichi Fujii)

their illuviation in the spodic B (Bs) horizon (Figs. 4.19 and 4.20). Podzols are typically seen in boreal coniferous forest, heathland, and some peatlands (Lundström et al. 2000). Acidity and the complexation of Al and Fe by organic acids can contribute to clay destruction and the illuviation of short-range-order (SRO) Al and Fe substances (e.g., imogolite and ferrihydrite) in the process termed podzolization (Lundström 1993). Despite the simple concept of podzolization, there is wide variation in the soil chemical properties and interpretation of the soil-forming processes of Podzols.

In Japan, Podzols constitute a minor fraction (ca. 2.1%) of the land area. Podzols are typically observed under subalpine coniferous forests in mountainous regions of Northeast and Central Japan (e.g., Hakkoda, Tateyama, Kiso, and Okuchichibu areas; Fig. 4.21). Some Podzols are located in coastal sand dune areas (e.g., Hamatonbetsu in the Hokkaido lowlands). Podzolization is ubiquitous in humid forest soils in Japan, with most of the forest soils affected by podzolization tending to fall into the categories of Andosols (Andisols), Regosols (Entisols), and Brown Forest soils (Inceptisols) due to the overwhelming effects of parent materials (esp., volcanic ash) and other soil-forming processes (e.g., erosion and andosolization).

Most Podzols occur on boundaries shared with Andosols, Regosols, and Brown Forest soils in Japan. Podzols share diagnostic soil properties with Andosols in terms of the accumulation of SRO Al- and Fe-containing substances. Since most Podzols in Japan are affected by the addition of volcanic ash as well as aeolian dust (silt to fine sand in size), the consistency of soil-forming process and the continuity of parent materials in the profile are always questioned. Additionally, some Podzols share similarities with Brown Forest soils, Ultisols, and Alfisols (Luvisols and Albeluvisols in the WRB classification) in terms of the acidified and clay-impoverished E horizon. Considering these complicated aspects, this section characterizes Podzols and podzolization in Japan in relation to pedogenetic processes and factors of the associated soil types.

4.4.2 Diagnosis of Podzols

Regarding soil classification worldwide, Podzols in the WRB classification and Spodosols in the USDA Soil Taxonomy are defined by a spodic B horizon with SRO substance accumulation ($\geq$0.5% acid oxalate extractable Al and Fe or $Al_o + 1/2Fe_o$, overlaid by a bleached and acidic E

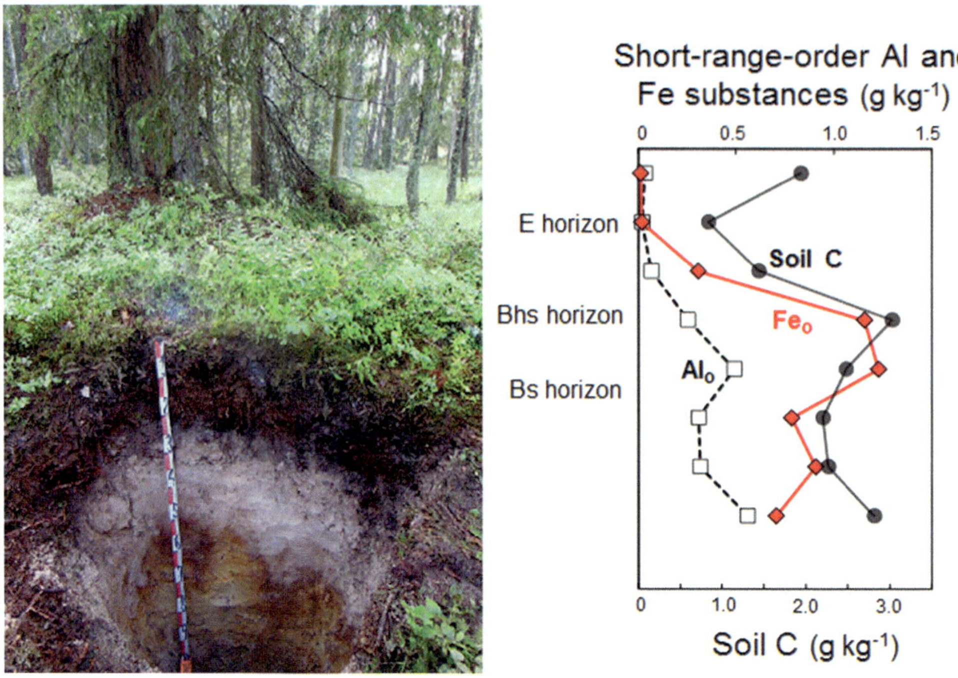

Fig. 4.20 Albic Podzol (Arenic) (Typic Haplorthods) on coastal sand dune under European red pine forest in Estonia. (Figure supplied by Kazumichi Fujii)

Fig. 4.21 Aquic Podzol (Andic Endoaquods) under subalpine cypress forest in Kiso Mountains (left) and Aquic Podzol under subslpine hemlock forest in Okuchichibu Mountains (right), Japan. (Figures supplied by Kazumichi Fujii)

horizon (pH < 5.9). Global classification systems adopt $Al_o + 1/2Fe_o$ as the diagnostic property, while the amount of pyrophosphate-extractable Al and Fe ($Al_p + 1/2Fe_p$) is used in some local classifications. The latter assumes migration of organo-mineral complexes (fulvic acid theory) while the former includes accumulation of inorganic SRO Al- and Fe-containing substances liberated from chelating organic acids (low-molecular weight organic acid theory) and the migration of sols in acidic condition (proto-imogolite hypothesis). The Bs horizon potentially consists of several B horizons: an upper Bh horizon rich in humus, followed by Bhir horizons rich in humus–Fe complexes, and deeper Bir horizons rich in inorganic Al and Fe oxides (e.g., imogolite and ferrihydrite) (Fig. 4.22). The heterogeneity of the spodic B horizon raises questions regarding podzolization and classification issues.

In the Soil Classification System of Japan, in a similar manner to global soil classification, Podzols can be classified by the morphological or chemical identification of the Bs horizon containing more than double the $Al_o + 1/2Fe_o$ compared to the acidic E horizon (Fig. 4.21). It is difficult to identify whether SRO substances which have accumulated in the B horizon are remaining substances that originated from the in situ weathering of volcanic ash or rather are illuviated SRO substances caused by podzolization. The former process, and a mixture of the two processes, can be mistaken for a well-developed Bs horizon, especially if the profile receives newly deposited materials (volcanic ash and aeolian dusts) (Ito et al. 1991). Podzolic Andosols (Fig. 4.22) are keyed out prior to Podzols in the Soil Classification System of Japan (Shoji and Ito 1990).

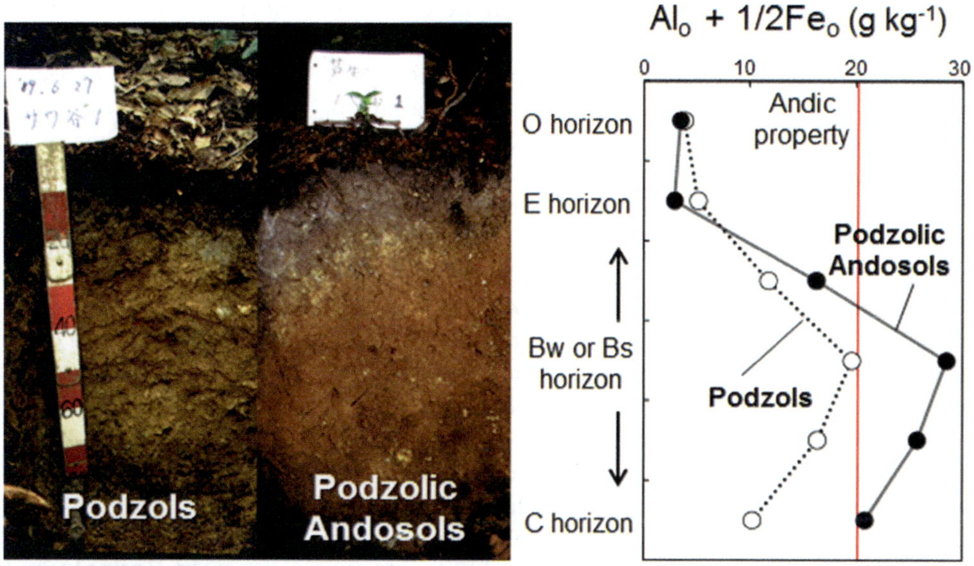

Fig. 4.22 Vertical distribution of SRO Al and Fe compounds in Podzol and Podzolic Andosol in northern Kyoto. *Data source* Funakawa (1993)

4.4.3 Podzolization: Eluviation and Illuviation of Short-Range-Order Al and Fe

To describe the uniqueness of Podzols in Japan, podzolization needs to be discussed in relation to existing theories and hypotheses.

Fulvic acid theory: Organo-mineral complexes coating the gains in the Bs horizons suggest that fulvic acids contribute to Al and Fe eluviation/illuviation (De Coninck 1980). Fulvic acids are typically a mixture of high molecular weight (>1000 D) dissolved organic matter (DOM) with carboxyl and phenolic functional groups with low charge density (e.g., 1 mol_c per 7 mol C from Ugolini and Sletten 1991). Organic acids do not simply work in mineral dissolution as proton sources (pH effect), but they also form complexes with Al and Fe through ligand reaction (Raulund-Rasmussen et al. 1998). The capture of Al and Fe by organic acids reduces inorganic Al concentrations in soil solution and leads to undersaturation in relation to aluminosilicate clays (e.g., vermiculite) and further dissolution of clays (e.g., vermiculite to smectite). Soil solution DOC concentration could decrease with depth due to sorption and mineralization. C/Al or Fe molar ratios of 30 are required for mobilization, but a C/(Al + Fe) ratio of 3 results in precipitation (Mokma and Buurman 1982; Jansen et al. 2005). The gradual increase of pH with depth can result in Al polymerization and the precipitation of SRO minerals. This can explain the accumulation of organo-mineral complexes in the upper part of the Bs horizon; however, fulvic acid theory fails to explain the accumulation of inorganic SRO Al and Fe oxides in the deeper part of the Bs horizon due to the recalcitrance of fulvic acids. An example of the measured mean residence time of fulvic acids (high-molecular-weight DOM) is 1.7 years (Qualls and Bridgham 2005), but complexation and precipitation can increase mean residence times to decades to hundreds of years (Harrison et al. 2000). The low capacity for chelation and low decomposability of fulvic acids cannot fully explain the illuviation of SRO minerals.

Fulvic acid-bicarbonate theory: To explain the accumulation of inorganic SRO minerals in the Bs horizon, fulvic acid-bicarbonate theory has been postulated based on soil solution studies collected by lysimeters (Ugolini and Dahlgren 1987). Fulvic acids contribute to Al- and Fe-leaching through congruent dissolution, while bicarbonate contributes to in situ weathering (incongruent dissolution) and the accumulation of inorganic SRO minerals in the higher-pH range of the Bs horizon. The pH rise in the B horizons is caused by proton consumption due to the mineralization and sorption of organic acids and nitrate uptake by plants. This theory has been developed by the study of Podzolized Andisols in Hakkoda (Japan) as well as the Washington Cascades (USA) (Ugolini et al. 1988; Ugolini and Sletten 1991).

Low-molecular-weight organic acid theory: Soil solutions extracted by laboratory centrifugation contain some low-molecular-weight organic acids (LMWOAs; <1000 D), such as citric, malic, and oxalic acids, with high charge density (1 mol_c per 1 to 2 mol C). Although the soil solution concentrations and percentages of LMWOAs in DOC are small (<50 μM and <10%, respectively), LMWOAs can nevertheless account for up to 50% of organically bonded Al and Fe due to the high charge density and chelating abilities of LMWOAs (Lundström et al. 2000). The LMWOAs can be exuded from roots, microorganisms, decomposed litters and humus, canopy leaching, lichen, and moss (Van Hees et al. 2005). In particular, some ectomycorrhizal fungi called "rock-eating fungi" can produce tunnels in the Podzol E horizon by releasing LMWOAs for the purpose of mining nutrients (esp., phosphorus) (Fig. 4.23). The Al and Fe which form complexes with LMWOAs are precipitated by microbial mineralization. The short mean residence time of LMWOAs (typically 0.1–24 h; Fujii et al. 2010) can account for the accumulation of inorganic Al and Fe oxides in the Bs horizons. The remaining issue of LMWOA theory is the short migration distances of a single LMWOA molecule, which can transport Al or Fe for only 10 nm downward within their lifetime (Van Hees et al. 2005). Despite this, cumulative effects of LMWOAs can contribute to the development of E horizons and the accumulation of SRO minerals, at least in the initial stage of podzolization.

Proto-imogolite theory: Imogolite and the other SRO aluminosilicates are assumed to be formed through neoformation from Si (H_4SiO_4) and Al in percolating soil solution of the Bs horizon, while migration can also occur in colloidal or hydroxyl sol forms. Positively charged SiO_2–Al_2O_3–Fe_2O_3 hydroxyl sols can be formed in the E horizons and precipitated in the B horizons (Farmer 1982). This raises controversy regarding whether imogolite formation is attributed to in situ weathering or the formation of proto-imogolite in the E horizon and its translocation downwards. This process can explain the formation of Bs horizons lacking organo-mineral complexes in Iron Podzols. LMWOAs can contribute to the initial stage of Al and Fe liberation and sol formation. However, no sol migration was observed in some Podzols derived from volcanic ash (Ugolini and Dahlgren 1991; Jansen et al. 2005). Sol mobility appears to be high in sandy soils or at the initial stage of podzolization.

Issues remaining in the classification of Podzols: The multiple hypotheses/theories are not exclusive, but there are some competing aspects which affect the decision of diagnostic criteria. When a central process in a Podzol is the

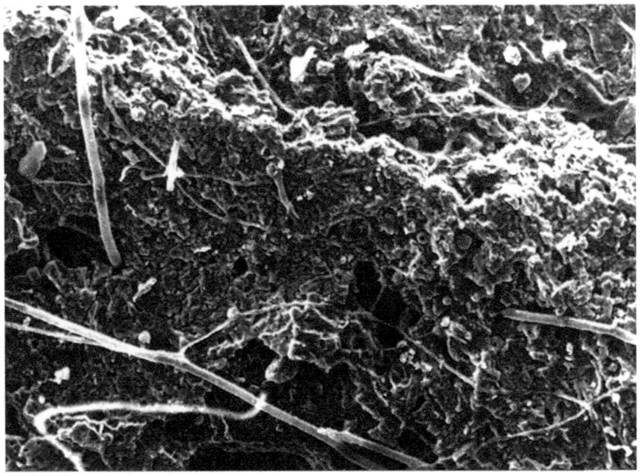

Fig. 4.23 Tunneling by "rock-eating" ectomycorrhizal fungi in the E horizon of a Podzol in Finland (Courtesy of Dr. Roger Finlay)

eluviation/illuviation of SRO minerals, and the accumulation of organo-mineral complexes in the Bs horizon is a secondary process, Podzols can be identified based on the amounts of acid oxalate extractable Al and Fe substances in soil, rather than the amounts of pyrophosphate-extractable Al and Fe. When the migration of organo-mineral complexes is a primary process, both indicators, $Al_o + 1/2Fe_o$ and $Al_p + 1/2Fe_p$, are informative and useful for Podzol identification. The vertical distribution of $Al_p + 1/2Fe_p$ can be useful in differentiating between Andosols and Podzols (Ito et al. 1990); however, the use of $Al_o + 1/2Fe_o$ that are shared between Andosols and Podzols risks the inclusion of Andosols into Podzols or vice versa in Andosol-dominated areas of Japan.

4.4.4 Dominant Process of Podzolization in Podzols in Japan

The presence of multiple podzolization hypotheses/theories is due to the wide variation in Podzols. The relative importance of the abovementioned processes to podzolization can vary depending on climate, vegetation, and parent materials.

The importance of fulvic acids or DOM in podzolization is supported by correlations between the DOC and Al concentrations found in soil solutions (Funakawa et al. 1992). The presence of bicarbonate in the higher-pH range of Bs horizons supports the fulvic acid-bicarbonate theory in Japan (Ugolini et al. 1988). On the other hand, the Al in soil solutions from Japanese cedar forest, northern Kyoto, is dominated by inorganic Al, not the organic form (Funakawa et al. 1992). This suggests that fulvic acids play roles in podzolization as sources of protons and counterions of Al in soil solution, rather than chelating agents, in some Podzols in Japan. Based on the fact that LMWOAs, potent weathering agents, are commonly present at high levels in both Podzols and Brown Forest soils (Fujii et al. 2010), the flux of leaching drivers (fulvic acids and nitric acid) may be important for Podzol formation.

Based on the study of soil solutions in beech forests in northern Kyoto, DOC fluxes are largest in the thick O horizon (Fig. 4.24). The proportion of DOC production relative to C input is consistently high in Podzols (Fig. 4.25); this supports considerable DOC leaching in acidic soils, including Podzols, although the causality between high DOC flux and Podzol formation could not necessarily be supported.

The importance of organic acids to podzolization is also supported by the analysis of proton budgets in soil (Fujii et al. 2008). The intensive acidification of the O horizon is contributed by the dissolution of organic acids, as well as by nitrification and cation uptake by dense root mats

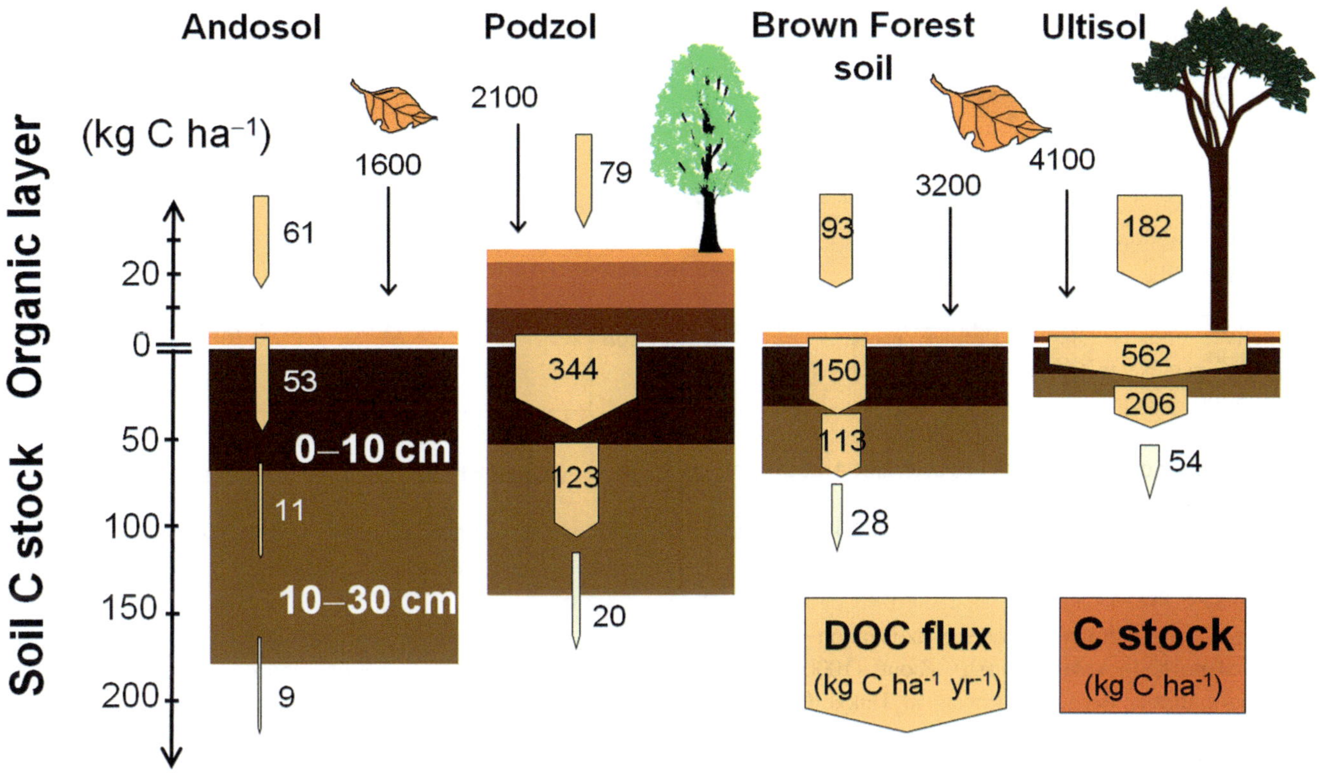

Fig. 4.24 Fluxes of dissolved organic C in an Andic Podzol (Andic Haplohumods) under beech forest in northern Kyoto, compared to Andosol, Brown Forest soil, and Ultisol profiles. *Data source* Fujii et al. (2011)

(Fig. 4.26). This acidity can induce the eluviation of Al and Fe from the E horizon. Proton consumption due to the mineralization or sorption of organic acids and nitrate uptake by plants in the Bs horizon results in the precipitation of Al and Fe and the illuviation of SRO minerals.

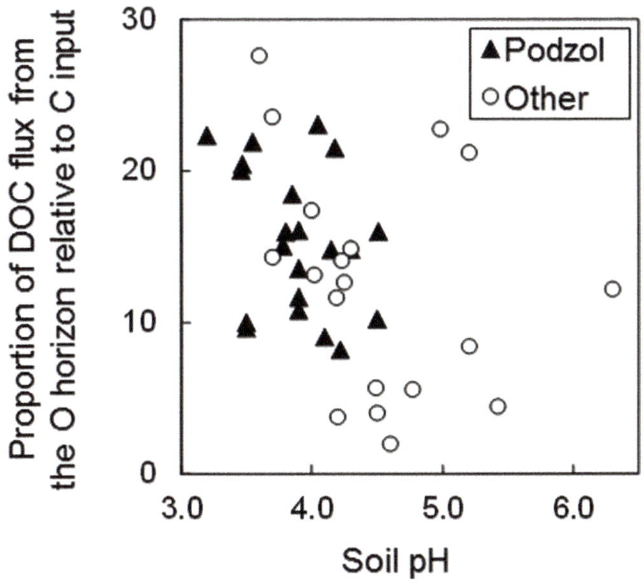

Fig. 4.25 Proportion of dissolved organic C fluxes in the O horizon relative to C input related with soil pH. The C input is the sum of litterfall C input and throughfall DOC input. *Data source* Fujii et al. (2009)

Podzolization also affects the composition of crystalline clay minerals. In Japan, hydroxyl-interlayered vermiculite is common in silicate clays produced through the weathering of mica (originating from bedrock or aeolian dust). Intensive acidification by organic acids drives the release of interlayered Al from the hydroxyl-interlayered vermiculite and the further weathering of vermiculite to smectite in the E horizon. In the Bs horizons, the high concentrations of Al in soil solution can support the Al-interlayering of vermiculite or the stability of hydroxyl-interlayered vermiculite (Funakawa et al. 1992; Kitagawa 2005).

4.4.5 Soil-Forming Factors Affecting Podzolization

Time: Most Podzols in the Northern Hemisphere were formed after the ice sheets began to retreat around ten thousand years ago. Among soil-forming processes, podzolization is relatively fast. Podzol formation requires at most hundreds of years in the case of sandy parent materials. Erosion and the input of volcanic ash and aeolian dust initializes soil formation. Podzols may have been formed from volcanic ash deposited

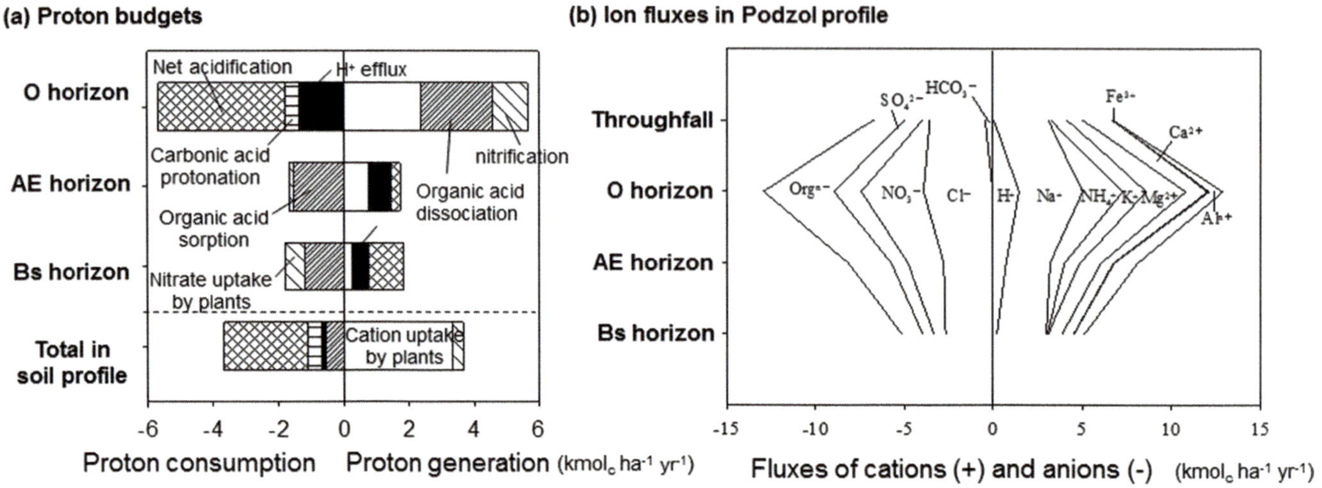

Fig. 4.26 Proton budgets **a** and ion fluxes **b** in a Podzol (Andic Haplohumods) under beech forest in northern Kyoto. *Data sources* Fujii et al. (2008)

5000 years ago (Chuseri), while podzolization restarted on newly deposited volcanic ash about 1000 years ago (Towada-a) under subalpine forests in Hakkoda, Northeast Japan (Shoji et al. 1982; Ugolini et al. 1988).

Climate: Podzols require conditions of intensive leaching (perudic to udic soil water regimes) where precipitation exceeds evapotranspiration. Most forest soils can satisfy this precondition, but leaching intensity is greater in alpine and northern areas of Japan. Acidic forest litters from conifers and a cool climate are favorable for podzolization; this is consistent with the rarity of Podzols in tropical and subtropical regions and the localization of Podzols in alpine or subalpine areas in Japan. The location of Podzols on a global scale and within Japan raises questions regarding the climate dependence (zonality) of Podzols due to the strong limitation of Podzol formation by parent materials (Fig. 4.27). Climate influences Podzol formation through various effects: the dominance of conifers in alpine and northern ecosystems in Japan, rates of mineral weathering (clay genesis), and microbial activities to produce and consume organic acids.

Landscape: Most of the Podzols in Japan are developed on well-drained positions on slopes, including ridges (Hirai et al. 1988). Podzols with an ortstein (iron pan) are not common in Japan because flat landscapes such as continental shields are required for their formation (Bockheim 2011). Since the resistance of Fe to acidity is stronger than that of Al, the clear albic E horizon is also rare in Japan, except in Podzols on sandy materials (e.g., coastal dunes). In seasonally flooded landscapes, and soils compacted by heavy snowfall (heavy snowpack pressure), a distinct E horizon can be formed, possibly due to seasonal Fe reduction.

Parent materials: At a global scale, the majority of Podzols are distributed in boreal forests in Northern Europe and the East Coast of America. During the Last Glacial Maximum (18,000 years ago), these regions were covered by ice sheets due to a relatively mild and humid climate (Fig. 4.28).

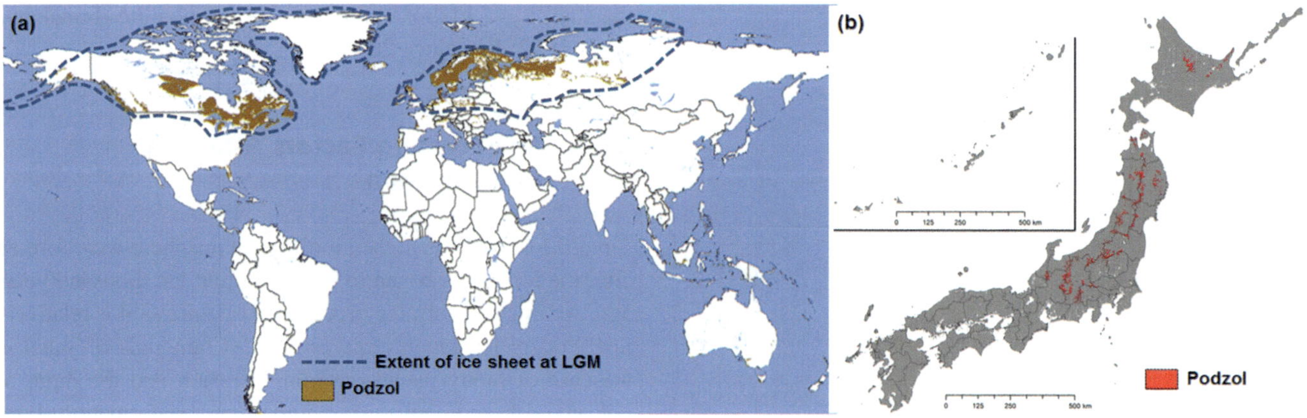

Fig. 4.27 Global distribution of Podzols in relation to extent of ice sheet at the Last Glacial Maximum (**a**) and Podzol distribution in Japan (**b**). *Data sources* University of Idaho (2019) and Kanda et al. (2018)

This contrasts with the dominance of permafrost-affected soils (Gelisols) in Siberia (Russia), the Northwest Territories (Canada), and Interior Alaska (USA) under a continental dry climate, where the absence of ice sheets promoted the development of permafrost. After the retreat of the ice sheets, Podzols formed on glacial sediments (till, loess, and glaciofluvial sands except for lacustrine clays). The formation of Podzols is favored by sandy parent materials that are poor in Fe, especially glaciofluvial sands or outwash. Since the mobility of organic acids can be limited by Fe oxides, podzolization is restricted by parent materials rich in Fe (>2% Fe_2O_3) (Duchaufour and Souchier 1978). Podzols can also be formed on clayey parent materials, including volcanic ash, in environments where the intensity of podzolization exceeds clay accumulation (sialitization).

Vegetation and microorganisms: Most Podzols in Japan occur in forests with a thick O horizon, which is a potential source of fulvic acids. However, a thick O horizon is not a prerequisite for Podzol formation; rather, Podzols tend to have a thick O horizon as a consequence of limited microbial and faunal activity (e.g., earthworms) under cool, acidic conditions (Hayakawa et al. 2014). In boreal regions of Europe and America, Podzols can also develop under a thin O horizon in open lichen-spruce woodlands. In Japan, Podzols are typically observed under subalpine coniferous forests dominated by fir (*Abies mariesii*), hemlock (*Tsuga diversifolia*), and cypress (*Chamaecyparis obtusa*). Some conifers which host ectomycorrhizal fungi are known to be "podzolizers," that is, to accelerate podzolization (Nielsen et al. 1999); however, Podzols also develop in forests of various non-ectomycorrhizal conifers, such as Japanese cedar (*Cryptomeria japonica*), and egg-cup Podzol under Kauri trees (*Agathis australis*) in New Zealand. Podzolization can also occur in some broad-leaved forests dominated by, for example, beech (*Fagus crenata*) in Japan (Fujii et al. 2008).

4.4.6 Boundary Between Andosolization and Podzolization

Andosols and Podzols are classified by the same diagnostic property (Al_o + $1/2Fe_o$). The vertical distribution of Al_o and Al_p in an Andosol suggests that organo-mineral complex was dominant in surface horizons (Fig. 4.28). The release of Al from volcanic ash can stabilize organic material. The presence of organic material and low pH conditions (<5) limit the formation of imogolite and allophane. In Andosols, in situ weathering by carbonic acids can result in the incongruent dissolution of minerals derived from volcanic materials and the accumulation of SRO minerals in the B horizon (Ugolini et al. 1988). This contrasts with podzolization characterized by eluviation and illuviation. However, as suggested by the fulvic acid-bicarbonate theory, the Podzol Bs horizon and Andosols share similarities in their weathering environments in terms of the dominance of non-complexing acids (e.g., H_2CO_3, HNO_3), low DOC concentrations, a pH range between 5 and 6.5 that promotes Al polymerization, and adequate Si concentrations that support the stability of imogolite (Fig. 4.29).

Acidity generated in the O horizon can be the determining factor of pedogenesis for Andosols and Podzols derived from volcanic ash (Shoji et al. 1982). The lower contribution of organic acids, as well as the lower soil acidification ratc, is responsible for andosolization (Fujii et al. 2008). This contrasts with podzolization caused by the intensive acidification of the O horizon.

4.4.7 Boundary Between Podzolization and Brunification

The formation of Brown Forest soils, known as "brunification," is characterized by the immobilization of organic matter and Fe oxides in the surface horizons, as suggested by the vertical distribution of Fe_o and Fe_p (Fig. 4.28) and higher saturation of the negative charge of organic acids by Al in soil solution in a Brown Forest soil (Fig. 4.29). Brown Forest soils in warmer broad-leaved forests are relatively acidic, but acidity derived mainly from plant uptake tends to distribute throughout the profile and be consumed in situ (Fujii et al. 2008). The dominance of non-complexing acids results in the accumulation of Fe oxides through incongruent mineral dissolution (Ugolini et al. 1990). Podzolization can be differentiated from soil acidification by the amounts of illuviated SRO minerals, although incipient podzolization can be involved in the formation of Brown Forest soils.

Large DOC fluxes from the O horizon can induce Al- and Fe-eluviation in the Ultisol profile of Southeast Asia (Fig. 4.24). On the other hand, the accumulation of SRO minerals in the B horizon is not significant (Fig. 4.28) and the Al concentrations of soil solution are lower than those for Brown Forest soil (Fig. 4.29). The clay-poor and acidic E horizon of Ultisols can cause weak eluviation of Al leaching; however, the leaching of Al and Fe appears to be within biological cycles of Al and Fe and the recrystallization of precipitated Al and Fe, except for in tropical Podzols on extremely sandy materials (Fujii et al. 2011).

Podzols can also be associated with Alfisols (Albeluvisols and Luvisols in the WRB classification) in temperate sub-humid regions. The clay migration and formation of the E horizons in Alfisols and Ultisols could provide a favorable condition for podzolization. In Japan, clay migration is not an active process due to the stability of colloids rich in organic matter, the low dispersibility of SRO clay minerals, and the absence of a distinct dry season. These factors also limit the wide distribution of Podzols in Japan.

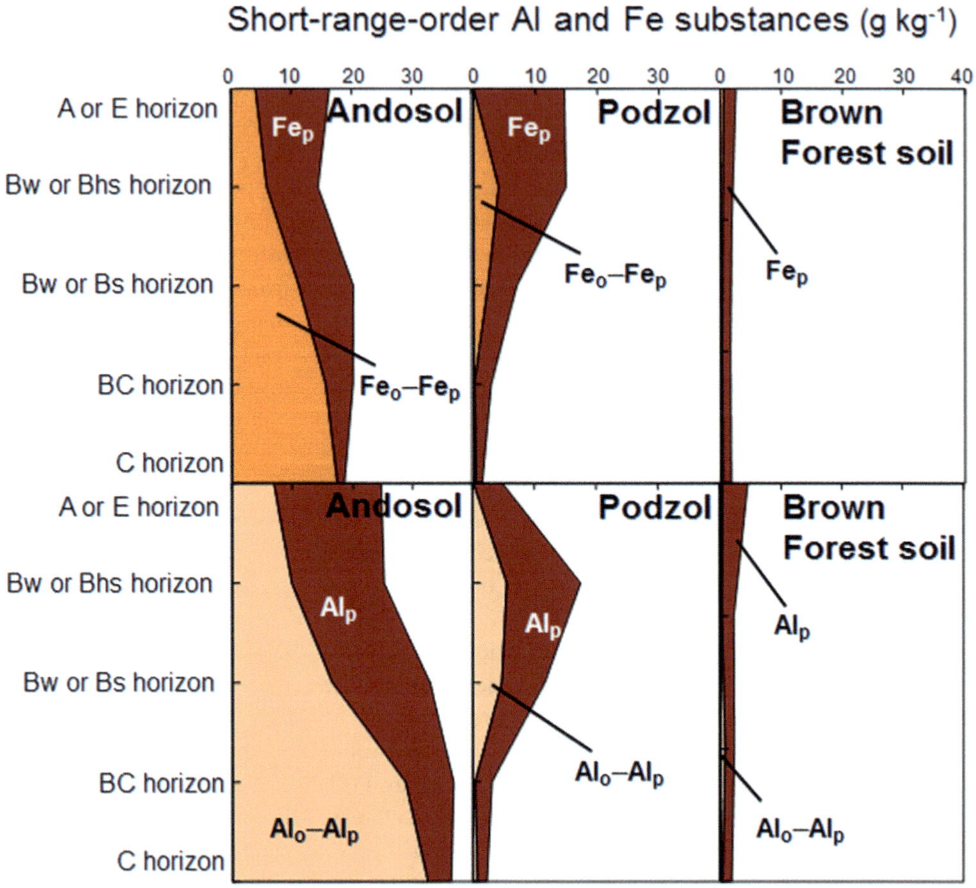

Fig. 4.28 Vertical distribution of SRO Al and Fe compounds in Andosol, Podzol, and Brown Forest soil, profiles. *Data source* Fujii et al. (2011)

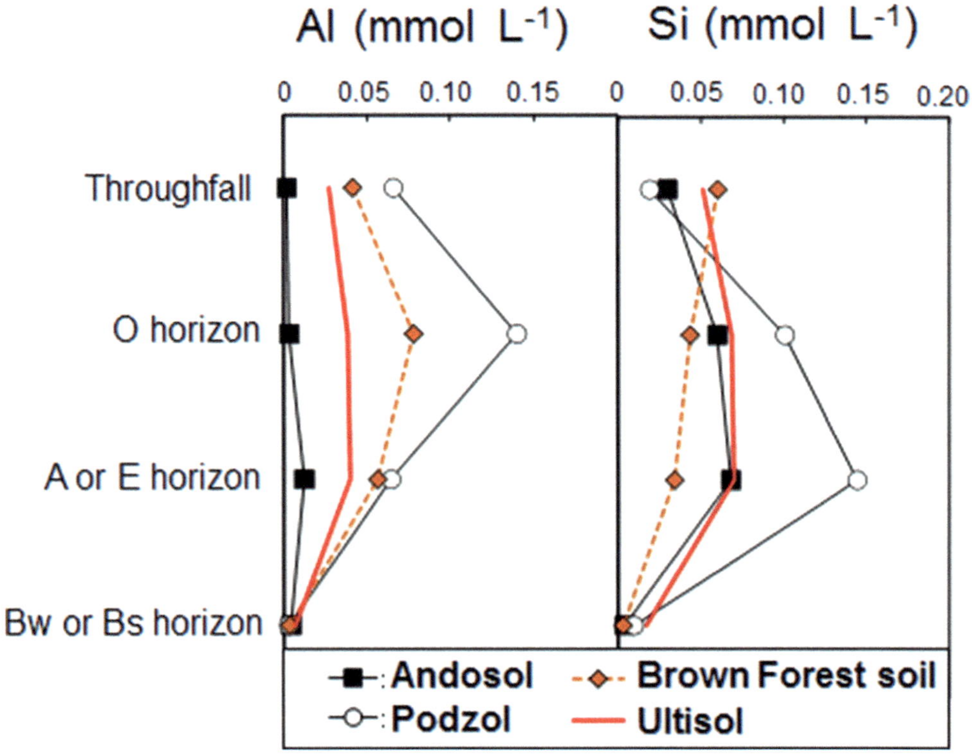

Fig. 4.29 Aluminum and silicon concentrations in soil solution of Andosol, Podzol, Brown Forest soil, and Ultisol profiles. *Data source* Fujii et al. (2008, 2009)

4.4.8 Management and Use of Podzols

In Japan, most Podzols are limited to forestry areas. In Podzols in the Kiso area (Fig. 4.21), forestry can produce excellent cypress timber. Due to steep slopes, alpine climate, and low productivity, Podzols are vulnerable to intensive use. The removal of the O horizon and erosion can cause a sharp decline in forest productivity or the failure of natural regeneration. Podzols are preserved or managed in national parks and non-private forests. This contrasts with the intensive potato production that occurs in Podzols in Canada (esp., Prince Edward Island). In Japan, Podzols provide habitats for rare alpine plants and ptarmigan nesting under *Pinus pumila*. Considering the uniqueness of Podzols, erosion management or preservation is required.

4.5 Fluvic Soils

4.5.1 Introduction

Fluvic soils are young soils developed from recent alluvial deposits. They occur on alluvial plains, river fans, valleys, and tidal marshes and occupy a significant proportion of the land surface (about 5200 km^2), amounting to approximately 14% of Japan's total land surface and 47% of its total crop field land (Fig. 4.30). Fluvic soils are usually fertile with weakly acidic to neutral soil reaction. These soils are used widely as paddy fields, and some well-drained soils are used for upland crops, vegetables, orchards, and grassland.

Fluvic soils are strongly affected by the variety of landform, relief, and moisture status of flood plains (Fig. 4.31). Natural levees consisting of sandy sediments occur adjacent to present or former river beds. Levees form the highest parts of the surrounding landscape and are freely drained. Brown Fluvic soils and Fluvic Paddy soils are widely found on levees. Old levees are used for upland crop fields and residential areas. Additionally, in suburban areas, many kinds of vegetables are grown on old levees.

Back swamp occurs behind the lower-lying levee area and consists of heavier-textured particles. In back swamps, the accumulation of organic matter is common, and Gley Fluvic soils, Gray Fluvic soils, and Peat soils are distributed. In Japan, back swamps are widely used for paddy fields, except in cold regions in northern and eastern parts of Hokkaido.

Fans (alluvial fans) are fan- or cone-shaped deposits of sediment built up by streams and are widely distributed in Japan. Fans are generally well-drained, have a coarse-grained texture with gravel and boulders and have gentle to medium slopes. On fans, mainly Brown Fluvic soils and Regosolic Fluvic soils are distributed. Fans are used for orchards, upland crops, and paddy fields with irrigation water. Some fans in steep mountainous areas are used for forest or left as wasteland.

Reclaimed land: Japan has a long history of marsh reclamation: since the 1600s (the Edo era), wide areas of paddy fields have been reclaimed from the sea. Gley Fluvic soils are the major soils appearing on reclaimed lands. Lands were reclaimed for paddy rice cultivation before the 1970s; however, in the 1970s, when national self-sufficiency of rice was attained, it was attempted to convert reclaimed lands into multipurpose crop fields.

4.5.2 Soil Classification of Fluvic Soils

Paddy fields are widely distributed in alluvial plains in Japan. Since rice is the most important cereal crop in Japan, many studies have been conducted on paddy soil, and a unique soil classification for paddy soil has emerged.

Paddy soils, both artificially and naturally, are submerged at least 3–4 months a year. During most of this period, the surface layer is kept in a reduced state, and the paddy soils have unique characteristics compared with the upland soils. However, these unique characteristics had not been recognized in the early soil classification system. For alluvial soils in Japan, Kamoshita (1940) adopted a scientific soil classification. He introduced the groundwater soil type (hydro-sequence system) developed in Germany for the preparation of soil survey reports. Groundwater soil types in this classification, i.e., Bog soil, Half-bog soil, Meadow soil, Gray Lowland soil, and Brown Lowland soil, were the basis of Japanese soil classification for alluvial soils under natural aquic conditions, but his soil classification system did not introduce any unique characteristics derived from the irrigation period specific to paddy fields. Uchiyama (1949) published the first soil classification referring to the characteristics of the subsurface layer of paddy soil derived from the submerged condition for a certain period. In his classification, aquic conditions resulting from continuous irrigation in well-drained paddy fields were found to promote the development of unique soil profiles, i.e., iron and manganese accumulation layers, in the subsurface layer. Following the classification of Uchiyama, several soil classifications were proposed for paddy field soils (Kanno 1957; Yamazaki 1960; Matsui et al. 1961). The characteristics developed under the influence of artificial irrigation of paddy fields were studied in detail, and a comprehensive classification framework including soil development under groundwater (natural) and artificial aquic conditions was proposed by Mitsuchi (1974). Although this framework was adapted to the soil classification system after Japan, some diagnostic horizons and properties were adopted in the USDA Soil Classification (1990) and World Reference Base for soil resources (IUSS Working Group 1998).

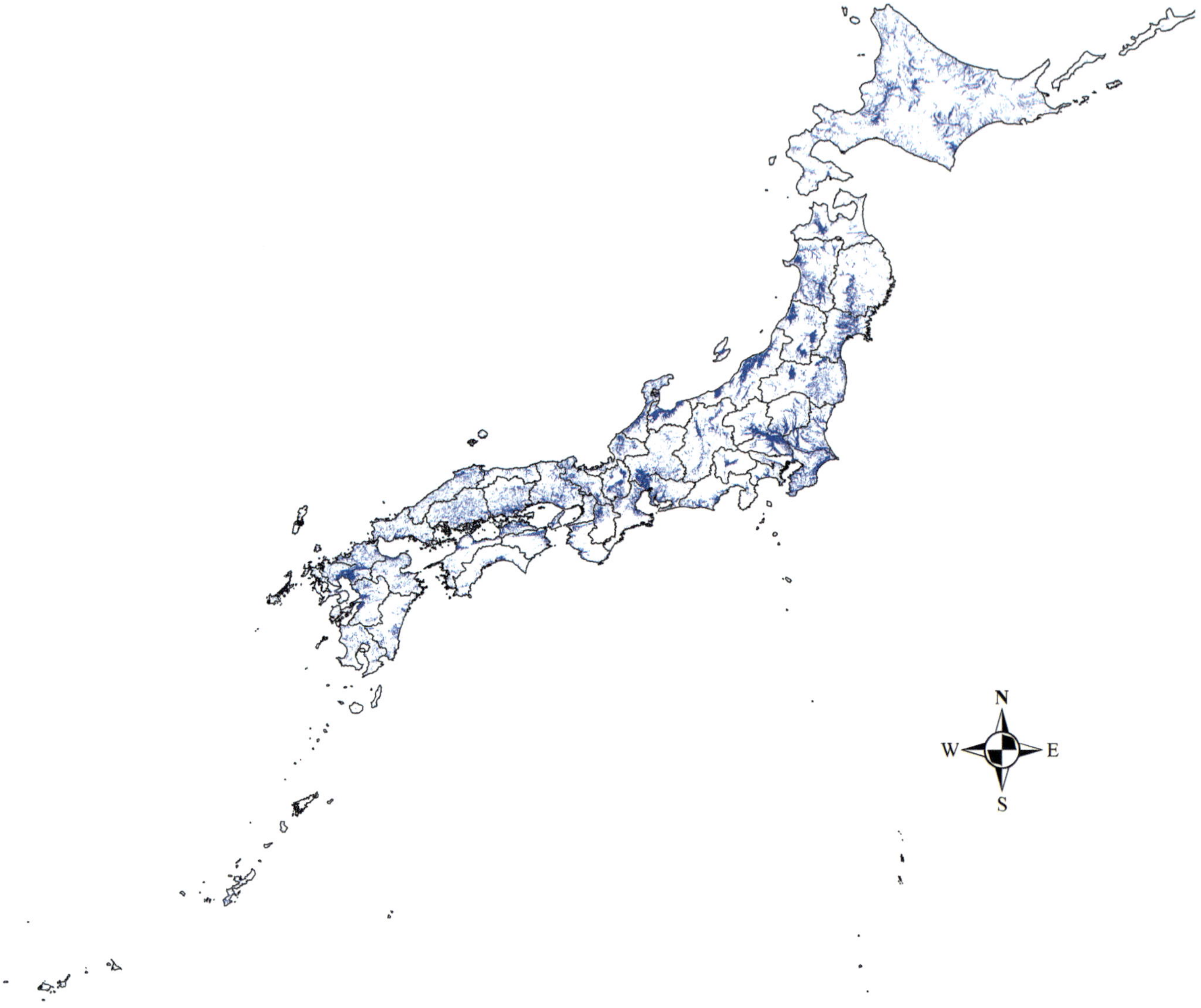

Fig. 4.30 Distribution of Fluvic soils. (Figure supplied by Hiroshi Obara)

For groundwater aquic soils, the relationship between groundwater dynamics and the diagnostic horizon (e.g., gley horizon, mottled horizon) was clarified (Hamazaki 1993).

In the new Soil Classification System of Japan (2017), Fluvic soils are subdivided into five soil groups by degree and type of aquic condition and degree of weathering. The aquic (reduced) condition forms the basis of the classification of Fluvic soils. Gley Fluvic soils are developed under permanently wet and strongly reduced conditions at shallow depth with high groundwater level. Brown Fluvic soils are developed in non-aquic or slightly aquic conditions. Gray Fluvic soils are intermediate between Gley and Brown Fluvic soils and are seasonally saturated with groundwater. Fluvic Paddy soils are characterized by anthraquic diagnostic horizons that developed under aquic conditions caused by irrigation water (Fig. 4.32). Regosolic Fluvic soils are young soils that show no changes after deposition; the gray color of sediments has not changed to brown even in non-aquic conditions.

(1) **Fluvic Paddy soils**

In the Japanese soil classification system (2017), irrigation-aquic soil and groundwater aquic soil are classified separately, and irrigation-aquic soil is first keyed out from the Fluvic soil group. Fluvic Paddy soils developed under

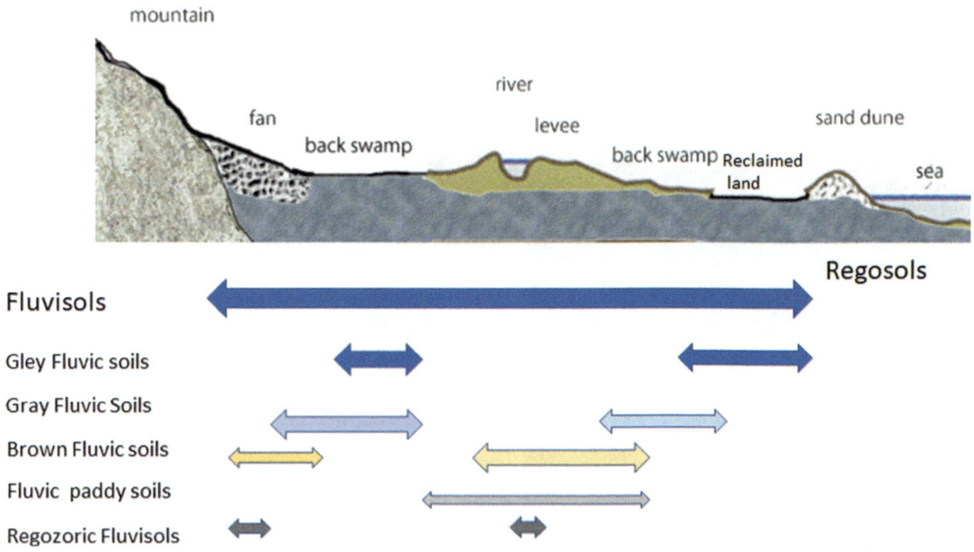

Fig. 4.31 Relationship between topography of alluvial plain and subgroups of Fluvic soils. (Figure supplied by Hiroshi Obara)

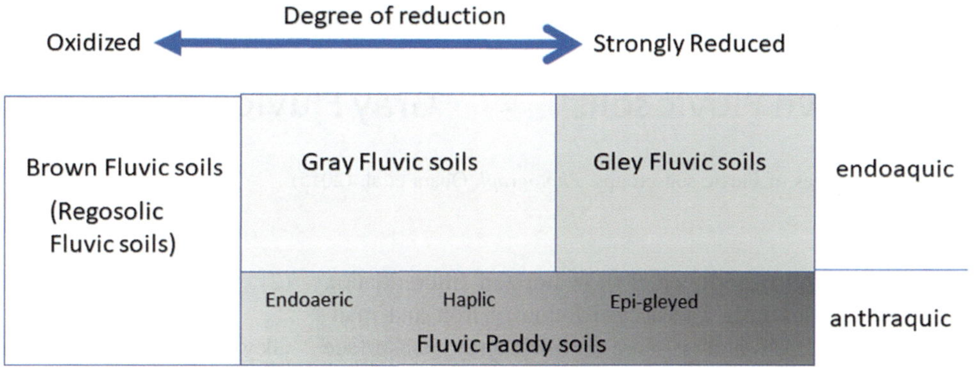

Fig. 4.32 Relationship among degree of reduction, type of aquic conditions and subgroups of Fluvic soils. Modified from Obara et al. (2014)

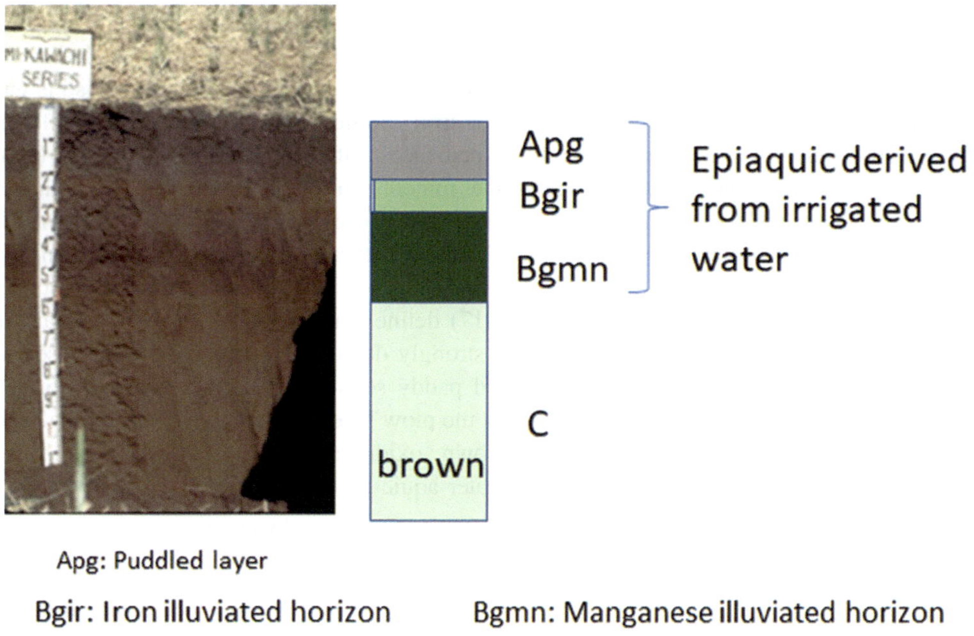

Fig. 4.33 An example of soils developed by irrigated water (anthraquic) (Endoaeric Fluvic Paddy soils). *Photograph* Obara et al. (2015)

Fig. 4.34 Profiles of Fluvic soil groups. *Photograph* Obara et al. (2015)

the influence of irrigation water show horizon differentiation as a result of leaching and accumulation of iron and manganese (Fig. 4.33) and/or have a thick gray subsurface horizon. In the Japanese soil classification system (2017), Fluvic Paddy soil is defined as "soils that have (1) an iron accumulation horizon (subsurface horizon 2 cm or more in thickness with plenty of thread-like and cloud-like iron mottles and two times or more the amount of iron (DCB extracted) contained in the plow layer); and/or (2) a grayed subsurface horizon with abundant cloud-like iron mottles and well-developed structures having shiny ped faces with its lower boundary being 50 cm below the mineral soil surface." These soils are generated by strong influence of artificial irrigation water and are typically found in natural levee areas and alluvial fans.

The Japan Soil Classification System (2017) defines the following subgroups: soils with surface soil strongly depleted of iron (Albic; generally called degraded paddy soils), soils having an inverted gley horizon beneath the plow layer (Epi-gleyed), soils having a yellowish-brown oxidizing subsoil (Endoaeric), soils having a groundwater aquic mottled subsoil (Aquic), and other soils (Haplic)

Fluvic Paddy soils are used for paddy rice and upland crops. These soils are easily used for upland crops due to their rather good drainage capacity.

(2) **Gley Fluvic soils**

Gley Fluvic soils were saturated almost all year round. It has gley horizon[1] and its upper boundary is within 50 cm of the inorganic soil surface. These soils are widely found in flood plains, deltas, tidal flats, and reclaimed land. The groundwater level is high, and the soil is generally poorly drained (Fig. 4.34).

The Japan Soil Classification System defines the following subgroups: soils with actual and potential acidity due to the presence of sulfuric acid (Thionic); soils with intervening organic layers (Peaty); soils with a humic or highly humic epipedone (Humic); soils with mottled surface gray layers (Epi—Gray); soils with iron mottles being confined to the upper 25 cm (Strong); and soils with thick gley horizons with iron mottles (Mottled).

Gley Fluvic soils have mostly been used for paddy fields. However, multipurpose uses of paddy fields are extending under the "*gentan*" policy (a governmental policy aimed at keeping rice prices high by reducing the supply of rice), and

[1]A gley horizons is defined as "a horizon that gives a positive reaction instantly to α,α'-dipyridyl, or is physically unripened and bluish-gray-colored." Gley horizons are separated into groundwater gley horizons, formed under the influence of groundwater, and stagnant-water gley horizons, formed under the influence of stagnant water.

Gley Fluvic soils are now also used for a wide range of crops with surface and underground drainage systems.

In recently reclaimed land, the drainage system develops from an early stage of reclamation. The soils are physically ripened after reclamation. In the first stage of ripening, gley horizons are dehydrated from the surface layer and are compacted, and cracks develop in deeper layers. At the same time, the upper part of the soil develops iron mottles in pores and crack surfaces. If the dehydration continues until soil matrices are unsaturated with water, the gley horizons should change to gray-colored mottled horizons. If a gray-colored mottled horizon develops at a depth of 50 cm or more from the surface, Gley Fluvic soils should change to Gray Fluvic soils. The change of a gley horizon to a gray-colored mottled horizon occurs rather quickly when the groundwater level drops. However, under traditional paddy rice cultivation, the change of a gley horizon to a gray-colored mottled horizon at greater depth takes place over a longer time. In the case of reclaimed land in Kojima, it took more than 70 years after reclamation for a gley horizon to change into a gray-colored mottled horizon at 50 cm depth.

(3) **Gray Fluvic soils**

Gray Fluvic soils have a mottled horizon formed under the seasonal saturation of groundwater and with its upper boundary being within 50 cm of the mineral soil surface (Fig. 4.34). Mottled horizons formed under the influence of groundwater are characterized by iron mottles along pores, such as tubular and root-like mottles occurring along root channels. These soils occur widely in nearly flat coastal plains, flood plains, delta plains, valley bottoms, and alluvial fans.

Seven subgroups are provided: Thionic, Peaty, Humic, Epi-gleyed (having an inverted gley horizon), Gleyic (having a groundwater gley horizon in the subsoil), Thapto-andic (having a buried Andosol in the subsoil), and Haplic (others).

In Gray Fluvic soils, the groundwater level is low relative to Gley Fluvic soils and soils are somewhat poorly drained in many cases. These soils are mostly used as paddy fields, and partly as upland crop fields.

(4) **Brown Fluvic soils**

These are soils which occur on alluvial plains. The subsurface horizon is yellowish-brown in color with no influence of groundwater within 50 cm of the surface and with no or weak influence of irrigation water (Fig. 4.34). Due to the earlier stage of weathering and the liberation of free iron oxides, soil materials of alluvial lowlands are normally dull yellowish-brown in color, with chroma ranging from 3 to 4. These soils occur mainly in lowlands with a low groundwater table, such as natural levees and alluvial fans.

Brown Fluvic soils are divided into four subgroups: soils with the upper boundary of a mottled horizon within 75 cm of the surface (Aquic), soils with a humic to highly humic epipedon (Humic), soils with a weak influence of irrigation water (Protoanthraquic), and others (Haplic). Brown Fluvic soils are used for upland crop fields or for residential areas.

(5) **Regosolic Fluvic soils**

Regosolic Fluvic soils are composed of unweathered clastic sediments in lowlands. These soils are normally gray in color (i.e., the color of the clastic materials is normally gray) and do not have rusty mottles due to the lack of free iron. Soils are generally sandy or sandy-skeletal in texture and are found in the present and past river beds. Part of these soils is seasonally saturated with underflow water, but the soil is not reduced.

Regosolic Fluvic soils are separated into two subgroups: seasonally saturated soils (Aquic) and others (Haplic). Because these soils are generally well-drained, they are mostly used for upland crop fields and orchards.

4.5.3 Relationships with Other Soil Great Groups

Fluvic soils are young soils developed from recent alluvial deposits and have only some diagnostic horizons related with aquic conditions, for example, gley horizons, mottled horizons, iron accumulation horizons, and grayed horizons. The important linkages of Fluvic soils are with other soils distributed in lowlands, namely Peat soils, Andosols, and Regosols.

Peat soils are usually distributed with Gley Fluvic soils. Gley Fluvic soils and Peat soils have similar wet conditions and only differ in their ratio of organic matter content and mineral sediment content. In some regions, Peat layers have been decomposed by drainage, and soils have been changed to Gley Fluvic soils or Gray Fluvic soils. Where Andosols are distributed on terraces or hills, materials with andic soil properties are transported by erosion or other processes, such as landfill, and are deposited in flood plains. These soils are classified as Gleyic Andosols when the thickness of layers with andic soil properties is at least 25 cm within 50 cm of the soil surface (Fig. 4.35). Gleyic Andosols are usually found in narrow valleys within terraces or hills that are covered by Andosols ("*Yachida*" in Japanese). These soils have been used for paddy rice cultivation since ancient times. The soils on slightly elevated lands along the sea coast such as sand dunes, sand bars, and sand banks are classified as Sandy Regosols.

Fig. 4.35 An example of Peaty Gley Andosols in small valley (Ibaraki prefecture). *Photograph* Obara et al. (2015)

4.6 Red-Yellow Soils

In Japan, red- and yellow-colored soils are widely distributed in the subtropical to warm temperate zone and are designated as Red-Yellow soils. These soils are closely related to Red-Yellow Podzolic soils as defined in the former USDA soil classification system; however, they have no characteristics of podzolization. Red- and yellow-colored soils are distributed adjacent to each other and share many similarities in physicochemical properties other than soil color. Therefore, both red- and yellow-colored soils were defined as "Red-Yellow soils" in Japan (Kanno 1961).

Here, "red" and "yellow" are defined following The Fifth Committee for Soil Classification and Nomenclature (2017), as follows: according to the Munsell color notation, a red color refers to one that is redder than hue 5YR, and that has value >3 and chroma ≥ 3, but excluding value/chroma 4/3 and 4/4, while a yellow color refers to one that is more yellow than hue 5 YR (not including 5 YR), and that has a value ≥ 3 and chroma ≥ 6, except value/chroma 3/6 and 4/6.

4.6.1 Genesis and Characteristics

(1) **Morphological and mineralogical properties of Red-Yellow soils**

Generally, the O horizon of Red-Yellow soils consists of an Oi horizon with fresh deciduous litter layer only. The A horizon is usually thin (less than 10 cm). The soil color of the A horizon is dark brown or reddish-brown, and its organic matter content is very low. The soil structure of the A horizon is strongly developed and has a granular structure. Below the A horizon is the B horizon, which is thick (several tens of centimeters) and has a brightly reddish-brown, orange, or yellow soil color. The organic matter content of the B horizon is much lower than that of the A horizon. The soil structure of the B horizon is subangular-blocky and relatively well-developed, and a clay film is often observed on the surface of the structure. Usually, the activity ratio of free iron oxide (Feo/Fed) is 0.4 or less, and the crystallinity ratio ([Fed-Feo]/Fet) is 0.5 or more (Nagatsuka 1975). The form of iron oxide is mainly hematite (Hm) or goethite (Gt), although yellow-colored soils sometimes contain lepidocrocite. The clay content of the B horizon is generally higher than that of the A horizon. This difference may be caused by: (1) an illuvial accumulation of clay; (2) the predominant pedogenetic formation of clay in the B horizon; (3) the destruction of clay in the A horizon; (4) the selective surface erosion of clay; (5) biological activity; or (6) a combination of two or more of processes (1)–(5) (IUSS Working Group WRB 2014). Clay minerals mainly consist of kaolin minerals and hydroxyl-interlayered vermiculite (HIV). In the lower horizon, iron mottles called "*Tora-han*" (tiger pattern) are sometimes present, and highly weathered gravels such as saprolites are often observed.

Schwertmann (1985) summarized the factors that predispose iron ions in soil to form Hm or Gt. When the rate of iron ion release from the parent materials is high, with the parent material having a high iron content, low organic matter contents, and slightly acidic to slightly alkaline conditions, ferrihydrite (Feh), which is a precursor of Hm, is more likely to be generated. Feh tends to dehydrate to Hm as the soil temperature increases and the soil moisture content decreases. On the other hand, Gt can be more produced under the such condition as the rate of iron ion release from the parent materials is low, with the parent material having a low iron content, relatively high organic matter contents, and strongly acidic conditions. It is important to note that Hm cannot be directly generated from Maejima et al. (2000) investigated the mineralogical composition of several red- and yellow-colored soils using differential X-ray diffraction analysis (DXRD) and revealed that the presence of Hm was associated with Gt in the red-colored soils, but detected no Hm in the yellow-colored soils.

(2) **Genesis of paleo-Red soils and the age of Red-Yellow soils**

There are two theories regarding the genesis of Red soils. One is the "zonal Red soil theory," whereby Red soil is considered to have formed under present climatic conditions.

The other is the "paleo-Red soil theory," whereby Red soil is deemed to be a relic of paleosols from a past geological age whose climate was warmer than the present climate. Ohmasa et al. (1955) first reported the existence of paleo-Red soils in Japan. Subsequently, Matsui and Kato (1961) proved that paleo-Red soils were distributed in Southwest as well as Northwest Japan. At the present, the paleo-Red soil theory is generally preferred, and research aiming to obtain a more accurate age for paleo-Red soils has begun. For example, Akagi et al. (2003) estimated the formation period of paleo-Red soils in southern Kyushu and found that paleo-Red soils were formed from 400–500 to 110–130 thousand years ago. Nagatsuka and Maejima (2001) also estimated the absolute age of soils on a raised coral limestone terrace in Kikai Island and stated that it would take at least 125,000 years for Red-Yellow soils to develop on coral limestone under a humid subtropical climate. However, Araki (1993) suggested that it is impossible to distinguish between paleo-Red soils and recent Red soils because the properties of Red-Yellow soils change continuously depending on whether they are located in a tropical, subtropical, or temperate zone. For instance, the CEC of red-colored B horizons in Northeast Japan is relatively high (usually exceeding 25 $cmol_c$ kg^{-1} clay) while those B horizons are in the early stage of the weathering process (Araki 1993).

4.6.2 Classification

The general properties of Red-Yellow soils are a red or yellow color, low accumulation of organic matter, low base saturation, and strong weathering. Generally, these soils are strongly or weakly acidic. Soils corresponding to the Red-Yellow soil great group satisfy all of the following requirements:

(1) An "Argic horizon" with "Red-Yellow properties" is present, having an integrated thickness of 25 cm or more and starting within 50 cm of the soil surface or the upper end of "Red-Yellow Cambic horizon," having an integrated thickness of 25 cm or more and starting within 35 cm of the soil surface;
(2) The subsurface horizon has a pH (H_2O) of less than 5.5 or a base saturation of less than 50%;
(3) Bedrock does not appear within 30 cm of the soil surface.

The Red-Yellow soil great group can be subdivided into two soil groups, "Argic" and "Cambic," depending on the presence of clay accumulation, in order to allow comparison between international and Japanese soil classification systems.

Argic Red-Yellow soils

These soils are Red-Yellow soils which have an argic horizon.

Cambic Red-Yellow soils

These soils are Red-Yellow soils which have a cambic horizon.

Takata et al. (2010) elucidated the differences in the physicochemical properties of argic and cambic horizons using database analysis and revealed that argic horizons develop on stable landscapes while cambic horizons develop on mountainous or hilly ranges. Therefore, higher categories (soil groups) of Red-Yellow soils are distinguished by the presence of argic or cambic horizons, and lower categories (soil subgroups) are distinguished based on soil color. This principle of division has been adopted in both Comprehensive Soil Classification System of Japan, First Approximation (Obara et al. 2011) and the Soil Classification System of Japan (The Fifth Committee for Soil Classification and Nomenclature 2017) (Fig. 4.36).

4.6.3 Correlation with International Classification Systems

Red-Yellow soils do not have a direct one-to-one correspond to soil definitions in international soil classification systems. That is, the correspondence relationship with international soil classification systems differs at higher level with and without clay accumulation for soils such as Acrisols, Alisols, Cambisols, and partly Stagnosols and Umbrisols (IUSS Working Group WRB 2014) and Udults or Udepts in the USDA Soil Taxonomy (Soil Survey Staff 2014). Because the

Fig. 4.36 Reddish Argic Red-Yellow soils (*left*) and Haplic Cambic Red-Yellow soils (*right*). Copyright by Dr. Y. Maejima

Table 4.3 Correlation of soil classification system for Red-Yellow soils

Soil Classification System of Japan (2017)		WRB (2014)	Soil Taxonomy (2014)
Red-Yellow soils		Acrisols, Alisols, Cambisols, Stagnosols, Umbrisols	Udults, Udepts
Argic		Alisols, Acrisols, Stagnosols, Umbrisols	Udults
	Anthraquic	Alic Stagnosols Acric Stagnosols	Anthraquic Paleudults Aquic Hapludults
	Albic	Albic Alisols Stagnic Albic Alisols	Typic Paleudults Aquic Paleudults Typic Hapludults Aquic Hapludults
	Psuedo-gleyed	Stagnic Alisols Stagnic Acrisols	Aquic Paleudults Aquic Hapludults
	Aquic	Gleyic Alisols Gleyic Acrisols	Aquic Paleudults Aquic Hapludults
	Humic	Alic Umbrisols Acric Umbrisols	Typic Haplohumults Humic Hapludults
	Reddish	Chromic Alisols Chromic Acrisols	Typic Paleudults Typic Hapludults
	Haplic	Haplic Alisols Haplic Acrisols	Typic Paleudults Typic Hapludults
Cambic		Cambisols, Stagnosols, Umbrisols	Udepts
	Anthraquic	Haplic Stagnosols	Anthraquic Eutrudepts
	Albic	Gleyic Cambisols Haplic Cambisols	Aquic Dystrudepts Typic Dystrudepts
	Psuedo-gleyed	Stagnic Cambisols	Aquic Dystrudepts Oxyaquic Dystrudepts
	Aquic	Gleyic Cambisols	Aquic Dystrudepts Oxyaquic Dystrudepts
	Humic	Cambic Umbrisols	Humic Dystrudepts
	Reddish	Chromic Cambisols	Typic Dystrudepts
	Andic	Dystric Cambisols	Andic Dystrudepts
	Haplic	Dystric Cambisols	Oxyaquic Dystrudepts Typic Dystrudepts

correspondence relationship with international soil classification systems is complicated, it is summarized in Table 4.3 as follows.

4.6.4 Distribution

The distribution area of Red-Yellow soils occupies about 7.6% of Japan's total land area. Red-Yellow soils are mainly distributed in middle-altitude to high Pleistocene terraces in Southwest Japan and in hills and low mountains in the Nansei Islands (also known as the Ryukyu Islands).

4.6.5 Land Use

Because Red-Yellow soils are distributed in hills and plateaus, they are often used as orchard, upland fields, and paddy fields in agricultural lands. Citrus, tea, grapes, and so on are cultivated in orchards, and vegetables and flower plants are cultivated in upland fields. Additionally, in recent years, greenhouse cultivation has also been carried out in Red-Yellow soils. Pineapple is cultivated in these soils on Okinawa Island.

For agricultural land use, the neutralization of acidity is necessary and is achieved by applying lime material such as carbonate lime, dolomite, and calcium silicate. Additionally, it is necessary to improve the physical and chemical properties of Red-Yellow soil by applying base cations and phosphorous via the application of fused magnesium phosphate and magnesium superphosphate and organic materials such as compost or manure. In many cases, deep tillage and pipe drainage are carried out in order to improve the drainage and expand the root area of crops because the subsurface soil is dense and has poor water permeability (Nagatsuka 1997).

4.7 Stagnic Soils

Stagnic soils are hydromorphic soils that are influenced by groundwater or surface water and have gley horizons or gray horizons with iron mottles in subsoils. While soils with hydromorphic features distributed in alluvial plains are classified as Fluvic soils, Stagnic soils are distributed in areas with poor drainage in terraces, hills, and mountainous areas. Stagnic soils can be divided into Stagnogley soils (mainly correlated with Gleysols; WRB 2006) and Pseudogley soils (mainly correlated to Stagnosols; WRB 2006) according to drainage conditions and groundwater level. While Stagnogley soils are strongly influenced by groundwater throughout the year or for most of the year, Pseudogley soils are influenced by surface water because of the presence of an impermeable layer that prevents the downward filtration of soil water.

4.7.1 Stagnogley Soils

(1) **Formation of Stagnogley soils**

Stagnogley soils are distributed in terraces, hills, and mountainous areas having a shallow groundwater table. Such areas are found in concave topography or at the bottom of slopes, where water gathers easily and drainage is poor. When a shallow groundwater table is present and water saturates a soil for a long time, a reductive condition develops in the subsoil. As a result, ferric ion is reduced to ferrous ion, the soil color changes to a blueish color, and a gley horizon forms. This soil formation process is known as "gleying." Gley horizons are easy to identify from field observation, because the soil has a blueish color and the α,α′-dipyridyl regent changes to a red color rapidly in gley horizons. The depth of appearance of gley horizons depends on the depth of the groundwater table and the length of time that the soil is water saturated. Stagnogley soils are characterized by the presence of gley horizons at shallow depths in the profile (Fig. 4.37). In Japanese soil classification systems, such as the Soil Classification System of Japan (The Fifth Committee for Soil Classification and Nomenclature 2017), Stagnogley soils are defined as having a gley horizon more than 10 cm in thickness that occurs within 50 cm of the soil surface. Under wet, cool conditions, plant debris does not decompose well and peat materials sometimes accumulate on the surface. This type of soil is classified as Epi-peaty Stagnogley soil.

(2) **Physical and chemical properties of Stagnogley soils**

Stagnogley soils are strongly acidic and have low base saturation because the bases have been leached out (Table 4.4). The soil texture of subsurface horizons is often clayey. As

Fig. 4.37 Soil profile of Stagnogley soil at Oshima, Niigata prefecture (Obara et al. 2015, modified by Kazunori Kohyama)

Stagnogley soils are saturated with water for a long time the compaction seen in Pseudogley soil is not recognized, and bulk density ranges from 0.8 to 1.2 Mg m^{-3}. As these soils have a shallow groundwater table, drainage is one of the most effective countermeasures for agricultural use.

Stagnogley soils are mainly distributed in Hokkaido and the Hokuriku region and are mainly used as paddy fields (Table 4.5). In the Hokuriku region, and particularly in Niigata Prefecture, Stagnogley soils are distributed on gentle sloping hills and are used as paddy fields. Land leveling is required for the construction of paddy fields, and accordingly, rice terraces, called "*Tanada,*" have been built and provide a unique landscape (Fig. 4.38). The original soils in the Hokuriku region were considered as Brown Forest soils that provide good drainage in hilly areas. However, as a result of long-term rice cultivation, the region later developed more aquic conditions and developed gleyic properties. In Hokkaido, the distribution area of Stagnogley soils is smaller than in the Hokuriku region. Here, these soils are

Table 4.4 Physicochemical properties of the Stagnic soils (unpublished data by Monolith data of NIAES)

Depth		BD	Permeability	pH		TC	TN	Exchangeable cation				CEC	BSat	PAbs
				H_2O	KCl			Ca	Mg	K	Na			
cm		g cm^{-3}	m s^{-1}			g kg^{-1}		$cmol_c$ kg^{-1}					%	
Stagnogley soil at Oshima, Niigata prefecture														
0	12	0.71	1.98 E−03	4.5	3.6	37	2.4	12.4	5.2	0.5	0.4	33.7	54.9	1170
12	19	0.96	2.26 E−04	4.5	3.6	29	2.0	13.6	5.5	0.4	0.2	33.6	58.6	1180
19	28	1.25	1.71 E−06	4.9	3.7	5	0.7	14.0	5.7	0.6	0.2	31.9	64.3	1140
28	54	1.25	2.00 E−06	5.0	3.5	2	0.5	17.7	7.9	0.9	0.2	34.6	77.2	1090
54	59			5.8	4.5	1	0.4	17.8	9.5	0.9	0.2	28.1	101.1	950
59	69	1.28	2.03 E−04	5.8	4.8	4	0.6	17.7	8.9	0.8	0.2	29.4	93.9	1210
Pseudogley soil at Mombetsu, Hokkaido														
0	14	0.56		4.8	3.8	83	3.6	1.3	1.5	0.7	0.5	26.2	15.3	1620
14	33	1.38	7.00 E−04	5.2	3.8	10	0.6	1.6	1.7	0.4	0.2	16.4	23.6	1130
33	53	1.34	1.00 E−05	5.3	3.7	4	0.3	3.5	2.7	0.2	0.4	17.7	39.0	820
53	83	1.60	4.00 E−06	5.5	3.5	1	0.2	5.4	4.4	0.2	0.5	17.1	61.1	510
83	100	1.71	2.00 E−06											

BD bulk density, *Bsat* base saturation, *PAbs* phosphorus absorption coefficient

Table 4.5 Distribution area of Stagnic soils in Japan (Obara et al 2016)

Region	Stagnic soils (km^2)		
	Total	Stagnogley soils	Pseudogley soils
Hokkaido	1500	6	1494
Tohoku	0	0	0
Kanto and Tosan	154	36	117
Hokuriku	748	682	65
Chubu	385	0	385
Kinki	150	0	150
Chugoku and Shikoku	8	0	8
Kyushu	1	0	1
Japan	2946	725	2221

mainly found in terraces accompanied by Pseudogley soils. In northern Hokkaido, Stagnogley soils are used as grassland and upland field because the climate is not suitable for rice cultivation.

4.7.2 Pseudogley Soils

(1) **Properties of Pseudogley soils**

Pseudogley soils are distributed in flat or gently sloped terraces, hills, and mountainous areas where water stagnates seasonally. Unlike Stagnogley soils, in Pseudogley soils the groundwater table is not always shallow. The soil profile of a Pseudogley soils is characterized by iron-mottle-rich horizons and compacted layers. The upper parts of the profile (Bg horizon) have mottled coarse blocky peds with bleached surfaces, while the lower parts (Cg horizon) have mottled plate-like blocky peds with bleached surfaces. As the peds of the Cg horizon are more compact than those of the Bg horizon, the Cg horizon becomes impermeable and water seasonally stagnates there; the stagnated water causes soil reduction and promotes the leaching of iron. In the dry season, the stagnated water disappears and the soil becomes drier. As a result, the redox potential of the subsoil increases, iron is oxidized, and free-iron mottles form (Fig. 4.39).

(2) **Physical and chemical properties of Pseudogley soils**

As bases are leached, Pseudogley soils show strong acidity and base saturation is low. The subsoil has a fine texture and its clay content often exceed 25%. The solid phase ratio is

Fig. 4.38 Landscape of rice terraces where Stagnogley soils distribute in Niigata prefecture. (Figure supplied by Kazunori Kohyama)

also high, and bulk density is 1.2–1.4 Mg m^{-3}. Due to the compactness of the subsoil, several physical conditions of Pseudogley soils are poor. The soils show a few coarse pores and low hydraulic conductivity (k = 10^{-6}–10^{-9} m s^{-1}). As a result, water stagnates with low rainfall and the water damage of crops occurs easily. Moreover, as the water-holding capacity of the soil is also small and root penetration is limited, droughts occur easily when rainfall is insufficient.

(3) **Distribution and land use**

In Japan, Pseudogley soils are mainly distributed in the Hokkaido, Chubu, and Kinki regions, and are used as paddy field. As flat topography and low permeability is suitable for maintaining irrigation water, Pseudogley soils are suitable for rice cultivation. In northern Hokkaido, which is not suitable for rice cultivation because of its cool climate, Pseudogley soil is used for upland and grassland. However, these uses require the soil's drainage condition to be improved. In Hokkaido, subsoiling and sand dressing are common practices used to improve the poor physical conditions of Pseudogley soils. Subsoiling with a pan breaker promotes the removal of stagnant water and the improvement of soil tilth by breaking and softening the dense subsoils. Underdrainage is also essential to remove stagnant water.

4.7.3 Topographic Sequence of the Terrace

The water condition depends on the microrelief of the terraces in which Stagnic soils are distributed. The water condition influences the soil type. For example, since the groundwater table is deep and the drainage condition is good in areas with convex topography, the soil is in an oxidized condition; Brown Forest soils, which are considered as drier soils, occur in such areas. On the contrary, in areas with concave topography or at the bottoms of slopes, surface water gathers easily, the groundwater table becomes shallower, and soil is in a reductive condition. Hydromorphic soils such as Stagnogley soils and Peat soils are distributed in such areas. Pseudogley soils are commonly found in the middles of slopes.

In the Soil Classification System of Japan (Japanese Society of Pedology), Brown Forest soils have two subgroups, Aquic Brown Forest soils and Haplic Brown Forest

Fig. 4.39 Soil profile of Pseudogley soil at Mombetsu, Hokkaido (Obara et al. 2015, modified by Kazunori Kohyama)

soils, according to the water condition; the former occurs in wetter conditions than the latter. Pseudogley soils are divided into three subgroups, Aeric Pseudogley soil, Haplic Pseudogley soil, and groundwater–aquic soil, according to the groundwater status. The Aeric subgroup is more oxidized than the Haplic subgroup, while the groundwater–aquic subgroup has a gley layer in the subsoil and is more reductive than the Haplic subgroup. Stagnogley soils are divided into two subgroups, Epi-peaty Stagnogley soil and Haplic Stagnogley soil, according to the water condition; the groundwater level of the Epi-peaty subgroup, which has a peat layer on the surface, is shallower than that of the Haplic subgroup (Fig. 4.40).

The ascending sequence from freely to poorly drained for the aforementioned soils is as follows: Haplic Brown Forest soils, Aquic Brown Forest soils, Aeric Pseudogley soils, Haplic Pseudogley soils, groundwater–aquic Pseudogley soils, Haplic Stagnogley soils, Epi-peaty Stagnogley soils, and Peat soils. These soils are distributed in a complex manner according to the water condition in the terrace.

4.8 Eutrosols

A "Eutrosol" is a soil with a high base saturation, that is, with a eutric condition. Eutrosol is a unique and uncommon soil name: As far as we know, this term is used only by the Japanese soil classification system.

4.8.1 Genesis and Characteristics

After the end of World War II, as national soil surveys progressed in Japan, the presence of dark red-colored soils and soils with high base saturation were reported. The term "*dark red soil*" was first reported by Takehara in 1961, who named the soil on the Ryukyu limestone in Iriomote Island as "*calcareous dark red soil*" (Takehara 1961). In 1966, in a soil survey in Kochi Prefecture, soils derived from serpentinite or limestone were morphologically named as "*dark red soil*," so as to distinguish them from "*red soil*." Dark red-colored soil was also found on hydrothermally altered igneous rocks. This type of soil is usually strongly acidic and has a low base saturation. After that, the "*dark red soil*" soil group was established in the "Classification of Forest Soil in Japan (1975)" (Forest Soils Division 1976) and was subdivided into three subgroups such as Eutric, Dystric, and Volcanogenous. The term "*dark red soil*" has been widely used since the "Classification of Cultivated Soils in Japan, Third Approximation" (Cultivated Soil Classification Committee 1995).

Soils termed "*dark red soil*" generally show high base saturation, although some do not. Furthermore, soils can be classified into this soil group even if they are not dark red in color; for example, this was the case for the aforementioned soil from the Ryukyu limestone. This causes confusion in the application of the soil classification system. The Fifth Committee for Soil Classification and Nomenclature of the Japanese Society of Pedology has been discussing changes in Soil Classification System of Japan and has reviewed the classification of "*dark red soil*" and related soils distributed in the country. Considering that these soils contain non-dark-reddish soil with a Eutric condition, the committee decided that it was better to prioritize not soil color but rather Eutric condition and did not adopt "dark red soil" as a soil name. Eutrosols were finally established as a new soil great group (The Fifth Committee for Soil Classification and Nomenclature 2017).

4.8.2 Classification

The general properties of Eutrosols are the presence of either a Cambic horizon or Argic horizon whose base saturation of the subsurface horizon is more than 50%. In the case of agricultural land, Eutrosols should not require acidity correction measures such as lime application. The criteria of

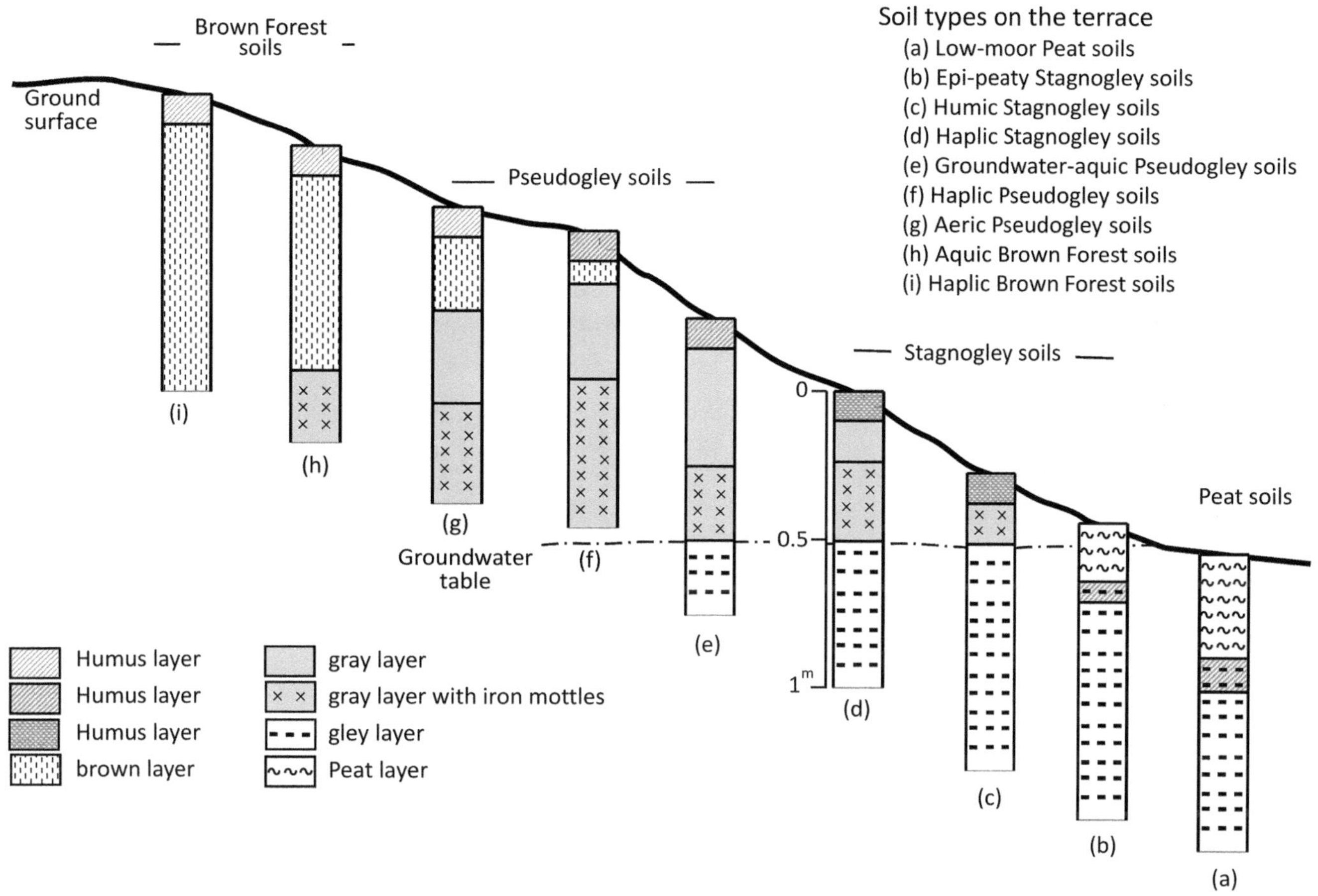

Fig. 4.40 Hydro-sequence of Stagnic soils and their related soils. (Figure supplied by Kazunori Kohyama)

both soil color and type of parent materials (parent rock) used in former classification systems (e.g., Obara et al. 2011) have been eliminated. Additionally, it was decided to key out the Eutrosol soil great group after Fluvic soils and Stagnic soils in order to exclude high-pH Fluvic soils and Stagnogley soils from Eutrosols. From the viewpoints of agricultural use and environmental conservation, this soil great group was subdivided into Magnesian and Calcareous groups. Nakao et al. (2011) reported a lower calcium/magnesium (Ca/Mg) ratio in serpentinitic soils and proposed the classification of "*dark red magnesian soils.*" Maejima et al. (2014) also investigated the genesis and classification of soils derived from serpentine and confirmed that exchangeable Ca/Mg ratio was an appropriate criterion for classifying serpentinitic soils. The exchangeable Ca/Mg ratio as a new diagnostic property was subsequently introduced to distinguish Magnesian Eutrosols from Calcareous Eutrosols (Fig. 4.41). Each group is further divided into Argic (clay accumulation) and Haplic (Cambic) subgroups according to the presence of clay accumulation in order to enable comparison between the international and Japanese soil classification systems.

The soils corresponding to the Eutrosol great group satisfy all of the following requirements:

(1) The top of the Argic horizon or Cambic horizon starts within 50 cm of the soil surface;
(2) All of the subsurface horizon has a pH (H_2O) of more than 5.5 or a base saturation of more than 50%;
(3) Bedrock does not appear within 30 cm of the soil surface.

The Eutrosol great group can be subdivided into two soil groups, Magnesian and Calcareous, depending on the exchangeable Ca/Mg ratio.

Magnesian Eutrosols
These soils are Eutrosols which have one or more subhorizons of subsurface horizons having an exchangeable Ca/Mg ratio of less than 0.1, or all of whose subhorizons have exchangeable Ca/Mg ratios of less than 1.

Calcareous Eutrosols
Other Eutrosols

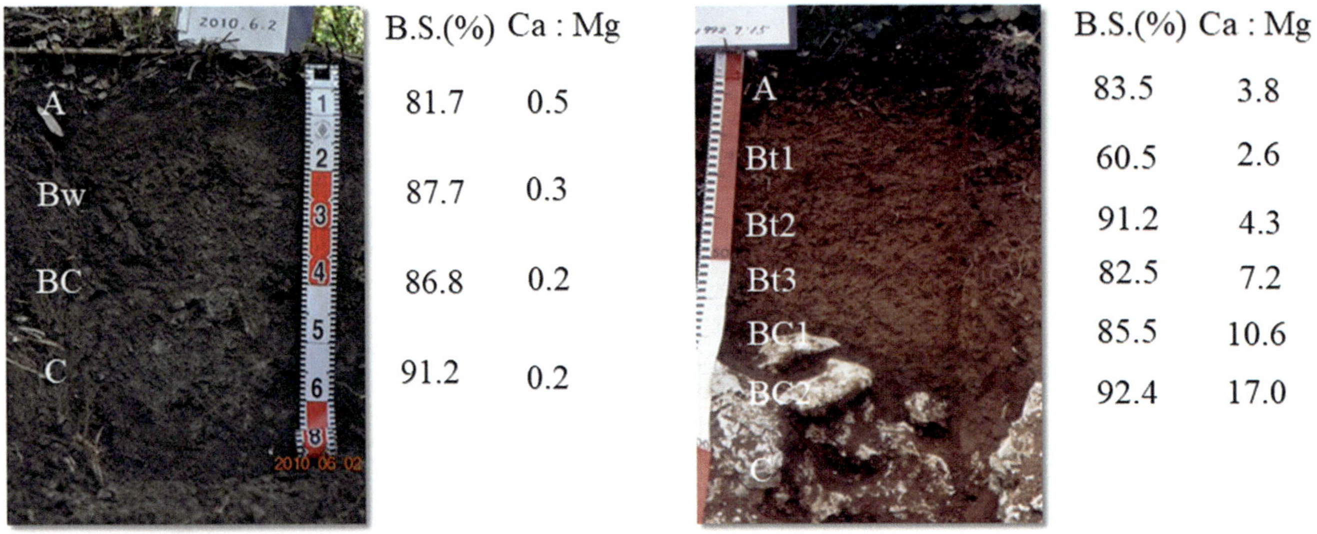

Fig. 4.41 Magnesian Eutrosols (*left*) and Calcareous Eutrosols (*right*). B.S.: base saturation, Ca:Mg: ratio of exchangeable Ca to Mg. *Data source* Maejima et al. (2014), Nagatsuka and Maejima (2001)

The Fifth Committee for Soil Classification and Nomenclature of the Japanese Society of Pedology examined the criteria of soil pH of Eutrosols and related soils in Japan, and the relationship between the pH (H_2O) and base saturation of the subsurface soils on limestone, serpentine, and basalt. The committee found that a soil pH of 5.5 or more was, for the most part, equivalent to a base saturation of 50% or more (Fig. 4.42). Therefore, the committee adopted the criterion of a soil pH of 5.5 or more for Eutrosols.

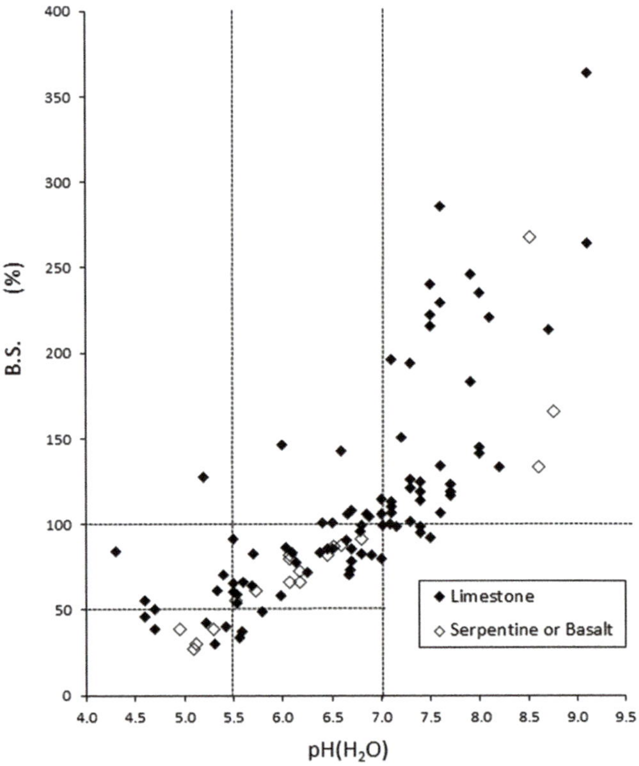

Fig. 4.42 Relationship between pH (H_2O) and base saturation of soils on limestone, serpentine, basalt. *Data source* Maejima et al. (2014), Nagatsuka and Maejima (2001)

4.8.3 Correlation with International Classification Systems

Eutrosols do not have one-to-one correspondence to soil definitions in international soil classification systems. The relationship between the Soil Classification System of Japan and international classifications differs at higher level with and without clay accumulation, such as for Luvisols or Cambisols (IUSS Working Group WRB 2014) and Udalfs or Udepts (Soil Survey Staff 2014). Correlation with international classification systems is summarized in Table 4.6.

4.8.4 Distribution

The distribution area of Eutrosols is extremely small (about 0.1% of Japan's land area) because the limestone and ultrabasic rocks (such as serpentinite and peridotite) which are the parent materials of these soils are locally distributed. In Japan, due to its temperate and humid climate, leaching of bases usually occurs, and soil reactions are acidic or weakly acidic in many cases. Therefore, the Eutrosols with a neutral or weakly alkaline pH are unusual.

Table 4.6 Correlation of soil classification system for Eutrosols

Soil classification system of Japan (2017)	WRB (2014)	Soil Taxonomy (2014)
Eutrosols	Luvisols, Cambisols	Udalfs, Udepts
Magnesian	Luvisols, Cambisols	Udalfs, Udepts
Argic	Leptic Luvisols (Chromic) Leptic Luvisols (Rhodic) Haplic Luvisols (Chromic) Haplic Luvisols (Rhodic)	Typic Paleudalfs Typic Rhodualfs Typic Hapludalfs
Haplic	Leptic Cambisols (Eutric) Haplic Cambisols (Eutric)	Lithic Eutrudepts Typic Eutrudepts
Calcareous	Luvisols, Cambisols	Udalfs, Udepts
Argic	Leptic Luvisols Haplic Luvisols	Typic Paleudalfs Typic Rhodualfs Typic Hapludalfs
Haplic	Leptic Cambisols (Eutric) Haplic Cambisols (Eutric)	Lithic Eutrudepts Typic Eutrudepts

Eutrosols exhibit various colors, such as chocolate, yellowish-brown, reddish-brown, and dark brown. Magnesian Eutrosols are generally distributed on serpentine and Calcareous Eutrosols are mainly distributed on limestone. Calcareous Eutrosols are one of the major soils on the Nansei Islands, which lie in a subtropical area where coral reef terraces are well-developed. It has been suggested that the Calcareous Eutrosols in this area formed from the residue of limestone (Nagatsuka et al. 1983). However, Inoue et al. (1993) estimated that the parent material of this soil was mainly eolian dust which was transported from continental China including East China Sea sediments in dry periods of glacial age. However, there is still no clear conclusion as to the parent material of Calcareous Eutrosols, and the elucidation of the potential contribution of eolian dust as a parent material is a subject to be studied in the future.

4.8.5 Land Use

In the case of Magnesian Eutrosols, due to the imbalance of base cations and the influence of heavy metals (nickel, cobalt, manganese), the growth of trees is usually not good, and unique vegetation with characteristic species of dwarf forests is observed in these soils. However, in some areas, these mineral-rich soils are suitable for agricultural use as orchards (mandarin oranges) and paddy fields.

On the other hand, the Calcareous Eutrosols occurring on limestone in the Nansei Islands are mainly used for sugar cane cultivation. The subsurface soil has a high clay content, is densic, and has low permeability. As a result, water percolates across the slope and flows out of sugar cane fields through cracks in the limestone bedrock, which makes the soils susceptible to drought.

4.8.6 Future Challenges

Volcanogenous Dark Red soil formed by hydrothermal metamorphism was classified as Dark Red soil in past soil classification systems (Forest Soil Division 1976; Cultivated Soil Classification Committee 1995). According to the Soil Classification System of Japan, Volcanogenous Dark Red soil is classified not into the Eutrosol soil great group, but rather into the Red-Yellow or Brown Forest soil great group. The formation process of this soil is different from that of Red-Yellow soils or Brown Forest soils, and therefore, we need to further consider how to classify acidic dark-reddish soils such as Volcanogenous Dark Red soil in the new classification system. Further data collection is needed to classify these soils.

4.9 Brown Forest Soils

In mountainous, humid parts of Japan, Brown Forest soils, which are relatively young soils, are widely distributed. In the Japanese archipelago, which stretches for over 3000 km in a north-to-southwardly direction and experiences climatic zones ranging from subarctic to subtropical, Brown Forest soils have a wide range of soil properties resulting from various soil-forming factors of other soil great groups, such as podzolization, including not only those having a cambic horizon but also those having an argic horizon. The difference between Brown Forest soils and Regosols is that they have a cambic horizon or argic horizon of sufficient thickness. Most Brown Forest soils are acidic or weakly acidic due to the leaching of base elements under a warm and humid climate condition. Most forest lands in Japan are covered by Brown Forest soils.

4.9.1 Genesis and Characteristics

Brown Forest soils are zonal soils under temperate to warm temperate humid climatic conditions in Japan. Based on the Classification of Forest Soils in Japan (Forest Soil Division 1976), the major forest soils in Japan were classified as "*brown forest soils*" and were distributed from Hokkaido to Kyushu ("Forest Soils of Japan" Editorial Committee 1983). These "*brown forest soils*" referred to soils distributed not only under forest vegetation but also under other land use types such as grassland and cropland; nevertheless, these soils were called "forest soils," because vegetation usually develops into forest under warm and humid environments in Japan, and so it represented brown-colored soils that develop under such circumstances. Under such forests, nutrient elements are taken up by the tree roots, which are widely and deeply developed in the soil, supplied to the ground surface as leaf litter, and returned to the soil by decomposition by small animals and microorganisms. As a result, a layer of fresh leaf litter is formed on the soil surface overlying the dark-colored "A horizon" with well-mixed humus. Underlying the A horizon is brown-colored Bw horizons without penetration of humus: The name of Brown Forest soils comes from the color of these Bw horizons. This brown color results from the color of the iron oxides that are present in large amounts in this type of soil. Furthermore, in the lower portion of the soil a C horizon is present, which is composed of physically weathered rocks and lacks the rock structure and color. As a result, Brown Forest soils have (O)-A-Bw-C horizons and show no eluviation or accumulation (Fig. 4.43).

Fig. 4.43 Soil profile of Thapto-red-yellow Brown Forest soil (Typic Dystrudept) under warm temperate cypress plantation in Sefuri Mountains, Japan. (Figure supplied by Akihiro Imaya)

In the Bw horizons, the chemical weathering of the parent rocks is advanced and free iron derived from rock minerals forms various iron oxides depending on the environment. Each variety of iron oxide shows a different color. In many cases, a brown color is due to the presence of goethite or maghemite, a yellow-brown color is due to goethite, an orange color is due to lepidocrocite, and a bright red color is due to hematite. In temperate zones, initially free iron ions precipitate to form brown-colored amorphous iron hydroxide; this is followed by a gradual progress of crystallization due to repeated wetting and drying to form goethite with a brown or reddish-brown color. In contrast, in warm temperate zones, iron hydroxide bonded to clay is crystallized due to partial dehydration and is strongly reddish in color. As described above, the kinds and proportions of iron oxides produced vary depending on climatic conditions, and thus the colors of the soils in which they occur are different even when all were classified into the same "*brown forest soils*". In other words, variations in the soil color of "*brown forest soils*" reflected differences in environmental and soil characteristics. Utilizing this feature, the difference between "*brown forest soils*" and "*red-yellow soils,*" which were distributed in temperate to subtropical climate zones, could be identified by the activity and crystallization index of free iron oxides, which indicates the form of free iron (Nagatsuka 1973).

The distribution of free iron in longitudinal section also differs among soil great groups. In the previous classification "*podzols,*" free irons are leached from the overlying layer due to dissolution by humic acid released from a thick organic layer and accumulate in the underlying layer. In "*red-yellow soils,*" the amount of free iron was found to be slightly larger in the underlying A horizon with a high crystalline ratio. In contrast, the "*brown forest soils*" had an almost constant free iron content throughout the soil profile (Nagatsuka 1993).

In most Brown Forest soils under the humid climate in Japan, some of the bases in the soil are leached. Therefore, acidic soils with a low base saturation are common. However, even under the environment where the Brown Forest soils were generated, Eutrosols (Soil Classification System of Japan 2017) which show 50% or more base saturation are derived from base-rich rocks such as limestone, calcareous sediments, peridotite, serpentine, andesite, agglomerate, and tuff.

The "*brown forest soils*" were derived from various parent materials, unlike Andosols, which are derived from a specific parent material such as volcanic ash. The Classification of Forest Soils in Japan (1976) emphasized classification based on soil genesis rather than classification based on parent materials (Kurotori and Ohmasa 1963; Mashimo 1977). For this reason, differences in soil chemical characteristics due to soil parent materials had not been systematically clarified. However, forest soils in Japan are generally young due to the country's climate and topographic conditions; as such, the influence of the parent material on soil chemical characteristics could not be ignored, and the necessity to set the soil class based on parent material was advocated (Endo 1969; Matsui 1977; Kato 1979; Wada 1986b). Differences based on the parent materials of "*brown forest soils*" were later somewhat clarified by Imaya et al. (2005).

Brown-colored soils which developed under temperate forest and which were derived from volcanic ash are classified as Andosols, not Brown Forest soils. Typically, Andosols have a black surface horizon; however, those which develop under forest vegetation have a dark brown surface horizon. Andosols with black horizons are developed under herbaceous vegetation, such as *Miscanthus sinensis,* contain large amounts of type A humic acid, and are strongly blackish in color due to advanced humification. Conversely, brown-colored Andosols are developed under deciduous broad-leaved forest, such as beech (*Fagus crenata*), and contain brown-colored humic acids of type B and type P, which do not lead to advanced humification (Kawamuro and Torii 1986). Andosols without a black-colored surface horizon were classified as "*brown forest soils*" in the Classification of Forest Soils in Japan (1976). In the Soil Classification System of Japan (The Fifth Committee for Soil Classification and Nomenclature 2017), the brown-colored Andosols as distinguished from black-colored Andosols and classified into the Brown-humic subgroup of the non-allophanic Andosols and Allophanic Andosols (Fig. 4.44).

In volcanic and mountainous areas of Japan, volcanic ash is mixed with weathered bedrock to form soil parent materials. This means that volcanic ash is a parent material not only of Andosols, but is also of Brown Forest soils. In the Soil Classification System of Japan (2017), Andosols and Brown Forest soils are divided according to: (1) the amounts of tephric material, which consists either of tephra, that is, unconsolidated, unweathered or only slightly weathered pyroclastic products of volcanic eruptions (including ash, cinders, lapilli, pumice, pumice-like vesicular pyroclastics, blocks, and volcanic bombs), or of tephric deposits, that is, tephra that has been reworked and mixed with material from other sources; and (2) the phosphate adsorption coefficient, degree of phosphate retention, or amounts of short-range-order Al and Fe oxides ($Al_o + 1/2Fe_o$). The values of Brown Forest soils vary depending on the degree of volcanic ash addition (Imaya et al. 2007; Fig. 4.45). Some Brown Forest soils which are especially influenced by volcanic ash are classified into the Andic subgroup (Fig. 4.46); these soils are classified using same diagnostic indicators as those of Andosols and Brown Forest soils (Fig. 4.47).

Fig. 4.44 Soil profile of Brown-humic Allophanic Andosol (Typic Hapludand) under deciduous broad-leave forest in Kyushu Mountains, Kumamoto, Japan *Photograph* A. Imaya

4.9.2 Classification

(1) **Great group of Brown Forest soils**

Brown Forest soils are listed ninth out of 10 soil great groups in the Soil Classification System of Japan (The Fifth Committee for Soil Classification and Nomenclature 2017). The main properties of these soils are a lack of andic and red-yellow soil properties, the presence of a cambic horizon or argic horizon, and no bedrock present shallower than 30 cm from the mineral soil surface. Brown Forest soils include soils which have a cambic horizon or argic horizon with red-yellow soil properties in the lower portion of the soil profile. These soils are distributed widely on hilly and mountainous landscapes throughout Japan and on the diluvial uplands in the Hokkaido and Tohoku regions. In hilly and mountainous landscapes, Brown Forest soils are derived from consolidated igneous rocks and partially consolidated or consolidated sedimentary and/or metamorphic rocks, and in the diluvial uplands are derived from unconsolidated sediments. In general, these soils lack rock structure and have somewhat high clay percentage, soil structure formation, and low base saturation.

Brown Forest soils are associated with other soil great groups. Soils distributed around the boundary between

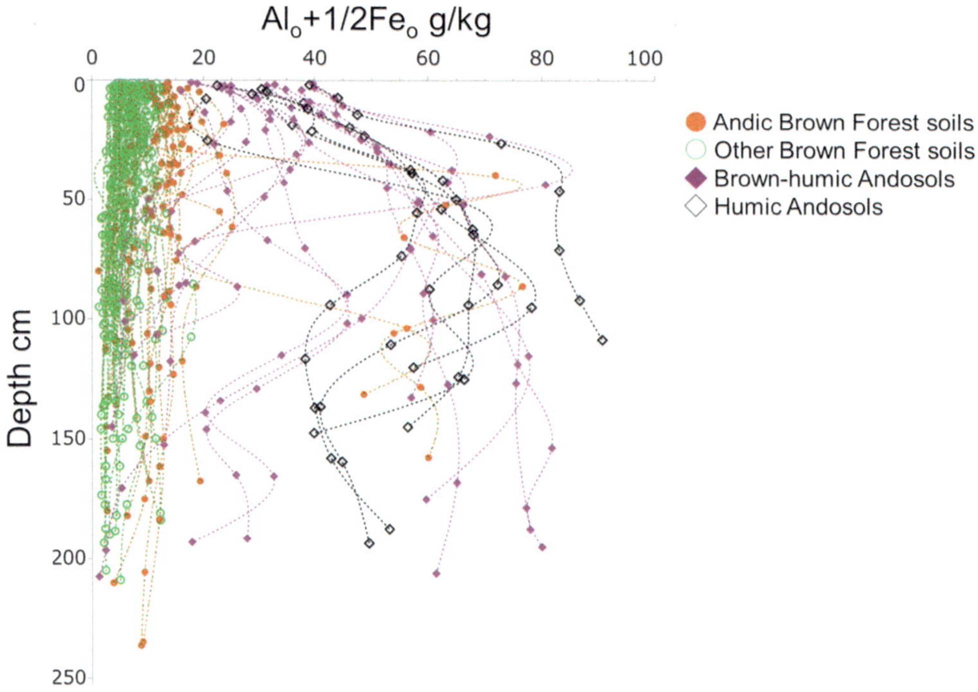

Fig. 4.45 Vertical distribution of SRO Al and Fe compounds in Brown Forest soil and Andosol profiles. *Data source* Imaya et al. (2007)

Fig. 4.46 Soil profile of Andic Brown Forest soil (Andic Dystrudept) under deciduous evergreen forest in Sefuri Mountains, Saga, Japan *Photograph* A. Imaya

Brown Forest soils and another soil great group have both or alternating characteristics of the soil great groups. Such soils are classified at the subgroup level of Brown Forest soils.

(2) **Soil group**

In the Soil Classification System of Japan (2017), the Brown Forest soil great group only consists of the Brown Forest soil group. Because the Japanese archipelago has a wide range of climatic conditions, the Brown Forest soil great group is considered to have various characteristics corresponding to climatic differences. Soils under warm temperate forest were regarded as "*warm temperate forest soils*" (Endo 1966) or "*yellow-brown forest soils*" (Nagatsuka 1975), which were different from the "*brown forest soils*" under cool temperate forest. The Unified Soil Classification System of Japan (first approximation) (Committee for Soil Classification and Nomenclature 1986) followed these ideas and classified soils developed under warm temperate forest as "*yellow-brown forest soils*" and soils developed under cool temperate forest as "*brown forest soils.*" A subsequent classification system (The Fourth Committee for Soil Classification and Nomenclature the Japanese Society of Pedology 2003) also keyed out "yellow-*brown forest soils*" from "*brown forest soils*" using soil color and organic carbon content as diagnostic criteria. However, these classifications did not consider the influence of volcanic ash on "*brown forest soils.*" Considering the influence of volcanic ash and comparing only soils with little or no influence from volcanic ash, the colors of subsurface

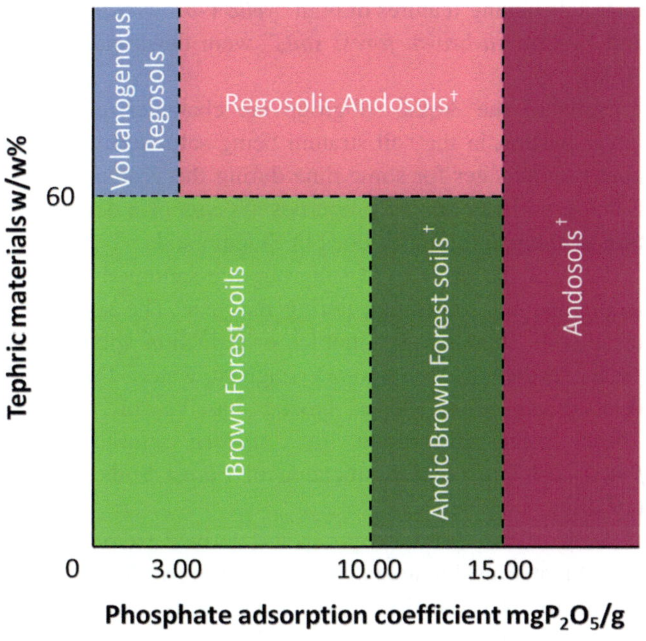

Fig. 4.47 Diagnostic criteria between Andosols, Regosols, and Brown Forest soils. (Figure supplied by Akihiro Imaya)

horizons of "*brown forest soils*" are in fact yellowish-brown, red, and yellow, rather than brown. As a result, the actual identification of "*brown forest soils*" was based on the organic carbon content, not the soil color. Nevertheless, the SOC contents of such soils were found to fluctuate depending on the extent of volcanic ash addition, so organic carbon contents may not be an appropriate indicator to divide "*yellow-brown forest soils*" and "*haplic brown forest soils.*" It has also been theorized that both of these soil groups could be classified according to the activity of free iron oxides (Nagatsuka 1973); however, the activity of free iron oxides was also found to fluctuate depending on the extent of volcanic ash addition. Therefore, it was uncertain whether the difference in the activity of free iron oxides between "*yellow-brown forest soils*" and "*haplic brown forest soils*" was due to climatic conditions or degree of volcanic ash addition. Because it was virtually impossible to distinguish the "*yellow-brown forest soils*" as a zonal soil, the Soil Classification System of Japan (2017) has only one soil group of Brown Forest soils in the Brown Forest soil great group.

(3) **Soil subgroup**

The soil group of Brown Forest soils has nine soil subgroups in the Soil Classification System of Japan (2017), namely the Anthraquic subgroup, having an anthraquic horizon; the Andic subgroup, having a layer that is ≥ 25 cm thick and starts ≤ 50 cm from the mineral soil surface, which has a phosphate adsorption coefficient ≥ 10.00 mgP_2O_5/g, a phosphate retention of $\geq 60\%$, or $\geq 1.2\%$ of $Al_o + 1/2Fe_o$; the Humic subgroup, having a humic epipedon; the Podzolic subgroup, having a spodic horizon (morphological) or spodic horizon (chemical); the Thapto-red-yellow subgroup, having a layer starting ≤ 75 cm from the mineral soil surface that has a cambic or argic horizon with red-yellow properties; the Aquic subgroup, having a layer with groundwater-aquic properties starting ≤ 75 cm from the mineral soil surface; the Epi-gleyed subgroup, having a layer with epiaquic properties starting ≤ 75 cm from the mineral soil surface; the Eutric subgroup, having a part of subsurface horizon with a $pH(H_2O) \geq 6.5$ or a base saturation of 50%; and the Haplic subgroup, which comprise all other Brown Forest soils.

The Anthraquic subgroup is developed in the rice paddy cultivation area in mountainous areas.

The Andic subgroup has a stronger influence of volcanic ash than other Brown Forest soils. (However, even Brown Forest soils that are not classified in the Andic subgroup are often affected by volcanic ash (Imaya et al. 2010b) to some extent because the Japanese archipelago is widely covered with volcanic ash due to its many volcanoes; accordingly, the influence of volcanic ash deposits cannot be ignored.)

The Podzolic subgroup can be associated with Podzols. These soils have weak podzolization under cool and humid climatic conditions. The Podzolic subgroup of Brown Forest soils is characterized by Al translocation by organic matter, resulting in the accumulation of active Al for phosphate absorption in the upper part of the subsurface horizon, and Fe-organic matter complexation in the surface horizons, leading to the increase of amorphous Fe. Andic-like properties in Brown Forest soils were considered to result from these two processes and not from the influence of volcanic ash (Hirai et al. 1990). Clay destruction in the Albic E horizon, which occurs in Podzols, is not seen in Podzolic Brown Forest soils. In the Classification of Forest Soils in Japan (1976), this Podzolic subgroup corresponds to dry slightly podzolic soils, wet iron slightly podzolic soils, and wet humus slightly podzolic soils. Dry slightly podzolic soils have a well-developed Oe horizon and an iron-rust-colored Bs horizon, and grayish-white eluvial spots are not recognized with the naked eye. Wet iron slightly podzolic soils have a thickly developed O horizon, in particular an Oa horizon and humus A horizon or Oa-A horizon, an E horizon with an unclear grayish-white portion, and a Bs horizon with rusty mottles. These soils are generally massive and often have longitudinal fissures contaminated by humus. Wet humus slightly podzolic soils have a black glossy Oa horizon, a dark-gray E horizon with unclear eluvial portion, a dark iron-rust-colored Bs horizon with rusty mottles, and the whole soil contains much humus and has a very dark color.

The Humic subgroup is characterized by the accumulation of soil organic matter in the deep horizon because organic matter slowly decomposes under cool climatic conditions. This subgroup includes soils that are classified as "*dark brown forest soils*" in the Classification of Forest Soils in Japan (1976). "*Dark brown forest soils*" were characterized as follows: "Blackish-brown, glossy Oa horizon or Oa-A horizon is recognized, the A horizon is blackish-brown, and the Bw horizon is dark brown (brightness and chroma are close to 3). An aggregate structure develops in the upper part of the A horizon or Oa-A horizon, and a massive structure is often seen in the Bw horizon and the lower part of the A horizon. The phenomenon of podzolization or gleization is cannot be recognized by the naked eye" (Forest Soil Division 1976).

The Thapto-red-yellow subgroup has thinner or deeper cambic or argic horizons with red-yellow properties compared to Red-Yellow soils. This soil subgroup corresponds to a part of "*reddish-brown forest soils*" and "*yellowish-brown forest soils*" in the Classification of Forest Soils in Japan (1976) and a part of "*yellow-brown forest soils*" in the Unified Soil Classification System of Japan (2003). Compared to the "*brown forest soil*" subgroup, the "*reddish-brown forest soils*" have an A horizon that is normally light-colored and thin, while the color of their Bw and C horizons are strongly reddish in color (the color of the Bw horizon is less reddish than almost 5YR 5/6 and more reddish than 7.5YR 5/8). These "*brown forest soils*" were formed from strongly reddish parent materials as a result of red-color weathering, and they are strongly acidic. Compared with the "*brown forest soil*" subgroup, the "*yellowish-brown forest soils*" had an A horizon that was normally light-colored and thin, and the color of their Bw and C horizons was strongly yellowish in color (the color of the Bw horizon is less yellowish than almost 10YR 6/6 and more yellowish than 7.5YR 6/8). These "*brown forest soils*" were formed from strongly yellowish parent materials as a result of yellow-color weathering, and they are strongly acidic. The soil group of "*yellow-brown forest soils*" was considered to be associated with a part of the yellowish and typical subgroups of "*brown forest soils*" in the Classification of Forest Soils in Japan (1976). The "*yellow-brown forest soils*" were distributed under a warm temperate zone, whereas the "*yellowish-brown forest soils*" were derived from paleo soil materials with yellow-color weathering and are also distributed under a cool temperate zone (Yambe and Yagi 1983). The "*yellow-brown forest soils*" and "*yellowish-brown forest soils*" were compared because the area with past yellow-color weathering greatly overlapped with the current warm temperate zone and the soils with yellow brunification derived from yellow-color weathering materials having features of both "*yellow-brown forest soils*" and "*yellowish-brown forest soils*" were developed (Imaya 2008).

Soils of the Aquic subgroup are characterized by the lower portion in the soil stratum being saturated with stagnant groundwater for some time during the year. A Stagnic horizon with endoaquic properties appears in a deeper portion in the solum than it does in Stagnic soils.

Soils of the Epi-gleyed subgroup are saturated with surface water for some time during the year. The surfaces of these soils contain rusty mottles or reduction spots caused by the presence of temporary stagnant water. These soils develop on upland slopes, having massive soil materials. These soils correspond to "*surface gleyed brown forest soils*" in the Classification of Forest Soils in Japan (1976).

Soils of the Eutric subgroup are derived from base-rich parent materials, but are not so much strongly affected by the parent materials as Eutrosols. These soils have a larger cation exchange capacity, higher pH, and higher base saturation than typical Brown Forest soils that are acidic or weakly acidic.

Since soil color has been a significant indicator in the previous soil classification systems in Japan, the dark red-colored acidic soils such as volcanic dark red soils derived from parent materials with thermal metamorphism were classified into the great group of "*dark red soils*"; however, most of these soils are now classified as Eutrosols having dark red color and derived from basic rocks such as limestone and serpentine. In the Soil Classification System of Japan, volcanic "*dark red soils*" are classified as Brown Forest soils or Red-Yellow soils, because the classification is based on soil physical and chemical properties rather than soil color and parent materials.

Brown Forest soils include soils having an argic horizon, although the majority of Brown Forest soils have a cambic horizon and may correspond to Cambisols. In order to classify soils having an argic horizon as an independent soil subgroup from ones having a cambic horizon, the information of these soils' characteristics is insufficient.

(4) **Further classification for practical use**

Brown Forest soils show various morphological features, such as degree of development of soil horizons and kinds of soil structure, resulting from various hydrological environments according to mountain slope topography (Ohmasa 1951). Hydrological conditions continuously change due to topographic position on the mountain slopes, where most of the forest lands in Japan are located. Additionally, their vegetation and tree growth performance also change relatively. Based on these facts, six soil types, from dry to wet, were established in the Classification of Forest Soils in Japan

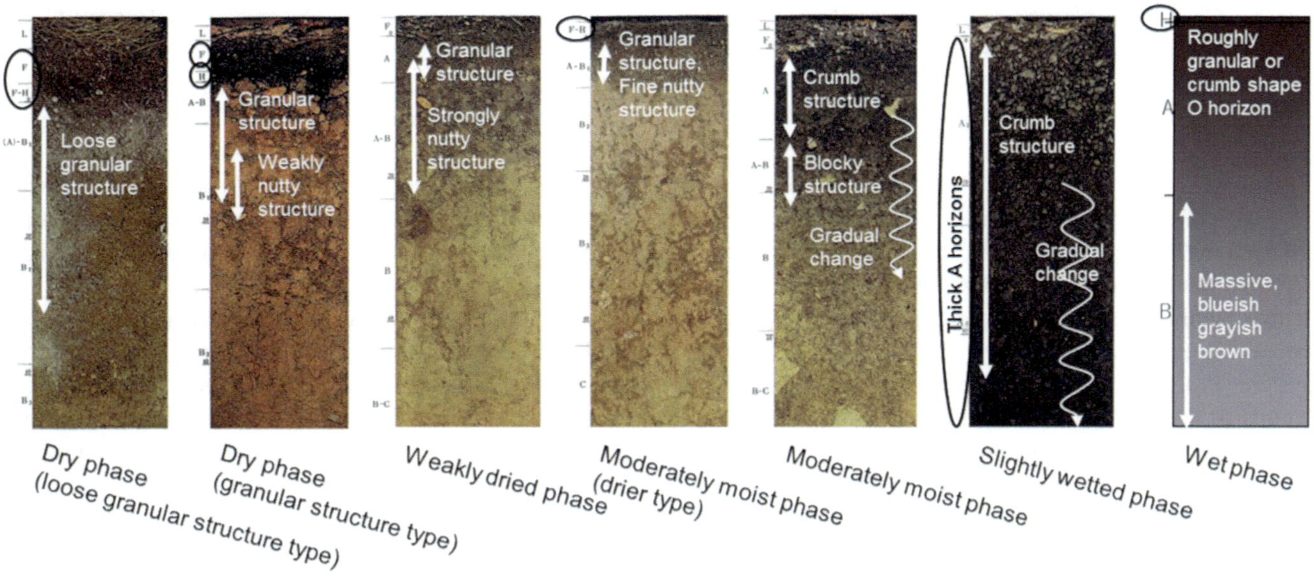

Fig. 4.48 Morphological characteristics of soil profiles of each soil phase. (Figure supplied by Akihiro Imaya)

(1976) and used for subdivisions of each soil subgroup. Soil type classification was used as an effective indicator for site evaluation for forest management aimed at timber production (Hashimoto 1969). Additionally, soil chemical properties showed differences depending on the soil types on a slope (Ohmasa 1951; Mashimo 1957; Kawada and Takami 1957). This idea of soil type classification is adopted as the forest soil phase in the Soil Classification System of Japan (2017). Brown Forest soils can use the following six forest soil phases (Fig. 4.48):

Dry phase (loose granular structure type): The O horizon is not thick. Oe or Oe/Oa horizons always develop, but the development of the Oa horizon is not significant. A black A horizon is generally thin, and the boundary with the B horizon is very apparent. A loose granular structure develops to a considerable depth in the A horizon and B horizon. This soil has rich mycelial strands and, in extreme cases, forms a mycelial net horizon. The color of the B horizon is generally light.

Dry phase (granular structure type): In this soil, a thick Oe horizon and Oa horizon develop, and a thin black A horizon or OA horizon is formed. A granular structure develops in the A horizon, and the boundary between the A horizon and B horizon is apparent. The color of the B horizon is normally bright. A granular or nutty structure develops above the B horizon, and a loose granular or fine nutty structure is often seen in the lower portion of the B horizon. A mycelial net horizon rich in mycelial strands is rarely formed.

Weakly dried phase: An Oe horizon and an Oa horizon are not well-developed. Humus penetrates relatively deeply; the humus is light in color and its profile is relatively hard and dense. A nutty structure frequently develops in the lower portion of the A horizon and in the upper portion of the B horizon. Mycelial strands are often recognized in the B horizon.

Moderately moist phase: This is the typical soil type. Oe and Oa horizons do not develop well. The A horizon is relatively thick and is rich in humus of a dark brown color. An aggregate structure develops in the upper portion of the A horizon, and a blocky structure is often seen in the lower portion. The B horizon is brown with a weak blocky structure but no other special structures are recognized. The transition from the A horizon to the B horizon is generally gradual.

Moderately moist phase (drier type): The soil profile shape is almost the same as that of the moderately moist phase, but these soils show somewhat drier characteristics, such as the occurrence of a granular structure in the upper portion of the A horizon or a nutty structure in the lower portion of the A horizon.

Slightly wetted phase: No O horizon develops. The A horizon is very rich in humus and very thick, with aggregate structures developed. The transition to the B horizon, which is brown to slightly dark gray in color, is gradual. No special structure is recognized in the B horizon.

Wet phase: A coarse granular or aggregate Oa horizon develops. The A horizon contains moderate levels of humus. The permeation of humus to the B horizon is low. The B horizon has massive soil materials, and is bluish-gray-brown in color. Rusty mottle is often recognized but a gley horizon is not recognized in the soil stratum within 1 m of the soil surface.

4.9.3 Correlation with International Classification Systems

Most Brown Forest soils with cambic horizons are classified as Cambisols in the WRB (IUSS Working Group WRB 2015) and as Inceptisols in the USDA Soil Taxonomy (Soil Survey Staff 2014). Brown Forest soils with argic horizons are classified as Acrisols, Alisols, Luvisols, or Lixisols in the WRB and as Ultisols in the USDA Soil Taxonomy. The soils of the Anthraquic subgroup are classified as Stagnosols in the WRB. As described above, in the Soil Classification System of Japan the Brown Forest soil great group corresponds to soils which are classified as different soil great groups in other soil classification systems.

In the Soil Classification System of Japan (2017), soils which correspond to Cambisols/Inceptisols include Cambic Red-Yellow soils and Brown Forest soils. Cambic Red-Yellow soils cannot be classified into soil orders which have argic horizons, such as Acrisols or Ultisols, because these soils do not have argic horizons in the control section. Brown Forest soils may have reddish cambic horizons while Cambic Red-Yellow soils have cambic horizons with red-yellow property, which are identified by soil color and SOC content. As is the case for Brown Forest soils, Red-Yellow soils contain an Argic group, which corresponds to Acrisols and Cambic group which corresponds to Cambisols in the WRB system. Soils having argic horizons and soils having cambic horizons coexist in the great groups of Brown Forest soils and Red-Yellow soils in the Japanese classification system, which strongly reflects the specificity of soil genesis in mountainous regions of Japan. On the slope, the movement of the clay fraction along the slope surface stronger than ones downward in the solum, even if the current climatic condition tends to cause clay illuviation. Therefore, both the overlying coarser-textured horizon and the underlying horizon with illuvial accumulation of clay are hard to develop in Japan. However, even for those soils without an argic horizon, weathering progressed. These soils have a reddish or yellowish color due to the presence of iron oxide consisting of goethite and hematite produced by the oxidation and dehydration of free iron by weathering. Moreover, these soils do not store significant amounts of soil organic matter and have a pale and thin A horizon as a result of an increase in the amount of low-activity clay minerals.

4.9.4 Distribution

(1) **Distribution of the Brown Forest soil group**

Brown Forest soils are distributed in a rather wide range of temperate and warm temperate zones in humid (high precipitation) climates. These are zonal soils which are formed between the zones of "*podzols*" and "*red-yellow soils*" (Forest Soils Division 1976). Brown Forest soils are mainly distributed in sloping land in hills and mountainous regions, between areas below the subalpine zone where Podzols are distributed and the alluvial fan, basin, and plain areas where Fluvic soils are distributed. However, on steeper slopes, Regosols are distributed more widely than Brown Forest soils.

Because Japan contains many volcanoes, the soils in most areas of Japan are affected by volcanic ash. The soils classified into the Andosol great group are distributed near volcanoes, and on flat and gently sloping land. In neighboring areas, Brown Forest soils are mixed with Andosols. From the eastern sides of volcanoes, the distribution ratio of Andosols increases, and the ratio of Brown Forest soils relatively decreases because the volcanic ash was blown by the westerlies and deposited in eastern sides from the volcano on the Japanese archipelago. The area classified into the Brown Forest soil group accounts for around 30% of the area in the Hokkaido, Tohoku, Tokai–Hokuriku, and Kyushu–Okinawa regions, and a little over 20% of the area in the Kanto region. In contrast, Brown Forest soils cover 47% of the area in the Kinki–Chugoku–Shikoku region (Table 2.1 in Sect. 2.5).

(2) **Distribution of each soil subgroup**

The soils of the Anthraquic subgroup are distributed in paddy fields in sloping lands in hilly and mountainous areas.

The soils of the Andic subgroup are distributed in marginal areas to the soils of the Andosol great group. In areas that do not contain Andosols, soils of the Andic subgroup are distributed in flat and gently sloping land affected by widespread tephra.

The soils of the Podzolic subgroup are distributed close to boundaries between Podzols and Brown Forest soils under a cold moist climate. The soils which correspond to dry slightly podzolic soils are distributed around locations which are susceptible to drying, such as crests, ridges, the upper parts of convex slopes, and the shoulders of plateaus. Such soils are derived from acidic parent materials, have a sandy texture, and are developed under forests of *hiba* (*Thujopsis dolabrata*) and Japanese umbrella pine (*Sciadopitys verticilata*) in subalpine and alpine zones. The soils which correspond to wet iron slightly podzolic soils are distributed around dull ridges, peneplains, and volcanic mudflow plateaus with heavy clayey and dense materials under natural forest vegetation (e.g., Sakhalin Spruce (*Picea glehnii*), *Abies mariesii*, Japanese white pine (*Pinus parviflora* var. parviflora), Japanese Thuja (*Thuja standishii*), Japanese cypress (*Chamaecyparis obtusa*), and Japanese Beech (*Fagus crenata*) from upper temperate to subalpine zones in north Hokkaido, backbone ranges on Honshu, and the Japanese Alps. Soils which correspond to wet humus slightly podzolic soils are distributed on gently sloping land under natural forest vegetation (e.g., Yezo spruce (*Picea jezoensis*), *Abies sachalinensis*, *Abies mariesii*, *Abies veitchii*, *Picea jezoensis* var. hondoensis, Japanese cypress (*Chamaecyparis obtusa*), Japanese thuja (*Thuja standishii*), *Betula ermanii*, and Japanese beech (*Fagus crenata*) from upper temperate to subalpine zones.

The soils of the Humic subgroup are distributed in the upper part of the zonal area of Brown Forest soils close to the zone of Podzols. These areas of Humic subgroup soils have a cold and humid climatic condition, which inhibits the decomposition of litter and soil organic matter. Consequently, these soils have Oa horizons and store large amounts of soil organic matter. A small amount of divalent iron is detected in the Oa and A horizons. This characteristic is similar to that of the wet humus podzolic soils in the Classification of Forest Soils in Japan (1976). Humic Brown Forest soils are widely distributed in mountainous areas on Hokkaido, the mountainous backbone of Honshu, and the Japanese Alps.

The soils of the Thapto-red-yellow subgroup are thought to correspond to a part of "*reddish-brown forest soils*" and "*yellowish-brown forest soils*" in the Classification of Forest Soils in Japan (1976) and to "*yellow-brown forest soils*" in the Unified Soil Classification System of Japan (second approximation) (The Fourth Committee for Soil Classification and Nomenclature, the Japanese Society of Pedology 2003). The distribution areas of "*reddish-brown forest soils*" overlapped with those of "*red soils.*" These soils are distributed mainly in the Okinawa Islands and are regularly interspersed in low altitude areas due to the topography throughout Japan (Kidachi and Ohmasa 1963). "*Yellowish-brown forest soils*" were distributed with the reddish subgroup on hilly landscapes in the western part of central Honshu. However, most "*yellow-brown forest soils*" are distributed under warm temperate forest in Southwest Japan.

The soils of the Aquic subgroup are distributed in areas with high groundwater level, such as around lakes and marshes, plateaus, flat lands, and the bottoms of slopes.

The soils of the Epi-gleyed subgroup are developed on plateaus on the Sea of Japan and Sea of Okhotsk coasts in Hokkaido and on the Sea of Japan coast in the Tohoku and Hokuriku regions.

The soils of the Eutric subgroup are distributed in close proximity to soils of the Eutrosol great group. These soils are derived from limestone, calcaric sediments, ultrabasic rocks such as serpentine, and their alternative sedimentary rocks such as sandstone.

The soils of the Haplic subgroup are generally distributed in sloping lands under temperate and warm temperate forests, unlike the above subgroups.

4.9.5 Land Use

In Japan, about 70% of the land area is covered by forest. Brown Forest soils have been used for tree plantation for timber production; Japanese cedar (*Cryptomeria japonica*), Japanese cypress (*Chamaecyparis obtusa*), and pine (*Pinus densiflora*) are the major planted tree species. Japanese cedar prefers relatively moist conditions such as those found in lower slopes and basins. In contrast, Japanese cypress prefers slightly dry conditions because this tree species is affected by Tokkuri disease, an abnormal overgrowth of tissues on the lower stem that occurs in moist conditions. In dry conditions, pine experiences less decline in growth than the Japanese cedar and Japanese cypress. In order to efficiently produce timber in this small country, it is necessary to select the optimal planting land for each tree species (Fig. 4.49). The Classification of Forest Soils in Japan (1976) classified soil types according to the hydrological environment of the sloping topography of mountain areas, which allowed the right tree to be planted in the right land type (Mashimo 1960; Fig. 4.50). The Soil Classification System of Japan (2017) incorporates this soil classification type as a forest soil phase. The soil distribution differs depending on the climatic and geological conditions, even on mountainous slopes under similar topography. For example, in hilly and sub-mountainous areas in Southwest Japan, dry soils occupy a large proportion of slopes. As a result, in these areas, the area which is suitable for the planting of Japanese cedar on sloping land is less than in northern parts of Japan and is limited to valleys.

Fig. 4.49 Japanese cedar and cypress plantation on the Brown Forest soil in Sefuri Mountains, Saga, Japan. (Figure supplied by Akihiro Imaya)

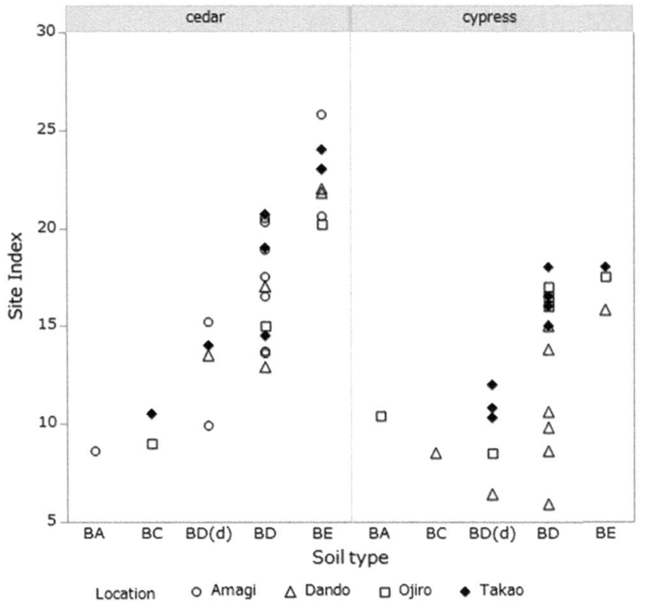

Fig. 4.50 Differences of tree growth due to each soil type in Classification of Forest Soils in Japan. Modified from Mashimo (1960), Copyright 1960, with permission from Forestry and Forest Products Research Institute

4.10 Regosols

4.10.1 Volcanogenous Regosols

Japan has one of the highest concentrations of active volcanoes in the world. Its 111 active volcanoes, which cover less than 0.3% of the global subaerial land area, account for 10% of the world's currently active volcanoes. The ejecta from volcanoes are important parent materials for regenerating soils in Japan. In recent years, several Japanese volcanoes have erupted, including Mt. Usu (2000), Miyakejima (2000), Shinmoedake (2011 and 2018), and Mt. Ontake (2014). Initial soil formation (early stage of pedogenesis) has begun in sites of recent ejecta deposition with vegetation recovery. A very weakly developed soil pedon observed in some ejecta deposits means that such deposits can be classified as Volcanogenous Regosols. The formation of active Al and Fe compounds in tandem with the cumulation of organic matter in volcanic ash soils is termed "andosolization" (Duchaufouur 1977). Soils having a developed A horizon (with regosolic andic soil properties) from andosolization are classified as Regosolic Andosols. In this section, the initial soil-forming process of Volcanogenous Regosols is reported.

In general, observations of the initial soil formation process are very difficult to obtain in the field. However, such observations are possible to obtain in Miyakejima (Miyake Island), a volcanic island. This island has erupted several times in the last 100 years, making it an ideal experimental field location for documenting soil formation from volcanic ejecta over time (Kato et al. 2005).

(1) Early-stage pedogenesis resulting from scoria

Miyakejima is a volcanic island on the Philippine Plate located approximately 180 km southwest of central Tokyo. The island is characterized by a humid warm temperate climate. The mean annual air temperature is 17.7 °C, and the mean annual precipitation is 2954 mm as of 2018. Miyakejima volcano erupted in 1874, 1940, 1962, 1983, and 2000. Activity prior to the 2000 eruption displayed an eruption style characteristic of basaltic magmas. As seen at Kilauea Volcano, Hawaii, basaltic magma released into the atmosphere fragments and forms scoria as dissolved gas escapes from individual globules. The accumulation of scoria forms a unique geographical surface, shaped like an inverted bowl, that is known as a "scoria cone." At Miyakejima, scoria cones of different eruption ages exist in isolation from one another. Each of these cones has the capacity to act as a growth medium for seeds dispersed from surrounding forests. Although local seed dispersal occurs on a yearly basis, no plants have ever grown on the youngest (1983) scoria cone. In contrast, perennial plants (*Fallopia japonica* var., *Miscanthus condensatus*) have formed communities on the 1940 and 1962 scoria cones, and deciduous/evergreen broad-leaved mixed forests (*Alnus sieboldiana* Matsum., *Machilus thunbergi*) have formed on the oldest (1874) scoria cone. Thus, plant succession can be observed at Miyakejima. In essence, Miyakejima was an experimental field area prior to the 2000 eruption, where soil formation factors, topography, parent material, and climate were similar, and only plants varied over time.

On Miyakejima, soils developed on the eruption products include a variety of profile morphologies (Fig. 4.51), a well-developed A horizon over the C horizon (A/C type) and lack an A horizon type (only consisted from C horizons). The particle size of scoriaceous material in these soils decreased through the initial soil formation process during the 125-year period. The decrease in the particle size of scoriaceous material is a result of soil weathering. Additionally, the length of plant roots and the thickness of the A layer both increase with an increase in the number of roots. Transition of mixed forests has also taken place on Miyakejima. However, even for deposits of the 1874 scoria, the development of a distinct soil B horizon is not observed. As plant succession proceeds in older deposits, the amount of soil organic matter, CEC, and the amount of exchangeable bases increases (Fig. 4.51). However, the formation of secondary clay minerals is not observed in this site. The increase in CEC is largely associated with the amount of soil organic matter accumulation. The CEC and the amounts of

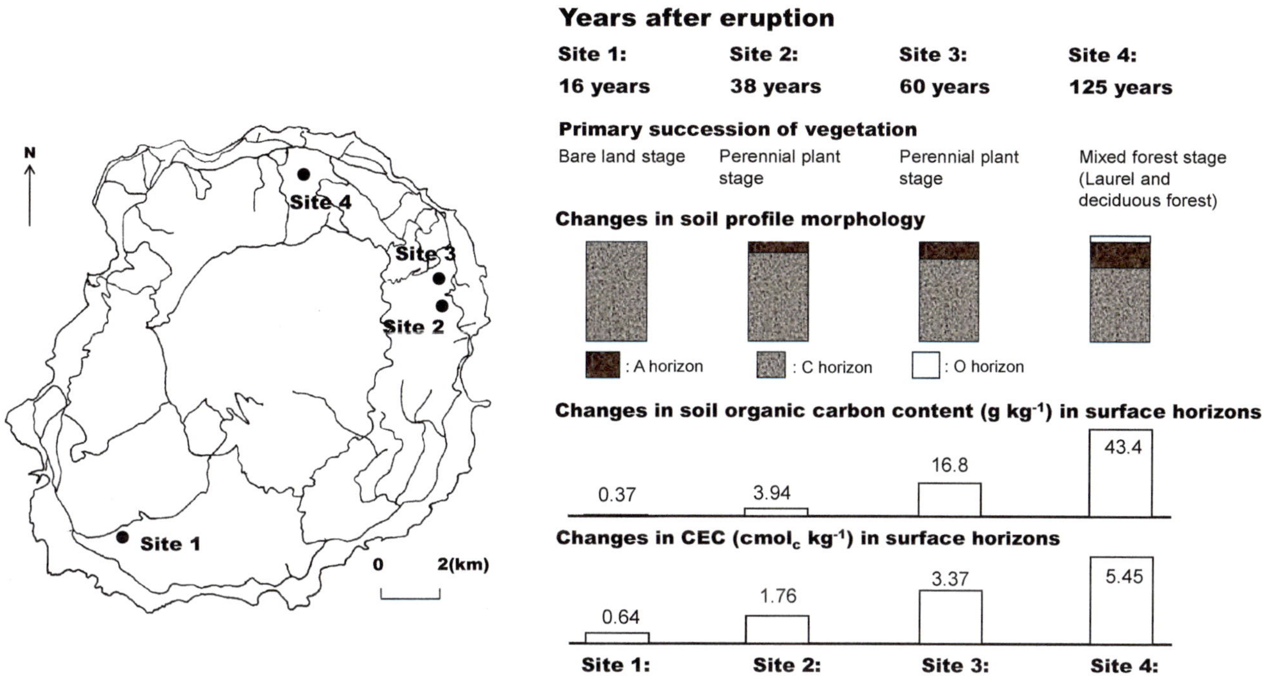

Fig. 4.51 Location of the soil pedons in Miyakejima, and the succession of vegetation, changes in soil profile morphology, and changes in physicochemical properties of soil surface during the 125 year period. Modified from Kato et al. (2005)

exchangeable bases increased at almost the same rate, resulting in a high degree of base saturation (100%) during the 125-year period of the initial soil formation processes of Volcanogenous Regosols. For 125 years after the 1874 eruption, the base saturation degree was always maintained at 100%. The increased amount of soil basic nutrients is thought to be limited by nutrient retention capacity. In young volcanic soils, the circumstances in which supply capacity exceeds nutrient retention capacity are maintained. If the soil nutrient amount reaches the prescribed value necessary for the establishment of a plant community "A" by the soil formation, the plant community "A" is established by the vegetation succession. Plant community "A" contributes to soil formation that shifts to the stage of the next soil nutrient state (the amount of soil nutrients necessary for the transition to plant community B). In the initial soil formation process, the nutrient state of the soil and the succession stage of the vegetation are closely related. In other words, in Miyakejima the initial soil formation process on the timescale of 125 years with scoria as a base material is closely related with vegetation transition.

(2) **Very early stage of pedogenesis derived from volcanic ash**

The eruption of Miyakejima in 2000 was characterized by the accumulation of several millimeters to several dozens of centimeters of volcanic ash over the entire island (Fig. 4.52). Variable ash deposition resulted from a magmatophreatic explosion at the summit, which was accompanied by the

Fig. 4.52 2000 eruption volcanic ash layer on the 1874 eruption sediments (Scoria). The gray layer to about 10 cm in depth is the volcanic ash layer which erupted in 2000. The Scoriaceous soil (Volcanogeneous Regosols) which erupted in 1874 is watched from 10 cm in depth. (Figure supplied by Taku Kato)

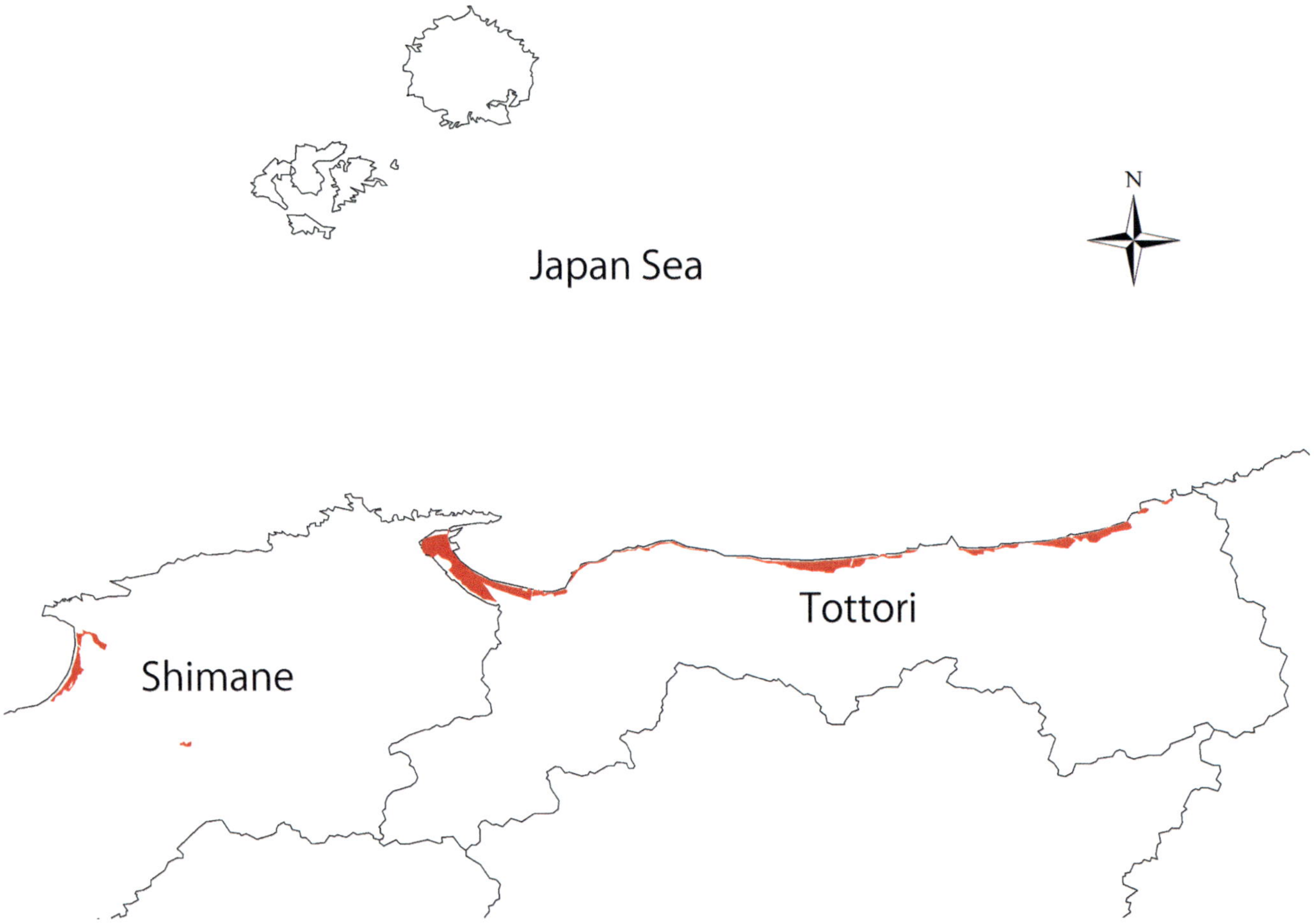

Fig. 4.53 Distribution of Sandy Regosols on the Japan Sea sides in Tottori and Shimane prefecture (Figure supplied by Hiroshi Obara)

progressive formation of a large caldera. In addition, volcanic degassing resulted in the emission of over 50,000 tons/day of sulfur dioxide (SO_2) at peak times. Although the volcanic ash retained a considerable amount of calcium sulfate ($CaSO_4$) shortly after the eruption, pH values sharply decreased from 4.0 to 3.1 due to the interaction with volcanic gas. This interaction resulted in the elution of large volumes of aluminum derived from primary minerals. The supply of organic materials from plant seedlings and litter was limited. Even under these harsh conditions, soil organic matter and the amount of microbial biomass increased year-on-year. The transfer and establishment of microbial communities appear to be more important in the very early stage of the pedogenic process than the establishment of subaerial plant communities (Fujimura et al. 2012). Fresh volcanic ejecta provides suitable conditions for the development of a microbial community and the formation of a new ecological system. Microbial groups with diverse metabolic abilities colonize soils before plants do (Guo et al. 2014). In particular, microbial species that are able to convert (fix) atmospheric carbon dioxide (CO_2) and nitrogen gas (N_2) into organic matter have a survival advantage. The first microbial species with these abilities which inhabit the soil are termed "pioneers" and contribute to soil evolution by supplying substances such as organic matter and ammonia that other species can directly use. Another condition required for the colonization of pioneer species in volcanic ash soil is the ability to produce energy without dependence on organic matter. Many heterotrophs obtain the energy necessary for their growth by decomposing organic matter. However, the energy sources available in volcanic ash soil in the early stages of pedogenesis, when limited organic matter is present, are sunlight and inorganic compounds. On Miyakejima, it has been observed that heterotrophic bacteria started to inhabit the soil approximately four years after the accumulation of volcanic ash, and that the proportions of chemolithoautotrophs and chemoorganotrophs were characteristically high compared with typical forest soils (Fujimura et al. 2012). Hydrogen-oxidizing bacteria and sulfur-oxidizing bacteria were also detected. However, iron-oxidizing bacteria that depend on bivalent iron (Fe^{2+}) as their energy source showed a significant increase in number. It has been suggested that a particularly predominant species of iron-oxidizing bacteria played the role of a pioneer, as it carried out nitrogen fixation (Sato et al. 2009). However, the population of these iron-oxidizing bacteria decreased in only

Fig. 4.54 Japanese shallot grown on Sandy Regosols in Tottori Sand dune. *Photograph* Copyright by Dr. H. Obara

six years. The most distinctive characteristic of the 2000 Miyakejima eruption was the emission of SO_2 gas that lasted for few years. As a result, the accumulated volcanic ash was acidified, which produced an optimal environment for iron-oxidizing bacteria. However, during the ensuing decade, the emission of SO_2 gas greatly diminished. This facilitated an increase in the pH value of sediments and an associated oxidation of Fe^{2+} (conversion to Fe^{3+}), which caused a reduction in the number of iron-oxidizing bacteria (Fujimura et al. 2016). In summary, on Miyakejima, pedogenesis on a timescale of a few years in soil of which volcanic ash was the precursor was strongly affected by the degree of change in the microbial ecosystem, which in turn was largely affected by SO_2 gas.

In Volcanogenous Regosols, the population of living organisms (plants and microbes) changes quite dynamically. If the soil is a survivable environment for living organisms, the dynamic transition of biota is interlinked with the dynamic change of the soil. If the ultimate form of the volcanic ash soil in Japan is Andosol, then the interpretation of changes over time of the Volcanogenous Regosols scattered around the country would reveal the complete picture of the soil formation process.

4.10.2 Sandy Regosols

These soils are coarse-textured soils which occur mainly in slightly elevated lands along the sea coast such as sand dunes, sand bars, sand banks, and spits. Sandy Regosols have a sand or loamy sand texture and gravel contents of less than 35%. Materials are mainly aeolian and partly marine deposits, and soils show a very poor horizon differentiation.

These soils are widely found along the coasts on the Sea of Japan and East China Sea, and on the Pacific Ocean coast (Fig. 4.53).

Sandy Regosols generally have high permeability and low water and nutrient storage capacity due to their sandy texture. On the other hand, Sandy Regosols offer ease of cultivation, rooting, and harvesting of root and tuber crops. The lands containing these soils are mostly idle, but some are used as wind breaks, upland crop fields, and for the growing of flowers and fruits. Additionally, in some places, such lands are intensively used for growing vegetables, for example, Japanese shallot (Fig. 4.54), radish, sweet potato, melon, and welsh onion, with irrigation systems such as sprinkler irrigation or drip irrigation.

Fig. 4.55 An example of Sandy Regosols (Shizuoka prefecture). *Photograph* Obara et al. (2015)

The Haplic subgroup is characterized by the absence of calcareous property and good to excessive drainage with a low groundwater table (Fig. 4.55). The Aquic subgroup comprises soils where the groundwater level is within 50 cm of the mineral soil surface for more than half the year. The Calcareous subgroup corresponds to soils having a calcareous property (more than 2.0% $CaCO_3$ equivalent) between 20 and 50 cm from the soil surface. The Calcareous subgroup occurs on the shores of the Nansei Islands.

On old, stable sand dunes, soil-forming processes can develop a dark-colored A horizon and somewhat developed B horizons. Podzols that have a light-colored E horizon and a rusty-colored spodic horizon (Bs) were reported in sand dunes of northern Hokkaido (Tomioka et al. 1973; Sasaki 1974). If the characteristics of the developed B horizon meet the Cambic B horizon requirements, the soils will change to Brown Forest soils or Red-Yellow soils.

4.10.3 Lithosols

These soils are residual soils that have bedrock with their upper boundaries within 30 cm of the mineral soil surface. They are widely distributed on strongly eroded sloping lands in mountainous and hilly areas throughout Japan. In such sites, landslides, creep, and other forms of erosion may remove surface materials from the site faster than soil material forms. On the other hand, some Lithosols can occur on limestone or fresh lava flows, where the supply of parent materials from rock weathering is not enough to develop a soil profile.

Lithosols are distributed in all regions of Japan; however, these soils occur in association with other soils and their area has not been exactly estimated. The lands where Lithosols occur are mostly forest or wasteland, but are partly used as upland crop fields and orchard.

Lithosols are divided into three subgroups by calcareous properties (Calcaric), endoaquic properties (Aquic), and others (Typic) (Fig. 4.56). Limestone-derived soils with bedrock appearing within 30 cm of the surface are classified as Calcaric Lithosols. If the bedrock appears at depths greater than 30 cm from the soil surface, the soils are classified as Calcaric Terrestrial Regosols or Eutrosols.

4.10.4 Terrestrial Regosols

Terrestrial Regosols are immature soils which occur on mountains, hills, and Pleistocene terraces. These soils are: (1) soils that have a gravel layer within 30 cm of the surface but do not have bedrock within 30 cm of the surface; or (2) unweathered soils having the color of the parent rock (or materials) due to the absence of free iron oxides liberated by weathering. In Japan, deeply weathered "*masa-do*" (weathered granite) is an example of (1), and "*jahgaru*" (semi-consolidated marl-derived soils) is an example of (2) (Fig. 4.57).

Terrestrial Regosols are distributed widely in mountainous, hilly, and plateau areas throughout Japan. Unstable landforms, the movement of soil materials, and the deposition of new soil material prevent the development of these soils. These soils are used for upland crops, orchard, grassland, and forest.

The parent materials influence the characteristics of these soils significantly. Thus, this soil group is subdivided based on parent rock into five subgroups, that is, soils derived from marl (Marlitic), calcareous type (Calcaric), soft rock type

Fig. 4.56 An example of Typic Lithosols (Miyagi prefecture). *Photograph* Copyright by Dr. H. Obara

Fig. 4.57 Examples of Terrestrial Regosols. "*Masa-do*" in left (1) and "*jahgaru*" in right (2). *Photograph* Obara et al. (2015)

(Paralithic), "*masa-do*" type (Granitic), and consolidated rock type (Typic).

Granitic Terrestrial Regosols are immature soils derived from weathered granite. Weathered granites contain coarse grains of primary minerals such as quartz, feldspar, and mica. Due to their coarse texture, these soils have no to slight stickiness, and high permeability. Accordingly, lands containing weathered granites have a risk of landslides when heavy rainfalls. Granitic Terrestrial Regosols have a low water and nutrient storage capacity. These soils are used for forest, orchard, tea fields, and upland crops. On stable land surfaces, these soils will change to Brown Forest soils or Red-Yellow soils when enough time is available for the development of a B horizon (Kanno et al. 1951; Imaya et al. 2005).

Marlitic Terrestrial Regosols in Okinawa Prefecture are gray-colored immature soils derived from marl. These are rather fertile soils that have alkaline to neutral soil reaction and high CEC (more than 20 cmol/kg). However, these soils have a clayey and silty texture and low to extremely low permeability. Therefore, the risk of water damage is one of the major problems in these soils for upland crop cultivation. These soils are used for upland crops such as sugarcane and vegetables. Marlitic Terrestrial Regosols will change to Eutrosols, Brown Forest soils, or Red-Yellow soils after cambic (Bw) or argillic (Bt) horizon development (Maejima et al. 2007; Kinjo et al. 2014).

References

Akagi I, Inoue Y, Nagatomo Y (2003) Pedological properties of Red-Yellow soils (paleo-Red soils) in southern Kyushu. Jpn J Soil Sci Plant Nutr 74:623–630 (in Japanese with English summary)

Araki S (1993) Comparative study of Kunigami Mahji and Feichisha Soils in Okinawa Islands, among Red-Yellow soils of the world. Pedologist 37:113–125 (in Japanese)

Bockheim J (2011) Distribution and genesis of ortstein and placic horizons in soils of the USA: a review. Soil Sci Soc Am J 75:994–1005

Buol SW, Southard RJ, Graham RC, McDaniel PA 2011: Histosols: organic soils. In: Buol SW et al (eds) Soil genesis and classification, 6th edn. Wiley and Blackwell, West Sussex, pp 307–320

Committee for Soil Classification and Nomenclature (1986) Unified soil classification system of Japan (1st Approximation). Pedologist 30:123–139

Cultivated Soil Classification Committee (1983) Classification of cultivated soils in Japan. Second Approximation Revised version, National Institute for Agricultural Technology (in Japanese)

Cultivated Soil Classification Committee (1995) Classification of cultivated soils in Japan, Third Approximation, Misc. Publ. Nat Inst Agro-Environ Sci 17:1–79 (in Japanese)

De Coninck F (1980) Major mechanisms in formation of spodic horizons. Geoderma 24:101–128

Duchaufour P, Souchier B (1978) Roles of iron and clay in genesis of acid soils under a humid, temperate climate. Geoderma 20:15–26

Edamura T, Kumagai K (2009) The dispersion of Jomon Ruins. Theory Appl GIS 17:63–72 (in Japanese)

Endo K (1966) Classification and nomenclature of the soils distributed in mountainous warm-temperate forest regions of Japan. Pedologist 10:2–10 (in Japanese with English summary)

Endo K (1969) Classification of the brown forest soils and related soils in Japan. Pedologist 13:43–46 (in Japanese)

FAO (1998) World reference base for soil resources, by ISSS–ISRIC–FAO. World Soil Resources Report No. 84. Rome

Farmer VC (1982) Significance of the presence of allophane and imogolite in podzol Bs horizons for podzolization mechanisms: a review. Soil Sci Plant Nutr 28:571–578

Food and Agriculture Organization of the United Nations (2015) World reference base for soil resources 2014. World Soil Resources Reports 106, Rome

Forest Soil Division (1976) Classification of forest soil in Japan (1975), Bull Gov For Exp Sta 280:1–28. (in Japanese with English summary)

Forest soils of Japan edit Committee (1983) Forest soils of Japan. Japan Forest Technical Association, Tokyo (in Japanese)

Fujii K, Funakawa S, Hayakawa C, Kosaki T (2008) Contribution of different proton sources to pedogenetic soil acidification in forested ecosystems in Japan. Geoderma 144:478–490

Fujii K, Uemura M, Funakawa S, Hayakawa C, Sukartiningsih Kosaki T, Ohta S (2009) Fluxes of dissolved organic carbon in two tropical forest ecosystems of East Kalimantan, Indonesia. Geoderma 152:127–136

Fujii K, Hayakawa C, Van Hees PAW, Funakawa S, Kosaki T (2010) Biodegrada-tion of low molecular weight organic compounds and their contribution to heterotrophic soil respiration in three Japanese forest soils. Plant Soil 334:475–489

Fujii K, Funakawa S, Shinjo H, Hayakawa C, Mori K, Kosaki T (2011) Fluxes of dissolved organic carbon and nitrogen throughout Andisol, Spodosol and Inceptisol profiles under forest in Japan. Soil Sci Plant Nutr 57:855–866

Fujimura R, Sato Y, Nishizawa T, Nanba K, Oshima K, Hattori M, Kamijo T, Ohta H (2012) Analysis of early bacterial communities on volcanic deposits on the Island of Miyake (Miyake-jima), Japan: a 6-year study at a fixed site. Microbes Environ 27:19–29

Fujimura R, Kim S, Sato S, Oshima K, Hattori M, Kamijo T, Ohta H (2016) Unique pioneer microbial communities exposed to volcanic sulfur dioxide. Sci Rep 6(Article number: 19687)

Fujita H, Takada M, Kaneko M (1997) Inventory of the present wetlands in Hokkaido, Japan. In: Report of the pro natura foundation. The Pro Natura Foundation, Tokyo, pp 3–14. (in Japanese)

Funakawa S, Hirai H, Kyuma K (1992) Soil-forming processes under natural forest north of Kyoto in relation to soil solution composition. Soil Sci Plant Nutr 38:101–112

Funakawa S (1993) Soil-forming processes under natural forest in northern Kyoto. Kyoto University, Kyoto, Japan.

Galbraith J (2018) Human-altered and human-transported (HAHT) soils in the U.S. soil classification system. Soil Sci Plant Nutr 64:190–199

GIO (Greenhouse Gas Inventory Office of Japan) (2012) National greenhouse gas inventory report of Japan. National Institute of Environment (NIES), Tokyo

Guo Y, Fujimura R, Sato Y, Suda Y, Kim S, Oshima K, Hattori M, Kamijo T, Narisawa K, Ohta H (2014) Characterization of early microbial communities on volcanic deposits along a vegetation gradient on the Island of Miyake, Japan. Microbes Environ 29 (1):38–49

Hamazaki T (1993) Moisture regime and its pedogenic roles in anthraquic paddy soils. Pedologist 37:15–27 (in Japanese and English summary)

Harrison AF, Harkness DD, Rowland AP, Garnette JS, Bacon PJ (2000) Annual carbon and nitrogen fluxes in soils along European forest transect, determined using 14C-bomb. In: Schulze ED (ed) Carbon and nitrogen cycling in European forest ecosystems. Springer, Berlin, pp 237–256

Hashimoto N (1969) Brown forest soils in the mountains of Japan. Pedologist 13:36–43 (in Japanese)

Hayakawa C, Funakawa S, Fujii K, Kadono A, Kosaki T (2014) Effects of climatic and soil properties on cellulose decomposition rates in temperate and tropical forests. Biol Fertil Soils 50:633–643

Hiradate S, Nakadai T, Shindo H, Yoneyama T (2004) Carbon source of humic substances in some Japanese volcanic ash soils determined by carbon stable isotopic ratio, δ13C. Geoderma 119:133–141

Hirai H, Araki S, Kyuma K (1988) Characteristics of brown forest soils developed on the Paleozoic shale in northern Kyoto with special reference to their pedogenetic process. Soil Sci Plant Nutr 39:157–170

Hirai H, Araki S, Kyuma K (1990) Characteristics of brown forest soils in northern Kyoto as compared with Andisols, Red- and/or Yellow-colored soils, and podzolic soils. Soil Sci Plant Nutr 36:623–632

Hokkaido Regional Development Bureau (1979) Survey for developing peatlands. In: Hokkaido Regional Development Bureau (ed) Specific soils in Hokkaido—formation, distribution and land improvement. Hokkaido Regional Development Bureau, Sapporo, pp 25–73. (in Japanese)

Honna T, Yamamoto S, Matsui K (1988) A simple procedure to determine melanic index useful to separation of melanic and fulvic Andisols. Pedolojist 32:69–78

Hosono M, Sase T (2015) The historical development of "Kurobo-kudo" layer: preliminary discussion from the perspective of human-made ecosystems. Quat Res 54:323–339 (in Japanese)

Imaya A (2008) Chemical properties and pedogenesis of yellowish/yellow brown forest soils in the submontane and mountain zones of Kyushu district, southwestern Japan. Soil Sci Plant Nutr 54:424–433

Imaya A, Ohta S, Tanaka N, Inagaki Y (2005) General chemical properties of Brown Forest Soils developed from different parent materials in the submontane zone of the Kanto and Chubu districts, Japan. Soil Sci Plant Nutr 51:873–884

Imaya A, Inagaki Y, Tanaka N, Ohta S (2007) Free oxides and short-range ordered mineral properties of brown forest soils developed from different parent materials in the submontane zone of the Kanto and Chubu districts, Japan. Soil Sci Plant Nutr 53:621–633

Imaya A, Yoshinaga S, Inagaki Y, Tanaka N, Ohta S (2010a) Volcanic ash additions control soil carbon accumulation in brown forest soils in Japan. Soil Sci Plant Nutr 56:734–744

Imaya A, Yoshinaga S, Inagaki Y, Tanaka N, Ohta S (2010b) Proposal for advanced classification of brown forest soils in Japan with reference to the degree of volcanic ash additions. Soil Sci Plant Nutr 56:454–465

Inoue K (1981) Implications of Eolian Dusts to 14 Å minerals in the volcanic ash soils in Japan. Pedolojist 25:97–118

Inoue K, Huang PM (1984) Influence of citric acid on the natural formation of imogolite. Nature 308:58–60

Inoue K, Satake H, Wakamatsu Y, Mizota C, Kusakabe M (1993) Eolian dust origin of parent materials of the red and yellow soil group in the Nansei syoto, Japan. Oxygen and carbon isotopic ratios of quartz, muscovite and calcite isolated from soils, bedrocks, and piston cores. Quat Res (Daiyonki-kenkyu) 32(3):139–155. (in Japanese with English summary)

Ito T, Shoji S, Masahiko S (1991a) Classification of volcanic ash soils from Konsen District, Hokkaido according to the last keys to Soil Taxonomy. Jpn J Soil Sci Plant Nutr 62:237–247 (in Japanese with English summary)

Ito T, Shoji S, Shirato Y, Ono E (1991b) Differentiation of a spodic horizon from a buried A horizon. Soil Sci Soc Am J 55:438–442

IUSS Working Group WRB (2006) World reference base for soil resources, 2006. World Soil Resources Reports No. 103. FAO, Rome

IUSS Working Group WRB (2014) World reference base for soil resources 2014. International soil classification system for naming soils and creating legends for soil maps. World Soil Resources Reports No. 106. FAO, Rome

IUSS Working Group WRB (2015) World reference base for soil resources 2014, update 2015 international soil classification system for naming soils and creating legends for soil map. World Soil Resources Reports No. 106. FAO, Rome

Jansen B, Nierop KGJ, Verstraten JM (2005) Mechanisms controlling the mobility of dissolved organic matter, aluminium and iron in podzol B horizons. Eur J Soil Sci 56:537–550

Japan Meteorological Agency (2017) Active volcanoes in Japan. http://www.data.jma.go.jp/svd/vois/data/tokyo/STOCK/kaisetsu/katsukazan_toha/katsukazan_toha.html

Kaila A (1956) Determination of the degree of humification in peat samples. J Agric Sci Finl 28:18–35

Kamoshita Y (1940) The soil types of the TSUGARU plain, Aomori prefecture, Nippon. J Imp Agr Exp Stat III-3:401–420. (in Japanese and English summary)

Kanda T, Takata Y, Wakabayashi S, Kohyama K, Obara H (2017) Development of the 1:50,000 digital soil map of cultivated soil Japan accordingly to the comprehensive soil classification system of Japan—first approximation. Jpn J. Soil Sci Plant Nutr 88:29–34 (in Japanese)

Kanda T, Takata Y, Kohyama K, Ohkura T, Maejima Y, Wakabayashi S, Obara H (2018) New soil maps of Japan based on the comprehensive soil classification system of Japan–first approximation and its application to the world reference base for soil resources 2006. JARQ 52:285–292–550

Kanno H. (1957) Fundamental soil profiles of inorganic paddy soils. Jpn J Soil Sci Plant Nutr 27:393–396. (in Japanese)

Kanno I (1961) Soils on genesis and classification of major soil types of Japan, No. 2, Red-yellow soil. Bull Kyushu Exp Stn 7:187–306 (in Japanese)

Kanno I, Nagai M, Arimura S (1951) The Red soils derived From the residua of Granitic Rocks, northern coast of the Inland Sea. Bull Kyushu Agric Exp Stat 22:41–45 (in Japanese with English summary)

Kanno H, Hirai H, Takahashi T, Nanzyo M (2008) Soil regions map of Japan based on a reclassification of the 1:1 million soil map of Japan (1990) according to the unified soil classification system of Japan—2nd approximation (2002). Pedologist 52(2):129–133 (in Japanese)

Kato Y (1961) Clay mineralogy of "Kuroboku" Soil at Nihondaira, Shizuoka Prefecture. J Sci Soil Manure Jap 32:328–332

Kato Y (1979) Significance of parent materials for soil genesis and soil classification. Pedologist 23:58–65 (in Japanese)

Kato T, Kamijo T, Hatta T, Tamura K, Higashi T (2005) Initial soil formation processes of Volcanogenous Regosols (Scoriacious) from Miyake-jima Island, Japan. Soil Sci Plant Nutr 51:291–301

Katsumi T (2015) Soil excavation and reclamation in civil engineering: environmental aspects. Soil Sci Plant Nutr 61:22–29

Kawada H, Takami M (1957) A study on correlations between chemical properties and types of forest soil. For Soils Jpn 8:81–124 (in Japanese with English summary)

Kawai N, Murata T, Watanabe M, Tanaka H (2015) Influence of historical manmade alterations on soil-forming processes in a former imperial estate (Shirogane-goryouchi), the Institute for Nature Study: development of a soil evaluation technique and importance of inventory construction for urban green areas. Soil Sci Plant Nutr 61:55–69

Kawamuro K, Torii A (1986) Difference in past vegetation between black soils and brown forest soils derived from volcanic ash at Mt. Kurohime, Nagano Pref., Japan. Quat Res 25:81–98 (in Japanese with English summary)

Kida K, Kawahigashi M (2015) Influence of asphalt pavement construction processes on urban soil formation in Tokyo. Soil Sci Plant Nutr 61:135–146

Kida K, Kawahigashi M (2017) The soil classification concept for soils influenced by artefacts; comparison between soil classification systems of Japan, Soil Taxonomy, and world reference base for soil resources. Pedologist 61:82–92 (in Japanese with English abstract)

Kidachi and Ohmasa (1963) Geological and mineralogical studies about pedogenesis of red soils in Japan (studies of red soils in Japan). Forest Soils Jap Rep 14:1–126 (in Japanese with English summary)

Kinjo K, Arasaki M, Asato M, Tokashiki Y (2014) Chemical and mineralogical properties of red-yellow soils in Shimajiri mudstone zone. J Soil Sci Plant Nutr 85:315–321 (in Japanese with English summary)

Kitagawa Y (2005) Characteristics of clay minerals in podzols and podzolic soils. Soil Sci Plant Nutr 51: 151–158

Kondo R (1997) Peat. In: Iwata S et al (eds) Environment area of soils. Fuji Technosystem, Tokyo, pp 234–246. (in Japanese)

Kondo R, Endo K (1993) Various indices for determining the degree of decomposition and some physico-chemical properties of peat soils. Pedologist 37:41–56 (in Japanese with English summary)

Kumada K (1987) Chemistry of soil organic matter, 1st edn. Jap Sci Soc Press, Elsevier, pp 17–56

Kurotori T, Ohmasa M (1963) Red soils and their accompanying black soils in Kyushu (studies of red soils in Japan). For Soils Jpn 13:1–88 (in Japanese with English summary)

Lundström US (1993) The role of organic acids in the soil solution chemistry of podzolized soil. J Soil Sci 48:115–120

Lundström US, van Breemen N, Bain D (2000) The podzolization process. A re-view. Geoderma 94:91–107

Machida H (2002) Volcanoes and tephras in the Japan area. Glob Environ Res 6(2):19–28

Maejima Y, Nagatsuka S, Higashi T (2000) Mineralogical compositions of iron oxides in Red- and Yellow-colored soils from southern Japan and Yunnan, China. Soil Sci Plant Nutr 46(3):571–580

Maejima Y, Nakaie K, Nagatuka S (2007) Genesis and classification of red soils derived from Marl in Miyako Island, Ryukyus. Pedologist 51-1:14–23. (in Japanese with English summary)

Maejima Y, Ohkura T, Kusumoto Y, Takata Y (2014) Genesis and classification of soils derived from serpentine and basalt on Mt. Mineoka, south Boso peninsula. Pedologist 58:2–16 (in Japanese with English summary)

Mashimo Y (1957) Moisture condition (pF) of forest soil. For Soils Jpn 8:43–66 (in Japanese with English summary)

Mashimo Y (1960) Studies on the physical properties of forest soil and their relation to the growth of Sugi (Cryptomeria japonica) and Hinoki (*Chamaecyparis obtusa*). Forest Soils Jap Rep 11:1–185 (in Japanese with English summary)

Mashimo Y (1977) On the 1975 system of forest soil classification. Jpn J For Environ 19:6–10 (in Japanese)

Matsui T (1977) On the revised classification system of forest soils (1975). Pedologist 21:139–146 (in Japanese)

Matsui T, Kato Y (1962) Notes on paleopedology of red soils in Japan. Quat Res (Daiyonki-kenkyu) 2:161–179. (in Japanese with English summary)

Matsui T, Kato Y, Kurobe T, Hamada R, Iwasa Y (1961) A tentative scheme of classification for paddy soils on the alluvial plain, with special reference to the soil survey around Shizuoka City, Pacific coast of central Japan. Pedologist 5:80–91 (in Japanese and English summary)

MLIT (Ministry of Land, Infrastructure, Transport and Tourism of Japan) (2014) White paper on land 2014. MLIT Tokyo. (in Japanese)

Ministry of the Environment (2017) https://www.env.go.jp/en/nature/nps/park/about/index.html

Mitsuchi M (1974) Pedogenic characteristics of paddy soils and their significane in soil classification. Bull Nat Inst Agric Sci B-25:29–115. (in Japanese and English summary)

Mizota C, Endo H, Um KT, Kusakabe M, Noto M, Matsuhisa Y (1991) The eolian origin of silty mantle in sedentary soils from Korea and Japan. Geoderma 49:153–164

Mokma DL, Buurman P (1982) Podzols and podzolization in temperate regions. International Soil Museum, No 1, p 131

Nagatsuka S (1973) On the features of free iron oxides in brown forest, yellow-brown forest and red soils of Japan. Pedologist 17:70–83 (in Japanese with English summary)

Nagatsuka S (1975) Genesis and classification of Yellow-Brown forest soils and Red soils in Southwest Japan. Bull Natl Inst Agric Sci Ser B. 26:133–257 (in Japanese with English summary)

Nagatsuka S (1993) Form of iron analysis. In: Japan Association for Quaternary Research (ed) A handbook of quaternary research, vol 2. University of Tokyo Press, Tokyo, pp 22–32

Nagatsuka S (1997) Ecological atlas of soils in Japan. Fuji Techno-System, Tokyo, p 218. (in Japanese)

Nagatsuka S, Maejima Y (2001) Dating of soils on the raised coral reef terraces of Kikai Island in the Ryukyus, Southwest Japan: with special reference to the age of Red-Yellow soils. Quat Res (Daiyonki-kenkyu) 40:137–147

Nagatsuka S, Kaneko S, Ishihara A (1983) Soil genesis on the raised coral reef terraces of Ishigaki- and Okinawa-Islands in the Ryukyu Islands, Japan. 1. Relations among soils and geomorphic plains and chronosequential changes of soil chemical properties, Soil Sci Plant Nutr 29(3):343–354

Nakao A, Watanabe T, Honda T, Hseu ZY, Funakawa S (2011) The calcium-magnesium ratio of serpentinitic soils in various topographic locations in Sekinomiya, Japan; a potential criterion for the classification of the Dark-red Magnesian soils. Pedologist 55:30–42

Nielsen KE, Ladekarl UL, Nornberg P (1999) Dynamic soil processes on heathland due to changes in vegetation to oak and Sitka spruce. For Ecol Manage 114:107–116

Obara H, Ohkura T, Takata Y, Kohyama K, Maejima Y, Hamazaki T (2011) Comprehensive soil classification of Japan—first approximation. Bull Natl Inst Agro-Environ Sci 29:1–73 (in Japanese with English summary)

Obara H, Maejima Y, Kohyama K, Ohkura T, Leon I, Ikeba M (2014) Characteristics and classification of paddy soils on the alluvial plain of KOKAI river in Ibaraki prefecture. Pedologist 58(1):17–29 (in Japanese and English summary)

Obara H, Kohyama K, Takata Y, Ohkura T, Maejima Y (2015) Photographs of soil. In: Natural Resources Inventory Center (ed), National Institute for Agro-Environmental Sciences, Tsukuba. (in Japanese)

Obara H, Takata Y, Kohyama K, Ohkura T, Maejima Y, Watabayashi S, Kanda T (2016) A new soil map of Japan based on comprehensive soil classification system of Japan first approximation. Bull Natl Inst Agro-Environ Sci 37:133–148 (in Japanese with English summary)

Ohmasa M (1951) Studies on beech forest soils. For Soils Jpn 1:1–243 (in Japanese with English summary)

Ohmasa M et al (1955) Studies of red soils of Japan I, Red soils of Niigata Prefecture. J Jpn Forest Soc 37:140–142 (in Japanese with English summary)

Ono K, Komoriya A, Tachibana R, Imaya A, Suzuki S, Noguchi K, Haginoe H (2018) Effects of row deep tillage for the growth base formed by piling up soil in damp lowlands behind coastal sand dunes to construct coastal disaster prevention forest belts on the Kujukuri coastline, Japan. Soil Sci Plant Nutr 64:168–180

Qualls RG, Bridgham SD (2005) Mineralization rate of 14C-labeled dissolved organic matter from leaf litter in soils of a weathering chronosequence. Soil Biol Biochem 37:905–916

Raulund-Rasmussen K, Borggaard OK, Hansen HCB, Olsson M (1998) Effect of natural organic soil solutes on weathering rates of soil minerals. Eur J Soil Sci 49:397–406

Saigusa M, Matsuyama N (1998) Distribution of allophanic Andosols and non-allophanic Andosols in Japan. Tohoku J Agric Res 48:75–83

Saigusa M, Shoji S, Takahashi T (1980) Plant root growth in acid Andosols from northeastern Japan: 2. Exchange acidity Y1 as a realistic measure of aluminum toxicity potential. Soil Sci 130:242–250

Sakaguchi Y (1974) Geology of Peatlands—search for environmental changes. Tokyo University Press, Tokyo (in Japanese)

Sasaki S (1974) Sand dune podsol in Hamatombetsu. Hokuno 464:40–45 (in Japanese)

Sase T (1986) Opal phytolith analysis of volcanic ash soils from recent Towada ashes. Pedologist 30:102–114 (in Japanese with English summary and figures)

Sato Y, Hosokawa K, Fujimura R, Nishizawa T, Kamijo T, Ohta H (2009) Nitrogenase activity (acetylene reduction) of an iron-oxidizing Leptospirillum strain cultured as a pioneer microbe from a recent volcanic deposit on Miyake-jima, Japan. Microbes Environ 24:291–296

Schad P (2018) Technosols in the world reference base for soil resources—history and definitions. Soil Sci Plant Nutr 64:132–137

Schwertmann U (1985) The effect of pedogenic environments on iron oxide minerals. Adv Soil Sci 1:172–200

Shindo H, Yoshida M, Yamamoto A, Honma H, Hiradate S (2005) δ13C values of organic constituents and possible source of humic substances in Japanese volcanic ash soils. Soil Sci 170:175–182

Shoji S, Ito T (1990) Classification of tephra-derived Spodosols. Soil Sci 150:799–815

Shoji S, Ono T (1978) Physical and chemical properties and clay mineralogy of Andosols from Kitakami, Japan. Soil Sci 126:297–312

Shoji S, Takahashi T (2002) Environmental and agricultural significance of volcanic ash soils. Glob Environ Res 6(2):113–135

Shoji S, Fujiwara Y, Yamada I, Saigusa M (1982) Chemistry and clay mineralogy of ando soils, brown forest soils, and podzolic soils formed from recent Towada ashes, northeastern Japan. Soil Sci 133:69–86

Shoji S, Ito T, Saigusa M, Yamada I (1985) Properties of nonallohanic andosols from Japan. Soil Sci 140:264–277

Shoji S, Ito T, Nakamura S, Saigusa M (1987) Properties of humus of Andosols from New Zealands, Chili and Ecuador. Jap J Soil Sci Plant Nutr 58:473–479 (in Japanese with English summary)

Shoji S, Takahashi T, Ito T, Ping CL (1988a) Properties and classification of selected volcanic ash soils from Kenai Peninnsula, Alaska. Soil Sci 145:395–413

Shoji S, Takahashi T, Saigusa M, Yamada I, Ugolini FC (1988b) Properties of Spodosols and Andisols showing climosequential biosequential relations in southern Hakkoda, northeastern Japan. Soil Sci 145:135–150

Shoji S, Nanzyo M, Dahlgren RA (1993a) volcanic ash soils-genesis, properties and utilization. Developments in Soil Science 21, Elsevier, Amsterdam

Shoji S, Nanzyo M, Shirato Y, Ito T (1993b) Chemical kinetics of weathering in young Andisols from northeastern Japan using soil age normalized to 10 °C. Soil Sci 155:53–60

Soil Survey Staff (1999) Soil taxonomy: a basic system of soil classification for making and interpreting soil surveys, 2nd edn. Natural Resources Conservation Service. U.S. Department of Agriculture Handbook 436

Soil Survey Staff (2014) Keys to soil taxonomy, 12th edn. United States Department of Agriculture Natural Resources Conservation Service, Washington DC, USA

Takahashi T, Dahlgren RA (2016) Nature, properties and function of aluminum-humus complexes in volcanic soils. Geoderma 263:110–121

Takahashi T, Shoji S (1996) Active aluminum status in surface horizons showing continuous climosequence of volcanic ash-derived soils in Towada district, northeastern Japan. Soil Sci Plant Nutr 42:113–120

Takata Y, Maejima Y, Ohkura T, Kohyama K, Hamazaki T, Obara H (2010) Locating the taxonomical position of argic and cambic horizons in Japanese Red-Yellow soils. Pedologist 54:11–20 (in Japanese with English summary)

Takehara H (1961) Secondary agricultural research report of Iriomote Island. Ministry of Agriculture, p 156. (in Japanese)

Tani M, Kondo R, Tsutsuki K (2001a) Iron forms in peat soils and their relation to indices for the decomposition degree of peat. Jpn J Soil Sci Plant Nutr 72:180–188 (in Japanese with English summary)

Tani M, Kondo R, Tsutsuki K (2001b) Dissolved organic matter and its interaction with dissolved iron in peat pore waters. Jpn J Soil Sci Plant Nutr 72:348–356 (in Japanese with English summary)

The Fifth Committee for Soil Classification and Nomenclature (2017). http://pedology.jp/img/SoilClassificationSystemofJapan.pdf

The Fourth Committee for Soil Classification and Nomenclature The Japanese society of pedology (2002). Unified Soil Classification System of Japan—2nd Approximation (2002). Jpn Soc Pedol. (in Japanese)

Tomioka E, Otowa M (1973) Soil survey of soya subprefecture, Soil Survey Report No. 22 p. 75, Hokkaido National Agricultural Experiment Staion, Sapporo, Japan. (in Japanese)

Uchiyama N (1949) Morphology of paddy soils. Chikyu shuppan, Tokyo (Japanese)

Uchiyama N, Masui J, Onikura Y (1960) Humus clay pan soil developed on a volcanic ash soil. Trans 7th Intern Congr, Soil Sci 4:443–450

Ugolini FC, Dahlgreen R (1987) The mechanism of podzolization as revealed by soil solution studies. In Podzols et podzolisation: table ronde internationale, Poitiers (France), 10–11 Apr 1986. INRA

Ugolini FC, Dahlgren RA (1991) Weathering environments and occurrence of imogolite/allophane in selected Andisols and Spodosols. Soil Sci Soc Am J 55:1166–1171

Ugolini FC, Dahlgren RA (2002) Soil development in volcanic ash. Glob Environ Res 6:69–81

Ugolini FC, Sletten RS (1991) The role of proton donors in pedogenesis as revealed by soil solution studies. Soil Sci 151:59–75

Ugolini FC, Dahlgren R, Shoji S, Ito T (1988) An example of andosolization and podzolization as revealed by soil solution studies, southern Hakkoda, north-eastern Japan. Soil Sci 145:111–125

Ugolini FC, Sletten RS, Marrett DJ (1990) Contemporary pedogenic processes in the Artic: brunification. Sci Edu Sol 28:333–348

University of Idaho (2019) https://www.uidaho.edu/cals/soil-orders/spodosols. Accessed on 17 May 2019

Uoi N, Kawahigashi M, Ono M, Murata T, Watanabe M (2013) Estimation of soil carbon storage in Shinjyuku Gyoen National Garden based on land use history and soil compactness profiles. Geogr Rep Tokyo Metrop Univ 48:13–24

Van Hees PA, Jones DL, Finlay R, Godbold DL, Lundström US (2005) The carbon we do not see—the impact of low molecular weight compounds on carbon dynamics and respiration in forest soils: a review. Soil Biol Biochem 37:1–13

Wada K (1985) The distinctive properties of Andosols. Adv Soil Sci 2:173–229

Wada K (ed) (1986a) Ando soils in Japan. Kyushu University Press, Japan

Wada K (1986b) An invitation to soil clay mineralogy. Clay Sci 26(1–1):1 (in Japanese with English summary)

Wada K, Gunjigake N (1979) Active aluminum and iron and phosphate adsorption in Ando soils. Soil Sci 128:331–336

Wada K, Higashi T (1978) The categories of aluminum- and iron-humus complexes in Ando soils determined by selective dissolution. J Soil Sci 27:357–368

Wada K, Okamura Y (1980) Electric charge characteristics of Ando A1 and buried A1 horizons soils. J Soil Sci 31:307–314

Yamazaki K (1960) Studies on the pedogenetic classification of paddy soils in Japan, Bulletin of Agricultural experimental station of Toyama prefecture, Special issue No. 1, 1–98 (in Japanese)

Yambe F, Yagi H (1983) Nature and genesis of main forest soils on the hilly region around Echigo Plain (I). Chemical properties and free iron oxides. Bull For For Prod Res Inst 324:125–139 (in Japanese with English summary)

Part III
Regional Features

Hokkaido Region

5

Toshiro Nakatsuji, Hiroyuki Shiga, Harunobu Takeuchi, Yasutaka Tsukamoto, Eiji Gotou, Yuji Watanabe, Michihiko Sakurai, Nobuhiko Fueki, Yuji Hikasa, Tetsuo Hayashi, Masami Sakaguchi, Toshiya Saigusa, Tetsuo Yagi, Osamu Sakai, Fuyuki Satoh, Hideaki Shibata, Kentaro Takagi, Makoto Kobayashi, Ryusuke Hatano, Takuji Sawamoto, Kanta Kuramochi, Yukio Hosobuchi, Osamu Nagata, Takehiko Matsumoto, Daiji Asaka, Hiroshi Nakamoto, Masayuki Onodera, Katsuhisa Niwa, Mariko Shimizu, Tomoyoshi Hirota, Ryo Ohtomo, Norikuni Oka, Rintaro Kinoshita, Jiwan Palta, and Masayuki Tani

Abstract

Hokkaido, the northernmost island in Japan, is located on the border between temperate and subarctic zones. Dominant soils distributed in Hokkaido have a wide variety such as Andosols, Fluvic soils, Brown Forest soils, and Peat soils. Paddy, upland, vegetable, dairy farming, and animal husbandry are being practiced on large scales according to the characteristics of soils and weather conditions in each region. The goal of agriculture style in Hokkaido is environmentally friendly agriculture. It can sustainably produce safe, high-quality, and high-yield agricultural products in harmony with the regional and global environments. To promote the environmentally friendly agriculture, soil fertility management techniques based on fertilization standards and soil diagnosis have been developed, with considering nutrient supply derived from applied organic matter. It has been demonstrated that practicing these techniques makes it possible to minimize the environmental impact such as nutrient leaching and greenhouse gas emissions from agricultural lands.

Keywords

Agricultural lands • Environmentally friendly agriculture • Fertilization standards • Organic matter • Soil fertility management

T. Nakatsuji (✉) · H. Takeuchi · Y. Tsukamoto · E. Gotou · Y. Watanabe · M. Sakurai · N. Fueki · T. Hayashi · M. Sakaguchi · T. Yagi · O. Sakai · Y. Hosobuchi · T. Matsumoto · D. Asaka · H. Nakamoto · M. Onodera
Agricultural Research Department, Hokkaido Research Organization, Sapporo City, Hokkaido, Japan
e-mail: nakatsuji-toshiro@hro.or.jp

H. Shiga · Y. Hikasa
HOKUREN Federation of Agricultural Cooperatives, Sapporo City, Hokkaido, Japan

T. Saigusa · T. Sawamoto
College of Agriculture, Food and Environment Sciences, Rakuno Gakuen University, Ebetsu City, Hokkaido, Japan

F. Satoh · H. Shibata · K. Takagi · M. Kobayashi
Field Science Center for Northern Biosphere, Hokkaido University, Sapporo City, Hokkaido, Japan

R. Hatano · K. Kuramochi
Research Faculty of Agriculture, Hokkaido University, Sapporo City, Hokkaido, Japan

O. Nagata · T. Hirota · R. Ohtomo · N. Oka
Hokkaido Agricultural Research Center, NARO, Sapporo City, Hokkaido, Japan

Present Address:
T. Hirota
Faculty of Agriculture, Kyushu University, Fukuoka City, Fukuoka, Japan

K. Niwa
Zukosha Co., Ltd, Obihiro City, Hokkaido, Japan

M. Shimizu
Civil Engineering Research Institute for Cold Region, Public Works Research Institute, Sapporo City, Hokkaido, Japan

R. Kinoshita · M. Tani
Research Center for Global Agromedicine, Obihiro University of Agriculture and Veterinary Medicine, Obihiro City, Hokkaido, Japan

J. Palta
College of Agricultural and Life Sciences, University of Wisconsin-Madison, Madison, USA

R. Hatano et al. (eds.), *The Soils of Japan*, World Soils Book Series,
https://doi.org/10.1007/978-981-15-8229-5_5

5.1 Outline of Agriculture and Soils

5.1.1 Agriculture

(1) **Early development up until World War II**

Until 1868, Hokkaido was a frontier island where few people, mostly the indigenous Ainu people, lived. At the beginning of 1869, just after the Meiji Restoration, the Japanese Government established the Development Commission, through which it launched a project to develop Hokkaido. Development was undertaken with the strong supervision and financial assistance of the central government, and the commission sent a contingent of colonial troops to develop and defend the island.

Unlike the other Japanese islands, Hokkaido is on the border between the temperate and subarctic zones. To adapt to the cool climate, modern technology from the USA and Europe was introduced. The Development Commission invited many foreign specialists, including Horace Capron, United States Commissioner of Agriculture, to lay the foundation of industries. During the early stage of development, the Commission undertook a strategy of agricultural development to introduce and extend the upland/dairy farming system on the advice of foreign agronomists.

The Hokkaido Government (established in 1886) initiated a survey to select suitable sites for cultivation and planned the placement of farmlands and urban areas partitioned into a grid by straight roads, forming a landscape that is particular to Hokkaido. At this time, development was led primarily by the private sector. After a law to dispose of undeveloped nationally owned land was enacted in 1897, many large farms were founded and accepted immense numbers immigrants from all over Japan. The peak period of immigration spanned almost 30 years from 1890 to 1920, and the population of Hokkaido surpassed 2 million in 1917.

Agricultural development in Hokkaido began with field crops combined with dairy farming, which were later followed by rice cultivation. Field crops first spread in lowlands with good drainage and then extended to uplands with volcanic ash soils and other soil types, furthermore, spread to the hilly lands along with livestock farming. In the 1890s, the system of agricultural work using horses and plows became widespread, making it possible to cultivate large areas. However, repeated cultivation without the use of fertilizer led to a decline in the fertility, and therefore the productivity, of the soil. During World War I, the price of beans, potato starch, and other products soared in overseas markets; as a result, exports increased rapidly, and the development of farmland also progressed. Mechanical soil improvement has been carried out since and has taken over after World War II (Sakuma 1987).

Although rice cultivation was initially considered inappropriate by foreign experts, most immigrants were farmers familiar with rice farming, and after migrating to Hokkaido they continued their efforts to establish rice cultivation on the island. Moreover, in 1892, the Hokkaido Government adopted policy to encourage the cultivation of rice. Paddy fields increased in number, replacing field crops in lowland areas and spreading throughout almost all of Hokkaido (excluding parts of the north and northeast) in the 1920s. In terms of technology, the expansion of rice cultivation was encouraged by the development and diffusion of cold-resistant rice varieties, the development of direct seed cultivation technology, the improvement of irrigation facilities, and so on. The Ishikari Plain and the Kamikawa Basin became areas of large-scale rice farming.

Regarding fertilization technology, the application of superphosphate started in the suburbs from around 1900. Its use spread a little later, but the average fertilization amount was small. From around 1925, in addition to superphosphate, fish meal, soybean meal, and the like were used for fertilization. However, the amount of fertilizer was drastically reduced at the end of World War II (Imura 1987).

The area of arable land in Hokkaido, which was 45,000 ha in 1890, had reached 983,000 ha in 1938 (Fig. 5.1). By the 1920s, the majority of the island's current agricultural land area, excluding pastures, had been developed. However, the decline in productivity due to exhausted soil fertility in the uplands became serious, and expansion to the marginal zone of rice cultivation worsened cold damage in the 1930s. Additionally, due to the Great Depression and declines in the labor force due to war, the area of arable land declined toward the end of World War II.

(2) **After World War II**

Immediately after the end of World War II, the development of agricultural land was carried out with urgency, with the

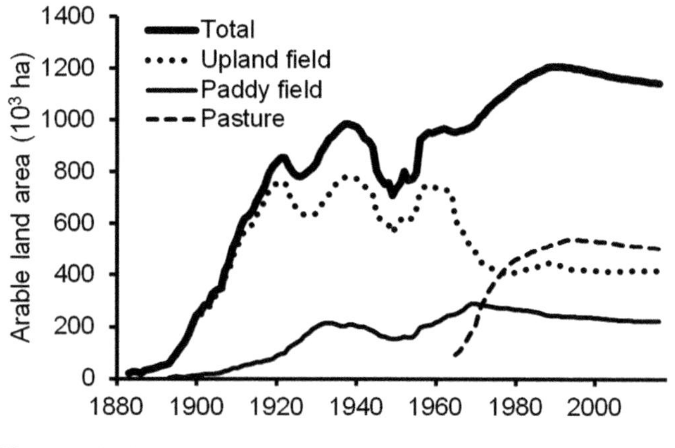

Fig. 5.1 Arable land area in Hokkaido. Figure supplied by Hiroyuki Shiga

aim of accepting surplus population and increasing food production. However, it was difficult for colonists to settle due to severe land conditions. In 1952, a comprehensive development plan was launched, involving the development of large-scale agricultural land, which positioned Hokkaido as a food production base in Japan. The development of paddy fields in the Ishikari peatlands and the construction of dairy pilot farms in the Konsen area were implemented, and by 1985 a total of 600,000 ha of farmland had been created. The area of cultivated land reached 1,209,000 ha in 1990 and then gradually declined to 1,146,000 ha in 2016.

Together with the "outline of the soils of suitable land for agriculture," which summarized the soil survey results of arable land in Hokkaido, the outline map of problem soils in Hokkaido (1:1,000,000 scale) was published in 1951 and was widely used as a basis for agricultural development and land improvement plans.

In Japan's period of high economic growth from around 1960, agricultural machinery (e.g. tractors) was introduced in response to the shortage of rural labor. With projects based on the Basic Law of Agriculture, large-scale field improvement, such as rezoning of areas, flattening, setting of underdrain, and so on, was advanced, and the efficiency of machine work and productivity were improved. With the development of dairy farms, pastures grew rapidly, and at the same time paddy rice production expanded. As a result, the area of uplands decreased. The agricultural land area per household increased due to the decrease in farming. However, the rice production adjustment policy that began in 1970 caused a sharp decline in the cultivation of paddy rice; as a result, rice cultivation in the eastern area of Hokkaido almost disappeared, and various field crops were introduced in paddy fields.

In Hokkaido, acidic soils, volcanic ash soils, peat soils, and heavy clay soils were known as "problem soils" before the war; however, after 1960, these soils were rapidly improved with the introduction of large machines: drainage and mineral soil-dressing were carried out in peatland; underdrainage and subsoiling were performed in heavy clay soils; and coarse volcanic soil, the improvement of which was delayed, could be improved by mixing tillage and so on (Nishimune and Sekiya 1999).

After World War II, the application of fertilizer increased rapidly both in paddy rice fields and upland farms, and yields increased accordingly. The application rate of nitrogen, phosphorus, and potassium per area almost doubled between 1950 and 1970, leading to a yield increase of 100% in upland crops and of 60% in paddy rice.

5.1.2 Soils

(1) Climate

Hokkaido is located in Northern Japan. The cold current "Oyashio" flows east side, and the warm current "Kuroshio" flows west side. The annual average temperature is 5–10 °C, while the annual average soil temperature at 0.5 m depth is estimated as 9.3 °C in Sapporo and 7.4 °C in Nemuro. The soil temperature regime is classified as mesic to frigid. Precipitation is in the range of 700–1200 mm per year, which is relatively low for Japan. Precipitation is low from February to June and peaks from August to September. The western area of Hokkaido receives large amounts of precipitation from October to January, causing heavy snowfall. Therefore, the soil drying season occurs in only a short period from spring to the beginning of summer. In Eastern Hokkaido, the temperature is very low in winter due to radiative cooling, and there is little snowfall. As a result, soil freezing occurs from December to April, to a depth of about 0.1–0.4 m. Recently, however, the soil freezing depth has changed to be shallow due to global warming.

(2) Soil parent materials

The geological basement of Hokkaido is composed of volcanic rocks (mainly andesite), metamorphic rocks, and sedimentary rocks such as sandstone and conglomerate of tertiary age. These rocks are the parent materials for soils such as brown forest soils in mountainous and diluvial plateau areas. Cultivated areas are scarce in these areas due to the harsh climate. In coastal and riverine areas, the land rose up in diluvium during the Quaternary Period (glacial period), forming terraces. The fine-textured sediment in these terraces has been dehydrated and compacted, forming "Terrace soil", which is characterized by its fine texture and hardness. In alluvial land around rivers, lowland soils are widely distributed. In particular, peat is formed in marshy areas in the lower reaches of large, calm rivers such as the Ishikari, Teshio, and Kushiro Rivers. These peat areas had generally been reclaimed by the end of the 1940s. Furthermore, as there are 20 active volcanoes on Hokkaido, volcanic ash or pyroclastic deposits widely cover the island. Some of these pyroclastic deposits were deposited 20 years ago or earlier, while others were deposited earlier in the Quaternary (Committee on Nomenclature of the Pyroclastic Deposit in Hokkaido 1974). These pyroclastic deposits provide parent material for volcanic ash soils. Volcanic ash soil formed from volcanic material which deposited less than a thousand year is not well weathered and is classified as Volcanogenous Regosol. In the southern half of Hokkaido, volcanic ash soils or Volcanogenous Regosols are generally distributed. The texture of the deposited volcanic ash typically changes according to the distance from volcanoes.

(3) **Soil distributions**

The cultivated area of Hokkaido is approximately 1,146,000 ha (2016 estimate). This accounts for 13.7% of the total area of Hokkaido (8,342,000 ha) and for 25.6% of the total cultivated area in Japan. The area of paddy field is relatively low, at 223,000 ha, as is the net area of cultivated paddy rice, at 105,000 ha. The area of upland field is 924,000 ha and is composed of ordinary upland field (416,000 ha), orchard land (3000 ha), and grassland (505,000 ha).

A 1:200,000 soil map, which includes non-cultivated area, was published by the Land Bureau (National Land Agency 1975–1979). However, in this map, the soil classification system is not consolidated for different land uses. Kanda et al. (2016) attempted to reclassify soil names under a single new soil classification system. Based on this system, Andosols occupy 43.1% of Hokkaido, brown forest soils 30.2%, and fulvic soils 11.1%. A 1:50,000 soil map of cultivated land was published in 1959–1979 based on as part of a national soil survey project and is widely used. Kanda et al. (2017) reclassified the original data from this map into the new soil classification system and reported that Andosols occupied 40.9% of the total cultivated area of Hokkaido and Fulvic soils 28.7%. Soil distribution areas can now be determined based on the new Soil Classification System of Japan (Fifth Committee for Soil Classification and Nomenclature of the Japanese Society of Pedology, 2017) (Table 5.1). Local governmental engineers use four divided soil types, namely volcanic ash soils, terrace soils, lowland soils, and peat soils, when developing new fertility techniques and leading farmers. The volcanic ash soil type in Hokkaido includes Volcanogenous Regosols.213. These relationships are shown in Table 5.1.

(4) **Soil improvement**

The soil drying period is shorter in Hokkaido than in the other areas of Japan. This fact means that upland crops can be cultivated without irrigation and that there is no risk of salt accumulation in surface soils. However, excessive rainfall on poorly drained soils usually results in a high water table or periodic flooding. This results in poor soil aeration, which may reduce crop growth. Therefore, drainage has been the most important countermeasure for upland crop fields. In paddy fields, drainage is important for producing high-quality rice or cultivating upland crops (under the paddy upland rotation). For these purposes, underground drainage pipes or open-channel drainage have been set in many fields for a long time. Thus, poorly drained soils have been converted to well-drained soils.

Soil improvement projects have been conducted in order to increase soil productivity. One major improvement program, involving soil-dressing, was carried out for paddy fields, which were covered by peat soils. After the 1980s, soil-dressing was widely carried out in upland fields with the objective of improving soil character—for example, increasing the depth of the plow layer or changing soil texture. More recently, soil-dressing has been carried out by national or local governmental land improvement projects. Mixing tillage, which mixed surface with subsoil, has been carried out in volcanic ash soil. Gravel removal has been performed in stony soils, which cover 12% of the cultivated area in Hokkaido. Subsoil breaking has also been carried out. Hardpan has formed directly under a surface layer with plowing or rotary tillage operations. Calcium or phosphate application has been continued. The inherent soil properties have been changed due to these improvement actions, now that appropriate fertilizations can be chosen based on the soil properties of individual fields.

5.2 Paddy Fields

5.2.1 Paddy Soil

(1) **Soil groups and characteristics of paddy fields in Hokkaido**

Of the total area of paddy fields in Hokkaido (ca. 223,000 ha), about 80% is located in the Ishikari, Sorachi, and Kamikawa areas. Gray Fluvic soils are most widely distributed in these areas, followed in order by Gley Fluvic soils, Peat soils, and Brown Fluvic soils. These four soil groups account for 80% of total paddy field area.

Gley Fluvic soils are mainly distributed in low wetlands and swampy areas of river basins, which are common in the Ishikari, Sorachi, and Kamikawa areas. These areas are characterized by a high groundwater level, clayey soils, and extremely low permeability. Gray Fluvic soils have a lower groundwater level than Gley Fluvic soils and slightly inferior drainage. These soils are mainly distributed in the middle and lower reaches of river basins and are common in the Sorachi and Kamikawa areas. Brown Fluvic soils are widely distributed in Hokkaido, mainly in natural levees and alluvial fans. These soils have good drainage, low groundwater level, and more oxidized condition. Brown Fluvic soils are widely used as uplands and grasslands, and in the Kamikawa

Table 5.1 Distribution rate of soil types in Hokkaido with the Soil Classification System of Japan and the relationship to Hokkaido agricultural diagnostic soil types

Classification System of Japan Great Groups Groups	Whole area (%)		Cultivated area (%)		Relationship to Hokkaido agricultural diagnostic soiltypes
Organic Soils		3.3		8.6	
Peatsoils	3.3		8.6		I. Peat soils
Andosols		41.8		40.9	
Podzolic Andsols	<0.1		0.0		-
Regosolic Andosols	7.0		7.8		II. Volcanic soils
Gleyed Andosols	<0.1		0.1		II. Volcanic soils
Wet Andosols	0.0		6.7		II. Volcanic soils
Non-allophanic Andosols	6.0		1.4		II. Volcanic soils
Allophanic Andosols	28.0		24.9		II. Volcanic soils
Podzols		1.2		0.0	
Podzols	1.2		<0.1		(not fixed)
Fluvic soils		10.3		28.7	
Fluvic Paddy soils	0.2		0.0		-
Gley Fluvic soils	1.4		5.5		III. Low land soils
Gray Fluvic soils	2.6		9.2		III. Low land soils
Brown Fluvic soils	5.3		13.8		III. Low land soils
Regosolic Fluvic soils	0.8		<0.1		III. Low land soils
Red-Yellow soils		0.9		1.9	
Argic Red-Yellow soils	<0.1		1.0		IV. Terrestrial soils
Cambic Red-Yellow soils	0.9		0.8		IV. Terrestrial soils
Stagnic soils		1.8		7.8	
Stagnogley soils	<0.1		0.5		IV. Terrestrial soils
Pseudogley soils	1.8		7.3		IV. Terrestrial soils
Eutrosols		<0.1	0.0		
Magnesian Eutrosols	<0.1		0.0		-
Calcareous Eutrosols	<0.1		0.0		-
Brown Forest soils		31.9		9.8	
Brown Forest soils	31.9		9.8		IV. Terrestrial soils
Regosols		8.1		2.3	
Volcanogenous Regosols	4.9		2.0		II. Volcanic soils
Sandy Regosols	0.6		0.3		(not fixed)
Lithosols	0.9		<0.1		-
Terrestrial Regosols	1.8		<0.1		-
Others		0.6		-	
Rocks and Stone	<0.1		-		
City and Water	0.6		-		
total		100.0		100.0	
total area (km^2)		81,488*		11,556**	

* official data of whole area is 83,424 (km^2)
** official data in 2010 of cultivated area
italic : soil group with no distribution

and Sorachi areas are used as paddy fields. Peat soils are mainly distributed in gentle streams of rivers and back-marshes, common in the Ishikari and Sorachi areas. These areas have a high groundwater level, poor drainage, and low bearing capacity.

(2) **Land consolidation projects in paddy fields**

In Hokkaido, many paddy fields are poorly drained and unsuitable for use as agricultural land. Therefore, the consolidation of drainage canals and subsurface drainage by public works has been actively carried out. From the late 1940s, with the assistance of loans from the World Bank, large-scale public works projects for irrigation and drainage canals, subsurface drainage, and soil-dressing were implemented, especially in peatland (Hasegawa and Tabuchi 1995). A total of 65,000 ha of peatlands was converted into paddy field in the 17 years from 1954 until 1970 (Matsushita et al. 1985).

Each year, public works projects improve subsurface drainage on 5000 ha of land through the installation of subsurface drainage systems. To increase the mechanization and efficiency of agricultural work, the land readjustment of large fields (1 ha or more) is being advanced. To promote subsurface drainage, drain pipes are installed at a depth of 60–80 cm with a spacing of 10 m, and filter material is filled from the drain pipe to the bottom of the surface layer. Furthermore, subsurface drainage is achieved by constructions that allow water to flow from irrigation canals into underdrain pipes for the purpose of cleaning and subirrigation. Rice husk has been used as a filter material, but it decays with time because of converted paddy field. Hence, the use of filter materials other than rice husk was examined; now, wood chips and inorganic materials such as gravel and volcanic gravel are used. The use of organic filter materials for improving subsurface drainage in paddy fields reduces the occurrence of methane, and a carbon storage effect is observed (Kitagawa and Tsukamoto 2014).

Clayey soil-dressing in paddy fields has been mainly conducted to increase the depth of the plow layer and the bearing capacity in peatlands. Recently, however, a project using sandy soil-dressing with a high silicon content has been conducted, mainly in the Ishikari area, to improve the taste of paddy rice.

(3) Changes in soil physicochemical properties in paddy fields

Table 5.2 shows the changes in the physicochemical properties of paddy fields in the past 50 years and a comparison of soil physicochemical values from 2010 compared with soil diagnosis reference values described in the "Hokkaido Fertilizer Recommendations 2015" (see Sect. 8.1).

The results show that sites with a soil hardness of the subsoil layer (measured by push cone-type penetrometer; "Yamanaka" method) of 20 mm or more occupied 39% of all sites. The total carbon and nitrogen content declined between the years 1970 and 1980; one reason for this result may have been the dilution of the plow layer by land readjustment during this time. Many converted paddy field sites require soil pH correction; 44% of sites had a pH of less than 5.5. Soil exchangeable potassium stopped increasing after 1990; sites that fall below the potassium soil diagnosis criterion account for 16% of cases, and sites that exceed the soil diagnosis criterion account for 17% of cases. The available phosphorus content in soil, as measured by the Bray-2 method, was on an increasing trend at the beginning of the survey, but has been flat since 2000; sites that exceed the soil diagnosis criterion for available phosphorus content account for 89% of cases. The difference in available phosphorus content between soil types is small. The recommended

Table 5.2 Changes in physicochemical properties of paddy field and evaluation of current conditions in Hokkaido

Properties	Changes during period from 1970 to 2010	Mean value in the year 2010	Percentage of sites (%)[a]			Total number of sites
			< criterion[a]	Meet the criterion[a]	> criterion[a]	
Soil hardness of subsoil layer	No change	18 mm	11	51	39	72
Total carbon	1970 ~ 1980: decrease	27 g kg^{-1}	NA[b]	NA	NA	74
Total nitrogen	1970 ~ 1980: decrease	2.3 g kg^{-1}	NA	NA	NA	74
pH(H_2O)	No change	5.5	44	44	12	74
Exchangeable K	1970 ~ 1990: increase	190 mg kg^{-1}	16	68	17	74
Available P	1970 ~ 2000: increase	212 mg kg^{-1}	3	8	89	74
Available N	1970 ~ 1980: decrease, 1980 ~ 2005: increase	133 mg kg^{-1}	NA	NA	NA	73
Available Si	2000 ~ 2005: increase	49 mg kg^{-1}	96	4	NA	73

[a]The value compared with the soil diagnosis criterion is the mean value in 2010. The soil diagnosis criterion of soil hardness of subsoil layer is 15 ~ 20 mm or less, pH(H_2O) is 5.5 ~ 6.5, exchangeable K is 125 mg-K kg^{-1}(150 mg-K_2O kg^{-1}) ~ 249 mg-K kg^{-1}(300 mg-K_2O kg^{-1}), available P (Bray No.2) is 44 mg-P kg^{-1}(100 mg-P_2O_5 kg^{-1}) ~ 87 mg-P kg^{-1}(200 mg-P_2O_5 kg^{-1}), and available Si (40-1 week incubation) is 75 mg-Si kg^{-1}(160 mg-SiO_2 kg^{-1}) or more

[b]NA Not available because of no criterion setting

amount of phosphorus fertilizer calculated from the actual field value of available phosphorus content in soil is estimated to be approximately 62% of the fertilization standards in Hokkaido. The available nitrogen content in soil has gradually increased, and in 96% of sites the available silicon (40 °C; 1 week incubation) is less than the soil diagnosis criterion.

5.2.2 Fertility Management

(1) **Nitrogen**

In the case of fertilizer incorporation in the plow layer, the fertilizing standards of nitrogen in the Hokkaido Fertilizer Recommendations 2015 (see Sect. 5.8.1) are 65–95 kg ha^{-1} in lowland soils, 60–85 kg ha^{-1} in terrace soils, and 50–75 kg ha^{-1} in peat soils.

The side-dressing application (applying fertilizer in 3 cm side line of a hill at a depth of 3–5 cm) has been developed for accelerating the initial growth due to high nitrogen concentrations in nearby rice roots. For the side-dressing method, the rate of nitrogen application is 30–40 kg ha^{-1} and the total rate of nitrogen fertilizer application is subtracted by 5 kg ha^{-1} from the standard. The side-dressed nitrogen is mainly absorbed by the plants before the flag leaf stage and is less distributed in polished rice. Therefore, this method was effective for producing low-protein rice.

It is recommended to apply about 2 kg ha^{-1} of nitrogen topdressing within seven days of the panicle formation stage, as this topdressing increases the number of ears and the number of unhulled rice grains per ear and can increase resistance to adverse weather conditions. Nitrogen topdressing after the flag leaf stage should be avoided, because during that time nitrogen is allocated to polished rice at high rate, which raises the protein content.

(2) **Phosphorus**

In cold regions, the enrichment of soil phosphorous has been emphasized, particularly as a countermeasure to cool summer damage, because phosphorous has the effect of accelerating initial plant growth. The fertilizing standard of phosphorous is determined at 35 kg P ha^{-1} (80 kg P_2O_5 ha^{-1}) in Hokkaido, but it is desirable to increase or decrease this based on soil diagnosis.

With regard to the method of fertilizer application, we should put emphasis on using the side-dressing method due to its effectiveness. It is reasonable that phosphorous absorption by rice plants depends on the soil condition in the latter half of the growing period. However, excessive amounts of fertilizer have been applied in many paddy fields. It is necessary to improve the phosphorous level in soil, because the amount of available phosphorous is very high; the mean value is 218 mg P kg^{-1} (500 mg P_2O_5 kg^{-1}).

(3) **Potassium**

The fertilizing standard of potassium was set at 66 kg K ha^{-1} (80 kg K_2O ha^{-1}) in the Hokkaido Fertilizer Recommendations 2015 (see Sect. 5.8.1), based on the degree of potassium absorption by rice plants. The standard concentration of exchangeable potassium in paddy fields is 125–249 mg K kg^{-1} (150–300 mg K_2O kg^{-1}). A total of 25 kg K ha^{-1} (30 kg K_2O ha^{-1}) of applied potassium fertilizer should be added to the fertilizing standard if the amount of exchangeable potassium is 125 mg K kg^{-1} or less, and 25 kg K ha^{-1} (30 kg K_2O ha^{-1}) of applied potassium fertilizer should be subtracted from the fertilizing standard if the amount of exchangeable potassium is 249 mg K kg^{-1} or higher.

Potassium is supplied to the soil by the return of rice straw, as 70% of the potassium in rice plants is distributed in the stems and leaves. We advise that potassium fertilizer should be reduced by 33 kg K ha^{-1} (40 kg K_2O ha^{-1}) when stems and leaves (10 t ha^{-1}) are applied and by 17 kg K ha^{-1} (20 kg K_2O ha^{-1}) when compost (10 t ha^{-1}) is applied.

(4) **Silicon**

When we studied rice plant growth in the paddy field for 10 years of successive calcium silicate (slag silicic acid manure) application, the yield was higher than in a control field. The rice plants absorbed a lot of silicon, which improved brown rice production efficiency and decreased the protein content in polished rice effectively. Accordingly, an index of available silicic acid of soil was formulated for the purpose of achieving a protein content of less than 80 g kg^{-1} of polished rice. Based on that index, soil with an available silicon concentration of below 47 mg Si kg^{-1} (100 mg SiO_2 kg^{-1}) is defined as lacking, soil with an available silicon concentration of between 47 and 75 mg Si kg^{-1} (100–160 mg $SiO_2^{-1}{}_{kg}$) is defined as slightly lacking, and soil with an available silicon concentration of above 75 mg Si kg^{-1} (160 mg $SiO_2^{-1}{}_{kg}$) is optimal. Half of paddy fields in Hokkaido require improvement; that is, they are diagnosed as lacking in available silicon. In particular, principal paddy field areas (Sorachi and Isikari areas) are low in silicon.

5.3 Upland Fields

5.3.1 Upland Soil

(1) **Characteristics and problems of upland soil**

In Hokkaido, problematic soils (see Sect. 5.1), which have serious problems in terms of crop productivity and soil management, are widely distributed, and the improvement of their physical and chemical properties according to the characteristics of each soil is required.

Volcanic ash soil generally has poor chemical properties, such as high phosphate retention capacity, and wet soil groups such as Gley Andosols and Wet Andosols have poor drainage. Terrace soil is highly viscous and often tight, and soil swelling and drainage measures are necessary. In lowland soil and fields converted from paddy field, drainage improvement measures are often required.

For these problem soils, soil improvement through public works, soil-dressing, open-channel drainage, subsurface drainage, subsoil breaking, phosphate fertilization, liming, and so on are highly effective. For large-scale countermeasures such as soil improvement projects, the reader is referred to Sect. 5.1.2, where we describe improvement technology that farmers can carry out within the range of usual farming work.

(2) **Subsoil breaking and drainage technology**

In order to maintain good physical properties of soil, it is important to suppress subsoil compaction and not to knead soil during tractor plowing. In the upland fields of Hokkaido, bottom plowing was typically done in autumn after the crop harvest. However, in recent years, it has been encouraged to perform a bottom plowing in spring where the soil is dry or a chisel plowing without kneading the soil (Onodera et al. 2004).

Hard plow pan is formed by bottom plowing and suppresses crop growth and water permeability. The hardness index value of plow pan is generally 1.5 MPa or more for a cone penetrometer or 20 mm or more for a Yamanaka-type soil hardness tester. The conducting of subsoil breaking is effective in fields which have a plow pan with these index values or greater (Nakatsu et al. 2004).

In recent years, wider types of construction, including a type that does not mix subsoil in the surface soil, a mole drain construction machine that does not use underdrain pipe (Kitagawa et al. 2008), and so on, have been developed in addition to the ordinary subsoil breaking instrument. These instruments are highly effective, and farmers can use them to improve the soil drainage. In order to maintain the function of underdrainage over a long period of time, we encourage the continual use of the subsoil breaking instrument.

Open-channel drainage is effective in fields where surface water occurs frequently after heavy or prolonged rain. Open-channel drainage is a farming technique involving digging a groove in a field or around a field and discharging the surface water outside the field.

The sloping of farm field is a farming technique which involves sloping a field with the assistance of a laser leveler (Wakasugi and Fujimori 2006). Farm field sloping promotes gravitational drainage and is effective for draining surface water.

In addition to the techniques mentioned here, the continuous application of organic material is very important for soil improvement. If it is difficult to obtain high-quality manure or compost, the cultivation of green manure is desirable.

5.3.2 Cropping Systems

(1) **Cropping system of upland field**

A characteristic cropping system is implemented in every area in Hokkaido according to the natural environment and the management scale. Tokachi area is characterized by large-scale farm management, and crop rotation of 3–4 years is conducted, mainly combining sugar beet, potato, wheat, and beans. However, the continuous cropping of wheat is increasing due to scale expansion and labor shortage in this area. Abashiri area is characterized by cool early summer temperatures, and hence, bean cultivation is not common. In this area, crop rotation of 2–3 years is conducted with sugar beet, potato, and wheat. In some fields, the continuous cropping of sugar beet and/or the alternating cropping of sugar beet and potato is also conducted. In the Shiribeshi area, where cultivated field is limited, Adzuki beans and potato are mainly cultivated. In Kamikawa area, wheat, beans and sugar beet are cultivated in the upland paddy rotational fields of the central part and in hilly land in the north and south.

(2) **The effect of crop rotation on soil fertility**

Plowing in crop residue is effective for the maintenance and improvement of soil fertility, as reflected by measures such as total carbon and total nitrogen. In Haplic Allophanic Andosols of the Tokachi area, the annual application rate of organic matter for maintaining the total carbon and total nitrogen of upland fields was 2.5 Mg ha^{-1} in dry matter. In the general 4-year crop rotation of sugar beet, soybeans, wheat, and potato, since the annual organic matter application rate is 2.4 Mg ha^{-1} in dry matter, it is presumed that the total carbon and total nitrogen in the soil can be maintained mostly by 4-year crop rotation. Additionally, when the application rate of organic matter was about 5.0 Mg ha^{-1} in dry matter, both the total carbon and total nitrogen in the soil rose by about 10% (Nakatsu and Tamura 2008). However, as with chemical fertilizers, it is necessary to pay attention to the fact that the excessive application of organic matter, such as compost, may cause excess nitrogen loading and nitrate contamination of groundwater.

The amount of nutrient input and output as fertilization, incorporating crop residue, and harvest depend on the crop species. In the case of sugar beet, the difference between nutrient input and nutrient output is greatly positive for nitrogen and potassium (Okumura et al. 1997). Because crop rotation combines multiple crops with different nutrient balances, it is also convenient for optimizing soil fertility.

(3) **Improvement of cropping systems**

Incorporating green manures into crop rotations is effective not only for avoiding replant disease but also for stabilizing the production of profitable crops and supplying organic matter to the soil. In Hokkaido, the proportion of succeeding crop green manures cultivated after the harvest of profitable crops is large. Oat, sunflower, and white mustard are mainly cultivated in winter wheat fields. The dry matter of green manure used is 2.6–0.6 Mg ha^{-1}, the amount of nitrogen reduction is 49–88 kg ha^{-1}, and the reduction in potassium is 102–158 kg K_2O ha^{-1}. Since fallow green manures cultivated by abandoning the cultivation of profitable crops are cultivated from spring to summer with good weather conditions, dry matter is 4.6–10.9 Mg ha^{-1}, the amount of reduction in nitrogen is 80–151 kg ha^{-1}, and the reduction in potassium is 158–457 kg K_2O ha^{-1}, more than for the succeeding green manures. The reduction in these elements in the field not only increases the amount of humus, nitrogen fertility, and cation exchange capacity (CEC), but also improves the overall physicochemical properties of the soil, for example, by decreasing soil hardness (Akashi et al. 2005).

5.3.3 Fertility Management

(1) **Current status of upland soils**

Physical properties: The topsoil depth and bulk density of the subsoil have both been increasing (Table 5.3). In Tokachi, a major upland area, as subsoil has been becoming harder and a stiff plow pan has been forming just under the plowed layer (topsoil), drainage degradation and reduced root elongation have been widely observed (Nakatsu et al. 2004). Therefore, many types of technology for subsoil breaking are employed frequently.

Chemical properties: The most controversial problem nowadays is the trend of increasing soil acidification (Table 5.3). The reason for this has been recognized as an avoidance of calcium application due to the fear of epidemics of soilborne diseases such as rhizomania of sugar beet and potato scab, which propagate in higher-pH soil. However, in fields with low-pH soils, growth disorders and yield losses of directly sown sugar beet have been reported (Fueki et al. 2002). Calcium application is an indispensable means to avert soil acidity, and a technology allowing the minimum fertilization of calcium by band application, which only raises soil pH around roots, is available (Furudate et al. 2000).

(2) **Soil diagnosis and fertilization**

Standard application amounts for nitrogen, phosphate, and potassium fertilizers are set up for each area and soil type. For example, for sugar beet, the standard amounts of fertilizer are 140–180 kg N ha^{-1}, 100–110 kg P_2O_5 ha^{-1}, and 140–160 kg K_2O ha^{-1}; these are the necessary fertilizer amounts required to obtain a standard crop yield at a standard soil fertility.

The soil criteria of phosphate and potassium are 100–300 mg P_2O_5 kg^{-1}, as determined by the Truog method and 150–300 mg K_2O kg^{-1} as exchangeable potassium (extracted by pH 7 ammonium acetate), respectively. For both phosphate and potassium, 1.3–1.5 folds of fertilization standard are applied when the analyzed soil conditions are lower than the aforementioned criteria, while 0.8–1.0 folds

Table 5.3 Transitional trend of upland soil properties in Hokkaido

Year	Topsoil depth	Bulk density of subsoil	Topsoil		Topsoil pH	Exchangeable cation of topsoil (mg kg^{-1})			Available phosphate of topsoil
			T-C	T-N					
	(cm)	(g mL^{-1})	(g kg^{-1})	(g kg^{-1})		CaO	MgO	K_2O	(mg kg^{-1})
1970	18.3	0.87	0.48	0.04	5.8	3280	280	270	100
1985	23.6	0.98	0.44	0.03	5.7	2950	450	540	240
2000	27.1	1.08	0.31	0.03	5.6	2440	390	420	310

of fertilization standard are applied when the analyzed conditions are higher than the criteria.

A total of 30–40 kg MgO ha^{-1} of magnesium is applied when the analyzed soil conditions (exchangeable magnesium extracted by pH 7 ammonium acetate) are within the range of 250–450 mg MgO kg^{-1}. A total of 1.3–1.5 folds of magnesium fertilizer are applied when the soil concentration of exchangeable magnesium is lower than 250 mg MgO kg^{-1}, and no magnesium fertilizer is applied when the concentration is higher than 450 mg MgO kg^{-1}.

For calcium fertilizer application, the soil pH criterion (5.5–6.5 as pH(H_2O), soil:water = 1:2.5) is regarded as more important than the concentration of exchangeable calcium; that is, calcium application is recommended to keep soil pH within the range 5.5–6.5.

(3) Fertilization with allowance for the application of organic matter

Application of organic matter is indispensable as a countermeasure against the degradation of soil fertility, and the application of 10 t ha^{-1} of manure every year is recommended in Hokkaido.

The use of fertilizers needs to be reduced, with allowance for the application of nutrient elements provided by applied organic matter. For example, 1 kg of N fertilizer can be reduced per 1 Mg of manure applied, because approximately 20% of N in manure could be taken up by plants, and the total amount of N per 1 Mg of manure is estimated as approximately 5 kg. When applying manure consecutively every year, the available N from manure is thought to increase due to the accumulation effect. When applying manure for 5–10 consecutive years, the amount of nitrogen fertilizer can be reduced by 2 kg of N fertilizer per 1 Mg of manure, and when applying manure over 10 consecutive years, the amount of fertilizer can be reduced by 3 kg of N fertilizer per 1 Mg of manure applied.

(4) Trace element fertilization

In Hokkaido, a deficiency of zinc (Zn) and copper (Cu) is frequently observed in typical commercial agricultural fields.

Zinc deficiency tends to occur when growing beans and corn (Fig. 5.2). A basic precaution against zinc deficiency is the application of zinc fertilizer to the soil. The criterion concentration of soil available zinc is set as 2–40 mg kg^{-1}, and 50 kg ha^{-1} of zinc sulfate (7-hydrate) is recommended to be applied to soil when a deficiency is observed, that is, when the concentration of soil available zinc is lower than the criterion.

Copper deficiency tends to occur in wheat (Fig. 5.3). The criterion concentration of soil available copper differs for different soil humus contents: The criterion is 0.7 mg kg^{-1} when soil humus content is less than 50 g kg^{-1}; 0.5 mg kg^{-1} when soil humus content is in the range of 50–100 g kg^{-1}; and 0.3 mg kg^{-1} when soil humus content is more than 100 g kg^{-1}. Between 20 and 40 kg ha^{-1} of copper sulfate (pentahydrate) is recommended to be applied to soil when a deficiency is observed, that is, when the concentration of soil available copper is lower than the criterion.

Fig. 5.2 Typical symptom of zinc deficiency of soybean (left) and azuki bean (right). Figure supplied by Nobuhiko Fueki

Fig. 5.3 Copper deficiency occurring in Tokachi region (Winter wheat maturity was delayed due to sterile symptom of copper deficiency). Figure supplied by Nobuhiko Fueki

5.4 Vegetable Fields

5.4.1 Vegetable Field Soils

(1) **Characteristics of the soil in vegetable fields**

Large amounts of fertilizer are often used for vegetable or flower fields in order to improve yield and quality. The contents of inorganic nitrogen, base, and phosphate are high in the soils of vegetable and flower fields, resulting in the accumulation of salt. Such soil is liable to lack micronutrients and may cause various physiological disorders in the plants.

Regarding soil in greenhouses, the stored water tends to move upward due to the absence of rainfall, making it easier for inorganic nitrogen and salt to accumulate on the surface. If such salt accumulation results in poor growth, a large amount of irrigation water is used to wash the salt down toward the subsoil as a corrective measure. However, as nitrate nitrogen can easily move toward the lower layers if the amount of irrigation water is great, this often results in a greater amount of additional fertilizer being required. Furthermore, as the movement of inorganic nitrogen and salt toward the subsoil caused by irrigation is only temporary due to the upward movement of the soil water, inorganic nitrogen and salt will immediately start to accumulate again; repeating this process will result in the discharge of nutrients, which is a burden on the environment.

(2) **Transitional trend of vegetable field soils**

In Hokkaido, a survey analysis of the physicochemical characteristics of agricultural soil was conducted every 5 years from 1979 to 2011, shedding light on the actual condition of the soil as well as on the direction of changes.

For example, the transition of the concentration of exchangeable potassium is shown in Fig. 5.4. Although significant differences can be seen in some years, it transitioned within the range of 410–520 mg kg^{-1}, that is, higher than the reference value of 150–300 mg kg^{-1}.

The main physicochemical characteristics of the soil are shown in Table 5.4. The depth of topsoil transitioned within the range of 20–23 cm, while there were no significant changes in the compactness or bulk density of the subsoil. However, 47% of agricultural soils had a compactness that exceeded the reference value, showing the necessity for regular crushing of the subsoil. While no significant change was seen in pH (H_2O), in 49% of agricultural fields the value was under 6.0, showing the need for acidity correction. There was no significant change concerning the concentration of exchangeable lime, with 17% of agricultural fields being under the reference value. Additionally, although the concentration of exchangeable magnesium continued to decrease until 2000, its fertilization must be reconsidered. Exchangeable potassium generally remained level, and the amount of fertilizer must be reduced, as 72% of agricultural fields were above the reference value. The Mg/K ratio

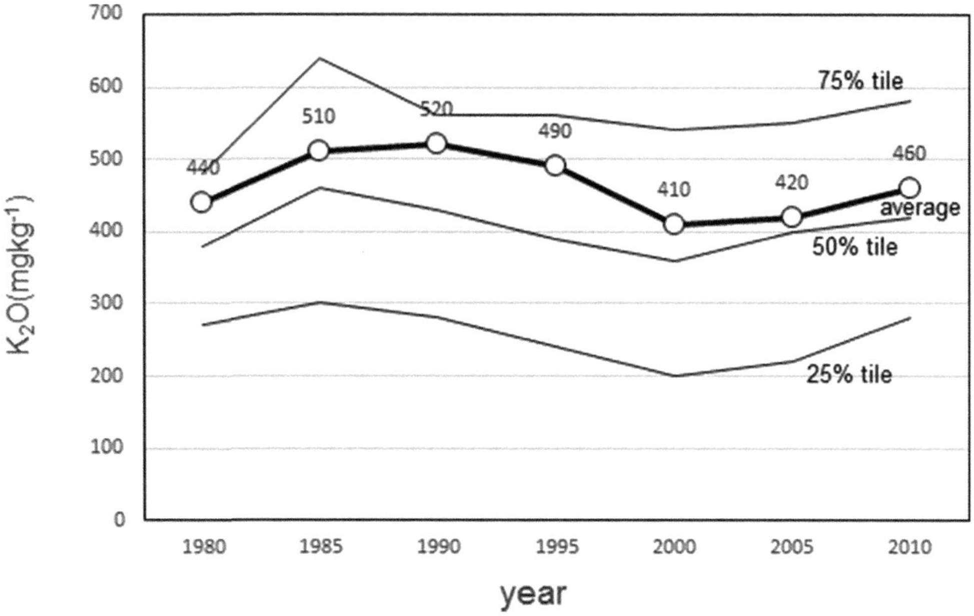

Fig. 5.4 Secular variation of soil exchangeable potassium in fixed points for vegetable fields in Hokkaido. Figure supplied by Yuji Hikasa

Table 5.4 Physical and chemical current status and evaluation of the soil in vegetable fields

Items	Average (2010) n = 123	Soil diagnostic Standard value	Number of points(%) Under	Within	Above
Depth of topsoil(cm)	20.4	20～30			
Compactness of subsoil(cm)	18.3	16～20	20	33	47
pH(H_2O)	6	6.0～6.5	49	35	16
Exchangeable lime($mgkg^{-1}$)	3540	Depends on soil[a]	17	43	40
Exchangeable magnesium($mgkg^{-1}$)	480	250～450	20	27	53
Exchangeable potassium($mgkg^{-1}$)	460	150～300	7	22	72
Mg/K ratio	3.2	2～	36	64	–
Available Phosphate($mgkg^{-1}$)	950	600～800[b]	35	19	46

[a]Coarse textured:1000～1700, midle textured:1700～3500, and fine textured:3500～4900
[b]Applying the standard for onions

generally remained level, with 36% of agricultural fields being below the reference value. No significant changes were seen in available phosphate. Even when applying the standard for onions, which have a high phosphate demand, 46% of agricultural fields exceeded the reference value (Hokkaido Research Organization 2014).

5.4.2 Fertility Management

(1) Open cultivation soil

Hokkaido is a cold region and receives much snow in winter. As such, most vegetable crops in open fields are cultivated only from May to September in this area. Therefore, the following points relating to the soil management of Hokkaido are listed.

(1) Because of the low temperature of the soil, available organic nitrogen accumulates easily in the soil and its mineralization is inactive. Additionally, the low temperature of the soil means that the degradation of organic matter applied to it is slow.
(2) Soil inorganic nitrogen is easily leached by the downward penetration of melting snow. This kind of nitrogen leaching is a common phenomenon in most open fields in Hokkaido. In general, however, vegetables require greater amounts of nitrogen, and their cultivation period is shorter than that of other field crops. Therefore, it is necessary to prevent nitrogen leaching after cultivation.
(3) It is important to enhance the initial growth of vegetable crops to complete the cultivation in a short period of time. In particular, it is necessary to raise soil phosphate content to increase the initial growth of crops in early spring, because phosphate is hard to elute from soil at low temperature.

Just before vegetable cultivation, there is insufficient soil inorganic nitrogen to grow the crops, due to the leaching of some inorganic nitrogen by snowmelt. Therefore, the amount of nitrogen fertilizer to be applied is decided based on the level of soil available organic nitrogen. In the cases of open vegetable fields in Hokkaido, organic nitrogen is tested as the soluble nitrogen extracted by autoclave. The amount of nitrogen fertilizer to be applied to each crop is determined using the Hokkaido Fertilizer Recommendations 2015 handbook (Sect. 8.1). Three standards for the soil available organic nitrogen content are set in this manual. The relations with the soil nitrogen and the nitrogen fertilization amounts vary based on the cultivation period and date of planting of each crop. Some nitrogen fertilization methods for major open vegetable fields in Hokkaido are shown in Table 5.5.

The amounts of phosphate and potassium fertilizer to be used are decided based on the level of the soil available phosphate and exchangeable potassium, respectively. The available phosphate is extracted by the Truog solution (pH 3.0) method in the case of open vegetable fields in Hokkaido. The amounts of phosphate and potassium fertilizer applied to each crop are determined from the abovementioned handbook. Five standards for each of the soil nutrients are set in this manual.

In open vegetable fields in Hokkaido, the standard annual application rate of cattle manure is determined to be 20 Mg ha^{-1} to maintain the physical and chemical properties of the soil. The required amounts of nitrogen, phosphate, and potassium fertilizer are reduced by partially evaluating the nutrients in cattle manure compared to those in chemical fertilizer.

Cattle manure contains approximately 5 kg of each of nitrogen and phosphate (P_2O_5), and 4 kg of potash (K_2O), per 1 Mg of fresh matter basis. The fertilization efficiency for single-year application is estimated to be around 20, 60, and 100% for nitrogen, phosphate, and potash, respectively. The chemical contents of cattle manure listed above allow the amount of chemical fertilizer to be reduced by 1 kg of nitrogen, 3 kg of phosphate, and 4 kg of potash per 1 Mg of fresh cattle manure applied to the soil. The nitrogen fertilization efficiency is particularly low in Hokkaido compared with other Japanese regions since it is the coldest region in Japan. The phosphate fertilization efficiency is the same as the value given in the British Nutrient Management Guide (RB 209) (Agriculture and Horticulture Development Board 2017), although it is somewhat lower than in other Japanese regions. Potash contained in manure is estimated to have the same content as that of chemical fertilizer since potash is almost completely water soluble.

Part of the nitrogen and phosphate contained in manure remains in the soil after application, and consequently they give load on the soil environment as surplus nutrients. Green manure cultivation at the site where vegetables were harvested is an effective method to collect surplus nitrogen mineralized in the soil and to prevent nitrogen leaching (Nakamoto 2007). The amounts of phosphate fertilizer applied are reduced based on soil diagnosis, and the accumulation of phosphate in the soil can be prevented since phosphate derived from manure gradually increases the available phosphate content in the soil.

The average values of soil available phosphate content in open cultivation vegetable fields observed in fixed points have been observed to be around three times the upper limit of the soil diagnostic criteria (from 150 to 300 mg P_2O_5 kg^{-1}) in the past 30 years (Fig. 5.5). Most vegetable farmers tend to excessive expectations for the fertilizing effect of phosphate, and the results of soil diagnoses are often not appropriately used to determine the appropriate amounts of phosphate fertilizer. Further educational activities are important to optimize the level of soil phosphate in vegetable fields in Hokkaido.

(2) **Greenhouse soils**

The procedure of fertilization design for greenhouse cultivation is basically based on fertilization based on soil diagnosis and the reduction in fertilizer application due to manure application. The evaluation method for soil nitrogen and the degree of fertilizer reduction due to manure application are different than in open cultivation.

In greenhouse cultivation, crops whose leaf color is required to be maintained until harvesting, such as spinach, and crops which require fertilization until the harvest stage, such as tomatoes, make up the majority. Therefore, in many cases, inorganic nitrogen remains in the soil before subsequent cropping. Nitrate nitrogen is used as an indicator of nitrogen fertility in greenhouses, because inorganic nitrogen is regarded as equivalent to fertilizer nitrogen and greatly affects crop growth. In nitrogen fertilization based on soil diagnosis, nitrogen fertility is classified into five levels based on soil nitrate nitrogen, 10 mg kg^{-1} of nitrate nitrogen is estimated to be equivalent to 10–15 kg N ha^{-1} of fertilizer

Table 5.5 Nitrogen fertilization amounts (kg ha^{-1}) upon soil available nitrogen standards for major open culture vegetable crops in Hokkaido

Crop	Cultivation dates[a]	Soil available nitrogen(mg kg^{-1})		
		~30	30~50	50~
Onion	May ~Sep.	180	150	120
Lettuce	Jun.~Jul.	140	120	100

[a]Cultivation after transplanted in fields

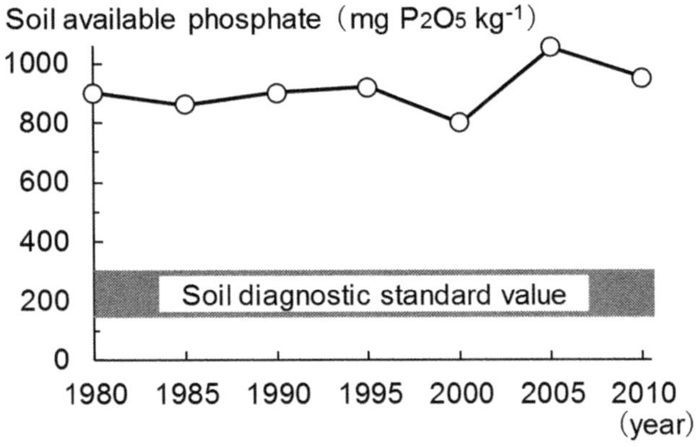

Fig. 5.5 Secular variation of soil available phosphate in fixed points for open culture vegetable crops in Hokkaido. Figure supplied by Tetsuo Hayashi

nitrogen, and the amount of nitrogen application at each level is determined (Table 5.6).

In aged greenhouses, nitrate nitrogen may remain in the subsoil (≥ 20 cm in depth) as well as in the plow layer (0–20 cm). In this case, even if ordinary soil diagnosis is carried out, the nitrogen nutrition of the crop becomes excessive, and it becomes difficult to control the plant vigor and internal quality of harvests. The root zones of deep-rooted vegetables, such as tomato, reach a depth of about 60 cm by the time of first additional fertilizer application, so nitrate nitrogen present at a depth of 20–60 cm is used for crops in addition to additional fertilizer nitrogen. By using subsoil diagnosis, it is possible to reduce the amount of additional fertilizer nitrogen used for tomato by evaluating the amount of nitrate nitrogen at a depth of 20–60 cm (Hayashi et al. 2004). In spinach cultivation, since the roots reach a depth of 40 cm or more by the time of harvesting, it is possible to lower the nitrate ion concentration of the plants by nitrogen fertilization based on a soil nitrogen diagnosis at 0–40 cm in depth (Hayashi and Nagao 2010).

Due to the development of diagnostic technology for inorganic nitrogen in the plow layer and subsoil and proper fertilization, the amount of residual nitrate nitrogen in greenhouse soils is decreasing. Along with this, the necessity of evaluating available nitrogen (autoclave-extractable nitrogen) in greenhouse soils is also increasing, and in some crops autoclave-extractable nitrogen is also used as an indicator of nitrogen fertility. In celery cultivation, nitrogen fertilization corresponding to the amount of autoclave-extractable nitrogen and nitrate nitrogen is established. In the greenhouse cultivation of mizuna and spinach, when the level of soil autoclave-extractable nitrogen is 100 mg kg^{-1} or more, and the amount of fertilizer can be further reduced by 30 kg N ha^{-1} from nitrogen fertilization corresponding to soil nitrate nitrogen content. Techniques for nitrogen fertilization which take autoclave-extractable nitrogen into consideration for the cultivation of tomato and other crops are presently being developed.

The application of organic matter, such as manure, is important for improving and maintaining the chemical, physical, and biological properties of soils. In Hokkaido, the physical properties of greenhouse soils have been improved by the successive application of 40 Mg ha^{-1} of cattle manure per year, but no further improvement was observed if the amount of applied cattle manure was increased. Also, because excessive application of cattle manure may cause groundwater pollution, the appropriate amount of cattle manure compost application for greenhouse cultivation is set at 40 Mg ha^{-1} per year (Hayashi and Hikasa 2017).

In order to avoid the excess accumulation of soil nutrients after manure application, it is necessary to reduce the amount of fertilizer component contained in manure. Table 5.7 shows the fertilizing effects of cattle manures in greenhouse soils. In greenhouse cultivation, the degree to which the application of nitrogen fertilizer can be reduced by manure application is larger than it is in open cultivation. This is because the soil temperature during the cultivation period is higher in greenhouse cultivation than in open cultivation, and the degradation of organic matter in the soil is also faster. The amount of nitrogen fertilizer reduction per 1 Mg of cattle manure is set as 2 kg for less than 4 years of successive application and as 3 kg for over 5 years of successive application. The nitrogen in manure is released over several years. Therefore, when compost is successively applied, nitrogen in the manure that was applied in the past several years is released in addition to the manure applied in the current year, and the reduction in the application of nitrogen fertilizer is therefore larger than in the case of single-year application.

Table 5.6 Nitrogen fertilization based on soil diagnosis (Greenhouse tomato)

Soil N fertility level	I	II (standard)	III	IV	V
Nitrate N (mg kg^{-1})	< 50	50 – 100	100 – 150	150 – 200	200<
N application rate (kg ha^{-1}) Basal	150	100	50	0	0
Additional (first)	40	40	40	0	0
Additional (second and after)	40	40	40	40	0

Source Hokkaido Government Agricultural Department (2015)
Note 1. Additional application time is the fruit development stage of each truss (excluding the two trusses below the pinching position)

Table 5.7 Fertile effects of cattle manures in greenhouse fields

Successive years	Nutrient contents (A, kg Mg^{-1} of manure)			Fertilizer efficiency index (B, chemical fertilizer = 1)			Amount to reduce chemical fertilizer (A × B, kg Mg^{-1} of manure)		
	T-N	P_2O_5	K_2O	T-N	P_2O_5	K_2O	T-N	P_2O_5	K_2O
1 – 4	5.0	5.0	4.0	0.4	0.6	1.0	2.0	3.0	4.0
5 <							3.0	3.0	4.0

Source Hokkaido Government Agricultural Department (2015)

5.5 Grassland

5.5.1 Grassland Soils

(1) **Dairy farming and sown grass in Hokkaido**

In Hokkaido, the dominant climate is subarctic humid, and forest is the climax vegetation. For this reason, almost all of Hokkaido's grasslands, which are used mainly for dairy farming, are artificial grasslands sown with temperate grasses. In 2014, the total area of grassland stood at 541,000 ha, representing 73% of Japan's total grassland area of 739,000 ha and 47% of Hokkaido's total cropland area of 1,148,000 ha. In the same year, the number of dairy cows being raised in Hokkaido stood at 792,400, representing 58% of the national total of 1,371,000.

Climatic conditions in Hokkaido vary according to region, with the eastern region in particular experiencing extremely cold winters during which the soil freezes, causing winter injury to even temperate grasses. Therefore, different grass species and cultivars are sown in different regions according to weather conditions. Currently, the most commonly sown species is Timothy-grass (*Phleum pratense* L.), but orchardgrass (*Dactylis glomerata* L.), meadow fescue (*Festuca pratensis* Huds.), perennial ryegrass (*Lolium perenne* L.), and others are also sown according to usage and weather conditions. Additionally, legumes, such as white clover (*Trifolium repens* L.), red clover (*Trifolium pratense* L.), and alfalfa (*Medicago sativa* L.), are often mixed in.

(2) **Characteristics of soils and respective fertilization improvement**

Grasslands in Hokkaido are located in volcanic ash soils, lowland soils, terrace soils, and peat soils at an estimated frequency of 50%, 22%, 21%, and 7%, respectively. Common characteristics in all of these grassland soils are the surface accumulation of nutrients and roots owing to surface fertilizer application and acidification owing to the leaching of minerals due to high precipitation.

(1) **Volcanic Ash soils**

Volcanic ash soil grasslands are widespread in Eastern and Southern Hokkaido. Although volcanic ash soils have good physical properties, their high phosphate retention capacity and the low nutrient content of their parent material have prompted research on the use of fertilizers to control the botanical composition of these soils. Interim results of an ongoing long-term grassland NPK field study launched in 1967 indicate that a good botanical composition can be maintained by low amounts of nitrogen, adequate amounts of phosphorus, potassium and magnesium, and also calcium to prevent soil acidification (Ohmura et al. 1985).

(2) **Terrace and Lowland soils**

Grasslands on terrace and lowland soils are widespread in Northern and Central Hokkaido. Although their parent material is rich in nutrients, clayey soils have inferior physical properties, including poor water retention, drainage, and crushability, and are accordingly susceptible to damage from excessive moisture and drought.

The yield of such grasslands depends on soil acidity and moisture content. Miki (1993) studied the effect of precipitation and soil pH on the process of accumulation and decomposition of soil organic matter. Soil acidification inhibits the growth of grass by solubilizing aluminum and also reduces the effectiveness of fertilizer owing to the increased absorption and fixation of phosphorus. Hojito (1994) conducted research on the acid resistance of grasses that quantified the effect of the above factors according to grass species.

(3) **Peat soils**

Land improvement for grasslands located on Peat soils must increase the bearing capacity of the soil through such means as soil-dressing, open ditch drainage, and underdrainage. Because excessive drainage promotes drought and nutrient runoff, maintaining a groundwater level of 30–50 cm is advised. Improving the drainage of peat soils promotes the drying and decomposition of peat together causing irregular subsidence.

Research has shown that maintaining high productivity in peat soils requires agricultural land improvement every few years.

5.5.2 Fertility Management

(1) **Approach to fertility management in the grasslands**

Field productivity is affected by botanical composition, and hence, the fertilizer recommendations for grasslands in Hokkaido are stipulated in Table 5.8 as "Fertilization Standards," considering the dominant grass species for each region and soil type (Hokkaido Fertilizer Recommendations 2015; see Sect. 8.1). Dominant species include Timothy-grass, as well as orchardgrass, perennial ryegrass, and alfalfa. The fertilization standards vary depending on the legume–grass ratio in the mixture. These fertilization standards indicate the total annual amount of fertilizer nutrients needed to obtain a standard yield when the chemical properties of the soil in the grasslands are within the ranges of the soil diagnosis criteria

(2) **Nitrogen**

As a general rule, the amount of nitrogen fertilizer is not dependent on soil analysis values, but is instead determined using the fertilization standards for the grass species, the climate zone, the soil type, and the percentage of legumes, as well as the fertilization process that takes into account the administration of organic material, as discussed in the next section. This is because soil analysis values are greatly affected by the percentage of legumes and the management of organic material in grasslands, making the conditions for setting the soil diagnosis criteria too complex.

The appropriate amounts of nitrogen fertilizer determined through such a process of fertilization are those which are at a level favorable for securing a standard yield and maintaining the percentage of legumes in the mixture (Kiso and Kikuchi 1988). As shown in Table 5.8, the fertilization standards for nitrogen in a meadow dominated by Timothy-grass on volcanic ash soil in the east of Hokkaido differ by four times depending on whether legumes are present. Increasing the nitrogen levels above the appropriate levels for the percentage of legumes present increases the vigor of the grasses, thereby reducing the percentage of legumes, whereas decreasing the nitrogen levels increases the vigor of legumes, thus increasing their share in the mixture.

(3) **Phosphorus**

The amount of phosphorus fertilization based on soil diagnosis is determined with respect to the soil type and the amount of available phosphorus in the soil (Table 5.9). The soil analysis value of phosphorus is calculated according to the Bray-2 method using the top 0–5 cm of soil. When testing a soil that varies greatly in phosphate absorption coefficient, it is not possible to uniformly evaluate the amount of absorbable phosphorus by herbage plants using this method, so a setting in different diagnosis criteria is needed for each soil type (Saigusa et al. 1990). The "fertilization rate" in Table 5.9 is a value relative to 100 that represents the fertilization standards.

(4) **Potassium**

Potassium fertilization based on soil diagnosis is stipulated based on the soil type and the amount of exchangeable potassium in the soil. The soil analysis value of potassium in the diagnosis is measured in the top 0–5 cm of the soil and is based on the amount determined by ion exchange extraction with 1 mol L^{-1} ammonium acetate solution.

The base material of volcanic ash soil has low potassium content, and hence, there is no need to take into consideration the amount of non-exchangeable potassium in the soil. The amount of potassium absorbed by the herbage plants can be controlled through the annual amount of potassium fertilizer applied and the amount of exchangeable potassium present in the top 0–5 cm of the soil in 1 ha (Saigusa and Noshiro 1990). If this total amount is 220 kg ha^{-1} of K_2O, legumes will be maintained favorably, and hence, a certain dry matter yield will generally be secured. For other

Table 5.8 Fertilization standards for meadow with dominant Timothy in volcanic ash soils of Eastern Hokkaido

Legume percentage category	Legume percentage	Timothy percentage	Standard yield	N	P	K
			Mg ha^{-1}	kg ha^{-1}		
1	≥30%	≥50%	45-50	40	44	149
2	15-30%	≥50%		60	44	149
3	5-15%	≥50%		100	35	149
4	<5%	≥70%		160	35	149

The forage legume seeded together cannot be alfalfa. The legume percentage category is based on the percentage wet weight of the first harvest. Assuming that the fertilizer is applied in two installments, the allocation between early spring and after the first harvest is in a ratio of 2:1

Table 5.9 Fertilizations for phosphorus based on soil diagnosis

Available phosphate content (mg-P kg^{-1})	Soil type		Less than standard values	Standard values	Above standard values	
	Volcanic ash soil	Regosolic	<131	131–262	262<	
		Ordinary	<87	87–218	218<	
		Cumulic	<44	44–131	131<	
	Lowland soil/terrace soil		<87	87–218	218–305	305<
Fertilization rate (%) relative to fertilization standards	Volcanic ash soil		150	100	50	50
	Lowland soil/terrace soil		150	100	50	0

The available phosphate content is measured by the Bray No. 2 method. The length of time for which reduced fertilization is possible is approximately 3 years

meadows, the amount of potassium fertilization based on soil diagnosis indicated in Table 5.10 is set for each soil type. The basis for the values in Table 5.10 is a study of potassium fertilization considering the yield standards in lowland soil and terrace soil (Miki et al. 1987) and the differing amounts of potassium supply based on the presence of dressing soil in peat.

5.5.3 Use of Livestock Manure

(1) Properties of livestock manure

The amount of urine and feces produced by dairy cows depends on their weight, the type and amount of feed provided, rearing methods, and environment, but a milking cow produces approximately 51 kg of feces and 13 kg of urine, totaling 64 kg of excrement, per day. The chemical composition of the feces is approximately 0.3% N, 0.1% P, and 0.1% K, while that of the urine is approximately 1.0% N, <0.01% P, and 1.0% K. The fraction of nitrogen in feces in the form of inorganic nitrogen (ammonium nitrogen and nitrate nitrogen), which has an immediate effect for grass growth, is low. Instead, most of the nitrogen is present in the form of organic compounds, and inorganic nitrogen is released only through the process of decomposition by microorganisms, which results in the feces having a slow-release effect when used as fertilizer. The nitrogen and potassium in the urine are in a similar form to that in chemical fertilizers, and thus, urine exhibits an immediate fertilizing effect. Therefore, assuming the average composition of manure or slurries available in Hokkaido, the quantities of fertilizer components in Table 5.11 were set to amounts that enable a reduction in chemical fertilization.

Table 5.10 Fertilizations for potassium based on soil diagnosis

Exchangeable potassium (mg-K kg^{-1})	Soil type		Less than standard values	Standard values	Above standard values	
	Volcanic ash soil	Regosolic	<58	58–75	75–249	249<
		Ordinary	<75	75–100	100–332	332<
		Cumulic	<83	83–108	108–374	374<
	Lowland soil/terrace soil		<125	125–166	166–415	415<
	Peat soil		<249	249–415	415–581	581<
Fertilization rate (%) relative to fertilization standards	Volcanic soil		125	100	75	50
	Lowland soil/terrace soil		110	100	50	0
	Peat soil	No soil admixture	125	100	75	50
		Soil admixture	110	100	75	0

Exchangeable potassium is calculated according to an exchange–extraction method using 1 mol L^{-1} ammonium acetate. The length of time for which reduced fertilization is possible is 1 year in volcanic ash soil and peat soil and 3 years in lowland soil/terrace soil

Table 5.11 Amounts of fertilizer components supplied to a grassland in the form of processed dairy cow manure in a field under maintenance management (kg M g^{-1})

	Type	Soil	Nitrogen (N)		Phosphorus (P)		Potassium (K)	
			This year	Second year	This year	Second year	This year	Second year
Solid manure	Farmyard manure	Volcanic ash soil	1.0	0.5	0.4	0	2.5	0
		Lowland/terrace soil	1.0	0.5	0.4	0	4.2	0.8
Liquid manure	Drainage from cattle bedded (primarily urine)		5.0	0	0	0	9.1	0
	Slurry, liquid derived from the mechanical separation of slurry Anaerobically digested slurry		2.0	0	0.2	0	3.3	0

Notes
(1) For application after use for the final seeding, it is notated as this year and year following application
(2) For continuous use, the quantities are added until the second year. In addition, if the soil analysis values are obtained in the second year of manure application, the fertilizing effect of the organic materials is considered to be reflected in the soil analysis values, and hence, fertilizer matching should be implemented based on only one source of information to avoid duplicate fertilizer reduction
(3) The values for urine fertilizer assume no dilution from rainwater, greywater, etc.

The contents of fertilizer components in manure, slurries, and the like vary significantly depending on their storage method, storage time, methods of secondary processing, and storage facility structure. Thus, it is recommended that manure and slurries be analyzed before they are applied to fields so that the application amount as well as the amount of chemical fertilizers applied at the same time can be determined based on the actual nutrient content.

(2) **Estimation for the fertilizer component of manure**

Chemical analysis is the most accurate method for estimating the amount of fertilizer components in manure, but because it involves specialized equipment, reagents, and technicians, this technique is time-consuming and costly. As a result, a method has been developed (Matsumoto et al. 2002) for estimating the total nitrogen (T-N), phosphorus, potassium, and ammonia nitrogen (NH_4-N) in manure by measuring the electrical conductivity (EC, mS/cm) and dry matter content (DM, weight%), which can be estimated relatively easily in Hokkaido. Additionally, in the case of slurries, it is possible to determine the dry matter content from the specific gravity as a substitute for the drying method, which requires a long time for measurement. Furthermore, the ammonia nitrogen content of manure can be measured at a high level of precision by using a simple reflective photometer (RQ flex, made by Merck).

(3) **Environmental-friendly application of livestock manure**

Figure 5.6 shows a procedure for determining a converted value for the actual amount of fertilizer applied to a grassland (Y, kg/t) from the fertilizer components of manure. If analysis values are not available, standard values can be substituted (Tables 5.11).

Table 5.12 Standard fertilizer conversion factors (Rs) for processed dairy cow manure

Type	Soil	Nitrogen (N)		Phosphorus (P)		Potassium (K)	
		This year	Second year	This year	Second year	This year	Second year
Compost	Volcanic ash soil	0.2	0.1	0.2	0.1	0.7	0.1
Urine fertilizer	Lowland/terrace soil	0.8	0	0	0	0.8	0
Slurries, separated liquid manure, methane-fermentation-digested-slurry		0.4	0	0.4	0	0.8	0

Notes
(1) For application after use for the final seeding, it is notated as this year and the year following application
(2) If both organic material application history and soil analysis values can be obtained, the fertilizing effect of the organic materials is considered to be reflected in the soil analysis values, and so fertilizer matching should be implemented based on only one source of information to avoid duplicate fertilizer reduction

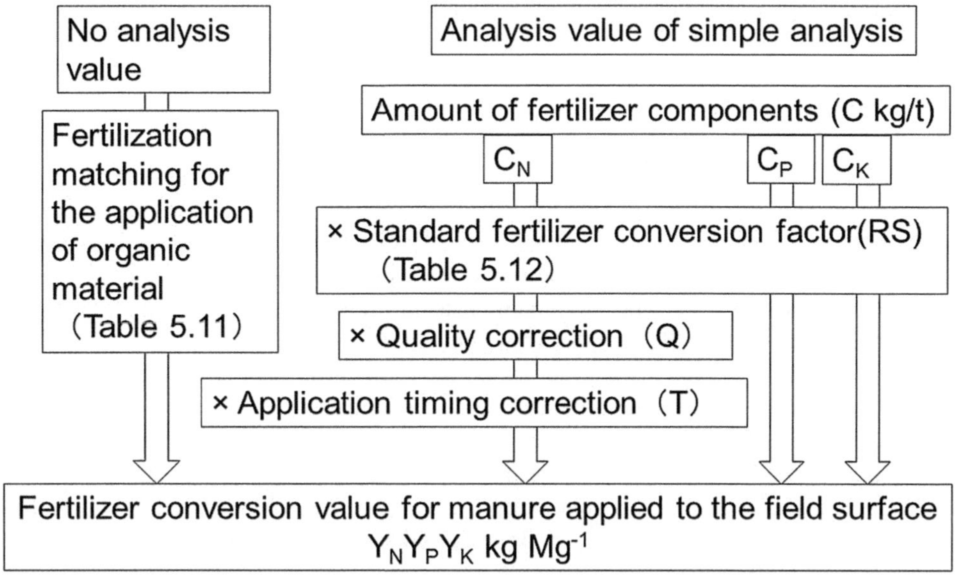

Fig. 5.6 Method for converting livestock manure to fertilizer. Figure supplied by Osamu Sakai

Below is an explanation of the case in which analysis values are available. The value Y for each of nitrogen, phosphorus, and potassium is determined by multiplying the fertilizer component content of the manure (C) with a fertilizer conversion factor (R) (Formula 5.1):

$$Y = C \times R \tag{5.1}$$

Here, the fertilizer conversion factor (R) signifies the fraction of the fertilizer components in the manure that is deemed to be equivalent to a chemical fertilizer (Table 5.12). This factor was determined through experimental results obtained in a number of laboratories in Hokkaido under varying conditions of weather, soil, grass species, and so on, and therefore, is applicable to grasses in Hokkaido in general. The fertilizer conversion factor for nitrogen varies depending on the fertilizer application time (T) and the manure quality (Q) (Matsumoto 2008), and thus correction factors are set for each.

5.6 Soils in Forests

5.6.1 Forest Soils

(1) **Forest management in Hokkaido**

Mixed forests of coniferous and deciduous trees are representative in Hokkaido. Most of the forests are affected to some extent by commercial logging operations. The presence of dwarf bamboo (*Sasa* spp.) on the forest floor is also a distinctive feature of forests in Hokkaido. *Sasa kurilensis* and *Sasa senanensis* occur in the west side of Hokkaido, where there is thick snow pack, while *Sasa nipponica* and *Sasa borealis* occur in the east side, where there is seasonal soil freezing in winter. Selective felling in order to induce the natural regeneration of trees after timber harvest is widespread, but the dense dwarf bamboo cover on the ground inhibits the recovery of trees as planned before cutting. Scarification treatment, which involves scratching the soil surface and removing dwarf bamboo roots, is performed using heavy machinery to encourage more rapid forest regeneration. As the method for controlling tree species has not yet been established yet in the regenerating site, birch trees usually tend to occupy the site, and young saplings are planted in order to produce artificial forest composed of coniferous trees.

When coniferous saplings are planted in a scarification site, they are forced to extend their roots into the nutrient-poor B horizon, because the mineral-rich O and A horizons have already been removed. The growth rate of the planted saplings is therefore so slow that cleaning operations to cut the dwarf bamboo and herbaceous plants around the planted trees are necessary for about 7 years, until the height of the planted trees exceeds that of the surroundings bushes.

In the case of artificial coniferous forest, thinning operations are additionally required to enlarge the stems of trees. However, many forest owners do not care for their forest due to the stagnant domestic wood price and the advancing age of forestry workers. This is the underlying problem for forest management in Hokkaido. Some measure should be taken to improve the current situation from the aspect of land conservation and global warming.

(2) **Chemical properties of forest soils in Hokkaido**

Since there is no input of substances other than atmospheric deposits and substances produced by litter decomposition, the nutrient supply in forest soils strongly depends on their natural environment (climate, parent material, vegetation, etc.). From the aspect of climate condition, Andosols and brown forest soils mainly appear in the forests of Central and Southern Hokkaido under temperate and humid climate regime. The soil pH in these forests is acidic (less than 6), and the concentration of soil solution is very low compared with that of greenhouse, upland, and grassland soils, for which the input of fertilizer is common as a means of soil management. Sodium and chlorine predominate in the soil solution in forest soils, indicating the strong influence of sea salts. As for the clay mineral compositions in the forest soils of Hokkaido, illite and montmorillonite containing potassium and calcium are not so significant, but rather vermiculite is the dominant mineral, being produced from the aforementioned clay minerals by the release of cations. The clay mineral stability diagram determined from soil solution shows that many forest soils are located in the stability area of vermiculite, which indicates the progression of the cation leaching process in Hokkaido (Satoh et al. 1982).

The evapotranspiration of Northern Hokkaido is very low because of the cool summers and lack of sunshine. In terms of soil geography, this area is a transition zone from acidified brown forest soils to Podzols due to the strong leaching potential. In the mountainous area of northwestern Hokkaido, heavy snowfall, which in some places exceeds more than 1000 mm water equivalent per year, produces thick accumulations of snow on the ground for more than 6 months of the year. The continuous downward movement of soil water, which is caused by snowmelt under the accumulated snow (about 0.5 mm/day) and during the spring melt season, accelerates the annual leaching of nutrients from the forest soil. Under such a strong leaching process, extensive soil acidification occurs, with the pH of forest soils in Northern Hokkaido becoming less than 4. In this soil, the concentration of aluminum in the soil solution increases, and Al-vermiculite is the dominant clay mineral in the B horizons due to the leaching of aluminum from surface horizons into the vermiculite clay sheets (Satoh et al. 1990).

(3) **Forest soils on volcanic ejecta**

Soils derived from volcanic ash that have strong phosphorus absorptivity are distributed widely in the forests of Central and Eastern Hokkaido. Because the input of fertilizer is not essential for these forest soils, trees growing on Andosols absorb the phosphorous supplied from the decomposition of leaf litter. However, this phosphorous absorption is very competitive in Andosols due to the presence not only of soil microorganisms but also of amorphous aluminosilicate minerals such as allophane with a high phosphorous-fixing ability. In the Podzols of the Scandinavian Peninsula of Northern Europe, the activity of soil microorganisms after adding phosphorous is reported to be affected by the presence of amorphous iron and aluminosilicate (Giesler et al. 2004).

Additionally, dense broad-leaved forests develop on the Volcanogenous Regosols in southwest Hokkaido. Despite the fact that these Regosols are very poor in nutrients, the growth of various broad-leaved trees is maintained by the decomposition of litter by the earthworms living in the O horizon.

(4) **Forest soils on ultramafic rocks**

Magnesian Eutrosols appear in the forests of Hokkaido on ultramafic peridotite and serpentine rocks. These soils are characterized by neutral soil pH, a high proportion of exchangeable magnesium, and a soil base saturation of close to 100%. These ultramafic soils are patchily distributed around the line connecting Cape Erimo (southernmost Hokkaido) with Cape Soya (northernmost Hokkaido), and spruce forest (*Picea glehnii*) is representative of them. The vegetation on serpentine rocks is generally considered to become steppe or bare land, and forests occurring on serpentine are thus very valuable. Large amounts of precipitation plays an important role in the development of the forests.

The reason why spruce trees precede other three species on ultramafic soils is that, despite not being able to compete with the other species, the spruce trees are able to tolerate harsh environments such as ultrabasic rock, peatland, and volcanic ash. An investigation into the mineral composition of trees growing on serpentine rock in Northern Hokkaido found that spruce did not absorb magnesium, unlike other tree species, suggesting that this tree has some magnesium excluding mechanism (Blandon et al. 1993). A symbiosis between spruce and mycorrhizal fungi has also been reported, which may also play a role in the tolerance of spruce on serpentine rocks (Kayama et al. 2006).

5.6.2 Material Cycling

(1) **Winter snow cover and material export**

In many forests in Hokkaido, snowfall dominates the annual precipitation. For example, in Northern Hokkaido about 50% of the annual precipitation is supplied as snowfall. Snowfall accumulates as snowpack in forests during winter and then is removed as water to streams during the short snowmelt season. Therefore, the leaching of material from soil, which is a significant component of material cycling, occurs during the snowmelt period. Since nutrient uptake by plants is generally negligible during the snowmelt period, many of the nutrients in the soil are leached out in the meltwater (e.g. Park et al. 2010). Further research is needed to clarify how material cycling in forests responds to variations in the winter climate, as recent changes in global climate may trigger significant changes in the snowfall and snowpack in Hokkaido (Park et al. 2010).

(2) **Soil freeze-thaw cycles and nitrogen dynamics**

As a porous medium, snowpack plays a role in insulating the soil. If the snowpack on the soil is sufficiently thick (e.g. <50 cm), the soil does not freeze and the soil temperature remains at around 0 °C. However, if the snowpack falls below certain thresholds (e.g. 50 cm in thickness), the soil freezes and thaws at variable frequencies and intensities as the ambient air temperature changes (e.g. Shibata 2016). Repeated soil freeze-thaw cycles physically crush litter, fine roots, and microbes, providing a substrate and an energy source for the microbes that survive and altering the nitrogen and carbon cycling during the thaw period. Recent studies have indicated that, as soil freeze-thaw cycles increase in frequency, the ammonium production from organic nitrogen via soil microbial reactions increases, and nitrification tends to decrease in forest soils (Shibata 2016). Hishi et al. (2014) indicated that slope topography (i.e. position and aspect) promotes the development of different microclimates, including the winter climate, and affects the soil nitrogen dynamics. Additionally, emissions of greenhouse gases from soil, such as nitrous oxide (N_2O), are enhanced by soil freeze-thaw cycles. Nitrogen produced in soil during winter is an important source of nutrients to streams and the atmosphere during the snowmelt period and also for plants and microbes during the following growing season (Shibata 2016).

(3) **Role of riparian wetlands**

Unlike forests in the southern islands in Japan, the topography is relatively gentle in the Hokkaido forests, and riparian wetlands in forested watersheds are generally located in areas adjacent to the stream channels. Organic matter accumulates in riparian wetlands because of the anaerobic conditions in the soil, resulting in emissions of the greenhouse gases, methane (CH_4), and N_2O. Emissions of N_2O from riparian soil indicate nitrate (NO_3^-) reduction via denitrification in soil and groundwater just before NO_3^- enters the stream water. The hyporheic zone, at the interface between the stream channel and the surrounding sediments where stream water and groundwater mix, plays an important role in buffering stream chemistry (Shibata et al. 2004). In natural forest watersheds in the northern part of Hokkaido, NO_3^- concentrations in stream water tend to be high in relatively steep watersheds; in contrast, low concentrations of NO_3^- and high concentrations of dissolved organic carbon (DOC) were observed in stream water because of microbial denitrification in DOC-rich soil and nutrient uptake by riparian plants in watersheds with relatively flat riparian wetlands in their lower reaches (Ogawa et al. 2006).

Further research should address how changes in global climate, including changes in the winter season, will impact forest ecosystems such as those in Hokkaido. Furthermore, there is little clarity about how increases in the magnitude and frequency of extreme climate events (e.g. typhoons and bomb cyclones) might influence material cycling in forests. There is thus a need for integrated research programs that assess the natural capital and ecosystem services of forest ecosystems in the uppermost reaches of watersheds and that pay special attention to both the local unique characteristics of Hokkaido and the coupling between natural and social systems in watersheds.

5.6.3 Carbon Cycle

(1) **Carbon cycle of forest soils in Hokkaido**

Andosols, brown forest soils, and Volcanogenous Regosols, which are distributed widely in forests in Hokkaido, contain large amounts of organic carbon in the top 1 m (Morisada et al. 2004). A major carbon input to the soil comes from dead forest vegetation, with leaves, branches, and fine roots being a stable and main source. The annual rate of carbon input from leaf and branch litter from trees ranges between 0.5 and 2 Mg C ha^{-1} in several forest types in Northern Hokkaido. Dwarf bamboo (*Sasa* spp.), a widely distributed undergrowth plant, supplies up to 1 Mg C ha^{-1} yr^{-1} of these litters (Watanabe et al. 2013). Additionally, fine roots are considered to supply a comparable amount of litter to the soil to that contributed by leaves. Coarse litter originated from trunks and coarse roots and root exudates need to be added to evaluate the total carbon input to the soil, however because of the large temporal and spatial heterogeneity of the coarse litter and the little quantitative evaluation for root exudates, large uncertainty exists in the estimation. For example, the long-term, large-scale averaged annual rate of carbon input from coarse litter (corresponding to plant mortality) ranges between 0.4 and 1 Mg C ha^{-1} yr^{-1} in several forest types in the Northern USA.

A major carbon output from the soil is the emission of CO_2 from the soil originating from the decomposition of soil carbon by microorganisms, which is known as microbial respiration. Root respiration also emits CO_2 from the soil surface, but this does not affect the soil carbon balance. The sum of the microbial (or heterotrophic) and root (or autotrophic) respiration is called soil respiration and is measured at many forests in Japan. Lee et al. (2006) report that the annual soil respiration rate in Japanese forests ranges from 2 to 14 Mg C ha^{-1} yr^{-1}, with an average rate of 6.47 ± 2.72 Mg C ha^{-1} yr^{-1}. The soil respiration rates of forests in Hokkaido lie within this range, although the soil respiration in forests with dense undergrowth of dwarf bamboo tends to be near the upper maximum (>10 Mg C ha^{-1} yr^{-1}) of the range (Aguilos et al. 2014). This reflects the high investment of the resources to the belowground organs of this plant (i.e. root biomass and respiration). In Hokkaido, soil respiration accounts for around 50–60% of total ecosystem respiration (10–15 Mg C ha^{-1} yr^{-1}) and microbial respiration for around 30–80% of the soil respiration, with the contribution of microbial respiration decreasing with an increase in the soil respiration.

Soil carbon balance and the carbon sink or source strength can be estimated by accounting for the aforementioned inputs and outputs. However, because the annual change in the soil carbon stock is considered to be less than a few percent of the stock, having large uncertainty in the estimation, many studies assume that carbon stock is constant during a certain period within several years.

(2) **Effects of forest management or disturbances**

Clear-cut harvesting decreases both the forest ecosystem photosynthesis and respiration rates. However, because the magnitude of the decrease is larger for the photosynthesis rates, such harvesting usually turns forest into a carbon source (Aguilos et al. 2014). Other serious disturbances, such as typhoons, windthrows, and fires, cause a similar change in the forest carbon balance. The change in the soil respiration and soil carbon balance caused by harvesting and these other disturbances strongly depends on the condition of the vegetation after such disturbances. In Northern Hokkaido, an observed increase in the soil respiration rate after clear-cut harvesting was mainly caused by an increase in the biomass and root respiration of the undergrowth dwarf bamboo resulting from better light conditions after the harvesting of overstory trees, while the increase in the biomass slightly decreased the soil surface temperature and suppressed the increase in soil carbon decomposition by microorganisms. Using a model and observed results, Aguilos et al. (2014) simulated the effect of the clear-cut harvesting on the forest carbon stock for each compartment and showed that the increase in the soil respiration can be explained by the increase in root respiration and in litter (residuals such as stumps and branches) decomposition after the harvesting. It is hard to recognize the change in the soil carbon stock during the course of management (Fig. 5.7). Soil carbon stock will decrease if such management or disturbance increases the soil temperature and enhances the microbial decomposition under poor undergrowth conditions; however, it is hard to find studies on this in Hokkaido. On the other hand, in Hokkaido, dense undergrowth and surface soils are often removed after harvesting for the plantation or natural regeneration of trees. Such practices have a large effect on the soil carbon stock and cause the loss of tens of Mg C ha^{-1}.

(3) **Effects of global warming**

Recent global synthesis studies show that ecosystems in cooler climates or with higher soil carbon stocks tend to lose a larger amount of soil carbon by CO_2 emissions in warmer conditions. Several soil warming experiments in black spruce and Scots pine forests in Northern Europe and the USA have shown that soil warming (3–6 °C temperature increase) enhances soil respiration by 11–45% for several years, with the limited temporal duration of the enhancement being due to the rapid decrease in soil labile carbon content as a result of warming. However, a soil warming experiment in a cool temperate forested peatland in Northern Hokkaido showed that warming (3.2 °C increase in the soil temperature at 5 cm depth) enhances the soil carbon decomposition by an average of 82% for the first 4 years (Aguilos et al. 2013), and that the enhancement rate continues to increase for more than 10 years. This suggests that soils with high substrate availability and without severe water stress would lose a huge amount of soil carbon in a future warmer environment, and that the effect would last for more than a decade.

5.6.4 Biodiversity

(1) **Importance and abundance of soil fauna in Hokkaido**

The organisms living in forest soils significantly influence biogeochemical cycles and, consequently, ecosystem function. In particular, soil fauna impacts biogeochemical cycles significantly via feeding on litter and coarse woody debris and also plays a role in the modification of soil structure as ecosystem engineers.

The study of soil fauna in the forest ecosystems of Hokkaido was begun by Yoshio Nakamura's group in the 1970s. Among various groups of soil mesofauna, which largely contribute to litter decomposition, *Collembola* (springtail) communities are the most abundant and *Oribatida* communities the second most abundant. For soil macrofauna, centipedes (predator) and *Enchytraeidae* (detritus feeder) are the

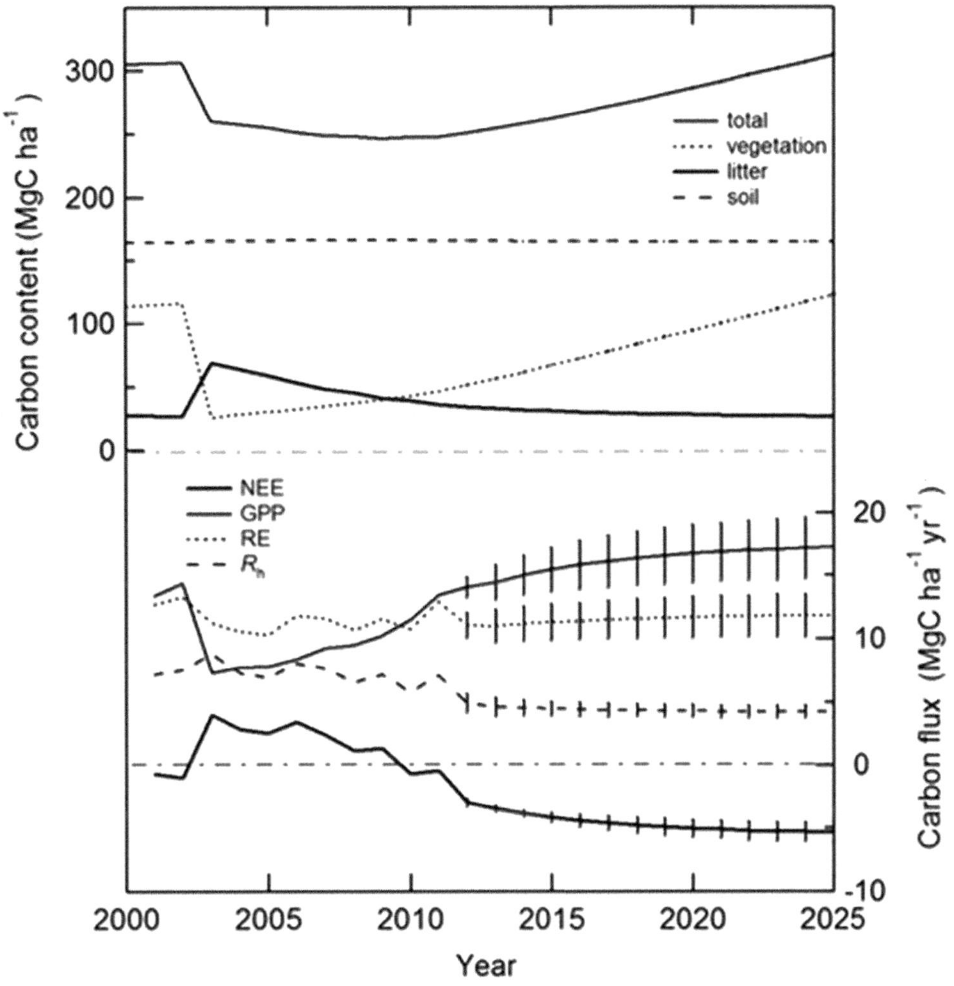

Fig. 5.7 Effect of clear-cut harvesting on the forest carbon cycle in a cool temperate mixed forest in Northern Hokkaido simulated using revised BIOME-BGC model and observed results. Reprinted from Agricultural and Forest Meteorology,197, Aguilos et al., Dynamics of ecosystem carbon balance recovering from a clear-cutting in a cool-temperate forest, 26-39, Copyright (2014), with permission from Elsevier. Clear-cut harvesting was in 2003, and results from 2012 are obtained using the same parameter set with that used during 2003–2011 and are shown as the average and the standard deviation (vertical bars) of 10 simulation runs each of which used the repeated annual variation of micrometeorology observed each year of the 10-year study period from 2002 to 2011 throughout the simulation, although the errors for the carbon contents were too small to identify. Total, vegetation, litter, and soil carbon contents and ecosystem photosynthesis (GPP), ecosystem respiration (RE), net ecosystem CO2 exchange (NEE), and soil heterotrophic respiration (Rh) rates are shown

most abundant groups, while millipedes show the second-highest abundance (Nakamura et al. 1970).

(2) **Environmental factors and soil fauna in Hokkaido**

Forest ecosystems establish on more complex abiotic environments than do other ecosystems, such as agro-environmental systems, and these complex environments influence the soil faunal distribution. For example, the abundance of earthworm communities is lower in alpine dwarf pine forests and subalpine broad-leaved forests compared with lowland forests. The distribution of soil fauna is influenced not only by the elevation, but also the topography. In Eastern Hokkaido, the density and biomass of earthworm communities are higher on north-facing slopes (Hishi et al. 2014, and related studies). Specifically, the abundance of Enchytraeidae communities is negatively correlated with the depth of the A0 layer and the soil C/N ratio, which vary with slope direction; this suggests that Encytraeidae prefers soil with a low C/N ratio, and also that they reduce the depth of the A0 layer by consuming organic matter. Soil type also influences the distribution of soil fauna. For instance, *Eisenia japonica* is abundant in Andosols, while *Dendrobaena octaedra* is abundant in organic soils.

The growth of *E. japonica* is known to be high when they feed on Calcareous Eutrosols with a high pH and calcium concentration (Kawakami and Makoto 2017). In North America, soil calcium concentration is positively correlated with the body weight of Lumbricidae earthworms. Together, these facts indicate that the difference in calcium concentration among soil types is an important determinant of the geographical variation of earthworm distribution. At a larger spatial scale, Uchida et al. (2004) reported that the density and biomass of earthworms in Central Hokkaido were far larger than in temperate forests in the Kanto region. However, little is known about the nationwide variation in the abundance of soil fauna and its diversity, or about the driver of the variation, and further nationwide comparative studies are therefore required.

In Hokkaido, many soil fauna groups show clear seasonal changes in their abundance. Their abundance peaks in the middle of autumn, and later the group of overwintering soil fauna moves to a deeper part of the soil profile to avoid the cold temperature (Nakamura et al. 1970). In Hokkaido, air temperature sometimes drops between −20 and −30 °C in the middle of winter. The deep snowpack in Northern Hokkaido insulates the soil from the cold air temperature, and soil does not freeze. As a result, some soil fauna groups can manage overwintering without suffering the damage caused by soil frost. In forests with a cold biome, the soil nitrogen availability for trees is the limiting factor affecting the primary production of forests (Aber 1989). In Hokkaido, the influence of earthworms is more predominant in summer compared with winter, which suggests seasonality in the relationship between soil fauna and its influence on biogeochemical cycles (Hishi et al. 2014). Species of Lumbricidae, which is the major earthworm group in boreal Eurasia, are more abundant in Northern Hokkaido than in Southern Hokkaido, which is consistent with the fact that forest vegetation in Northern Hokkaido is in the transient zone between temperate forests to boreal forests.

(3) **Human activity and soil fauna**

Artificial factors such as land use change can influence the soil faunal community. In Central Hokkaido, the soil faunal abundance was found to be greater in natural mixed-conifer broad-leaved forests than in artificial forests of *Abies sachalinensis* (Nakamura et al. 1970). Among artificial forests, those of non-native *Picea abies* have a lower abundance of soil faunal than that of native *A. sachalinensis* because of the poor understory vegetation in *P. abies* forests. This study suggests that the proper management of understory vegetation can maintain the diversity and abundance of the soil faunal community in artificial forests. On the other hand, little is known about the influence of silvicultural practices other than plantation (e.g. scarification of understory dwarf bamboo), and further studies are necessary.

In high-latitude areas such as Hokkaido, the influence of climate warming on organisms is predicted to be predominant. Based on experimental studies and latitude gradient studies (e.g. Fujii et al. 2018), climate warming is predicted to change the soil faunal contribution to litter decomposition and consequently the plant growth. Although winter climate influences the structure and function of forest communities, the magnitude of change in temperature has been greater in winter than in summer. In Northern Hokkaido, the snow cover period has become shorter due to the warming climate. In future studies, the influence of characteristic climate change in Hokkaido (e.g. advancing snowmelt) on the biodiversity and the associated biogeochemical cycle should be clarified by utilizing the long-term monitoring data obtained by the Japan Long-term Ecological Research (JaLTER) and Monitoring 1000 research programs being conducted by the Japanese Ministry of the Environment.

5.7 Agriculture and Environment

5.7.1 Environmental Burden and Conservation

As a result of the material cycling through the soil, agricultural environmental burden extends to the aquasphere and atmosphere. In Hokkaido, in particular, the NO_3-N contamination of groundwater is a problem. Figure 5.8 shows the results of an investigation of the concentration of NO_3-N in well water across the whole of Hokkaido (Hokkaido 2013). A large number of wells were contaminated with NO_3-N in upland fields of Andosols and livestock areas, especially in the Abashiri area. The environmental standard (10 mg N L^{-1}) was exceeded in 5.7% of cases (for details, see Sect. 5.7.2). Based on the results of this survey, it is thought that the groundwater pollution was derived from nitrogen fertilization, and efforts have been made to improve fertilization methods. In particular, the upper limit of the nitrogen fertilizer application for producing crops is set in accordance with the nitrogen environmental capacity which is the sum of the amount of nitrogen taken up by the crop and the allowable remaining amount of NO_3–N in soil (Matsumoto and Tou 2006). The nitrogen environmental capacity (kg ha^{-1} yr^{-1}) was 157 ± 19 for paddy fields 218 ± 39 for grasslands and 169 ± 30 for upland fields in Hokkaido. The groundwater NO_3–N concentration increased in proportion to the difference between the nitrogen fertilizer application rate and the nitrogen environmental capacity in all plots, and in upland fields where the nitrogen fertilizer application rate was above the nitrogen environmental capacity, the groundwater NO_3–N concentration was higher than 10 mg N L^{-1}. It is recommended to account for surplus nitrogen in soil in previous cultivation by implementing a three-crop rotation, for example, sugar beet, azuki bean, and spring wheat. This is because surplus nitrogen easily exceeds

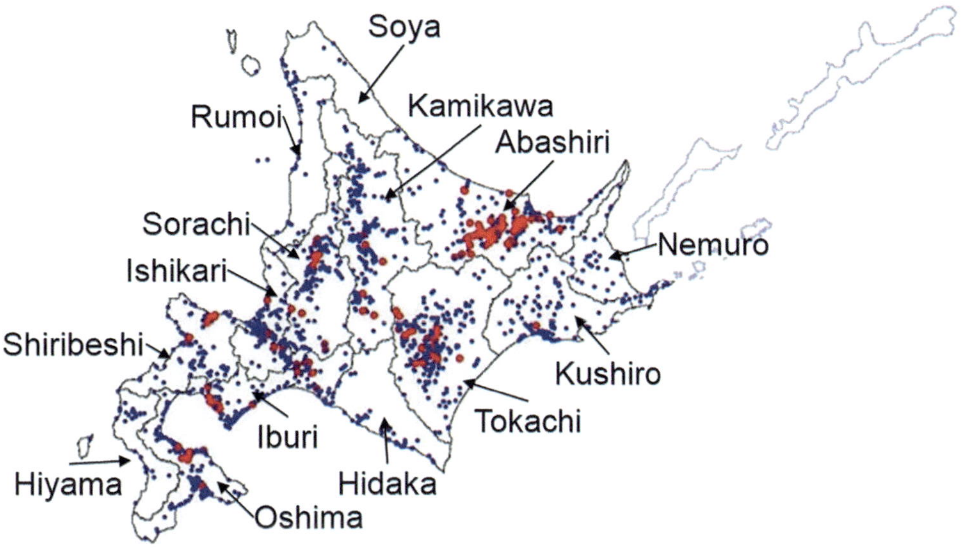

Fig. 5.8 Nitrate nitrogen concentration in groundwater in Hokkaido. *Source* (Hokkaido 2013). Blue → within environmental standard (less than 10 mgN L^{-1}) Red → beyond environmental standard (more than 10 mgN L^{-1})

the nitrogen environmental capacity when a crop with a high nitrogen demand is successively cultivated. In Hokkaido, the reduction in soil organic matter levels is continuing. Based on soil survey data from about 6000 points conducted between 1959 and 2003, the carbon content of upland field soils in Hokkaido decreased by 35% between 1970 (4.8%) and 2000 (3.9%). The Hokkaido Clean Agriculture System is promoted to increase soil carbon levels and reduce groundwater contamination. This system is based on reducing the amount of inorganic nitrogen derived from compost (assuming 1 kg N per ton of compost) from the application amount of nitrogen chemical fertilizer by applying more than 10 tons of compost per hectare per year. Also, since N_2O emission from upland fields strongly depends on the application rate of nitrogen chemical fertilizer (Shimizu et al. 2013), reducing the application rate of nitrogen chemical fertilizer contributes to environmental conservation. Methane emissions from paddy fields can be reduced by 43% by intermittent irrigation, and a reduction of over 30% is possible by the incorporation of rice straw into soil in autumn (Goto et al. 2004).

5.7.2 Greenhouse Gas Emissions, Nutrient Loads, and Heavy Metal Contamination

(1) **Impact on the atmosphere**

(1) **Carbon dioxide (CO_2)**

The flows of carbon and CO_2 in cultivated lands are shown in Fig. 5.9. Crops absorb CO_2 from the atmosphere (gross primary production [GPP]) and emit CO_2 as aboveground respiration (AR) and root respiration (RR). The difference, GPP—(AR + RR), is called the net primary production (NPP). Some of the NPP is moved outside of the field (OUT) during harvesting, and the rest is put back into the soil as residue. Organic matter is input as manure (IN). In addition, CO_2 is emitted in the atmosphere by soil organic matter decomposition (OMD). The net CO_2 emission from cultivated land is calculated by AR + RR + OMD + OUT − GPP − IN, or, OMD − (NPP − OUT) − IN, which shows that the decrease in the soil organic matter content is equal to the net CO_2 emission.

The area of cultivated land in Hokkaido is 1,145,000 ha (25.8% of the total area of cultivated land in Japan), in which Andosols, which have a high soil organic matter content, are widely distributed. Reducing the decomposition of soil organic matter and increasing soil carbon sequestration through the appropriate application of manure is needed both for the maintenance of soil fertility and as a measure against global warming.

(2) **Methane (CH_4)**

Under anaerobic conditions, such as in rice paddy fields, CH_4 is produced by methanogens in the soil and emitted through rice stems to the atmosphere. The factors controlling methane emissions are water management, the application of organic matter and nitrogen (N), and so on.

The area of rice paddy field in Hokkaido is 222,000 ha (9.2% of the total rice paddy area in Japan). Under regional conditions (e.g. a very slow decomposition rate of paddy straw due to low temperature in winter), some measures have been proposed to mitigate CH_4 emissions by maintaining stable rice production in Hokkaido.

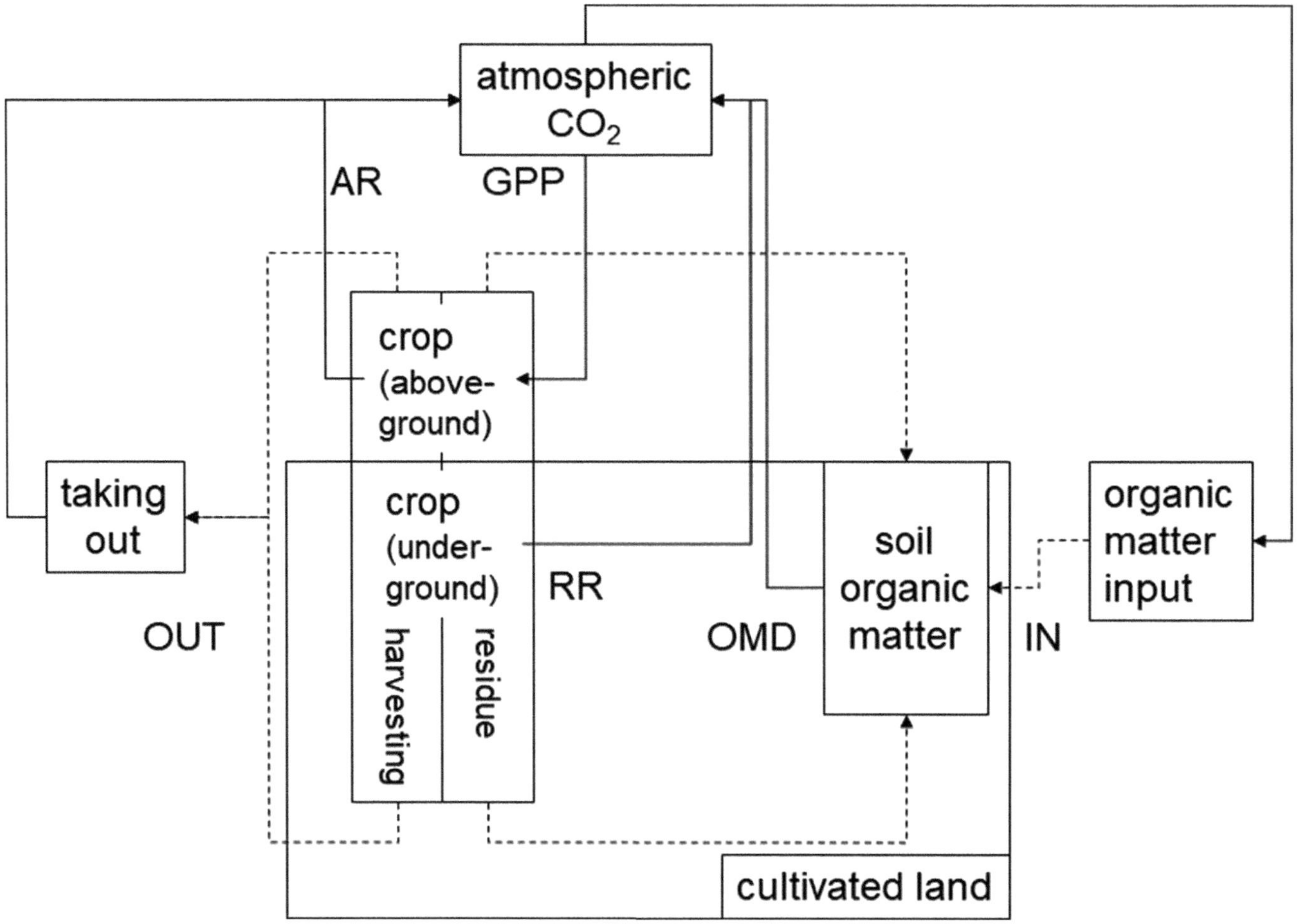

Fig. 5.9 Carbon flows in cultivated land. Solid and dotted lines denote the flows of CO_2 and organic matters, respectively. See the text for the abbreviations. Figure supplied by Takuji Sawamto

(3) **Nitrous oxide (N_2O)**

In cultivated lands, N_2O is produced by soil microbial processes (nitrification and denitrification) and is directly emitted to the atmosphere (Fig. 5.10). Nitrous oxide produced in soil is also indirectly emitted through soil water drainage (Fig. 5.10). The factors controlling N_2O production are soil moisture, soil temperature, mineral N content in the soil, N fertilization rate, and so on.

The area of upland field (for crops and vegetables) and grassland in Hokkaido is 416,000 ha (36.5% of the total upland field area in Japan) and 503,000 ha (83.8% of the total grassland area in Japan), respectively. Emission factors (kg N_2O-N/kg N) for direct N_2O emission are estimated to be 0.0062 ± 0.0048 for all the fields (excluding rice paddy fields) in Japan and are relatively higher for poorly drained fields (Akiyama et al. 2006). In Hokkaido, high N_2O emissions have been observed in summer and autumn, probably due to the denitrification process triggered by abundant rainfall (Kusa et al. 2002; Katayanagi et al. 2008). Some measures have been proposed to mitigate N_2O emission by the efficient application of N without reducing crop production.

(4) **Ammonia (NH_3)**

In Hokkaido, it is estimated that as much as 112 Gg N (36.5% of the total in Japan) is emitted annually through livestock animal waste, which is equal to a load of 96 kg N per hectare of cultivated land.

Ammonia volatilization from N in chemical fertilizer is negligible in Hokkaido because alkali soils are not widely distributed there. Thus, almost all ammonia volatilization in Hokkaido originates from livestock manure. The volatilization rate can be controlled by the manure application method, manure pH, type of manure, wind speed, temperature, and soil moisture. Mitigating ammonia volatilization is important for environmental conservation. Management for enhancing the nutrient utilization is also essential because the NH_3 remaining in the manure may lead to higher nitrate leaching and N_2O emissions.

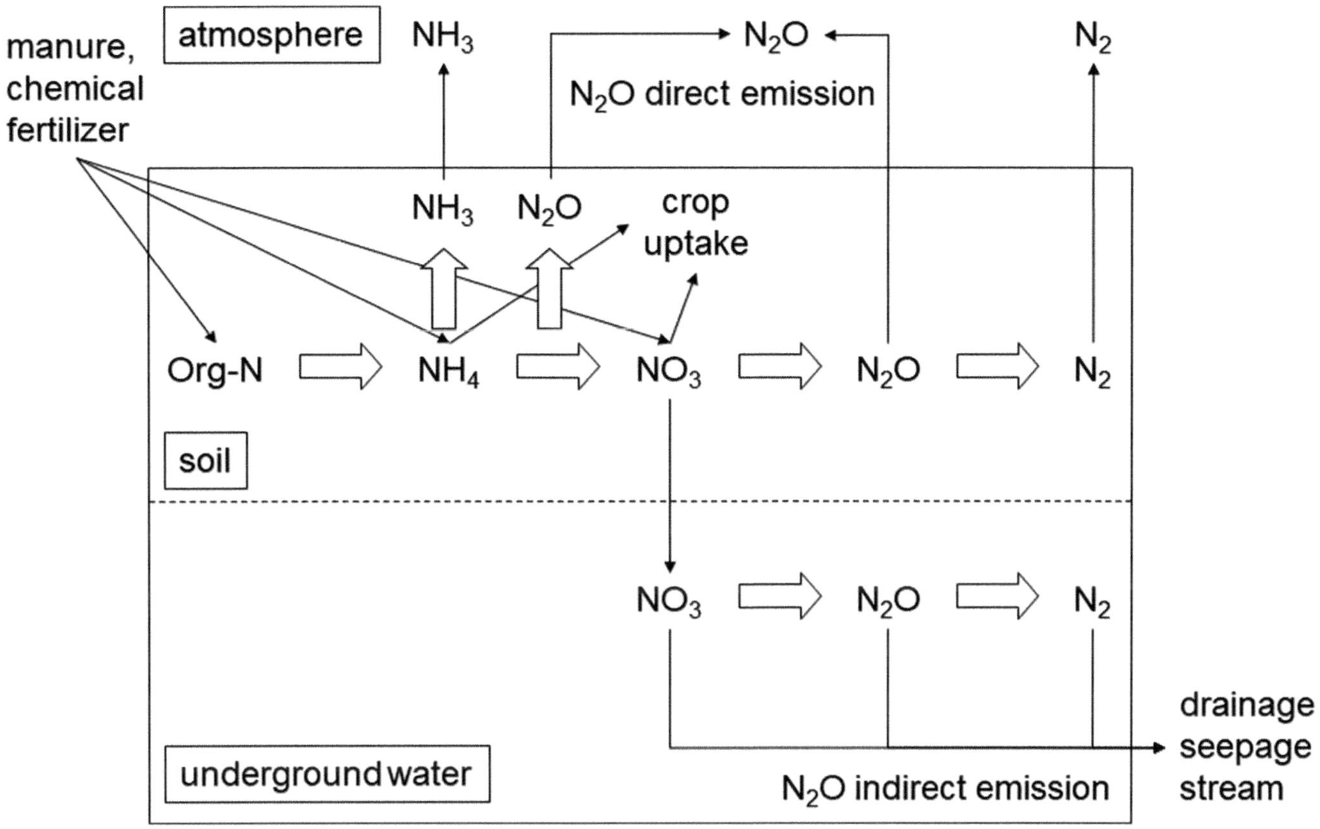

Fig. 5.10 Nitrogen flows and its transformations in the atmosphere, soil, and underground water. Figure supplied by Takuji Sawamto

(2) **Impact to the water system**

(1) **Rivers**

For this analysis, we used data from the water quality hydrology database of the Ministry of Land, Infrastructure and Transport (2017). Concentrations of total nitrogen (TN), total phosphorus (TP), and nitrate nitrogen (NO_3-N) in major rivers in Hokkaido are summarized in Table 5.13. Concentrations of TN and TP are average values for a single observation point in the most downstream area measured over 16 years from 2001 to 2016. The concentrations of NO_3-N are averages of multiple observation points in the mid-downstream area measured over 10 years from 2007 to 2016.

For various rivers, TN concentrations follow the order of Tokoro > Tokachi > Ishikari = Abashiri = Rumoi > Kushiro = Teshio > Saru. Concentrations were high in the upland field areas (Tokoro and Tokachi Rivers), except for the Abashiri River, had medium values in the paddy field areas (Ishikari and Rumoi Rivers), and were low in the areas of dairy farming and racehorse production (Kushiro, Teshio, and Saru Rivers). Concentrations of TP were less than 0.09 mg P L^{-1} in all rivers. In the upland field, the TP concentration in the Tokoro River was the highest and was moderate in the Tokachi River—0.09 mg P L^{-1}. This is probably due to the fact that the Tokachi River basin is covered by Andosols with a high phosphate absorption coefficient. Except for the Teshio River, TP concentrations were low in Andosol areas (Abashiri, Kushiro, Tokachi, and Saru Rivers), high in non-Andosols areas (Tokoro, Rumoi and Ishikari Rivers). Concentrations of NO_3-N were in the range of 0.1–1.4 mg N L^{-1}, in the order of Tokoro > Tokachi > Abashiri > Kushiro = Teshio > Ishikari = Saru = Rumoi. As for TN, NO_3-N concentrations were the highest upland fields, intermediate in paddy fields, and the lowest in dairy farm areas.

Woli et al. (2002) measured the concentration of river water NO_3-N in various watersheds in Hokkaido and found a positive linear relationship between the land use ratio in the watersheds and the NO_3-N concentration. They showed that the slopes of the linear regression lines were large for upland crops, open vegetable fields, and intensive livestock farming, and small for areas of dairy farming and racehorse production. This indicates that agriculture affects river water quality to a larger extent in upland fields, open vegetable fields and intensive livestock farming, and to a smaller extent in areas of dairy farming.

Table 5.13 Comparison of water quality of major rivers in Hokkaido

	Ishikari R.	Rumoi R.	Teshio R.	Tokoro R.	Abashiri R.	Kushiro R.	Tokachi R.	Saru R.
TN (mg N L^{-1})	1.21c	1.06c	0.60d	2.21a	1.06c	0.72d	1.95b	0.36e
TP (mg P L^{-1})	0.074a	0.076a	0.033c	0.085a	0.061b	0.057b	0.052b	0.035c
NO_3- N (mg N L^{-1})	0.21e	0.11e	0.39d	1.44a	0.58c	0.39d	1.19b	0.21e

Source Ministry of Land, Infrastructure, Transport and Tourism (2017)
TN and TP concentrations are average values of single observation point in the most downstream for 16 years from 2001 to 2016. And NO_3-N is the average of multiple points in the mid-downstream for 10 years from 2007 to 2016
Different letters in the table indicate that there is a significant difference (Tukey, $p < 0.05$, n = 90–385)

In November 1999, the "Act on Proper Management and Promotion of Use of Livestock Manure" was enacted, and it came into full force in November 2004. Under this law, it was obliged to thoroughly manage the livestock manure that could be a point source. By comparing concentrations of biochemical oxygen demand (BOD)—an indicator of organic pollution—before and after the enforcement of this law, we evaluated the impact of livestock manure point sources (Ministry of Land, Infrastructure and Transport 2017). Four rivers were selected in Northern and Eastern Hokkaido, which are dairy areas. The observed trend varied according to area (Table 5.14). In Northern Hokkaido, BOD concentrations decreased significantly in all rivers after the enforcement of the law, whereas in the eastern part, BOD concentrations in the two rivers decreased significantly, but others showed no significant difference, or increased significantly. Whether these changes were due to farming practices or natural conditions could not be clarified. However, by the enforcement of this law, some degree of water quality improvement was achieved. Rivers in Hokkaido have been negatively affected by agriculture; however, the impact is low, and the point source pollution from livestock manure is declining.

(2) **Groundwater**

A detailed investigation of the NO_3-N concentration in groundwater in Hokkaido was performed in the 3 years from 1999 to 2001 (Table 5.15). Of a total of 9528 groundwater measurement sites (wells), 546 cases (5.7%) exceeded the environmental standard value of 10 mg N L^{-1}. However, in the Abashiri area, 30.7% of measurements exceeded this standard. By contrast, in the Sorachi, Tokachi, and Iburi areas (upland crop areas), this figure was around 3%, and it was almost 0% in areas mainly comprising dairy farming or horse production (Rumoi, Soya, Nemuro, Kushiro, and Hidaka).

The high NO_3-N concentration in groundwater in the Abashiri area is due to the fact that it contains upland field and receives little annual precipitation. While the average annual rainfall for the whole of Hokkaido is 1100 mm, zones that receive an average annual rainfall of less than 900 mm exist from the Tokachi area to the Abashiri area. The average annual rainfall in Abashiri is less than 750 mm. If precipitation is low, the amount of groundwater recharge will be less, but will not exceed the annual evapotranspiration. In agricultural lands, as water migrates it dissolves nitrogen loaded by fertilizer, compost, and manure, and the recharged water therefore contains high concentrations of NO_3-N. Thus, groundwater with a high concentration of NO_3-N is present in the Abashiri area. Matsumoto and Tou 2006 calculated the surplus water volume by subtracting the amount of evapotranspiration from rainfall and residual nitrogen tolerance that the leachate concentration did not exceed 10 mg N L^{-1} for each municipality. Furthermore, they defined the nitrogen assimilation capacity taking into consideration the amount of nitrogen absorbed by crops. The determined nitrogen assimilation capacities for various municipalities ranged from 94 to 308 kg N ha^{-1}, with an average value of 183 kg N ha^{-1}. The highest value was observed in grassland (225 kg N ha^{-1}), followed by mixed upland field/grassland (200), upland field production

Table 5.14 Comparison of average of BOD concentration (mg O_2 L^{-1}) of 8 rivers in dairy farming area before and after "law about promotion of adequacy of the management and utilizing of the animal excrement" enforcement

	The eastern part				The northern part			
	1	2	3	4	5	6	7	8
1995–1999	0.78	0.85	0.49	1.09	1.06	0.84	1.09	0.48
2012–2016	0.54	0.68	0.58	1.04	0.73	0.49	0.62	0.25
p-value	<0.01	<0.05	<0.05	0.315	<0.01	<0.01	<0.01	<0.01

Source Ministry of Land, Infrastructure, Transport and Tourism (2017)
1995–1999: Before enforcement of the law, 2012–2016: After enforcement of the law, *p*-value: *t* test (n = 60–135)

Table 5.15 Results of groundwater NO_3-N concentration survey in each area of Hokkaido

Area	Total number of well	>10 mg N L^{-1} number of well	Excess rate (%)	>30 mg N L^{-1} number of well	>50 mg N L^{-1} number of well
Ishikari	933	22	2.4	1	
Oshima	1136	34	3.0		
Hiyama	247	0	0		
Shiribeshi	370	7	1.9		
Sorachi	771	28	3.6	1	
Kamikawa	1435	18	1.3		
Rumoi	140	0	0		
Souya	85	0	0		
Abashiro	1089	334	30.7	39	18
Nemuro	197	0	0		
Kushiro	421	1	0.2		
Tokachi	1452	48	3.3	4	
Hidaka	265	0	0		
Iburi	987	54	5.5	2	
Hokkaido	9528	546	5.7	47	18

Source (Hokkaido 2013)

(170) and paddy fields (159). Additionally, the average amount of nitrogen input from municipal fertilizer, manure, and so on was found to be 185 kg N ha^{-1} for the whole of Hokkaido. The highest average nitrogen input was found in mixed upland crop/dairy areas (224 kg N ha^{-1}), followed by grassland (213), upland crop (190), and paddy field (151). The excess amount of nitrogen, which is the difference between nitrogen input and nitrogen assimilation capacity, was on the order of 2 kg N ha^{-1} on average for the whole of Hokkaido. The largest amount of excess nitrogen was observed in mixed upland field/dairy areas (24 kg N ha^{-1}), followed by upland farming (20), paddy field (−8), and grassland (−12). Furthermore, the authors considered the relationship between the NO_3-N concentration in the groundwater and the excess nitrogen amount with reference to the above well measurement result. No significant relationship was found for the grassland and paddy field areas, whereas a significant positive relationship was found for the upland field areas. This indicates that agriculture affects groundwater NO_3-N concentrations in upland field areas.

(3) **Impact of heavy metals**

(1) **Heavy metal concentrations of soils and soil amendments**

The cadmium concentration in non-polluted soils in Japan is 0.30 mg kg^{-1} (Asami et al. 1988), and the median value of typical domestic soil is 0.27 mg kg^{-1} (Yamazaki et al. 2009). Compared to these, the cadmium concentrations in soils in Hokkaido are lower (Table 5.16). Comparing heavy metal concentrations of cultivated soils and non-cultivated soils in Hokkaido, most of the levels are higher in cultivated soils, especially orchard soils (Table 5.16). This can be attributed to the loading of heavy metals from the application of agricultural materials.

The concentrations of heavy metals in organic materials used for agriculture and fertilizers have been determined (Table 5.17). Zinc and copper are present in high amounts in pig manure, sewage sludge, and compost made from these materials. Copper is also contained in pesticides such as Bordeaux mixture. The main loads of cadmium are due to phosphate fertilizer, compost, and sewage sludge. Large amounts of squid and scallop are fished in Hokkaido, and a large amount of fishery waste is generated; this waste includes the digestive glands of squid and mid-gut glands of scallop, which have high cadmium concentrations. Starfish, which are caught together with scallops, also contain high levels of cadmium. As such, compost and fertilizer which use such fishery wastes as raw materials also have high cadmium concentrations (Furudate and Otobe 2009). The load of lead (Pb) is due to paint and leaded gasoline. Lead concentrations in orchard soil may have increased as a result of the application of lead-containing pesticide in the past. Arsenic (in the form of lead arsenate, calcium arsenate, and organoarsenic compounds) was also used in pesticides in the past. Currently, it seems that the arsenic load derived from pesticide is small. However, organoarsenic compounds are

Table 5.16 Heavy metal concentrations in cultivated and non-cultivated soils in Hokkaido

Soils	Zn	Cu	Cd (mg kg^{-1})	Pb	As
Paddy	5.2	4.7	0.11	1.2	1.6
Upland	3.4	1.0	0.08	0.7	0.5
Vegetable	5.3	0.4	0.06	0.7	1.2
Orchard	16.6	5.8	0.20	5.1	2.6
Glassland	4.7	0.5	0.02	0.5	0.5
Whole cultivated	4.7	1.6	0.09	0.7	0.8
Non-cultivated	4.3	0.6	0.05	0.7	0.4

Zn, Cu, and Cd were extracted with 0.1 mol L^{-1} HCl
Pb was extracted with 1 mol L^{-1} ammonium acetate (pH4.5)
As was extracted with 1 mol L^{-1} HCl

Table 5.17 Heavy metal concentrations in organic materials for compost and fertilizers

		Zn	Cu	Cd	Pb	As
				(mg kg^{-1})		
Compost Material (dry weight basis)	Rice straw	83	30	0.24	7.8	0.72
	Bark	225	40	0.28	11	1.5
	Cattle manure	95	21	0.25	7.6	1.2
	Pigs manure	738	244	1.0	9.5	1.4
	Layers manure	218	34	0.59	12	0.30
	Horses manure	135	32	0.21	10	2.0
Sewage sludge (dry weight basis)	Lime-treated sludge	777	125	0.80	54	11
	Polymeric sludge	1139	203	1.6	58	19
Fisheries waste (fresh weight basis)	Digestive gland of squid	83	141	23		
	Mid-gut gland of scallop	28	5.8	18		
	starfish	43	1.4	2.8		
	Holdfast of kelp					19[a]
Chemical Fertilizer (fresh weight basis)	Ammonium sulfate			0.01[b,d]		
	Super-phosphate	95[c]	11[c]	4.0[d]		
	Double super-phosphate	110[c]	10[c]	6.7[b]		
	High analysis compound	69[c]	15[c]	3.0[d]		
	Potassium sulfate			0.20[b]		

[a]Jin et al. (2012) Rep. Hokkaido Inst. Pub. Health, 62, 21-
[b]Hosobuchi et al. (2011) Jpn. J. Soil Sci. Plant Nutr., 82, 207-
[c]Mishima et al. (2005) Soil Sci. Plant Nutr., 51, 437-
[d]Mishima et al. (2004) Soil Sci. Plant Nutr., 50, 263-

contained in the residue of marine algae that is present in fishery waste, and compost with this waste may consequently have a high arsenic concentration.

(2) **Estimation of cadmium load and removal in vegetable soils**

Cadmium concentrations in cultivated soils in Hokkaido (0.09 mg kg^{-1}) were higher than those in non-cultivated soils (0.05 mg kg^{-1}) (Table 5.16). Therefore, Hosobuchi et al. (2011) compared the cadmium load introduced to soil by fertilizer and compost and its removal by crops in vegetable fields. The cadmium loads introduced to soils that cultivated tomato, cabbage, and Japanese radish were found to be mostly derived from the application of compost and phosphate fertilizer, with the load being the highest for compost (Table 5.18). When comparing the cadmium load with the cadmium removal in tomato, which had a large cadmium uptake and for which both the edible and residual parts were removed from the field, the cadmium removal was found to be large irrespective of whether compost was applied. However, for cabbage and Japanese radish, the

Table 5.18 Cadmium load and removal in vegetable soils

Soils	Compost	Cd load[a] (g ha^{-1})	Cd removal[b] (g ha^{-1})	Difference[c] (g ha^{-1})	Cd in plowed soil[d] (g ha^{-1})
Tomato	×	2.57	36.4	−33.8	1584
	○[e]	6.76	33.6	−26.8	1584
Cabbage	×	2.41	0.47	1.94	1334
	○	4.50	0.63	3.87	1334
Japanese radish	×	1.33	0.92	0.41	1334
	○	3.45	0.95	2.50	1334

[a]Applied in compost and fertilizers
[b]Cd uptake by edible and residual parts from the tomato soil and by edible part from the cabbage and Japanese radish soils
[c]Load–Removal
[d]Up to 20 cm deep
[e]Applied 40t ha^{-1} for tomato soil and 20t ha^{-1} for cabbage and Japanese radish soils

cadmium loads were larger, because the cadmium uptakes of these plants were lower and only the edible parts of the plants were removed from the field. Thus, there was a difference in the cadmium removal depending on crop type: Soil cadmium levels tended to decrease for tomato and to accumulate for cabbage and Japanese radish. The magnitude of the decrease and the accumulation in a single year were slight compared to the amount of cadmium accumulation in the soil, but if similar cultivation is continued for many years the cadmium in the soil is likely to change significantly.

5.7.3 Soil Management Strategies

(1) **Paddy fields**

(1) **Atmospheric environment**

Paddy fields are the major agricultural source of methane in Hokkaido. One anthropogenic factor that affects the generation of methane is the soil type. Other anthropogenic factors affecting methane generation are field management, organic matter treatment, and cultivation, and the amount of methane generated is influenced by the combination of these factors. Methane emission increases in proportion to the amount of residual rice straw left in the field, with about 50% of the carbon content in the rice straw residue being emitted as methane (Naser et al. 2007).

The plowing of rice straw in autumn reduces methane emissions by 55–70% compared to plowing in spring (Goto et al. 2004). Furthermore, by adding fertilizer and decomposition materials in autumn before rice straw plowing, an additional 10% reduction in methane emissions is achieved (Goto et al. 2004).

By performing midsummer drainage in late June, methane emissions are reduced by about 50% compared to conditions with continuous flooding. Performing intermittent irrigation after late June (after the heading stage) reduces methane emissions by about 40% compared to continuous flooding (Goto et al. 2004).

Soil type is a factor affecting methane emissions. In Hokkaido, methane emissions are the lowest in Brown Fluvic soils, intermediate in gray fluvic soils, and the highest in gley fluvic soils. Furthermore, when a field is managed without tillage or without puddling, methane emissions are reduced as a result of the suppression of soil reduction which accompanies the improvement in drainage.

(2) **Aquatic environment**

When paddy fields are flooded, the supply of oxygen to the soil ceases and denitrification by microorganisms occurs. Due to this phenomenon, paddy fields have the function of purifying nitric acid. However, wastewater from paddy fields contains excess fertilizer components such as nitrogen and phosphorus, and there is concern about the adverse effects of these components on the surrounding aquatic environment. Since the nitrate concentration and water volume of rivers become markedly higher in the middle of May, the catchment area mainly composed of rice paddies in Hokkaido has the highest nitric acid outflow at this time. Nitrate nitrogen tends to flow out from paddy fields to rivers until the middle of June; however, this discharge is suppressed after the middle of June. In order to prevent water pollution, it is important to take measures against wastewater at the time of puddling.

At the time of transplant from puddling, wastewater containing a large amount of clay causes river water pollution. Comparing the nitrogen concentration of wastewater from paddy fields, the concentration of organic nitrogen in the surface drainage water before transplantation is remarkably high. This organic nitrogen is derived from suspended matter such as clay contained in surface drainage water. Since the concentration of suspended matter tends to increase as the degree of puddling increases, reducing the degree of puddling is an effective way to prevent water pollution. Furthermore, because the concentration of suspended matter tends to increase as the amount of water at the time of puddling increases, saving water at the time of puddling is also an effective means of mitigating clay outflow.

Lime-based sewage sludge compost is classified as ordinary fertilizer, but a fertilization standard (Urban Sewage Sludge Agricultural Land Application Standard) is set because such sewage sludge contains more heavy metals than compost, and it easily increases the pH of soil. In Hokkaido, composting is carried out, and the application amount per hectare per year is indicated as 10 t dry matter (Hokkaido Fertilizer Recommendations 2015; see Sect. 8.1). In rice cultivation, if the application amount of lime-based sewage sludge compost is 1 t ha^{-1} or less, and the reduction amount of chemical fertilizer nitrogen and phosphoric acid per 1 ton of compost is about 5 kg, the yield and quality are the same as in conventional rice cultivation and heavy metals do not accumulate in white rice and soil (Sugikawa et al. 2009).

(2) **Upland**

(1) **Soil management for mitigating greenhouse gas emissions**

In general, plowing or rotary tillage increases carbon dioxide emissions by promoting the decomposition of soil organic matter. For example, in Haplic Allophanic Andosols in a 5-year crop rotation system in Eastern Hokkaido, the decreases in soil carbon content over 20 years for minimum tillage (rotary tillage once in spring) and for conventional tillage (rotary tillage twice in spring and plowing once in autumn) were 1.04 Mg C ha^{-1} y^{-1} and 1.34 Mg C ha^{-1} y^{-1}, respectively. This shows that it is possible to mitigate the degradation of soil organic matter by performing only minimum tillage.

However, efforts to positively increase soil carbon stock are also important. The application of organic materials (including the incorporation of crop residue) is effective for this purpose. Nakatsu and Tamura (2008) investigated the effect of 30 years of continuous crop residue incorporation and/or cattle manure application on the soil carbon content of Haplic Allophanic Andosols under a 4-year crop rotation system in Eastern Hokkaido. The results showed that soil carbon content clearly decreased without the application of crop residue and manure. In contrast, soil carbon content did not change over the 30-year period with the application of 2.5 Mg ha^{-1} of organic materials (dry matter basis), equivalent to 1 Mg C ha^{-1}. Furthermore, soil carbon content increased by 10% from its initial value with the application of 5 Mg ha^{-1} of organic materials, equivalent to 2 Mg C ha^{-1}. These results suggest that the annual decrease in soil carbon is about 1 Mg C ha^{-1}, which is nearly the same as the value determined in the abovementioned case of 5-year rotational cropping (1.04–1.34 Mg C ha^{-1}). We can also estimate the ratio of the carbon accumulated in the soil to the applied carbon to be about 8%, assuming that the organic material was incorporated within the top 25 cm of the soil and that the soil bulk density was 800 kg m^{-3}.

Emissions of nitrous oxide associated with nitrogen fertilization and/or the incorporation of crop residues are inevitably associated with crop production. Until recently, it was considered difficult to balance the improvement of crop productivity and nitrous oxide emission control at the same time. Now, however, the following practical emission control techniques have been proposed.

One such technique is the use of controlled-release fertilizers such as coated fertilizer, chemically synthesized slow-release fertilizer, and fertilizer containing nitrification inhibitor (Akiyama et al. 2009). The application of these fertilizers suppresses the rapid increase in soil solution nitrogen concentration, resulting in lower nitrous oxide emissions than when using usual chemical fertilizers.

Additionally, nitrous oxide emissions accompanied by the incorporation of crop residues are mainly caused by denitrification in anaerobic soil conditions. Accordingly, basic soil management, such as ensuring drainage and air permeability, is also effective for the control of nitrous oxide emissions.

(2) **Soil management for preventing nitrate pollution in groundwater**

As described in Sect. 5.7.1., the concept of "nitrogen environmental capacity" is used as an index for the risk assessment of nitrate pollution of groundwater in agricultural lands in Hokkaido. This value is also considered as the upper limit of nitrogen input to keep the nitrate nitrogen concentration in groundwater below the environmental criterion (10 mg NO_3-N L^{-1}).

In Hokkaido, the fertilization standards (see Sect. 8.1 for details) for various crops are defined as the recommended rate of fertilizer application for environmentally friendly agriculture practice, and the observance of fertilization standards ensures the prevention of nitrate pollution in groundwater. If fertilization management based on the fertilization standards is practiced in upland fields, the risk of nitrate pollution in groundwater is low, even without paying special attention to the concept of nitrogen environmental capacity (Suzuki et al. 2004).

The available nitrogen released from cattle manure, which is the most commonly used type of manure in Hokkaido, increases cumulatively with continuous application. Therefore, a guideline annual upper limit for the application of cattle manure is indicated and for upland fields is 30 Mg ha^{-1} (about 150 kg N ha^{-1} as total nitrogen). This limit was determined in view of both changes in the nitrate nitrogen concentration in soil solution and the yield or quality of crops under cattle manure application. From the same point of view, the annual upper limit of cattle manure application for open cultivation vegetable fields is 25 Mg ha^{-1} for crops with one cultivation per year such as onion and 50 Mg ha^{-1} for crops with two cultivations per year. During the combined use of manure and chemical fertilizer, it is important to reduce the amount of chemical fertilizer by the correct amount by considering the nitrogen derived from the manure; the amount of chemical nitrogen that can be reduced per 1 Mg of applied manure is estimated to be 1 kg in the case of less than 4 years of continuous application, 2 kg for 5–10 years, and 3 kg for over 10 years.

(3) **Grassland**

(1) **Air**

Air pollution caused by dairy farming includes the generation of unpleasant odors and greenhouse gases. The former mostly occurs in the spring and autumn, which are the primary times for the application of dairy cow manure and liquid manure (slurry). One countermeasure to this problem is an improved method of slurry application. The amount of ammonia volatilized from slurry that is shallowly injected into grassland is less than the amount that is dispersed in the customary farming method of using an impact plate. However, the working efficiency of the shallow injection method is inferior to that of the customary method, and as such it is not widely adopted.

Regarding the issue of the release of greenhouse gases from dairy farming, there is an increase in nitrous oxide emissions when manure is applied to the surface of grassland together with chemical fertilizers, compared with fertilization using chemical fertilizers alone. However, carbon derived from the manure accumulates in the grassland, which is considered to suppress the greenhouse effect of the nitrous oxide emissions.

(2) **Water**

The main causes of water pollution generated from grasslands in dairy farming regions are the improper management and use of cow manure. Representative examples of the former are the creation of outdoor manure piles and the storage of slurries in simple pit lagoons, while representative examples of the latter are the excess application of manure and the application of manure during the winter.

In Hokkaido, a research project for promoting the appropriate management and use of livestock manure began in 1994, and its results were summarized in the "Handbook on the Treatment and Use of Livestock Manure, 2004." The project proposed the "manure-focused fertilization method" for grassland, which involves the evaluation of fertilizer nutrient contents supplied by dairy manure, as detailed in Sect. 5.5.3. As a result of attempts for the local introduction of the manure-focused fertilization method, the Konsen Agricultural Experiment Station and the Northern Kushiro Agricultural Extension Center, in Hokkaido, succeeded in improving fertilization, for example by greatly reducing phosphate concentrations, which had markedly accumulated in the soil in the target area. Additionally, promoting the introduction of this method demanded the establishment of a farm support system to enable the implementation of manure-focused fertilization.

In 2009, Matsunaka et al. (2009) developed the AMAFE2006 software program (Decision Support System for Application of Manure and Fertilizer to Grassland and Forage Corn Field based on Nutrient Recycling) to support the "manure-focused fertilization method."

Furthermore, the maximum amount of processed dairy cow manure that could be applied for grassland renovation was studied from the perspective of water pollution caused by the leaching of nitrogen and maintaining the botanical composition of grassland with mixed seeding. An upper limit of 60–50 Mg ha^{-1} was proposed for the application of manure in mineral soils and volcanic ash soils, respectively. For dairy slurries, the limit was a value equivalent to 40 kg N ha^{-1} of the fertilizer nitrogen conversion amount for the second year of renewal.

(3) **Heavy metals**

Analysis of the necessity for the application of zinc and copper in fields maintained on volcanic ash soils revealed no decrease in grass yield if these heavy metals were not supplemented, and a soil concentration level at which the yield was reduced could not be found (Saigusa et al. 1997, 1999). Additionally, it was concluded that, as measures to increase the content of these heavy metals in grass considering the needs of cattle, rather than applying fertilizer, it is preferable to increase the fraction of forage legumes, to harvest earlier and to apply dairy manure. This is because most of the applied heavy metals accumulate in the surface of grassland, and therefore, damage from the excess accumulation of these metals resulting from the continuous application of dairy manure cannot be ruled out.

5.8 Environmentally Friendly Agriculture (Clean Agriculture)

5.8.1 Fertilizer Recommendations

(1) **Outline of the Hokkaido Fertilizer Recommendations**

Appropriate fertilizer management is important for crop production. Fertilizer management in Hokkaido is implemented based on the "Hokkaido Fertilizer Recommendations" published by the Agricultural Department of the Hokkaido Government. The Hokkaido Fertilizer Recommendations aim to communicate a standard of fertilizer application to farmers, agricultural improvement extension centers, agricultural cooperative staff, and so on. The recommendations give the standard amount of fertilizer (fertilization standard) by area and soil type for major crops and also detail the application of fertilizer based on soil diagnosis and crop nutrition diagnosis. These recommendations were initially published in 1957 as the "Hokkaido Fertilization Standard" and have since been revised almost every 7 years. The "Hokkaido Fertilizer Recommendations" were published in 2002, integrating with the separately published "Soil and Plant Diagnosis Criteria" and "Fertilizations based on Soil Diagnosis." Revisions were made to the recommendations in 2010–2015, which included the addition of new research results. Currently, the "Hokkaido Fertilizer Recommendations 2015" (Hokkaido Government Agricultural Department 2015) are used. The number of target crops has increased from 10 in 1957 to 73 at present.

The Hokkaido Fertilizer Recommendations 2015 define standard fertilizer application rates for nitrogen, phosphate, potassium, and other elements, corresponding to the reference yield set for each region (and soil), and classification for targeted crops (item, cropping type) has been performed. An appropriate fertilizer application amount is calculated by increasing or decreasing the amount of fertilizer based on soil diagnosis and crop nutrition diagnosis and increasing or decreasing the amount of fertilizer depending on the kind and the application amount of organic matter such as compost and manure.

The Hokkaido Fertilizer Recommendations 2015 comprehensively judge the natural conditions, such as weather conditions and topography, of each region of Hokkaido, and divide Hokkaido into 18 regions. Additionally, soil types are classified into four types, namely lowland soil, volcanic ash soil, terrace soil, and peat soil, according to sedimentary pattern and geological base material.

The "standard yield" is set at a yield level that is achievable through proper cultivation control under the standard weather and soil conditions and is set from the past statistical yield.

The "standard amount of fertilizer (fertilization standard)" is the amount of fertilizer (nitrogen, phosphate, potassium, etc.) necessary to secure the standard yield. For soil conditions, nitrogen has a moderate fecundity level, and phosphate, potassium, and so on are within the range of the soil diagnostic reference value.

Figure 5.11 shows a schematic diagram of the relationship between nutrient supply to crops and the yield, quality,

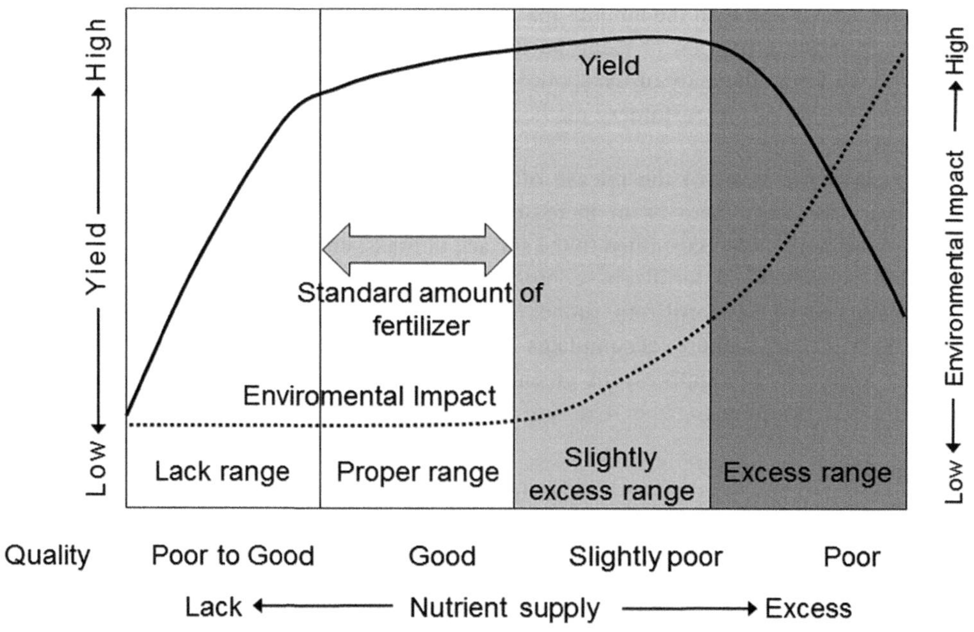

Fig. 5.11 Schematic diagram of the relationship between nutrient supply to crops and yield, quality, and environmental impact. Figure supplied by Daiji Asaka

and environmental impact. When nutrient supply is lacking, the yield/quality improves as the nutrient supply increases, the yield/quality is maintained at a high level in the proper range, and the yield/quality decreases and the environmental impact increases in the excess range. These relationships tend to be clear, especially for nitrogen supply. From the viewpoint of crop yield, it is desirable to have nutrient supply of a "slightly excess range" that can stably obtain high yields. However, from the viewpoint of environmental impact, quality improvement, and production cost reduction, the "standard amount of fertilizer" aims to supply nutrients within the "proper range".

The "soil diagnostic criteria" indicate the criteria for the soil chemistry and physical properties which are desirable for crops to achieve good growth and yield. The "crop nutrition diagnostic criteria" indicate a measure of the nutrient content in the plant body when the crop grows normally; physiological disorders such as nutrient deficiency or excess are expected when conditions greatly deviate from the diagnostic criteria. The "soil diagnostic criteria" and "crop nutrition diagnostic criteria" are measured via the method described in "analytical method for soil and crop nutrition diagnosis 2012" (Hokkaido Research Organization 2012).

For the sustainable production of crops, it is important to keep the soil in good condition. For each crop, we have specified the amount of manure application and fertilizer reduction necessary to maintain soil fertility. For example, 10 Mg ha^{-1} of manure for rice plants and 20 Mg ha^{-1} of manure for open field vegetables are applied per year, and the amount of fertilizer (nitrogen, phosphate, potassium, etc.) is reduced according to the type of manure and the application period.

(2) **Standard fertilizer application procedure**

Figure 5.12 shows the standard fertilizer application procedure in the Hokkaido Fertilizer Recommendations 2015. First, the standard amount of fertilizer required for a standard yield, corresponding to crops, variety, cropping type, regional zone, and soil type is determined. Next, the amount of nutrients supplied based on the result of the soil diagnostic, increasing or decreasing the amount of fertilizer application, is estimated. Then, the amount of fertilizer is increased or decreased according to the applied amount of organic matter (e.g. manure), the kind of organic matter (e.g. green manure or the residue of previous crops), and the amount of plowing in that has been performed. When calculating the amount of fertilizer required, it is assumed that a certain amount of nitrogen fertilizer is applied as "starter nitrogen" in order to ensure good initial growth.

The idea of calculating the amount of nutrients necessary for crops, including the amount of nutrients supplied by organic matter and the residue of previous crops, is also useful from the viewpoint of environmental conservation. This reflects the basic idea of "clean agriculture", which is Hokkaido's environmental conservation agricultural technology, as explained in the next section. The Hokkaido Fertilizer Recommendations have been revised almost every 5 years, and the next revision is scheduled to be carried out in 2020.

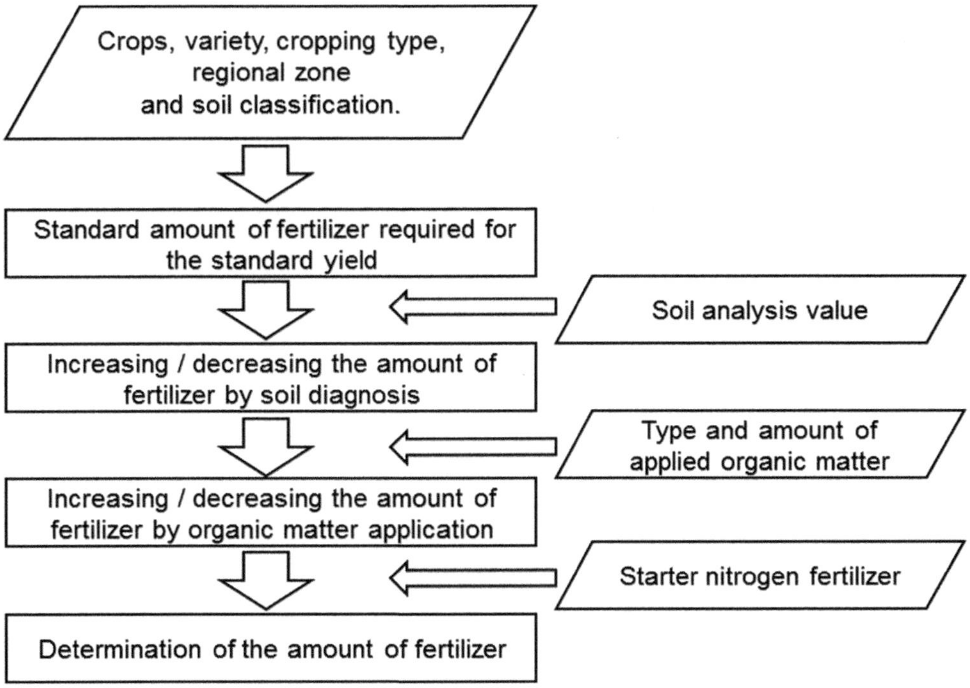

Fig. 5.12 Standard fertilizer application procedure in "Hokkaido Fertilizer Recommendations, 2015". Figure supplied by Daiji Asaka

5.8.2 Soil Management

(1) **Labeling the clean agriculture**

"Clean agriculture" is a mode of environmental conservation agriculture, which was first advocated in Hokkaido, in 1991. It involves producing safe, high-quality agricultural products in harmony with the environment, using methods such as soil preparation by the application of organic materials and cultivation using minimal amounts of chemical fertilizers and chemically synthesized pesticides.

In 2000, the Hokkaido Clean Agriculture Promotion Committee established the "Certification Mark System for Northern Clean Agricultural Products" through which the YES! Clean mark (Fig. 5.13) is displayed on products that meet predetermined fertilizer application and pest control standards, present cultivation information and have been promoting the spread of the system and the development of registered production groups.

(2) **Fertilizer application standards for certification system**

The system not only obligates soil diagnosis and sets the upper limit for the amount of chemical fertilizer application, but also sets the upper limit for the total amount of nitrogen application, which is the total amount of nitrogen components supplied from fertilizers, including compost and organic fertilizers, corresponding to the soil nitrogen fertility level (Table 5.19). Additionally, for the application of organic materials such as compost, the lower limit for the application amount of organic material including compost and the upper limit for the application amount of compost are set to ensure the maintenance and improvement of soil fertility and the mitigation of the nitrogen load on the environment. Details of soil diagnosis and fertilizer application standards are presented below.

For fields registered in the Certification Mark System for Northern Clean Agricultural Products, it is necessary to conduct soil diagnosis using analytical parameters predetermined for different kinds of crops. The nitrogen fertility of respective fields is evaluated using the diagnosis results. The upper limit of total nitrogen application described below is determined based on this nitrogen fertility. "Wet incubation nitrogen," "autoclave-extractable nitrogen," and "raw soil incubation nitrogen," which are analytical parameters used for paddy rice, field crops, and vegetables grown outdoors, show stable values for a long period. Therefore, a value from analysis in the last 3-year period is useful. Performing soil diagnosis every 3 years is sufficient after registration. For "nitrate nitrogen", which is an analytical parameter used for greenhouse vegetables, an analysis value within the last 1-year period must be used. Soil diagnosis is conducted every year after registration.

Total nitrogen application consists of the integrated amount of nitrogen component derived from organic materials including compost, chemical fertilizers, and organic fertilizers used in the period relevant to each crop. The standard fertilizer application corresponding to the soil nitrogen fertility level is fundamentally applicable to the upper limit of total nitrogen application. However, the standard fertilizer application might vary for some crops by zone category or soil category. The maximum value of the standard fertilizer application was used for such crops.

The upper limit of chemical fertilizer application is determined using the remaining nitrogen amount obtained by subtracting the nitrogen supply when the lower limit of organic matter including compost is applied from the upper limit of the total nitrogen application. However, fertility enhancement by the application of organic matter including compost is required in a field with soil nitrogen fertility classified as "low." Therefore, the value of "middle" soil nitrogen fertility was applied for the upper limit of chemical fertilizer application.

The lower limit of application of organic materials including compost represents the amount of organic material which must be applied annually to fields using the "amount of compost application to maintain soil fertility" shown in the Hokkaido fertilizer application guidelines. The value is 10 Mg ha^{-1} for paddy rice and field crops, 20 Mg ha^{-1} for vegetables, flowers, and ornamental plants grown outdoors and fruit trees, and 40 Mg ha^{-1} for vegetables, flowers, and ornamental plants grown in greenhouses.

However, for cases in which only a small amount of compost is available, or when quality benefits are achieved in crops more easily without compost application, the concept described below is used for the application of organic materials including compost: Organic matter that can supply nitrogen component or dry weight equivalent to compost is useful as "organics equivalent to compost." Organic materials such as fish meal and fermented chicken manure can be converted into compost with an equivalent nitrogen component (Table 5.20). Till in, green manure can also be converted into compost based on dry weight. Therefore, the standard can be met using by applying appropriate organic material instead of using compost.

Nitrogen mineralization occurs in compost during periods when no crop is grown. It is therefore feared that the application of large quantities of compost might become a source of environmental pollution such as groundwater contamination. For that reason, an upper limit of compost application amount is set for compost containing a high percentage of nitrogen-rich cattle manure, such as cattle manure/straw compost and cattle manure/bedding compost.

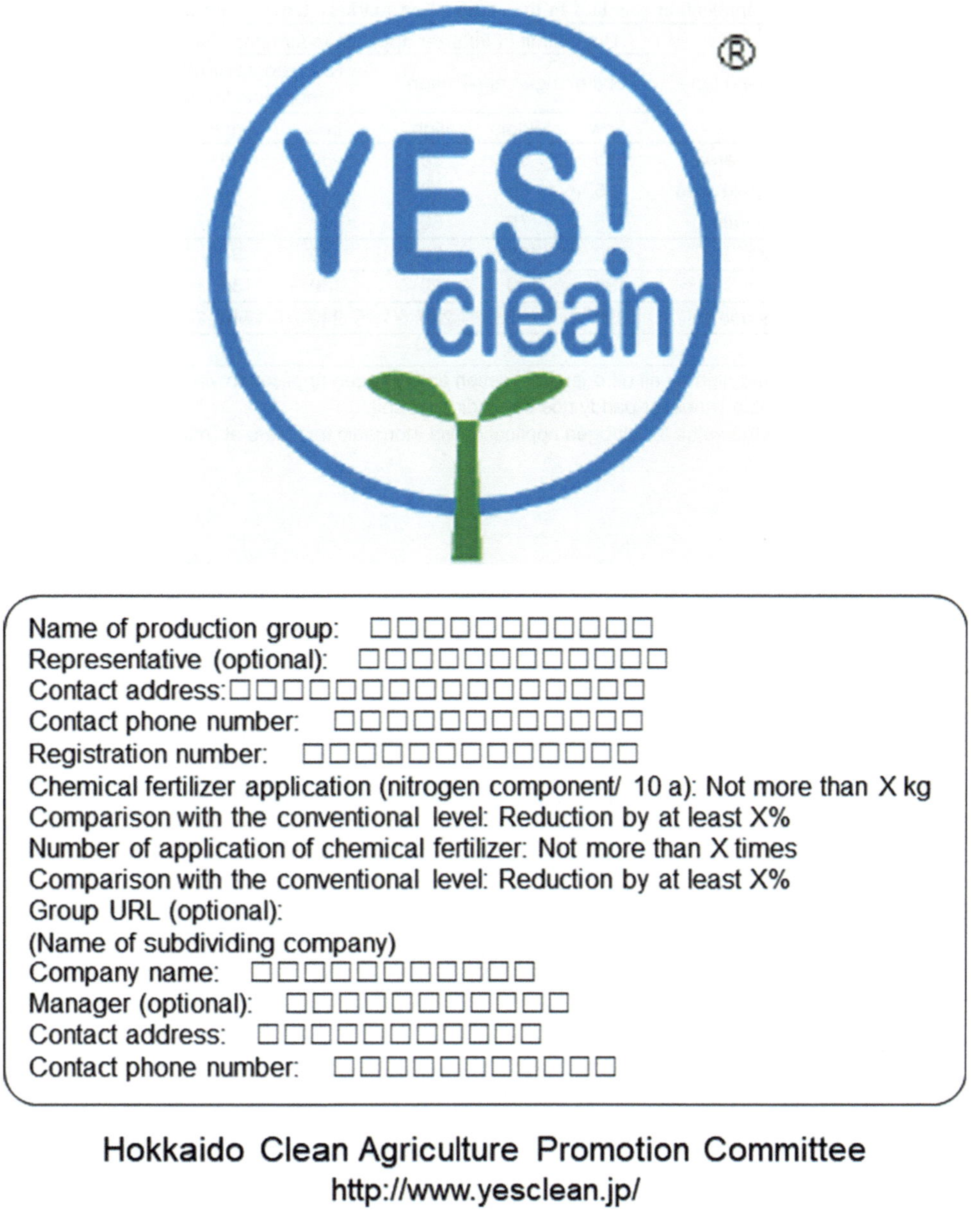

Fig. 5.13 Display example of the YES! Clean mark. *Source* (Hokkaido clean Agriculture promotion committee, 2017b)

The annual upper limit of compost application amount is 30 Mg ha^{-1} for field crops and vegetables grown outdoors (single harvest per year) and 50 Mg ha^{-1} for vegetables grown outdoors (two harvests per year).

5.9 Recent Topics on Soil Management

5.9.1 ICT for Soil Management

Currently, ICT is utilized in large-scale upland fields in Hokkaido. For example, wheat harvesting is carried out based on early-late maps of wheat maturity created using satellite images, and tractors are equipped with automatic steering.

Under such circumstances, variable fertilizer application using ICT is expected as a new soil management technology. Hara et al. (2015) constructed a system for the variable application of nitrogen fertilizer on wheat by mounting an optical sensor in a tractor. This system can perform variable fertilizer application in real time based on variations in crop growth (which corresponds to N absorption) obtained from the sensor. The experimental results of a study using the above system in eight fields showed that the average wheat yield with variable fertilizer application increased by 4.0% compared with conventional uniform fertilizer application.

Table 5.19 Fertilizer application standard in the certification mark system for northern clean agricultural products (abstract)

Crop	Cropping type	Upper limit of fertilizer application standard (kg ha^{-1}): Total nitrogen application			Nitrogen application via inorganic fertilizers			Conventional level of nitrogen application via inorganic fertilizers	Reduction rate of nitrogen from inorganic fertilizers
		Low	Middle	High	Low	Middle	High	(kg ha^{-1})	(%)
Paddy rice	High-yield area	95	90	80	80	80	70	100	20.0
	Middle-yield area	85	80	70	70	70	60	100	30.0
	Low-yield area	75	70	60	60	60	50	100	40.0
Potato	Outdoors	120	100	80	90	90	70	110	18.2
Onion	Outdoors	180	150	120	130	130	100	200	35.0
Tomato	Greenhouse	350	300	250	240	240	190	260	7.7
Apple	Outdoors	–	70	–	–	50	–	70	28.6

Note) "Low, middle and high" in all nitrogen application and nitrogen application via inorganic fertilizers are the level of soil nitrogen fertility. The standard value also varies for paddy rice depending on soil.
The reduction rate is the value for nitrogen application via inorganic fertilizers at "middle" soil nitrogen fertility level compared to the conventional level.

Table 5.20 Examples of nitrogen and dry weight equivalents of organic materials such as compost

Type	Unit (Mg)	Nitrogen equivalent (kg/unit)	Dry weight (kg/unit)
Cattle manure/straw compost and cattle manure/bedding compost	10	10	3000
Bark compost[a] and rice husk compost	10	0	4000
Cattle manure slurry	10	13	800
Horse manure compost	10	5	3000
Pig manure compost[b]	10	20	4000
Dried chicken manure	1	13	900
Fermented chicken manure	1	13	800
Rice straw compost	10	10	3000
Garbage compost	10	10	4000
Rapeseed meal	1	30	850
Fish meal	1	50	950

Source (Hokkaido Clean Agriculture Promotion Committee, 2017b)
[a]Bark compost with a high percentage of manure shall be in accordance with cattle manure/straw compost and cattle manure/bedding compost
[b]Pig manure compost with a high percentage of manure shall be in accordance with cattle manure/straw compost and cattle manure/bedding compost

We have also constructed a variable fertilizer application system that utilizes images of the surface soil taken by unmanned aerial vehicle drone (UAV drone). In this subsection, the outline and effect of our system are described.

In Hokkaido, hot water extractable N is used for the evaluation of N fertility, and a method for estimating the appropriate N fertilizer application rate based on hot water extractable N is established in upland crops such as sugar beet, potato, and the like (Hokkaido Fertilizer Recommendations 2015; see Sect. 8.1).

We demonstrated that the variation of hot water extractable N within an upland field can be accurately determined from visible images of the surface soil taken by UAV drone and constructed a system for variable fertilizer application based on the map of hot water extractable N.

The procedure of our system is as follows (Fig. 5.14).

(1) A technique for variable fertilizer application was designed based on the hot water extractable N map, the above described method of estimating appropriate N fertilizer application rate, and the application history of organic matter such as crop residues and compost with reference to the method described by Fueki et al. (2010). When blended fertilizers are used, variable

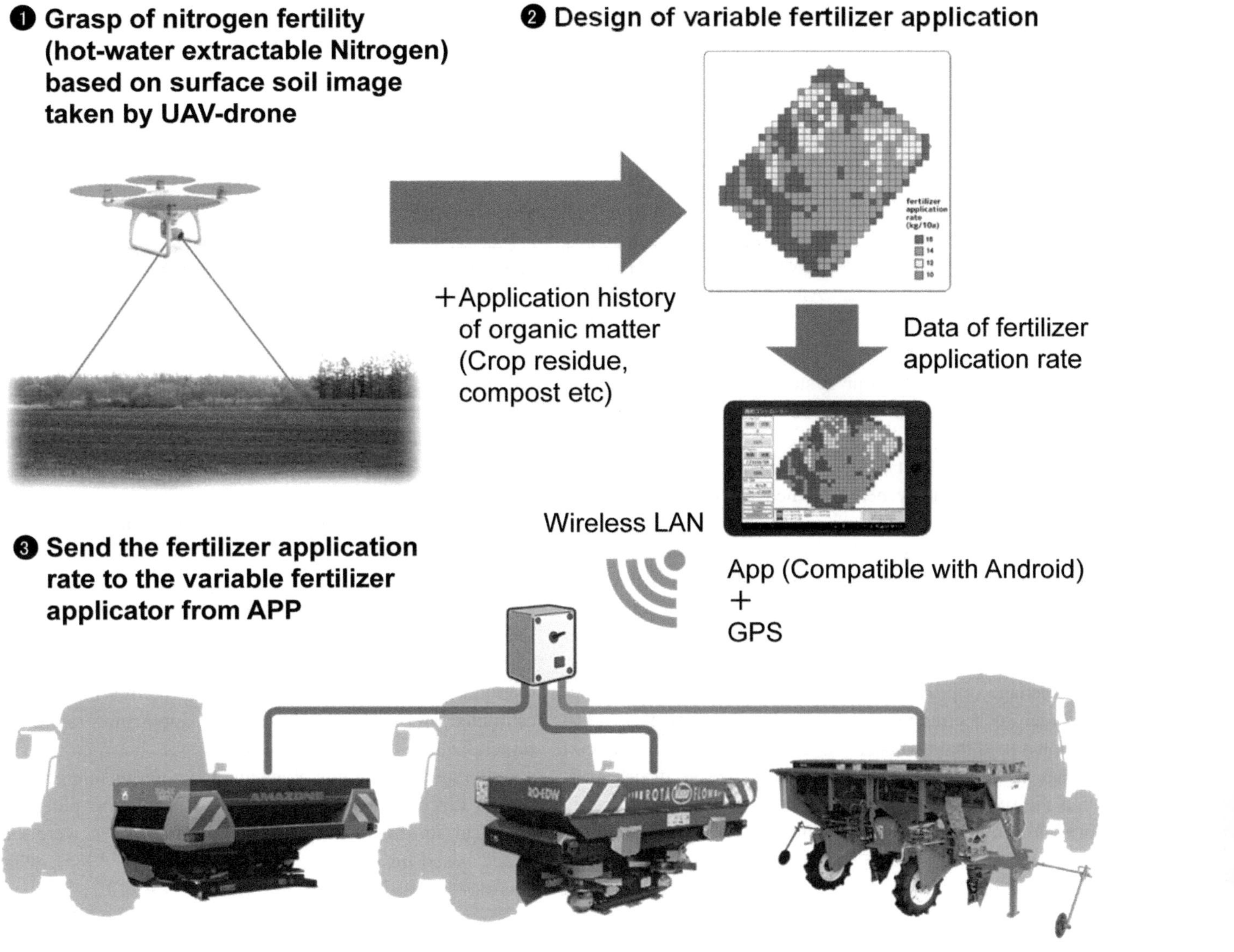

Fig. 5.14 Outline of variable fertilizer application system. Figure supplied by Katsuhisa Niwa

fertilizer application can be designed within a range for which the amounts of phosphate and potassium application are not reduced below the required rates.

(2) The data of variable fertilizer application rates are input into an application (App), compatible with the Android mobile operating system, that was developed by our company (Zukosha co., Ltd). This app can determine the present location using GPS and send the fertilizer application rate of the present location to the fertilizer applicator via a wireless LAN.

(3) By using a variable fertilizer applicator that can receive fertilizer information from the app, variable fertilizer application can be performed automatically.

The effect of our system was investigated in sugar beet in three Andosol fields in the Tokachi area of Hokkaido in 2015. Experimental plots were established in each field. The areas of each plot were more than 1 ha. Conventional fertilizer application (CF) and variable fertilizer application (VF) treatments were applied to the plots, and the fertilizer application rates and sugar yields for both treatments were compared.

For the VF treatment, the nitrogen fertilizer application rate was reduced by an average of 25% in each field compared with the CF treatment. Moreover, the sugar yield was higher in each field for the VF treatment than for the CF treatment, with an average of 9% higher yields (Table 5.21).

Table 5.21 N application rate and sugar yield for each treatment in three experiment fields

Field	Average N application rate[a]			Sugar yield[b]		
	VF (kgN ha^{-1})	CF (kgN ha^{-1})	Ratio (%; VF/CF)	VF (t ha^{-1})	CF (t ha^{-1})	Ratio (%; VF/CF)
A	158	184	14.1	13.3 ± 0.7	12.2 ± 1.1	9.0
B	68	151	55.0	14.6	13.2	10.3
C	108	114	5.3	12.5 ± 0.6	11.7 ± 0.5	6.7
Average	111	150	24.8	13.5	12.4	8.7

Source (Niwa et al. 2016)
VF: Variable fertilizer application, CF: Conventional fertilizer application. [a] VF: 120–200, 40–175, 80–140 kgN ha - 1 in field A, B, C, respectively, CF: Uniform fertilizer application. [b]Field A, C: Mean values for five replicates ± standard deviation

Following these results, our system was introduced in an area of upland field of about 120 ha in 2016 and about 160 ha in 2017.

Currently, the effect of our variable fertilizer application system is continuously being studied for wheat and potato crops.

In Hokkaido, the growth of field crops is restricted by poor drainage, and pipe drains have therefore been constructed as a countermeasure. However, when three typhoons passed through Hokkaido in 2016, enormous water damage occurred in upland fields. As such, it is considered that drainage improvement is still insufficient.

Under such circumstances, a request was made to photograph a field that had been partially flooded due to typhoon damage using a UAV drone. The drone image clearly captured the partially flooded area (Fig. 5.15). After geometric correction, a flood map was created by visual interpretation and provided to the client. The map was used for the design of a pipe drain system, which was then densely constructed in the flooded area.

In this way, ICT can be utilized in various soil management techniques. From now on, it is important to apply ICT not only to variable fertilizer applications, but also to soil management for land improvement such as drainage improvement.

5.9.2 Farmyard Manure Application

Applying livestock manure can allow the amount of chemical nitrogen fertilizer to be decreased and can increase the soil carbon sequestration. Additionally, the application of chemical N fertilizer and manure can decrease the methane (CH_4) uptake in soils and increase nitrous oxide (N_2O) emissions. This section reports on the effect of the application of farmyard manure on the net ecosystem carbon balance (NECB), emissions of CH_4 and N_2O, and global warming potential (GWP) at four managed grasslands on Andosols under different climatic regions ranging from warm to cool temperate (Shimizu et al. 2014).

(1) **Evaluation method of the effect of farmyard manure application**

Two treatments were performed: one treatment with chemical fertilizer (fertilizer plot) and another treatment using cattle manure and chemical fertilizer (manure plot). In the manure plots, the nutrient supply rate from manure was estimated by multiplying the manure application rate by the mineralization rate, and the difference between nutrient supply rate from manure alone and the nutrient supply rate in the fertilizer plots was supplied to the manure plots with chemical fertilizer. The N mineralization rates were estimated based on Uchida's model (Shiga et al. 1985), which was developed in Japan, and amounted to 13.2, 7.02, and 5.53% of total manure N in the first, second, and third years after manure application, respectively. The application of manure allowed the application rate of chemical fertilizer N to be decreased by 12–35% of that in the fertilizer plots. In both plots (fertilizer plot and manure plot) at each site, net ecosystem exchange (NEE) was measured by the eddy covariance method (Hirata et al. 2013). The NECB in hay meadow grasslands, which includes C input through manure application and C output through crop harvest as well as NEE, was estimated using the following equation:

$$\mathrm{NECB} = \mathrm{NEE} - \mathrm{C\,applied\,in\,manure} + \mathrm{harvested\,C}$$

Positive NECB represents C loss from the ecosystem. The fluxes of CH_4 and N_2O were measured by dark chamber methods. (Relative to CO_2, the GWP values with a 100-year time horizon are 25 times greater for CH_4 and 298 times greater for N_2O.)

Fig. 5.15 UAV drone image of a field flooded partially due to typhoon damage. Figure supplied by Katsuhisa Niwa

(2) **Effect of manure application on net ecosystem carbon balance (NECB)**

The amounts of harvested C in the fertilizer and manure plots were 4.3 ± 0.8 and 4.1 ± 0.6 Mg C ha^{-1} yr^{-1}, respectively, and there was no significant difference in harvested C among the sites or between the fertilizer and manure plots. However, the NEE varied between the fertilizer and manure plots, being −2.4 ± 1.1 and −1.6 ± 0.7 Mg C ha^{-1} yr^{-1}, respectively, and significant differences in NEE were also found between the treatment plots ($p < 0.05$). This indicates that more net CO_2 uptake occurred in the fertilizer plots. However, there was no significant difference in grass production between fertilizer plots and manure plots because harvested C in both plots was in the same range. The difference in NEE between treatments was 0.9 ± 0.9 Mg C ha^{-1} yr^{-1} on average, and the rate of C loss as CO_2 from total amount of C applied in manure in 3 years was estimated as 25 ± 37%. Therefore, the differences in NEE between the plots was likely caused by organic matter decomposition.

NECB was significantly higher in the fertilizer plots (1.9 ± 0.9 Mg C ha^{-1} yr^{-1}) than in the manure plots (−1.8 ± 1.8 Mg C ha^{-1} yr^{-1}), indicating that grassland soils of all sites experience C loss without manure application.

(3) **The effect of manure application on CH_4 and N_2O emissions**

The CH_4 emissions ranged from −0.8 to 3.9 kg C ha^{-1} yr^{-1}. CH_4 was emitted from poorly drained grassland because soil moisture promotes microbial methanogenesis. There was no significant difference in CH_4 emissions among plots (Shimizu et al. 2013).

The N_2O emission from the manure plots was 6.2 ± 3.7 kg N ha^{-1} yr^{-1} and that from the fertilizer plots was 3.6 ± 3.2 kg N ha^{-1} yr^{-1} ($p < 0.05$). High available C from the applied manure and mineral N from the chemical fertilizer may have enhanced denitrification and N_2O production in the manure plots. Emissions of CO_2 and N_2O from arable soils would increase with increasing soil C and N stocks as a result of the regular application of farmyard manure for many years. However, the continuous application of manure can allow the supplementary application rates of chemical fertilizer to be reduced over the years in the manure plots because of increasing soil nitrogen availability. This may mitigate N_2O emissions (Mori and Hojito 2012).

(4) **The effect of manure application on GWP**

The GWP was greater in fertilizer plots (2.3 ± 1.1 Mg CO_2 eq.-C ha^{-1} yr^{-1}) than in manure plots (−1.0 ± 2.1 Mg CO_2 eq.-C ha^{-1} yr^{-1}) (Fig. 5.16), indicating that manure application can mitigate greenhouse gas emissions from grassland. The major contributors to GWP were components of the NECB, followed by N_2O emission. The difference in GWP between the manure and fertilizer plots had a negative relationship with the application rate of manure C ($y = -4.45\ \ln(x) + 2.84$; $R2 = 0.85$; $p < 0.01$) (Fig. 5.17) and was negative for application rates of manure C of more than 1.9 Mg C ha^{-1} yr^{-1}. This indicates that increasing the application rate of manure C can mitigate global warming at least at the site scale.

5.9.3 Volunteer Potato Weed Control

The depths of soil frost have been decreasing in Eastern Hokkaido, due to the development of thick snow cover in early winter that insulates the ground (Hirota et al. 2006). After these decreases in soil frost depth, we have observed unexpected and undesirable effects of soil frost depth reduction in agriculture, namely weed problems from volunteer potatoes (Fig. 5.18). The soil in this region used to freeze up to a depth sufficient to kill leftover potato tubers, thereby providing a natural mechanism of weed control. However, the decreasing soil frost depth has facilitated the survival of unharvested tubers, causing a large outbreak of volunteer potatoes as a weed in the subsequent year of the crop rotation. These must be removed to prevent competition with the current year's crop, in order to prevent an invasion or outbreak of pest or vermin such as potato cyst nematode, and to prevent the contamination of different crops or cultivars. Weeding volunteer potatoes using herbicides, rotary tillage, or manual labor is time-consuming, onerous work, because the sprouts emerge aboveground sporadically, and not simultaneously. Therefore, farmers are forced to undertake repeated sessions of manual removal of volunteer potatoes over a long time (several tens of hours per hectare per person) across the main farming season (Hirota et al. 2011).

(1) **Snow plowing (yukiwari)**

To overcome the problem of the reduction in soil frost depth, local farmers have devised a countermeasure to the changing conditions. Some farmers have started snow plowing ("*yukiwari*" in Japanese), which temporarily removes snow cover within a field, allowing the frost to penetrate to a greater depth and thereby killing volunteer potatoes. This method consists of three steps as shown in Fig. 5.19.

This method of volunteer potato elimination appears to be effective, although it is based on trial-and-error, relying on the producer's experience and intuition. In some years, soil frost is inadequate, allowing the survival of volunteer potatoes, while in other years, excessive frost depth delays planting.

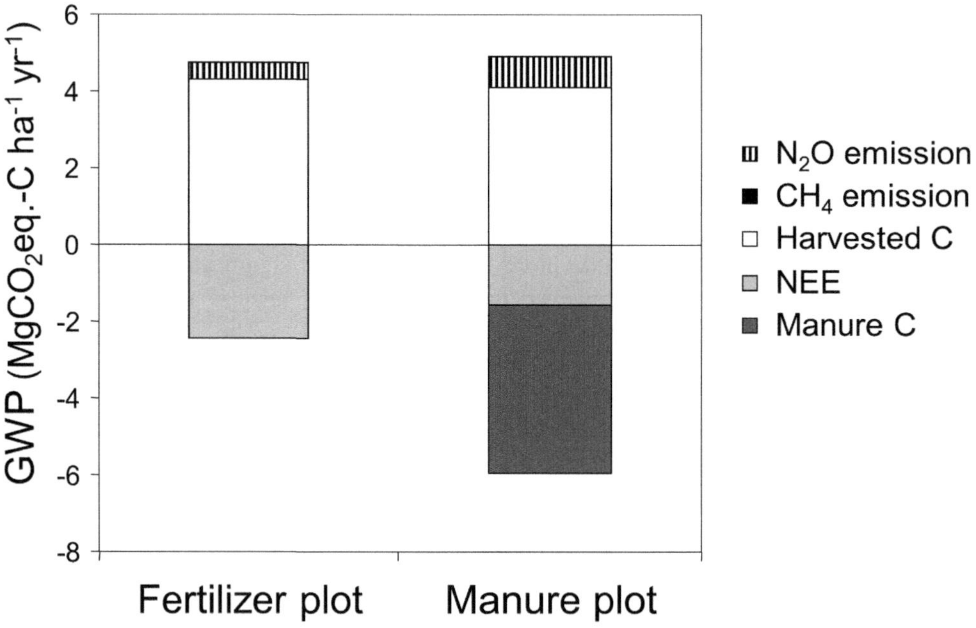

Fig. 5.16 Global warming potential (GWP) for the fertilizer and manure plots. Reprinted by permission from Springer Nature: Springer, Shimizu et al. 2014, COPYRIGHT (Copyright Springer Nature Singapore Pte Ltd. 2015)

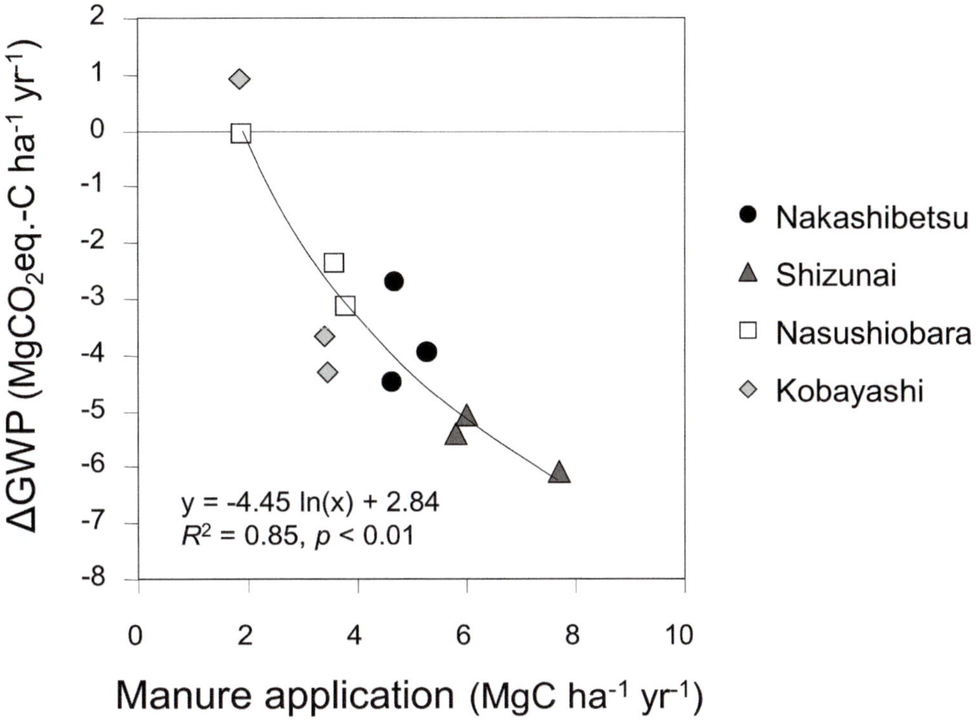

Fig. 5.17 Difference in global warming potential (GWP) between the manure and fertilizer plots (manure plots–fertilizer plots) compared with the application rate of manure C. Reprinted by permission from Springer Nature: Springer, Shimizu et al. 2014, COPYRIGHT (Copyright Springer Nature Singapore Pte Ltd. 2015)

Fig. 5.18 Photographs showing (**a**) potatoes left after harvest, (**b**) volunteer potatoes in a sugar beet field, (**c**) volunteer potatoes in a wheat field, and (**d**) volunteer potatoes in a bean field. Reprinted by permission from Springer Nature: Springer, Hirota et al., 2011, Copyright c 2011, Springer Science Business Media B.V.

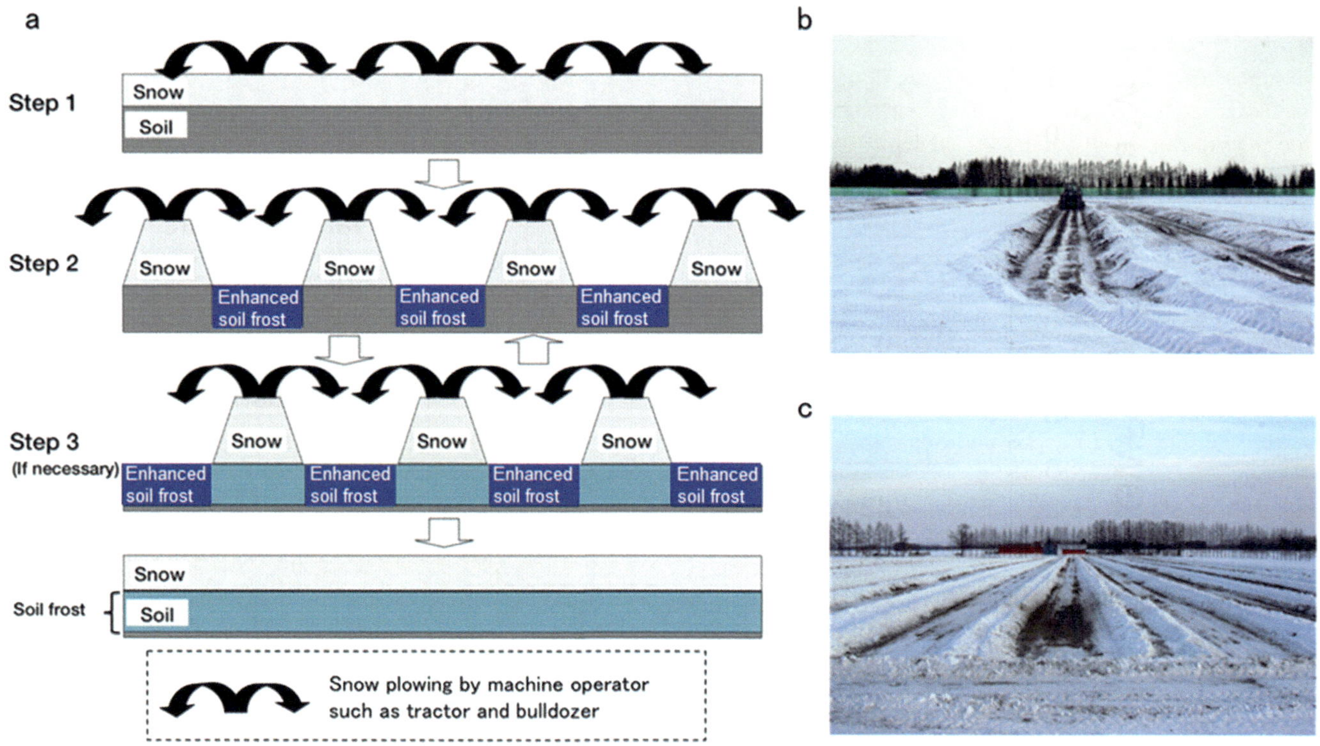

Fig. 5.19 Schematic diagram of (**a**) the operation sequence of Yukiwari (snow plowing) and photographs of (**b**) a tractor plowing snow and (**c**) a field after plowing. Reprinted by permission from Springer Nature: Springer, Hirota et al., 2011, Copyright c 2011, Springer Science Business Media B.V.

(2) **Soil frost control**

This method needs to balance (i.e. optimize) the following two conflicting objectives: (1) allow sufficiently deep frost penetration to kill the potato tubers and (2) keep the frost depth shallow enough so that it does not cause a delay in seeding or negatively impact the growth of overwintering crops. To solve this optimization problem, a method was developed to manipulate soil frost depths, by artificially controlling snow cover thickness; this was guided by a simple numerical model that simulates the soil freezing-thawing processes using daily mean air temperature and snow cover thickness as input variables. The optimum soil frost depth was founded to be 0.3 m, which is a compromise between the elimination of volunteer potatoes and a minimum frost depth to prevent negative effects on agriculture. Soil frost depth can be controlled by adjusting the timing and duration of thick snow cover that insulates the soil surface during cold periods. To achieve the optimal frost depth, the duration of treatment (i.e. snow removal and subsequent re-deposition or natural accumulation) was predicted using a simple numerical soil temperature model that required only daily mean air temperature and snow depth as input variables. The soil frost control remarkably decreased potato tuber survival to less than 0.5% of unharvested tubers. This soil frost control method is expected to be most effective in regions which have a mean air temperature during December to February of between −12 and −5 °C. In the Tokachi area, a Web-based decision-making system for soil frost control for volunteer potato management has been disseminated. Snow plowing has been used widely by producers in Tokachi, and the large-scale operation of this practice can now be optimized by the real-time prediction of frost depths using the efficient and robust soil frost model (Hirota et al. 2011; Hirota et al. 2013; Yazaki et al. 2013).

5.9.4 Use of Arbuscular Mycorrhizal Fungi

Arbuscular mycorrhizal fungi (AMF) colonize plant roots as obligate symbionts and enhance the water and nutrient uptake of their host. The Japanese Government has designated AMF inoculum as a soil improvement material to improve soil function by providing phosphate nutrition to plants. The maximum utilization of AMF in agriculture is expected.

In soybean or field corn cultivation in Hokkaido, it has been recognized that yield of the crop grown after AMF host plants is generally higher than that grown after non-AMF host plants. This phenomenon is known as the previous crop effect (Arihara and Karasawa 2000; Fig. 5.20). Using this previous crop effect, Oka et al. (2010) indicated that the

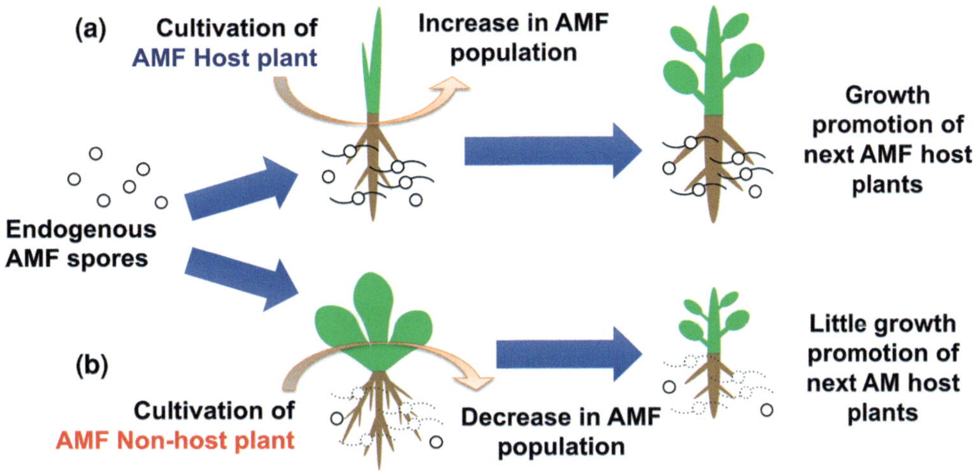

Fig. 5.20 Mechanism of previous crop effect. When host plants of arbuscular mycorrhizal fungi (AMF) are cultivated (**a**), endogenous AMF population increase which leads growth promotion of succeeding AMF host plants. Conversely, monoculture of AMF non-host plant or laying land fallow (**b**) leads decrease in the population of endogenous AMF, and poor growth promotion of succeeding AMF host plants would be expected. Figure supplied by Ryo Ohtomo and Norikuni Oka

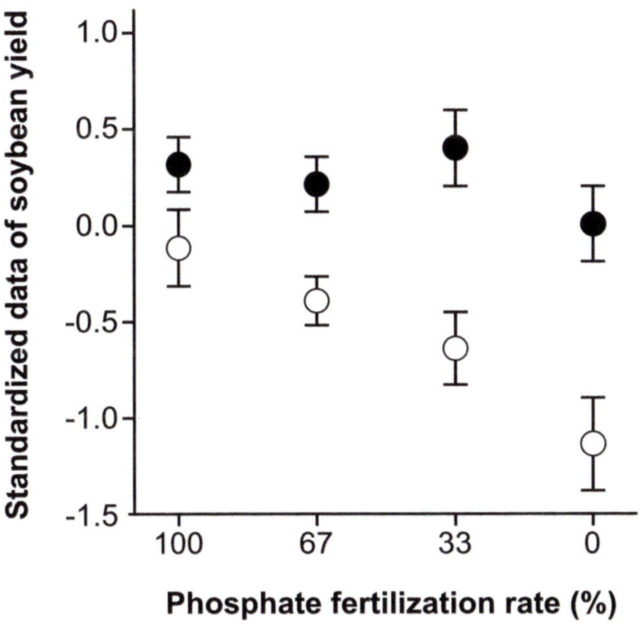

Fig. 5.21 Trends in soybean yield following arbuscular mycorrhizal fungal (AMF) host and non-host plants. Closed symbols (●) indicate yield following AMF host plants, and open symbols (○) indicate yield following AMF non-host plants or fallow fields. Data standardized within each year were averaged over years and shown with standard error. Figure supplied by Ryo Ohtomo and Norikuni Oka

application of phosphate fertilizer to soybean can be reduced to 30–70% of the recommended amount at that time without reduction in the yield (Fig. 5.21). To verify whether this reduced fertilizer rate is applicable to farmer's field, two surveys were conducted, one involving the AMF colonization of soybean in farmer's field, and another involving an on-site fertilizer reduction test at farmer's field. Finally, the recommended application level of phosphate fertilizer in Hokkaido soybean cultivation was revised considering these results together.

AMF colonization is higher in Andosols than in other soil types and lower in phosphate-rich fields or fields with higher phosphate application (Table 5.22). Although the level of AMF colonization in soybean of farmer's field was lower than that in experimental field observed by Oka et al. (2010), the difference between the previous crop type (i.e. AMF host plant or non-AMF host plant) was significant. Therefore, we considered that the reduction in the amount of phosphate fertilizer considering previous crop effect would be applicable to soybean production in farmer's field.

5.9.5 Calcium Fertilization for Potato Cultivation

(1) **Functions of calcium in plants**

The addition of essential plant nutrients through the application of fertilizer is crucial to improving crop yield and quality. In particular, the application of calcium (Ca) is important for adequate crop growth and the quality of the harvested crop. Ca is abundant in cell membranes and cell walls and forms gel structures in the cell wall by complexing with pectin, thereby maintaining plant structure and protecting cell structure. Ca also reacts as a messenger to changes in the cell growth environment, including light and heat conditions. Ca is therefore an essential nutrient for

Table 5.22 Arbuscular mycorrhizal fungal infection to soybean in Hokkaido, Japan (2011-2013)

Items	Class	All			After hosts		After non-hosts	
		AM (%)[b]		n	AM (%)	n	AM (%)	n
Previous crop	AMF host[d]	34	c	(75)				
	AMF non-host[d]	22		(23)				
Soil type[a]	Andosols	42	a	(29)	50	(17)	30	(12)
	Peat soils	28	ab	(22)	30	(19)	11	(3)
	Others	26	b	(47)	28	(39)	14	(8)
Available phosphate Troug, mg P/kg (mg P_2O_5/100 g)	Less than 43.6 (10)	51	a	(7)	48	(6)	69	(1)
	43.6 (10) to 131 (30)	30	b	(57)	33	(42)	21	(15)
	More than 131 (30)	28	ab	(34)	31	(27)	18	(7)
Phosphate fertilization (relative to recommendation)	More than 70%	28	c	(77)	31	(64)	12	(13)
	Less than 70%	41		(21)	48	(11)	34	(10)
0–0.3 m average soil hardness	Less than 1.5 MPa	38	c	(40)	43	(28)	26	(12)
	More than1.5 Mpa	13		(5)	14	(4)	5	(1)

Source (Ohtomo et al. 2015)
[a]Soil type was classified according to Soil Classification System, 2017
[b]Numbers indicate average AMF infection rate to soybean, and numbers with parenthesis indicate sample number (n). Symbols in "All" indicate statistical differences which was evaluated within each categories, i.e. [c]indicate significant difference between two (Mann–Whitney's U test, $p < 0.05$), and different letters indicate significant differences (Steel–Dwass, $p < 0.05$)
[d]AMF host: wheat, potato, oats, sunflower, soybean, and kidney bean. AMF non-host: buckwheat, white mustard, and beet

growing strong and healthy crops that can adapt to pests, disease, and numerous environmental pressures.

(2) **Status of soil calcium content in Hokkaido**

In Japan, acidic soils are widespread due to high rainfall leaching base cations including potassium (K), magnesium (Mg), and Ca. The application of liming materials, including calcium carbonate (limestone) and calcium magnesium carbonate (dolomite), has been widely practiced in order to increase soil pH. However, these materials have extremely low solubilities and supply little Ca to plants. For potato production, high pH has been considered as the cause of common scab, and the application of liming materials has been avoided (Mizuno 2001). A regional survey of 170 potato-growing farms in the Tokachi area (n = 90; primarily Andosols) and Kamikawa area (n = 80; primarily brown forest soils and pseudogley soils) showed that the mean Ca saturation, the fraction of exchangeable soil Ca to CEC, was 33%. The Hokkaido Fertilizer Recommendations 2015 advises that growers apply inputs to maintain the Ca saturation at above 40%; however, about 80% and 70% of the surveyed farms in the Tokachi and Kamikawa areas, respectively, had Ca saturations below this threshold (Fig. 5.22). This shows that the supply of Ca to crops from soils is not sufficient in these areas.

(3) **Calcium fertilization in the USA**

Research carried out by the University of Wisconsin–Madison has shown that the incidence of black bruise, internal brown spots, and hollow heart can be suppressed when the tuber Ca concentration exceeds 250 mg kg^{-1} by the application of water soluble calcium salt (calcium nitrate or calcium chloride) on top of ridges during the tuber bulking period. Our survey of 170 farms in Hokkaido found that the mean tuber Ca concentration was 122 mg kg^{-1}, and that no locations exceeded 250 mg kg^{-1}.

(4) **Calcium fertilization in Hokkaido**

The type of soil and the method of ridging need to be considered when determining the method of Ca fertilization in potato cropping systems in Japan. The soils of Japan have high organic matter content, and the side-dressing of calcium nitrate therefore results in excess nitrogen and causes a lower yield and specific gravity. The early ridging method that is widely practiced in Hokkaido and the mulching method that is practiced in Honshu make the application of Ca-containing fertilizer on top of the ridges challenging.

A collaborative research project between Calbee Potato Inc. and the Obihiro University of Agriculture and Veterinary Medicine has been carrying out the application of

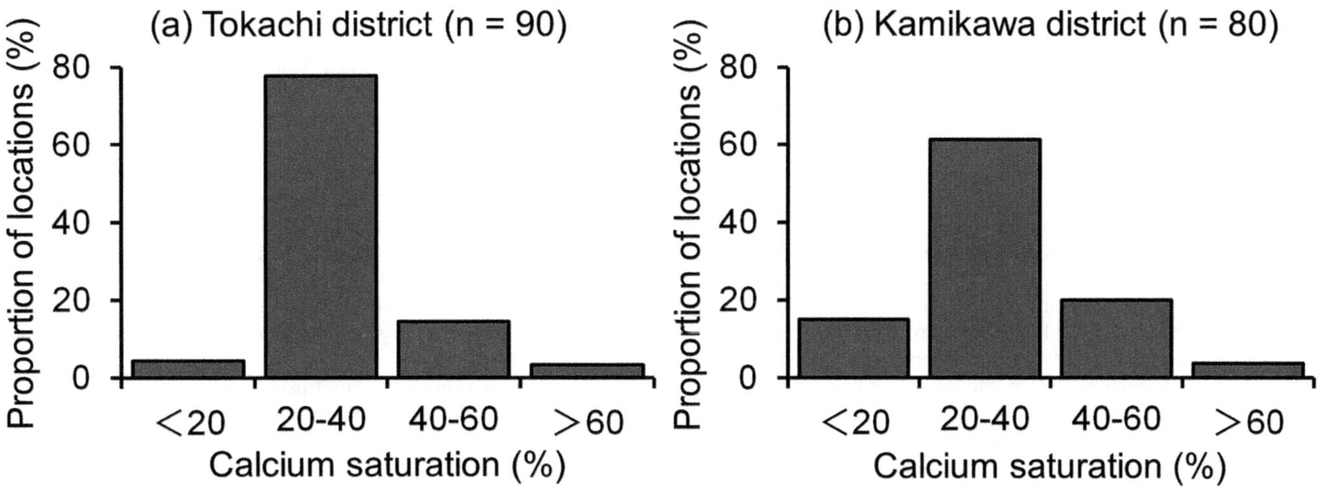

Fig. 5.22 Histogram of surface soil calcium saturation in potato growing farms of Tokachi and Kamikawa districts. Surface soil samples were collected in potato growing farms in Tokachi (n = 90) and Kamikawa (n = 80) district in 2013 and 2014 growing seasons. Calcium saturation was determined by cation exchange capacity by the Schollenberger method and the analyzed exchangeable calcium content. Figure supplied by Rintaro Kinoshita, Jiwan Palta and Masayuki Tani

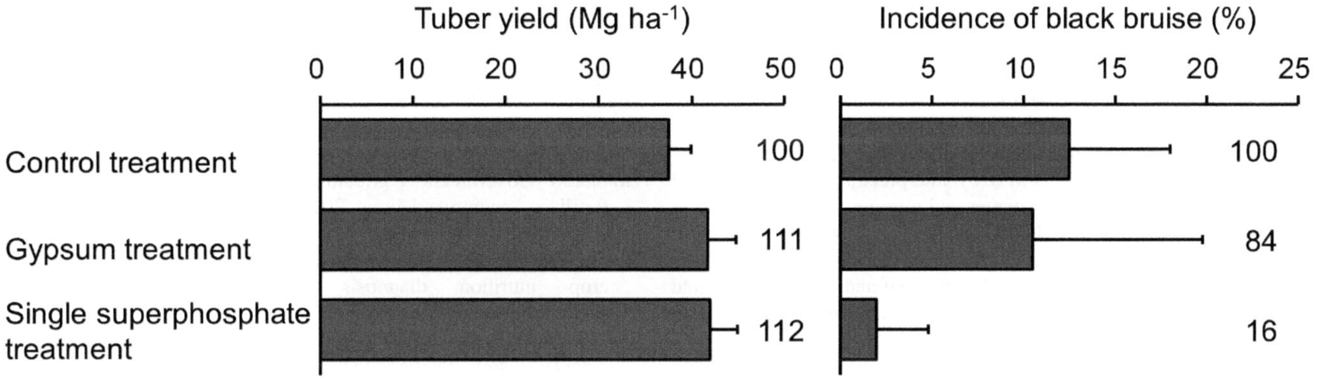

Fig. 5.23 Effects of broadcasted gypsum and single superphosphate on tuber yield of potatoes and the incidence of black bruise. Data extracted from an experimental result from 2015 growing season (Cultivar: Toyoshiro, location: Memuro town, soil type: Low-humic Andosols, Typic Hapludands). Values indicated in the figure show relative values to control set at 100. Figure supplied by Rintaro Kinoshita, Jiwan Palta and Masayuki Tani

gypsum (calcium sulfate), single superphosphate (SSP), or triple superphosphate (TSP) by broadcasting after potatoes are planted. An experiment was conducted involving (1) control treatment following the standard fertilizer rate for potatoes from the Hokkaido Fertilizer Recommendations 2015 (60 kg N ha^{-1}, 200 kg P_2O_5 ha^{-1}, 110 kg K_2O ha^{-1}) by applying ammonium sulfate, ammonium phosphate. and potassium sulfate in the basal soil; (2) gypsum fertilizer treatment by applying the same fertilizer as the control and an additional broadcasted gypsum after planning at 500 kg ha^{-1}; and (3) SSP treatment by applying fertilizer at the standard rate but replacing ammonium phosphate with SSP. In the gypsum and SSP treatments, both tuber starch content and tuber yield increased by more than 10%. In the SSP treatment, the incidence of black bruise decreased (Fig. 5.23).

References

Agriculture and Horticulture Development Board (AHDB) (2017) Nutrient management guide (RB209) Section2 p. 17. https://ahdb.org.uk/documents/RB209/RB209_Section2_WEB_2017-12-01.pdf (March, 2018)

Aguilos M, Takagi K et al (2013) Sustained large stimulation of soil heterotrophic respiration rate and its temperature sensitivity by soil warming in a cool-temperate forested peatland. Tellus B 65:20792. https://doi.org/10.3402/tellusb.v65i0.20792

Aguilos M, Takagi K et al (2014) Dynamics of ecosystem carbon balance recovering from a clear-cutting in a cool-temperate forest. Agric For Meteorol 197:26–39

Akashi K, Konno K, Otsuka H (2005) Agronomic characters on green manures in Abashiri district of Hokkaido. Hokunou 72:30–35 (in Japanese)

Akiyama H, Yan X, Yagi K (2006) Estimations of emission factors for fertilizer-induced direct N2O emissions from agricultural soils in Japan: summary of available data. Soil Sci Plant Nutri 52:774–787

Akiyama H, Yan X, Yagi K (2009) Evaluation of effectiveness of enhanced-efficiency fertilizers as mitigation options for N2O and NO emissions from agricultural soils: meta-analysis. Glob Change Biol. https://doi.org/10.1111/j.1365-2486.2009.02031.x

Arihara J, Karasawa T (2000) Effect of previous crops on arbuscular mycorrhizal formation and growth of succeeding maize. Soil Sci Plant Nutr 46:43–51

Asami T, Kubota M, Minamizawa K (1988) Natural abundance of cadmium. Antimony, bismuth and some other heavy metals in Japanese soils. Jpn J Soil Sci Plant Nutr 59:197–199 (in Japanese)

Blandon MZD, Satoh F et al (1993) The mineral condition of soils and tree species in serpentine and non-serpentine areas of northern Hokkaido. Res Bulletin Hokkaido Univ Forest 51:1–13

Committee on nomenclature of the pyroclastic deposit in Hokkaido (1974) Distribution of the late quaternary pyroclastic deposit in Hokkaido. Miscellaneous publication of the Hokkaido national agricultural experiment station, Japan

Fifth Committee for Soil Classification and Nomenclature of Japanese Society of Pedology (2017) Soil classification system of Japan. (in Japanese)

Fueki N, Nakatsu S et al (2002) Cause of disorders in early stage of sugar beet growth cultivated by paper-pot planting and direct sowing in Hokkaido. Jpn J Soil Sci Plant Nutr 73:373–382

Fueki N, Sato K, Nakatsu S (2010) Interpretation of soil mineral nitrogen by scoring organic matter and nitrogen management as an 'N-score' in the fields of Hokkaido before sugar beet planting. Soil Sci Plant Nutr 56:750–759

Fujii S, Cornelissen JHC et al. (2018):Tree leaf and root traits mediate soil faunal contribution to litter decomposition across an elevational gradient. Functional Ecology, in press

Furudate A, Otobe Y (2009) Application effects of cattle feces/marine waste mixture composts and sludge composts on cadmium contents of timothy (Phleum pretense L.) in Northern Hokkaido, Japan. Jpn J Soil Sci Plant Nutr 80:506–510 (in Japanese)

Furudate A, Yamagami M et al (2000) The effect of furrow application of liming materials on the growth of transplanted sugar beets. Hokunou 67:14–19

Giesler R, Satoh F et al (2004) Microbialy available phosphorous in boreal forests: effects of aluminium and iron accumulation in the humus layer. Ecosystems 7:208–217

Goto E, Miura S et al. (2003) The present condition of chemical properties of paddy soils in Hokkaido. Jpn J Soil Sci Plant Nutr 74:475–483

Hara K, Suda T, Watanobe K (2015) Evaluation of sensor-based variable rate nitrogen fertilization for winter wheat production. JSAM 77(6):485–493 (in Japanese with English summary)

Hasegawa S, Tabuchi T (1995) Well facilitated paddy fields in Japan. In: Tabuchi T, Hasegawa S (eds) Paddy fields in the World. Japanese Society of Irrigation, Drainage and Reclamation Engineering, Tokyo, pp 103–124

Hayashi T, Hikasa Y (2017) Fertile effects of cattle manures to tomato cultivations in greenhouse fields in cold regions. Hokunou 84:252–260 (in Japanese)

Hayashi T, Nagao A (2010) A diagnostic method for measuring nitrogen in the subsoil to decrease the nitrate accumulation in Spinach. Jpn J Soil Sci Plant Nutr 81:263–266 (in Japanese)

Hayashi T, Hikasa Y, Nakazumi H (2004) Improved nitrogen application method based on the subsoil test under greenhouse conditions. Jpn J Soil Sci Plant Nutr 75:617–621 (in Japanese)

Hirata R, Miyata A et al (2013) Carbon dioxide exchange at four intensively managed grassland sites across different climate zones of Japan and the influence of manure application on ecosystem carbon and greenhouse gas budgets. Agric For Meteorol 177:57–68

Hirota T, Yazaki T et al. (2013) Soil frost control: its application to volunteer potato management in a cold region. In Imai R, Yoshida M, Matsumoto N (eds) Plant and microbe adaptations to cold in a changing world. Springer, New York, pp 51–61

Hirota T, Iwata Y et al (2006) Decreasing soil-frost depth and its relation to climate change in Tokachi, Hokkaido. Jpn J Meteorol Soc Jpn 84:821–833

Hirota T, Usuki K et al (2011) Soil frost control: agricultural adaptation to climate variability in a cold region of Japan. Mitig Adapt Strateg Glob Change 16:791–802. https://doi.org/10.1007/s11027-011-9296-8

Hishi T, Urakawa R et al (2014) Seasonality of factors controlling N mineralization rates among slope positions and aspects in cool-temperate deciduous natural forests and larch plantations. Biol Fertil Soils 50:243–356

Hojito M (1994) Studies on grassland acidification caused by fertilizer application and its effects on grass growth. Rep Hokkaido Prefectural Exp Stations 83:1–106 (in Japanese with English summary)

Hokkaido (2013) Contamination of groundwater with nitrate and nitrite nitrogen. http://www.pref.hokkaido.lg.jp/ks/jss/khz/mizukankyo/syou.htm

Hokkaido Clean Agriculture Promotion Committee (2017) Guidelines for the certification mark system for Northern clean agricultural products. Forms and formats. http://www.yesclean.jp/crop/d00.html

Hokkaido Government Agricultural Department (2015) Hokkaido fertilizer recommendations 2015. http://www.pref.hokkaido.lg.jp/ns/shs/clean/sehiguide2015.htm

Hokkaido Research Organization (2012) Analytical method for soil and crop nutrition diagnosis 2012. https://www.hro.or.jp/list/agricultural/center/bunseki2012/index.html

Hosobuchi Y, Hayashi T, Nakazumi H (2011) Estimation of cadmium balance in soils: load applied in fertilizer and compost and removal in edible and uneaten parts of vegetables. Jpn J Soil Sci Plant Nutr 82:207–213 (in Japanese with English summary)

Imura K (1987) Development of agriculture in Hokkaido and problems of soil and plant nutrition. In: Soil science and plant nutrition in Hokkaido agriculture, 1987, Hokunoukai, Sapporo, pp 1–8 (in Japanese)

Kanda T, Takata Y et al (2016) Soil map of Hokkaido developed with the comprehensive soil classification system of Japan, first approximation: change of andosols distribution area in forest. JSSPN 87:184–192 (in Japanese with English summary)

Kanda T, Takata Y et al (2017) Development of the 1:50,000 digital soil map of cultivated soil in Japan according to the comprehensive soil classification system of Japan—first approximation. JSSPN 88:29–34 (in Japanese with English summary)

Katayanagi N, Sawamoto T et al (2008) Nitrous oxide and nitric oxide fluxes from cornfield, grassland, pasture and forest in a watershed in Southern Hokkaido. Jpn Soil Sci Plant Nutri 54:662–680

Kawakami T, Makoto K (2017) Does an earthworm species acclimatize and/or adapt to soil calcium conditions? The consequence of soil nitrogen mineralization in forest soil. Ecol Res 32:603–610

Kayama M, Choi DS et al (2006) Comparison of growth characteristics and tolerance to serpentine soil of three ectomycorrhizal spruce seedlings in northern Japan. Trees Struct Funct 20:430–440

Kiso S, Kikuchi K (1988) Optimizing the application rates of nitrogen fertilizer on the basis of legume ratio in timothy (Phleum pratense

L.) dominant meadows. Jpn J Grassl Sci 34:169–177 (in Japanese with English summary)
Kitagawa I, Tsukamoto Y (2014) Evaluation of technologies for sequestering carbon in agricultural land by using organic filter underdrains. Paddy Water Environ 12(Supp. 2):287–294
Kitagawa I, Takeuchi H et al (2008) The new drainage technology "Cutting Drain" of digging type mole drain. Jpn J Soil Sci Plant Nutr 79:313–316 (in Japanese)
Kusa K, Sawamoto T, Hatano R (2002) Nitrous oxide emissions for 6 years from a gray lowland soil cultivated with onions in Hokkaido. Jpn Nutr Cycl Agroecosys 63:239–247
Land bureau, National land agency (1975–1979) Land classification map 01 (Hokkaido I ~ VIII) (in Japanese)
Lee M-S, Mo WH, Koizumi H (2006) Soil respiration of forest ecosystems in Japan and global implications. Ecol Res 21:828–839
Matsumoto T (2008) Studies on the fertility management of timothy sward based on use of dairy cattle manures evaluated as available nutrients. Report of Hokkaido Pref. Agric Exp Stn 121:1–61 (in Japanese with English summary)
Matsumoto T, Tou S (2006) Risk assessment of nitrate pollution in groundwater based on the environmental nitrogen-assimilation capacity of agricultural lands in Hokkaido. J Soil Sci Plant Nutr 77:17–24 (in Japanese)
Matsumoto T, Tamura T et al (2002) Simple estimation method for nutrients content in dairy cattle manure. Jpn J Soil Sci Plant Nutr 73:169–173 (in Japanese)
Matsunaka T, Saigusa T et al (2009) Decision support system for application of manure and fertilizer to grassland and forage corn field based on nutrient recycling. Jpn J Soil Sci Plant Nutr 80:177–182 (in Japanese)
Matsushita K, Igarashi Y, Ueda Y (1985) Genesis and transfiguration of the Ishikari peat-land in Hokkaido. Rep Geol Surv Hokkaido 57:71–88
Miki N (1993) Study on the effective application of nitrogen fertilizer to cutting sward based on evaluation of soil nitrogen release in Tenpoku district. Rep Hokkaido Prefectural Exp Stations 79:1–98 (in Japanese with English summary)
Miki N, Higashida S et al (1987) Potassium application to orchardgrass sward corresponding with soil potassium fertility in heavy clay soil of tenpoku district. Jpn J Soil Sci Plant Nutr 58:758–761 (in Japanese)
Ministry of Land, Infrastructure, Transport and Tourism (2017) Ministry of land, infrastructure, transport and tourism, water quality hydrology database. http://www1.river.go.jp
Mizuno N (2001) Studies on soil characterization of conductive soil and suppressing method for potato common scab. Jpn J Soil Sci Plant Nutr 72:327–330 (in Japanese)
Mori A, Hojito M (2012) Effect of combined application of manure and fertilizer on N_2O fluxes from a grassland soil in Nasu. Jpn Agric Ecosyst Environ 160:40–50
Morisada K, Ono K, Kanomata H (2004) Organic carbon stock in forest soils in Japan. Geoderma 119:21–32
Nakamoto H (2007) Assessment for risk of groundwater contamination caused by nitrate and reduction measures for reduction on onion fields etc. Agric Hortic 82:482–488 (in Japanese)
Nakamura Y, Fujikawa T et al (1970) Distribution and dynamics of some forest soil animals in Hokkaido. J Jap For Soc 52:80–88 (in Japanese with English summary)
Nakatsu S, Tamura H (2008) Effects of thirty years continuous application of organic materials (bark manure and crop residues) on total carbon, total nitrogen and physical characteristics of upland field soil in light colored andosol in Hokkaido. Jpn J Soil Sci Plant Nutr 79:139–145 (in Japanese with English summary)
Nakatsu S, Higashida S, Sawazaki A (2004) Easy estimation method for plow pan and effect of improved pan breaking. Jpn J Soil Sci Plant Nutr 75:265–268
Naser HM, Nagata O et al (2007) Methane emissions from five paddy fields with different amounts of rice straw application in central Hokkaido. Jpn Soil Sci Plant Nutr 53:95–101
Nishimune A, Sekiya S (1999) Hokkaido agriculture in the expansion phase. In: Soil science and plant nutrition in Hokkaido agriculture, 1999. Hokunoukai, Sapporo. (in Japanese). pp 1–5
Niwa K, Yokobori J et al (2016) The monitoring of soil information within an agricultural field using low altitude remote sensing. Water Land Environ Eng 84:749–752 (in Japanese)
Ogawa A, Shibata H et al (2006) Relationship of topography to surface water chemistry with particular focus on nitrogen and organic carbon solutes within a forested watershed in Hokkaido. Jpn Hydrol Process 20:251–265
Ohmura K, Kiso S, Sekijyo T (1985) Influences of manuring management on changes in productivity and nutritional environment of volcanic ashy grassland. Bulletin Hokkaido Prefectural Agric Exp Stations 52:65–76 (in Japanese with English summary)
Ohtomo R, Sakai O et al (2015) Reduced phosphate fertilization to soybean cultivation in Hokkaido considering crop rotation sequence. Jap J Soil Sci Plant Nutr 86:549–552
Oka N, Karasawa T, Okazaki K, Takebe M (2010) Maintenance of soybean yield with reduced phosphorus application by previous cropping with mycorrhizal plants. Soil Sci Plant Nutr 56:824–830
Okumura M, Yamagami M, Higashida S (1997) Rotation term and sequence in upland crops for effective soil management. Jpn J Soil Sci Plant Nutr 68:331–335 (in Japanese)
Onodera M, Nakamoto H, Okumura M (2004) Soil management based on microbial activity (α-glucosidase activity) in clay soil of vegetable field. New research results Hokkaido. (in Japanese). pp 113–119
Park JH, Duan L et al (2010) Potential effects of climate change and variability on watershed biogeochemical processes and water quality in Northeast Asia. Environ Int 36:212–225
Saigusa T, Noshiro M (1990) K and P application on grassland corresponding to the properties of andosols popular in eastern Hokkaido, Transactions of the 14th international congress of soil science, vol IV, pp 714–715
Saigusa T, Hojito M, Noshiro M (1997) Effects of Zinc and Cupper application on timothy grasslands on andosols in Konsen district. Jpn J Soil Sci Plant Nutr 68:453–457 (in Japanese)
Saigusa T, Hojito M, Noshiro M (1999) Effects of Zinc application on timothy (Phleum pratense L.) grasslands in Konsen district. Jpn J Soil Sci Plant Nutr 70:10–18
Sakuma T (1987) Agricultural zone and land use. In: Soil science and plant nutrition in Hokkaido agriculture, 1987. Hokunoukai, Sapporo. (in Japanese). pp 1–8
Satoh F, Imai H, Okajima H (1982) Ionic composition of soil solutions separated from forest soils and the stability of clay minerals in terms of chemical equilibrium. Jpn J Soil Sci Plant Nutr 53:219–226 (in Japanese with English summary)
Satoh F, Sakuma T, Okajima H (1990) A toposequence of fine textured soils in the hilly area of the northernmost part of Hokkaido, I. Morphological Phys-Chem Prop Soil Sci Plant Nutr 36:9–21
Shibata H (2016) Impact of winter climate change on nitrogen biogeochemistry in forest ecosystems: a synthesis from Japanese case studies. Ecol Indic 65:4–9
Shibata H, Sugawara O et al (2004) Nitrogen dynamics in the hyporheic zone of a forested stream during a small storm, Hokkaido. Jpn Biogeochem 69:83–104
Shimizu M, Hatano R et al. (2014) Chapter 33—Mitigation effect of farmyard manure application on greenhouse gas emissions from

managed grasslands in Japan. In Hartemink AE, McSweeney K (eds) Soil carbon. Springer. pp 313–325

Shimizu M, Hatano R et al (2013) The effect of fertilizer and manure application on CH_4 and N_2O emissions from managed grasslands in Japan. Soil Sci Plant Nutr 59:69–86

Sugikawa Y, Goto E, Asaka D (2009) Application standard of lime-treated sewage sludge compost for paddy rice in Hokkaido. Jpn J Soil Sci Plant Nutr 80:530–533 (in Japanese with English summary)

Suzuki K, Shiga H (2004) The maximum permissible amount of nitrogen input into an andisol upland field in Abashiri area assessed by percolate nitrate concentration. Jpn J Soil Sci Plant Nutr 75:45–52 (in Japanese with English summary)

The Hokkaido Research Organization, Agricultural Research Department, Central Agricultural Experiment Station, etc. (2014) Status and changes of the properties and the storage amount of carbon of cultivated soil in Hokkaido (1959 ~ 2011). (in Japanese)

Uchida T, Kaneko N et al (2004) Analysis of the feeding ecology of earthworms (Megascolecidae) in Japanese forests using gut content fractionation and δ15 N and δ13C stable isotope natural abundances. Appl Soil Ecol 27:153–163

Wakasugi K, Fujimori S (2006) Developed of efficient construction of sloped farm field. Tech Rep Natl Inst Rural Eng Jpn 204:195–202 (in Japanese with English summary)

Watanabe T, Fukuzawa K, Shibata H (2013) Temporal changes in litterfall, litter decomposition and their chemical composition in Sasa dwarf bamboo in a natural forest ecosystem of northern Japan. J For Res 18:129–138

Woli KP, Nagumo T, Hatano R (2002) Evaluating impact of land use and N budgets on stream water quality in Hokkaido, Japan. Nutr Cycl Agro Ecosyst 63:175–184

Yamasaki S, Kimura K et al (2009) Cadmium concentrations in soils of Japan. Jpn J Soil Sci Plant Nutr 80:30–36 (in Japanese with English summary)

Yazaki T, Hirota T et al (2013) Effective killing of volunteer potato (Solanum tuberosum L.) tubers by soil frost control using agrometeorological information—An adaptive countermeasure to climate change in a cold region. Agric For Meteorol 182–183:91–100

Tohoku Region

6

Hiroshi Fujii, Shizuka Mori, Yumi Matsumoto, Yuka Sasaki, Chiharu Ito, Shinpei Nakagawa, Tadashi Takahashi, Nobuhiko Matsuyama, Mizuhiko Nishida, Yoshihiro Kaneta, Haruki Fujisawa, Norimasa Tanikawa, Tadashi Ando, Hiroyuki Shiono, Teruo Shima, Masakazu Aoyama, Mikio Morioka, Takayuki Ando, Keitaro Tawaraya, Takumi Sato, Fumiaki Takakai, Takashi Sato, Tomoki Takahashi, Masashi Ito, Weiguo Cheng, Miyuki Nakajima, Toyoaki Ito, Hisashi Nasukawa, Toru Uno, Ryousuke Tajima, Tomonori Abe, Takuro Shinano, Takashi Saito, Shokichi Wakabayashi, Shigeto Fujimura, Hisaya Matsunami, Takashi Hirayama, Katashi Kubo, Takeshi Ota, Masanori Saito, Tetsuya Katagiri, and Kazuto Ando

Abstract

The Tohoku region is about 210 km east to west and 510 km north to south, with the Ohu Mountains and the Nasu Volcanic in the central area, the Kitakami and Abukuma Mountains on the east, and the Dewa Mountains and Chokai Volcano on the west. Therefore, because the Sea of Japan side and the Pacific side is divided into

H. Fujii (✉) · S. Mori · Y. Matsumoto · Y. Sasaki · K. Tawaraya · W. Cheng
Faculty of Agriculture, Yamagata University, Yamagata City, Yamagata Prefecture, Japan
e-mail: hfujii@tds1.tr.yamagata-u.ac.jp

C. Ito · S. Nakagawa · M. Ito
Akita Prefectural Agricultural Experiment Station, Akita Prefectural Government, Akita City, Akita Prefecture, Japan

Tadashi Takahashi · Y. Kaneta · F. Takakai · Takashi Sato
Faculty of Bioresource Sciences, Akita Prefectural University, Akita City, Akita Prefecture, Japan

N. Matsuyama · M. Aoyama
Faculty of Agriculture and Life Science, Hirosaki University, Hirosaki City, Aomori Prefecture, Japan

M. Nishida · H. Nasukawa · T. Uno · R. Tajima · M. Saito
Graduate School of Agricultural Science, Tohoku University, Sendai City, Miyagi Prefecture, Japan

H. Fujisawa · N. Tanikawa
Aomori Prefectural Industrial Technology Research Center, Kuroishi City, Aomori Prefecture, Japan

Tadashi Ando · M. Morioka
Yamagata Integrated Agricultural Research Center, Yamagata Prefectural Government, Yamagata City, Yamagata Prefecture, Japan

H. Shiono
Yamagata Prefecture Murayama Area General Branch Administration Office, Yamagata City, Yamagata Prefecture, Japan

T. Shima
Iwate Agricultural Reserch Center, Kitakami City, Iwate Prefecture, Japan

Takayuki Ando
Horticultural Experiment Station, Yamagata Integrated Reserch Center, Yamagata Prefectural Government, Yamagata City, Yamagata Prefecture, Japan

Takumi Sato
RIKEN BioResource Research Center, Tsukuba City, Ibaraki Prefecture, Japan

Tomoki Takahashi · M. Nakajima · S. Wakabayashi
Tohoku Agricultural Research Center, NARO, Morioka City, Iwate Prefecture, Japan

T. Ito
Faculty of Agriculture, Niigata Agro-Food University, Tainai City, Niigata Prefecture, Japan

T. Abe
Furukawa Agricultural Experiment Station, Miyagi Prefectural Government, Furukawa City, Miyagi Prefecture, Japan

T. Shinano · S. Fujimura · H. Matsunami · K. Kubo · T. Ota
Agricultural Radiation Research Center, Tohoku Agriculture Research Center, NARO, Morioka City, Iwate Prefecture, Japan

Present Address:
T. Shinano
Graduate School of Agriculture, Hokkaido University, Spporo City, Hokkaido, Japan

R. Hatano et al. (eds.), *The Soils of Japan*, World Soils Book Series,
https://doi.org/10.1007/978-981-15-8229-5_6

the mountainous backbone, the soil and the climate are very different. The soil is widely distributed in fertile soil based on tuff on the Japan Seaside. In the northern part of the Pacific Ocean, volcanic ash soil (Andosols) is distributed, and in the southern part, soil consisting of granite is distributed. In the Tohoku region, while overcoming cold and snow damage, paddy farming centered on rice farming, livestock farming using rich feed resources in the middle and mountain areas, resource crops utilizing the cool climate, production of vegetables and fruits, etc. A variety of agriculture has been developed and played a role as a food supply base for Japan. In the future, as agriculture becomes more internationalized, people's interest in the stable supply of safe and secure food and environmental conservation will be further heightened. Under such circumstances, Tohoku agriculture is strongly required to switch to a combined system combining field crops such as wheat, soybeans, vegetables, flowers, etc. with livestock, and to expand the scale of management, while putting rice farming on a basic basis. The labor-saving and low-cost production of high-quality, and high-value-added agricultural products that meet safety needs, stable production of food by overcoming weather disasters, development of natural circulation agriculture in harmony with the environment, promotion of technology development that responds to agriculture, etc. are being carried out.

Keywords

Andosols • Paddy field • Environmentally-friendly agriculture • Fluvic soils • The great east Japan earthquake

T. Saito
Hama Agricultural Regeneration Research Centre, Fukushima Prefecture Agricultural Technology Centre, Soma City, Fukushima Prefecture, Japan

T. Hirayama
Fukushima Prefecture Agricultural Technology Centre, Fukushima Prefectural Government, Koriyama City, Fukushima Prefecture, Japan

T. Katagiri
KONICA MINOLTA, INC., Tokyo, Japan

K. Ando
Yanmar HeliAgri CO., LTD, Osaka City, Osaka Prefecture, Japan

6.1 Outline of Soils in Tohoku

6.1.1 Soils in Tsuruoka City, Yamagata Prefecture

Yamagata Prefecture is divided into four municipal areas: Murayama, Mogami, Okitama, and Shonai. Of these, the Shonai area has the largest area of arable land and paddy fields and the highest rice yield (Ministry of Agriculture, Forestry, and Fisheries [MAFF] 2015a, b). Paddy field accounts for 90% of the arable land in the Shonai area (MAFF 2015a, b). The Shonai area is surrounded by mountains from the north, east, and south until the southwest, and is separated from the Sea of Japan by sand dunes in the west, with a plain in the center.

Tsuruoka City is located in the south of the Shonai Plain, a fluvial alluvial plain whose soils are derived from and accumulated by rivers from the surrounding mountains (Ariga 1984), including the Akagawa River system and the Ohoyama River system. When we analyzed the clay mineral composition of 35 paddy soils in Tsuruoka City, all soils contained smectite as the main clay mineral and kaolin and 2:1–2:1:1 intergraded minerals. The mountains surrounding the Shonai Plain are located in a green tuff region (Kitamura 1986). Green tuff contains a substantial amount of smectite (Masui 1959a) and its sediment paddy soil is dominated by smectite and kaolin mineral (Masui 1959b). Therefore, the clay mineral composition of paddy soils in Tsuruoka City is highly affected by green tuff from geological features distributed in the upper reaches of the Akagawa River system and the Ohoyama River system.

Some paddy soils in Tsuruoka City contain chlorite. These paddy soils were divided into three types based on their chlorite content (Fig. 6.1). The soils free of chlorite are distributed in the Ohoto River basin (Fig. 6.1a). The soils containing small amounts of chlorite are distributed in the Yazawa River basin and in a part of the Ohoto River basin (Fig. 6.1b). The soils containing significant amounts of chlorite are distributed in the Akagawa River, Yazawa River, and Ohoyama River basins (Fig. 6.1c).

Granite of geological features is distributed in the upper reaches of the Akagawa River and Yazawa River, but not in the Ohoto River (Jinbo 1960). Thus, the chlorite weathered from granite (Kanno et al. 1957) could be contained in the paddy soil sediment following transport by the Akagawa River and Yazawa River. The soil sediment deposited by the Akagawa River may contain more materials derived from granite than that deposited by the Yazawa River. The variation in the chlorite contents of different types of soil may reflect the distribution of the flood basin of each river, that is, the soils containing high levels of chlorite indicate the flood basin of the Akagawa River, the soils containing lower

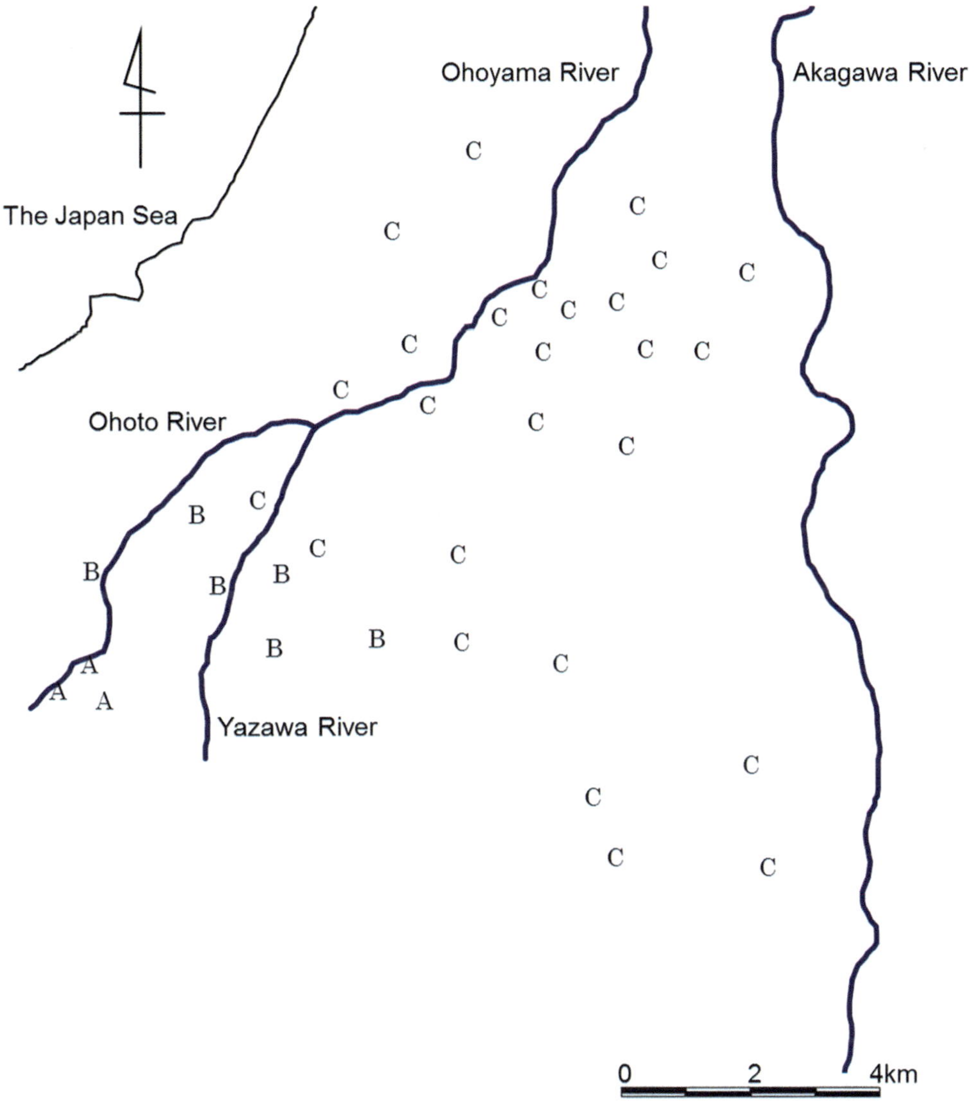

Fig. 6.1 Spatial distribution of clay mineral composition in paddy soils of Tsuruoka city. **A** Clay mineral composition free from chlorite; **B** Clay mineral composition slightly contain with chlorite; **C** Clay mineral composition obviously contain with chlorite. *Source* Figure provided by Yuka Sasaki

levels of chlorite indicate the flood basin of the Yazawa River and the soils free from chlorite indicate the flood basin of the Ohoto River.

We measured soil chemical properties of 261 paddy fields collected by 0.5-km mesh in Tsuruoka City. The spatial distribution of the soil chemical properties was divided into three features. The first feature showed lower values of soil chemical properties in the Akagawa and Ohoyama River basins and higher values in the Yazawa and Ohoto River basins. The specific properties and their lower and upper ranges were as follows: total nitrogen (N), lower range 0.9–1.9 g kg^{-1}, upper range 2.9–5.3 g kg^{-1}; phosphate absorption coefficient (PAC), lower range 3.0–7.5 g P_2O_5 kg^{-1}, upper range 12.8–19.5 g P_2O_5 kg^{-1}; cation exchange capacity (CEC), lower range 8–14 cmol(+) kg^{-1}, upper range 25–40 cmol(+) kg^{-1}; exchangeable calcium (Ca), lower range 0.2–1.0 g kg^{-1}, upper range 2.0–4.3 g kg^{-1}; and exchangeable magnesium (Mg), lower range 0.05–0.19 g kg^{-1}, upper range 0.62–1.19 g kg^{-1}. Organic matter in soil forms complexes with clay and becomes stabilized (Yonebayashi 1997). Additionally, higher ratios of 2:1 clay mineral in soil lead to the accumulation of higher amounts of soil organic matter (SOM; Yonebayashi 1997). The soils in the Yazawa and Ohoto River basins contain relatively higher levels of smectite than the Akagawa River basin because of the lower influence of granite from geological features in the upper reaches of these rivers comparing to the Akagawa River. Therefore, these soils can accumulate higher amounts of organic matter, resulting in higher total N, PAC, and CEC values. Moreover, higher CEC facilitates the absorption of higher amounts of exchangeable Ca and Mg in these soils.

The second feature showed lower values of soil chemical properties in the left bank of the Yazawa–Ohoyama River and higher values in the right bank of the Yazawa–Ohoyama

River until the Akagawa River. The specific properties and the lower and upper ranges were as follows: exchangeable potassium (K), lower range 0.01–0.09 g kg^{-1}, upper range 0.22–0.43 g kg^{-1}; and exchangeable sodium (Na), lower range 0.03–0.06 g kg^{-1}, upper range 0.14–0.39 g kg^{-1}. Paddy fields in Tsuruoka City receive irrigation water from the Akagawa, Ohoyama, or Ohoto River. Each river distributes irrigation water to the fields located on the left bank of the river (Yazawa River land improvement district 1986, 1988; Shoryuji River land improvement district 1998)—that is, the paddy fields on the right bank of the Yazawa–Ohoyama River until the Akagawa River receive irrigation water from the Akagawa River. Granite contains the highest amounts of K and Na among igneous rocks (Naganori 1984). Thus, the irrigation water in the area from the right bank of the Yazawa–Ohoyama River until the Akagawa River may contain higher amounts of K and Na due to the granitic geological features that are present in the upper reaches of the Akagawa River, resulting in higher amounts of exchangeable K and Na in soils.

The third feature was not clear. The spatial distributions of pH (4.6–6.3), available P (0.03–0.45 g P_2O_5 kg^{-1}), and base saturation percentage (19–97%) had no definite trend. These chemical properties may not be affected by geological features in the upper reaches of many rivers, but rather by cultivation practices such as fertilizer and organic matter management in each paddy field.

The following is a summary of the above discussion. The clay mineral composition of alluvial paddy soils in Tsuruoka City, Yamagata Prefecture, has a spatial characteristic that can be affected by geological features, namely, green tuff that is distributed in the upper reaches of all the Akagawa River and Ohoyama River systems, and granite that is distributed in the upper reaches of the Akagawa River and the Yazawa River, flow into the plain of Tsuruoka City. The spatial characteristics of the chemical properties of the paddy soils in Tsuruoka City can be divided based on three features. One is the spatial distribution of clay mineral composition and organic matter content of the soil that is derived from different geological features in the upper reaches between the Akagawa River and the Yazawa and Ohoto Rivers, such as total N, PAC, CEC, and exchangeable Ca and Mg. The second is affected by geological features in the upper reaches of rivers that provide irrigation water, such as exchangeable K and Na. The third is not affected by geological features in any upper reaches, such as pH, available P, and base saturation percentage.

6.1.2 Soils in the Hachirogata Land Reclamation Area

(1) **Soil properties of the Hachirogata land reclamation area**

The Hachirogata land reclamation area is located to the east of the Oga Peninsula, which is situated in the central part of Akita Prefecture. A crossing point between latitude 40° N and longitude 140° E is located within this area. Hachirogata was once the second-largest lake in Japan, with a total area of 22,024 ha. However, after a severe food shortage after World War II, the lake underwent extensive reclamation in 1966 under the technical supervision of Dutch experts.

The central land reclamation area occupies 15,640 ha (71% of the lagoon area), and the entire area is situated below sea level (average elevation, −4 m above sea level [a.s.l.]). An extremely moist and soft sludge-like heavy clay soil covers approximately 80% of the land reclamation area. This soil is characterized by the following unique properties that are not commonly present in general farming soils:

(1) It contains an average clay content of 51% (72% at maximum), half of which comprises minute particles (diameter < 1 μm). The primary clay mineral is an expansible 2:1 smectite.
(2) It exhibits relatively favorable chemical properties, specifically a large base-exchange capacity. It naturally contains large quantities of nitrogen, potassium, phosphate, silicate, and other compounds. The most abundant phosphate form is the Al type, followed by the Fe and Ca types.
(3) Soil pH mainly ranges from neutral to slightly alkaline, which is uncommon among Japanese soils. This contributes to the loss of fertilizer nitrogen through phenomena such as ammonia volatilization. Accordingly, the soil requires the application of large quantities of nitrogenous fertilizer despite its high nitrogen content.
(4) In surface layers, the presence of calcium derived from seashells significantly affects the soil alkalinity. At deeper layers, however, marine compounds, such as sodium, magnesium, chlorine, and sulfur, exert a greater influence on soil properties, such as soil pH, as well as on the development of soil structures and the elongation of crop roots.

(5) Following the land reclamation, the soil has become highly consistent, with a plasticity index exceeding 100%. Since the soil's water content is equal to the liquid limit, it is rather fluid. Additionally, due to its high clay content, the soil contracts and expands significantly through repeated drying and wetting cycles. A soil mass that is dry and consolidated in summer may crumble and revert to mud when exposed to water accumulated during winter.

(6) The soil has an extremely low saturated hydraulic conductivity of 10^{-9} cm s^{-1}, perhaps due to the presence of few coarse pores. A large volume of water is retained at a low suction pressure of −0.1 MPa or below, and a considerable proportion of water is strongly adsorbed into the soil and is hence unavailable to crops.

These properties are attributed to the presence of the expansible clay mineral 2:1 smectite and the fact that this reclamation land area was once a brackish lake with a lakebed sedimentary environment highly susceptible to marine water.

(2) **Productivity of paddy rice in heavy clay soil**

Using the data from the nitrogen–phosphorus–potassium ratio (NPK ratio) experiments conducted on paddy rice over three decades at the Ogata farm of the Akita Agricultural Experiment Station (hereinafter referred to as "Ogata") and the Niida farm of the former Akita Agricultural Experiment Station (hereinafter referred to as "Niida"), we evaluated the productivity of paddy rice at Ogata.

Table 6.1 summarizes the effects of the nutrient deprivation and the application of fertilizer on paddy rice yield. Among the nitrogen, phosphorus, and potassium, the cropping index significantly decreased due to nitrogen deprivation during both the first (the 3rd to 11th year of the examination start) and second (the 23rd to the 31st year of the test start) periods at Niida. In contrast, the yield dramatically increased following fertilizer application. While the deprivation of phosphate and potassium did not affect the yield as clearly as did nitrogen deprivation, potassium deprivation exhibited a larger effect during the second period. Additionally, at Ogata, the effect of nitrogen deprivation was the most significant among all three elements (nitrogen, phosphate, and potassium) during both the first and second periods. Although the deprivation of both phosphate and potassium slightly affected the yield, no significant difference was observed for these NPK ratios. Moreover, the increase in the yield following fertilizer application was negligible in both the first and second periods. Despite smaller amounts of fertilizer being applied in Ogata than in Niida, all fractions in Ogata yielded a higher harvest yield than those in Niida, and the difference in the harvest yield between the potassium- and nitrogen-deprived fractions was particularly significant.

Table 6.1 Effect of lack of elements and compost application rate on the yield of paddy rice

Item	Experiment plot	Ogata		Niida	
		First periods	Second periods	First periods	Second periods
Brown rice yield (gm^{-2})	Non-nitrogen	382	434	372	340
	Non-phosphoric acid	590	542	527	523
	Non-potassium	587	555	516	436
	Three elements	626	553	541	509
	Compost	632	567	577	545
Yield index	Non-nitrogen	61	78	69	67
	Non-phosphoric acid	94	98	98	103
	Non-potassium	94	100	96	86
	Three elements	(100)	(100)	(100)	(100)
	Compost	101	102	107	107

Note 1
In the previous term, the 3rd to 11th year of the examination start
Late term, beginning of trial 23 to 31 years
Late term, the average from the 23rd to the 31st year of the test start
The value in the lower row: the index with the yield of the three-element groups as 100
Note 2
Nitrogen, phosphoric acid, potassium: Ogata: 7 kg in the previous term, 5 kg of plowing
Niida: 6 kg in both the first and second term
Compost application amount in manure district: Otaga: 1.2 tons, Niida: 1 ton
Note 3
Soil conditions: Ogata: Gley Lowiand soil. Niida: Gray Lowland soil

Table 6.2 Influence of lack of elements and compost application rate on yield component of paddy

Yield component	Ogata					Niida				
	Non-nitrogen	Non-phosphoric acid	Non-potassium	Three elements	Compost	Non-nitrogen	Non-phosphoric acid	Non-potassium	Three elements	Compost
Panicle number	328	422	431	442	443	273	414	456	444	468
(m-2)	74	95	98	100	100	61	93	103	100	105
Number of panicle	64.8	67	67.1	67	69.4	64.4	71	59.4	65.4	70.8
	97	100	100	100	104	99	109	91	100	10.8
Percentage of ripened grains	90.1	87.6	87.2	87.8	86.2	92.2	88.1	81.1	86.5	81.8
(%)	103	100	99	100	98	107	102	100	100	94
1000-grain weight	22.3	21.9	21.9	21.9	21.8	21.8	21	20.8	20.8	20.7
(g)	102	100	100	100	100	105	101	100	100	99

Table 6.2 summarizes the effect of the nutrient deprivation and the application of fertilizer on different yield components. Compared with the yield components for the NPK ratio at Niida, the NPK ratio at Ogata developed an almost equal number of ears, but the number of unhulled grains per ear, ripening rate, and thousand-kernel weight were all slightly higher at Ogata. While the deprivation of nitrogen and phosphate resulted in a decrease in the number of ears in both fields, the impact was less significant at Ogata than at Niida. At Niida, the impact of potassium deprivation was reflected in the number of spikelets per ear and the ripening rate. In contrast, potassium deprivation did not have any noticeable effect on the yield component at Ogata. Both the number of ears and the number of spikelets per ear increased with fertilizer application at Niida. In contrast, at Ogata, the number of spikelets per ear increased, although the number of ears did not increase, with fertilizer application.

These results indicate that, compared with general Gray Fluvic soil, the Gley soil paddies at Ogata supply more nitrogen and potassium from the soil. While the harvest level was generally high, the degree to which the increase of harvest yield was attributable to fertilizer application remains unclear. Fertilizer application at Ogata contributed more to the number of spikelets per ear than to the number of ears, suggesting that nitrogen fertilization was more effective in the latter stages of growth.

(3) **Cultivation management for developing soil structures**

In paddy–upland rotations of meadow heavy clay soil, the Gley layer thins every year during the upland period. However, when the crop field was reconverted to a rice paddy, the Gley layer thickened by 10 cm per year during puddling cultivation. Three years after the rice paddy reconversion, the second layer immediately below the surface soil had significantly thinned and the soil structure had degraded into a wall-shaped structure. Therefore, although the paddy was reverted into a crop field, the second layer remained thin, and consequently, the soil structure did not develop.

Crops that are sensitive to excessively wet soil conditions, such as soybeans, require transplantations through non-tillage and non-puddling techniques to avoid moisture damage in the first year of crop field conversion. In non-tillage and non-puddling cultivation, the soil structure and acidity of the underlying soil layer that is developed through paddy–upland rotations could be maintained. Moreover, the improved drainage results in increased soil pulverization at the surface, which contributes to increases in soybean production.

Therefore, in meadow heavy clay soil, techniques to promote crop field conversion by cultivation management for the development of soil structures and the maintenance of oxidized soil during paddy rice cultivation are required.

In the initial stage after land reclamation, it was aimed to achieve large-scale farming using large machinery on the Ogata reclamation land area; however, this was significantly impeded by the physical properties of the soft, heavy clay soil. Recently, improved drainage and soil drying have stabilized the plow layer immediately beneath the surface soil, making the soil-bearing capacity less problematic. Furthermore, techniques have been developed to increase the productivity of paddy rice and other crops, through fertilizer application methods to increase the availability of fertilizer nitrogen and green manure cultivation aimed at improving the physical properties and nitrogen fertility of the soil.

Fig. 6.2 Active volcanoes in Tohoku district Active volcanoes in Tohoku region. *Data Source* Japan Meteorological Agency (2017)

Table 6.3 Percentage of Andosols in upland fields and paddy fields (%)

Prefecture	Upland field	Paddy field
Aomori	81	23
Iwate	61	33
Miyagi	26	7
Akita	51	10
Yamagata	33	13
Fukushima	31	10
Tohoku district	54	15
Japan	47	13

The compression and re-plowing of the plowed field using large machinery have contributed to the thickening of the soil. Compressed and thickened soil suppresses the elongation of crop roots and promotes excessive wetness, thus hampering the growth and harvest of crops and the conversion of rice paddies into other crop fields. At the Hachirogata reclamation area, the physical properties of the soil remain a major factor determining the adaptability of machines to fields as well as the growth rates and yields of crops. Therefore, it is essential to continue to develop soil structures and improve the physical properties of the soil.

6.1.3 Andosols in the Tohoku Region

(1) **Formation and distribution of Andosols**

(1) Distribution of Andosols in the Tohoku region

A total of 111 active volcanoes are identified in Japan (Japan Meteorological Agency 2017). Among them, 18 volcanoes are located in the Tohoku region (Fig. 6.2). As a result, Andosols—whose parent material is volcanic ejecta—are the most widely distributed soils in the Tohoku region (Fig. 6.2.); Andosols cover 36% of its total land territory, followed by Brown Forest soils (28%), Fluvic soils (14%), and other soil types.

Because volcanic ejecta tend to be carried towards the eastern sides of volcanoes by westerlies in Japan, Andosols are concentrated in areas on the Pacific Ocean side. For instance, in Aomori Prefecture, the Towada volcano (Lake Towada) has produced repeated huge explosions, and Andosols are widely distributed on the stable Sanbongihara Plateau. Additionally, in Iwate Prefecture, many volcanoes in the Ohu Mountains and the Kitakami peneplain also contributed to the formation of Andosols.

(2) Distribution of Andosols in arable lands

In Japan, Andosols are mainly used for upland fields. In the Tohoku region, Andosols comprise 54% of total upland fields (common upland plus orchard), as detailed in Table 6.3, which is higher than the value for the whole of Japan (47%). By prefecture, the abundance of Andosols in upland fields is 81% in Aomori Prefecture and more than 50% in Iwate and Akita Prefectures. As described earlier, the high coverage of Andosols in Aomori and Iwate Prefectures is explained by their stable terrain and the abundant supply of volcanic ejecta. In Akita Prefecture, the abundance of Andosols is due to the preservation of volcanic ejecta in the faces of volcanic terraces. On the other hand, the abundance of Andosols in Yamagata Prefecture is low (33%) reflecting the small supply of volcanic ejecta there. The low Andosol abundances in Miyagi and Fukushima Prefectures (26% and 31%, respectively) are due to the progress of the development of terrain in upland fields.

The abundance of Andosols in paddy fields in the Tohoku region is 15%, which is higher than the value for the whole of Japan (13%). Especially high abundances are observed in Iwate (33%) and Aomori (23%) Prefectures.

(3) Distinctive formation and distribution of Andosols in the Tohoku region

As mentioned above, soils in the Tohoku region are largely influenced by volcanic ejecta. Many studies concerning these soils have been conducted and published. Here, two distinctive studies of Andosol formation in the Tohoku region are shown below.

(i) Climosequence and biosequence of Podzolic Andosols–Brown Andosols–Melanic Andosols

Volcanic ejecta from Towada volcano (Lake Towada), which mainly consists of Towada-a ash (with an age of ca. 1100 years ago) and Chuseri ash (ca. 6300 years ago), widely cover the eastern part of Aomori Prefecture. Reflecting on the difference in climate (elevation) and vegetation, different types of Andosols are formed from the same volcanic materials (Shoji et al. 1988). Under mixed Maries' fir (*Abies mariesii*)–Japanese beech (*Fagus crenata*) vegetation at elevations of 800–1200 m, Podzolic Andosols with eluvial and illuvial horizons are formed. In Japanese beech forest at elevations of 400–800 m, Brown Andosols are formed. At similar elevations (400–600 m), Melanic Andosols appear in areas strongly influenced by grass vegetation such as *Miscanthus sinensis*.

On the Shimokita Peninsula of Aomori Prefecture, different types of Andosols are formed depending on vegetation under similar climatic conditions (Takahashi et al. 1989). Brown Andosols are observed in the Japanese beech forest, while Podzolic Andosols are formed under *Thujopsis dolabrata* (cypress family; conifer), showing that this tree is a strong podzolizer.

(ii) Distribution pattern of Allophanic Andosols and Non-allophanic Andosols

Andosols can be divided into allophanic and non-allophanic types based on their dominant clay mineralogical composition. In Allophanic Andosols, the clay fraction is dominated by allophane and imogolite, and the potential for Al toxicity is low. Conversely, in Non-allophanic Andosols, the clay fraction is dominated by 2:1 minerals and soils are strongly acidic. In the Tohoku region, Allophanic Andosols cover 856,000 ha, which is equal to 48% of the region's total Andosol land area, while Non-allophanic Andosols cover 935,000 ha, equal to 52% of its total Andosol land area. Allophanic Andosols preferentially form in thick Holocene tephra deposits, while Non-allophanic Andosols form in areas with minimal Holocene tephra deposition (Saigusa and Matsuyama 1998).

(2) Characteristics of Andosols and their use for agriculture

(1) Soil properties relating to productivity

Andosols have several unique properties, such as thick black horizons, high phosphorus retention, and low bulk density (Shoji et al. 1985). The high humus content of Andosols is associated with low bulk density and high water retention. These unique physical properties provide good conditions for upland crops. One of the most important factors limiting crop growth in Andosols is their low phosphorus availability, which is due to the presence of high amounts of active Al. Consequently, the heavy application of phosphate fertilizer is necessary to achieve a high yield of upland crops in Andosols. Because Non-allophanic Andosols have large amounts of exchangeable Al, they often cause symptoms of serious aluminum toxicity in sensitive crops. It is important that liming be carried out appropriately in both types of Andosols, due to the soil acidity and the requirement of calcium for crop nutrition.

(2) Utilization and management of Andosols

As already noted, Andosols are mainly used for upland crops in Japan. The Towada area in Aomori Prefecture is a large production center for root crops (Chinese yam, edible burdock, Japanese radish, and carrot). In particular, the Towada region is Japan's largest producer of Chinese yam (contributing about 40% of the total national production) and edible burdock (about 35% of the total national production). The reason that Towada is such a large producer of these crops is mainly due to the soil properties of Allophanic Andosols (i.e., a thick A horizon, low bulk density, friable consistency, and weakly acidic pH). On the other hand, in the Tsugaru area of Aomori Prefecture, apple is widely cultivated in Allophanic Andosols, with the production being the highest in Japan (about 60% of the total national production of apples). Andosols are considered to be unsuitable soils for apple cultivation due to their high water retention, low temperature, and slow nitrogen mineralization pattern; however, the development of a unique cultivation technique for Andosols has made such cultivation possible in the Tsugaru area (Sekiya 1982).

(3) Soil degradation by agriculture

In the Tohoku region, about 10% of total arable land is estimated to be at risk of soil erosion. Soil erosion has been a severe problem for crop growth in Andosols due to their low bulk density. Therefore, several agronomic countermeasures, such as minimum tillage, mixing of soil layers, improvement of drainage, and application of compost, are recommended for agriculture.

Since Allophanic Andosols are originally weakly acidic, aluminum toxicity does not occur frequently in plants. However, Allophanic Andosols can become strongly acidic following the heavy application of chemical fertilizer. Strongly acidic Allophanic Andosols with an accumulation of acidic materials dissolves a part of the active Al fraction in the soils, which causes Al toxicity and thereby leads to the shallow rooting of Al-susceptible crops. Additionally, the soil productivity of strongly acidic Allophanic Andosols is lower than that of the original weakly acidic soil; for example, the number of bacteria decreases (e.g., from 160×10^6 cfu g^{-1} for weakly acidic soil to 10×10^6 cfu g^{-1} for strongly acidic soil) as do the levels of readily mineralizable soil nitrogen (Matsuyama et al. 2005).

As the cultivation of specific crops continues, diseases related to continuous cropping become a more severe problem. In the Towada region of Aomori Prefecture (an area of Allophanic Andosols), the production of Chinese yam is reduced by Chinese yam root rot (Rhizoctonia rot) even for only a few years of continuous cropping (Oikawa and Arai 1996). Therefore, cultivation techniques, such as soil disinfection, the application of green mature crop and fully fermented compost, soil improvement, disinfection of seed tubers, and crop rotation with vegetables, are very important in this upland soil.

(3) **Environmental and cultural significance of Andosols**

(1) Natural parks and Andosols

Among the 18 active volcanoes in the Tohoku region (Fig. 6.2), 10 volcanoes and their surrounding areas including Andosols are contained within National Parks: the Hakkodasan, Towada, Hachimantai, Akita-Yakeyaka, Iwatesan, and Akita-Komagatake volcanoes are situated in the Towada-Hachimantai National Park; the Azumayama, Adatarayama, and Bandaisan volcanoes belong to the Bandai-Asahi National Park; while Hiuchigatake volcano is located in the Oze National Park. In addition, the areas of six active volcanoes (Osorezan, Iwakisan, Chokaisan, Kurikomayama, Naruko, and Zaozan) are designated as Quasi-National Parks. The unique morphology of volcanic mountains and changes in vegetation according to elevation create attractive landscapes.

(2) Ancient human activity and Andosol formation

It has been revealed that there is a close relationship between the distribution of Jomon ruins (ca. 15,000–2300 years ago) and the distribution of Melanic Andosols in Japan (Edamura and Kumagai 2009). For example, in the Tohoku region, many ruins of the early to late Jomon Period have been discovered in areas of volcanic ash soils. It is thought that the Jomon people resided in permanent settlements and created a society in the volcanic areas with hunting and fishing. Moreover, they might have practiced farming (Sato et al. 2003a, b). Thus, Melanic Andosols are thought to have been created under ecosystems largely influenced by human beings (Hosono and Sase 2015). The Melanic Andosols could be recognized as a soil containing important ruins.

6.2 Management of Paddy Soils

6.2.1 Organic Material Application

The effects of applying organic material to paddy rice (*Oryza sativa* L) fields have been investigated at many agricultural research centers in the Tohoku region. At the Tohoku Agricultural Research Center, National Agriculture and Food Research Organization (TARC, NARO, N39°29′, E140°30′, altitude 30 m a.s.l.) located in Daisen, Akita Prefecture, the effects of organic materials have been examined in long-term field experiments with rice straw compost (RSC) and livestock manure compost (LMC) (Ohyama 1982; Ohyama et al. 1983; Sumida et al. 2002; Nishida et al. 2007). In one long-term field experiment using RSC, compost was applied at the rate of 0, 1, 2, 3, or 4 kg m^{-2} $year^{-1}$ from 1968. In all the treatments, N, P_2O_5, and K_2O were applied at 8 g m^{-2} as chemical fertilizer. In a long-term field experiment using LMC, compost was applied at the rate of 0 or 3.6 kg m^{-2} $year^{-1}$ from 1973. The soil was classified as fine-textured Gray Fluvic soil (Fluvisol). Results from the relatively early period of these experiments were reported by Ohyama (1982) and Ohyama et al. (1983). In the long-term field experiment using RSC, the highest yield was observed in the 2 kg m^{-2} plot, which had a 10% higher yield than the plot without RSC based on the average yield for the 13 years following the initiation of the experiment. In the 3 and 4 kg m^{-2} plots, the yields were lower than those in the 1 and 2 kg m^{-2} plots due to rank growth. In the long-term field experiment using LMC, the beneficial effect of the repeated application of LMC on rice yield was clear in the plot with small amounts of chemical N fertilizer application. In the plot with LMC and large amounts of chemical N fertilizer application, however, yield decreased due to the occurrence of rice blast. On the other hand, the application of calcium silicate alleviated the rice blast in the plot with the LMC and large amounts of chemical N fertilizer, resulting in higher rice yield than in the plot with LMC without calcium silicate. Nitrogen supply to the rice plants from the RSC gradually increased over time, averaging 1.0 g m^{-2} for the 1 kg m^{-2} application rate for the 13 years following the initiation of the experiment. It was estimated that 42.8% of the N in the RSC was retained in the soil, 14.7% of which was taken up by the rice plants, and 42.5% of which was lost. The N supply to rice plants from LMC was 1.1 g m^{-2} for the 1 kg m^{-2} application rate of LMC. It was estimated that 48% of the N in the LMC was retained in the soil, 23.6% of which was absorbed by the rice plant, and 28.4% of which was lost. It is noted that the rice plant recovery of the N in the LMC increased by 4% following the application of calcium silicate.

After the above investigations (Ohyama 1982; Ohyama et al. 1983), when the effect of repeated compost application had become stable, the nitrogen budgets of these long-term field experiments was examined (Sumida et al. 2002). The N balance in the plot without any N application indicated that about 3–4 g m^{-2} of N every year was supplied through irrigation water, wet and dry deposits, and biological fixation. Chemical fertilizer N applied at 8 g m^{-2} was scarcely retained in the plow layer, and about 4 g m^{-2} of that was taken up by the rice plants, and about 4 g m^{-2} was lost. About 2 g m^{-2} of N from RSC applied at 11 g N m^{-2} was retained in the topsoil, about 3 g m^{-2} of which was absorbed by the rice plants, and the remainder (about 6.5 g m^{-2}) was lost. When the application rate of RSC was higher than 2 kg m^{-2}, the rice yield plateaued (at a yield 20% higher than that without RSC), and the N absorbed by the rice plants decreased and then N loss increased. About 5 g m^{-2} of the N from LMC applied at 17 g N m^{-2} was retained in the topsoil, about 6 g m^{-2} of was absorbed by the rice plants, and the remainder (about 6 g m^{-2}) was lost.

Temporal changes in the natural abundance of ^{15}N (as expressed by $\delta^{15}N$), the ratio of ^{15}N to ^{14}N in the soils and in the RSC and LMC used in these long-term field experiments were investigated by Nishida et al. (2007). The $\delta^{15}N$ values of LMC composted prior to 1983 were around +6–7‰. The values of LMC composted between 1984 and 1997 were approximately +11–12‰ and those after 1998 were about +17–18‰. The $\delta^{15}N$ values of RSC were almost constant and averaged +5.5‰. The variation in the $\delta^{15}N$ values of LMC reflects the degree of maturity of composting and the type of livestock manure composted. The LMC produced before 1983 was created by piling cattle manure on a concrete pad in the open air without turning and was not well composted. However, the LMC produced from 1984 to 1997 was created by piling cattle manure in a compost house and turning it, and was relatively better composted than LMC produced before 1983. After 1998, LMC was produced using three types of livestock manure, namely, cattle manure, swine manure, and poultry manure, which were stirred with a rotary mixer under forced aeration in a compost house. This composting method ensured sufficient maturity of composting. The addition of swine and poultry manure promoted ammonium volatilization which decreased the amount of ^{14}N. The stability of the $\delta^{15}N$ values of RSC might have resulted from the rice straw being collected from nearby and the consistent composting method, which involved piling and occasional turning in a compost house. In soils under the successive application of LMC, $\delta^{15}N$ values increased, whereas the $\delta^{15}N$ values in soil without compost tended to decrease (Fig. 6.3). The upward trend of the $\delta^{15}N$ values in soil treated with LMC was also observed in the $\delta^{15}N$ values of the LMC itself, which increased. Soils treated with chemical N fertilizer tended to show lower $\delta^{15}N$ values than those not treated with chemical N fertilizer regardless of whether LMC was applied. In soils under the successive application of RSC (Fig. 6.4), the $\delta^{15}N$ values of the soils remained nearly the same. However, a downward trend was observed in the $\delta^{15}N$ values of soils not treated with RSC, and the trend was more pronounced when chemical N fertilizer was applied. Interestingly, in paddy soil without compost and without chemical N fertilizer, the $\delta^{15}N$ values of the soil decreased. This can be attributed to the natural input of N by biological fixation and deposition with rain, which have lower $\delta^{15}N$ values. The documentation of the transition of the $\delta^{15}N$ values of soils in long-term paddy field experiments suggests that the $\delta^{15}N$ value of paddy soil could potentially be lowered by natural N supply, and that the $\delta^{15}N$ value of paddy soil is determined by the balance between the $\delta^{15}N$-lowering effect of natural N input, the amount and $\delta^{15}N$ value of applied nitrogen materials, and the behavior of their nitrogen.

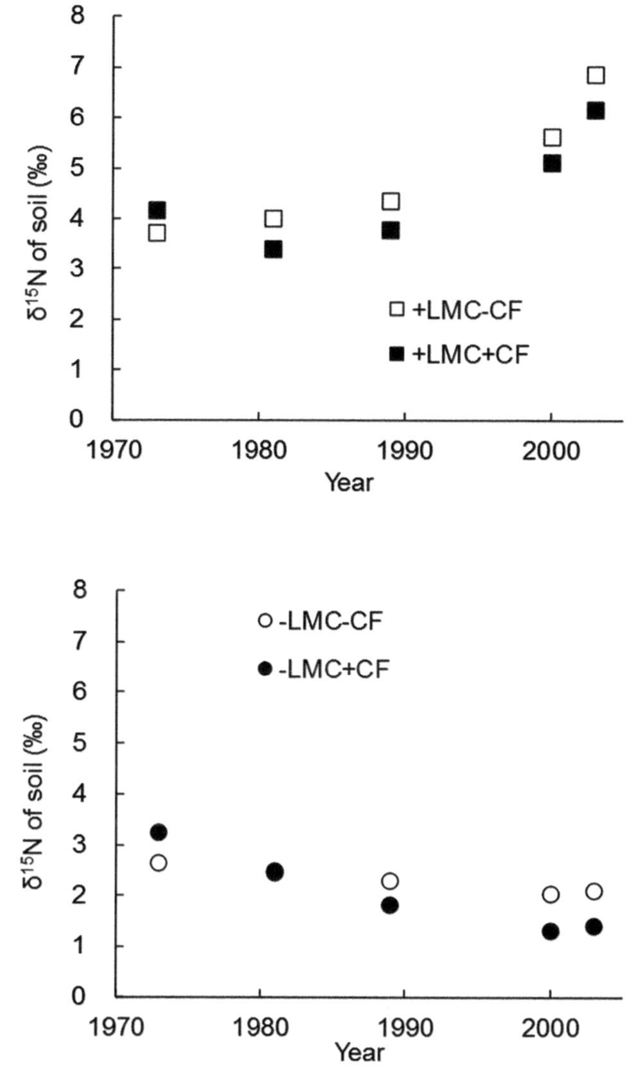

Fig. 6.3 Temporal changes in $\delta15$ N values of soil in a long-term field experiment with repeated application of livestock manure compost conducted at Tohoku Agricultural Research Center, National Agriculture and Food Research Organization located in Daisen, Akita, Japan. +LMC: livestock manure compost applied, −LMC: livestock manure compost not applied, +CF: chemical N fertilizer applied, −CF: chemical N fertilizer not applied. *Data source* Nishida et al. (2007)

6.2.2 Single Application of Fertilizer

In heavy clay paddy fields, drainage is easily reduced through the destruction of soil structure by plowing and

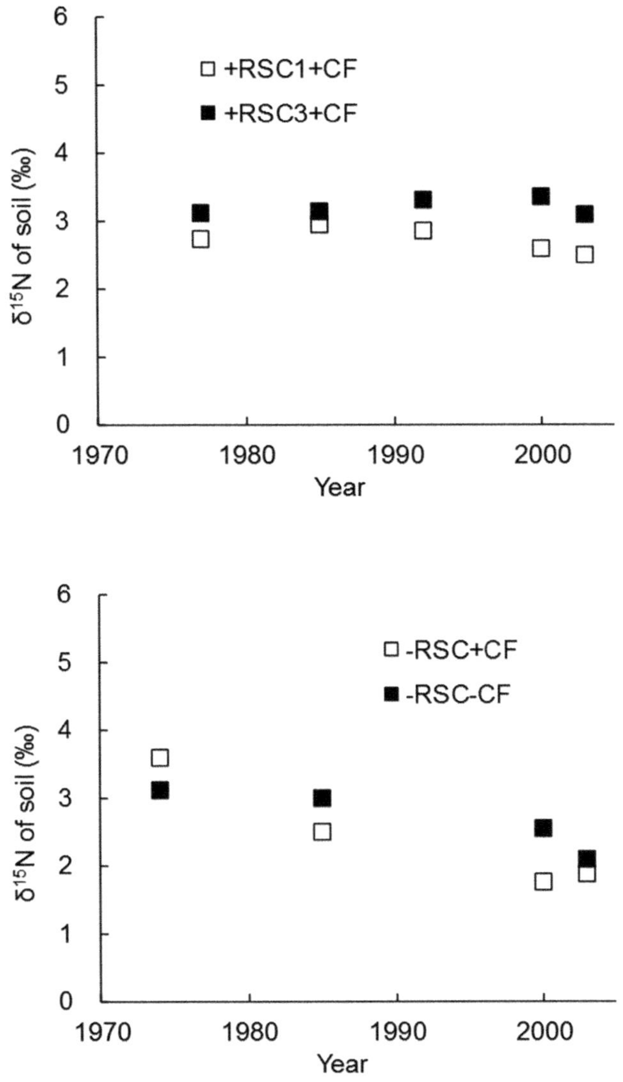

Fig. 6.4 Temporal changes in $\delta^{15}N$ values of soil in a long-term field experiment with repeated application of rice straw compost conducted at Tohoku Agricultural Research Center, National Agriculture and Food Research Organization located in Daisen, Akita, Japan. +RSC: rice straw compost applied, −RSC: rice straw compost not applied, +CF: chemical N fertilizer applied, −CF: chemical N fertilizer not applied. *Data source* Nishida et al. (2007)

puddling. In the background of these unfavorable soil characteristics, no-tillage transplantation of rice was conducted. The recovery rate of fertilizer nitrogen was remarkably low due to the surface fertilization of basal dressing in the no-tillage cultivation system. Improving the method of fertilization was necessary to address this problem.

(1) **For no-tillage cultivation**

The single application of fertilizer in a nursery box was first reported by Sato and Shibuya (1991), who used a controlled availability fertilizer (CAF). In this CAF, the sigmoidal release properties of nitrogen were accurately controlled by a resin film coating the fertilizer granules. By a single application of fertilizer in a nursery box, an amount of fertilizer nitrogen was delivered in a controlled manner over time corresponding to the requirements of rice plants throughout an entire growing season, which makes the application of nitrogen fertilizers to the field unnecessary.

The CAF in nursery boxes was applied to paddy fields at the transplanting stage, where it came into contact with the roots of rice seedlings. The CAF used in the study was a sigmoid-type slow-release nitrogen fertilizer (N content: 40%). Negligible release of nitrogen was observed for about 30 days after the application of CAF to the nursery box; thereafter, approximately 80% of the total nitrogen content leached into the water at 25 °C within 70 days of the lag period. Readily available fertilizer (e.g., ammonium sulfate and urea) and CAF of the normal-type release, which elutes immediately after application, cannot be used, because abnormally high nitrogen concentration is present in the nursery soil. Currently, the fertilizer used in our investigation (product name: "Naebako Makase") is sold only for the single application of fertilizer in a nursery box. The elution during the seedling period (approximately 35 days in the middle seedling stage) is insignificant. There is no abnormally high nitrogen concentration or spindly growth of seedlings. Moreover, the topdressing of nitrogen fertilizer to the nursery box during the seedling period can be omitted. The single application of fertilizers in a nursery box can supply the fertilizer through contact with the rootlets of the seedlings. Therefore, the recovery rate of nitrogen fertilization applied to the field with the no-tillage transplanting of rice was improved by the single application of fertilizer in a nursery box (Kaneta et al. 1994a; Kaneta 1995). At first, the CAF was completely mixed with the soil in a nursery box. Then, we devised a method in which fertilizers were applied in layers in the nursery box without mixing (Kaneta et al. 1994b).

(2) **Characteristics of nitrogen release from fertilizer**

In Akita Prefecture in the Tohoku region, the cumulative soil temperature in the nursery boxes during the 34 days of the rice-growing period was about 615 °C. The cumulative nitrogen release from the CAF during this period was 2.8% of its total nitrogen content. The nitrogen released during the seedling period was exceedingly low, and thus nitrogen topdressing to the nursery box was unnecessary. The effect of temperature change on nitrogen release was minor. After transplantation, the rate of nitrogen release from the CAF increased during the tilling stage (cumulative soil temperature of 1000 °C) and the reduction division stage. After that stage, the nitrogen release was extremely slow until the

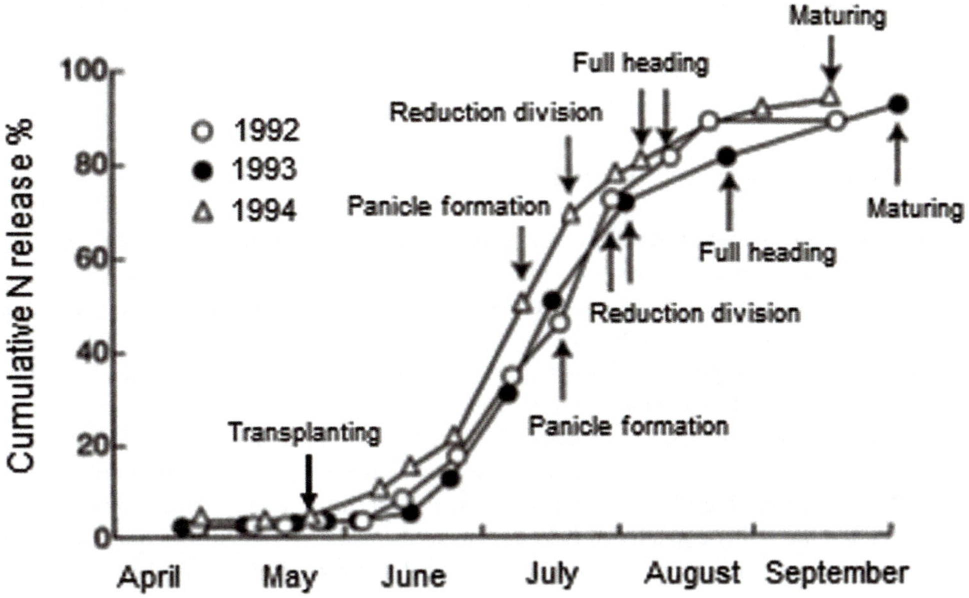

Fig. 6.5 Nitrogen release pattern of CAF (sigmoid 100 days type) in the nursery greenhouse and the following paddy field (5 cm soil depth, 1996). CAF: (controlled availability fertilizer) modified by Kaneta et al. (1997). *Source* Figure provided by Yoshihiro Kaneta

cumulative temperature reached approximately 3500 °C. The cumulative nitrogen release from the CAF amounted to about 80% of its total nitrogen content at a cumulative temperature of 2500 °C, and 92% at the maturing stage at about 3500 °C.

As can be seen in Fig. 6.5, nitrogen release after the end of July 1993, which was characterized by low soil temperature, was lower than that in 1992. Furthermore, the high temperature in 1994 induced a faster nitrogen release until the end of the year than in 1992. However, the amounts of nitrogen released in each rice growth stage were largely similar among the 3 years (Kaneta and Tsuchiya 1997a). Therefore, the nitrogen release from CAF used in paddy fields is well adapted to the growth stage of rice, even in different weather conditions, and interannual variation in the nitrogen release was low.

(3) **Nitrogen recovery rate**

We investigated the effect of the fertilizer placement of the basal dressing on rice nitrogen recovery rate when using ammonium sulfate and CAF (coating urea of sigmoid 100 days type) containing heavy nitrogen (^{15}N) as a tracer. As shown in Fig. 6.6, the nitrogen recovery rate of ammonium sulfate was 9% after surface dressing and 33% after side dressing. On the other hand, when the ammonium

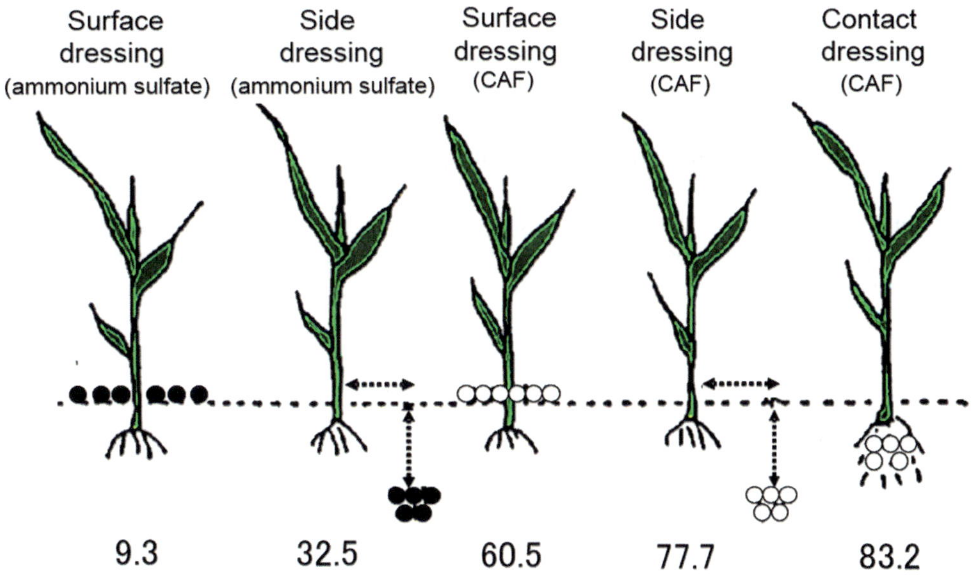

Fig. 6.6 The effect of the fertilizer placement of the basal dressing on rice nitrogen recovery rate (Unit: %, Cultivar: *Akitakomachi*) CAF: controlled availability fertilizer (sigmoid 100 days type) modified by Kaneta (1995). *Source* Figure provided by Yoshihiro Kaneta

sulfate was applied by direct contact with the seedlings, an abnormally high concentration occurred, but there was still no injury by the CAF (coating urea), and the nitrogen recovery rate improved significantly by 83% (Kaneta 1995).

(4) **Characteristics of growth and yield**

In the single application of fertilizer in a nursery box, since the initial absorption of nitrogen by plant roots just after transplantation is reduced, the number of tillers per plant is lower by 10–20% compared to that in the conventional method of fertilizer application. However, the panicle number is equivalent to that obtained in conventional fertilizer use because of the percentage of productive tillers increases. Therefore, excessive dressing should be avoided when using a single application of fertilizer in a nursery box. As a result, the yield and the esthetic quality of rice are almost the same as those obtained in the conventional method of fertilizer application, even if the amount of fertilizer is decreased for the single application of fertilizer in a nursery box (Kaneta 1996; Kaneta and Tsuchiya 1997b). In the future, the spread of technologies allowing labor-saving and low-cost, stable, and high-quality yields, such as the single application of fertilizer in a nursery box, is expected.

6.2.3 Effective Silicate Application Technique

In recent years, annual fluctuations in the yield and quality of paddy rice have been increasing, and climate change is one of the factors responsible for such fluctuations. The presence of silicate reduces various stresses on rice plants, and it can be expected to reduce the stress caused by climate change. Here, we report the absorption characteristics of silicate in paddy rice, the application technique for silicate material based on its characteristics, and the effect of the application of silicate material on one effect of climate change (salt-adhesion damage).

(1) **Silica application and adsorption of rice plant**

This analysis of the silicate absorption rate was based on the amount of silicate absorbed by rice plants over several growth phases. The study was conducted in paddy fields with soils containing different amounts of silicate in the Shonai area of Yamagata Prefecture.

Considering the silicate absorption rate of paddy rice at each growth stage, the absorption rate increases after the panicle formation stage. The silicate absorption rate at each stage is arranged in descending order as follows: Panicle formation stage–full heading stage> Full heading stage–maturing stage> Bearing tiller stage–panicle formation stage> Early stage of tillering–bearing tiller stage> Transplantation stage–early stage of tillering

The silicate absorption rate of the paddy rice increases after the panicle formation stage. This is thought to correspond to the formation of rice husks, which are the accumulation site of silicate, and also to the time when the amount and number of roots reaches a maximum. Additionally, the amount of absorbed silicate increases after the panicle formation stage, as the amount of silicate supplied from the soil increases. Thus, if the supply of silicate after the panicle formation stage is insufficient (i.e., in fields where the content of silicate in the soil is low), it is critical to efficiently supply silicate after this stage.

(2) **Silicate application technique during the panicle formation stage**

In farmland where the supply of silicate from the soil after the panicle formation stage is insufficient, an efficient silicate application technology is needed that includes the topdressing of silicate material. A topdressing application of silicate material is necessary as it is difficult to spray a large amount of such material via basal fertilization.

The amount of silicate absorbed by the stem and leaf during the mature stage in the plot where topdressing was applied during the panicle formation stage was 65 g m^{-2} (110% the value of the silicate-free plot), 63.5 g m^{-2} in the plot where calcium silicate was applied during the entire growth phase (107% the value of the silicate-free plot), and 59.3 g m^{-2} in the silicate-free plot. The amount of silicate absorbed by the panicles of rice plants was 35.3 g m^{-2} in the plot where topdressing was applied during the panicle formation stage (116% the value of the silicate-free plot), 33.2 g m^{-2} in the plot where calcium silicate was applied during the entire growth phase (109% the value of the silicate-free plot), and 30.4 g m^{-2} in the silicate-free plot.

Table 6.4 Silicate absorption rate and nitrogen absorption rate in rice plant

Plot	Silicate absorption rate (gm^{-2})			Nitrogen absorption rate (gm^{-2})		
	Shoot	Panicle	Total	Shoot	Panicle	Total
Control	59.3	30.4	90(100)	3.2	6.7	9.9(100)
Calcium silicate	63.5	33.2	97(108)	3.4	6.9	10.2(104)
A topdressing	64.6	35.3	100(111)	3.4	7.3	10.7(108)
B topdressing	65.4	35.2	101(112)	3.2	7.1	10.3(104)

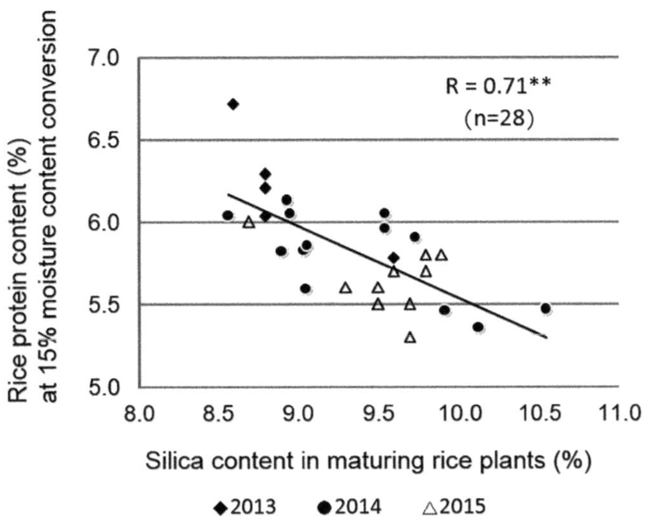

Fig. 6.7 Relationship between rice protein content and silica content in maturing rice plants. *Source* Figure provided by Haruki Fujisawa

Although the timing of application is later during the panicle formation stage and the amount of calcium silicate needed is lower than that needed for the entire growth phase, the amount of silicate absorbed was observed to be equal to or greater than that for calcium silicate application over the entire growth phase (Table 6.4).

The amount of nitrogen absorbed at the mature stage of crop growth was higher in the silicate application plot than in the silicate-free plot, and the utilization rate of the topdressed nitrogen increased during the panicle formation stage due to application of silicic acid (full layer application and surface application). The increase in the number of spikelte rice achieved with silicate application is thought to be related to the increase in the utilization rate of topdressed nitrogen and the increased nitrogen absorption during the panicle formation stage (Table 6.4).

6.2.4 Application Standard of Silicate Fertilizer

(1) **Background on formulation of application standards**

A particularly tasty variety of rice, '*Seitein no Hekireki*', is currently cultivated in Aomori Prefecture. This rice has a rice protein content (a metric negatively related to the taste of rice required to meet market standards) of 6.4% or less, achieved with the conversion of 15% moisture content ratio. An increase in silica content tends to decrease the protein content of rice (Fig. 6.7). By suppressing the rise in rice protein content, the preparation of soils with silicate fertilizer has been noted as a technology for producing tasty varieties of rice. In Aomori Prefecture, the soil amendment standard for bioavailable silicate is 150 mg kg^{-1} soil. Shortages in silicate supply in any given year can be caused by a variety of factors, such as the insufficient application of silicate fertilizer to cropland, silicate deficiencies in soils, and silicate uptake by rice plants. In order to determine the amount of silicate application needed to offset deficiencies, one must also take into account the amount of silicate supplied by irrigation water. In fact, the silicate content of irrigation water has never been quantitatively established, but it has historically been assumed that irrigation water supplies about 300 kg Si ha^{-1}. When cultivating *Seiten no Hekireki*, rice, the required amount of silicate fertilizer, and its frequency of application are determined based on soil analysis. However, rather than estimating the amount of silicate supplied by irrigation water, we investigated the application quantity that appeared to be consistent with on-site situations to promote the effective application.

(2) **Survey of silicic acid concentrations in irrigation and river water**

Silicic acid concentrations in irrigation and river water were measured twice (at the same time) in both July and August of 2015 and 2016, when silicate uptake by rice was maximal.

Silicic acid concentrations were measured by absorption spectroscopy of the yellow complex formed by adding molybdic acid reagent. The survey sites ($n = 42$) were located on the Tsugaru Plain, where the cultivation of *Seiten no Hekireki* rice is concentrated. The average silicic acid concentrations over the two studied years at these sites are shown in Fig. 6.8. Silicic acid concentrations in irrigation water ranged from 6.6–41.1 mg SiO_2 L^{-1}, with an average value of 22.9 mg SiO_2 L^{-1} (Fig. 6.8). The Iwaki River is the main river draining the Tsugaru Plain, and tributaries flow into this river from the surrounding mountains. Irrigation water derived from tributaries draining the Hakkoda mountain range and Tsugaru highlands on the eastern drainage boundary of the plain tends to have high concentrations of silica (19.9–41.1 mg SiO_2 L^{-1}), whereas tributaries draining the Shirakami Mountains (including the Iwaki River) on the western drainage boundary tend to have lower silicic acid concentrations (6.6–28.4 mg SiO_2 L^{-1}). The relationship between the silicic acid concentration in river and irrigation waters is represented by the regression equation, $y = 0.952x + 0.663$ ($r = 0.92$, $P < 0.01$; Fig. 6.9). This equation indicates that there is a strong relationship between silica concentrations in river water and silica concentrations in irrigation water. From this relationship, we inferred that the silica concentration of irrigation water could be estimated from the silica concentrations in nearby (source) rivers.

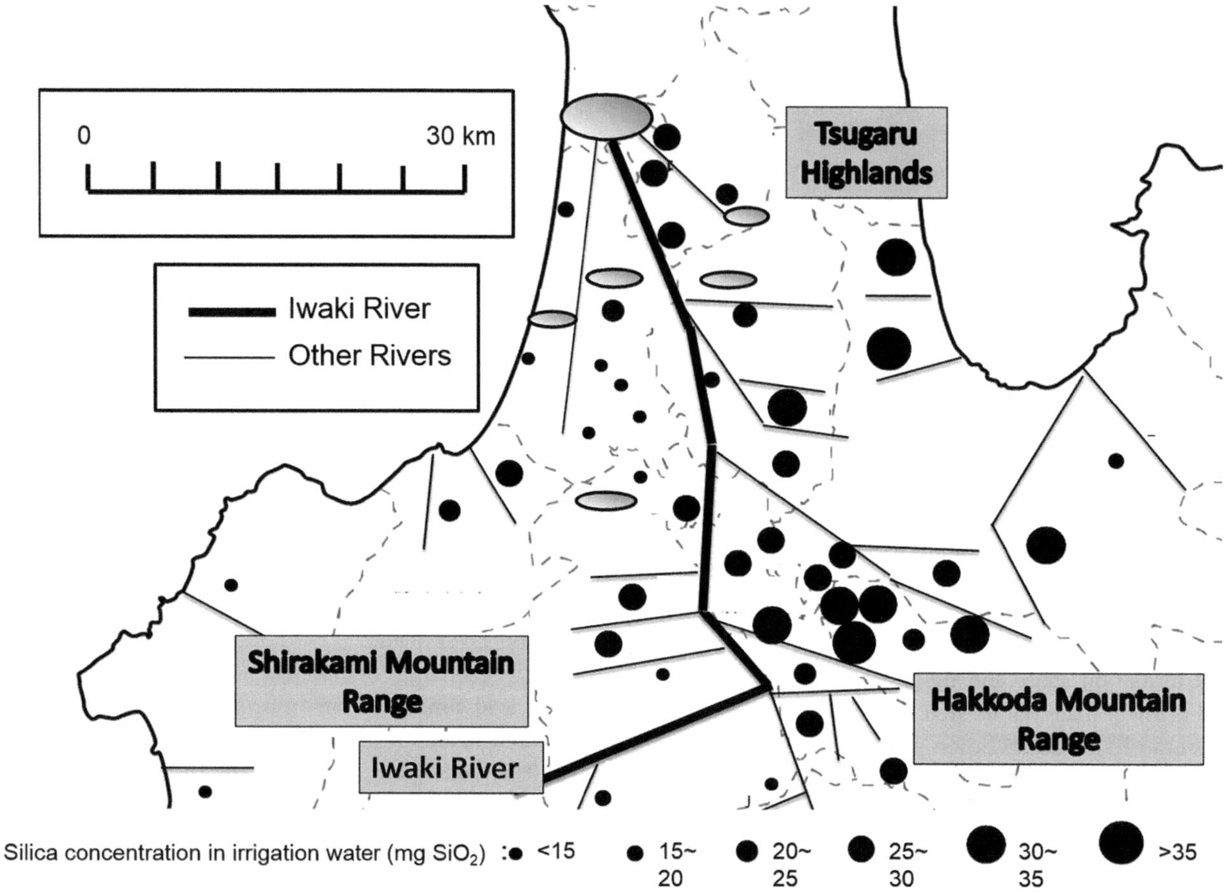

Fig. 6.8 Distributions of silica concentrations in irrigation and river water. *Source* Figure provided by Haruki Fujisawa

(3) **Estimates of silica uptake in rice**

Between 2015 and 2016, a survey was performed to determine the influence of the amount of available silicate in irrigation water on silicate uptake by cultivated rice plants. Samples of paddy soils were collected at the same 42 survey sites where water samples were collected (see Sect. 5.2.4.2). The amount of available silicate in the soils was determined by the flooded-thermal-incubation method, while the silicate uptake by maturing rice plants was determined using the gravimetric method. The cultivation of *Seiten no Hekireki* rice was limited (eight sites); therefore, we also surveyed locations where the rice varieties '*Tsugaru-roman*' (20 sites), '*Masshigura*' (11 sites), and other varieties (three sites) were being grown. Figure 6.10 shows the relationship between amount of available silicate and silicate uptake in maturing rice plants; an increase in available silicate concentrations in soils tended to increase the amount of silicate uptake by growing rice plants (Fig. 6.10). Our multiple regression analysis, which used the concentration of silicic acid in irrigation water and the available silicate content of soils as explanatory variables and the amount of silicate uptake in maturing rice plants as the dependent variable, revealed a multiple correlation coefficient of $r = 0.98$. The amount of silicate uptake (kg ha^{-1}) in maturing plants is equal to [$4.10 \times$ available silicate amount (mg kg^{-1})] + [$10.7 \times$ silicic acid concentration (mg SiO_2 L^{-1}) in irrigation water] (correlation coefficient: $r = 0.98^{**}$). Based on this equation, we surmise that silicate uptake in maturing rice plants can be reasonably estimated for cultivated fields from measurements of silicic acid concentration in irrigation water and the available silicate content of soils.

(4) **Development of fertilizer application standards**

The amount of silicate amendment required in a particular soil is determined based on the amount needed to supplement a shortage in silicate supply in a given year. This shortage is determined by the target amount of silicate needed by rice plants for growth, minus the concentration of silicate available in the soil, and the amount provided by irrigation water (if the difference is negative). We were able to estimate the amount of silicate absorption from the irrigation water silicic acid concentration and available silicate content, and by

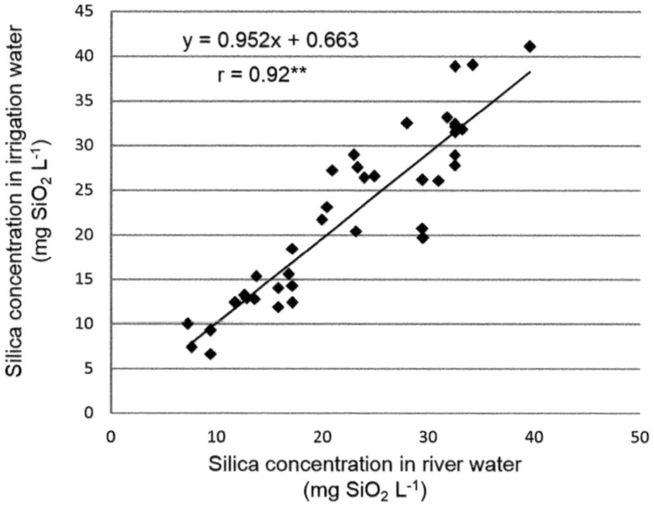

Fig. 6.9 Relationship between silica concentration in irrigation and river waters (based on averages for 2015 and 2016). *Source* Figure provided by Haruki Fujisawa

setting the target silicate absorption quantity, the following equation was formulated: Silicate supply shortage = target silicate absorption quantity—estimated silicate absorption quantity. The soil silica deficiency variable enabled us to determine the silicate fertilizer application standard by setting the target silicate absorption quantity using the relationship depicted in Fig. 6.10. For example, an available silicate reference value of 150 mg Si kg^{-1} soil would yield a silicate uptake of 920 kg ha^{-1} in plants. Therefore, the silicate application standard is equal to 920 kg Si ha^{-1} minus (4.10 × bioavailable silica concentration) minus (10.7 × silicic acid concentration in irrigation water) plus the level of soil silicate deficiency. Using this relationship, we were able to calculate an application standard corresponding to the silicic acid concentration in irrigation water.

6.2.5 Fertilizer Calculation System

In Aomori Prefecture, we aim to increase profitability by linking the production, distribution, and sales of prefecture-made agricultural, forestry, and fishery products and processed products that are safe, secure, and superior from the consumer's point of view. We have named and promoted "active agricultural, forestry, and fisheries" as agricultural promotion measures that are focused on sales.

As part of this effort, in order to build a system that can produce and supply high-quality agricultural products stably, safely, and securely, all producers in the prefecture aim to address "healthy soil making". This effort is known as "Japan's healthiest soil-making campaign", which was initiated in 2007.

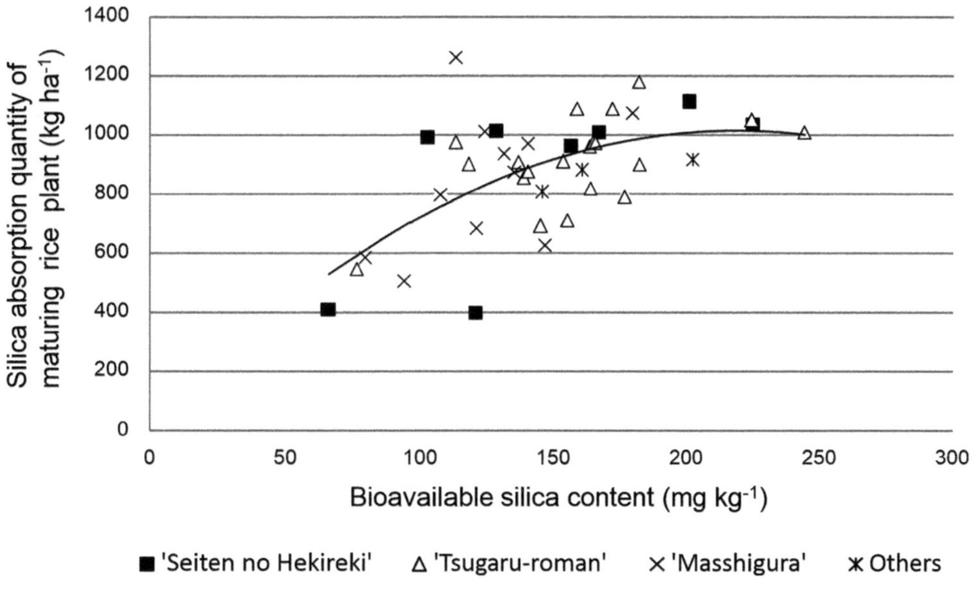

Fig. 6.10 Relationship between silica uptake in maturing rice plants and bioavailable silica content in soils (based on averages for 2015 and 2016). *Source* Figure provided by Haruki Fujisawa

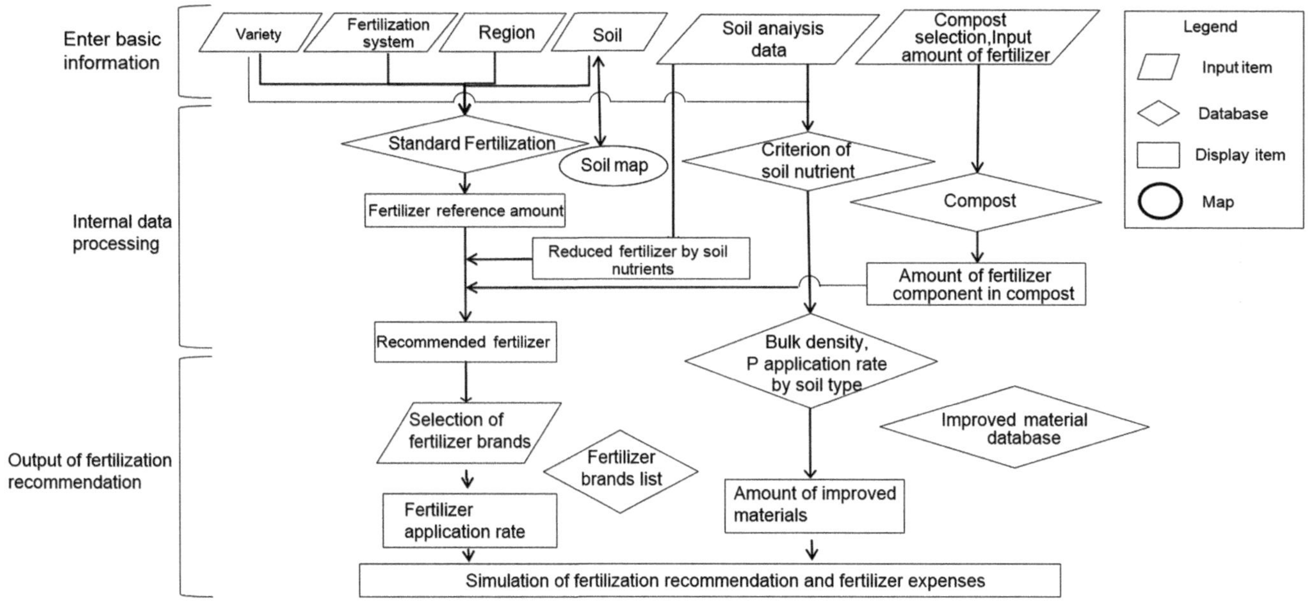

Fig. 6.11 Schematic diagram of fertilization design system. *Source* Figure provided by Norimasa Tanigawa

Aomori Prefecture has defined "healthy soil making" as follows (Aomori Prefecture 2017): An appropriate application of organic soil improvement materials, such as composts and composted animal manures, on the basis of soil analysis for achieving soil with balanced physical, chemical, and biological properties by conducting appropriate soil management strategies that combine deep tillage with crop rotation.

In 2016, the Aomori Prefecture Industrial Technology Center developed an internet application, "Sehi Navi", which can be used to easily calculate the amounts of fertilizer and soil improvement materials that need to be applied to soil. This application takes into account the appropriate amount of soil improvement materials to be applied according to the soil analysis results and the amount of plant nutrients that are already present in the soil and present in the soil improvement materials to be applied. Sehi Navi has been released on our homepage at the following URL and can be used free of charge: http://www.aomori-itc.or.jp/sehisekkei/.

A schematic diagram of the fertilizer application scheme for the Sehi Navi application is shown in Fig. 6.11. Crops for which Sehi Navi can be used to calculate fertilizer requirements are transplanted lowland rice, field crops, and outdoor-grown vegetables. For transplanted lowland rice only, the calculation of silicic acid-containing material to be applied, as determined based on the amount of available silicic acid in the soil, is also incorporated into the fertilizer calculation scheme. For field crops and outdoor-grown vegetables, the required level of fertilizer reduction is based on the amount of nitrate-nitrogen in the soil. Because of this difference, several fertilizer scheme screens are present in the application.

Target nutrient values for improving the chosen crop are selected from the target value database and are displayed; any excess or deficiency of the nutrients in the soil is determined by comparing the target values with the nutrient values obtained from the soil analysis results. If a deficiency is noted, the appropriate amount of soil improvement material to be applied is calculated based on the deficit of each component (phosphoric acid, silicic acid, magnesium, calcium, or potassium), the depth of soil cultivation, the provisional specific gravity, and magnifying the application of phosphoric acid registered for each type of soil. The individual type of soil improvement material to be used can then be automatically selected, and any brand name of fertilizer can possibly be designated. A graph shows the predicted amounts of individual nutrient concentrations and other soil analysis parameters that are improved by applying a given amount of soil improvement materials.

A standard amount of fertilizer relevant to the crop, variety, fertilizer system (or cropping type), field area, and soil type is selected from the database and displayed by the application. The recommended amount of fertilizer is calculated by subtracting from the standard fertilizer amount the amount of nutrients that are already present in the soil, as calculated based on the soil analysis results, and the amount of nutrients contained in the soil improvement materials to be applied, including composts and composted manures. This final recommended amount of fertilizer is displayed by the application. When the user selects the desired fertilizer application, quantity, and fertilizer brand name for each ingredient with reference to the recommended fertilizer amount, the amount of fertilizer to be applied for the area of the field and the fertilizer costs are calculated and displayed.

A database of the nutrient composition of the 26 main livestock manures and composts distributed in Aomori Prefecture has been registered, making it is possible to calculate the level of fertilizer reduction required based on the amount of nutrients in the soil and in the soil improvement materials applied.

In Aomori Prefecture, all producers are involved in the "healthy soil making" initiative, using appropriate (and rationalized) amounts of inorganic and organic soil improvement materials on the basis of soil analysis, and hence reducing the need for (and cost of) fertilizers. Using safe and reliable amounts of agricultural products as a selling point, we aim to raise the added value of the prefectural agricultural products. An effective tool to achieve this is the Sehi Navi internet application, which as described can compute the proper application rates of inorganic soil improvement materials, composts, composted manures, and fertilizers, in a simple operation.

6.2.6 Usage of Excess Accumulated Phosphate

Being susceptible to the effect of cold winds, called "*Yamase*" in Japanese, Iwate Prefecture has experienced a number of seriously cold weather events in the past. Additionally, volcanic ash-derived Andosols, which efficiently immobilize phosphate, are widely distributed in the paddy field soils of this prefecture. Phosphate application is considered to promote the initial growth of paddy rice, and multiple applications are actively conducted to fertilize the soil and counteract low soil temperatures. As a result, large amounts of phosphate have accumulated in the paddy field soils of Iwate Prefecture.

In an effort to minimize over-application, we developed a fertilizer application method that monitors the phosphate which accumulates in the soil.

To evaluate the relationship between the available phosphate content and the amount of phosphate applied to each type of soil, we investigated numerous test cases to obtain a yield increase index. Figure 6.12 summarizes the influence of phosphate fertilizer application on paddy rice yield; the indicator used here is based on the gross brown rice amount set as 100 from the same or adjacent fields when phosphate is applied at levels of 0 or 30 kg ha^{-1}. The results revealed that when the level of available phosphate was below 60 mg kg^{-1} the increase in rice yield was substantial, but that when level of available phosphate was 60–300 mg kg^{-1} there were large annual fluctuations in yield. When the level of available phosphate was more than 300 mg kg^{-1}, phosphate application did not influence yield. Additionally, at available phosphate levels of 300 mg kg^{-1} or below, tillering was delayed in years when low temperatures were experienced at the start of the growth period. Table 6.5 shows the cases in which yields were affected. In 1993, the yield was considerably lower because of very cold weather, which is apparent in the graph.

Considering the abovementioned facts, the levels of available aqueous phosphate in paddy rice, regardless of soil type, are summarized in Table 6.6.

It was confirmed that when the level of available phosphate was more than 300 mg kg^{-1}, yields were unaffected, even when no additional phosphate was applied; however, discontinuing phosphate application would likely decrease available phosphate. Thus, periodic soil analysis is required.

6.3 Management of Upland Soils

6.3.1 Organic Matter Application

In the Tohoku region, upland fields are often located on Andosols generated from volcanic products. Additionally, livestock farming is concentrated in Iwate Prefecture and on the Pacific Ocean side of Aomori Prefecture. Therefore, animal manure is mainly applied to the Andosol upland fields. Although Andosols are characterized by a high organic matter content, the application of organic matter on these soils nevertheless has a positive effect on the soil properties and crop growth. This section describes the dynamics of applied organic matter in Andosols and its effect on the formation of macroaggregates.

Generally, the level of SOM tends to decrease over time when the soil is used as an upland field. Thus, it is necessary to apply an appropriate amount of organic matter to the field in order to maintain the level of SOM.

We investigated the effects of the continuous application of cattle manure on the quantity and quality of SOM in an Andosol (Aoyama and Kumakura 2001). Over a period of 20 years, surface soil samples were collected periodically from plots treated with NPK and NPK + manure, with application rates of 80, 160, and 320 Mg ha^{-1} yr^{-1}, at the Tohoku Agricultural Research Center in Morioka. Particulate organic matter (POM, > 53 μm) and mineral-associated organic matter (MAOM, < 53 μm) fractions were separated from the soil samples by sieving after mechanical dispersion.

In the soil treated with only NPK, a significant decrease in the concentration of total organic carbon (C), reaching up to 10%, was observed after 20 years (Fig. 6.13). Although the application of only NPK had no significant effect on the amount of POM-C, the amount of MAOM-C decreased over time. This indicates that the decrease in the concentration of organic C in the soil treated with only NPK was due to a decrease in the amount of MAOM.

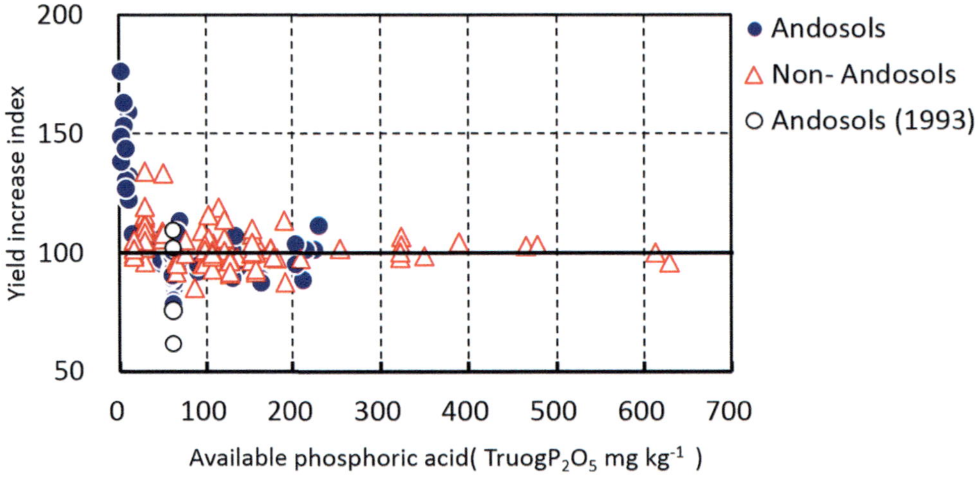

Fig. 6.12 Application effect of phosphoric acid by available phosphoric acid level since yield in black soil in 1993 is lower due to cold weather, it is excluded from analysis. *Source* Figure provided by Teruo Shima

Table 6.5 Effect of the tiller and panicle number of rice plant by phosphate application rate in cool

Year	Spot	Phosphate application rate*1	Middle of June tiller number*2	End of June tiller number*2	Beginning of July tiller number*2	Middle of July tiller number*2	Maturity panicle number*2	Brown rice yield t/ha	Note
1993	Esashi	P-120	345	–	–	847	584	6.71	Brown lowland soil
		P-50	308	–	–	793	587	6.69	TruogP_2O_5 120 $mgkg^{-1}$
		P-0	289	–	–	724	567	6.33	
1996	Sawauchi	P-160	–	99	275	397	381	4.62	yellow soil
		P-0	–	98	210	393	378	4.61	TruogP_2O_5 620 $mgkg^{-1}$

*1Numbers behind *P* Phosphate application amount
*2Tiller number · pniclenumber (m-2)

Table 6.6 Phosphate application rate by available phosphate level

Available phosphate level $mgkg^{-1}$	Phosphate application rate $kgha^{-1}$
< 60	70 + Phosphate improvement
60–300	70
300 ≦	0
	(phosphate: P_2O_5)

The concentration of organic C in the soil that was amended with 80 Mg ha^{-1} yr^{-1} of cattle manure did not change significantly. However, manure application at rates of 160 and 320 Mg ha^{-1} yr^{-1} significantly increased the concentration of organic C and POM-C, while there was a significant decrease in the amount of MAOM-C. This indicates that most of the organic matter derived from the cattle manure accumulated in the form of POM and was then decomposed and transformed into MAOM. The preferential accumulation of POM was also observed in Andosol upland fields amended with compost and crop residues (Aoyama and Kumakura 2002).

The application of organic matter altered SOM not only quantitatively but also qualitatively. In the field plots described above, the application of manure lowered the humification degree of humic acid in the soil and increased the proportion of large-molecular-size humic acid. The qualitative changes in the organic matter of the Andosol upon manure application were mainly due to the accumulation of manure-derived POM.

An adequate supply of water and oxygen to plant roots is essential for the normal growth of plants in upland fields. The supply of water and oxygen is related to the presence of soil aggregates. Thus, we investigated the effects of the

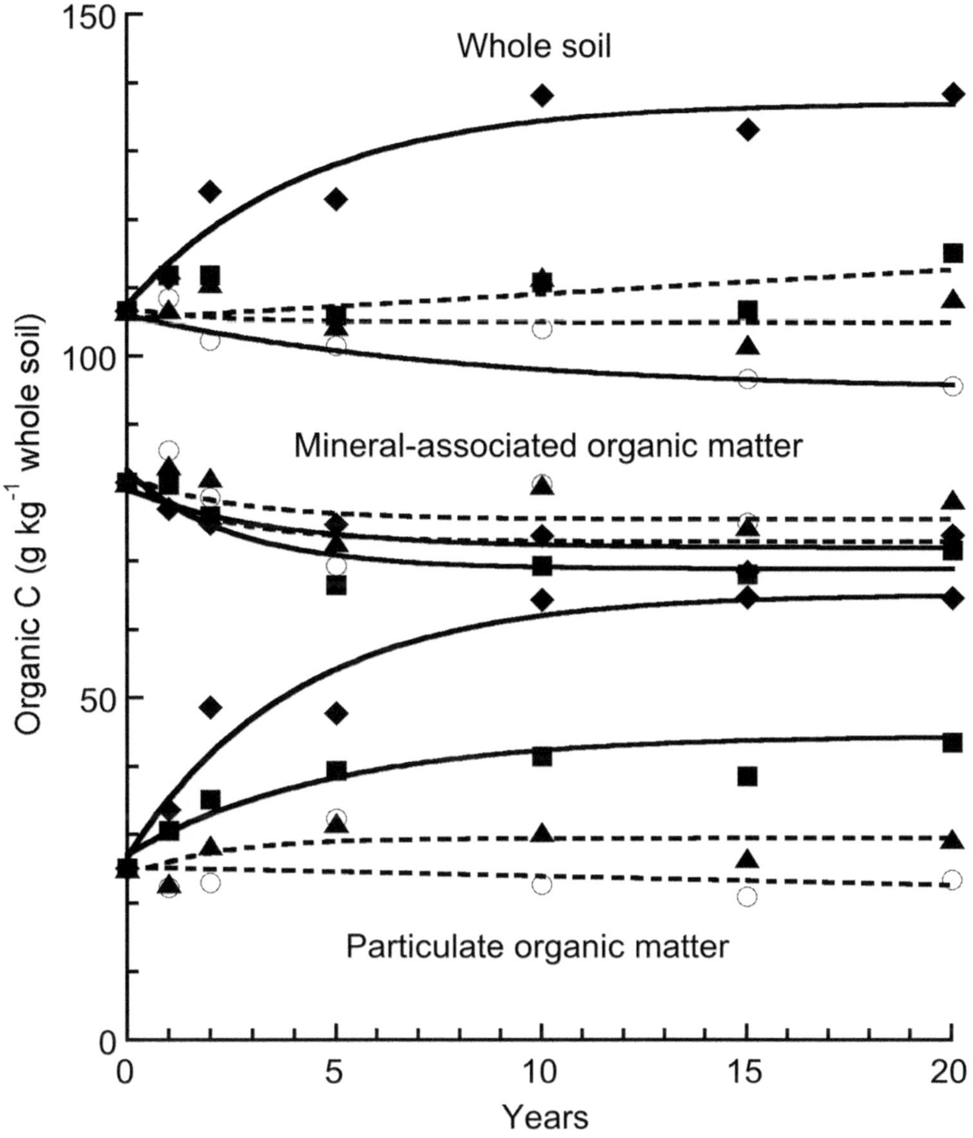

Fig. 6.13 Changes in the amounts of organic C in the whole soil sample and the particulate and mineral-associated organic matter fractions. Regressions shown by a solid line are statistically significant at the 0.05 level, whereas those shown by a dashed line are not significant. ○, 0 Mg ha^{-1} yr^{-1}; ▲, 80 Mg ha^{-1} yr^{-1}; ■, 160 Mg ha^{-1} yr^{-1}; ◆, 320 Mg ha^{-1} yr^{-1}. Reprinted from Aoyama and Kumakura 2002

application of manure on the aggregate composition in an Andosol upland field (Aoyama and Taninai 1992). Aggregates of different sizes were separated by wet sieving from soil samples collected from field plots that were amended with cattle manure at rates of 0 to 200 Mg h^{-1} for 4 years. The results showed that the proportion of macroaggregates increased with an increasing manure application rate. Furthermore, the application of manure increased the amount of organic C in the macroaggregates. This indicates that manure-derived organic matter accumulates as POM, which contributes to the formation of macroaggregates. Macroaggregates are formed by the association of microaggregates and POM (Aoyama et al. 1999). This association is considered to be due to the binding action of microbial mucilage and fungal hyphae, with the POM acting as a substrate for microorganisms (Aoyama 2015).

6.3.2 Watermelon Cultivation on Andosols

The cultivation of watermelon (*Citrullus lanatus*) in Yamagata Prefecture is mostly concentrated in its northeastern part, in the Kitamurayama area. Here, 850 ha are under watermelon cultivation (as of 2016) and 33,700 tons of watermelons are grown annually. Located in the Murayama Basin, which is surrounded by the Ōu Mountains and Dewa Hills, Kitamurayama experiences a wide range of temperatures during summer and relatively heavy snowfall during winter. Kitamurayama also receives relatively less rainfall during the rainy season and is covered with volcanic soils (Andosols) that are suitable for watermelon cultivation.

Watermelon cultivation in this area began in the 1960s. Since then, technical advancements, such as row covers and plastic mulch, together with the local soil characteristics and climatic factors, have helped to make a major producer of good-tasting watermelons in Kitamurayama area (Kikuchi 2009).

There are two main methods of watermelon cultivation in Kitamurayama: "row cover/vine repositioning" and "row cover/cover repositioning". The vines are planted in late April and in May and are harvested in late July and in August.

Nevertheless, in both methods, it is important that the soil is kept warm after planting out and that the vines are pollinated when their flowers open. Plastic mulch and row covers perform an important role in cultivation by preserving soil heat and protecting the vines from rainfall during flowering. In the Kitamurayama area, snow covers the ground for around 120 days of the year and can reach 1–2 m in depth. Since the snow does not melt until April, there may not be enough time in spring to complete all the work required for watermelon cultivation. Moreover, because the soil temperature in spring does not increase once the ground is tilled and covered with plastic mulch, there is no guarantee that the planted vines will achieve the necessary initial growth. In light of these constraints on watermelon cultivation, "autumn mulching" (Fig. 6.14), a method in which the soil is fertilized, tilled, formed into ridges, and then covered with plastic mulch during autumn (when there is plenty of time to work) is also gaining popularity (Ishiyama et al. 2000).

Kitamurayama is covered by Non-allophanic Andosols with low specific gravity and high humus content, which easily lend themselves to being formed into ridges. Covering the soil with plastic mulch during autumn can allow the snow to compact the soil to a certain extent during the time between the first snowfall and planting, thereby helping the soil to retain its warmth and moisture.

Fig. 6.14 Laying a plastic mulch for watermelons during autumn. *Source* Figure provided by Mikio Morioka

Watermelon cultivation in the Kitamurayama area makes good use of the local soil characteristics and climatic conditions, despite the difficulties associated with agriculture in cold, snowy regions.

6.3.3 Sweet Cherry Cultivation

(1) Production overview of sweet cherry in Japan

The production volume of sweet cherry (*Prunus avium* L.) in Japan is just under 20,000 tons, 75% of which is grown in Yamagata Prefecture. In 2015, the production value of sweet cherry in Yamagata Prefecture was 33.7 billion yen (Production Agriculture Income Statistics in FY2015), accounting for roughly half of the total fruit production value in Yamagata Prefecture (67.3 billion yen; Production Agriculture Income Statistics in FY2015). Yamagata Prefecture is the best-known prefecture for sweet cherry production in Japan. The main variety of sweet cherry grown in Yamagata Prefecture is "Satonishiki", which accounts for 72% of sweet cherry cultivation, followed by "Benisyuho" (14%) (Production Dynamics Survey of Special Fruit FY2015). During production, sweet cherries can be damaged by low temperature, precipitation, and frost, both before and after the flowering period in early spring. For this reason, yield can fluctuate greatly (Fig. 6.15). Consumers are always seeking larger, higher quality fruit.

(2) Actual nitrogen content in the annual fertilizer amount

The amount of fertilizer to be applied in a year is decided by consulting the fertilizer application record from the previous year, assessing current tree vigor (which is the result of the fertilizer management of previous years), the degree of fruiting, and the results of soil analysis. If tree vigor is high, the amount of fertilizer application can be reduced from the amount used in the previous year; if tree vigor is weak,

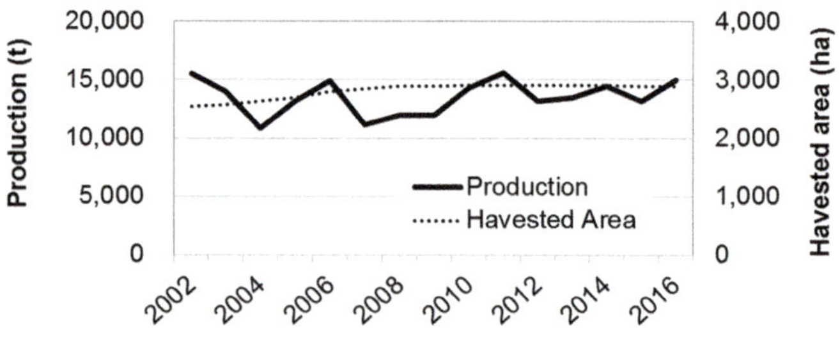

Fig. 6.15 Trends in cherry production and harvested area in Yamagata Prefecture. *Source* Figure provided by Takayuki Ando

fertilizer application should be increased and tree vigor should be maintained by trimming (pruning), disbudding, and thinning the fruit. A farm that does not have satisfactory tree vigor even when fertilizer application is increased may have problems with drainage or depth of effective soil, and these factors should be improved. The nitrogen concentration of the fertilizer applied to the chief cherry-producing district is approximately 100 kg N ha^{-1}.

(3) **Method of split application of nitrogen**

The period from harvest to leaf fall is long. Therefore, it is important to promote early growth and full (strong) flower bud formation in the following year. To increase the next year's harvest, topdressing is applied immediately after harvest to restore exhausted trees and maintain leaf color.

We have been investigating the effect of nitrogen fertilizer on tree vigor and fruit quality at the Yamagata Horticultural Experiment Station. We found that, in the case of highly productive varieties that are prone to reduced tree vigor, an increased amount of fertilizer topdressing should be applied after harvest to recover tree vigor, maintain dark-colored leaves, and increase both yield and production. The nitrogen utilization rate in trees increases when fertilization is performed earlier, and nitrogen is distributed mainly in the younger parts of the plant, such as the leaves. Post-harvest topdressing contributes to the storage of nutrients, including assimilated nutrients. It has been suggested that sufficient nitrogen absorption in the tree immediately after harvest is beneficial and allows the effects of fertilization to last longer. For this reason, the amount of topdressing applied after harvest is typically 20% or more of the annual fertilizer application amount. However, depending on specific factors, such as tree vigor or yield, the proportion of fertilizer that is applied as topdressing may need to be adjusted.

(4) **Application of the whole amount of fertilizer immediately after harvest**

It is possible to maintain tree vigor by increasing the proportion of fertilizer that is applied as topdressing after harvest. Furthermore, labor can be reduced if all the fertilizer is applied at once after harvest. Therefore, we compared the conventional fertilizer application method with the "whole-volume" fertilizer application method in which the entire amount of fertilizer is applied immediately after harvest for "Satonishiki" cherry. For the whole-volume method, the soil inorganic nitrogen content increased, leaf color was maintained (Fig. 6.16), and comparable growth was observed in the following year. Moreover, the fruit quality and yield were equal to or better than those produced using the conventional fertilization method (Table 6.7; Fig. 6.17). Therefore, the method of applying all the fertilizer at once after harvest is beneficial in terms of both yield and labor reduction (Ando and Shiono 2019).

6.3.4 Use of Arbuscular Mycorrhizal Fungi

(1) **Arbuscular mycorrhizal fungi**

Arbuscular mycorrhizal (AM) fungi belong to the Glomeromycota and form a symbiotic relationship with the roots of a host plant. Most crop plants are host plants of AM fungi; these include plants of families of Poaceae, Fabaceae, Solanaceae, Liliaceae, Cucurbitaceae, Asteraceae, Rosaceae, and Rutaceae. The fungi receive sugars from the plant and supply phosphate to the plant. External hyphae extend into the soil far from the root surface where *P* is depleted. Therefore, AM-colonized plants take up more soil phosphate than non-colonized plants.

Under soil conditions where *P* is a limiting factor of plant growth, AM colonization can increase plant growth. External hyphae can connect different host plant species. Growth improvement of mixed croppings such as that consisting of Poaceae and Fabaceae is due to a hyphal network of AM fungi in soil. The growth response of plants to AM colonization is different among crop species and cultivars (Tawaraya 2003).

(2) **Arbuscular mycorrhizal fungi for horticultural plants**

Inoculum of the AM fungus *Glomus* R-10 (Idemitsu Kosan Co. Ltd., Tokyo, Japan) was mixed with sterilized soil (Tawaraya et al. 2012). Seeds of Welsh onion (*Allium fistulosum* L. cv. Motokura) were sown in paper pots containing the soil. Seedlings were grown in a glasshouse for 58 days. The paper pots were transplanted into plot plants were harvested 131 days after transplanting.

The shoot length and stem diameter of the inoculated plants were higher than those of the non-inoculated plants at soil P concentrations of 300 and 600 mg P_2O_5 kg^{-1}. The shoot *P* contents of inoculated plants were also higher than those of non-inoculated plants at 300 and 600 mg P_2O_5 kg^{-1}. The yield (shoot fresh weight) of non-inoculated plants was higher at 1000 and 1500 mg P_2O_5 kg^{-1} than at concentrations of 300 and 600 mg P_2O_5 kg^{-1} (Fig. 6.18). The yield of inoculated plants was not different among the four P levels. The yield of inoculated plants at 300 mg P_2O_5 kg^{-1} was the same as the yield of non-inoculated plants at

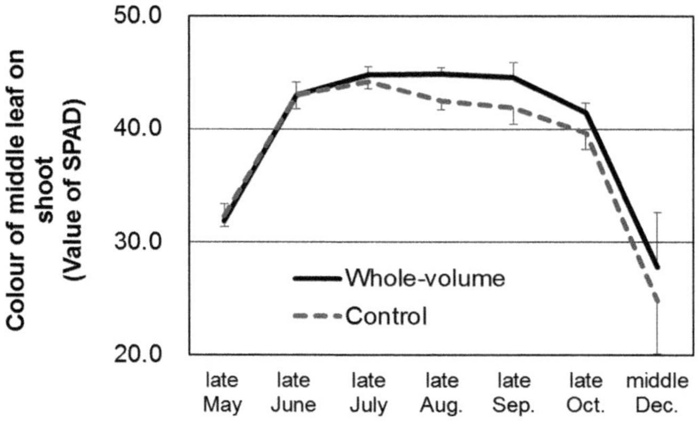

Fig. 6.16 Five-year average of annual changes in the color of middle leaves on shoots of "Satonishiki" cherry. *x*: Whole-volume fertilizer application after harvesting *y*: Conventional fertilizer application. Reprinted with permission from Ando and Shiono 2019. The original publication is available at https://doi.org/10.17660/actahortic.2019.1235.43

Table 6.7 Number of fruit set per spur

Treatment	Number of fruit set per (bouquet) spur				
	2011	2012	2013	2014	2015
Whole-volume	3.1 a*	2.9 a	3.2 a	2.3 a	3.2 a
Control	2.7 a	2.6 a	1.9 b	0.9 b	1.6 b

*Values with a different letter are significantly different ($p < 0.05$) between treatments

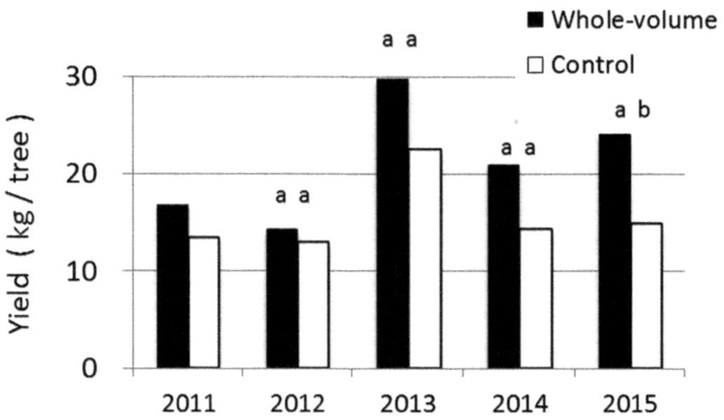

Fig. 6.17 Changes in the "Satonishiki" cherry yield per tree * Values with a different letter are significantly different ($p < 0.10$) between treatments. Reprinted with permission from Ando and Shiono 2019. The original publication is available at https://doi.org/10.17660/actahortic.2019.1235.43

1000 mg P_2O_5 kg^{-1}. These results indicate that the inoculation of plants with AM fungi can reduce the required amount of P fertilizer applied to Welsh onion under field conditions.

The population and *P* uptake ability of indigenous AM fungi can affect the growth improvement achieved by introduced AM fungi. AM fungi should be introduced if the population of indigenous AM fungi is small and/or the uptake ability of indigenous AM fungi is low. If there is a population of AM fungi with a high uptake ability, the heavy application of *P* fertilizer and fungicide should be avoided in order to maintain the population of indigenous AM fungi.

(3) **Arbuscular mycorrhizal fungi in living mulch system**

Living mulch is a cropping system in which a cover crop is planted either before or together with the main crop and maintained as a living ground cover throughout the growing season. Living mulch prevents soil erosion, increases SOM, and suppress weeds without the application of chemical herbicides. If the main crop and cover crop are host plants of AM fungi, both plants are colonized with AM fungi, and improved *P* nutrition can be expected. We used white clover (*Trifolium repens* L.) living mulch for the cultivation of

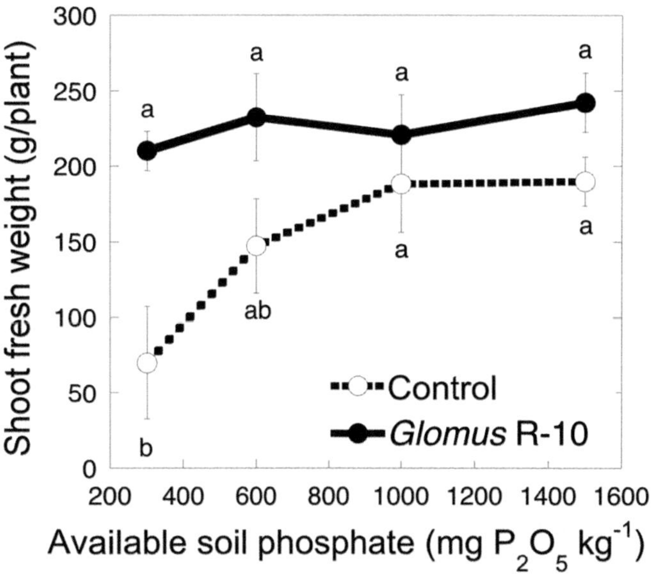

Fig. 6.18 Influence of inoculation of Glomus R-10 for the yield of Welsh onion. *Source* Figure provided by Keitaro Tawaraya and Takumi Sato

maize (*Zea mays* L.) as a main crop. AM colonization by indigenous fungi increased the P uptake and growth of corn (Deguchi et al. 2012). Furthermore, the yield and total digestible nutritional yield of the corn with living mulch and with no *P* application was comparable to the maximum yield of corn without living mulch and with *P* application, suggesting that white clover living mulch reduced the need for the application of *P* fertilizer to silage corn under field conditions (Deguchi et al. 2017).

6.4 Paddy–Upland Rotation

6.4.1 Nitrogen Fertility

In Japan, rice supply has exceeded domestic demand and rice production has been adjusted accordingly for about 40 years. A typical adjustment is the rotation of paddy rice (*Oryza sativa* L.) in paddy fields and upland crops in drained paddy fields, which has been promoted to improve the self-sufficiency rates in upland crops. Soybean (*Glycine max* (L.) Merr.) is a major crop in this rotation sequence in the Tohoku region. In leading soybean-producing countries, namely, the United States and Brazil, the yield of soybean has increased to almost 300 g m^{-2}, whereas the yield of soybean in drained paddy fields in Japan remains as low as about 150 g m^{-2}. The principal problem might be that soybean plants can easily suffer from water damage in the poorly drained paddy fields. A decline in the fertility of paddy soil, where the crop rotation of paddy rice and upland soybean (paddy–upland rotation) has been continued, has recently received attention as another reason for the stagnation of soybean yield.

In a long-term field experiment conducted at TARC, NARO (N39°29′, E140°30′, altitude 30 m a.s.l.), a decline in the soil available nitrogen (N), which was mineralized from air-dried soil under submerged conditions at 30 °C for 4 weeks, was observed (Sumida et al. 2005; Nishida 2016). This field experiment with paddy–upland rotation consisted of three treatments for upland frequency. The upland frequency was defined as the ratio of the number of soybean cropping years to total cropping years after initiation of the paddy–upland rotation. The experimental treatments were short-term upland rotation, medium-term upland rotation, and continuously irrigated paddy. In the short-term upland rotation, soybean and paddy rice were planted with a cycle of approximately 1 year of upland (soybean cultivation) and 2 years of paddy (rice cultivation) (upland frequency: about 35%). In the medium-term upland rotation, a cycle of approximately 3 years of upland and 1 year of paddy was adopted (upland frequency: about 75%). Sub-treatment, with or without the repeated application of organic matter, was performed in each treatment for upland frequency. The soil was classified as fine-textured Gray Fluvic soil (Fluvisol, United Nations Food and Agriculture Organization 2006). The level of soil available N declined in the paddy–upland rotations, regardless of upland frequency, with a greater decrease in soil available N being observed in the medium-term upland rotation than in the short-term upland rotation. The decline in soil available N was alleviated by the application of organic materials. The level of total carbon in the soil also decreased in the paddy–upland rotation.

An investigation of the soil fertility in farmers' fields with paddy–upland rotation was also conducted (Nishida et al. 2013). The relationship between levels of soil available N and upland frequency in four different farmers' fields where

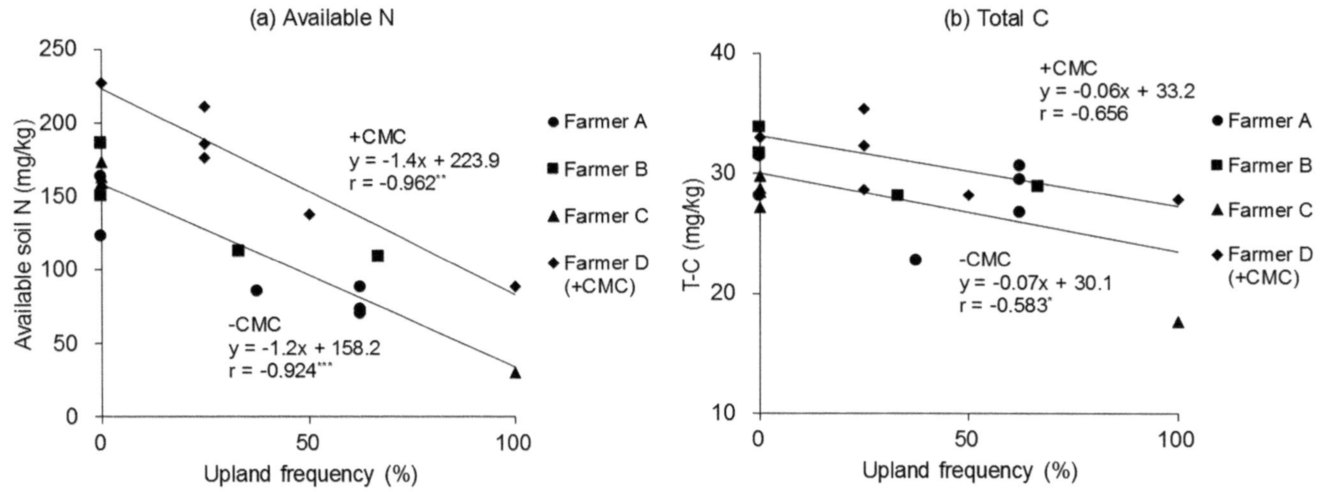

Fig. 6.19 Relationship between upland frequency and available soil nitrogen (**a**), and total soil carbon (**b**) in four different farmers' fields in Daisen, Akita, Japan. ***$P < 0.001$, **$P < 0.01$, *$P < 0.05$. +CMC: cattle manure compost applied; −CMC: cattle manure compost not applied. Modified from Nishida (2016). *Source* Figure provided by Mizuhiko Nishida

paddy–upland rotation had been carried out in Daisen, Akita Prefecture, is shown in Fig. 6.19a. Regardless of cattle manure compost (CMC) application, a negative correlation was observed between the soil available N and upland frequency. It was demonstrated that soil N fertility, as represented by soil available N, tended to decrease in paddy–upland rotations under different management practices in this region and that the decrease in soil N fertility was quantitatively associated with the upland frequency. The level of soil available N was maintained at a level approximately 60 mg kg^{-1} higher in soils under repeated CMC application at a rate of 2–3 kg m^{-2} than in soils without CMC application where only crop residues were returned. The soil total carbon also declined in the paddy–upland rotation (Fig. 6.19b). These results indicate that SOM is decomposed in paddy–upland rotations, leading to the decline in the soil available N.

A decline in soil N fertility causes a decrease in soybean yield. Yield data were comprehensively collected from the experimental paddy–upland rotation plots at TARC, NARO, during the period from 1991 to 2009 (Nishida 2016). Soybean yield decreased as soil available N decreased, and a significant quadratic relationship was found between soil available N and soybean yield. According to the quadratic equation, when target yields were set to 300 g m^{-2} for soybean and 600 g m^{-2} for rice, soil available N needs to be more than 80 mg kg^{-1} to obtain the target soybean yield. A significant relationship was not found between soil available N and rice yield in the paddy–upland rotation. However, rice yield decreased, due to lodging, when the level of soil available N was over 200 mg kg^{-1}, indicating that soil available N should be lower than 200 mg kg^{-1} in the paddy–upland rotation. Rice yield tended to be greater than 600 g m^{-2} when the soil available N was between 80 and 200 mg kg^{-1}. Consequently, the suitable range of soil available N for paddy–upland rotation is 80 to 200 mg kg^{-1}, which corresponds with the optimal range identified by MAFF (2009) for paddy rice.

The keys to controlling soil N fertility in paddy–upland rotation are the temporal balance of paddy rice and soybean cultivation (i.e., the upland frequency), and the application of organic materials. As stated above, the suitable range of soil available N level for paddy–upland rotation is 80–200 mg kg^{-1}. Figure 6.19a clearly displays the management strategies required to maintain the minimum level of soil available N for paddy–upland rotation. According to the regression line, in fields without the application of organic materials except crop residues, soil available N levels were lower than 80 mg kg^{-1} when the upland frequency was higher than approximately 60%. Therefore, the upland frequency needs to be lower than 60% to maintain the minimum level of soil available N (80 mg kg^{-1}) in cases without the application of organic materials except crop residues. However, under the repeated application of CMC at a rate of 2–3 kg m^{-2}, soil available N can be maintained at more than 80 mg kg^{-1} even with an upland frequency of 100%. With the repeated application of CMC to paddy rice and soybean cultivations performed at the same frequency (i.e., upland frequency of 50%), soil available N can be maintained at the same level as that in continuous paddy fields without the application of organic materials except for crop residues (upland frequency: 0%). Soybean cultivation can be conducted more frequently in fields to which CMC is applied than in fields to which no organic material is applied, as soil available N can be maintained within a suitable range.

6.4.2 Nitrogen Budget

Soybeans require a large amount of N due to the large accumulation of N in their seeds. Although soybeans obtain some of their total N accumulation from symbiotic N_2 fixation in their root nodules, they also take up significant amounts of soil N to meet their high N requirements. Therefore, there is a possibility that the N output from soybean-cultivated fields can exceed the N input to the field, and thus the N budget could be negative, indicating N loss from the field. The N loss from the field could thus cause a decrease in soil available N in rotated paddy–upland fields. Although significant N loss from rice paddy fields is not thought to have occurred, a detailed N budget in a rice-cultivated rotated paddy field has not yet been well established. Therefore, in order to maintain soil N fertility in rotated paddy fields, it is essential to evaluate their N budget during both the soybean and rice cultivation period.

(1) **Nitrogen budget in an upland field with soybean cultivation**

The annual N budget in a soybean-cultivated upland field was evaluated for 3 years in an experimental lysimeter field filled with Gray Fluvic soil (Fulvic paddy soil), a soil type which is typically found in the Sea of Japan side of Northern Japan. Input (seed, fertilizer, bulk N deposition, and symbiotic N_2 fixation) and output (harvested grain, N leached via drainage water and nitrous oxide (N_2O) emission) flow of N were measured, and the field N budget was estimated by subtracting the N output from the N input (Fig. 6.20, Takakai et al. 2010, 2017a). The grain yield of soybean for the 3 years ranged from 291 to 410 g m^{-2}, with an average of 341 g m^{-2}.

The annual N budgets in an upland field with soybean cultivation for 3 years were consistently negative (i.e., there was a net N loss from the field), and ranged from −10.9 to −7.9 g N m^{-2} y^{-1}, with an average of −9.6 g N m^{-2} y^{-1}. The validity of the estimation was confirmed by its agreement with the amount of N loss estimated based on changes in soil N storage (Takakai et al. 2017a). The major component of total N input (20.1 g N m^{-2} y^{-1}) was symbiotic N_2 fixation, which accounted for 83% of total input, and the major components of total N output (29.8 g N m^{-2} y^{-1}) were harvested grain and N leaching, which accounted for 74% and 25% of total N output, respectively. Based on the cultivation guidelines for this region, to avoid the excessive growth of soybean, chemical fertilizer was not applied for the first year of soybean cultivation after field conversion. The amount of chemical fertilizer application for the following 2 years (second and third years) was 2 g N m^{-2}, which was applied as basal fertilizer. Bulk N deposition tended to increase in the fallow season, including in winter. The percentage of soybean N accumulation that was derived from N_2 fixation remained relatively constant (60–69%) over the 3 years.

(2) **Nitrogen budget in a paddy field with rice cultivation**

Following the 3 years of soybean cultivation under upland condition, the annual N budget during rice cultivation under flooded paddy conditions was also evaluated for 3 years. The input and output N flows were measured and compared with those measured during soybean cultivation (Table 6.8; Takakai et al. 2017a). The yield of brown rice for the 3 years ranged from 422 to 608 g m^{-2}, with an average of 519 g m^{-2}.

The three-year average of N budget during rice cultivation was negative (–2.3 g N m^{-2} y^{-1}), indicating a net N loss, although this loss was less than that observed during soybean cultivation. The N budget indicated a significant loss of N, unlike previous reports in continuous paddy fields which indicated no change in N budget or a slight N gain (e.g., +1.28 g N m^{-2}; Katayanagi et al. 2013). Additionally, the N budget measured during rice cultivation varied greatly among the 3 years (–5.7 to +0.2 g N m^{-2} y^{-1}) due to the difference in fertilizer application rates, which accounted for 63% of the total N input over the 3-year period. In paddy fields converted from upland fields, the reduction of basal N fertilization is recommended to avoid excessive growth and lodging due to the increased nitrogen uptake. Chemical fertilizer was applied at a rate of 0 or 6 g N m^{-2} to all plots as a basal fertilizer in the first or second and third years, respectively. A total of 2–3 g N m^{-2} of chemical fertilizer was also applied as topdressing in late July. The N loss from the paddy fields in this study could be significantly higher in the first year after conversion with low N fertilization. Consequently, in the paddy–upland rotation system, considerable loss of N may occur in both the soybean and rice cultivation periods.

The major components of the total N output were harvested grain and N leaching, which accounted for 49 and 29% of total N output, respectively. The annual N loss via leaching during soybean cultivation was lower than that during rice cultivation.

In order to maintain the soil N fertility in a rotated paddy field with soybean cultivation, techniques for the mitigation of N loss, such as the application of organic matter (e.g., green manure such as hairy vetch and manure compost; Sato T, in this chapter; Nishida M, in this chapter; Nishida et al. 2013) may be essential. However, because the application of organic matter to rotated paddy fields can significantly change the N flows, its effects on the N budget

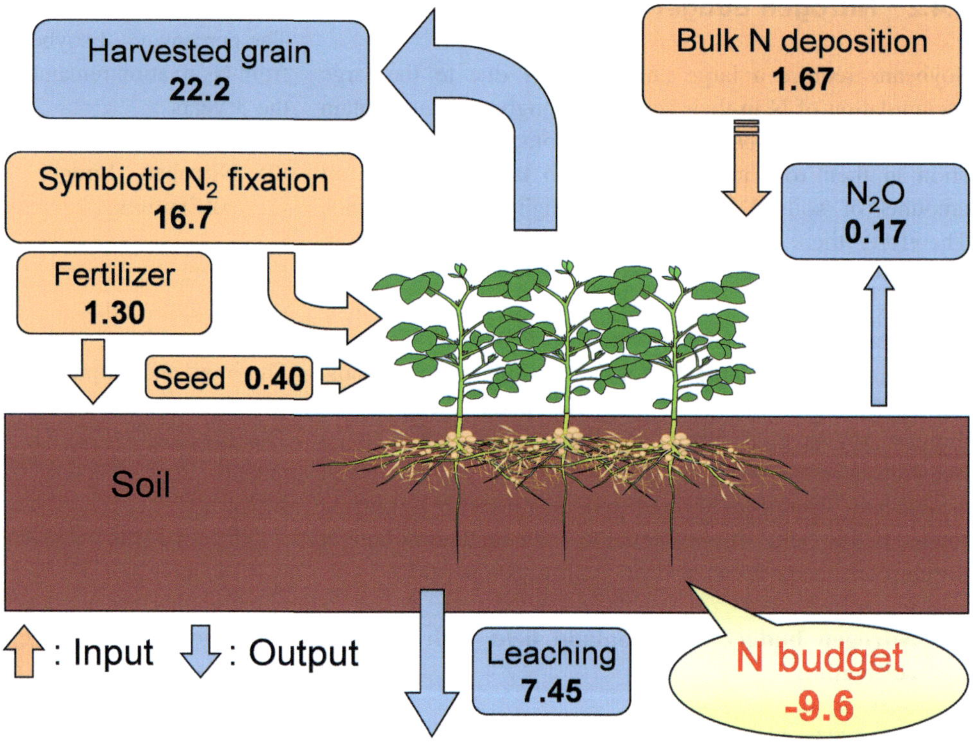

Fig. 6.20 Annual nitrogen (N) flows and a budget in an upland soybean field converted from a rice paddy field (Three years average, unit: g N m^{-2} y^{-1}). N_2O, nitrous oxide. N_2 emission via denitrification was not considered in this study. Modified from Takakai et al. (2017a), Copyright 2017, The Authors licensed under CC BY 4.0. *Source* Figure provided by Fumiaki Takakai

Table 6.8 Nitrogen (N) flows and budgets in soybean and rice-cultivated field [unit: g N m^{-2} y^{-1}, modified from Takakai et al. (2017a)]

		Upland soybean	Paddy rice
Input	Bulk N deposition	1.67	1.60
	Irrigation	–	0.29
	Seed (soybean) or Seedlings (rice)	0.40	0.12
	N_2 fixation	16.7	2.0
	Fertilizer	1.3	7.0
	Total	20.1	11.1
Output	Harvested grain	22.2	6.51
	Surface drainage	–	0.64
	Leaching	7.45	3.94
	N_2O emission	0.17	0.01
	NH_3 volatilization	ND	0.10
	N_2 emission via denitrification	ND	2.19
	Total	29.8	13.4
N budget		−9.6	−2.3

Positive and negative values indicated N input and output, respectively. The N budget was calculated by subtracting N output from input. All values were expressed as annual value (three years average). N_2 fixation, NH_3 volatilization, and N_2 emission via denitrification were estimated from literature value. NH_3, ammonia; N_2, dinitrogen; N_2O, nitrous oxide

should also be evaluated. Furthermore, the application of organic matter to rotated paddy fields could also increase environmental loads such as greenhouse gas emissions and N leaching (Takakai et al. 2017a, b). Consequently, the careful management of organic matter application is required.

6.4.3 Green Manure

The cultivation of green manure crops has been used as a soil amendment for increasing soil fertility (Cherr et al. 2006). Legumes have been utilized as green manure crops in agriculture because they can accumulate a large amount of nitrogen through N_2 fixation in their nodules. Accordingly, incorporating a leguminous green manure crop provides a large amount of nitrogen and carbon to the soil for subsequent crops. Hairy vetch (*Vicia villosa* Roth) is one such leguminous cover crop that is used as green manure, as well as for weed management. Its symbiosis with *Rhizobium leguminosarum* bv. viciae can fix 100–200 kg ha^{-1} $year^{-1}$ of atmospheric nitrogen. Hairy vetch has recently been used to improve the physical properties of the soil in the Hachirogata Polder in Akita Prefecture (Sato et al. 2007). Succeeding soybean yield was found to increase when hairy vetch was planted in an upland field converted from a paddy field (Sato et al. 2011).

In the Tohoku region, hairy vetch seeds are generally sown at a density of 3 g m^{-2} (approximately 200 seeds per square meter) in September to October. The hairy vetch in the field germinates two weeks after sowing and grows until the winter. The hairy vetch can overwinter in this state and grows vigorously from April to June the next year. During the period of maximum growth, in June, the plant height reaches about 180 cm, and the plants cover the entire field (Fig. 6.21). The average dry weight of the aboveground portion of the plants is about 400 g m^{-2} and the average nitrogen content is about 4.0%. It is therefore predicted that plowing in the aboveground portion of the hairy vetch would supply about 16 g of nitrogen per square meter (Sato et al. 2007). Finally, the aboveground part of the hairy vetch is chopped up and plowed into the ground at a depth of 10 cm prior to soybean cultivation (Sato et al. 2007).

The soil structure in the first few centimeters from the surface is made up of small granules left by the hairy vetch planting. The 10 cm layer below this level has a lumpy consistency due to the presence of small clods. The surface has a cracked structure, with the cracks being up to 50 cm deep and up to 2 cm wide. The roots of the hairy vetch are often observed 15 cm below the surface, following the crack lines and extending to a depth of up to 50 cm (Fig. 6.21). The hairy vetch roots grow along the soil crack line, which means that it can be desirable to promote a crack structure (Sato et al. 2007).

Fig. 6.21 Photo of hairy vetch shoot (left) and soil structure (right). Modified from Sato et al. (2007). *Source* Figure provided by Takashi Sato

The planting of hairy vetch provides favorable conditions for root growth and the modulation of soybean by improving soil permeability and maintaining suitable water conditions. Improving soil effectiveness also promotes nitrogen fixation and absorption in the roots of soybean (Sato et al. 2007). The root system of soybean in the field with hairy vetch planting is broader and the roots grow deeper and longer compared with the field without hairy vetch planting (Sato et al. 2011). Thus, soybean growth is promoted by favorable soil conditions in the field with hairy vetch planting (Fig. 6.22).

The soybean yields obtained in the field with hairy vetch planting were approximately 30% higher than those obtained by conventional cultivation in both 2005 and 2006. The number of pods and the number of seeds per pod were significantly higher in the field with hairy vetch planting (Table 6.9). The soil improvements caused by hairy vetch planting may have promoted vegetative growth, leading to an increased number of nodes and therefore higher rates of pod formation (Sato et al. 2011).

6.4.4 Effect of Soil Properties

One of the major factors which decrease the yield of upland crops in upland field converted from paddy field is water damage, which can be caused by soil physical conditions such as poor field drainage, low air diffusivity, bad soil tilth, and a shallow plow layer. In this chapter, we introduce the physical properties of soils that are used as upland crops and paddy rice rotational uses.

The dominant soils in Japanese paddy fields are non-volcanic ash soils, in which it is easy to form a low-permeability layer that can maintain flooded conditions. In particular, clayey alluvial soils are widespread in Japanese paddy fields. This is one of the reasons why water damage is the most significant problem in upland fields converted from the rice paddy field. Table 6.10 shows the percentage distribution of each soil type in paddy fields in the Tohoku region. A total of 70% of paddy fields in the Tohoku region are in Fluvic soil, with the frequency being higher on the Sea of Japan side than on the Pacific side. The percentage distribution of fine-textured Fluvic soil is the highest, with such soil accounting for 45% of paddy fields. Fine-textured Fluvic soil has low water permeability and bad soil tilth. Consequently, this soil often leads to water damage in upland crops.

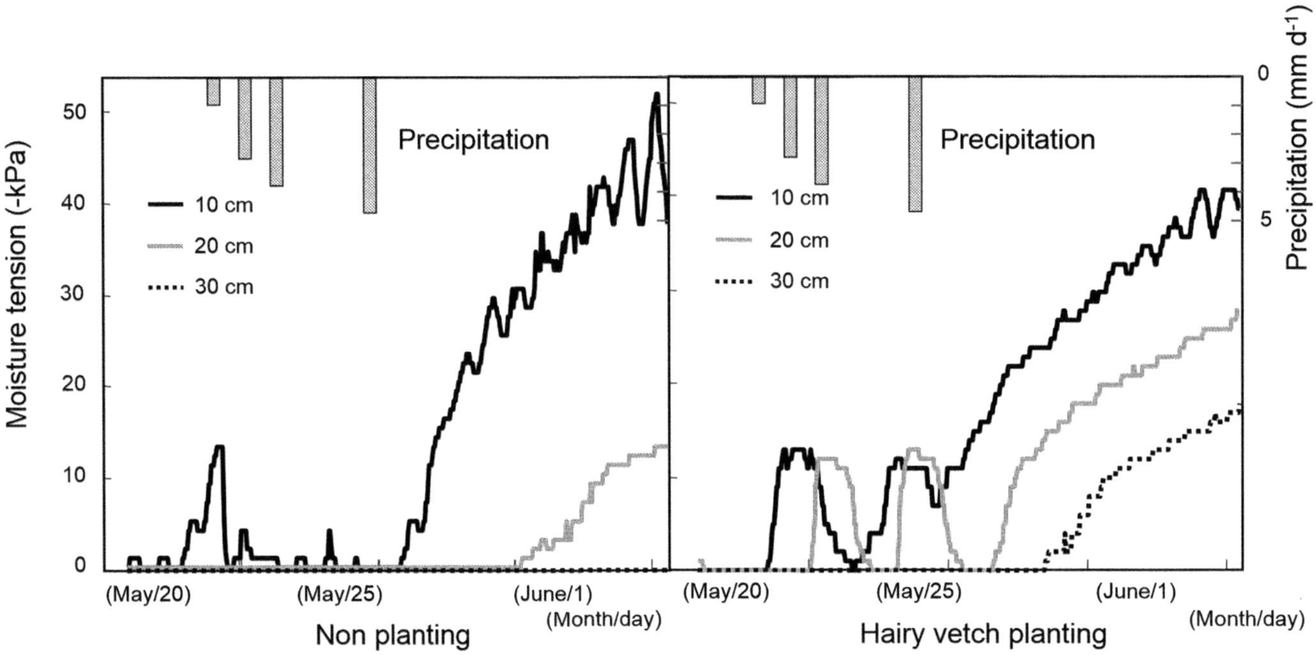

Fig. 6.22 Changes in soil moisture tension in each depth after the precipitation in 2005. The data of the precipitation was obtained from automated meteorological data acquisition system (AMeDAS) setting up at Ogata village. Modified from Sato et al. (2007). *Source* Figure provided by Takashi Sato

The second factor that decreases the yield of upland crops is the existence of a plow pan. The depth of the plow layer in paddy fields is about 11–13 cm, and below this there is a low-permeability plow pan which maintains flooded conditions. During upland cropping, this plow pan often leads to the presence of excess water in the field, causing water damage to crops.

(1) **Process of changing to upland soil**

The general profile of paddy soils shows a massive structure that indicates very weak development of soil structure, which results from periodic paddling and the lack of a continual soil wetting–drying cycle due to flooding.

The conversion of paddy field to upland field forces strong drying. Capillary forces developed by the drying process compress soil, decreasing the volume of micropores and amount of water retention. Thus, soil develops hydrophobic properties by the conversion to upland field. On the contrary, cracking that occurs due to the drying process increases the water content of soil micropores to greater than −6.2 kPa (Nakano 1978). Consequently, soil structure that is unique to upland soil forms gradually (Fig. 6.23).

In terms of field management, it is important to change the water potential of field capacity. Upland fields converted from rice paddy field show a low tillability and it is hard to obtain good soil tilth. When the conversion period to upland field increases, field capacity becomes close to the plastic

Table 6.9 Yield and yield components in soybean (2005 and 2006)

Year	Field	Stem length (cm)	Stem diameter (mm)	Branch (number plant^{-1})	Pod number (number plant^{-1})	Total seed number (number plant^{-1})	1000 seeds weight (g 100 seeds^{-1})	Yield (g m^{-2})
2005	Non-planting	49.7a	9.6a	6.2a	103.2a	200.4a	27.6a	350.7a
	Hairyvetch planting	59.0b	10.1a	6.4a	107.6a	211.0a	27.2a	369.3a
2006	Non-planting	45.9a	8.0a	4.5a	45.3a	81.9a	28.0a	276.2a
	Hairyvetch planting	61.9b	8.8a	5.7b	70.5b	136.4b	28.1a	393.3b

Means followed by a common letter are not significantly different at $p < 0.05$ ($n = 4$) level by t-test
Modified from Sato et al. (2011)

Table 6.10 Ratio of soil types (%) in paddy fields spread in Tohoku region

	On Pacific Ocean				On Japan sea		Total
	Aomori	Iwate	Miyagi	Fukushima	Akita	Yamagata	
High-humic Andosols	7	0	1	1	0	4	2
Humic Andosols	17	35	3	10	7	6	13
Low-humic Andosols	0	4	0	0	1	2	1
Subtotal (Andosols)	23	38	4	11	8	12	16
Fine-textured Fluvic soils	25	29	35	56	63	52	45
Medium-textured Fluvic soils	25	8	20	16	12	22	17
Coarse-textured Fluvic soils	7	1	6	1	8	3	4
Subtotal (Fulvic soils)	56	39	61	73	83	76	66
Fine-textured upland soils	1	16	2	7	3	3	5
Other upland soils	0	5	1	0	0	0	1
Subtotal (Upland soils)	1	21	2	7	3	3	6
Organic soils	20	2	33	9	6	9	13
Total, ha	96,932	109,682	110,684	101,327	135,246	119,190	673,061

Prepared by the author based on the data in Soil fertility conservation survey in Aomori Prefecture (1978), Agricultural Experimental Station in Iwate (1978), Agricultural Research Center in Miyagi (1978), Agricultural Experimental Station in Akita (1978), Agricultural Experimental Station in Yamagata (1978), and Agricultural Experimental Station in Fukushima (1978). The categorizing of soil types is obeyed by Takimoto et al. (2017). Andosols group are categorized by soil organic matter (SOM) contents; High-humic Andosols contain >10% of SOM, humic Andosols contain 5–10%, and others are grouped in low-humic Andosols. Fluvic soils group is categorized by clay contents; fine-textured lowland soils contain >15% of clay, medium-textured lowland soils contain <15% of clay, and <65% of sand, or >40% of fine sand and <45% of coarse sand. Other Fluvic soils are categorized into coarse-textured Fluvic soils. Upland soils contained >15% of clay contents are categorized by fine-textured upland soils and others are other upland soils

limit (Fig. 6.23). The soil tillability improves as a result of this process because the dominant yield pattern changes from plastic deformation to brittle fracture in the ordinal conditions of field water content.

As mentioned above, the drying history of soil drives the change in its physical properties by conversion. The "uplandization index" is an index used to quantitatively express the progress of the change of soil physical properties toward that of a completely upland field; its value ranges from 0 to 1, with a value of 1 indicating that the subjected field has completely upland properties. The uplandization index uses the sediment volume of the paddy field and upland field conditions. Sediment volume is defined as the volume of well-puddled soil in 40 mmol L^{-1} of NH_4Cl solution, which is considered as an index of soil microstructure. It is known that the sediment volume is decreased by conversion to upland field and increased by conversion to the paddy field (Fig. 6.24). The uplandization index is calculated by the following equation (Naganoma and Moroyu 1983):

$$\text{Uplandization Index} = \frac{V_p - V}{V_p - V_u}$$

where V, V_p, and V_u indicate the sediment volume of the subjected field, the sediment volume of the continuous paddy field near the subjected field, and the sediment volume of the continuous upland field near the subjected field or of the air-dried soil in the subjected field.

What is the nature of *uplandization index*? The sediment volume decreases irreversibly due to soil drying to a water potential exceeding −1.5 MPa (Katou et al. 1985). The sediment volume is related to the properties of well-puddled soil. This implies that the soil microstructure is changed by drying exceeding −1.5 MPa, and this change is not easily restored by puddling. Additionally, the soil microstructure does not disappear in the rewetting process, because sediment volume is measured in solution. In conclusion, the uplandization index shows the soil microstructure that results from drying exceeding −1.5 MPa and that is not

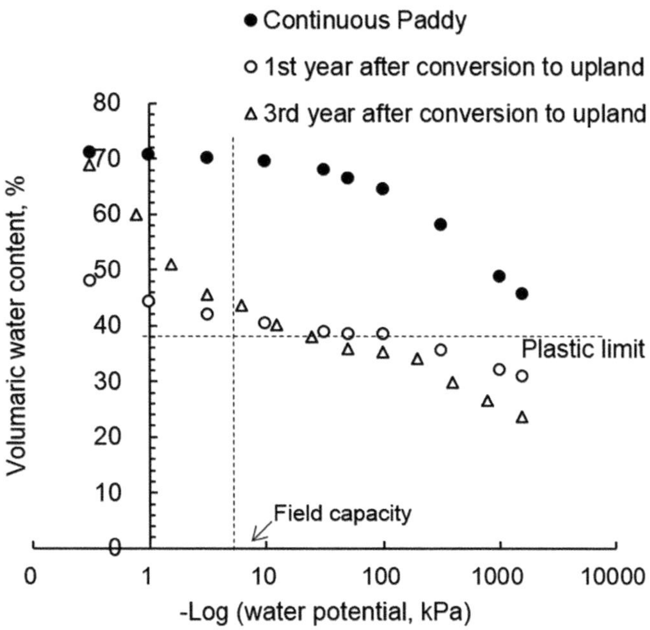

Fig. 6.23 Changes of water retention characteristics by conversion of field from paddy to upland. Dashed vertical line indicates field capacity (−6.2 kPa) and dashed horizontal line shows the plastic limit. *Source* Nakano (1978)

altered by the rewetting process. In other words, it shows the history of soil drying

(2) **Process of changing to paddy soil**

The soil structure gradually restores to a massive state in paddy fields that have been reconverted from upland fields. This process is the opposite of the process of change that occurs in the conversion to upland field. In particular, in the first year after reconversion to paddy field, puddling

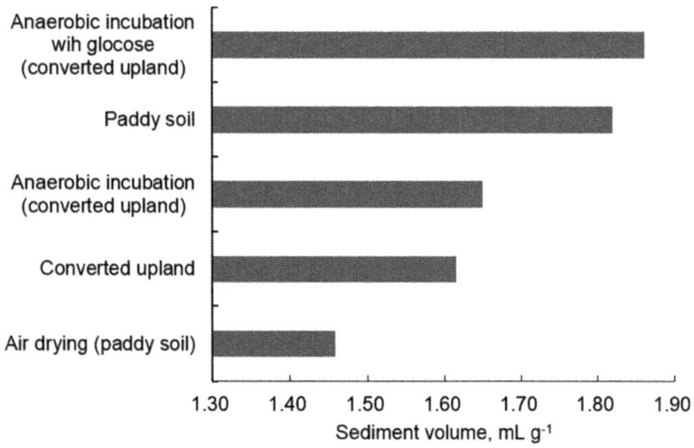

Fig. 6.24 Changes of sediment volume by various treatments for paddy soil and converted upland soil. *Source* Naganoma and Moroyu (1983)

sometimes results in the hardening of the soil, which has a negative impact for the transplanting of seedlings and the trafficability of machinery (Kitagawa et al. 1988). This phenomenon is called "*itsuki*" in Japanese. *Itsuki* is considered to result when an upland-like soil microstructure remains. Increasing soil micropore by paddling decreases water potential if the soil water content is constant, due to the movement of soil water to smaller pores. *Itsuki* is considered to be caused by increasing soil micropore by puddling.

Soil reduction is a factor that changes soil microstructure in reconverted paddy fields. The reduced soil has a greatly increased sediment volume (anaerobic incubated data in Fig. 6.24). Takahashi and Toriyama (1999) showed that free iron oxides are a key material for increasing sediment volume through reduction and puddling processes. They considered that free iron oxides bind soil particles together during the drying process, and the breaking of this binding by reduction results in an increase in sediment volume by puddling. The degree and timing of the increase of soil micropore by puddling would affect the *Itsuki* phenomenon.

6.5 Environmentally-Friendly Agriculture

6.5.1 Countermeasures for Cadmium-Contaminated Paddy Soils

Cadmium (Cd) contamination in agricultural soil in Akita Prefecture is the result of past mining activities. Akita was once one of the most active mining areas in Japan, with 248 modern mines documented. During the period of peak activity, the prefecture accounted for approximately half of the domestic copper output and was also one of the top producers of nonferrous metals, including lead, zinc, and silver. In particular, the northern part of the prefecture operated many large-scale mines equipped with refineries, around which numerous mid- to small-scale mines were distributed. Similarly, many large-scale and satellite mines also operated in the southern part of the prefecture, primarily producing black ore ("*kuroko*").

The operations of these mines peaked in the first half of the twentieth century (shortly before the Pacific War). At that time, the prevention of mine pollution was virtually ignored, with miners haphazardly leaking Cd, a heavy metal contained in mine drainage water and soot, into surrounding farmlands and other areas.

In many cases, Cd contamination was spread unnoticed, because drainage water discharged directly into rivers from the mines drifted into rice paddies during irrigation and because contaminated riverbed soil seeped into and was deposited in farmlands following the collapse of tailings dams and the flooding of rivers.

Table 6.11 Cadmium concentration in grain brown rice by experiment soil dressing (mg kg^{-1})

Thickness of layer of soil dressing	Year			Average
Plot name (cm)	1979	1980	1981	
0	0.48	1.27	0.41	0.72
10	0.28	0.18	0.12	0.29
15	0.11	0.12	0.07	0.10
20	0.15	0.12	0.03	0.10
25	0.11	0.08	0.16	0.12
30	0.18	0.16	0.01	0.12
35	0.14	0.13	0.06	0.11

Method of soil dressing adding on surface layer
Reprinted from Bunro OGAWA (1994)

Although the downstream basins of the mines once prospered, due to the mining boom these regions have been forced to take measures against farmland pollution since the latter half of the twentieth century. This is the negative legacy of the mining industry.

(1) Measures and challenges

In 1970, the Japanese government enacted the "Act to Prevent Soil Contamination on Agricultural Land", stipulating that contaminated rice paddies producing Cd-contaminated rice (the reference concentration was then 1.0 mg kg^{-1} in paddy rice, but was later revised to 0.4 mg kg^{-1} in 2011 in accordance with the Codex standards) be decontaminated through "soil dressing". Soil dressing refers to a civil engineering technique whereby non-contaminated soil from mountains and other places is transported to the contaminated site, where the contaminated soil is either replaced by or covered with the non-contaminated soil.

Akita Prefecture implemented soil dressing over 1630 ha of farmland from the 1970s through 2015, effectively decontaminating all high Cd-contaminated lands in the prefecture. The majority of the decontamination was achieved through coverage soil dressing, with the thickness of the soil dressing layer being set at 27.5 cm based on the results of an experiment conducted to reduce Cd concentrations in brown rice (Table 6.11), and through the decrease of soil layer thickness by compression. Assuming a soil bulk density of 1.0 kg L^{-1}, this corresponds to a soil volume of 2750 Mg ha^{-1}.

The effect of soil dressing on the contaminated rice paddies was outstanding, such that today the Cd concentrations in the soil and brown rice produced in this area are negligible, 40 years after the decontamination. However, the scarcity of humus, organic compounds, and nutrients in the dressed soil places a heavy burden on farmers in terms of soil development and management.

Moreover, around the areas that were subjected to decontamination operations, there are many low-contaminant-concentration farmlands containing concentrations of Cd that are slightly higher than the naturally occurring levels. Performing soil dressing over all these areas would require large amounts of non-contaminated soil to be transported, replaced, and added, which is cost-prohibitive, time-consuming, and thus impracticable. To prevent Cd-contaminated rice from being produced in such farmlands, since the 1970s Akita Prefecture has been engaged in the development and dissemination of cultivation techniques for suppressing Cd absorption by crops.

As shown in Fig. 6.25, there is no correlation between the Cd concentrations in rice and those in the farming soil. This is because the configuration of Cd compounds in the soil

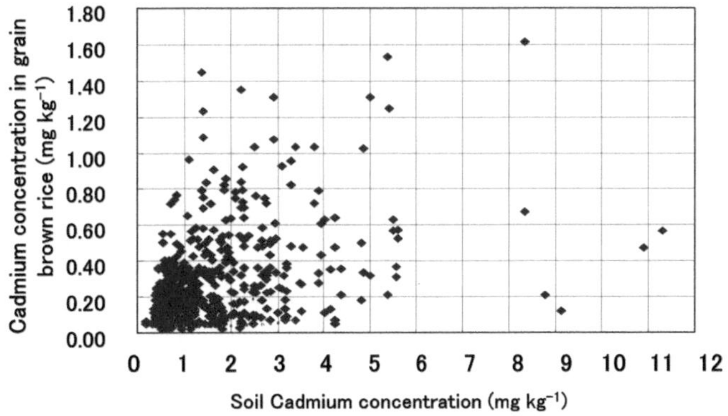

Fig. 6.25 Relationship between soil and Cadmium concentration in brown rice. *Source* Figure provided by Masashi Ito

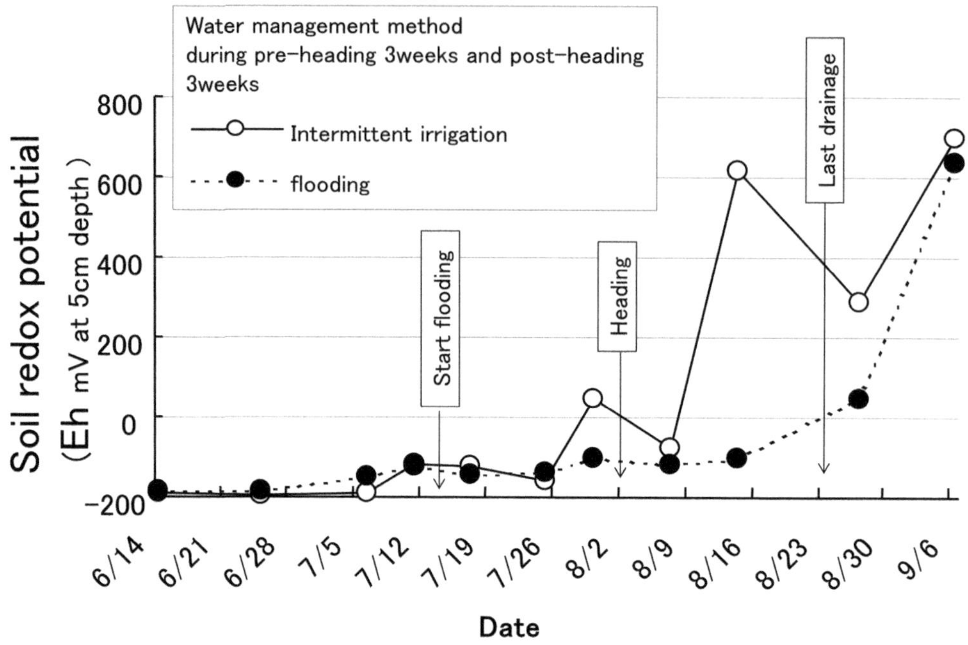

Fig. 6.26 Change in redox potential at 5 cm soil depth. *Source* Figure provided by Masashi Ito

Edited by Ministry of Agriculture, Forestry and Fisheries Agriculture,Forestry and Fisheries Research Council secretariat ,"Development of technology for suppression of Cadmium absorption by crops in arable soils", Agriculture,Forestry and Fisheries Research Council secretariat,2005,65p,(Kenkyuseika 434). Agriculture,Forestry and Fisheries Research Council

varies depending on the cultivation and management conditions, with different forms of Cd compounds having different rates of absorption into the rice. Studies have also revealed that rice absorbs Cd most abundantly during the booting stage and the milk-ripe stage (three weeks before and after the heading stage). The absorption suppression technology aims to make Cd in the soil unavailable during this peak absorption period, thereby suppressing Cd absorption into the rice. In 1978, researchers in Akita Prefecture developed a technique for suppressing Cd absorption through the controlled flooding of rice paddies. Controlled flooding reduces the oxidation-reduction potential of the soil, and reduced sulfide ions strongly bond to residual Cd compounds in the soil, making them insoluble and unabsorbable. The level of reduction is determined based on the oxidation-reduction potential (Eh), with the reference range being below −100 mV.

Figure 6.26 shows changes in the oxidation-reduction potential in a controlled flooding plot and intermittent flooding plot (three weeks before and after the heading stage). (conventional management). The oxidation-reduction potential fluctuated more at lower levels in the former plot than in the latter. As shown in Table 6.12, the Cd levels in brown rice were significantly suppressed in the controlled flooding plot.

Although the absorption suppression technique utilizing oxidation through controlled flooding is highly effective, when the water disappears from the paddy surface the oxidation-reduction potential immediately exceeds 500 mV (oxidized state), thus Cd form in paddy soil change easily absorption form into paddy rice.

In Akita Prefecture, this controlled flooding method has been implemented to suppress Cd absorption in areas around the already decontaminated zone, ensuring the production of safe rice containing Cd levels below the reference range. Rice can only be distributed after its safety is confirmed by Cd concentration examinations performed for every production batch.

Table 6.12 Cadmium contents by water management

Water management method during pre-heading 3 weeks and post-heading 3 weeks	Cadmium contents in grain brown rice (mg kg^{-1})
Flooding	0.09
Intermittent irrigation (normal cultivation)	0.23

Soil cadmium concentration 2.1 mg kg^{-1} by 0.1 M HCl extraction

Source Development of technology for suppression of cadmium absorption by crops in arable soils, Kenkyuseika 434(2005), Agriculture, Forestry and Fisheries Research Council secretariat

6.5.2 Methane Emission from Paddy Fields and Its Mitigation Strategy

The Tohoku region is a relatively cold temperate area in Northeast Japan, having a mean annual temperature of about 10 °C. Since fields are covered by snow during the long winter season in this region, rice is generally grown from early May to mid-October by single cropping (Fig. 6.27). For the single-cropping system, rice paddies were often left under aerobic (unflooded) conditions after harvest and during the whole winter fallow season. Promoting the aerobic degradation of rice straw in fields during the fallow season can decrease CH_4 emission during the next rice growth season. In previous studies, many strategies have been investigated for improving the aerobic degradation of rice straw during the rice off-season, such as placing the rice straw on the surface of the field, incorporating the rice straw into the soil top layer (0–5 cm), and adding N fertilizer to promote rice straw decomposition. However, the decomposition rate of rice straw has large and obvious variation and is affected by the cumulative temperature. Meteorological factors such as temperature, rainfall, and snow strongly affect rice straw decomposition during the fallow season, especially in high-latitude temperate regions. Lab incubation experiments have shown that modeling the aerobic decomposition of rice straw during the rice off-season was lower than about 15% when the daily temperature changed from −5 to 5 °C, which was used to model freeze-thaw cycles which occur during the winter season (Nakajima et al. 2016; Tang et al. 2016). Actual measurements carried out by Kumagai et al. (2000) and Eusufzai et al. (2011) showed that rice straw application increased CH_4 emissions by 2–5 times compared with treatment without rice straw in Yamagata and Iwate prefectures. Although the Tohoku region contains 25% of Japan's total area of rice paddies, it contributed 54% of Japan's total CH_4 emissions in 1990 based on a process-based biogeochemical model estimation (Hayano et al. 2013). The authors concluded that the retardation of the decomposition of organic matter in the fallow season due to the cool climate, and the relatively lower drainage, in this region were the reasons for the relatively higher CH_4 production in the rice-growing season. The mitigating effects of autumn shallow tillage (lower than 5–8 cm depth) with straw incorporation on the CH_4 emission during the following rice-growing season were evaluated in an Inceptisol paddy field in Yamagata City, Yamagata Prefecture and an Andisol paddy field in Morioka City, Iwate Prefecture by Shiono et al. (2016) and Nakajima et al. (2017), respectively. However, the results were inconsistent. Additionally, Shiono et al. (2016) showed that shallow tillage in autumn with the incorporation of rice straw into the soil in Yamagata City reduced CH_4 emissions; however, Nakajima et al. (2017) showed that CH_4 emissions were not different between conventional (15 cm depth) and shallow tillage (7 cm depth)

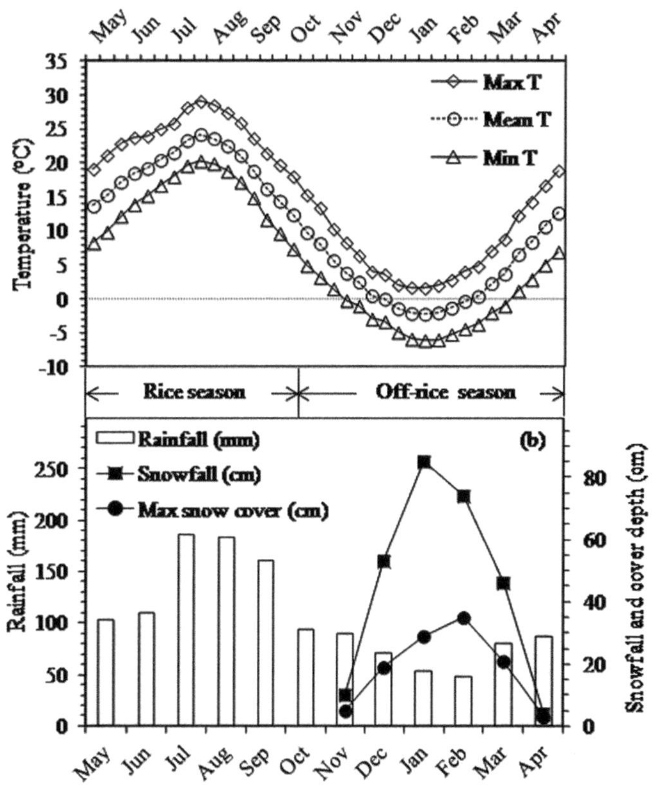

Fig. 6.27 The averages of maximum, mean, and minimum air temperatures (*T*) for every ten days from May to April (**a**); the monthly rainfall, snowfall, and maximum snow cover in the winter season (**b**) in Morioka, Tohoku region, Japan. Data are the average values for the period 1981–2010 and provided by the Japan Meteorological Agency. *Source* Figure provided by Weiguo Cheng and Miyuki Nakajima

Table 6.13 Generation amount of methane

Experimental year	Plot	Methane (CH_4) emission			
		(g CH_4 m^{-2})		(kg CO_2-eq m^{-2})	Control ratio
2011	Shallow tillage in autumn	49.9	a	1.40	80
	Control(tillage in spring)	62.6	a	1.75	100
2012	Shallow tillage in autumn	19.9	a	0.56	44
	Control(tillage in spring)	45.6	a	1.28	100
2013	Shallow tillage in autumn	85.6	a	2.40	64
	Rice straw without application	51.6	a	1.45	38
	Control(tillage in spring)	134.1	a	3.75	100
ANOVA					
	Treatment(T)	*			
	Year(Y)	**			
	(T) × (Y)	n.s			

$*p < 0.05$, $**p < 0.01$,n.s:not significant. Analysis of variance (ANOVA) test
CO_2-eq(kg m^{-2}) = methane emission(kg m^{-2}) × GWP
GWP(Global Warming Potential) of CH_4: 28 (IPCC 2013)

in Morioka City. This inconsistency could be due to the differences in environmental conditions (mostly temperature and moisture), as well as soil properties, between the two paddy sites. Compared with rice straw, rice straw compost clearly decreased CH_4 production and emission from a rice paddy in the Tohoku region (Kumagai et al. 2000; Cheng et al. 2016). However, composting rice straw requires much cost and labor. Developing a simpler method for composting rice straw is therefore desirable.

In Japan, the area of rice cultivation decreased from 3.3 million ha in the 1960s to 1.6 million ha in 2015, while rice consumption decreased from a peak of 118 kg per person per year in 1962 to 55 kg in 2015 (Cheng et al. 2018). The paddy–upland rotation cultivation system (see Sect. 6.4), which involves cycling between two crops, such as rice and soybean, for various years at a time, is adopted in many areas and is increasingly used in the Tohoku region. Since CH_4 emissions from rotated rice paddy fields have decreased remarkably in the Tohoku region, as confirmed by many studies (e.g. Eusufzai et al. 2010), this practice should be an effective strategy to decrease CH_4 emissions from Japanese rice paddy fields in the future. The water management of midseason drainage also is an effective strategy to decrease CH_4 emissions from rice paddies, as confirmed by Itoh et al. (2011) in the cities of Tsuruoka and Yamagata, Yamagata Prefecture.

6.5.3 Shallow Fall Tillage for Mitigating of Methane Emissions

Shallow fall tillage promotes the aerobic decomposition of rice straw while maintaining the workability of the soil. Furthermore, the production of organic acids, which is associated with the anaerobic decomposition of rice straw, decreases, because of the more aerobic conditions associated with fall-shallow tillage; this effect reduces the growth inhibition which is caused by the production of organic acid. Consequently, an increase in early rice growth is expected by fall-shallow tillage. To test these hypotheses, we investigated the influence of fall-shallow tillage in paddy fields in Yamagata Prefecture on methane generation and the subsequent growth of paddy rice crops. Trials were conducted in a field site (soil type: Gray Fluvic soil) in the Yamagata Integrated Agriculture Research Center from 2011 to 2013, using the "*Hae-nuki*" japonica rice cultivar.

The measured methane emissions varied depending on the year, being the lowest in 2012 and highest in 2013 (Table 6.13). In 2012, methane generation in the fall-shallow tillage treatment decreased by 44–80% compared with the control treatment (spring tillage treatment). In the no-straw treatment, the measured methane emission was 51.6 g m^{-2}, which was the lowest rate among the treatment plots set up that year.

Table 6.14 shows the growth and yield of paddy rice. With respect to plant height, no significant difference was observed between plants after the fall-shallow tillage treatment and after the spring tillage treatment. However, significant differences were found between these two treatments in terms of the number of tillers counted on June 20 and July 10: The number of tillers was significantly greater in the fall-shallow tillage plots than in the control plots (spring tillage treatment). In 2013, the no-straw treatment had the highest number of tillers, followed by the fall-shallow tillage treatment, with the number of tillers being lowest in the control treatment (rice straw application and spring tillage).

Table 6.14 Growth, yield, and quality of paddy rice

Experimental year	Plot	Plant length, culm length (cm)						
		June 20 (active-tillering stage)	July 10 (maximum tiller number stage)	September 10 (maturity stage)	June 20 (active-tillering stage)	July 10 (maximum tiller number stage)	September 10 (maturity stage)	Brown rice yield (Mg ha-1)
2011	Shallow tillage in autumn	39.2	72.9	82.5	578	549	467	6.78
	Control (tillage in spring)	36.4	69.0	82.1	440	499	451	6.51
2012	Shallow tillage in autumn	31.7	53.1	79.9	480	724	545	6.25
	Control (tillage in spring)	32.0	51.9	80.1	414	684	549	6.31
2013	Shallow tillage in autumn	36.2	58.5	78.2	437	529	434	6.78
	rice straw without application	38.5	61.4	81.1	485	588	472	6.95
	Control (tillage in spring)	35.5	54.9	75.3	357	447	385	6.26
ANOVA								
	Treatment (T)	n.s	n.s	n.s	*	**	n.s	n.s
	Year (Y)	**	*	**	*	**	**	n.s
	(T) × (Y)	n.s	n.s	n.s	n.s	n.s	n.s	n.s

*$p < 0.05$, **$p < 0.01$, n.s: not significant. Analysis of variance (ANOVA) test

Table 6.15 Decomposition and composition of rice straw

Season	Plot	Decomposition rate (%)		T–C	T–N	C/N
		Dry weight	T–C	(%)	(%)	
2012 October	Before installation	–	–	39.6	0.64	61.5
2013 April	Shallow tillage in autumn	100.0**	100.0*	37.4	0.77	48.6
	Control (tillage in spring)	100.0	100.0	37.8	0.78	48.7

* $p < 0.05$,**: $p < 0.01$. Analysis of variance (ANOVA) test

There was no significant difference in the number of panicles per plant and in brown rice yield among the different treatments.

In the Murayama area of Yamagata Prefecture, the crop index values relative to the average yield (crop index = 100) for the trials from 2011, 2012, and 2013 were 102, 101, and 105, respectively. For this reason, although the number of tillers increased in response to fall-shallow tillage, this was not reflected in an increase in the number of panicles per plant or the brown rice yield in the control plots (rice application and spring tillage), indicating that stem density in the control plots was already optimal for maximum yield. However, stressful weather conditions, such as a cold summer, could result in sub-optimal tillers populations in the control treatment. Furthermore, the increased tillers number achieved following shallow fall tillage could result in higher and more stable grain yields as a consequence of avoiding the growth suppression associated with spring tillage.

The rice straw decomposition rate in the fall-shallow tillage plots, when aerobic decomposition occurred, was higher than in the control plots (spring tillage treatment) when anaerobic decomposition occurred on the soil surface under snowfall and subsequent snowmelt (Table 6.15). The no-straw treatment in 2013 had the largest number of tillers of all three treatments trialed. This suggests that fall-shallow tillage promotes the aerobic decomposition of rice straw during the fallow period, decreasing the amount of methane and growth-inhibiting substances generated, which leads to a decrease in methane emissions and an increase in the number of tillers. Thus, applying the fall-shallow tillage technique after the spreading of rice straw on the land is expected to contribute to an improvement in the growth of rice plants and a reduction of methane emissions from paddy fields in cold climates where there is considerable snowfall. In Yamagata Prefecture, the fall tillage technology has been selected as an objective of the environmentally protective conservation-type agriculture; local officers are attempting to spread the use of this new tillage strategy to farmers, and farmers who adopt such technology are entitled to a direct payment grant.

6.6 Great East Japan Earthquake

6.6.1 Calcium Silicate Application in Tsunami-Affected Soils

The Great East Japan Earthquake and tsunami disaster of 11 March 2011 claimed many human lives. The tsunami inundated regions along the Pacific coast and caused severe damage to lowland farmlands. About 15,000 ha of paddy fields were damaged in Miyagi Prefecture, Northeast Japan. Most of the tsunami-affected farmland has since been desalinated by irrigation. Although water-soluble salts were effectively removed from the plow layer soils, some fields had a poor basic cation balance. Exchangeable calcium (Ca) ions in soils were replaced by sodium (Na) ions derived from seawater, and Ca ions leached downward during the desalination process. Some Na ions remained at the cation exchange sites, and exchangeable Ca contents were reduced.

In soils with a high exchangeable Na concentration, rice (*Oryza sativa* L.) sometimes shows poor growth due to the excess uptake of Na (Gong et al. 2006; Matoh et al. 1986). To mitigate Na toxicity and restore soil productivity, it is essential to optimize the basic cation balance in soils that have been desalinated. In soils with an exchangeable sodium percentage (ESP) of above 20%, rice yield may begin to decrease (Dobermann and Fairhurst 2000). Rice yield was found to decrease by half in soil with an ESP of over 80% (Gupta and Sharma 1990). When wheat takes up excessive Na, the uptake of potassium (K) and Ca is suppressed and plant growth is limited (Kinraide 1999). In this study, we examined the effectiveness of applying calcium silicate materials (fertilizers made from steelmaking slag) to solve the basic cation balance problem that leads to Na toxicity in rice.

Figures 6.28 and 6.29 show the temporal changes in the Ca and Na concentrations in soil solutions collected from the plow layer in the four treatments. In the control treatment, the Na concentrations were about 2–6 times the Ca concentrations. The addition of slag or gypsum increased the Ca concentration; the Ca concentration was highest for the

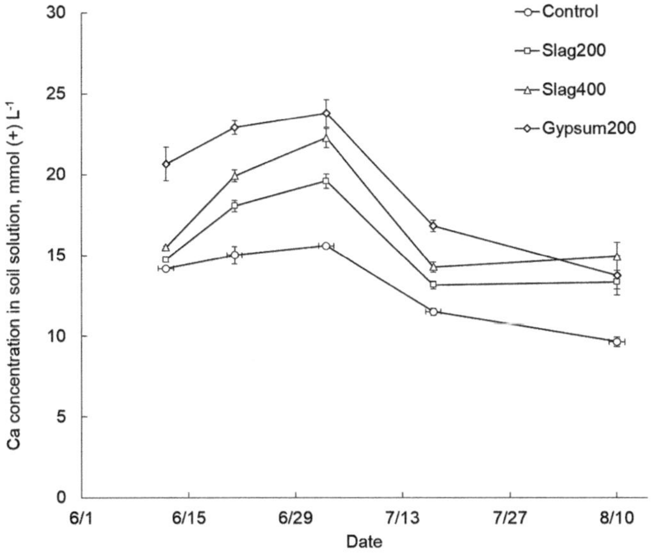

Fig. 6.28 Temporal pattern of calcium concentration in the soil solution. *Source* Figure provided by Toyoaki Ito, Hisashi Nasukawa, Toru Uno, Ryousuke Tajima, and Masanori Saito

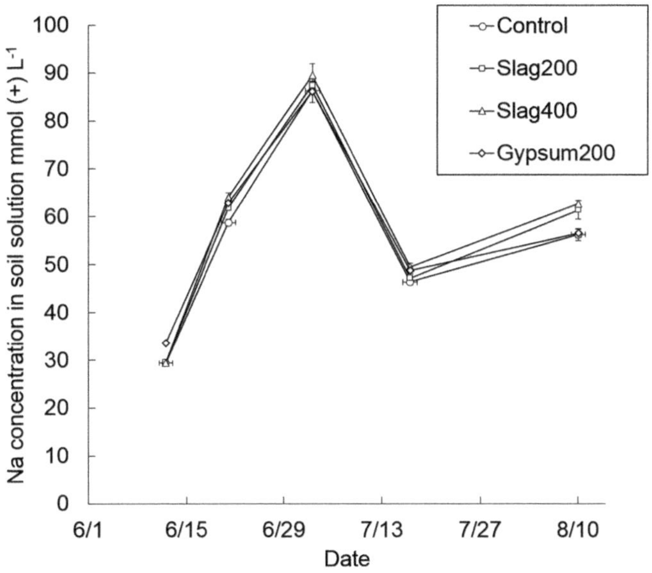

Fig. 6.29 Temporal pattern of sodium concentration in the soil solution. *Source* Figure provided by Toyoaki Ito, Hisashi Nasukawa, Toru Uno, Ryousuke Tajima, and Masanori Saito

gypsum treatment, even though the 200 g m^{-2} slag treatment had the same Ca input rate. This finding indicates that the steelmaking slag dissolves more slowly than gypsum. On the other hand, there were no differences in Na concentrations among the four treatments. The increase in the Na concentration in the soil solution from 12 June to 3 July is considered to be due to the progression of ferrous iron formation under submerged conditions and ion exchange between ferrous iron in solution and Na ions at the soil exchange sites. These results indicate that the desalinated soil supplies a large amount of Na to rice plants and that the application of slag and gypsum can increase plant-available Ca in desalinated soils.

Relative to the control treatment, the slag treatments with application rates of 200 and 400 g m^{-2} increased brown rice yields by 10% and 21%, respectively, and the latter significantly (Fig. 6.30). Gypsum treatment had little effect on the rice yield, despite having a greater effect on increasing the Ca and K contents and decreasing the Na content of rice straw than did the 200 g m^{-2} slag treatment (Figs. 6.31–6.33). This result suggests that hydrogen sulfide was generated by the reduction of sulfate derived from gypsum and inhibited rice root elongation and/or nutrient uptake and reduced rice yield. The application of slag or gypsum increased the Ca and K concentrations in rice straw by 9–20% and 9–16%, respectively, relative to the control (Figs. 6.31, 6.33), but decreased the Na concentrations by 11–16% (Fig. 6.32). The slag treatments significantly enriched the silicate contents of rice straw at maturity (Fig. 6.34). These findings suggest that Ca-containing

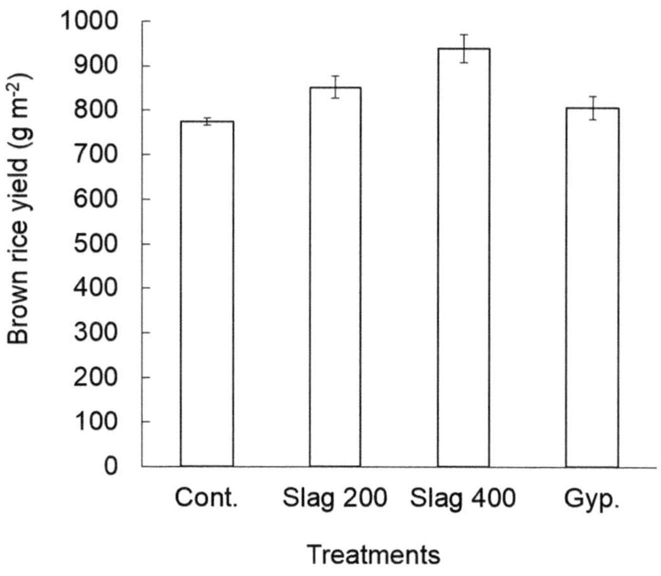

Fig. 6.30 Mean (±SE) brown rice yield in each treatment Bars with the same letters are not significantly different ($P < 0.05$) according to the Tukey–Kramer test. *Source* Figure provided by Toyoaki Ito, Hisashi Nasukawa, Toru Uno, Ryousuke Tajima, and Masanori Saito

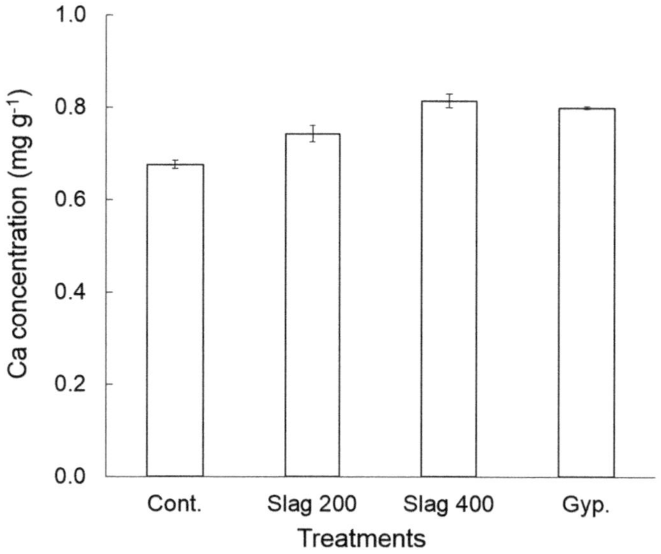

Fig. 6.31 Mean (±SE) calcium concentration in rice straw in each treatment Bars with the same letters are not significantly different ($P < 0.05$) according to the Tukey–Kramer test. *Source* Figure provided by Toyoaki Ito, Hisashi Nasukawa, Toru Uno, Ryousuke Tajima, and Masanori Saito

materials such as steelmaking slag and gypsum suppress Na uptake and improve the K nutrient condition of rice plants by promoting Ca absorption.

It is well known that the supplementation of Ca reduces Na absorption and enhances K absorption (Khan et al. 1992; Song et al. 2006). Additionally, silicate is taken up by rice plants, which increases photosynthetic capacity, resistance to insects, and salt resistance (Ma 2004). Silicate is deposited in the leaves and restricts moisture loss from the leaf cuticle, thus improving water use efficiency. Furthermore, the enhanced physical strength of leaves and stems after silicate accumulation improves the plant's ability to remain upright, as well as its light absorption. Together, these physical and physiological changes result in enhanced photosynthetic capacity. Moreover, silicate application reduces Na uptake in rice and thus mitigates Na toxicity (Gong et al. 2006; Matoh et al. 1986). The application of slag and gypsum did not accelerate Na leaching from plow layer soils, but did increase the content of plant-available Ca (exchangeable Ca).

Our findings indicate that fertilizer made of steelmaking slag is more effective than gypsum in restoring the productivity of desalinated tsunami-affected soils containing high Na concentrations. Slag can supply Ca and Si to rice plants and is effective in alleviating Na toxicity without the risk of increasing hydrogen sulfide injury to rice roots.

6.6.2 Green Manure Application in Tsunami-Affected Soils

The tsunami resulting from the Great East Japan Earthquake (11 March 2011) deposited seawater and sand into fields in

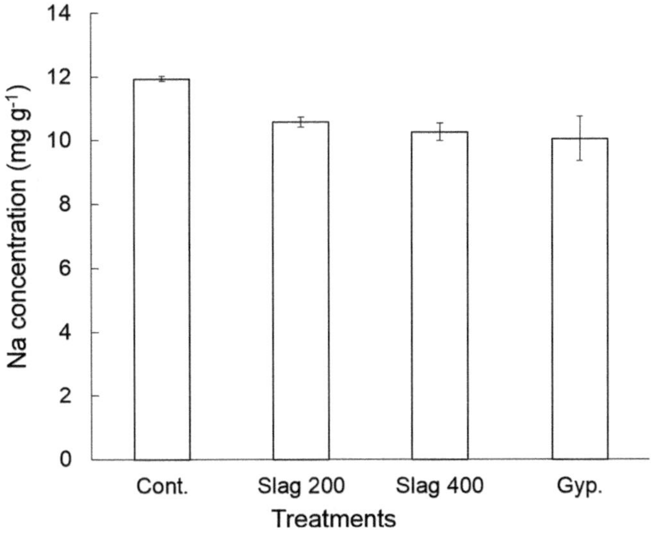

Fig. 6.32 Mean (±SE) sodium concentration in rice straw in each treatment Bars with the same letters are not significantly different ($P < 0.05$) according to the Tukey–Kramer test. *Source* Figure provided by Toyoaki Ito, Hisashi Nasukawa, Toru Uno, Ryousuke Tajima, and Masanori Saito

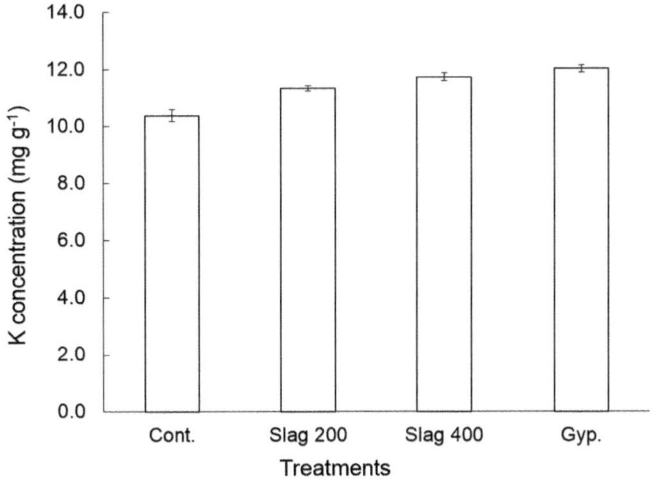

Fig. 6.33 Mean (±SE) potassium concentration in rice straw in each treatment Bars with the same letters are not significantly different ($P < 0.05$) according to the Tukey–Kramer test. *Source* Figure provided by Toyoaki Ito, Hisashi Nasukawa, Toru Uno, Ryousuke Tajima, and Masanori Saito

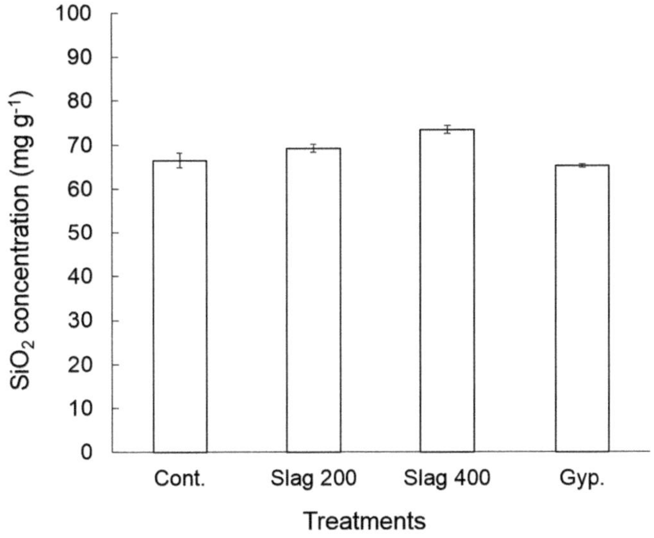

Fig. 6.34 Mean (±SE) silicate concentration in rice straw in each treatment Bars with the same letters are not significantly different ($P < 0.05$) according to the Tukey–Kramer test. *Source* Figure provided by Toyoaki Ito, Hisashi Nasukawa, Toru Uno, Ryousuke Tajima, and Masanori Saito

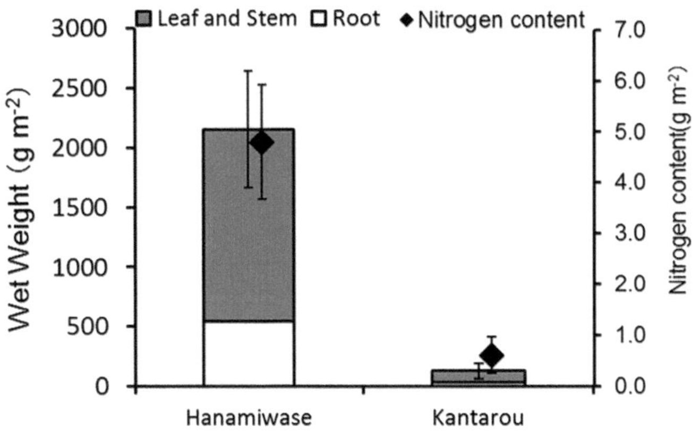

Fig. 6.35 The wet weight of green manure and the nitrogen content in green manure.(Abe and Honda 2015). The sampling of green manure was conducted on 23 April just before the green manure was plowed back. The Hanamiwase and the Kantarou were taken from 0.054 m^2 and 0.25 m^2, respectively in each field. The root samples were taken from 0 to 15 cm layer in the soil of each plot. The error bar showed a standard deviation (n = 3). The statistical analysis was conducted by Tukey test between the Hanamiwase and the Kantarou about nitrogen content. *Source* Figure provided by Tomonori Abe

coastal areas of Miyagi Prefecture. In some paddy fields, soil fertility was decreased after the removal of salt and debris, and the subsequent rice growth was poor. Generally, in fields where the soil fertility has declined, CMC is applied to increase soil fertility. However, in coastal areas from Ishinomaki City to Yamamoto Town, there are few livestock farmers, and farmers often do not have the machinery necessary for manure spraying. Therefore, in these areas, another method to increase soil fertility is required instead of cattle manure application.

We examined the improvement of soil fertility that can be achieved by introducing green manure to paddy rice cultivation in low-fertility paddy fields that were damaged by the tsunami (Abe and Honda 2015). The test field was located in Higashi-Matsushima City and had an area of about 1 ha. Green manure was sown in October 2012, in the second year after the tsunami disaster, after cropping paddy rice. The green manure used was Italian ryegrass ("Hanamiwase" variety) and hairy vetch ("Kantarou" variety). A green manure-free seeded area was also set up. The respective plots names were "HANAMIWASE", "KANTAROU", and "No Seeding". In the test field, due to the ground subsidence caused by the earthquake, there is some agricultural water reflux from the drainage to the field from late April when the agricultural water flows into the irrigation canal. For this reason, the growing period of green manure was limited. The ground part weight of the ryegrass just before plowing-in was secured adequate growth amount of 1620 g m^{-2} with fast-growing early, whereas the slowly growing hairy vetch had a very small ground part weight of 90 g m^{-2} (Fig. 6.35). A large amount of CH_4 or H_2S gas was generated during rice cultivation in June in the HANAMIWASE plot. It is considered that the soil abnormally reduced in this plot. As a result (Fig. 6.36), the initial growth of the paddy rice in the ryegrass-containing plot was temporarily suppressed; there was a tendency for the number of tillers to be smaller than that of the other plots. However, the number of stems was larger than in the other plots (KANTAROU and No Seeding) after the panicle initiation stage, when the gas was generated in low amounts. The amount of nitrogen mineralization of green manure was measured by an indoor experiment. The amount of nitrogen mineralization in the first 4 weeks, in the KANTAROU plot, was more than 4–8 weeks. The amount of mineralization in the HANAMIWASE plot continued to increase for eight weeks. It was inferred that continuous nitrogen mineralization improved the late growth of rice in the HANAMIWASE plot (Fig. 6.37).

The number of rice grains per square meter in the KANTAROU plot was equal to that in the No Seeding plot. There was also no significant difference in yield between these two plots (Table 6.16). However, the number of ears

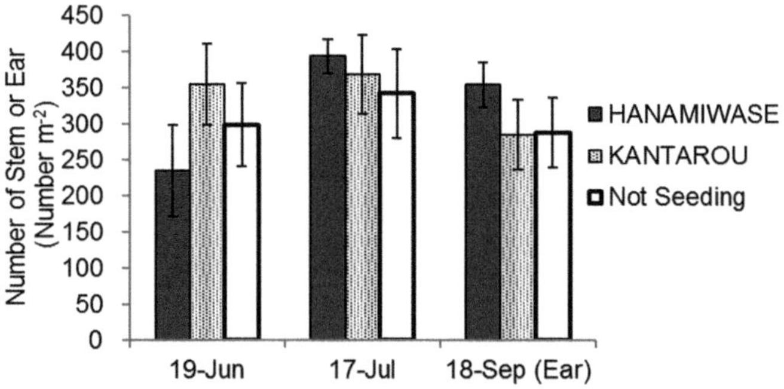

Fig. 6.36 Comparison of number of stems or number of ears in paddy rice (Abe and Honda 2015) The error bar showed standard deviation (n = 3). By Tukey's method, a significant difference test was conducted for each time period. Different alphabets indicate that there is a significant difference at the 5% risk level. *Source* Figure provided by Tomonori Abe

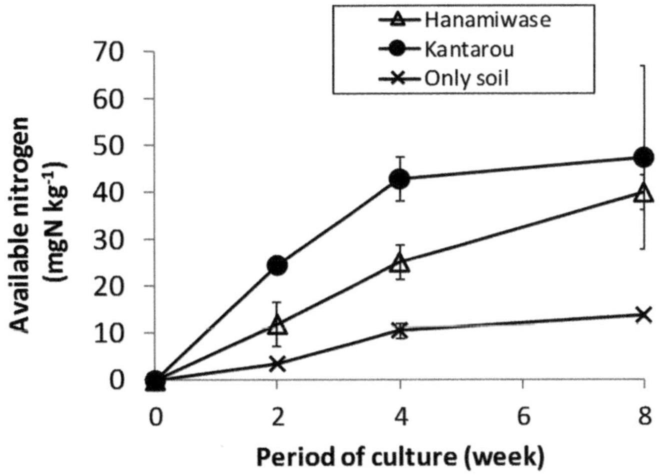

Fig. 6.37 Nitrogen mineralization pattern of soil with or without green manure (Abe and Honda 2015) Fresh green manure, 3 g of the aboveground portion and about 0.1 g of the underground portion, was added to raw soil equivalent to 15 g of dry soil. And they were cultivated in a flooded state at 30 °C. The error bar showed a standard deviation ($n = 3$). *Source* Figure provided by Tomonori Abe

per square meter, and the number of grains per ear were higher in the HANAMIWASE plot than in the No Seeding plot. Furthermore, the number of grains per square meter was 1.5 times larger in the HANAMIWASE plot than in the No Seeding plot. The yield in the HANAMIWASE plot was 477 g m^{-2}, which was significantly larger than that in the No Seeding plot, and the whole grain rate was 88.2%, which was higher than that of the other plots.

However, no significant difference was observed in the amount of soil available nitrogen of the plots after cultivation (data collected in April 2014). Although only one green manure cultivation and plowing alone had an effect on rice cultivation immediately afterward, it was inferred that the amount of available nitrogen from green manure was small in the second year. Therefore, in order to raise the level of available nitrogen in the soil, it was considered necessary to consecutively apply at least green manure.

6.6.3 Tsunami-Deposited Sediment

Following the disaster, tsunami-affected fields could not be used for farming because of the deposition of salts from seawater, the deposition of sediment from downstream sources, the incorporation of rubble and gravel, and the destruction of inflow and drainage facilities. Planting began in May 2011 in a limited number of fields that had only limited seawater intrusion and sediment deposition. This planting was made possible by the removal of surface sediment and soil and desalination by irrigation and rain. In most fields, however, heavy machinery was required to remove saline- and rubble-mixed deposited sediment and the repair and refurbishment of field sections and waterways was critical. Surface soils were transported from other regions and applied to farmlands; however, securing appropriate soil dressing was difficult, as these soils were mixed with rubble with a size greater than 10 cm, which made plowing immediately after treatment challenging. At the same time, in Rikuzentakata—the area most severely affected by the tsunami in Iwate Prefecture—the processing of rubble-mixed sediment deposited on disaster-affected farmlands was another challenging issue.

Therefore, there was a request from the local Farmland Maintenance Department to determine whether soil extracted from rubble-mixed sediment could be used as surface soil for restoring paddy fields. Originating mostly from farmland regions, this tsunami-deposited sediment was confirmed not to contain pollutants such as heavy metals. We conducted tests as part of the efforts to restore disaster-affected farmland.

Using soil obtained from the local rubble-processing facility for deposited sediment, the following procedures were conducted: (1) determination of soil chemical properties (e.g., pH, electric conductivity (EC), total carbon (T–C), total nitrogen (T–N), and exchangeable bases) as measured using conventional methods; (2) assessment of the residual amount of rubble in processed sediment using sieve separation (5 mm mesh) and weighing the soil and rubble separately; and (3) field cultivation of rice (frame: 1.2 m × 0.8 m × 15 cm surface soil) in a paddy field by replacing a portion of the original surface soil with rubble-extracted soil and also a rubble-extracted soil/mountain soil mixture.

The method used in the processing facility for sorting rubble from sediment that was collected from disaster-affected fields comprised sifting the load through a

Table 6.16 The growth on maturing stage and yield

Plot		Ear length (cm)	Number of ear (m^{-2})	Number of grains per ear	Number of grains (m^{-2})	Yield ($g\ m^{-2}$)	Thousand-kernel-weight (g)	Percentage of ripened grains (%)	Percentage of Whole grain (%)	Percentage of protein (%)
HANAMIWASE	72.2	17.5	355	63.4	22.6	477*	21.9	96.4	88.2	5.1
KANTAROU	68.3	17.3	286	52.9	15.2	319	22.0	94.8	84.2	5.0
Not Seeding	67.3	16.3	288	52.2	15.1	306	21.1	96.0	74.3	4.6

Table 6.17 Characteristic of the rubble-extracted soil

Sample	pH (H_2O)	EC dS m^{-1}	CEC (cmol(+) kg^{-1})	Exchangeable base mg kg^{-1}				Phosphate absorption coefficient	Available phosphate mg kg^{-1}	Free ion oxide g kg^{-1}	T–C g kg^{-1}	T–N g kg^{-1}
				Calcium (CaO)	Magnesium (MgO)	Potassium (K_2O)	Sodium (Na_2O)					
Extracted soil (dry classification*)	7.5	0.42	14.3	852	35.6	11.0	3.0	320	355	12.0	26.4	1.44
Extracted soil (wet classification**)	8.0	0.32	9.2	853	23.6	8.6	2.6	250	206	8.6	15.1	0.64
【control】 Kitakami paddy soil	5.4	0.06	21.1	114	27.7	19.8	1.0	1710	74	32.6	34.2	2.11
Reference 1: mountain soil A (for mixing)	6.7	0.15	14.7	280	58.4	5.6	1.9	520	71	14.4	7.4	0.45
Reference 2: mountain soil B (for mixing)	6.3	0.02	7.8	39	5.1	2.0	1.5	160	116	6.5	2.5	0.12

*Dry classification
**Wet classification
(Reference) Two kinds of mountain climate mixed in sorting soil

Fig. 6.38 Amount of residual rubble in rubble sorting soil Photo on the right (Residual rubble in 5 kg of rubble sorting soil) Upper left: glass, Upper right: wood, Bottom left: Shell, Bottom right: plastic. *Source* Figure provided by Teruo Shima

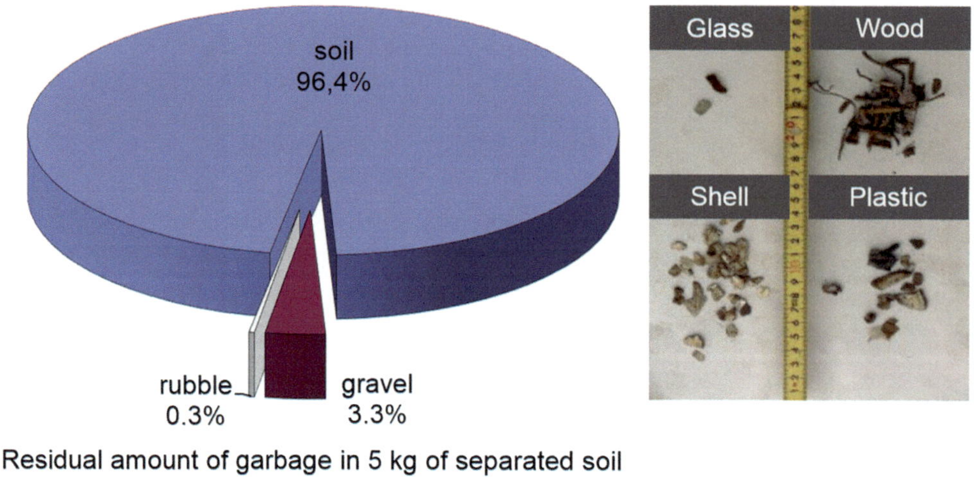

Table 6.18 Growth of rice plants and brown rice yield in frame test using sorted soil

Plot no	Experiment plot mixing rate	Soil pH	Plant length (cm)		Tiller		number (m^2)	Culm length (cm)	Panicle length (cm)	Panicle number (m^{-2})
6–21	7–4	6–21					Brown rice yield gm^{-2}	7–4		
1	Separated soil: mountain soil *A* = 2:1	7.3	42.3	52.8	475	630	85.5	19.5	456	755
2	Separated soil: mountain soil *A* = 1:1	7.4	41.3	51.5	490	647	85.7	19.4	470	750
3	Separated soil: mountain soil *A* = 1:2	7.1	41.8	51.1	418	612	85.3	19.3	462	746
4	Separated soil	7.5	43.3	51.7	414	588	85.0	19.7	429	726
5	【control】 kitakami top soil	5.4	38.7	49.4	278	466	82.1	19.5	383	689

series of sieves (dry classification). A second method, whereby rubble was separated by adding water to loosen sediment clods (wet classification) was also planned, but for the field cultivation test we used only soil that was sorted by dry classification.

The results indicated that the rubble-extracted soil had a higher exchangeable lime content (pH = 7.5–8.0) than the paddy field soil that is typically found in Iwate Prefecture, despite the amount of exchangeable sodium originating from seawater being only slightly higher than typical. Furthermore, the rubble-extracted soil had a sufficient exchangeable magnesium and available phosphate content, and higher total carbon and total nitrogen contents than the mountain soil, which presumably contains both carbon and nitrogen. These properties indicated that the rubble-extracted soil was appropriate for rice cultivation in paddies. The exchangeable potassium content was somewhat low, and electrical conductivity was slightly elevated, but not at levels that were deemed to require desalination (Table 6.17).

In the residual rubble of sorted sediment by dry classification, fragments of glass, wood, plastic, and the like were occasionally observed. However, by separating them out, these items were considered not to be detrimental for agricultural operations (Fig. 6.38).

In cultivation tests, when rubble-extracted soil was used as the agricultural surface soil, no abnormalities in rice growth were observed; both growth and yield were comparable to or higher than that of the control (Table 6.18). To adjust the high pH of the rubble-extracted soil and to ensure sufficient soil dressing, mixed rubble-extracted/mountain soils were tested. The mixed soil also provided similar growth and yield to those of the rubble-extracted soil (Table 6.18).

Although the data is not shown, in pot experiments, rice growth was higher in soil mixed with rubble-extracted soil than in mountain soil. These results indicate that rubble-extracted soil can be employed as surface soil to restore paddy fields, as was done in some regions.

In 2016, rice growth and yield in the fields that used rubble-extracted soil as soil dressing were higher than in the fields containing only mountain soil. We are yet to confirm whether or not growth and yield are influenced by soil type after the second year of cultivation. Weeds that were not present prior to the tsunami disaster now occur extensively. It, therefore, seems that in addition to growth yield, other factors need to be investigated.

Although a planting period of several years has passed since the restoration of disaster-affected farmlands, various issues persist. We believe that continuous support for the farmers is critical.

6.6.4 Decontamination of Radioactive Cesium

As early as 2011, the Nuclear Emergency Response Headquarters announced their decontamination methods for agricultural fields, based on the radioactivity of the soil and the purpose of the field. The purposes of decontamination are (1) to reduce the radioactivity of agricultural products; and (2) to reduce the external radiation exposure of farmers who are living in the area. Ever since the accident at the Fukushima Daiichi Nuclear Power Plant (FDNPP) operated by the Tokyo Electric Power Company (TEPCO), cultivation has not been allowed when the soil radioactivity is high. Soil radioactivity of 5000 Bq kg^{-1} has been considered to be threshold for the cultivation of rice based on a previous observation of the transfer of radiocesium from soil to plants after global fallout since 1960, which showed that the transfer factor (radiocesium content of brown rice/soil) was slightly higher than 0.1 at maximum. Notably, some fields had transfer factor values that were higher than 0.1 and exceeded the provisional regulation value for radioactivity (500 Bq kg^{-1}). The reason for this result is that the potassium level in the soil plays a critical role in regulating the transfer factor. An agricultural field recovery has been performed through a combination of decontamination and potassium application.

After the topsoil was removed, non-contaminated soil was added. Because the required amount of soil was too large, the mountainous soil around the area was collected and used. This soil was primarily weathered granite and/or soil containing a large amount of sand, and the fertility was therefore poor. Even after decontamination, agricultural activities have been very limited, and thus the manure supply from animal husbandry is non-existent. Several methods were examined to improve and maintain the soil fertility, as shown below.

(1) **Physical decontamination methods**

Three methods to decontaminate and/or decrease the radioactivity of topsoil were developed in 2011, and are summarized in Table 6.19. Topsoil removal without the use of hardener is widely applied to paddy fields and upland crop fields. If the soil contamination level was not high, reverse tillage was performed over a large pasture field area. Topsoil removal with the use of hardener has not been performed, since it requires extra labor. Stirring cleaning methods were not used, as their efficiency is largely dependent on the clay content of the soil, and such methods are therefore applicable only to paddy fields with sufficient available water.

Although topsoil removal effectively removes contaminated soil from the field, it also removes the most fertile soil that has accumulated in the field. Based on a survey of the applied dressed soil (Fig. 6.39), the fertility of the dressed soils is very low, especially in terms of total nitrogen (<0.5%) and total carbon (<0.5%) in four different areas (Yoshino et al. 2015). For this reason, it is recommended to apply a sufficient amount of chemical fertilizer, zeolite, and/or manure to the field before cultivation. Decontamination does not lead to the complete removal of contaminated soil; the field is not labeled, and some (although a very limited amount) of the radioactive materials move downwards from the surface. As a result, it is highly necessary to continue using countermeasures to mitigate the transfer of radiocesium from soil to plants until the soil radioactivity decreases sufficiently, by applying an adequate amount of potassium fertilizer and/or potassium-containing resources to the soil.

(2) **Soil fertility recovery using green manure crops in farmland after decontamination**

We examined the soil fertility recovery and radiocesium uptake effects in the soil after cropping by plowing green manure into a test field. The field was located in Futaba

Table 6.19 Comparison of decontamination methods

Decontamination method	Characteristics	Radioactivity (Bq/kg)		Dose rate (μS/h)		Amount of waste (mainly soil) (t/ha)
		Before	After	Before	After	
Topsoil removal	Remove the few cm top soil (4 cm)	10,370	2599	7.14	3.39	400
	Remove the topsoil using hardener (3 cm)	9616	1721	7.76	3.57	300
	Remove the topsoil with grass (3 cm)	13,600	327	–	–	400
Stirring cleaning method	Remove clay fraction selectively	16,052	9859	7.50	6.48	12–15
Reverse tillage	Turn over more than 30 cm	–	–	0.66	0.30	0

Created by the author based on the trial in Iitate village in 2011 (Ministry of Agriculture, Forestry and Fisheries 2011)

Fig. 6.39 Decontaminated paddy field with dressed soil taken from a nearby mountain. *Source* Figure provided by T. Saito, S. Fujimura, H. Matsunami, T. Hirayama, K. Kubo, T. Ota, T. Shinano

Table 6.20 Physicochemical properties and ^{137}Cs concentrations in the experimental field soil (0–15 cm)

pH (H_2O)	EC	T–C	T–N	Av–P	Ex–Ca	Ex–Mg	Ex–K	CEC	Cs-137
	mS/cm	(g kg^{-1})	(g kg^{-1})	(mg kg^{-1})	(mg kg^{-1} DW)			(cmol$_c$ kg^{-1})	(Bq kg^{-1} DW)
6.1	0.05	13	1.1	237	1946	342	140	9.0	713

Part of the data is based on Farming resumption demonstrated technical information (Fukushima Agricultural Technology Centre 2019)

town, Fukushima Prefecture, which was designated as an area in preparation for the lifting of the evacuation order in May 2013. In the test field, the decontamination work to remove the top 5 cm of radiocesium-contaminated surface soil, which was topdressed with mountain sand, involved plowing to a depth of approximately 15 cm in April 2016. We set up three plots, each of which was 10 m in length and 7 m in width. At first, *Sesbani*, *Crotarralia*, and buckwheat were sown in each plot, respectively, in June 2016. After sampling these green manures, we plowed them into the soil in September 2016. Then, Italian ryegrass was sown in all of the plots in October 2016. We sampled the Italian ryegrass in April 2017. The chemical properties and ^{137}Cs concentrations in the experimental field soil are shown in Table 6.20. The ^{137}Cs radio activities in the soils of this field ranged from 490–1170 Bq kg^{-1} dry weight after decontamination.

The aboveground dry weights of *Sesbania*, *Crotarralia*, and buckwheat were 5.2, 2.2, and 1.1 t ha^{-1} (Table 6.21). It is clear that *Sesbania* cultivation is suitable in farmland with poor drainage. There was no difference in the total carbon (T–C) between these green manures. The total nitrogen (T–N) contents of *Sesbania* and *Crotarralia* were higher than those of buckwheat. In *Sesbania*, the amounts of T–C, and T–N that were plowed into the soil were higher than for other green manure crops. However, the plowing amount was not influenced by the T–C and T–N in the soil.

The ^{137}Cs radioactivities of *Sesbania*, *Crotarralia*, and buckwheat were 88, 369, and 121 Bq kg^{-1} dry weight (Table 6.21). The ^{137}Cs concentration of *Crotarralia* was higher than those of *Sesbania* and buckwheat. For *Sesbania* and *Crotarralia*, the amount of ^{137}Cs that was plowed into the soil was higher than it was for buckwheat.

The dry matter weights of Italian ryegrass in the *Sesbania* and *Crotalaria* plots were higher than those in other plots (Table 6.22). It was thus clear that the effect of soil fertility recovery by plowing green manure crops into the soil is high in farmland following decontamination. The ^{137}Cs concentration of the Italian ryegrass in the *Sesbania* plot was higher than that in the other plots. However, plowing green manure into the soil had little effect on the ^{137}Cs uptake since the ^{137}Cs concentration of various green manure crops is low.

Seven years after the FDNPP accident, a large area has been decontaminated and agriculture has been restarted in many places. As previously mentioned, "decontamination" does not indicate the complete removal of radiocesium from the soil; it is still necessary to continue using countermeasures to mitigate the transfer of radiocesium from soil to plants through additional applications of potassium fertilizer.

Table 6.21 Aboveground plant dry weight, concentration and the plowing amount of various green manure crops

Experimental plot	Aboveground part dry weight	T–C		T–N		^{137}Cs	
	(t ha^{-1})	(%)	(t ha^{-1})	(%)	(kg ha^{-1})	(B q kg^{-1})	(kBq ha^{-1})
Sesbania	5.2	44.4	229	2.1	107	88	467
Crotarralia	2.2	42.5	95.2	1.4	33	369	841
Buck wheat	1.1	44.3	49.6	0.4	3.4	121	140

Part of the data is based on Farming resumption demonstrated technical information (Fukushima Agricultural Technology Centre 2019)

Table 6.22 Above-ground part dry weights and ^{137}Cs concentration of Italian ryegrass

Experimental plot (Previous crops)	Above-ground part dry weight (t ha^{-1})	^{137}Cs (Bq kg^{-1})
Sesbania	4.6	71
Crotarralia	2.4	36
Buckwheat	1.1	29
No cultivation	0.8	24

Part of the data is based on Farming resumption demonstrated technical information (Fukushima Agricultural Technology Centre 2019)

However, more than 2000 ha of agricultural fields remain untouched in a difficult return zone. For further decontamination and subsequent agricultural usage, it is important to investigate the behavior of radiocesium in these areas after 7 years.

6.6.5 Potassium Fertilizer Application for Mitigation of Crop Radiocesium Uptake

Because of their long half-lives, there is concern that radiocesium isotopes will remain on the surface of agricultural land and persist for a long time. Therefore, from March 2011 we started monitoring the radiocesium in the soil and agricultural products that have been collected from agricultural land in Fukushima Prefecture, and we have investigated the distribution of radiocesium in farmland. Based on these data, the Nuclear Emergency Response Headquarters has indicated that there are rice planting areas in all the regions of the prefecture, except in the 20-km exclusion zone and the deliberate evacuation zone (DEZ). However, the brown rice produced in some areas of Fukushima Prefecture exceeded the provisional regulation radioactivity level for agricultural crops at that time (>500 Bq kg^{-1}). We investigated the radiocesium content of soil and brown rice and the exchangeable potassium (Ex-K) content of soil. There was a high correlation between the radiocesium in the brown rice and the Ex-K in the soil (Fig. 6.40), while no correlation was observed between the radiocesium in the soil and the radiocesium in the brown rice from these fields (Fig. 6.41).

(1) **Effect of potassium fertilizer application timing on the root uptake of radiocesium in brown rice**

We investigated the optimum timing and amount of K application needed to reduce the ^{134}Cs and ^{137}Cs uptake by brown rice. A field was contaminated with radiocesium; the brown rice produced in this field in 2011 exceeded the provisional regulation value for radiocesium and radiostrontium at that time (500 Bq kg^{-1}), and the planting of rice in this area was restricted in 2012.

The measured ^{137}Cs radioactivities in brown rice were 32 Bq kg^{-1} without K application, 5.0 Bq kg^{-1} with 8.0 g m^{-2} of basal K, 15 Bq kg^{-1} with K applied at 50 days after transplanting, and 36 Bq kg^{-1} with K applied at 80 days (Fig. 6.42). Therefore, the basal application of K fertilizer decreased the ^{137}Cs uptake the most. The ^{137}Cs radioactivity of brown rice was 32 Bq kg^{-1} without K application, 4.0 Bq kg^{-1} with 8.0 g m^{-2} of K, and 2.0 Bq kg^{-1} with 16 g m^{-2} of K (Fig. 6.43). The application of K at a level above 16 g m^{-2} had no further effect. Thus, increasing the K rate to 16 g m^{-2} decreased the ^{137}Cs concentration in brown rice. We focused on exchangeable radiocesium as a proxy for available radiocesium because there was no correlation between total radiocesium in paddy soil and that in brown rice (Saito et al. 2012, 2015) (Table 6.23). Heavy applications of K decreased the mean radioactivity of exchangeable ^{137}Cs in the soil from 234 to 25 Bq kg^{-1} dry weight. Hence, heavy applications of K fertilizer during the early growing period could decrease the ^{137}Cs uptake by rice plants and the concentration of exchangeable ^{137}Cs in the soil.

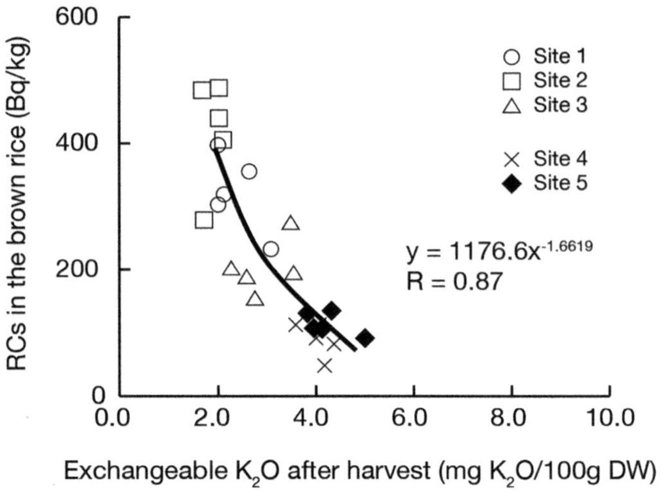

Fig. 6.40 Relationship between radioactive cesium concentration in rice and exchangeable potassium in soil. Saito et al. (2012)

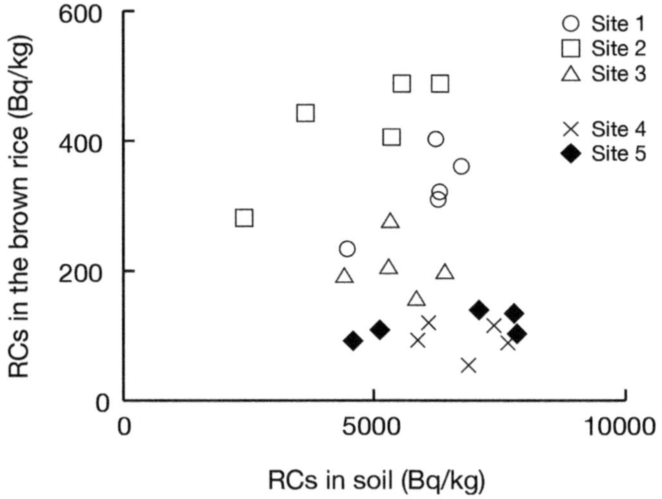

Fig. 6.41 Relationship between radioactive cesium concentration in rice and soil. Saito et al. (2012)

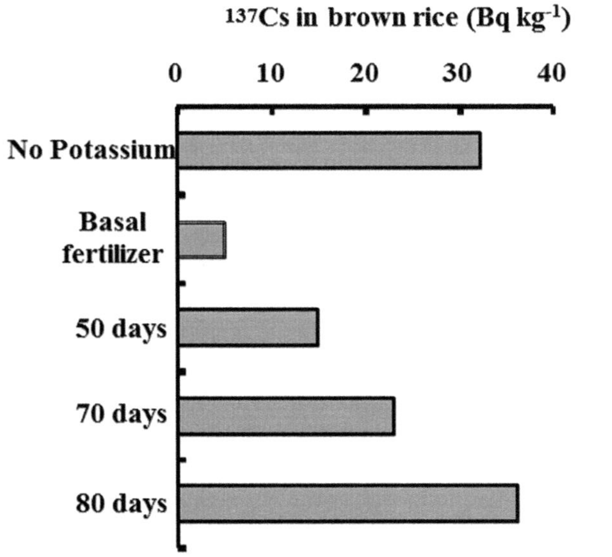

Fig. 6.42 Concentrations of ^{137}Cs in brown rice by timing of K fertilizer application. Saito et al. (2015)

(2) **Effect of rice straw application for potassium supply**

It is also important to consider the balance of the input and output of K throughout the crop growth period because rice straw contains approximately 2% K (weight basis), and incorporating rice straw is one of the sources of K for rice grown in paddy fields. Approximately 75% of the rice straw produced in Fukushima Prefecture is plowed into the soil before the next cultivation season, and other uses include compost, mulch, forage, and bedding (Fukushima Prefecture 2015). The soil Ex-K level was significantly lower in paddy fields without rice straw incorporation (n = 146; 62 ± 33 [SD] mg K kg^{-1}) than those with rice straw incorporation (n = 22; 95 ± 36 [SD] mg K kg^{-1}; $p < 0.01$ by t-test). The quantity of supplied K was analyzed for 42 paddy fields in Fukushima Prefecture using data obtained by the MAFF in 2014 and 2015. The amount of K in 2015 was defined as the sum of (1) soil Ex-K at harvest in 2014, assuming that the bulk density was one Mg m^{-3} and the plow layer was 15 cm in depth; (2) the K content of rice straw if the straw was

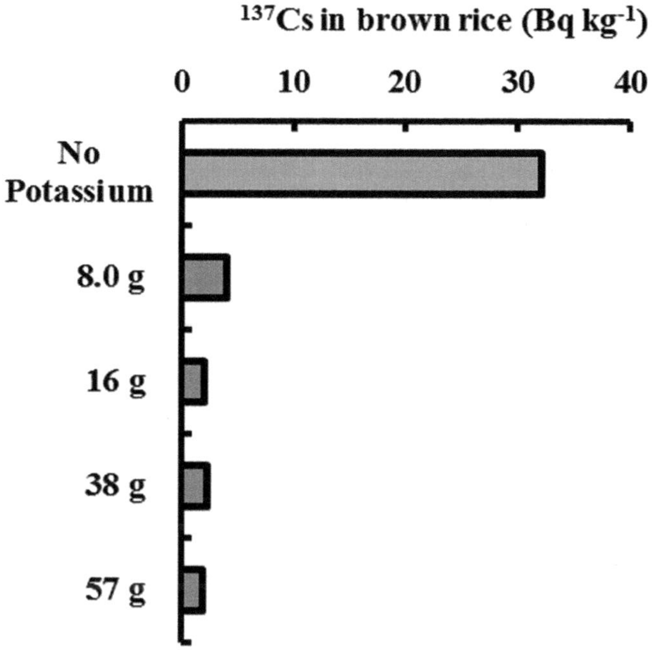

Fig. 6.43 Concentrations of ^{137}Cs in brown rice by amount of basal K fertilizer application. Saito et al. (2015)

Table 6.23 Relationship between exchangeable K, ^{137}Cs in soil

Timing of potash fertilizer	Amount of potash fertilizer (g m^{-2})	Exchangeable K (mg kg^{-1} dw)	Exchangeable ^{137}Cs (mg kg^{-1} dw)
No potassium	0	95.4	234
Basal fertilizer	8.0	107	209
Basal fertilizer	16	318	43.8
Basal fertilizer	38	356	31.0
Basal fertilizer	57	395	25.0

Created by the author based on Saito et al. 2015 with permission from Springer

incorporated into the soil, assuming that the rice straw contained 2% K and the weight of the rice straw was 1.3-fold that of the brown rice yield, according to Fukushima Prefecture (2015); (3) the K from normal fertilization; and (4) the K contained in countermeasure fertilization to prevent radiocesium uptake. The soil Ex-K, rice straw K, usual fertilization K, and countermeasure fertilization K, accounted for 50%, 18%, 7%, and 26% of the K supply, respectively, in the 32 paddy fields in which straw was incorporated. These results suggested that rice straw incorporation was a significant countermeasure for increasing the soil Ex-K content and preventing radiocesium uptake by rice.

(3) **Countermeasure to mitigate the transfer of radiocesium from soil to upland crops by potassium management**

During soybean cultivation, the target value of soil Ex-K is set to 25 mg K_2O 100 g^{-1} (207 mg K Kg^{-1}), except in areas where high levels of contamination in soybean grain were observed. In high contamination cases, approximately 50 mg K_2O 100 g^{-1} (415 mg K Kg^{-1}) is recommended, which is applied by amending the K fertilizer (Ministry of Agriculture, Forestry and Fisheries, National Agriculture and Food Research Organization, National Institute for Agro-Environmental Sciences 2015a). The most effective timing for K applications is a basal application (Hirayama and Igarashi 2017). The K application does not change the taste of the food (Hirayama et al. 2018). In multiple soybean fields where the radiocesium radioactivity of the soybean grain exceeded 100 Bq kg^{-1}, the grain radioactivity decreased with an increase in the soil Ex-K level, with the tendency being different among soil types. In some soils, the radiocesium transfer factor is significantly higher than in the soils of other areas with the same Ex-K level, or the increase in Ex-K due to the application of K fertilizer is limited (Hirayama et al. 2018). A study found that radiocesium that was attached to the soil was gradually fixed to the soil solid phase over time (Takeda et al. 2013) and the transfer factor to soybeans decreased year by year (Fig. 6.44). As in the field with a relatively high radiocesium transfer factor, the

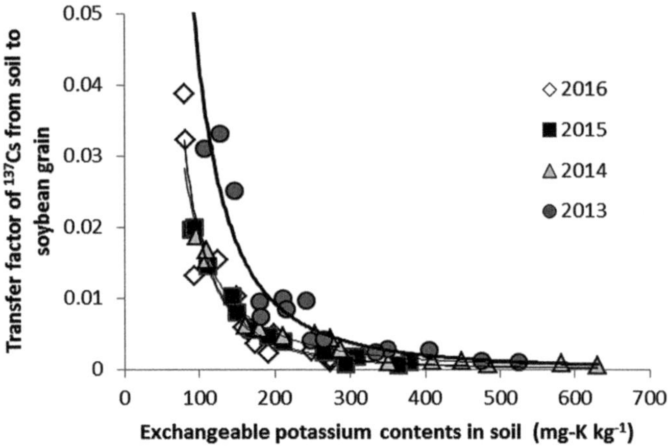

Fig. 6.44 Transfer factor of radioactive cesium to soybean and exchangeable potassium contents in soil. Unpublished data were obtained from Ministry of Agriculture, Forestry, and Fisheries. *Source* Figure provided by T. Saito, S. Fujimura, H. Matsunami, T. Hirayama, K. Kubo, T. Ota, T. Shinano

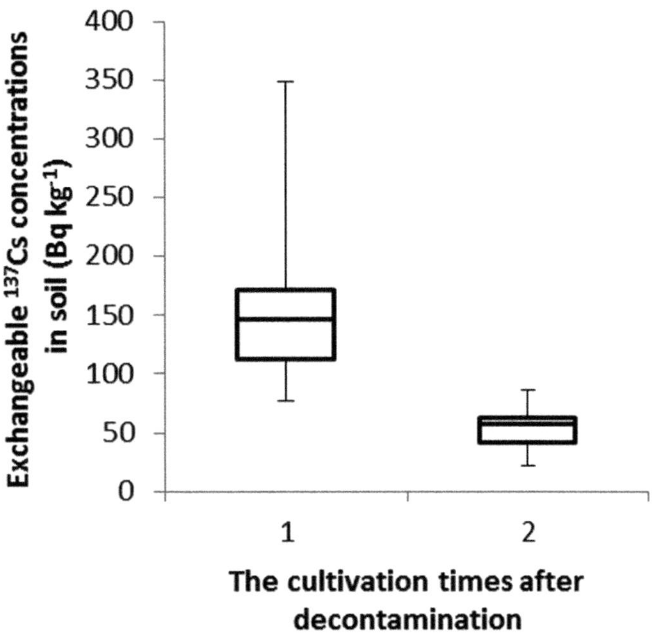

Fig. 6.45 The successive change of the percentage of exchangeable radioactive cesium to radioactive cesium in the soil in a formerly restricted area. Based on Radiation-related support technology information Fukushima Agricultural Technology Centre (2017). *Source* Figure provided by T. Saito, S. Fujimura, H. Matsunami, T. Hirayama, K. Kubo, T. Ota, T. Shinano

continuous cultivation of soybeans dramatically decreased the seed radiocesium concentration; thus, it is suggested that this fixation to the soil seems to be slow in these fields compared with other fields (Hirayama et al. 2018). In the field where the Ex-K level was not increased sufficiently, additional K fertilization at the V6 stage (sixth trifoliolate leaf developmental stage) is effective (Ministry of Agriculture, Forestry and Fisheries, National Agriculture and Food Research Organization, National Institute for Agro-Environmental Sciences 2015b). The soil is thought to have a strong capacity to fix applied K to make it insoluble (Hamamoto et al., in the press, Kubo et al. 2018). Topsoil removal and soil dressing have been performed in formerly restricted areas, and it is highly necessary to increase the K fertilization, since the level of exchangeable radiocesium is high enough to increase the radiocesium transfer from the soil to plants, especially during the initial cultivation after decontamination in some fields (Fig. 6.45).

In 2012, some buckwheat samples were found to exceed the standard limitation value for radiocesium concentration in the area which had been affected by the fallout of radiocesium after the FDNPP incident. The soil Ex-K level is the most important factor for determining the grain radiocesium, as was observed in other crops. As is shown in Fig. 6.46, it is necessary to maintain the soil Ex-K level at more than 30 mg K_2O 100 g^{-1} (249 mg K Kg^{-1}) at the harvest to have sufficiently low radiocesium levels in buckwheat grain (Ministry of Agriculture, Forestry and Fisheries, National Agriculture and Food Research Organization, National Institute for Agro-Environmental Sciences 2014; Kubo et al. 2015). According to Kubo et al. (2017), buckwheat takes up radiocesium vigorously by the flowering stage (Fig. 6.47), and no clear effect was obtained when increasing the Ex-K level in the soil using additional K fertilizer at the flowering stage (unpublished data). From these results, it is recommended to apply K fertilizer as basal fertilization.

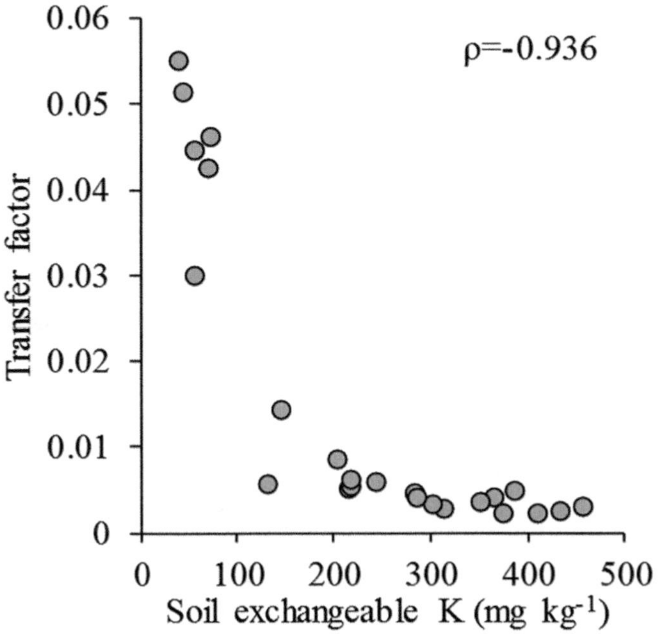

Fig. 6.46 Relationship between soil exchangeable potassium content and transfer factor of radioactive cesium from soil to buckwheat. Modified from Kubo et al. (2015), Copyright 215, with permission from Elsevier

The presence of radiocesium in brown rice produced in Fukushima Prefecture has been monitored by testing the entire volume and all the bags (over 10 million bags per year and 30 kg of brown rice per bag) of rice produced since the 2012 cropping season (Nihei et al. 2015), which have been confirmed to be within the standard limit. Following decontamination and countermeasures to reduce the radiocesium transfer, there were no reports of the standard limit level having been exceeded in later years. On the other hand, it is difficult to remove all the radiocesium from the field by decontamination. And further decrease of radiocesium concentration in the soil will be very slow, because of the long half-life of ^{137}Cs. Because there is still a high risk of increasing the transfer factor if the soil Ex-K level decreases by a substantial amount, continuous countermeasures against radiocesium transfer to crops should be performed. Further investigation will be required to elucidate what level of soil Ex-K is needed to prevent radiocesium transfer to rice based on the soil type, that is, the K adsorption ability, the natural K supply of the soil, and the soil radiocesium concentration.

6.6.6 Behavior of Radiocesium in Japanese Cedar Forest Soil

Both agricultural lands and forests in the Tohoku region were extensively contaminated with radioactive fallout, mostly radiocesium (^{137}Cs), derived from the accident at the FDNPP in 2011. The proportion of forest in the contaminated area was as high as 60% in Miyagi Prefecture and as high as 70% in Fukushima Prefecture. Between 30 and 50% of the contaminated forest was a man-made planted forest, most of which was the Japanese cedar forest (Forest Agency 2013). When the incident happened, in March, it was still cold in the north of the Tohoku region, and some parts of the contaminated forest were still covered with snow. In the deciduous forest, trees were free of leaves, while evergreen coniferous trees held their leaves. These differences between deciduous and coniferous tree species affected the distribution of radiocesium in forest environments; in deciduous forest, the radiocesium fallout directly precipitated on the forest floor, while in the coniferous forest of Japanese cedar, a large part of the radiocesium fallout precipitated on aboveground parts of trees (Forestry Agency 2011).

Thus, tree species and their phenology not only largely affect the distribution of precipitated radiocesium, but may also affect its dynamic behavior, such as leaching to soil. To mitigate the risk of environmental radiocesium contamination, it is important to understand the behavior of radiocesium in the forest, over the years, after the incident. Therefore, we surveyed changes in radiocesium distribution

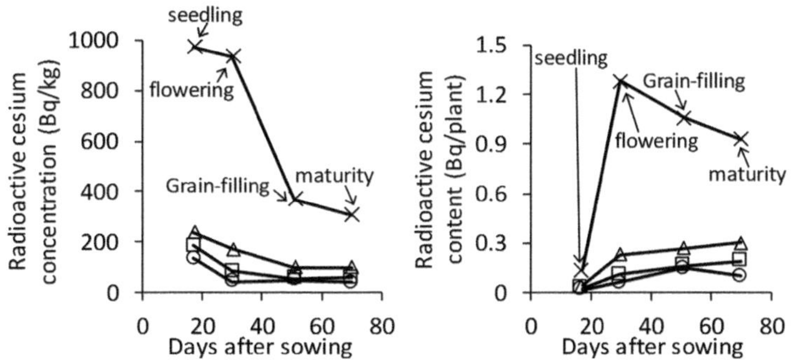

Fig. 6.47 Changes in radioactive cesium concentration (left) and content (right) in whole buckwheat plants in different soil exchangeable potassium content. Modified from Kubo et al. (2017), with permission from Taylor & Francis Online. × 66 mgK kg^{-1} (no potassium fertilization), Δ 207 mgK kg^{-1}, □ 374 mgK kg^{-1}, O 540 mgK kg^{-1}

Fig. 6.48 Scenes of the understory of the experimental forest in Field Science Center, Tohoku University. (Left) The unthinned plot, (Right) Thinned plot. *Source* Figure provided by Masanori Saito

on the forest floor and in the soil in a forest of Japanese cedar (*Cryptomeria japonica*) with different thinning intensity.

The forest belongs to the Field Science Center of Tohoku University and is located 150 km north of the FDNPS. Following the 2011 incident, the forest was contaminated with radiocesium fallout. The radiocesium fallout in this area was estimated at 30–60 kBq m^{-2} based upon aerial survey (MEXT 2011) and ground survey of neighboring pastures (Ogura et al. 2014). A part of the forest was thinned by up to 67% and the rest was kept unthinned. In the unthinned part, the tree density was about 12,000 trees/ha, and almost the whole area was covered with Japanese cedar canopy and minor understory vegetation. In the thinned part, the tree density was about 4000 trees/ha, and various broad-leaved shrubs grew well as understory among cedar trees (Fig. 6.48). The soil is Non-allophanic Andosol.

Samplings of litter and soil were conducted in June 2011 and June 2014. Litter samples collected from the forest floor (O layer) were classified into a twig, cedar leaves, broad leaves, and humified litter. Soil was collected from depths of 0 to 30 cm. radiocesium in these samples was measured with a gamma counter. The measured ^{137}Cs values were expressed on an area basis. The total litter weight per area was not different between in the thinned and the unthinned plots, while the proportion of "broad leaves" litter was much higher in the thinned plot than in the unthinned plot. Soil carbon content and C/N ratio were slightly higher in the unthinned plot due to the thick accumulation of Japanese cedar litter. No changes were found between the two sampling times.

The change in the distribution of ^{137}Cs between June 2011 and June 2014 is shown in Fig. 6.49. In 2011, the total concentration of ^{137}Cs on the forest floor and in the soil in the unthinned plot was just less than half of those in the thinned plot. Most of the ^{137}Cs in the soil were found in the surface layer (0–15 cm depth), while a small proportion of the ^{137}Cs was detected below 15 cm depth. In 2014, the total ^{137}Cs on the forest floor and in the soil in both the thinned and unthinned plots had increased, and no difference was observed between the two plots. In both plots, the ^{137}Cs concentration on the forest floor decreased remarkably, while the ^{137}Cs concentration in the soil increased. The increase in ^{137}Cs was found only in the surface soil layer (0–15 cm).

These results suggest that in 2011, ^{137}Cs remained in the aboveground parts of Japanese cedar and that the ^{137}Cs was gradually washed out from the canopy to the forest floor and then moved into the surface soil layer, although not into the deep soil layer below 15 cm depth. This movement of ^{137}Cs from the tree canopy to the soil through the litter layer was greatly affected by thinning intensity. In the unthinned plot, a large part of ^{137}Cs fallout was first trapped within the aboveground parts of cedar trees and remained there for years. In the thinned plot, the ^{137}Cs directly precipitated onto the forest floors. The ^{137}Cs deposited in the litter was washed out and moved down to the soil layer. In soil, the clay fraction may fix ^{137}Cs.

A prolonged survey of forests in Fukushima Prefecture (Forestry Agency 2017) indicated that in 2016 more than 90% of the total ^{137}Cs in forests were distributed in the soil layer and may be immobilized with soil particles. The same trend of ^{137}Cs dynamics may be found in the Japanese cedar forest in Miyagi Prefecture and elsewhere. Further continuous surveys are needed.

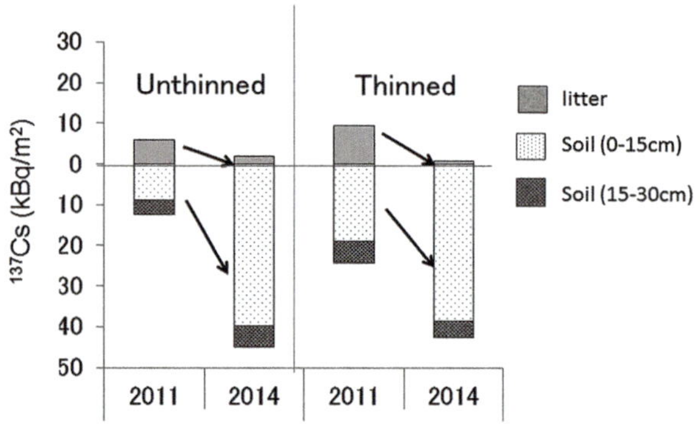

Fig. 6.49 Change of distribution of ^{137}Cs in forest floor and soil in a Japanese cedar forest with different intensity from 2011 to 2014. *Source* Figure provided by Masanori Saito

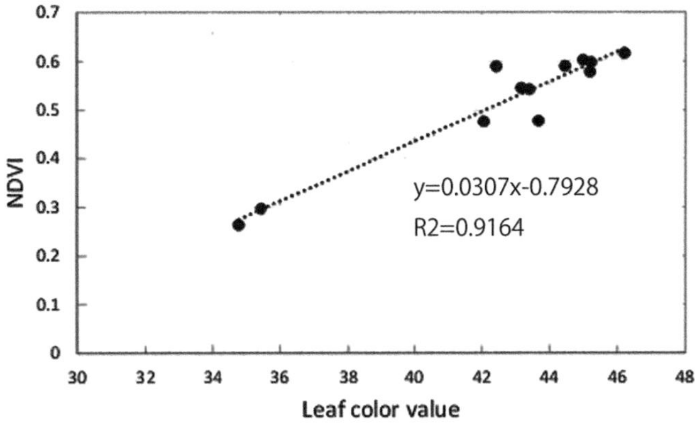

Fig. 6.50 NDVI and leaf color value. A strong correlation appears between the NDVI value and the leaf color value measured with a chlorophyll meter. *Source* Figure provided by Hiroshi Fujii, Shizuka Mori, Yumi Matsumoto, Tetsuya Katagiri, and Kazuto Ando

6.7 Information and Communication Technology

6.7.1 ICT for Fertilizer Management

(1) **Aerial photographic system using a drone equipped with a multispectral camera**

In an effort to evaluate the growth conditions of the entire studied farmland over a wide range and at high speed, we used an aerial photographic system equipped with a multispectral camera mounted on a drone. The system consists of the drone, the multispectral camera, a GPS, and an azimuth meter. The multispectral camera was equipped with a bandpass filter that selects only visible/near-infrared light (i.e., making it a monochrome camera). The camera and GPS were controlled by a tablet PC, and the accumulated images were output by flash memory. The maximum flight time of the drone was eight minutes, during which it was possible to capture images over approximately 2 ha. The ground resolution at a flight altitude of 30 m was 3 cm. Up to 40–50 images could be captured from a 30-ha field.

The vegetation rate and the normalized difference vegetation index (NDVI) were obtained from the acquired images. The vegetation rate is an index corresponding to the number of rice stems and is used to calculate the proportion of soil and leaves in the image. Both indicators show a high correlation with field survey results of leaf color and the number of stems measured at agricultural testing sites in various places (Figs. 6.50 and 6.51).

(2) **Value provided**

The growth map obtained from the data processing system can be viewed with *geographic information system* (GIS) software. Figures 6.52 and 6.53 both show examples of NDVI/vegetation rate maps of about 100 farmlands. In Farmland X (Fig. 6.52), the color scale changes greatly from blue to yellow. Due to the merging of multiple farmlands, differences in soil productivity between the fields appeared as a growth difference. In Field D, NDVI was low and the vegetation rate was high, and, while the number of stems was high, the leaf color was thin. In other words, although the growth immediately after rice planting was good, it is believed that a nitrogen deficiency occurred in the subsequent growth phase. As a countermeasure, it is necessary to improve soil productivity by applying organic matter such as

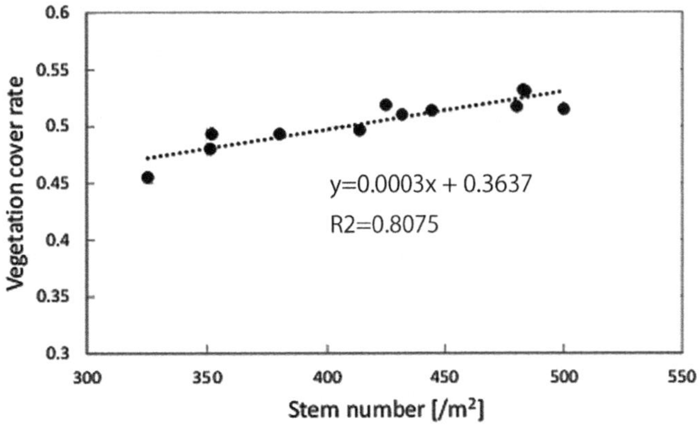

Fig. 6.51 Vegetation cover rate and stem number. A strong correlation appears between the vegetation cover rate calculated from the multispectral image and the stem number counted in the rice fields. *Source* Figure provided by Hiroshi Fujii, Shizuka Mori, Yumi Matsumoto, Tetsuya Katagiri, and Kazuto Ando

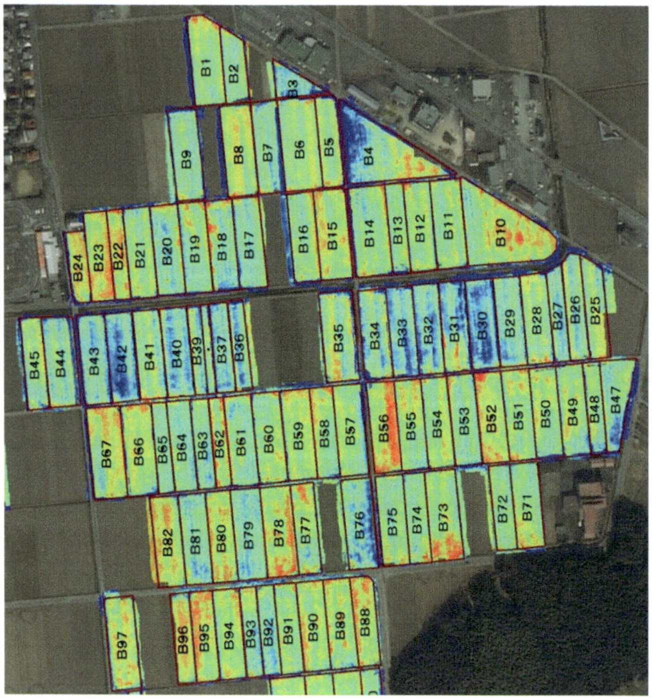

Fig. 6.52 NDVI map. In field *X*, there is a large variation in color from blue to red, indicating a large variation of nitrogen in the soil. The fields in row *D* are generally blue, indicating a lack of fertilizer. *Source* Figure provided by Hiroshi Fujii, Shizuka Mori, Yumi Matsumoto, Tetsuya Katagiri, and Kazuto Ando

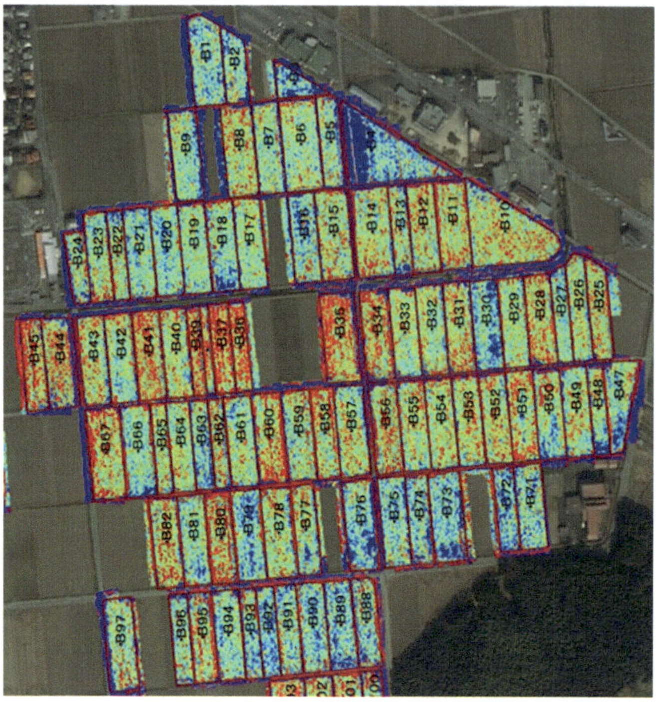

Fig. 6.53 NDVI map. In field *X*, there is a large variation in color from blue to red, indicating a large variation of nitrogen in the soil. The fields in row *D* are generally blue, indicating a lack of fertilizer. *Source* Figure provided by Hiroshi Fujii, Shizuka Mori, Yumi Matsumoto, Tetsuya Katagiri, and Kazuto Ando

compost and preparing the soil with fertilizer. These variations in rice growth occur due to variations in soil productivity, and efficient soil improvement is possible by "visualizing" growth in farmland.

A growth map is converted into a map of areas that require nitrogen, and by controlling the fertilizing amount (variable fertilization) for each part of the field with a fertilizer spraying machine, such as an unmanned helicopter, it is possible to minimize growth variation and stabilize the quality and quantity of the crop. In the demonstration experiment described above, yield stabilization and improved protein content of the crop were confirmed by variable fertilization. In farmland with poor growth, as identified by a low NDVI and vegetation rate, additional

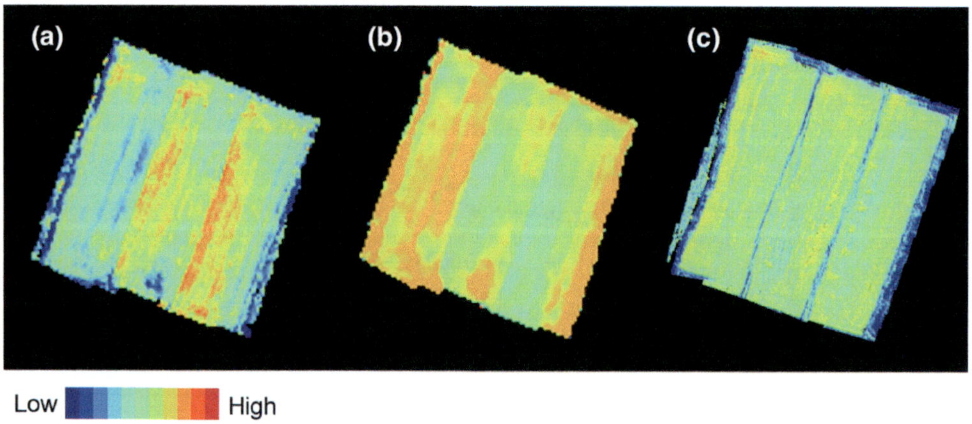

Fig. 6.54 Visualized effects. **a** NDVI map. **b** fertilizer map indicating the amount of fertilizer required. **c** NDVI map after the application of fertilizer. After fertilization, the NDVI variation seen in map a has now been reduced in map **c**. *Source* Figure provided by Hiroshi Fujii, Shizuka Mori, Yumi Matsumoto, Tetsuya Katagiri, and Kazuto Ando

nitrogen fertilizer can be applied to increase yield. On the other hand, in regions where the NDVI is high, the leaf color is thick, and the growth is good, the taste (protein content) of the crop can be improved by reducing topdressing with nitrogen. Figure 6.54a shows an NDVI map of a field and Fig. 6.54b shows an example of a corresponding topdressing amount map. In the farmland with low growth on the left side of the NDVI map, the topdressing application was increased.

Figure 6.54c shows the NDVI map after topdressing. It reveals that growth variability was improved by the variable topdressing application. Fertilizer was applied by an unmanned helicopter equipped with GPS, with which it is possible to change the amount of fertilizer to be sprayed in different places based on the topdressing map.

Variations in a growth map occur not only due to soil properties but also due to uneven work. After analyzing the images, it was speculated that tractor work was the cause of the variations. By interviewing the farmers, it was found that a fertilizer shortage occurred in places where some work was omitted. By feeding back this unevenness to the operator, the process can be improved. By evaluating variations over time using the work process management system, the degree of improvement can be quantified. In Japanese agriculture, field consolidation has been proceeding; where previously the managers themselves were the workers, now the manager and the worker are different individuals, so the management system is becoming more divided. When managers, who are capable of observing the appearance of farmland, instruct, and educate on-site staff, they express a subjective opinion on growth. However, by quantifying growth using the growth map, it is possible to share information promptly and accurately.

References

Abe T, Honda S (2015) Measures of desalinization in tsunami-hit farmland. 10. Measures to improve soil fertility in tsunami-affected paddy fields in Miyagi Prefecture. Jap J Soil Sci. Plant Nutr 86:441–442 (in Japanese)

Ando T, Shiono H (2019) Effects of whole-volume fertilizer application after harvesting on 'Satonishiki' sweet cherry growth and yield. Acta Hortic 1235:313–318 https://doi.org/10.17660/ActaHortic.2019.1235.43

Aomori Prefecture (2017) Second phase promotion plan of "Japan's healthiest soil-making plan" Aomori Prefectural Industrial Technology Research Center. https://www.aomori-itc.or.jp/sehisekkei/ Accessed on 17 May 2019. (in Japanese)

Aoyama M (2015) Functional roles of soil organic matter. Humic Subs Res 12:21–28

Aoyama M, Kumakura N (2001) Quantitative and qualitative changes of organic matter in an Ando soil induced by mineral fertilizer and cattle manure applications for 20 years. Soil Sci Plant Nutr 47:241–252

Aoyama M, Kumakura N (2002) High and low molecular weight components of soil humic acids as affected by plant residue applications. In: Proceedings of the international humic substances society twentieth anniversary conference, humic substances: nature's most versatile materials. Northeastern University, Boston, pp 25–27

Aoyama M, Taninai Y (1992) Organic matter and its mineralization in particle size and aggregate size fractions of soils with four-yeara pplication of farmyard manure. Jpn J Soil Sci Plant Nutri 63:571–580 (in Japanese with English summary)

Aoyama M, Angers DA, N'Dayegamiye A (1999) Particulate and mineral-associated organic matter in water-stable aggregates as affected by mineral fertilizer and manure applications. Can J Soil Sci 79:295–302

Ariga T (1984) Geomorphological development of Shonai coastal plain, northeastern Japan, The behavior of alluvial fan's front since latest Pleistocene. Annal Tohoku Geographycal Asocciation 36:13–24 (in Japanese)

Cheng W, Padre AT et al (2016) Changes in the soil C and N contents, C decomposition and N mineralization potentials in a rice paddy after long-term application of inorganic fertilizers and organic matters. Soil Sci Plant Nutr 62:212–219

Cheng W, Kimani SM et al (2018) Forage rice varieties Fukuhibiki and Tachisuzuka emit larger CH_4 than edible rice Haenuki. Soil Sci Plant Nutr 64:77–83

Cherr CM, Scholberg JMS, McSorley R (2006) Green manure approaches to crop production: a synthesis. Agron J 98:302–319

Deguchi S, Uozumi S et al (2012) Arbuscular mycorrhizal colonization increases phosphorus uptake and growth of corn in a white clover living mulch system. Soil Sci Plant Nutr 58(2):169–72. 10.1080/00380768.2012.662697

Deguchi S, Uozumi S et al (2017) White clover living mulch reduces the need for phosphorus fertilizer application to corn. Eur J Agron 86:87–92. https://doi.org/10.1016/j.eja.2017.03.006

Edamura T, Kumagai K (2009) The dispersion of Jomon Ruins. Theory Appl GIS 17:63–72 (in Japanese)

Eusufzai MK, Tokida T et al (2010) Methane emission from rice fields as affected by land use change. Agr Ecosyst Environ 139:742–748

Eusufzai MK, Tokida T et al (2011) Effect of rice straw application on CH_4 emission in continuous and recently converted paddy fields. J Agri Meteorol 67:185–192

FAO (Food and Agriculture Organization of the United Nations) (2006) World reference base for soil resources 2006. FAO, Rome

Forestry Agency (2011) Survey of distribution of radio-compounds in forests. www.rinya.maff.go.jp/j/press/hozen/pdf/111227_2-01.pdf Accessed on 1 May 2019

Forestry Agency (2013) Forest rate and artificial forest rate in prefectures in 2013. http://www.rinya.maff.go.jp/j/keikaku/genkyou/h24/1.html. Accessed on 1 May 2019

Forestry Agency (2017) Survey of distribution of radio-compounds in forests. http://www.rinya.maff.go.jp/j/kaihatu/jyosen/attach/pdf/H28_jittaihaaku_kekka-1.pdf. Accessed on 1 May 2019

Fukushima Agricultural Technology Centre (2017) Reduction of radioactive cesium transfer to soybean by addition potassium application and aging effect in decontamination agricultural land. In: Radiation related support technology information. http://www.pref.fukushima.lg.jp/uploaded/attachment/255860.pdf. Access on 23 Apr 2019

Fukushima Agricultural Technology Centre (2019) Soil fertility remediation effect by cultivation and incorporation of leguminous green manure crops in agricultural land after decontamination. In: Farming resumption demonstrated technical information. http://www.pref.fukushima.lg.jp/uploaded/attachment/315540.pdf. Access on 23 Apr 2019

Fukushima Prefecture (2015) Annual reports for rice, soybean, wheat and buckwheat productions. https://www.pref.fukushima.lg.jp/uploaded/attachment/165778.pdf. Access on 22 Apr 2019

Hayano M, Fumoto T et al (2013) National-scale estimation of methane emission from paddy fields in Japan: database construction and upscaling using process-based biogeochemistry model. Soil Sci Plant Nutri 59:812–823

Hirayama T, Igarashi Y (2017) The best time for application of potassium fertilizers used as absorption inhibitor for radiocesium of soybean. Tohoku J Crop Sci 60:49–50 (in Japanese)

Hirayama T, Takeuchi T et al (2018) Effect of application of many potassium fertilizers to decrease transfer of radiocesium in soil to soybean plants and its verification of seed quality. Bull Fukushima Agric Technol Centre. (in Japanese with English abstract, accepted for publication)

Hirayama T, Takeuchi T et al (2018) Effects of decreasing radiocesium transfer from the soil to soybean plants and changing the seed nutrient composition by the increased application of potassium fertilizer. Bull Fukushima Agri Tech Cent 9:1–10 (in Japanese with English abstract)

Hosono M, Sase T (2015) The historical development of "Kurobo-kudo" layer: Preliminary discussion from the perspective of man-made ecosystems. Quat Res 54:323–339 (in Japanese)

Ishiyama S et al (2000) Watermelon basics–management, harvest and shipment, agricultural technology encyclopedia, *Vegetables*, **4**, Edition No. 25: 159–234, Rural Culture Association (Tokyo). (In Japanese)

Itoh M, Sudo S, Mori S, Saito H, Yoshida T, Shiratori Y, Suga S, Yoshikawa N, Suzue Y, Mizukami H, Mochida T, Yagi K (2011) Mitigation of methane emissions from paddy fields by prolonging midseason drainage. Agr Ecosyst Environ 141:359–372

Japan Meteorological Agency (2017) Active volcanoes in Japan. http://www.data.jma.go.jp/svd/vois/data/tokyo/STOCK/kaisetsu/katsukazan_toha/katsukazan_toha.html

Jinbo N (1960) Geological map of yamagata prefecture. Mining section, Yamagata Prefecture (in Japanese)

Kaneta Y (1995) Inovation of fertilizer application by using controlled-release fertilizers (Part 2): The single application of fertilizer in nursery box to non tillage transplanting rice. Jpn J Soil Sci Plant Nutr 66:176–181 (in Japanese)

Kaneta Y (1996) Single application of fertilizer in a nursery box for rice. Agri Horti 71:802–806 (In Japanese.)

Kaneta Y, Tsuchiya K (1997a) Relationship between variation of weather and recovery rate of rice plants in non-tillage culture by single application of fertilizer in nursery boxes. Jpn J Soil Sci Plant Nutr 68:112–115. (In Japanese)

Kaneta Y, Tsuchiya K (1997b) Effects of single application of fertilizer in nursery boxes on growth and nitrogen uptake of rice in wet paddy fields with rice straw plowed. Jpn J Soil Sci Plant Nutr 68:185–188. (In Japanese)

Kaneta Y, Awasaki H, Murai Y (1994a) Non-tillage rice culture by single application of fertilizer in a nursery box with controlled-release fertilizer. Jpn J Soil Sci Plant Nutr 65:385–391 (In Japanese)

Kaneta Y, Awasaki H, Murai Y (1994b) Single application of fertilizer in a nursery box with controlled-release fertilizer (Part 1): stratified application of controlled-release fertilizer. Tohoku Agric Res 47:115–116 (In Japanese.)

Kanno I, Honjyo Y, Tokudome S, Arimura S (1957) Clay minerals of red-yellow soils derived from granitic rocks (Part 2), Weathered materials of granite from Tamana, Kumamoto prefecture. Jpn J Soil Sci Plant Nutr 28:55–58 (in Japanese)

Katayanagi N, Ono K et al (2013) Validation of the DNDC-Rice model to discover problems in evaluating the nitrogen balance at a paddy-field scale for single-cropping of rice. Nutr Cycl Agroecosyst 95:255–268

Katou H, Nakaya N, Maeda K (1985) Changes in sedeminnt volume, liaquid limit, and plastic limit of alluvial soils upon drying. Soil Sci Plant Nutr 31:215–226

Kikuchi R (2009) Climatological background on the watermelon cultivation in Obanazawa-basin, Yamagata prefecture, faculty of liberal arts review. Tohoku Gakuin Univ 154:29–50 (in Japanese)

Kitagawa Y, Okayama K, Tojo S (1988) Hardening of puddled soil in rotational paddy field (part I) Occurrence of hardening of puddled soil in rotational paddy fields. Jpn J Soil Sci Plant Nutri 59:149–155

Kitamura N (1986) Neogene stratum of Tohoku district. Urban Kubota 25:26–31 (in Japanese)

Kubo K, Nemoto K et al (2015) Analyses and countermeasures for decreasing radioactive cesium in buckwheat in areas affected by the nuclear accident in 2011. Field Crops Res 170:40–46. https://doi.org/10.1016/j.fcr.2014.10.001

Kubo K, Fujimura S et al (2017) Effect of soil exchangeable potassium content on cesium absorption and partitioning in buckwheat grown in a radioactive cesium-contaminated field. Plant Prod Sci 20:396–405. https://doi.org/10.1080/1343943X.2017.1355737

Kubo K, Hirayama T et al (2018) Potassium behavior and clay mineral composition in the soil with low effectiveness of potassium application. Soil Sci Plant Nutr 64:265–271. https://doi.org/10.1080/00380768.2017.1419830

Kumagai K, Konno Y, Togashi M (2000) Effect of water management and rice plant growth on methane emission from rice paddy fields in a cold district of Japan. Jpn J Soil Sci Plant Nutr 71:643–650 (in Japanese)

Masui J (1959a) Clay mineral of neogene pumice tuff soil in Matsushima environs, Miyagi prefecture. Jap J Soil Sci Plant Nutr 30:361–366. (in Japanese)

Masui J (1959b) Clay mineral of paddy soils derived from neogene pumice tuff rocks in Matsushima environs, Miyagi prefecture. Jap J Soil Sci Plant Nutr 30:443–448. (in Japanese)

Matsuyama N, Saigusa M et al (2005) Acidification and soil productivity of allophanic andosols affected by application of fertilizers. Soil Sci Plant Nutr 51:117–123

Ministry of Agriculture, Forestry and Fisheries (2011) Outline of proved decontamination techniques. http://www.affrc.maff.go.jp/docs/press/pdf/110914-02.pdf. (in Japanese) Accessed on 22nd April 2019

Ministry of Agriculture, Forestry and Fisheries (2015a) Agricultural statics of Yamagata in 2014

Ministry of Agriculture, Forestry and Fisheries (2015b) Production agriculture income statistics

Ministry of Agriculture, Forestry and Fisheries, National Food and Agriculture Research, Natural Institute for Agro-Environmental Sciences (2015b) Factors and countermeasure of soybean with high radioactive cesium concentration. Investigation of factors and result of trial examination (Outline, the third edition). http://www.maff.go.jp/j/kanbo/joho/saigai/pdf/youin_daizu_3.pdf (In Japanese). Accessed on 22 Apr 2019

Ministry of Agriculture, Forestry and Fisheries, National Food and Agriculture Research, Natural Institute for Agro-Environmental Sciences, (MAFF, NARO, NIAES) (2014) Factors and counter-measure of buckwheat with high radioactive cesium concentration. Investigation of factors and result of trial examination (Outline, the third edition). http://www.maff.go.jp/j/kanbo/joho/saigai/pdf/h25soba_yoin.pdf (In Japanese)

Ministry of Agriculture, Forestry and Fisheries, National Food and Agriculture Research, Natural Institute for Agro-Environmental Sciences, (MAFF, NARO, NIAES) (2015a) Factors and counter-measure of soybean with high radioactive cesium concentration.

Investigation of factors and result of trial examination (Outline, the third edition)

Ministry of Education, Culture, Sports, Science and Technology (MEXT) (2011) Aerial monitoring of radioactive compounds (Concentration of radio Cs in soil). 2011. https://radioactivity.nsr.go.jp/ja/contents/5000/4931/24/1940_0830_1.pdf. Accessed on 1 May 2019

Naganoma H, Moroyu H (1983) Index for uplandization degree of paddy field as the soil microstructure, Studies for Upland cropping techniques for upland converted from rice paddy fields. Agri Res Center Tsukuba 1:46–53 (In Japanese)

Naganori Y (1984) II. Soil composition, 2. Mineral component. In: Yamane I, Hamada R, Yoshinaga N, Asami T, Matsuda Kei-ichiro, Sakuma T, Kobayashi M, Yumura Y (eds) Soil science. Buneido, Tokyo, pp 30–43. (in Japanese)

Nakajima M, Cheng W et al (2016) Modeling aerobic decomposition of rice straw during off-rice season in an Andisol paddy soil in a cold temperate region, Japan: effects of soil temperature and moisture. Soil Sci Plant Nutr 62:90–98

Nakajima M, Cheng W et al (2017) Shallow autumn tillage does not reduce CH_4 emission from an Andisol paddy field in Morioka, a cold region in Japan. J Agri Meteorol 73:91–98

Nakano K (1978) Changes in soil physical properties of clayey soil by conversion from ill-drained paddy field into upland field. Bull Hokurikku Natl Agric Exp Stn 21:63–94 (In Japanese with English Summary)

Nihei N, Tanoi K et al (2015) Inspections of radiocesium concentration levels in rice from Fukushima Prefecture after the Fukushima Dai-ichi Nuclear Power Plant accident. Sci Rep 5:8653. https://doi.org/10.1038/srep08653

Nishida M (2016) Decline in fertility of paddy soils induced by paddy rice and upland soybean rotation, and measures against the decline. JARQ 50:87–94

Nishida M, Iwaya K et al (2007) Changes in natural 15 N abundance in paddy soils under different long-term soil management regimes in the Tohoku region of Japan. Soil Sci Plant Nutr 53:310–317

Nishida M, Sekiya H, Yoshida K (2013) Status of paddy soils as affected by paddy rice and upland soybean rotation in northeast Japan, with special reference to nitrogen fertility. Soil Sci Plant Nutr 59:208–217

Ogawa B (1994) Studies on the actual situation of cadmium pollution to paddy rice in Akita prefecture and its control. Bull Akita Agric Exp stn 35

Ogura S-I, Suzuki T, Saito M (2014) Distribution of radioactive cesium in soil and its uptake by herbaceous plants in temperate pastures with different management after the Fukushima Dai-Ichi Nuclear Power Station accident. Soil Sci Plant Nutr 60:790–800

Ohyama N (1982) Decomposition and accumulation of some kinds of organic matters applied to the paddy field and their effects on growth of rice plant. Tohoku Agric Res 30:57–72 (in Japanese)

Ohyama N, Satoh T et al (1983) Relationship between amount of manure and nitrogen application effecting on rice yield. Tohoku Agric Res 33:49–50 (in Japanese)

Oikawa K, Arai S (1996) Occurrence of rhizoctonia rot on tubers of Chinese yam in the crop-rotation field based on the Chinese yam (Dioscorea opposite Thumb. cv. Nagaimo). Ann Rept Plant Prot North Japn 47:55–58 (in Japanese)

Saigusa M, Matsuyama N (1998) Distribution of allophanic Andosols and non-allophanic Andosols in Japan. Tohoku J Agric Res 48:75–83 (In Japanese.)

Saito T, Ohkoshi S et al (2012) Effect of potassium application on root uptake of radiocesium in rice.KUR Research Program for Scientific Basis of Nuclear Safety, pp 165–169

Saito T, Takahashi K et al (2015) Effect of application timing of potassium fertilizer on root uptake of 137Cs in brown rice. J Radioanal Nucl Chem 303:1585–1587

Sato T, Kaneta Y et al (2003b) Effect of soil physical properties on soybean nodulation and N_2 fixation at the early growth stage in heavy soil field in Hachirougata polder, Japan. Jpn Soil Sci Plant Nutr 49:695–702

Sato T, Shibuya K (1991) One time application of total fertilizer at nursery stage in rice culture. Rep. Tohoku Br. Crop Sci Soc Jpn 34:15–16 (In Japanese.)

Sato Y, Yamanaka S and Takahashi M (2003a) Evidence for Jomon plant cultivation based on DNA analysis of chestnut remains. Senri Ethnol Stud 63:187–197

Sato T, Yoshimoto S et al (2007) Effect of hairy vetch planting on changes in soil physical properties and soybean early growth in a heavy clayey soil field. Jpn J Soil Sci Plant Nutr 78:53–60 (in Japanese)

Sato T, Yoshimoto S et al (2011) Effect of planting with leguminous green manure plant hairy vetch on growth and yield of succeeding crop soybean in a heavy clay soil field converted from paddy field. Jpn J Soil Sci Plant Nutr 82:123–130 (in Japanese)

Sekiya K (1982) Characteristic and productivity of orchard soil. In: Chiba T (ed) Soil management and fertilizer application on orchard. Hakuyusha, Tokyo, pp 38–43. (in Japanese)

Shiono H, Saito H et al (2016) Shallow tillage in autumn reduced methane and nitrous oxide emissions from a paddy field with incorporated rice straw in a cold region of Japan. Jpn J Soil Sci Plant Nutr 87:101–109 (in Japanese with English summary)

Shoji S, Ito T et al (1985) Properties of nonallophanic Andosols from Japan. Soil Sci 140:264–277

Shoji S, Takahashi T et al (1988) Properties of Spodosols and Andosols showing climosequential and biosequential relations in southern Hakkoda, northeastern Japan. Soil Sci 145:135–150

Shoryuji River land improvement district (1998) Operation and maintenance scheme drawing in Shoryuji River land improvement district (in Japanese)

Sumida H, Kato N, Nishida M (2002) Fate of fertilizer nitrogen and compost nitrogen in long-term experiments on gray lowland soil in a cool region. Bull Natl Agric Res Cent Tohoku Reg 100:49–59 (in Japanese with English summary)

Sumida H, Kato N, Nishida M (2005) Depletion of soil fertility and crop productivity in succession of paddy rice-soybean rotation. Bull Natl Agric Res Cent Tohoku Reg 103:39–52 (in Japanese with English summary)

Takahashi T, Toriiyama K (1999) Role of iron oxides in change in soil microstructure with drying and reducing treatments. J Jpn Soc Soil Phys 78:35–43 (In Japanese)

Takahashi T, Shoji S, Sato A (1989) Clayey Spodosols and Andosols showing a biosequential relation from Shimokita Peninsula, northeastern Japan. Soil Sci 148:204–218

Takakai F, Kikuchi T et al (2017a) Changes in the nitrogen budget and soil nitrogen in a field with paddy–upland rotation with different histories of manure application. Agriculture 7:39

Takakai F, Nakagawa S et al (2017b) Net greenhouse gas budget and soil carbon storage in a field with paddy–upland rotation with different history of manure application. Agriculture 7:49

Takakai F, Takeda M et al (2010) Effects of preceding compost application on the nitrogen budget in an upland soybean field converted from a rice paddy field on gray lowland soil in Akita, Japan. Soil Sci Plant Nutr 56:760–772

Takeda A, Tsukada H, Nakao A et al (2013) Time-dependent changes of phytoavailability of Cs added to allophanic Andosols in laboratory cultivations and extraction tests. J Environ Radioact 122:29–36

Takimoto T, Takata Y, Kuwagata T (2017) Development of a soil physico-environmental database on prefectural scale for Japanese agricultural land for evaluating soil temperature and moisture. Jpn J Soil Sci Plant Nutri 88:309–317 (in Japanese with English summary)

Tang S, Cheng W et al (2016) Simulating the effects of soil temperature and moisture in the off-rice season on rice straw decomposition and subsequent CH_4 production during the growth season in a paddy soil. Biol Fertil Soils 52:739–748

Tawaraya K (2003) Arbuscular mycorrhizal dependency of different plant species and cultivars. Soil Sci Plant Nutr 49:655–68

Tawaraya K, Hirose R, Wagatsuma T (2012) Inoculation of arbuscular mycorrhizal fungi can substantially reduce phosphate fertilizer application to Allium fistulosum L. and achieve marketable yield under field condition. Biol Fertil Soils 48:839–843. https://doi.org/10.1007/S00374-012-0669-2

The Ministry of Agriculture, Forestry and Fisheries (2009) Basic direction for improvement of soil fertility. http://www.maff.go.jp/j/seisan/kankyo/hozen_type/h_dozyo/pdf/chi4.pdf (February 2018, in Japanese)

Yazawa River land improvement district (1986) Irrigation system map in Yazawa River land improvement district

Yazawa River land improvement district (1988) Operation and maintenance equipment map in Yazawa River land improvement district

Yonebayashi K (1997) Organic matter of soil. In: Kyuma K (ed) Updated soil science. Asakura Publishing, Tokyo. (in Japanese)

Yoshino N, Kobayashi H et al (2015) Effects of soil stripping and dressing for decontamination of radioactive materials on soil fertility of agricultural land. J Soc Remed Radioact Contam Environ 3:145–152 (in Japanese with English summary)

Kanto-Koushinetsu Region

7

Hiroaki Sumida, Yusuke Takata, Masataka Komatsu, Kumiko Baba, Yutaka Shiratori, Hiroshi Takesako, Kenji Kanazawa, Masayuki Kawahigashi, Masaharu Ikeba, Naruo Miyazaki, Nobuyuki Kanuma, Kazuko Someya, Kunio Maruoka, Takashi Tsukamoto, Kosuke Homma, Akira Noguchi, Fumie Shinmachi, Hiroko Yamaya-Ito, Reiji Takahashi, Takayuki Kobayashi, and Syunrokurou Fujiwara

Abstract

The area of Kanto-Koushinetsu occupies 16% (63,041 km^2) of the land area of Japan. This region includes the Tokyo metropolis and the prefectures of Ibaraki, Tochigi, Gunma, Saitama, Chiba, Kanagawa, Yamanashi, Nagano, Niigata. Significant depletion in agricultural land has been progressed from 1972 to 2011 due to the population concentration in the urban area of Tokyo metropolis and the prefectures of Saitama, Chiba, Kanagawa. The area around the population concentration zone plays an important role in supplying vegetables, and the national share of items is 33.8% for vegetables and 25.6% for a flowering plant, and these values are the top of the Japan. In the Kanto region, the plain area occupies a large area, and agricultural activities are being developed under good water use and the existence of a populated area. The characteristics of geology, topography, climatic conditions and soil characteristics and the effects of volcanic ash soil in this area are described. The relationship between the recent agricultural production of Kanto-Koshinetsu area and the soil characteristics of each area was clarified. On the other hand, anthropogenic soils are increasing in Tokyo metropolis and problems in current soil classification system emerged from these soils has been pointed out. As a topic, an action on restoration of habitat for the reintroduction of the crested ibis in Sado Island (Sado City, Niigata Prefecture) for selected as Globally Important Agricultural Heritage System

H. Sumida (✉) · A. Noguchi · F. Shinmachi · H. Yamaya-Ito · R. Takahashi · T. Kobayashi
College of Bioresource Sciences, Nihon University, Fujisawa City, Kanagawa Prefecture, Japan
e-mail: sumida.hiroaki@nihon-u.ac.jp

Y. Takata
Institute for Agro-Environmental Sciences, NARO, Tsukuba City, Ibaraki Prefecture, Japan

M. Komatsu
Nagano Prefecture Farmers Academy, Nagano Prefectural Government, Nagano City, Nagano Prefecture, Japan

K. Baba
Yamanashi Prefectural Agricultural Technology Center, Yamanashi Prefectural Government, Kai City, Yamanashi Prefecture, Japan

Y. Shiratori
Niigata Agricultural Research Institute, Niigata Prefectural Government, Nagaoka City, Niigata Prefecture, Japan

H. Takesako · S. Fujiwara
Former Kurokawa Field Science Center, Meiji University, Kawasaki City, Kanagawa Prefecture, Japan

K. Kanazawa
Central Region Agricultural Research Center, NARO, Tsukuba City, Ibaraki Prefecture, Japan

M. Kawahigashi
Department of Geograpy, Tokyo Metropolitan University, Hachioji City, Tokyo, Japan

M. Ikeba
Ibaraki Agricultural Center, Ibaraki Prefectural Government, Mito City, Ibaraki Prefecture, Japan

N. Miyazaki
Tochigi Prefectural Agricultural Experiment Station, Tochigi Prefectural Government, Utsunomiya City, Tochigi Prefecture, Japan

N. Kanuma · K. Someya
Gunma Agricultural Technology Center, Gunma Prefectural Government, Isesaki City, Gunma Prefecture, Japan

K. Maruoka
Saitama Prefectural Honjo Agriculture and Forestry Promotion Center, Saitama Prefectural Government, Kumagaya City, Saitama Prefecture, Japan

T. Tsukamoto
Chiba Prefectural Agriculture and Forestry Research Center, Chiba Prefectural Government, Chiba City, Chiba Prefecture, Japan

K. Homma
Faculty of Agriculture, Field Center for Sustainable Agriculture and Forestry, Niigata University, Niigata City, Niigata Prefecture, Japan

R. Hatano et al. (eds.), *The Soils of Japan*, World Soils Book Series,
https://doi.org/10.1007/978-981-15-8229-5_7

(GIAHS) was introduced. Furthermore, examples of agricultural production and soil environment conservation were introduced.

Keywords

Geography and geology • Climate condition • Andosols • Anthropogenic soils • Agricultural production and soil properties • GIAHS

7.1 Outline of the Kanto-Koushinetsu Region

7.1.1 Perspective of the Kanto-Koushinetsu Region

Kanto-Koushinetsu is a geological area shared between the island of Honshu and other nearby islands (Fig. 7.1). The area has an extremely large population (more than 48 million) and large amounts of industry and includes the Tokyo metropolis and the prefectures of Ibaraki, Tochigi, Gunma, Saitama, Chiba, Kanagawa, Yamanashi, Nagano, and Niigata. The Kanto-Koushinetsu region is usually divided into three sub-regions: (1) the Kanto region, which includes the Tokyo metropolis and the prefectures of Ibaraki, Tochigi, Gunma, Saitama, Chiba, and Kanagawa; (2) the Koushin region, which includes Yamanashi and Nagano Prefectures; and (3) Echigo, which includes Niigata Prefecture. In general, the Koushin region and Echigo are together referred to as "Koushinetsu."

A total of 40% of the Kanto region consist of mountainous areas, which is lower than the national average (61%). The region also contains the Kanto Plain, the largest plain in Japan. The proportion of cultivated land in the Kanto region (18.6%) is higher than the national average (11.5%), due to agricultural activities on the region's wide flat areas, such as the Kanto Plain, that are suitable for cultivation. Highly profitable urban agriculture, such as vegetable production, is conducted while taking advantage of proximity to big cities. In contrast, however, the distribution of paddy fields in the Kanto region is lower than the Japanese average.

The Koushin region is characterized by high mountainous areas. The average land altitude in Japan is 378 m above sea level (a.s.l.), but in Nagano and Yamanashi Prefectures, the averages are 1033 m and 913 m a.s.l., respectively; these prefectural averages are the highest and the second highest in Japan. Furthermore, the top five highest mountains in Japan (Mt. Fuji, Mt. Kita, Mt. Okuhodaka, Mt. Mano, and Mt. Yarigatake) are located in these prefectures. Moreover, the area ratio of mountains in Nagano Prefecture (86%) and Yamanashi Prefecture (85%) is much higher than the national average. These two prefectures have unique geological conditions and are famous for their fruit and highland farming. Yamanashi Prefecture produces the largest amount of grapes and peaches of all Japanese prefectures, while Nagano Prefecture produces the largest amount of highland lettuce and the second largest amount apples.

Niigata Prefecture is located in the northern part of the Kanto-Koushinetsu region and is the only prefecture in the region that faces the Sea of Japan. The Echigo Mountains are located in the south of Niigata Prefecture and divide the prefecture from the Kanto region. The Echigo Mountains are known to receive among the heaviest snowfall of any area in the world, with an average snow depth of 3 to 4 meters. As a result of this large snowfall, the mountains are the source of several rivers, which flow into the Sea of Japan. In the lower reaches of these rivers, the Echigo Plain extends to the coastal area of Niigata Prefecture. This plain is the fourth largest in Japan, and is also the largest rice production area, using the abundant supply of snowmelt from the Echigo Mountains to irrigate rice paddy.

7.1.2 Geography and Geology of the Kanto-Koushinetsu Region

(1) Geography and geology of the Kanto region

The Kanto region has the largest plain in Japan (the Kanto Plain, with an area of about 17,000 km^2). This plain is bordered by 10 active volcanoes (Mt. Nasutake, Mt. Takahara, Mt. Nikko-Shirane, Mt. Nikko-Nantai, Mt. Akagi, Mt. Haruna, Mt. Kusatsu-Shirane, Mt. Asama, Mt. Fuji, and Mt. Hakone) located in the north and southwest of the region, by the Kanto Mountains in the west, and by the Pacific Ocean to the south and east (Fig. 7.1). Large quantities of tephra from the surrounding volcanoes—Mt. Asama, Mt. Haruna, and Mt. Akagi to the north and Mt. Hakone and Mt. Fuji to the south—have been deposited on the Pleistocene terraces of the Kanto region, thereby maintaining the Kanto Plain as the largest plain in Japan. The Pleistocene terraces in this region are the most important geomorphological plains in Japan from an agricultural and pedological viewpoint, because Andosols (which are mainly used for upland crops) are widely distributed on these plains (see Sect. 7.1.3).

The Kanto Mountains consist of three major tectonic belts: the Sambagawa belt, in the north, and the Chichibu and Shimanto belts, in the south. The Sambagawa belt is constituted chiefly of crystalline schist that is composed largely of quartz, graphite, piedmontite, and stilpnomelane. This schist is thought to have been originally formed from Jurassic to Cretaceous accretionary deposits. The Chichibu belt consists mainly of mélanges of pelagic, hemipelagic, and trench-fill

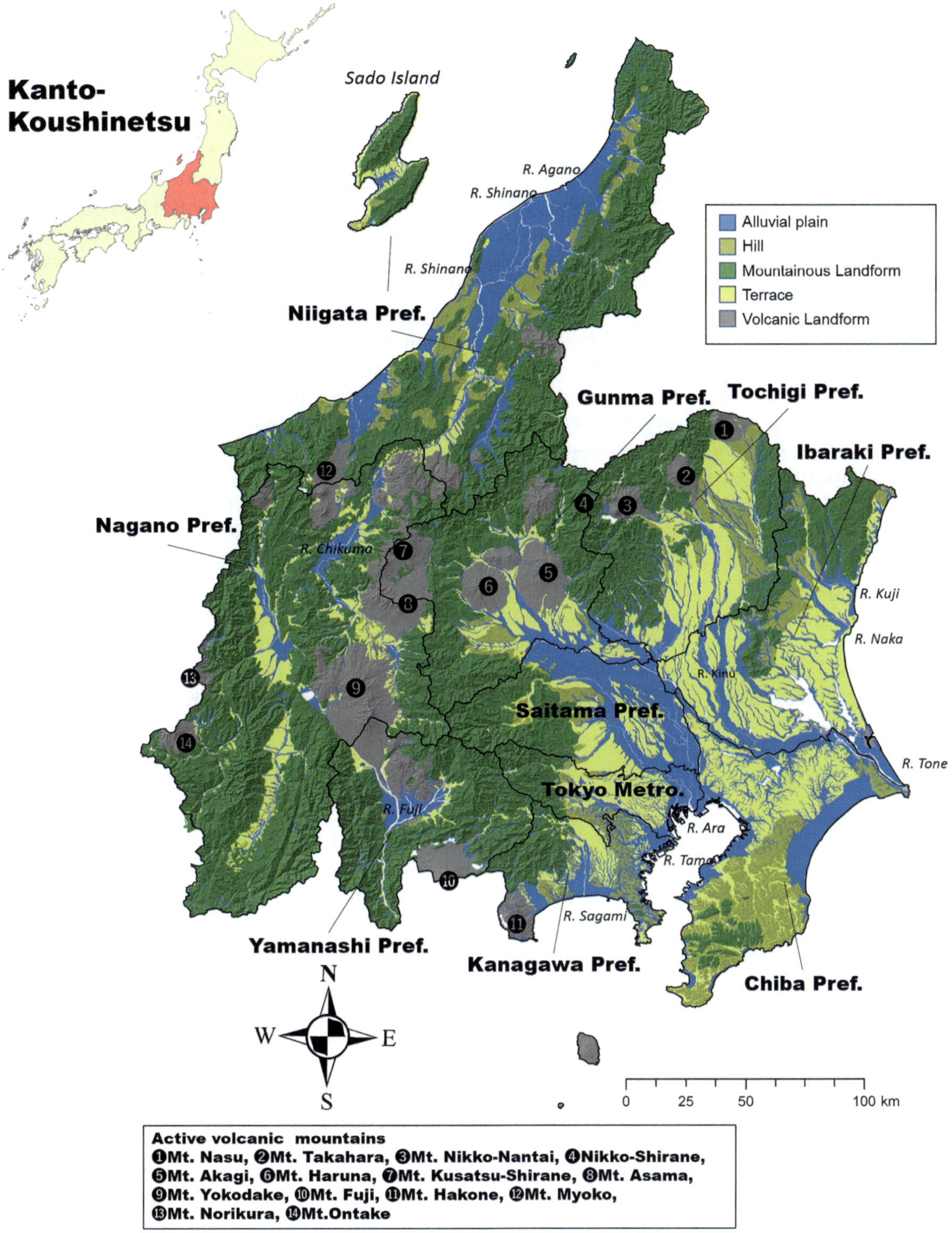

Fig. 7.1 Geomorphological map of Kanto-Koushinetsu region. (Figure supplied by Yusuke Takata)

deposits. Most of the rocks of this belt are also thought to be Jurassic to Cretaceous accretionary deposits. The Shimanto belt consists mainly of alternating beds of sandstone and mudstone and occasionally intercalated mélange units. The rocks of the Shimanto belt are interpreted to be Cretaceous to Paleogene accretionary deposits (Takahashi 2000).

(2) **Geography and geology of the Koushin region**

There are many mountains in Nagano and Yamanashi prefectures whose peaks are higher than 3000 m in altitude, while land that is over 1000 m in altitude accounts for more than 50% of the area of these prefectures. There are some flatlands between mountains, such as the Nagano (Zenkohji) Basin, Saku Basin, Matsumoto Basin, Suwa Basin, Kofu Basin, and Ina Basin. Additionally, the prefectures contain several active volcanoes, such as Mt. Fuji, Mt. Asama, Mt. Yatsugatake, and Mt. Ontake. Relatively small alluvial plains have been formed by rivers in these prefectures, such as the Sai, Chikuma, Tenryu, and Kiso Rivers. These plains are used as paddy field.

The Itoigawa–Shizuoka Tectonic Line runs from north to southeast through the middle of Nagano Prefecture, dividing an area of the prefecture into two geomorphological regions. The western side of the line consists mainly of igneous rocks formed from the Paleozoic Era to the Mesozoic Era. (In fact, all the mountains of the northern, central, and southern Japanese Alps are made of igneous rocks.) To the east of the Itoigawa–Shizuoka Tectonic Line is the Fossa Magna ("Great Rift Belt"), where the basement rocks are mainly marine strata formed in the Neogene and some kinds of igneous rocks. Quaternary eruption products of volcanoes such as Mt. Asama, Mt. Yatsugatake, and Mt. Ontake have accumulated and covered these basement rocks.

(3) **Geography and geology of Niigata Prefecture**

Niigata Prefecture, which faces the Sea of Japan side of central Honshu, is an elongated area that stretches for about 240 km along the Sea of Japan coast from the northeast to the southwest, with an area of 12,584 km^2 (making it the fifth largest prefecture in Japan). Along the prefectural border are mountains ranging from 1500 to 3000 meters in height, and the Agano River and Shinano River, which originate from these mountains (Fig. 7.1). The Echigo Plain, the largest plain in the prefecture, lies between these mountains and coastal dunes. The Kashiwazaki Plain lies in the downstream part of the Sabaishi River, and the Takada Plain lies in the downstream part of the Seki River. Other areas in this prefecture are hilly and mountainous areas. On Sado Island, the Kuninaka Plain spreads between the Oosado Mountains and the Kosado Hills.

The geological features of Niigata Prefecture can be divided into four parts: (1) in the eastern part of the Kasiwazaki–Chiba Tectonic Line, rigid rocks (sedimentary rocks and granitoids) of Paleozoic and Mesozoic formations are mainly distributed and are widely covered by a green tuff layer of Neogene strata; (2) the central part of Niigata Prefecture contains an area called Fossa Magna between the Kasiwazaki–Chiba and Itoigawa–Shizuoka Tectonic Lines, where sedimentary rocks (soft stones such as sandstone and mudstone) of Neogene and Quaternary strata are thickly distributed; (3) in the western part of the Itoigawa–Shizuoka Tectonic Line, sedimentary and metamorphic rocks of Paleozoic and Mesozoic formations are distributed; and (4) in the mountains of Sado Island, volcanic rocks and volcanic sedimentary rocks (green tuff), mainly deposited in Paleogene to Neogene formations, are distributed.

Coastal sand dunes are connected to the plains of Niigata Prefecture along the coastline. About 9000 to 7000 years ago, after the Holocene glacial retreat, sediments were carried by rivers, and these coastal dunes are estimated to have formed about 1000 years ago. As the mouth of the rivers was blocked by the sand dunes, a vast wetland zone was formed between the mountains and the coastal dunes, which became the current Echigo plain. Alluvial plain deposits (unsolidified sediments) are thickly distributed in the plains of Niigata Prefecture, and in the Niigata Plain the thickness reaches 160 m. The area close to the coast is mainly composed of sand and viscous soil, and the ground in the upstream area close to the mountains is mainly composed of gravel and sand.

7.1.3 Climate Condition of the Kanto-Koushinetsu Region

(1) **Climate condition of the Kanto region**

The climate of Japan is influenced by a monsoonal flow that carries moist air from the Indian Ocean and Pacific Ocean. In general, Japan has four distinct seasons, namely, spring (March to May), summer (June to August), autumn (September to November), and winter (December to February). In the Kanto region, it is usually warm year-round, being rainy in the summer and dry in the winter with dry winds. The summer rain is due to seasonal rains (called "*Tsuyu*") which occur from mid-June to mid-July), as well as typhoons.

Mean annual precipitation in the Kanto region ranges from 1000 to 3000 mm (in Tokyo the value is 1528 mm), with the Kanto Plain receiving relatively little precipitation. The southwest area of the Kanto region is a common pathway for typhoons, and this area therefore receives especially high rainfall. The annual mean air temperature is 16.3 °C in Tokyo, and is higher in southern coastal areas than in mountainous zones.

The Kanto Plain is drained by the Kuji, Naka, Tone, Ara, Tama, and Sagami river systems (Fig. 7.1). The Tone river system has the largest drainage area of any river system in Japan (16,840 km^2) and is also the country's second longest river system, at 322 km in length. The Tone river system divides the Kanto region into northern and southern parts. Its annual gross discharge is 4500–7000 × 10^6 m^3. The Naka river system basin has an area of 3270 km^2, and its annual gross discharge is 1500–3000 × 10^6 m^3. The areas of the basins of the Ara, Sagami, Kuji, and Tama Rivers are 2940, 1660, 1490, and 1240 km^2, respectively, and all of these four rivers have annual gross discharges of 500–1500 × 10^6 m^3 (Takamura et al. 1981).

(2) **Climate condition of the Koushinetsu region**

In the Sea of Japan area of the Koushinetsu region, northwestern winter winds bring heavy snowfall. In the summer, the region receives less rain than the Pacific area, though it sometimes experiences extremely hot temperatures because of the foehn wind. The annual average temperature in Niigata Prefecture is 13.6–13.9 °C, and the annual average precipitation is 1500–2800 mm. In winter, there is snowfall of about 1 m on the plains and of over 3 m in the mountains due to the influence of westerly wind from the Sea of Japan. There is more precipitation in winter than summer, and from spring to autumn, there is a relatively high number of sunshine hours.

The climate of the Koshin region varies greatly with altitude, with annual mean temperatures ranging from 6.4–15.0°C and mean annual precipitation ranging from 900–2500 mm. The annual mean air temperature in the foothills of the high mountains is approximately 10 °C. The climate in the northern part of Nagano Prefecture belongs to the climate type of the Sea of Japan, and that in the south of the Koshin region belongs to the typical inland humid continental climate, with large temperature differences between summer and winter and between day and night.

7.1.4 Soils of the Kanto-Koushinetsu Region

(1) **Soil map of the Kanto-Koushinetsu region**

The soil map of the Kanto-Koushinetsu region is shown in Fig. 7.2. In terms of soil great group, Andosols are distributed over the largest area (40% of the total land area) in this region, covering the terrace of the Kanto Plain and the mountainous area in the northeast. Allophanic Andosols (Silandic Andosols) account for approximately 75% of these Andosols, making them one of the most representative soils in the Kanto-Koushinetsu region. Regosolic Andosols (Vitric Andosols) have the second largest distribution area of the Andosol great group in the region, and cover volcanic landforms. Non-Allophanic Andosols, whose distribution area is limited to the western part of the Kanto-Koushinetsu region, are the third most widely distributed Andosol in the region.

Soils belonging to the Brown Forest soil great group have the second largest distribution by area in this region. These soils are mainly distributed in the Koushinetsu region, which is characterized by a temperate climate and hilly-mountainous areas without volcanic ash deposition. The third most widely distributed soil great group by area in this region is Fluvic soil, which mainly covers alluvial plains in the Kanto region and Niigata Prefecture. In the Japanese Soil Classification System, Fluvic soils are subdivided by the condition of the groundwater table into Gley Fluvic soils, Gray Fluvic soils, and Brown Fluvic soils. Gley Lowland soils and Gray Lowland soils are mainly used for rice cultivation in this region.

(2) **Volcanic ash soil in the northern Kanto region**

The northern Kanto region, which includes Ibaraki, Tochigi, and Gunma Prefectures, is located near Tokyo, which has a huge consumer market for agricultural products. Eight active volcanoes (Mt. Nasu, Mt. Takahara, Mt. Nantai, Mt. Nikko-Shirane, Mt. Kusatsu-Shirane, Mt. Akagi, Mt. Haruna, and Mt. Asama) are located in the western part of this region (Fig. 7.1). These volcanoes are located on the volcanic front line and accordingly are among the most active volcanoes in this area. Therefore, an enormous amount of volcanic ash has been generated from these volcanoes, which has covered the plains in the northern Kanto region with layers of tephra.

Tephra layers, both fallout-deposited and flow-deposited, play an important role in characterizing soils and landscapes in the northern Kanto region. In the early stage of stratigraphic studies in Japan, tephra sequences were studied in conjunction with geomorphic surfaces and paleontological and geomagnetic studies. For instance, the Pleistocene terraces in Utsunomiya City, in the central part of the northern Kanto region, had been divided into the Hoshakuji Terrace, the Takaragi Terrace, and the Tawara Terrace (Fig. 7.3). The higher terrace (i.e., Hoshakuji) contains thicker deposits of aeolian volcanic ash (so-called loam) lying on top of fluvial and/or marine deposits. These aeolian weathered Pleistocene volcanic ash layers were divided into three stratigraphic formations, namely the Tawara loam, the Takaragi loam, and the Hoshakuji loam. Developments in characterization techniques for tephra identification and dating methods revealed the presence of "marker" tephras spread across the whole of Japan, and have placed regional tephra stratigraphies on a more solid framework (Machida 1999). Based on recent results, this type locality has been divided into eight geomorphic surfaces (Suzuki and Koike 2000), namely, the Tobiyama-Kamiketsu surface (terrace age = 370–380 ka),

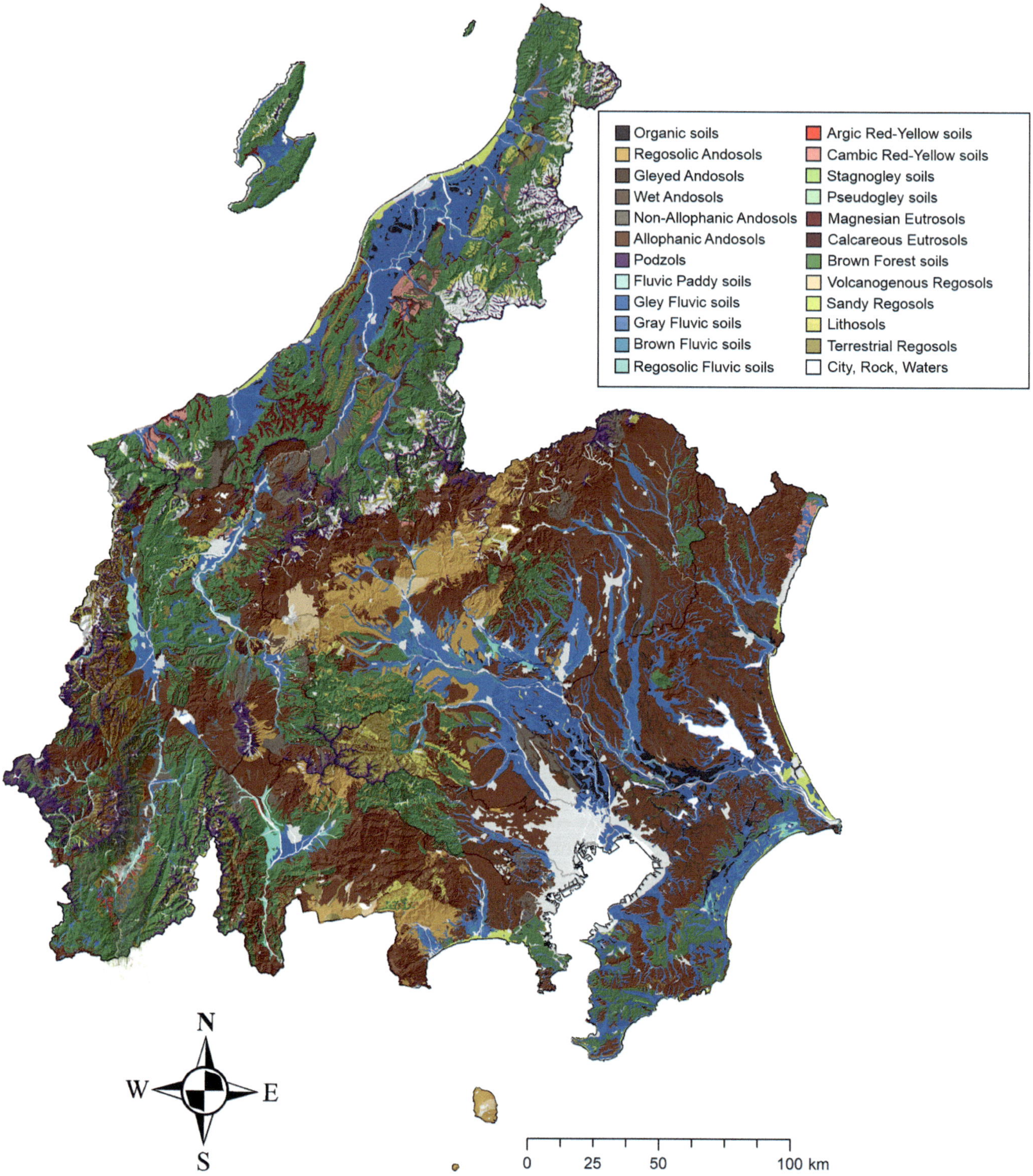

Fig. 7.2 Soil map of Kanto-Koushinetsu region. (Figure supplied by Yusuke Takata)

the Hoshakuji surface (250–260 ka), the Kanuma surface (145–155 ka), the Okamoto-Ohwada surface (100–110 ka), the Takaragi surface (40–65 ka), the Minemachi surface (35–40 ka), the Tawara surface (20–27 ka), and the Kamasusaka surface (13–16 ka). The compilation of a tephra catalogue for Japan has provided fundamental data such as

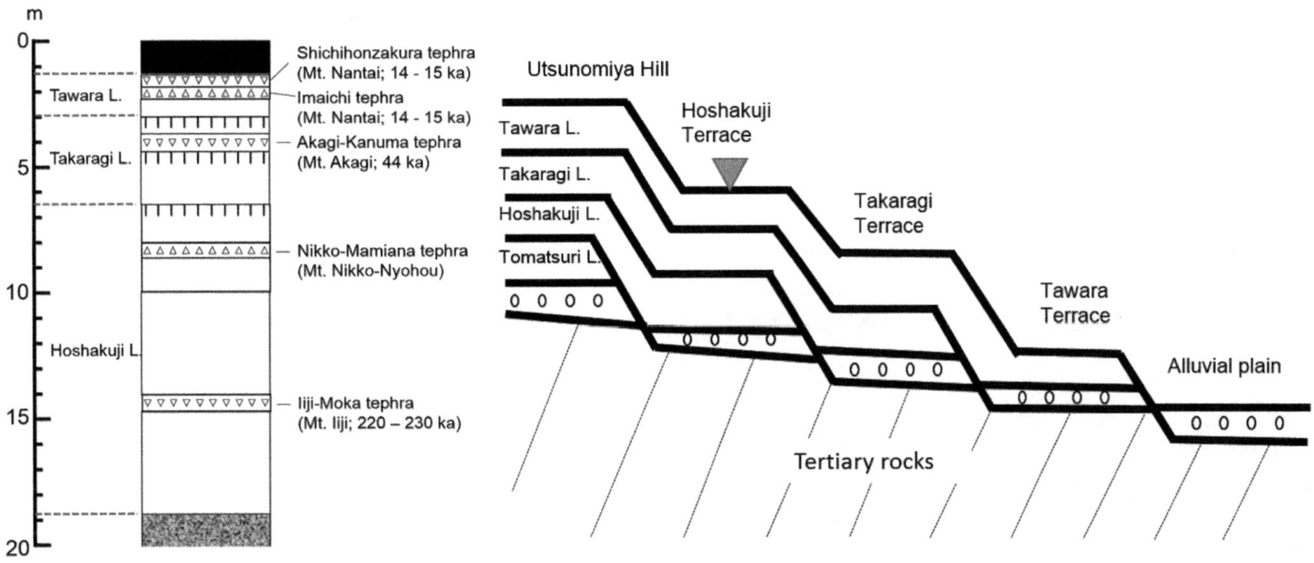

Fig. 7.3 Historical classification of the Pleistocene terrace and stratigraphic profile in Utsunomiya city, northern part of the Kanto region. Stratigraphic profile was modified from Akutsu (1957). Age and origin of tephra were compiled from Yamamoto (2007) and Suzuki (2011)

characteristics of tephra for identification, tephra age, source volcano, volcanic parameters, and tephra chronostratigraphic position (Machida and Arai 2003).

A stratigraphic profile of tephra on the Hoshakuji Terrace in Utsunomiya City (Fig. 7.3) is traditionally divided into three parts, namely, Tawara loam, Takaragi loam, and Hoshakuji loam. The Tawara loam includes the Shichihonzakura pumice and the Imaichi pumice and scoria. The Shichihonzakura tephra, which was ejected from Mt. Nantai, has been weathered to a yellow color, and its main mineral is allophane. The Imaichi tephra is found just below the Shichihonzakura tephra layer and is distributed across almost the same area as that layer. The Imaichi tephra consists of pumice and scoria and is also regarded as ejecta from Mt. Nantai. The tephra has been weathered to a reddish-brown color, and its main minerals are allophane and imogolite. The allophane is found mainly in scoria particles, while the imogolite is found as films covering the surface of pumice grains and filling their interstices. These tephras are widely observed in Nikko City (Fig. 7.4) near Mt. Nantai. Figure 7.4 shows a typical pedon (Cumulic Allophanic Andosol) in the Tawara surface of Nikko City. The Shichihonzakura tephra layer (yellow) and Imaichi tephra (reddish-brown) are observed at depths of 90–124 cm and 124–200 cm, respectively.

The Takaragi loam includes the Akagi-Kanuma pumice (ejecta from Mt. Akagi; age = 45 ka). This pumice has been weathered to a yellow color and to a relatively soft texture, and its main clay mineral is allophane. The Hoshakuji loam consists of thick deposits of volcanic ash materials, including the Mamiana scoria (origin and age are unknown) and the Iiji-Moka pumice (Mt. Iiji; age = 220–230 ka).

The soils generated from volcanic ash are classified as Andosols (Podzolic Andsols, Regosolic Andosols, Gleyed Andosols, Wet Andosols, Non-Allophanic Andosols, Allophanic Andosols). As shown in Fig. 7.2, Andosols cover 63% of the total area of the northern Kanto region (Ibaraki, Tochigi, and Gunma Prefectures). Of these soils, Allophanic Andosols are the dominant Andosol soil group, while Regosolic Andosols are observed mainly in Gunma Prefecture, where fresh volcanic ash (derived from Mt. Asama and Mt. Haruna) is widely distributed.

(3) Volcanic ash soil in the southern part of the Kanto region

In southern Kanto, pyroclastic fall materials derived from Mt. Fuji, are typically andesitic-basaltic in composition, except beginning material of the Hoei tephra (rhyolic) were deposited almost due east direction by the prevailing westerlies. Tephra layer thicknesses and grain sizes thus decrease eastward. Because smaller grains have a larger specific surface area, weathering is enhanced; this is indicated by various properties including Al/Si ratio, the formation of different clay minerals, and the degree of humification. We examined six pedons formed by upbuilding pedogenesis derived from Holocene pyroclastic fall deposits ejected from mainly Mt. Fuji at different distances (km) and azimuths (degrees) from the volcano (Fig. 7.5): (1) the Yamanakako pedon (18 km; 27°); (2) the Ohkura pedon (40 km; 6°); (3) the Nishi-Ohtake pedon (48 km; 1°); (4) the Fujisawa pedon (67 km; 2°); (5) the Sagamihara pedon (64 km; 30°); and (6) the Tachikawa pedon (72 km; 33°). At each site, the Holocene deposits were divided pedostratigraphically into four

Fig. 7.4 Cumulic Allophanic Andosols in Nikko City. There are approximately 100,000 hectares of this soil series. The age of the deposits is 0–70 cm 7 ka; 90–124 cm is referred to as Shichihonsakura pumice, 14 ka; 124–200 cm Imaichi pumice 15 ka. Vegetation is *Quercus serrata*, *Sasaella ramosa* and *Pinus densiflora*. (Figure supplied by Institute for Agro-Environmental Sciences, NARO)

horizons according to eruption age. The uppermost scoria material, the Hoei tephra, was deposited in AD 1707. The material comprising the second horizon was deposited around 1200 years before present (BP), corresponding to the Medieval Warm Period. The materials comprising the third horizon were deposited 1200–4000 years BP. The fourth horizon is a composite of materials that are grouped into a horizon named the "Fuji Black Horizon", which is characterized by relatively high clay and humus contents; these materials formed 4000–10,000 years BP, corresponding to the Jomon transgression and the Holocene Climate Optimum.

With increasing distance from Mt. Fuji, the textures of the soil horizons in each pedons become finer. These textures reflect the original (primary) volcanic particles, and different amounts of secondary particles (clay-sized minerals) that formed to a greater or lesser extent according to the primary particle sizes, which in turn relate to the distance from the source volcano (Mt. Fuji). The Yamanakako soil is a Udivitrand with around 1.8 m of coarse-textured (sandy loam) materials, including scoria (particle size > 2 mm) formed from the Hoei tephra to the base of the third horizon. The texture of the fourth horizon (the Fuji Black Horizon) in this soil is light clay, being finer than those of the upper horizons but coarser than the corresponding horizons in the other five pedons. The soil at Ohkura is a Udivitrand, and the texture of the first horizon (the Hoei tephra) is loamy sand to sand. This soil has the coarsest overall texture of the six pedons because the thickness of the Hoei tephra exceeds that in other pedons. The Nishi-Ohtake soil is a Melanudand and has different textures (sandy clay loam to heavy clay) than the Ohkura soil. The Fujisawa soil is a Melanudand, and the textures of its horizons range from light clay (first horizon; Hoei tephra) to heavy clay (fourth horizon; Fuji Black Horizon). This soil has a thick (2 m) humic horizon and the highest humus content of the six pedons. The Sagamihara soil is a Melanudand, and the textures range from silty loam (first horizon; Hoei tephra) to heavy clay (fourth horizon; Fuji Black Horizon). The Tachikawa soil is a Typic Melanudand. It has not share the characteristics of the other five pedons. This soil has both of lower clay and humus content (despite being far from Mt. Fuji) because it located on a flat terrace with the lowest horizons formed on well-drained alluvium, so, the soil have been received wind erosion as effect on inhibitory action for Fuji black horizon formation.

The contents of SiO_2, CaO, MgO, K_2O, and Na_2O in soil are decreased by eluviation, whereas the contents of Al_2O_3, Fe_2O_3, TiO_2, and MnO_4 are increased because they are residual. Therefore, the values of weathering indices such as silica–alumina molar ratio (SiO_2/Al_2O_3) decrease in the Yamanakako to the Tachikawa soil with increasing distance from Mt. Fuji. The value of (Alo-Alp)/Sio ratio indicating of types of allophane, it increase from Yamanakako (1.3) to Fujisawa (2.0), reflecting the concomitant formation of Si-rich to Al-rich allophane (Fig. 7.6).

The thickness of the Hoei tephra in the Yamanakako and Ohkura soils is 20 and 60 cm, respectively. Both horizons are lightly weathered, and hence have high Si/Al ratios and small amounts of Si-rich allophane, halloysite, and opaline silica, indicating a Si-rich weathering environment. Beyond the Fujisawa area, about 27 km southeast of Ohkura, the soils contain Al-rich allophane and gibbsite, indicating an Al-rich weathering environment. In contrast, the Nishi-Otake soil, located 8.2 km southeast of Ohkura, contains allophane but not gibbsite or halloysite, suggesting that this soil is located in a transitional zone rich in both Si and Al.

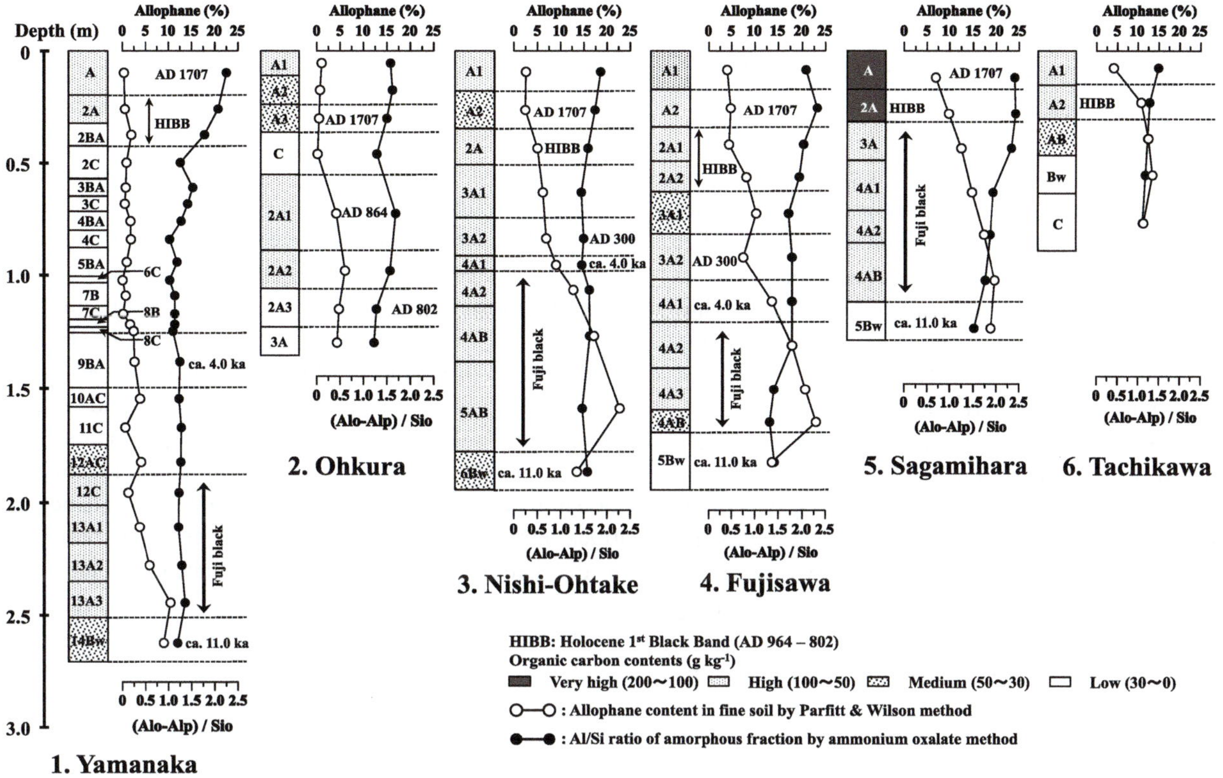

Fig. 7.5 Soil columnar section and vertical distribution of allophane content and silica–alumina molar ratio in fine soil fraction at 6 site in the southern part of Kanto region. (Figure supplied by Hiroaki Sumida and Hiroshi Takesako)

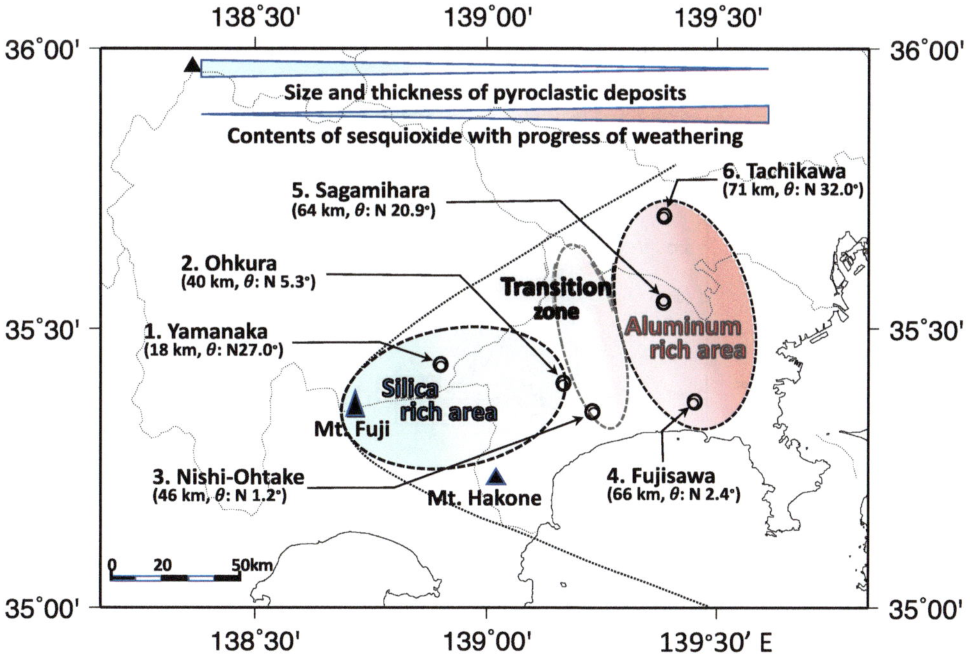

Fig. 7.6 Weathering style of pyroclastic deposits derived from Mt. Fuji in the southern part of the Kanto region. The numbers in parentheses indicate the distance from the central crater of Mt. Fuji to the east (km); θ, showed the angle toward from the east to the north-south. (Figure supplied by Hiroaki Sumida and Hiroshi Takesako)

7.2 Agricultural Features and Soil Properties

7.2.1 Production Status of Major Vegetables and Livestock Products

Japanese agricultural production has undergone various trends in the past six decades. The total production rapidly expanded until about 1980, and then gradually declined after around 1995. The crop cultivation sector experienced a similar trend, but in the livestock sector, production has started to rise again since around 2010 after decreasing during 1985 and 2005 (Fig. 7.7a). Regarding categories of agricultural production, a remarkable decrease in rice production has taken place recently. Furthermore, vegetable production had been declining until 2005, but it has started to rise again recently (Fig. 7.7b). In all livestock categories production has started to rise again in recent years, following a decrease which took place until around 2005 (Fig. 7.7c), although there are some differences in degree and timing between categories.

When calculating Japan's agricultural statistics, the country is often divided into 10 local regions and prefectures, one of which is the Kanto region (see the horizontal axis in Fig. 7.8b). In terms of the recent agricultural production status in the Kanto region, between 2011 to 2015, the region's average agricultural production accounted for 23.0% of the national production, which was the largest share of the 10 local regions and prefectures; during this period, the Kanto region accounted for 25.6% of the crop cultivation in Japan, representing the largest contribution of any region or prefecture, and for 17.8% of the livestock sector, representing the third largest contribution, after the Kyushu region and Hokkaido Prefecture. This fact shows that the Kanto region is one of the major agricultural areas of Japan. The high productivity of the Kanto region is especially clear in the crop cultivation sector, considering that the arable land area of the region is 16.3% of Japan's total, the third largest after Hokkaido Prefecture and the Tohoku region. The production of vegetables, pigs, and flowering plants in the Kanto region is higher than the national average (Fig. 7.8a). Furthermore, the vegetable production of the Kanto region is the highest in Japan (33.8% of the total national production), its pig production is the second highest (27.2%), after the Kyushu region, and its production of flowering plants is the highest (25.6%). The region also produces the second highest amount of fruits (22.2% of the total national production), rice (19.2%), and dairy cows (17.1%). In order to identify the regional features, the difference between the statistics for various agricultural categories in each region or prefecture and the national statistics were calculated, as shown in Fig. 7.8b. For example, in the Hokuriku region, rice production accounted for 58.9% of the total agricultural production of the region (2.9 times the national average value of 20.3%). Of the 10 regions and prefectures, the agricultural activities of the Kanto region most closely match the national average composition, although the production of vegetables is higher, as stated above.

The Kanto region is located in Central Japan and has a moderate climate. On wide plains, there is easy access to water resources for irrigation including rivers, natural and/or agricultural ponds, lakes, and so on, and in megacities in the region, where there is a huge demand for food, agricultural activities are ongoing with a high activity. In these circumstances, the unique local feature of the Kanto region is

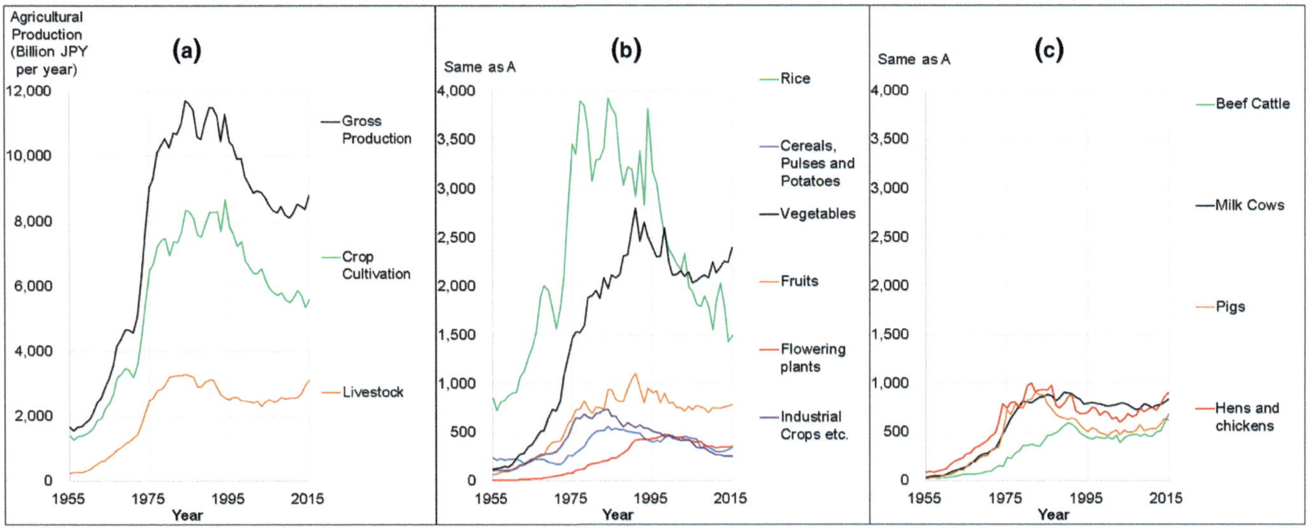

Fig. 7.7 Long-term trends of Japan's agricultural production (Billion Japanese Yen (JPY) per year) **a** total amount and sum of agricultural sectors **b** categories in crop cultivation sector **c** categories in livestock sector. (Figure supplied by Kenji Kanazawa)

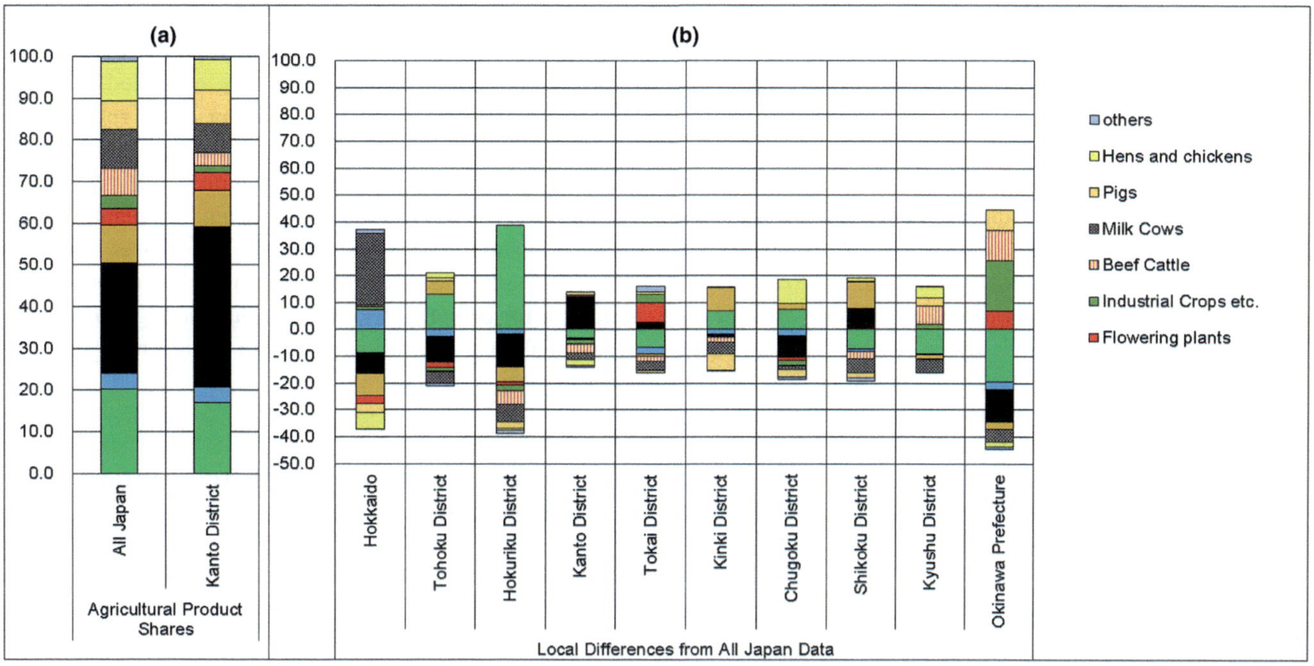

Fig. 7.8 Local differences from all Japan data in agricultural product composition during 2011–2015 **a** composition of agricultural products by categories in all Japan and the Kanto region **b** local differences from all Japan data in agricultural product composition. (Figure supplied by Kenji Kanazawa)

obviously its vegetable production, especially of leaf and stem vegetables (which are not suitable for long-term storage), although the Kanto region resembles the whole of Japan in terms of the composition of its agricultural production at the category level.

7.2.2 Anthropogenic Soils in Reclaimed Areas

(1) **Increase in anthropized area accompanied by urbanization**

The Kanto region is a large, highly populated area with a high population density. The region is widely covered with concrete and asphalt in the form of buildings and road constructions. These urbanized areas have been extended by the conversion of agricultural areas that took place during Japan's period of high economic growth after World War II. The rate of urbanization of agricultural areas was highest in 1992 (MLIT 2007). Although the conversion rate has decreased recently, the decrease in agricultural land continues. The area of agricultural land reduced by 40% between 1965 and 2005, while the area of urbanized land increased by 50% in the same period (MLIT, 2007). However, since the beginning of the twenty-first century, the rate of increase in the area of urbanized land has been reducing due to decreasing population. The amount of greenery areas is also increasing along with the increase in the urbanized area in the Kanto region from 2000 to 18,000 ha in this half century. The vegetation in these greenery areas stands on human-made soils.

(2) **Human-made soils under greenery areas**

In the Tokyo metropolitan area, urban green spaces with large areas have long land-use histories of anthropogenic management. Shinjuku Gyoen Park, located in central Tokyo, has been managed since the late sixteenth century, when it was the private property of the Daimyo (Japanese feudal lords). The area has been developed by small- to large-scale civil engineering works for housing and agriculture both before and after WWII. Artificial plantation has also been established on the land through large-scale civil engineering works. Civil engineering processes can be identified on soil profiles. A drastic change in soil color, structure, and texture indicates a process of cut and bank at the ground surface of the park. Frequent small-scale civil engineering works can also be recognized by changes in soil softness (Uoi et al. 2013). Despite the soil disturbance, these soils can be classified as Haplic or Humic Allophanic Andosols (Allophanic Andosols Reformed Phase) according to the Soil Classification System of Japan. Additionally, soils containing many artifacts in the soil profile are sporadically found in the park, and are classified as "Mineral Artifactual Soils."

The Institute for Nature Study, located in the Minato Ward in Tokyo, has also been the site of civil engineering processes since the eighth century. The presence of a large-scale earthwork constructed through cut and banking to create a residential area has also been confirmed. Disturbed soils through the earthwork have already started soil formation processes affected by the development of vegetation. Accumulated humus in the soil profile indicates vegetation succession by stable carbon isotope analysis, indicating that soils on the earthworks were naturally processed over several centuries (Kawai et al. 2015).

Andosols have potentially occupied a large area in the Tokyo metropolitan area. Soils taken by historical civil engineering works demonstrate typical andic properties, as indicated by a high content of non-crystalline minerals, high phosphate retention, and high cation exchangeable capacity. However, the physical properties of these soils have been drastically altered, becoming highly compacted through civil engineering processes. A larger amount of artifacts in the soils is also a remarkable feature of their profile after anthropogenic management and accordingly they are classified as human-made soils in most classification systems. It can be recognized from the soil profiles that even the soils under greenery areas are strongly affected by large-scale civil engineering.

(3) **Soils on human-made islands in the Tokyo Bay**

The Tokyo Bay area has already been fully occupied by human-made islands, except for in the sea lane, which is used for transportation. The extension of the sea shore along the Tokyo Bay had already been started in the Edo period as waste landfill. The reclamation of excessively growing urban waste for management purposes has been frequently conducted during the period of high economic growth between late 50s and 1973 when Oil shock occurred in Japan. The human-made islands in the bay area have been used for industrial and urban purposes as well as for greenery areas. The basements of the islands have been constructed with domestic and construction wastes. Even in planting layers for vegetation, construction debris is often present in large amounts, which has resulted in soil alkalinization. Subsoils are highly compacted by heavy machinery used to construct the planting bases, which results in the limitation of vertical root extension to shallow depths. Transported Andosols have been dressed to create a planting surface horizon accompanied by buried construction debris artifacts. Large amounts of construction debris in the planting horizons define the soil as Artifactual soil.

(4) **Soils out of the scope of the soil classification**

The center of the urbanized area in the Tokyo metropolitan area has been highly saturated with buildings, which has caused soils on the ground surface to become sealed as a result of constructions such as roads and buildings. Soils beneath constructions are outside the scope of the Japanese Soil Classification System because they are sealed by technic hard materials for a long time. Soils in green spaces along road constructions, in open spaces between buildings, on building walls and rooftops using tree pits, in planters with a bottom, and in vertical hanging media for plants are not classified using the system because in most cases those soils are completely separated from the natural soil profile and lack time for pedogenesis and material cycling. Urban areas in the Tokyo metropolitan area are represented by white spaces on the soil map.

7.3 Status and Changes of Agricultural Soil Properties

7.3.1 Kanto Region

(1) **Ibaraki Prefecture**

There are 77,000 ha of paddy field in Ibaraki Prefecture, a total of 70% of which are occupied by either Gley Fluvic soils, Gray Fluvic soils, or Fluvic Paddy soils.

The distribution area of upland fields is 74,000 ha, the main soils of which are Allophanic Andosols and Wet Andosols.

(a) Actual condition of paddy soil

In recent years, a reduction in the application rate of organic matter and improvements to the drainage systems in paddy fields have resulted in a decline in the soil organic matter content of these fields, and there are concerns that soil productivity will decline as a result.

The current soil nutrient status in paddy fields in Ibaraki Prefecture is characterized by a slightly high amount of magnesium and potassium, although these are almost at the same level as the prefectural recommendations. However, the level of available nitrogen in the soils is decreasing, and there is concern that farmers are less willing to maintain levels of soil available nitrogen soil. The distribution frequencies of available phosphate and exchangeable potassium in paddy soils are shown in Fig. 7.9. In Ibaraki Prefecture, available phosphate was deficient in 23%, and in surplus in 14%, of paddy fields, while exchangeable potassium was deficient in 4% and in surplus in 28%.

(b) Upland field soil

Allophanic Andosols are the main soil group in upland fields in Ibaraki Prefecture. Liming is one of the most common soil

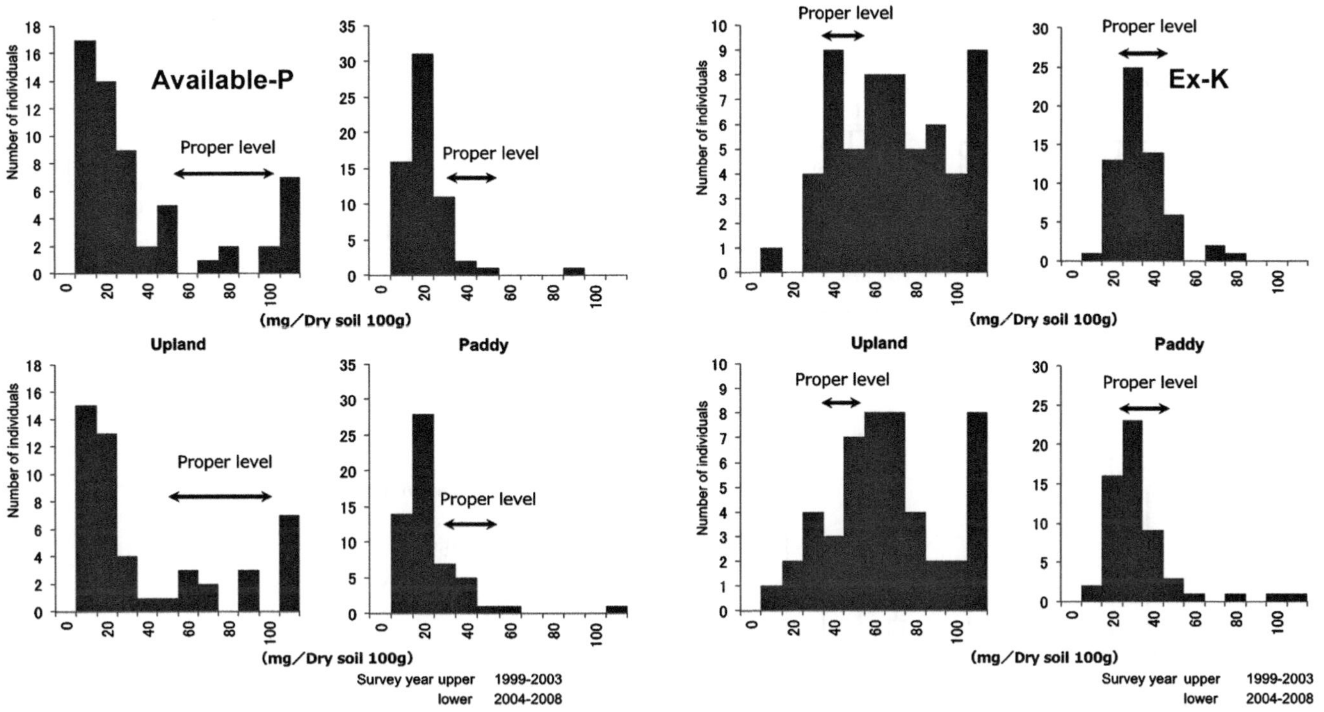

Fig. 7.9 Current status and changes of soil available phosphate (P_2O_5) and exchangeable potassium (K_2O) in Ibaraki prefecture. (Adapted from National Farmland Soil Guidebook 2012)

management techniques to neutralize soil acidity in this area. However, the base saturation level of upland soil has not been increased. Potassium is excessively accumulated in upland soils, and the ratio of magnesium to potassium is low. The level of available phosphate in upland soils tends to be higher than the prefectural recommendation. To improve these nutritional imbalances, the appropriate use of soil amendments based on soil diagnosis is desired. Furthermore, the level of available nitrogen is still low in upland soils. The distribution frequency of available phosphate and exchangeable potassium in upland soil are shown in Fig. 7.9.

Available phosphate was deficient in 57%, and in surplus in 24%, of the upland fields, while exchangeable potassium was deficient in 6% and in surplus in 80%.

(2) **Tochigi Prefecture**

The area of agricultural land in Tochigi Prefecture is 124,200 ha, which accounts for about 20% of the total area of Tochigi Prefecture (6408 km^2). Of this, paddy field covers 96,800 ha, and upland field 27,100 ha, which includes 2190 ha of orchard and 2570 ha of grassland. Wet Andosols occupy 44% of paddy field soils, Gray Fluvic soils 39%, Gleyed Andosols 9%, and Gley Fluvic soils 6%. Allophanic Andosols occupy 81% of upland field and orchard soil and Brown Forest soils 7%.

(a) Paddy fields

The plow depths (Fig. 7.10) of half of the monitoring points in Tochigi Prefecture were less than the prefectural criterion (15–20 cm). Almost half of the points were within the appropriate value for soil pH (6.0–6.5) (Fig. 7.10). Many points exceeded the appropriate value for soil Truog-P_2O_5 (100–150 mg P_2O_5 kg^{-1}) (Fig. 7.10), but there were points of low value. The soil Truog-P_2O_5 value increased a little between 1979 and 2003, but decreased a little between 2004 and 2013.

(b) Upland fields

Approximately 75% of monitoring points had plow depths that were less than the prefectural criterion (20 cm). Half of the points had soil Truog-P_2O_5 values of between 200 and 600 mg P_2O_5 kg^{-1}, with values that were lower and higher than this range also being observed. The same tendency of increasing soil Truog-P_2O_5 values has continued from 1979 to the present.

(3) **Gunma Prefecture**

The total area of agricultural soil in Gunma Prefecture is 76,300 ha. This consists of paddy (28,400 ha) and upland

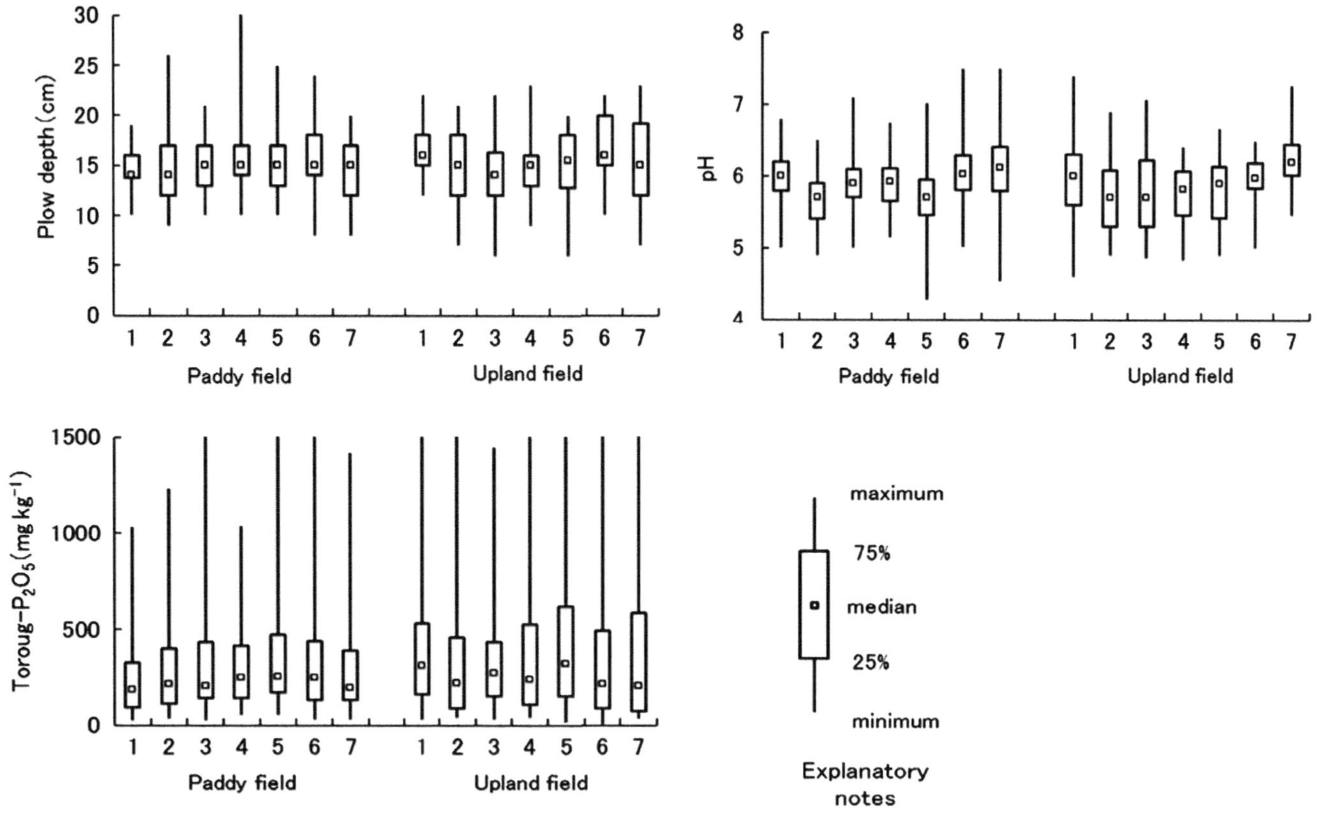

Fig. 7.10 Current status and changes of plow depth, soil pH and soil available phosphate (P_2O_5) in Tochigi prefecture. (Adapted from National Farmland Soil Guidebook 2012)

(47,900 ha, including orchard soil) fields. The most dominant soil group in paddy fields is Fluvic Paddy soil (40% of paddy fields) followed by Gray Fluvic soils (19%), Wet Andosols (16%), and Gley Fluvic soils (10%). In the upland fields, Allophanic Andosols are predominant (38%), followed by Regosolic Andosols (35%), Brown Fluvic soils (7%), and Brown Forest soils (5%) (Statistical Yearbook of the Ministry of Agriculture 2009). Figure 7.11 shows the actual conditions of soil in Gunma Prefecture based on the results of a survey from the first round (1979) to the fourth round (1994–1998) of the "Soil Environmental Monitoring Project", which was carried out in four regions of the prefecture (north, central, western, and eastern).

(a) Paddy soil

The depth of the plow layer is small in the eastern part of Gunma Prefecture where Gley Fluvic soils are dominant. However, a significant change in depth is not observed in most years. Soil base saturation tended to slightly decrease throughout the region, and in 60% of paddy fields, soil pH is below 6.0. There is no remarkable trend in the content of plant-available phosphate, although about 30% of paddy fields contain relatively high amounts (≧0.3 g kg^{-1}). Soil total nitrogen and plant-available nitrogen are almost the same over the region and throughout the year.

(b) Upland soil

The depth of the plow layer tends to be shallow as a whole. In the western and eastern parts of Gunma Prefecture, soil base saturation and soil pH tended to decrease. The level of plant-available phosphate in the soil tended to increase, particularly in the northern area (≧1 g kg^{-1}).

In Konjac field, the plow layer depth and base saturation decreased. However, the plant-available phosphate content of the soil is extremely high, leading to an increase in the average plant-available phosphate content in the northern upland soil.

(c) Orchard soil

Significant change in soil pH has not been observed in recent years. On the other hand, plant-available phosphate content tended to increase, and 80% of soils exhibit a relatively high available phosphorous content (≧0.3 g kg^{-1}). The lime saturation depends on the location: it tended to increase in the northern part of Gunma Prefecture but decreased in the

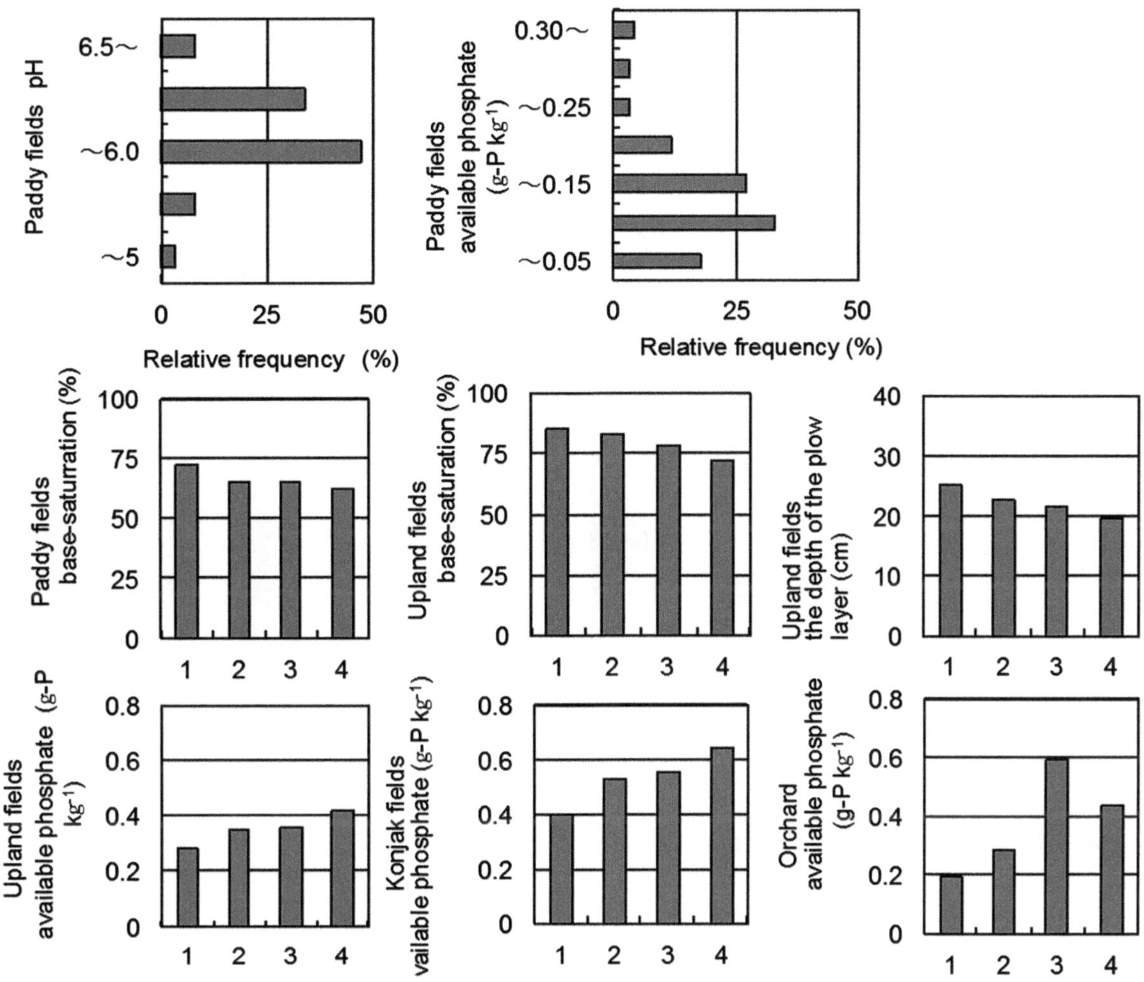

Fig. 7.11 Current status and changes of the cultivated soil properties. The side bar graph is the relative frequency (1994 ~ 97). A bar graph is an annual change (1:1979 ~ 82,2:84 ~ 87,3:89 ~ 92,4:94 ~ 97). (Adapted from National Farmland Soil Guidebook 2012)

western part. Although no significant change in Mg and K contents has been observed, the proportion of soils with deficiencies in both elements is relatively high.

(4) **Saitama Prefecture**

The total area of agricultural land in Saitama Prefecture is 75,000 ha. This mainly consists of paddy (42,000 ha), upland (30,000 ha), and orchard (3000 ha) soils. The most dominant types of paddy soil are Gray Fluvic soils and Gley Fluvic soils, which occupy almost the same area. Peat soils are distributed in the southeastern region. Allophanic Andosols are the dominant soil (34%), followed by Gray Fluvic soils (17%). In orchard fields, Allophanic Andosols are the most dominant soil group in the eastern plateau area, and Brown Forest soils are the most dominant soil group in the Chichibu area.

The results of the "Soil-Environment Monitoring Project" are described below. The average values in the plow layer at monitoring sites (65 paddy soils, 26 upland soils, and 16 grove soils) from the fifth (1999–2003), sixth (2004–2008), and seventh (2009–2013) surveys are presented in Fig. 7.12.

(a) Paddy soil

The concentrations of available phosphate and exchangeable potassium meet a prefectural soil diagnostic criteria, as shown in Fig. 7.12. In recent years, the application of slow-release fertilizer based on prefectural cultivation standards and varietal characteristics of rice has been spreading.

(b) Upland and orchard soil

Generally, fertilizer components tend to accumulate in the upland soil. In particular, exchangeable Ca accumulated in the plow layer of upland fields due to liming and the exchangeable Ca/Mg ratio tends to be higher there. However, the ratio of exchangeable Mg to exchangeable K tends

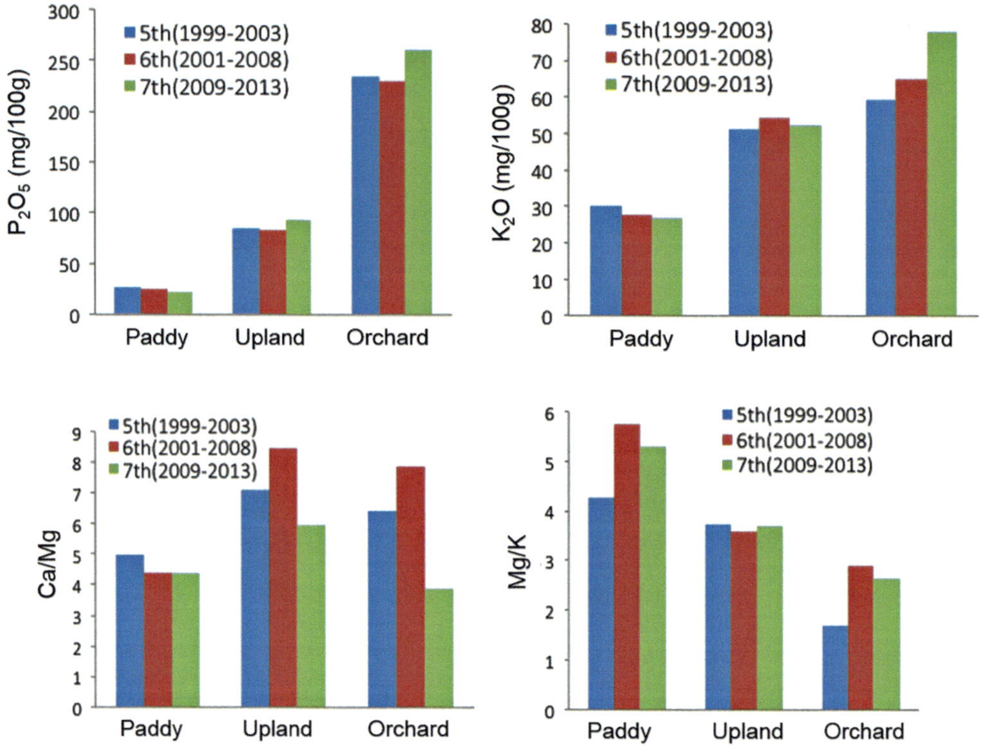

Fig. 7.12 Current status and changes of soil available phosphate (P_2O_5), exchangeable K, Ca/Mg, Mg/K in Saitama prefecture. (Adapted from National Farmland Soil Guidebook 2012)

to be lower (Fig. 7.12). In orchard fields, the levels of soil available phosphate and exchangeable K are higher than in the other land-use types (Fig. 7.12). In tea fields, the amount of liming is low, and the ratio of exchangeable Mg to exchangeable K tends to be lower.

(5) Chiba Prefecture

The total area of paddy field in Chiba Prefecture is approximately 75,000 ha. Among soil groups, Gley Fluvic soils occupy most of this area, 66%, while Gray Fluvic soils occupy 19% and Peat soils occupy 6%. The area of upland field in the prefecture is approximately 54,000 ha. Among soil groups, Allophanic Andosols are the most dominant soil group in upland fields, occupying 68% of the total area; Brown Fluvic soils, which are mainly distributed on the Kujukuri coast, occupy 13%, while Brown Forest soils, which are distributed in the southern hilly area, occupy 14%.

(a) Paddy soil

The appropriate levels of various soil parameters are as follows: soil pH: 5.5–6.5; exchangeable Ca: 2250–3650 mg CaO kg^{-1}; exchangeable Mg: 400–800 mg MgO kg^{-1}; exchangeable K: 100–500 mg K_2O kg^{-1}; available P_2O_5: 50–200 mg P_2O_5 kg^{-1}; and available SiO_2: 100–250 mg SiO_2 kg^{-1}. The measured levels of these parameters were within the appropriate range in most of the surveyed fields (Fig. 7.13).

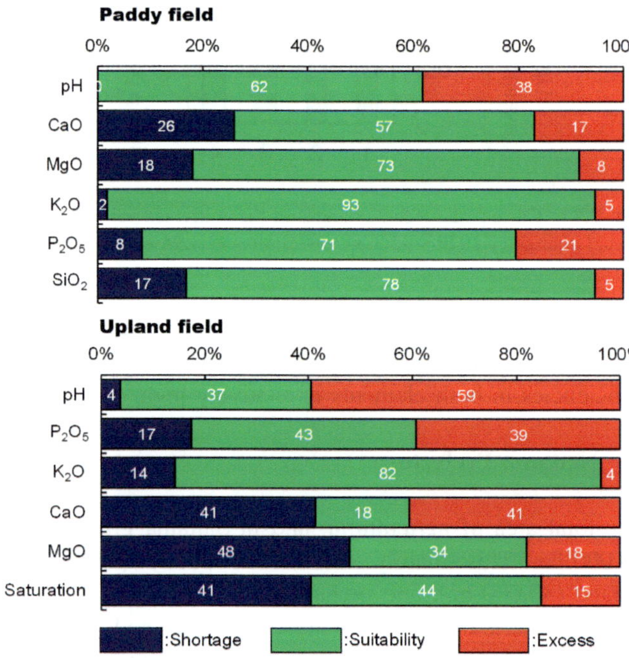

Fig. 7.13 Results of soil diagnostic in paddy field and upland field at Chiba prefecture. Shortage: < criteria value, Suitability; within the criteria, Excess: > criteria value. (Adapted from National Farmland Soil Guidebook 2012)

The concentration of soil available N tended to decrease in the paddy fields with decreasing amounts of N fertilization. The standard amount of N fertilizer application was necessary for maintaining nitrogen fertility.

(b) Upland field soil

Soil pH tended to increase in the upland fields (Fig. 7.13), and 59% of upland fields had soil pH values in excess of pH 6.5. Levels of soil available P_2O_5 were approximately appropriate, although levels were excessive in the Brown Fluvic soils and the Brown Forest soils. Levels of soil exchangeable K were appropriate in most fields. A total of 41% of fields were deficient in exchangeable Ca, and 41% had excess levels. There was large difference among the fields. A total of 48% of fields were deficient in exchangeable Mg. Mg was especially deficient in the Brown Fluvic soils and the Brown Forest soils. Levels of soil available N tended to decrease in the upland fields (data not shown).

7.3.2 Koushin Region

(1) **Yamanashi Prefecture**

The total area of agricultural land in Yamanashi Prefecture is 24,900 ha. Of this, orchard (11,600 ha) is the most dominant land-use type.

Allophanic Andosols are the most dominant agricultural soil group in this prefecture, occupying 31% of the total area of agricultural land, followed by Brown Forest soils (30%), Regosolic Fluvic soils (12%), and Brown Fluvic soils (11%). In orchard fields, Brown Forest soil is the most dominant soil group (43%), followed by Allophanic Andosols (21%) and Brown Forest soils (19%). These three soil groups are all characterized by well-drained conditions and are therefore suitable for fruit cultivation.

(a) Paddy fields

About 60% of paddy fields have a shortage of exchangeable potassium relative to the soil diagnostic criterion, and about 20% of paddy fields have a shortage of available phosphoric acid (Fig. 7.14). Allophanic Andosols are distributed in paddy fields, which have shortage of available phosphoric acid. On the other hand, about 40% of the paddy fields, mainly those covered by Fluvic soil groups, exceeded the criterion of available phosphoric acid (Fig. 7.14). Soil humus content tends to decrease due to the reduction of manure application and the dryness of paddy fields.

In the rice–vegetable rotation fields, there is an excess of soil available phosphoric acid due to residual nutrients from vegetable cultivation.

(b) Upland fields

Upland fields are mainly located in mountainous areas or on the Pleistocene plateau. Allophanic Andosols are mainly distributed in these areas. In about 70% of upland fields, the level of exchangeable potassium exceeded the soil diagnostic criterion (Fig. 7.14), while the level of available phosphoric acid is also excessive in about 80% (Fig. 7.14). Nutrient imbalance is caused by the excessive application of chemical fertilizer, that is, that which is performed without taking into consideration the nutrients remaining from the

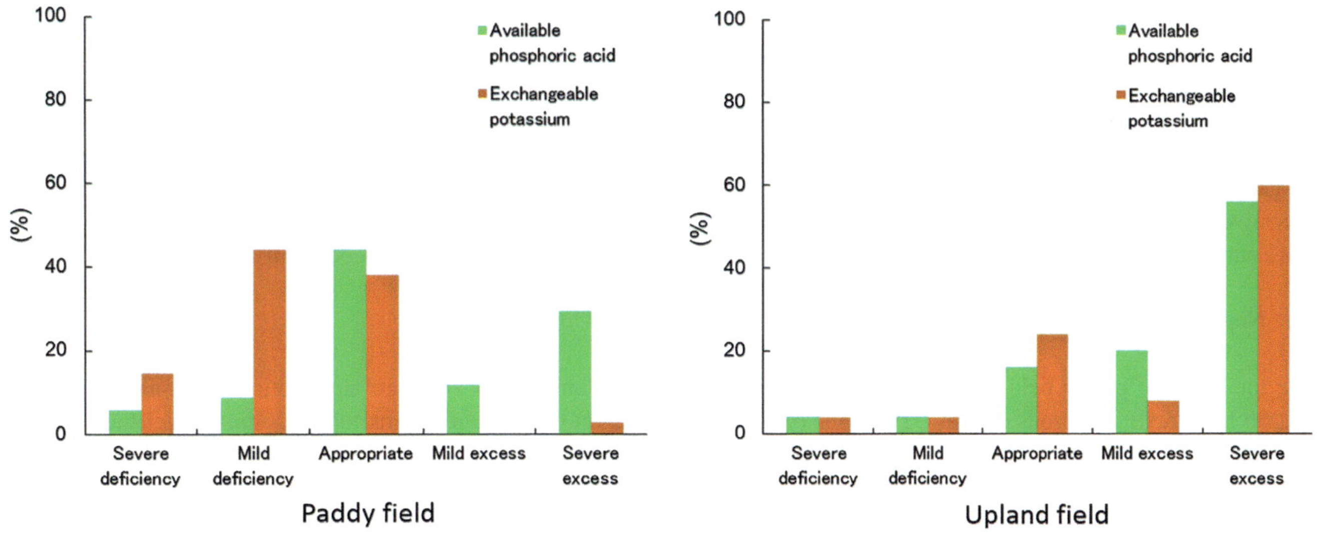

Fig. 7.14 Results of soil diagnostic in paddy field and upland field at Yamanashi prefecture. (Adapted from National Farmland Soil Guidebook 2012)

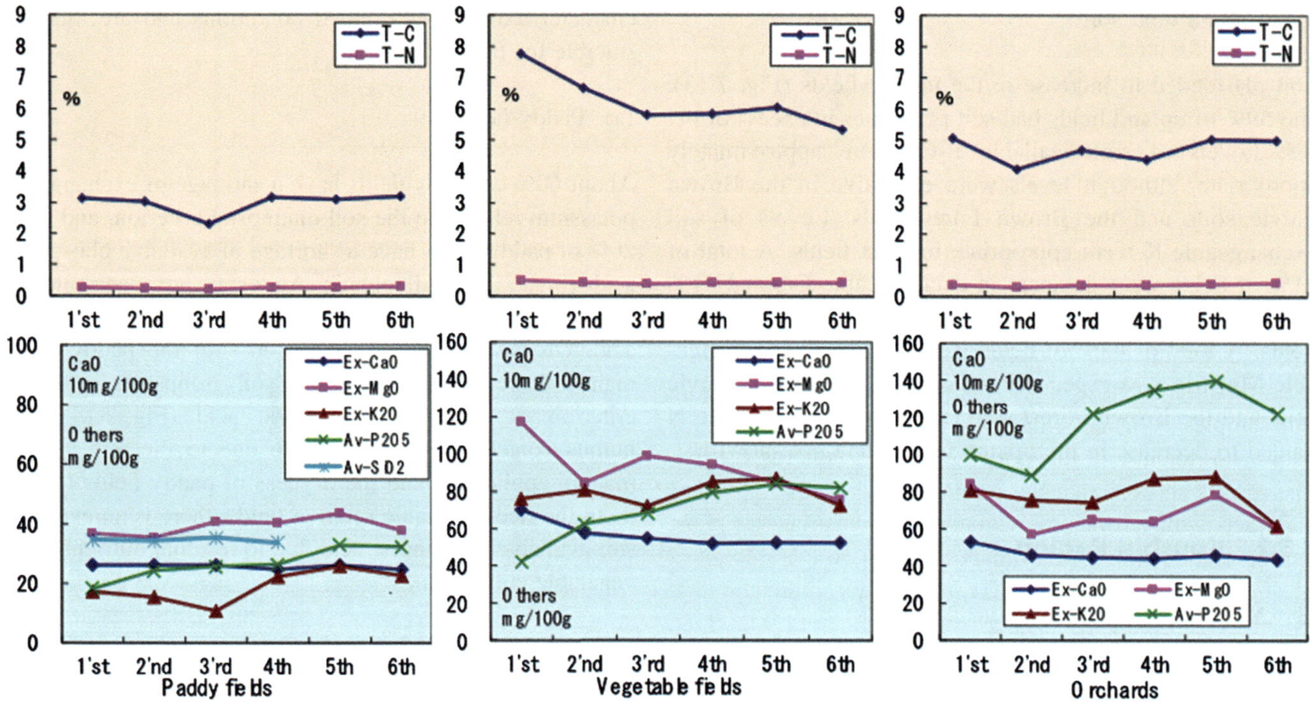

Fig. 7.15 Current status and changes of T–N, T–C, Av.–P and Ex-cations in Nagano agricultural soils. Survey turn; 1st: 1979–1983, 2nd: 1984–1988, 3rd: 1989–1993, 4th: 1994–1998, 5th: 1999–2003, 6th: 2007–2008. (Adapted from National Farmland Soil Guidebook 2012)

previous cultivation nor nutrients supplied from soil organic matter. Decreasing plow layer depth is also a trigger for nutrient accumulation in the surface layer.

(c) Orchard fields

Concentrations of exchangeable calcium and potassium exceed the soil diagnostic criterion in 70% to 80% of the orchard fields in Yamanashi Prefecture. Levels of available phosphoric acid are also excessive in most of the orchard fields due to the decrease in the plow depth. The levels of soil humus are increasing with the spread of herbage cultivation. The soil pH of tea gardens is lower than the soil diagnostic criterion due to the excessive application of N fertilizer. Moreover, in the tea gardens, nitrate-nitrogen tends to be lower than in other prefectures.

(2) **Nagano Prefecture**

The total area of paddy field in Nagano Prefecture is about 70,000 ha. Gray Fluvic soils occupy 21% of the paddy fields and Brown Fluvic soils account for 19%. In upland fields, Allophanic Andosol is the most dominant soil group, accounting for 39% of fields, and the second most dominant soil group is Brown Forest soils, which account for 17%.

(a) Paddy fields

The fertilizer application rate in paddy fields is at an almost appropriate level based on the prefectural guidelines. The amount of available silicon in paddy field soil has tended to decrease. In order to improve the soil chemical fertility of paddy fields, the application of integrated amendments, including minor elements, has been recommended.

The contents of soil total nitrogen and soil total organic carbon have been relatively stable during the survey period (1979–2008), but those of exchangeable cations have been unstable (Fig. 7.15). The content of available phosphate in paddy field soils has tended to increase (Fig. 7.15), and accordingly the amount of phosphate fertilizer applied to paddy fields shall be reduced from now on.

(b) Upland fields

The amount of manure application in upland fields of Nagano Prefecture seems to be moderate, yet the total carbon content of soils in vegetable fields has likely decreased annually over the survey period (Fig. 7.15); as such, it is suggested that the annual amount of manure application be decreased.

The mean contents of minerals in soils of upland vegetable fields have either increased significantly, been relatively stable, or decreased significantly. For instance, the content of exchangeable magnesium has experienced a great decline since the third round of surveys. Additionally, the content of available phosphate has increased remarkably since the first round of surveys, and although no increase was observed between the fifth and sixth round of surveys, its content is still high, and it will therefore be advisable to dress a reasonable amount of fertilizers.

7.3.3 Niigata Prefecture

The total area of agricultural land in Niigata Prefecture is about 174,000 ha, about 90% of which is occupied by paddy fields. Approximately 70% of paddy field soils in the prefecture are classified as Gley Fluvic soils, and about 15% are classified as Gray Fluvic soils. Large areas of the alluvial plains which constitute the flatland (such as the Niigata Plain, Kashiwazaki Plain, Takada Plain) are occupied by heavy clay soils with poor drainage.

Table 7.1 shows the changes in the condition of soil management in paddy fields in Niigata Prefecture for about 30 years. The application of nitrogen fertilizer has been decreased greatly in order to produce good-tasting rice, while the application of phosphorus and potassium fertilizers have been greatly decreased to reduce fertilization costs. However, no significant fluctuation has been seen in the rice yield. The application of rice straw on fields has been greatly increased in order to prevent smoke pollution from incineration and to improve soil fertility. Rice straw is applied to most paddy fields in the prefecture. Compost is difficult to obtain because the livestock industry in the prefecture is partly concentrated and the number of livestock is decreasing. The soil nitrogen content temporarily decreased to less than 5%, but is now 11.9% due to the improvement of soil fertility and the promotion of environmental conservation agriculture. Additionally, improvements to the land have increased the number of multipurpose paddy fields. As a result, it seems that the gradual transformation of soil into well-drained paddy soils due to the lowering of the groundwater level is progressing.

Table 7.2 shows the change in soil physical and chemical properties of rice paddies in the prefecture. There were no changes in the plow layer depth or bulk density. However, the compactness of soils decreased due to rice straw application. Levels of total carbon and total nitrogen were thought to have been maintained by rice straw application, but levels of available nitrogen increased. Levels of available phosphorus increased due to the application of phosphorus fertilizer, which accumulated in soils. Levels of exchangeable calcium and exchangeable magnesium decreased due to the decrease in the application of soil amendments. On the other hand, levels of exchangeable potassium increased due to the application of potassium fertilizers and rice straw. Additionally, although no change was observed in free iron oxide content, the contents of available silicon and easily reducible manganese decreased due to a decrease in the application of soil amendments and reformation into well-drained paddy soils.

In recent years, the deterioration of rice quality due to climate change—for example, high or low temperature, heavy rain, and strong typhoon winds, which are thought to be caused by global warming—has been a problem in Niigata Prefecture. Furthermore, "*Akiochi*" (an autumnal decline in plant vigor) and brown spot rice disease have been spreading, mainly in degraded paddy fields which have well-drained paddy soils. It is well known that these degraded paddy soils are lacking in free iron oxide and have low content of silica, and manganese with low cation exchange capacity. In future, it will be necessary to pay

Table 7.1 Changes in the conditions of soil managements in paddy fields for about 30 years

		1979-83(a)	2009-13(b)	b/a
Amount of nitrogen fertilizer	kg-N ha^{-1}	74	50	0.67
Amount of phosphorous fertilizer	kg-P ha^{-1}	37	19	0.52
Amount of potassium fertilizer	kg-K ha^{-1}	73	39	0.53
Crop yield (brown rice)*	Mg ha^{-1}	4.95	5.29	1.07
Area ratio of rice straw application**	%	34.4	93.9	2.73
Area ratio of compost and barnyard manure application**	%	20.6	11.9	0.58
Construction area of multipurpose paddy field***	%	11.3	48.5	4.29

*; Crop statistics (Source: Ministry of Agriculture, Forestry and Fisheries: 2017)
**; Values in 1980 (Source: Niigata Prefecture 2007 Agriculture, Forestry and Fishery Industry in Niigata Prefecture)
**; Values in 2013 (Source: Niigata Prefecture 2016 Agriculture, Forestry and Fishery Industry in Niigata Prefecture)
***; Values in 1980 and 2015 (Source: Paddy field improvement in Niigata Prefecture: 2017)

Table 7.2 Changes in soil physical and chemical properties of rice paddies in the prefecture for about 30 years

	1979-83(a)	2009-13(b)	b/a
Plow layer depth (cm)	14.1	14.1	1.00
Compactness of soil* (mm)	10.1	7.4	0.73
Bulk density (Mg m^{-3})	0.84	0.83	0.99
pH(H_2O)	5.5	5.4	0.99
Total carbon (g C kg^{-1})	31.1	30.9	0.99
Total nitrogen (g N kg^{-1})	2.47	2.50	1.01
Available nitrogen (mg N kg^{-1})	186	240	1.29
Available phosphorous (Troug P) (mg P kg^{-1})	56.7	75.1	1.32
Exchangeable (Ca $cmol_c$ kg^{-1})	8.35	7.42	0.89
Exchangeable (Mg $cmol_c$ kg^{-1})	3.92	2.85	0.73
Exchangeable (K $cmol_c$ kg^{-1})	0.37	0.53	1.42
Available silicon mg (Si kg^{-1})	52.3	41.1	0.79
Free iron oxide (g Fe kg^{-1})	12.6	13.2	1.05
Easily reducible manganese (mg Mn kg^{-1})	100	92	0.92

*Yamanaka's soil hardness tester index

close attention to changes in the soil environment due to the reformation into well-drained paddy soils and to conduct periodic soil diagnosis to manage plant nutrient practice.

7.4 Globally Important Agricultural Heritage Systems

7.4.1 The "Satoyama" of Sado Island

In 2011, Sado Island (Sado City, Niigata Prefecture) was approved to be a Globally Important Agricultural Heritage System (GIAHS) site by the United Nations Food and Agriculture Organization (FAO), the first such approval in a developed country. GIAHS sites tend to be recognized only as heritage sites with traditional agricultural legacies, as suggested by the name. However, Sado Island was approved due to a combination of traditional ecological knowledge and applications of modern technology (FAO 2017).

The subject of the GIAHS of Sado Island is "*satoyama* in harmony with the Japanese crested ibis (*Nipponia Nippon*)." A "*satoyama,*" as defined by the Japanese Ministry of the Environment, is a landscape complex composed of landscape elements such as villages, rice paddies, agricultural ponds, coppices, and grasslands, located between wilderness and the city (Homma 2012) (Fig. 7.16). Satoyama ecosystems have been one of the most important types of ecosystem in East Asian countries, including Japan, because approximately 50% of endangered species are distributed there, even though it is secondary nature.

The Japanese crested ibis, the most famous endangered bird in Japan, is a symbolic species of satoyama (Fig. 7.17). It feeds on loaches, frogs, aquatic insects, and grasshoppers in rice paddies, and roosts in coppices or planted coniferous forest that surround their feeding grounds. The Japanese crested ibis was widely distributed in the Far East (China, Taiwan, South Korea, North Korea, Eastern Russia, and Japan) until the nineteenth century. However, the population of the species decreased abruptly in each country from the late nineteenth century to the early twentieth century due to overhunting, which caused it to become extinct in most distribution areas except Sado Island, Japan, and Xansi Province, China.

After World War II, the satoyama ecosystems in Japan were changed drastically by land consolidation, agricultural mechanization, and the use of herbicide, fungicide, and pesticide. Satoyamas in mountainous regions was mostly abandoned, because their production efficiency was not high enough compared with that of lowland satoyamas and because fuel wood and charcoal were no longer in demand as energy sources. Thus, as a result of the environmental change, the crested ibis had been extinct in the wild in Japan since 1981. Around the year 2000, an artificial breeding technique for the crested ibis was established in China and Japan, and a project for the restoration of satoyama environments and the reintroduction of the crested ibis in the wild was started in Sado Island in 2002.

Habitat restoration for the reintroduction of the crested ibis is segmented into two approaches. One is to improve the environment of modern rice paddies, and another is to restore abandoned rice paddies and coppices. These projects

Fig. 7.16 Typical satoyama landscape in Sado Island (Iwakubi village). (Figure supplied by Kosuke Homma)

Fig. 7.17 Japanese crested ibis individual, released in the wild in 2008. (Figure supplied by Kosuke Homma)

started in 2002, and have continued for more than 15 years. In 2008, the reintroduction of the crested ibis was started in Sado Island, and fortunately more than 300 individuals now live in restored satoyama.

In order to promote ibis-friendly agriculture and to improve biodiversity in rice fields, many approaches have been attempted in order to modify traditional farming techniques:

(1) Creating a catch drain or a swale ('*e*' in Japanese) approximately 20–30 cm in depth from the soil surface of the rice paddy in order to provide a refuge for aquatic organisms when the fields are drained in early summer (Fig. 7.18).
(2) Irrigating paddy fields in winter to create habitats for aquatic organisms to survive in winter.
(3) Creating fish passes to connect paddy fields with drainage to minimize elevation gaps in order to allow fish to migrate.

These efforts are intended to convert temporal aquatic areas in modern rice fields to permanent wetlands functionally. From the viewpoint of ecology,

Fig. 7.18 Example of the swale in the rice paddy fields. (Figure supplied by Kosuke Homma)

connectivity among landscape elements has been much improved spatiotemporally by reconstructing aquatic areas. The approaches of performing irrigation in winter and creating catch drains are based on traditional agricultural techniques of this region.

(4) Creating biotopes utilizing fallow rice fields and abandoned coppices to maintain the biodiversity and biomass of aquatic organisms all year-round.
(5) Reducing the application of chemical fertilizer, pesticide, and fungicide to less than 50% of the conventional application amount.
(6) The complete prohibition of herbicide application on levees.
(7) The certification of certain farmers as "Eco-Farmers" by Niigata Prefecture.
(8) Monitoring species diversity and the quantity of aquatic organisms in rice paddies every year.

Creating and maintaining biotopes is sometimes too heavy a burden for farmers, because fallow rice fields sometimes spread in mountainous areas that are difficult to access, and preventing this takes a lot of time and money. Therefore, the activities of non-profit organizations and volunteers are very important. The monitoring of aquatic organisms is important for stakeholders in order to recognize the effects of their farming methods and to form continuous motivations.

The Japanese crested ibis is one of the top predators in the food web and acts as an "umbrella species." Attempts for the reintroduction of the crested ibis have improved the satoyama environment and the feeding grounds of the species. However, it has also emerged that there are lots of ecological and/or agro-economic problems in the modern agricultural ecosystem to be solved. Thus, the Sado model, authorized as a GIAHS, realistically shows the direction of eco-farming in the near future.

7.5 Topics of Interest in Suburban Agricultural Land

7.5.1 Concerning "Kanpyo," a Popular Sushi Item

Few people might know of "*kanpyo,*" a food item which is used for *norimaki-sushi* and is loved by many people. Kanpyo is made from the immature fruits of the cucurbitaceous *yugao* (a white-flowered gourd, *Lagenaria siceraria* (Molina) Standl. var. hispida (Thunb.) H.Hara).

The industrial crop production of kanpyo in Japan was about 319 tons in 2014. A total of 99% of this were produced in Tochigi Prefecture, and especially in Shimotsuke City, Mibu Town, and Kaminokawa Town, which are located in the southern part of the prefecture.

The cucurbitaceous yugao belongs to the *Cucurbitaceae* family and is indigenous to the tropical regions of North Africa to India. Although "*hyotan*" (a bottle gourd) is also made from the fruit of the cucurbitaceous yugao, it contains large amounts of cucurbitacin and is thus inedible. On the other hand, kanpyo has been determined to have a low cucurbitacin content. The fruits are almost spherical in shape, are 30 to 40 cm in diameter, and 7 to 8 kg in weight.

The roots of the cucurbitaceous yugao distribute at depths of up to 5 cm from the soil surface under the area over which the lateral branches spread. However, in the case of deep soil plowing, the roots reach 40 cm in depth and the plant tolerates drought. Therefore, land which has a deep surface soil layer and good drainage should be suitable for cultivating cucurbitaceous yugao. Due to the fact that the southern part of Tochigi Prefecture is covered by soil of the Kanto loam layer, which has good drainage and a light weight, cucurbitaceous yugao roots can grow well there. As this is an inland area, the temperature tends to be higher and evening rain showers often occur in midsummer, thus providing the necessary temperature and rainfall required for growing cucurbitaceous yugao. Due in part to the suitable soil and weather conditions, the cultivation of cucurbitaceous yugao became popular in the area.

Mature fruits of the cucurbitaceous yugao which have reached the size mentioned above are harvested in the early morning from July to August. Harvested fruits are immediately set on a lathe-like machine; after removing the skin with the edge of a blade, the pulp is then peeled to produce a long, thin belt (3 cm in width, 3 to 4 mm in thickness, 2 m in length). The peeled fruit belts are then hung on a pole outside or inside a hothouse to dry. Most farmers start to harvest the fruits before dawn in order to finish the harvesting early and allow more time for drying. In order to avoid the oxidation-browning of kanpyo, dried kanpyo belts are normally fumigated with sulfur dioxide prior to packing.

7.5.2 New Aspects on Nitrifying Microorganisms

Nitrification is a process that is performed by ammonia-oxidizing bacteria (AOB) and nitrite-oxidizing bacteria (NOB). Ammonia oxidation is the first step of nitrification. Since the presence of ammonia-oxidizing archaea (AOA) was confirmed in 2005, these organisms are always found to be present in various natural environments, and it is clear that the contribution of AOA to nitrification is very large.

We here describe the new findings and topics regarding nitrifying microorganisms that were clarified in Japanese soil environments over the past decade; we present studies that target soil, compost, and coastal soil environments.

In non-cultivated soil, Onodera et al. (2010) revealed the seasonal variation of the vertical distribution and nitrification activity of the AOA and AOB in temperate forest soil. They suggested that AOB contributed to the net nitrification of the top layer of such soil.

Fujii et al. (2010) investigated the annual fluctuations in the abundance of AOB and AOA living in paddy field soil and the number of bacteria in both groups. A difference in the seasonal variation was shown. They suggested that in Japanese paddy field soil there was difference in the succession of abundances between the *amoA* genes of AOB and AOA.

Hayatsu et al. (2017) isolated an acid-adapted AOB from acidic agricultural soil. This strain, TAO100, belongs to the *Gammaproteobacteria* class, and the optimum pH for its growth was found to be pH 5.0–7.5. However, TAO100 survived under highly acidic conditions (down to pH 2) by forming agglomerates and was not halophilic phenotypically. The authors proposed TAO100 as a novel species of a new genus, *Candidatus Nitorosoglobus terrae*.

Yamamoto et al. (2010) investigated the relative diversity and abundance of AOB and AOA in cattle manure composting. They revealed a change in AOB community structure during composting by denaturing gradient gel electrophoresis analysis and real-time PCR. The results showed that both AOB and AOA played active roles in nitrification in the composting system.

Yamamoto et al. (2011) also analyzed the dynamics of an archaeal community during composting. Their results revealed that the methanogenic archaea and AOA may be the dominant microorganisms in the composting process. In their analysis of the archaeal *amoA* gene, the dominant *amoA* gene sequence showed 99% homology with *Candidatus Nitrososphaera gargensis*. These results suggest that in the composting procedure, AOA, may play a role in ammonia oxidation in spite of the high ammonia concentration.

Nakagawa and Takahashi (2015) isolated AOB that were tolerant to high ammonia concentration from composted cattle manure. The isolated strain belongs to the genus *Nitrosomonas* and grew at a high ammonium concentration of 1000 mM. The authors proposed a provisional taxonomic assignment of *Nitrosomonas stercoris* and registered the strain as a new species.

Ando et al. (2009) investigated the number of bacteria and nitrification activity in the bottom sand of an eelgrass zone in Tanoura Bay, Shizuoka, Japan. They observed a seasonal change in the abundance of AOA and AOB derived from the sand. These results suggest that these microorganisms contributed to nitrification in the sediment of the eelgrass zone.

Matsutani et al. (2011) highly enriched a new marine ammonia-oxidizing crenarchaeote from the coastal sand of the eelgrass zone. This strain (NM25) belonged to *Candidatus Nitrosopumilus* and grew in an ammonium concentration of 15 mM in a medium, and showed the existence of AOA adapted to an environment with a high ammonium concentration.

Ishii (2017) enriched AM1 strains from the coastal sand in an eelgrass zone and examined their physiological characteristics; the AM1 were non-marine cold-adapted NOB belonging to the genus *Nitrotoga*. The abundance of AM1 increased to about 80% of the total bacterial population, and AM1 was the only detectable NOB in the bacterial community. The authors showed that the enriched culture of *Nitrotoga* could adapt to low temperature and also to relatively high concentrations of ammonia.

7.5.3 Drins in Upland Soils and Agricultural Crops

Persistent organic pollutants (POPs) are synthetic chemicals that: (1) pose a risk of causing adverse effects to human health; (2) persist in the environment; (3) bioaccumulate through the food web; and (4) can be transported across international boundaries far from their sources. Currently, there are 24 POPs that are controlled under the Stockholm Convention on POPs. POPs can be classified into the following three broad categories:

- Pesticides such as dichloro-diphenyl-trichloroethane and drins
- Intentionally produced industrial chemicals such as polychlorinated biphenyls
- Unintentional by-products such as dioxin and furans.

In this section, the distribution and contamination of drins (Fig. 7.19) in upland fields and agricultural crops is reviewed. Additionally, effective methods for mitigating drin contamination in agricultural crops are described. Drins consist of three compounds, dieldrin, aldrin, and endrin, as shown in Fig. 7.19. These chemicals are still detected in upland fields as well as certain agricultural crops, even though the use of drins was prohibited in Japan more than four decades ago.

(1) **Distribution of drins in upland fields**

Drins were widely used as pesticides on arable land in upland fields from the 1950s to the 1960s. Due to their persistence and toxic properties, the Japanese government banned the use of drins in 1975. However, dieldrin, which is more stable than the other two drins, still remains in soil. According to Hashimoto (2005), who examined the concentration of drins in 814 surface soils (0–15 cm depth) from agricultural fields in Tokyo, dieldrin residue was detected in 85 soil samples, in concentrations ranging from 0.01 to 2.6 mg kg^{-1}, whereas endrin was detected in only three soils and aldrin was not detected in any.

(2) **Drin contamination in agricultural crops**

It is known that drins are readily taken up by plants of the cucurbit family, such as cucumber, pumpkin, and squash, while such accumulation has not been found in most non-cucurbit crops. Hashimoto (2007) reported that dieldrin was not detected in 90% of 351 analyzed cucumber fruit samples from Tokyo, while the remaining 10% contained dieldrin, nearly half of which (4.8% of the 351 samples) were beyond the tolerance level (0.02 mg kg^{-1} on a fresh weight basis) set by the Food Sanitation Law of Japan. Therefore, dieldrin contamination in cucurbit crops, and particularly in cucumber, is a serious concern.

(3) **Method for the suppression of dieldrin uptake by cucumber fruits**

Currently, there is no official threshold for soil dieldrin concentration. However, the recommended soil dieldrin concentration for cucumber production is below 0.01 mg kg^{-1}; if the dieldrin concentration is beyond this value, the cucumber farmer must take measures to prevent dieldrin contamination in the cucumber fruits. Although discarding contaminated fields and removing contaminated soil are options, the following environmentally friendly, low-cost treatments are also expected for mitigating dieldrin uptake:

(1) Changing crops from cucumber to non-cucurbit crops
This is the most effective and affordable option to prevent dieldrin contamination in crops. However, appropriate crop choice is necessary because, as

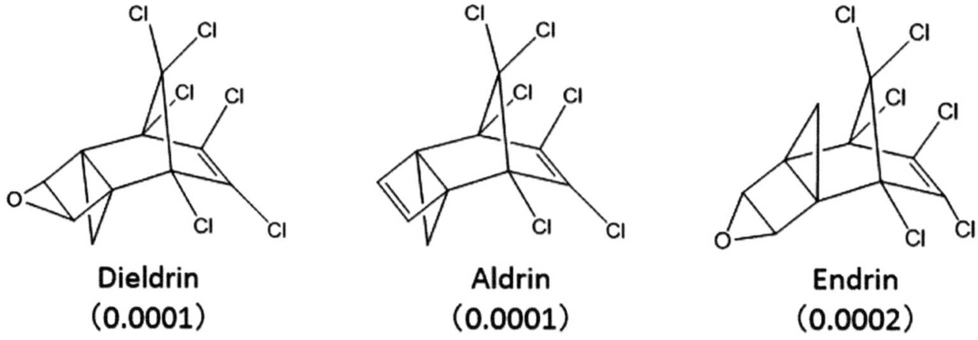

Fig. 7.19 Chemical structure of drins (Number in the bracket is acceptable daily intake; mg kg^{-1} body weight). (Figure supplied by Takayuki Kobayashi)

reported by Yamamoto et al. (1973), dieldrin contamination occurs in root vegetables such as carrot and radish.

(2) Application of low-dieldrin-uptake rootstock into cucumber cultivation
Otani and Seike (2007) proposed that selecting low-dieldrin-uptake rootstock is a promising practical technique to significantly reduce dieldrin concentration in cucumber fruits grown in contaminated fields.

(3) Application of carbonaceous adsorbents
Organic pollutants, including POPs, is strongly adsorbed to carbonaceous materials such as activated carbon and biochar. Saito et al. (2011) demonstrated that the application of activated carbon to dieldrin-contaminated soil is an effective and practical technique for reducing dieldrin concentration in cucumber fruits.

7.5.4 Recycling of Organic Resources

The application of organic matter is an essential practice in local agricultural production. Cattle manure compost is the most commonly used type of organic matter, but organic waste garbage from cities and vegetable waste from fields is produced in massive quantities. This organic waste is a valuable resource. However, it is not used effectively because it comes in many forms and compositions and because of its high water content. If it could be used as fertilizer for agricultural production instead of being disposed of as waste, partnerships could be formed between cities and the farming industry.

Although methane fermentation technology and technologies for converting organic waste into livestock feed and fertilizer have been established to facilitate the effective use of organic waste with high moisture content, most agricultural waste is deposited in landfills and urban waste is disposed of by incineration. The ingredients of animal feed must be fresh. Methane fermentation poses the problem of the need to process residual liquid. Converting organic waste to fertilizer and compost offers the advantage of being able to process such waste completely on farmland. However, converting organic waste to compost requires expansive facilities and lengthy processing; it also poses environmental problems such as foul odors and is not suitable for urban areas. Efforts are therefore being made to develop a way to produce liquid fertilizer by thermal decomposition to convert organic waste with high moisture content to fertilizer in narrow spaces in a short period of time.

When water is placed in an airtight container and heated, the water is converted to steam and expands; the resulting increase in pressure then causes the steam to convert back to water. Raising the temperature to 374 °C and the pressure to

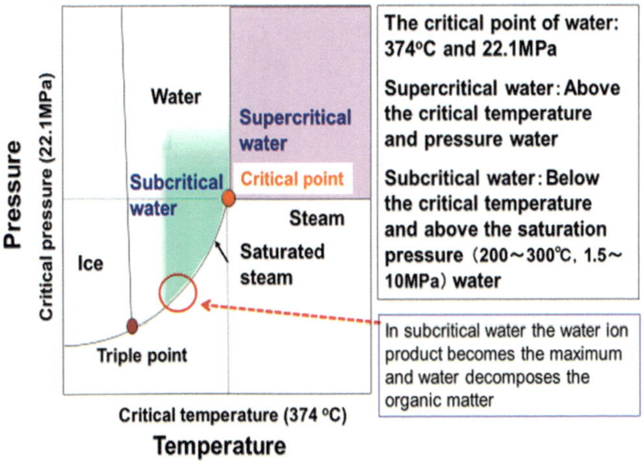

Fig. 7.20 Subcritical water. (Figure supplied by Syunrokurou Fujiwara)

22.1 MPa equalizes the density of water and water vapor; this condition is referred to as the "critical point" of water (Fig. 7.20). When the temperature or pressure exceeds that of the critical point, the water enters a state known as "supercritical," and conversely, when either is below the critical point the water enters a state known as "subcritical."

Ions of subcritical water take up more space and enhance hydrolytic capacity while reducing the dielectric constant, thereby making the water act like an organic solvent. In other words, subcritical water offers both the hydrolytic capacity of water and the affinity of oil. Because its hydrolytic capacity is enhanced, especially in the area of subcritical water, the reaction that occurs here is referred to as "hydrothermal decomposition."

Researchers from Meiji University set up a "hydrothermal decomposition unit" equipped with a 200-L decomposition tank in order to convert organic waste to liquid fertilizer on a practical scale in the Kurokawa Field Science Center (Fig. 7.21). This equipment performs hydrothermal decomposition by introducing high-temperature/high-pressure water vapor from a high-temperature/high-pressure boiler into a 200-L decomposition tank with an agitator. The information collection and control of the entire system can be controlled by a touch screen on the control panel. This equipment was manufactured in 2013 by Fujimura Event Inc. with financial assistance from the Ministry of Education, Culture, Sports, Science and Technology.

The equipment was used to study how to convert vegetable scraps into liquid fertilizer in order to make effective use of agricultural waste. Vegetable scraps from the field science center were thermally decomposed for 30 min at temperatures of 170 to 200 °C to produce liquid fertilizer. It was found that liquid fertilizer could be produced at 170 °C, but that a temperature of 200 °C was more suitable for vegetables with abundant fiber.

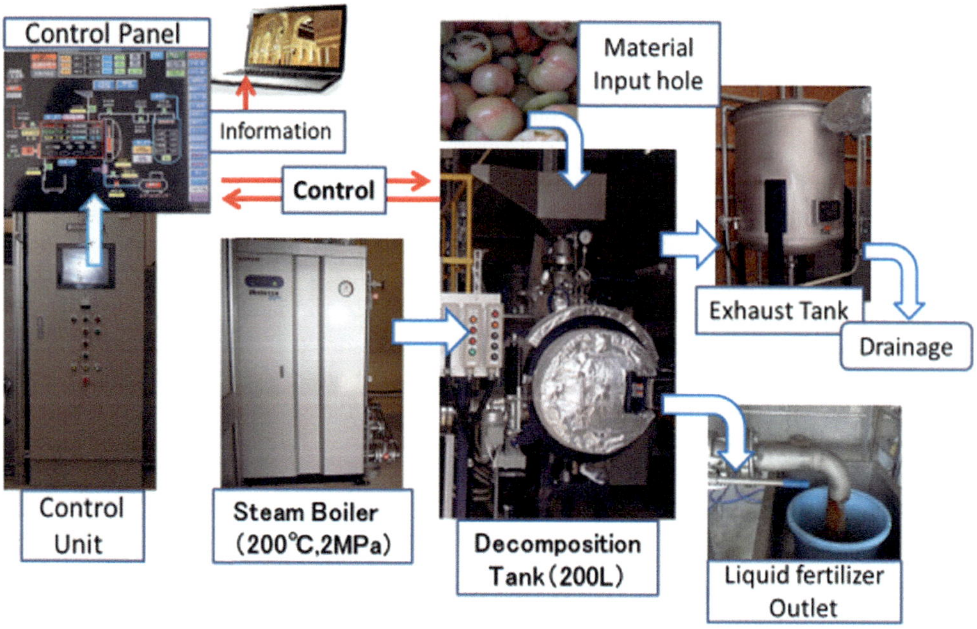

Fig. 7.21 Structure of subcritical water hydrolysis system. (Figure supplied by Syunrokurou Fujiwara)

Fig. 7.22 Cucumber cultivation test of hydrothermally decomposed liquid fertilizer. (Figure supplied by Syunrokurou Fujiwara)

By repeating the vegetable cultivation test using thermal decomposition liquid fertilizer various times with various crop types, such as leafy vegetables and fruits (Fig. 7.22), it was found that growth inhibition can be avoided by either spreading thermal decomposition liquid fertilizer on the soil and decomposing the organic acid before planting the crops or fertilizing after the crops have taken root.

The thermal decomposition of water for agricultural use is a new technology, and there are still many problems that must be solved before it can be practically applied. However, because organic waste can be converted into liquid fertilizer quickly using just water, such conversion is an extremely effective way to use organic waste in urbanized areas without contaminating the environment. In addition, because it can detoxify dead or decayed plants, this technique is a beneficial technology for developing countries that have a shortage of fertilizer but may face environmental pollution by organic waste.

References

Akutsu J (1957) On the Kanto Loam (Ash) Formation in the environs of Utsunomiya city. Tochigi Prefecture, Chikyu Kagaku 33:1–11 (in Japanese with English summary)

Ando Y et al (2009) Seasonal changes in abundance of ammonia-oxidizing archaea and ammonia-oxidizing bacteria and their nitrification in sand of an eelgrass zone. Microbes Environ 24:21–27

FAO (2017) http://www.fao.org/giahs/giahsaroundtheworld/designated-sites/asia-and-the-pacific/sados-satoyama-in-harmony-with-japanese-crested-ibis/en/. Accessed on 2 Apr 2017

Fujii C et al (2010) Succession and community composition of ammonia-oxidizing archaea and bacteria in bulk soil of a Japanese paddy field. Soil Sci Plant Nutr 56:212–219

Hashimoto Y (2005) Dieldrin residue in the soil and cucumber from agricultural field in Tokyo. J Pestic Sci 30:397–402

Hashimoto Y (2007) Reduction of dieldrin concentration in cucumber fruits using Cucurbita rootstocks and activated carbon. J Pestic Sci 32:229–234

Hayatsu M et al (2017) An acid-tolerant ammonia-oxidizing γ-proteobacterium from soil. The ISME J 11:1130–1141

Homma K (2012) Re-introduction of Japanese Crested Ibis. In: Miyazono M, Uchida K and Homma K (eds) Sadogashima Kankyo Taizen (Reviced Version). pp 145–151. Sado City. (in Japanese)

Ishii K (2017) Enrichment and physiological characterization of a cold-adapted nitrite-oxidizing Nitrotoga sp. from an eelgrass sediment. Appl. Environ. Microbiol., 83:e00549–17

Japanese Soil Conservation Research Project Nationwide Council (2012) Japanese soil conservation research project Nationwide council (ed) National Farmland Soil Guidebook,. Japan Soil Association, Tokyo (in Japanese), pp 36–57

Kawai N, Murata T et al (2015) Influence of historical manmade alterations on soil-forming processes in a former imperial estate (Shirogane-goryouchi), the Institute for Nature Study: development of a soil evaluation technique and importance of inventory construction for urban green areas. Soil Science and Plant Nutrition 61:55–69

Machida H (1999) Quaternary Widespread Tephra Catalog in and around Japan: Recent Progress, The Quat. res. (Daiyonki-Kenkyu), 38, 194–201 (in Japanese with English summary)

Machida H, Arai F (2003) Atlas of Tephra in and around Japan (Revised edition), University of Tokyo Press, pp 336

Matsutani N et al (2011) Enrichment of a novel marine ammonia-oxidizing archaeon obtained from sand of an eelgrass zone. Microbes Environ 26:23–29

Ministry of Agriculture, Forestry and Fisheries 2017, Crop statistics (2017) http://www.maff.go.jp/e/data/stat/90th/. Accessed on 4 Aug 2018

Nakagawa T, Takahashi R (2015) Nitrosomonas stercoris sp. nov., a chemoautotrophic ammonia-oxidizing bacterium tolerant of high ammonium isolated from composted cattle manure. Microbes Environ 30:221–227

Niigata Prefecture (2007) Agriculture, Forestry and Fishery Industry in Niigata Prefecture (in Japanese)

Niigata Prefecture (2016) Agriculture, Forestry and Fishery Industry in Niigata Prefecture (in Japanese)

Niigata Prefecture (2017) Paddy field improvement in Niigata Prefecture (in Japanese)

Onodera Y et al (2010) Seasonal change in vertical distribution of ammonia-oxidizing archaea and bacteria and their nitrification in temperate forest soil. Microbes Environ 25:28–35

Otani T, Seike N (2007) Rootstock control of fruit dieldrin concentration in grafted cucumber (Cucumis sativus). J Pestic Sci 32:235–242

Saito T, Otani T et al (2011) Suppressive effect of soil application of carbonaceous adsorbents on dieldrin uptake by cucumber fruits. Soil Sci Plant Nutr 57:157–166

Suzuki T (2011) Quaternary tephrostratigraphy in the Kinu River Lowland, Northeast Japan: eruptive history of Quaternary volcanoes and landform development. J Geol Soc Japan 117:121–133 (in Japanese)

Suzuki T and Koike K (2000) Kinugawa Lowland, In Regional Geomorphology of the Japanese Islands, Vol 4, pp 172–183 University of Tokyo Press, Tokyo (in Japanese)

Takamura H et al (1981) The study on mapping of hydrologic environment, grants in aid for scientific research of the 1980, miniatry of education, culture and science, Japan

Uoi N, Kawahigashi M et al (2013) Estimation of soil carbon storage in Shinjyuku Gyoen National Garden based on land use history and soil compactness profiles. Geogr Rep Tokyo Metrop Univ 48:13–24

Yamamoto T (2007) Tephrostratigraphy of Middle Pleistocene Iiji volcano, Niigata Prefecture: geochronological relationship between the Iiji-Moka tephra and the sea-level change at MIS 7 in the northern Kanto region NE Japan. Bull Geol Surv Japan 58:117–132 (in Japanese with English summary)

Yamamoto N et al (2010) Diversity and abundance of ammonia-oxidizing bacteria and ammonia-oxidizing archaea during cattle manure composting. Microb Ecol 60:807–815

Yamamoto N et al (2011) Archaeal community dynamics and detection of ammonia-oxidizing archaea during composting of cattle manure using culture-independent DNA analysis. Appl Microbiol Biotechnol 90:1501–1510

Chubu Region (Hokuriku/Tokai)

8

Naoto Ogawa, Susumu Asakawa, Jun Murase, Akira Watanabe, Hirotatsu Murano, Hidetaka Sasaki, Kouichi Hosokawa, Masashi Saitou, Minako Kanda, Fumio Uno, Jun Koike, Hitoshi Watanabe, Masahiro Kasuya, Yuhei Hirono, and Masayuki Hara

Abstract

This chapter presents characteristics of soils of the Chubu region, which encompasses between the Sea of Japan and the Pacific Ocean in the middle of Honshu. In the region, there is Mt. Fuji, the active volcano and the highest mountain in Japan. Not only that, the geology of the region is diverse. The region has several mountain ranges, diverse climates and vegetation. So, you can see a lot of different type of soils in the region. In this chapter, we will first give an overview of the soil formation factors and the existing soils in the region. The chapter then describes the agriculture and forestry for the three distinct regions of the Chubu region. The northern part of the Chubu region, called Hokuriku, centers on paddy production on wet and clayey soil by virtue of snow-meltwater and summer high temperature. For the central part of the Chubu region consisting of highland, the sustainable soil management for forestry is described. As for the southern part of the Chubu region, Tokai, characteristics of agriculture for cultivation of vegetables and tea on mainly Red-Yellow soil under mild climate are described, including efforts toward effective use of fertilizers.

Keywords

Paddy field • Pelleting of livestock manure • Red-Yellow soil • Restoration of cadmium-contaminated soil • Sustainable soil management • Tea • Winter flooding of paddy fields • Vegetable

N. Ogawa (✉)
Graduate School of Integrated Science and Technology, Shizuoka University, Shizuoka City, Shizuoka Prefecture, Japan
e-mail: ogawa.naoto@shizuoka.ac.jp

S. Asakawa · J. Murase · A. Watanabe
Graduate School of Bioagricultural Sciences, Nagoya University, Nagoya City, Aichi Prefecture, Japan

H. Murano
Faculty of Agriculture, Meijo University, Nagoya City, Aichi Prefecture, Japan

H. Sasaki
Fukui Public Works Office, Fukui Prefectural Government, Fukui City, Fukui Prefecture, Japan

K. Hosokawa
Regional Agriculture Division, Fukui Prefectural Government, Fukui City, Fukui Prefecture, Japan

M. Saitou
Fukui Agricultural Experiment Station, Fukui Prefectural Government, Fukui City, Fukui Prefecture, Japan

M. Kanda
Reinan Regional Bureau, Fukui Prefectural Government, Fukui City, Fukui Prefecture, Japan

F. Uno
Ishikawa Agriculture and Forestry Research Center, Ishikawa Prefectural Government, Kanazawa City, Ishikawa Prefecture, Japan

J. Koike
Toyama Prefectural Agricultural, Forestry and Fisheries Research Center, Toyama Prefectural Government, Toyama City, Toyama Prefecture, Japan

H. Watanabe
Gifu Prefectural Research Institute for Forests, Gifu Prefectural Government, Mino City, Gifu Prefecture, Japan

M. Kasuya
Aichi Agricultural Research Center, Aichi Prefectural Government, Nagakute City, Aichi Prefecture, Japan

Y. Hirono
Institute of Fruit Tree and Tea Sciences, NARO, Shimada City, Shizuoka Prefecture, Japan

M. Hara
Mie Prefectural Agricultural Research Institute, Mie Prefectural Government, Matsusaka City, Mie Prefecture, Japan

R. Hatano et al. (eds.), *The Soils of Japan*, World Soils Book Series,
https://doi.org/10.1007/978-981-15-8229-5_8

8.1 Soils of the Chubu Region

8.1.1 Soil-Forming Factors

1. **Topography and geology**

The Chubu region is located approximately in the center of Honshu, facing the Sea of Japan in the north and the Pacific Ocean in the south (Fig. 8.1). Highlands, including the Hida Mountains, Kiso Mountains, and Akaishi Mountains, spread inland, and the elevation of the region ranges from 0 m above sea level (a.s.l.) on the coastal plain to 3000 m at the mountain tops. The region contains two large faults, namely the Itoigawa–Shizuoka Tectonic Line and the Median Tectonic Line, and its geology is largely divided by these two faults. Eastward from the Itoigawa–Shizuoka Tectonic Line, rocks from the Neogene Period are mainly distributed. West of the Itoigawa–Shizuoka tectonic line, the region is divided north and south by the Median Tectonic Line. In the northern part (inner zone), plutonic rocks such as granite, low-pressure metamorphic rocks, and middle Paleozoic rocks are mainly distributed, and in the southern part (outer zone), high-pressure metamorphic rocks and superbasic rocks are distributed (Kato 2002). Most of the Hida and Kiso Mountains located in the inner zone consist of granitic rocks. Mt. Norikura and Mt. Ontake, two volcanoes located on the south side of the Hida Mountains, consist of Quaternary andesitic volcanic rocks. On the southern side of these volcanoes, along the boundary between Gifu and Nagano Prefectures, a mountainous area (the Atera Mountains) that extends up to Aichi Prefecture consists of felsic volcanic rocks of the Late Cretaceous (Nohi Rhyolite). The Akaishi Mountains, which are part of the outer zone, consist mainly of metamorphic rocks (the Sanbagawa belt) and sedimentary rocks (the Chichibu belt) (Sakai 2007). From Mt. Fuji to the Izu Peninsula, volcanic ejecta derived from the activity of Holocene volcanoes such as Mt. Fuji, Mt. Hakone, and Mt. Amagi are widely distributed, including volcanic ash, pumice stone, and scoria. Additionally, many volcanic rocks erupted to the ocean floor in the Neogene Period are distributed in the Izu Peninsula (Wakasawa and Takahashi 2012).

The depositional area of Holocene tephra erupted from the active volcanoes in the Chubu region (e.g. Mt. Ontake, Mt. Haku, Mt. Tateyama) is not very clear, but the amount of Holocene tephra in the region could be relatively lower than in other regions. Machida (1999) reported that the Chubu region is also affected by nationally distributed Holocene marker tephras, such as the U-Oki tephra (ca. 9.3 ka) from the Ulleung Island volcano and the K-Ah tephra (ca. 6.3 ka) from Kikai, a major caldera largely submerged off the coast of southern Kyushu. However, these Holocene volcanoes are far from the Chubu region. Machida (1999) also reported that several nationally spread Quaternary tephras, such as the Aso-4 tephra (ca. 84–89 ka) from the Aso Caldera in southern Kyushu and the AT tephra (ca. 24–25 ka) from the Aira Caldera, are observed in the Chubu region. Furthermore, on the Sea of Japan side of the region, the DKP tephra (ca. 43–55 ka), erupted from Daisen volcano, is distributed.

In the Chubu region, alluvial fans are distributed between the mountains and the plain (Fig. 8.2) (Yoshikawa et al. 1981). The Kurobe River alluvial fan, in the eastern part of Toyama Prefecture, approaches the ocean with the uplifting Hida Mountains behind (Kamishima and Takeuchi 2016). From the Middle to Late Pleistocene, alluvial fans and terraces were extensively formed in the northwestern part of Mie Prefecture (Yoshida 1983). In the west, from Cape Omaezaki in Shizuoka Prefecture, hilly lands, diluvial uplands, and alluvial fans developed at the southern foot of the mountains belonging to the outer zone (Wakasawa and Takahashi 2012).

2. **Climate**

The Chubu region is affected by the Asian monsoon and is struck by typhoons in the summer months. Due to these effects, the climate of the Chubu region is divided into the following three distinctive regions (Fig. 8.3): Hokuriku region, on the Sea of Japan side; the Tokai region, on the Pacific Ocean side; and the Hida region, in the Central Highlands.

(1) Hokuriku region: During the winter months, moist monsoon wind blowing from the Sea of Japan, which is stopped by high mountains, brings large amounts of precipitation. The precipitation usually falls as snow, making this one of the snowiest areas in Japan. The snow melts in spring.
(2) Central Highlands: These contain 3000-m-high mountains (the Hida Mountains and Kiso Mountains). In winter, the temperature is low due to lower amounts of atmospheric water vapor, and the temperature difference between day and night is large. This region has a so-called continental climate.
(3) Tokai region: In winter, dry and sunny days continue due to foehn wind. This area is warm even in winter.

3. **Vegetation**

The evergreen broad-leaved forest zone extends to around 800 m a.s.l. on the Pacific Ocean side (400 m a.s.l. on the Sea of Japan side) and includes plains, hills, and low mountainous areas. The summer green broad-leaved forest zone extends up to around 1500 m a.s.l. The subalpine coniferous forest zone extends up to an altitude of around 2400 m a.s.l. The alpine zone spreads beyond the subalpine zone (Fig. 8.4) (Numata

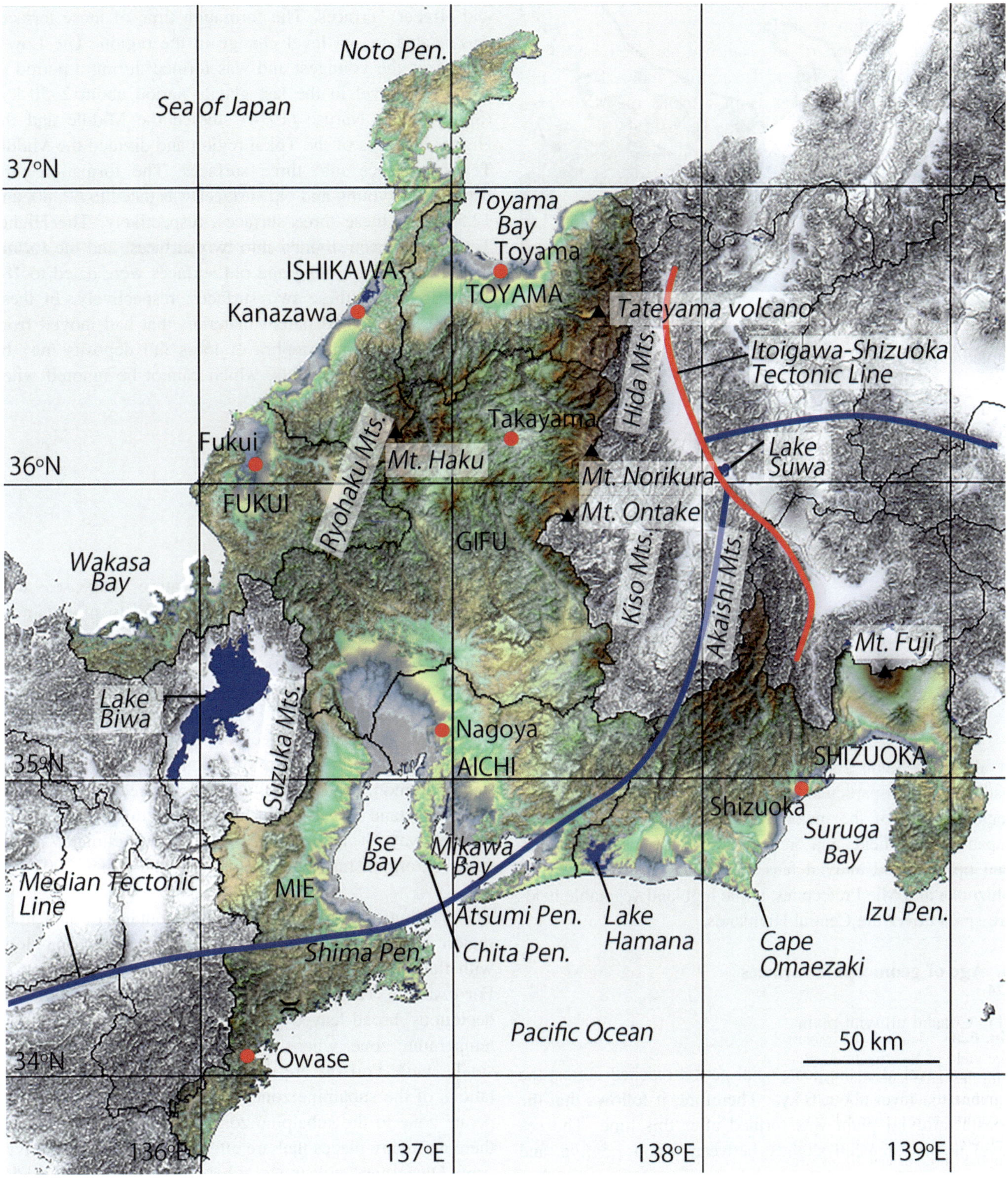

Fig. 8.1 Map of the Chubu Region. The figure was created by Hirotatsu Murano using the free software KASHMIR 3D ver. 9.3.1 (http://www.kashmir3d.com/)

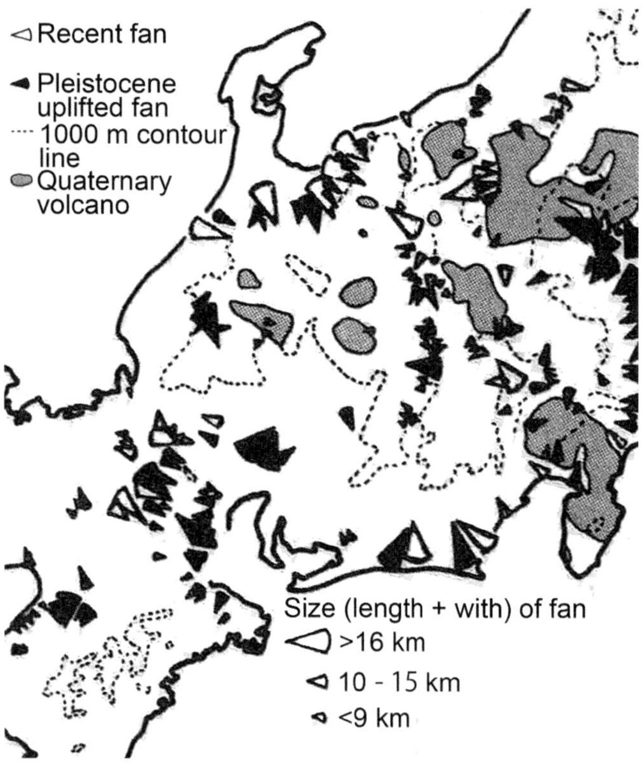

Fig. 8.2 Distribution of alluvial fan in Chubu region Adapted in part Yoshikawa et al. (1981). The figure in Yoshikawa et al. (1981) was reproduced from Toya et al. (1971). The figures were used with permission from the University of Tokyo press and Kokon Shoin publishers

et al. 1972; Sakai 2007). Alluvial plains are often used as paddy fields. In particular, the Sea of Japan side of Hokuriku region is one of the major paddy rice production areas in Japan. Upland fields and orchards are distributed in the diluvial uplands and alluvial fans. Tea fields are distributed in Shizuoka and Mie Prefectures, while highland vegetable fields are spread over the Central Highlands.

4. **Age of geomorphic surfaces**

(1) Coastal alluvial plain

The sea level after the postglacial period reached almost the current sea level about 6 kya. Therefore, it follows that the coastal alluvial plain was formed after this time. The sea level in this area fell slightly between 6 and 1.5 kya, and therefore, many of the alluvial plains are younger than 1.5 ka (Kato 1975).

(2) Terraces

Terraces in the Tokai region on the Pacific Ocean side of the Chubu region are mainly divided into the Lower, Middle, and Higher Terraces. The formation time of these terraces was related to sea level change in the region. The Lower Terrace is the youngest and was formed during a period of falling sea level in the last glacial period about 2–50 kya (Sakai 1963). Naruse (1974) studied the Middle and the Higher Terraces of the Tokai region and divided the Middle Terrace surface into three surfaces. The formation time between the young and old surfaces was dated to 60, 80, and 12.5 ka for these three surfaces, respectively. The Higher Terrace has been divided into two surfaces, and the formation times of the young and old surfaces were dated to 180 and 206 ka for these two surfaces, respectively. In these terraces, redeposited parent materials that had moved from above the slope and tephra or loess fall deposits may be mixed in certain amounts, which cannot be ignored when considering soil formation.

8.1.2 Soils Distribution

(1) **Soils in mountainous areas**

In the Chubu region, the temperature decreases with increasing elevation, from coastal lowland to mountainous areas; the vegetation zone also changes from warm temperate to alpine (Fig. 8.5). For this reason, climate-series changes can be easily observed in the forest soil of this area from Brown Forest soils to Podzols. Furthermore, weather conditions are region-specific, for example, abundant snowfall occurs on the Sea of Japan side, while rainfall is more common in Kiso area, which is located further inland. To understand the soil distribution in this region in an orderly way, it is necessary to consider climate, parent material, organisms, topography, and time (Sakai 2007).

The distribution of forest soils in the mountains of the Chubu region basically changes with increasing elevation, along with the change in climate zone (Fig. 8.6) (Kondo 1967). For example, Haplic Brown Forest soils are distributed in deciduous broad-leaved forests of the mountainous cold temperature zone, where the influence of tephra deposits is small, while Podzols are distributed in conifer forest plantations of the subalpine zone. However, from the cool temperate zone to the subalpine zone of the mountainous area, there are many places that are affected by the tephra derived from Quaternary volcanoes, where Non-Allophanic Andosols are often observed. In particular, there is a dominant Non-Allophanic Andosol distribution area on the Sea of Japan side. However, at the foot of active volcanic mountains, such as Mt. Haku, Mt. Ontake, and Mt. Norikura, Allophanic Andosols are distributed mainly in the cold temperature zone (Kanda et al. 2016).

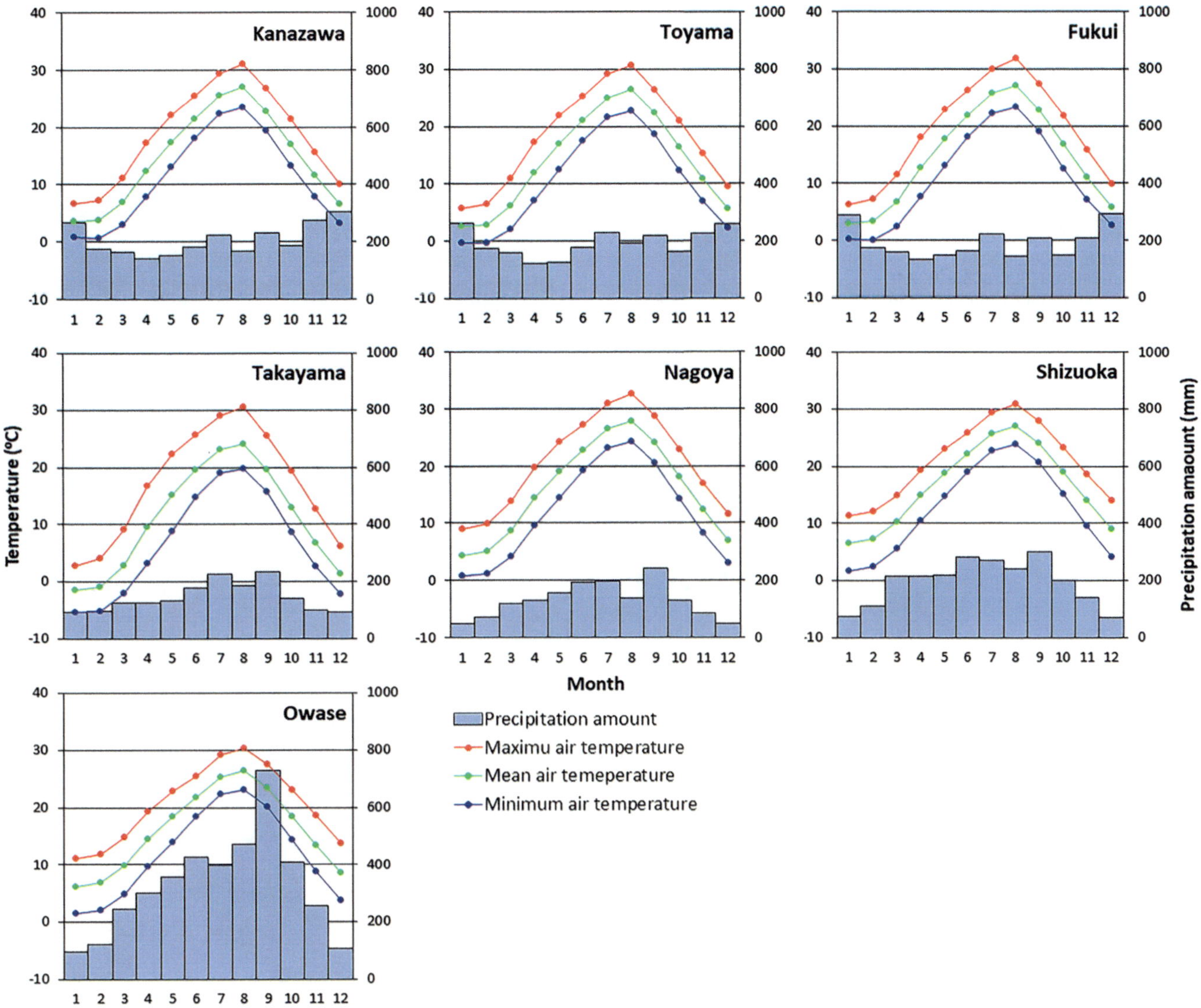

Fig. 8.3 Climographs for seven representative sites within the Chubu region. The figures were created by Hirotatsu Murano based on 1976-2015 data from the Japan Meteorological Agency (2017)

Peat soils and Epi-Peaty Podzols are observed in flat areas or on gentle slopes with low drainage from the upper cool temperate zone to the subalpine zone. Places where Peat soil is distributed in a coherent area are Mt. Hakuba, located to the east of Tateyama volcano (Fig. 8.7), and Mt. Norikura in the Hida Mountains. These peat layers sometimes contain marker tephra that were spread nationwide by eruptions of south Kyushu caldera volcanoes and volcanic tephra from the Chubu region. In the Ryohaku Mountains, the AT tephra and eruption products from Mt. Haku are intercalated in the peat surrounding this volcano (Higashino, 2006). The K-Ah tephra, U-Oki tephra, and AT tephra are observed in the Ohnohara peat layer located about 50 km east-southeast of Nagoya (Arai et al. 1988).

(2) **Soils on hills and terraces**

Reddish Red-Yellow soils are distributed from the foot of the hills to the Higher Terrace of the oldest generation (Fig. 8.8), Red-Yellow soils to Thapto-Red-Yellow Brown Forest soils are distributed on the Middle Terrace surface, and Brown Forest soils are distributed on the Lower Terrace of the youngest generation. This shows the time sequence of soil development. The Reddish Red-Yellow soils are more weathered than the surrounding soils under the temperate humid climate. Most of the silicate and base is leached from the soil, which is rich in iron and aluminum, showing red and yellow layers (Matsui and Kato 1962). The Red-Yellow soils in the Chubu region are considered to have formed

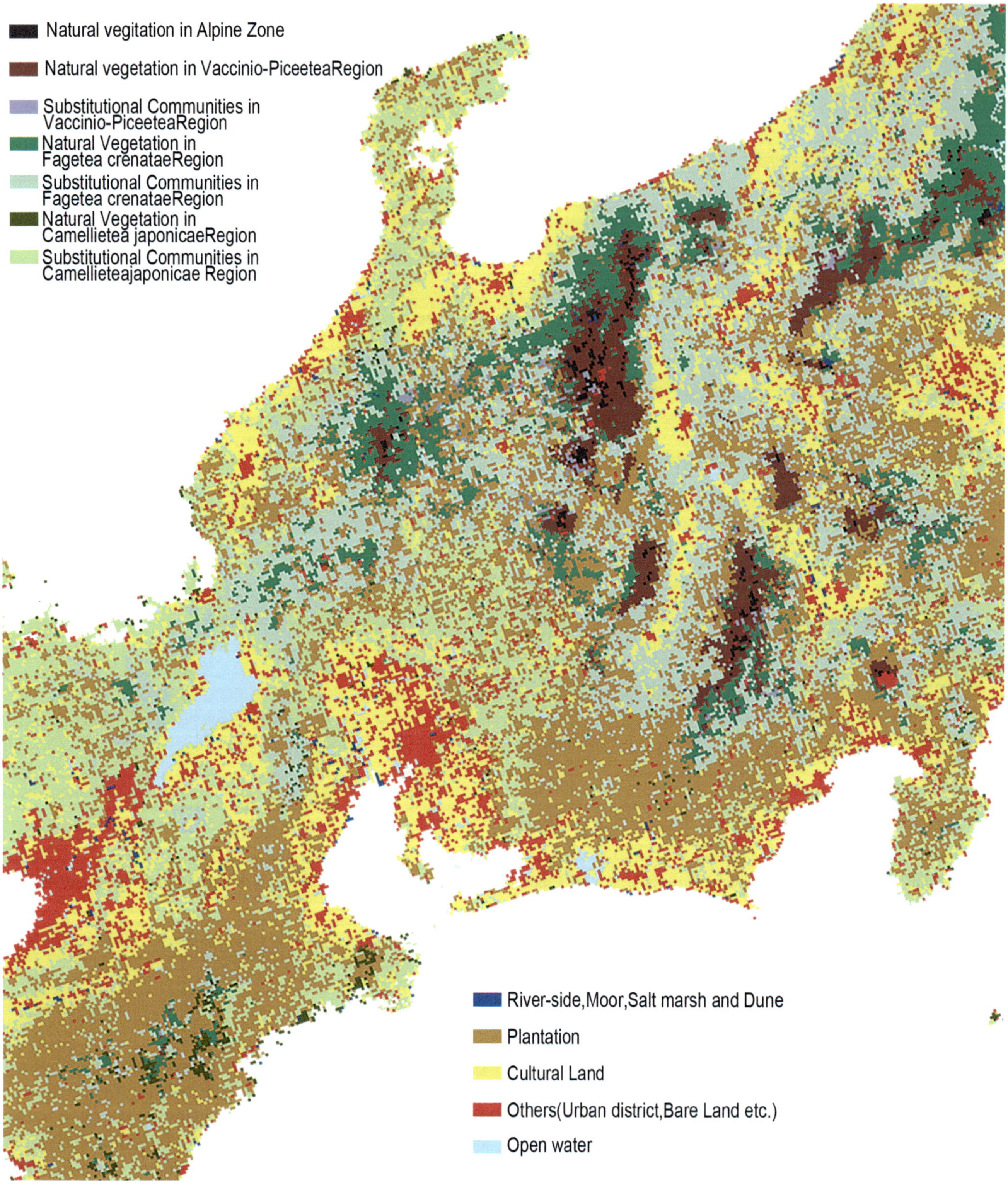

Fig. 8.4 State of vegetation obtained in the fourth national surveys on the natural environment conducted from 1989 to 1993. Figure created by Hirotatsu Murano based on state fourth national surveys (Biodiversity Center of Japan 2018)

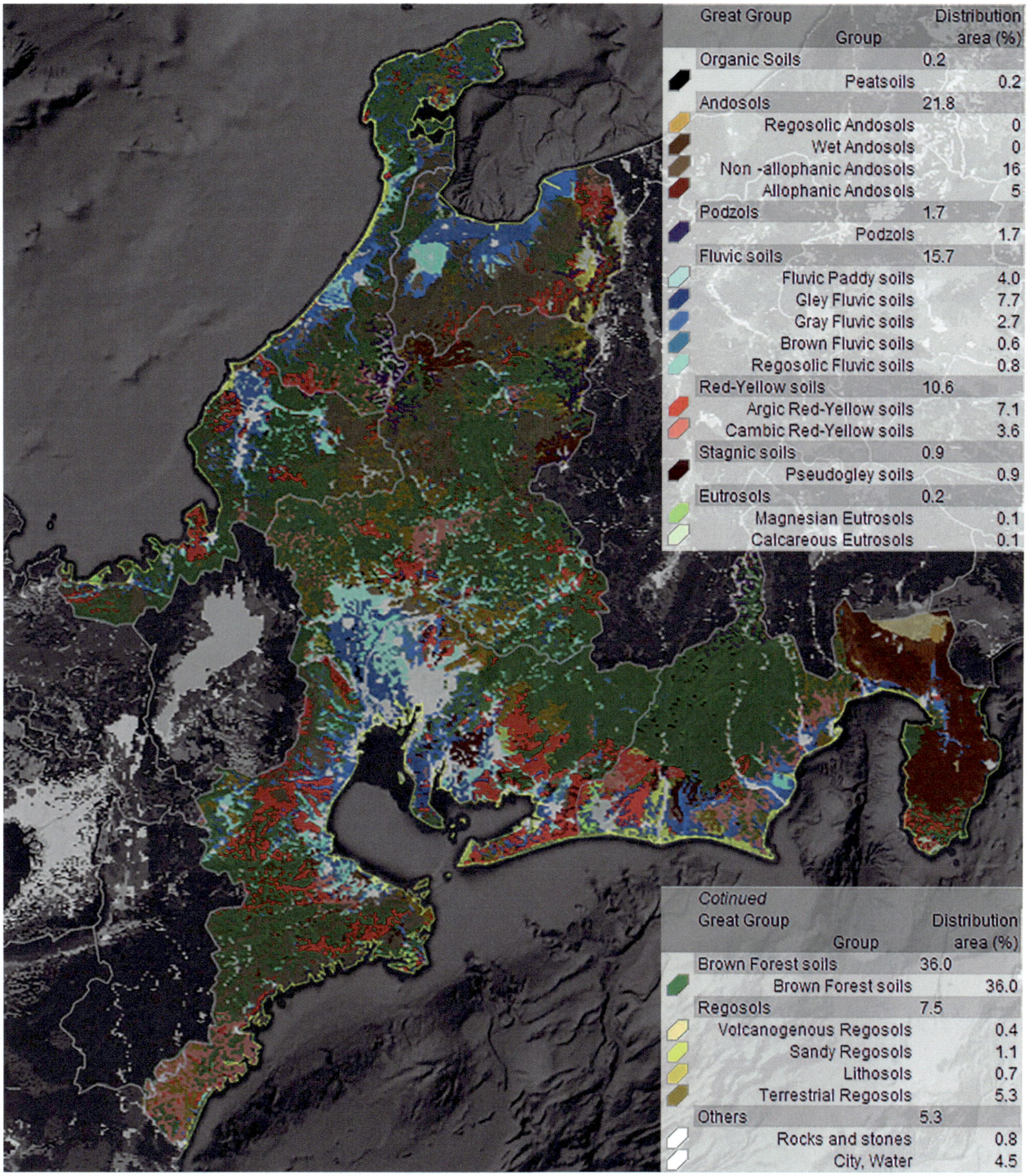

Fig. 8.5 Soil orders of Chubu region. The figure was created by Hirotatsu Murano based on soil classification system of Japan

when the Higher or Middle Terraces were formed (Matsui and Kato 1962). At that time, the Chubu region was in a subtropical humid climatic zone with a mean annual temperature of 20°C or more, annual precipitation of 1500–3000 mm, and an annual cumulative temperature of more than 5000°C. Kimura (1973a, b) studied the geomorphic surfaces and geology of the middle part of the west coast area of Ise Bay. The "Highest Terrace," "Higher Terrace,"

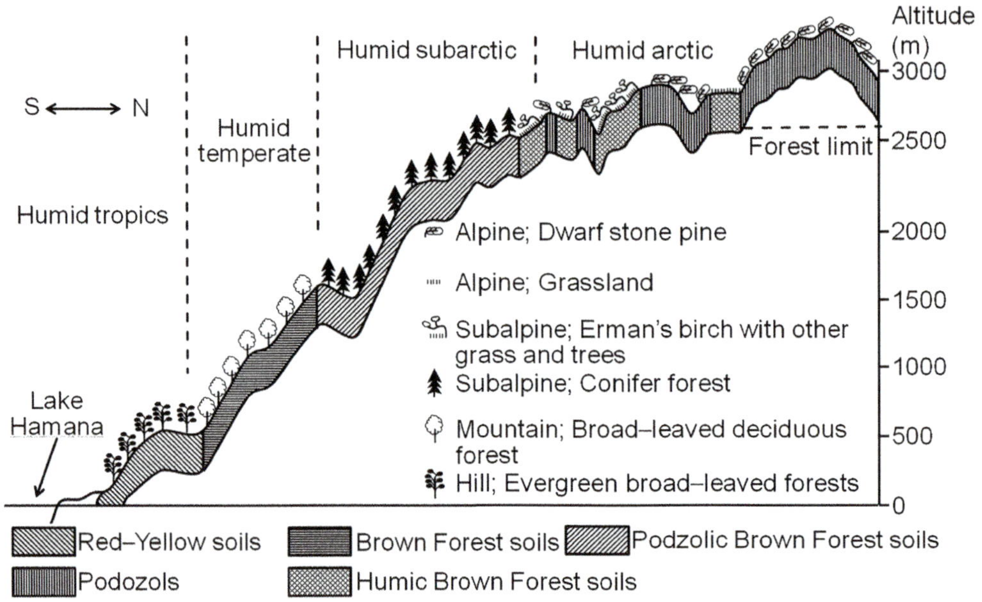

Fig. 8.6 Vertical zonality of soils in the Chubu region Modified from Kondo (1967) with permission from Japanese Society of Pedology

Fig. 8.7 Soil profiles of the Chubu Region. **a**. Peat soil at Midagahara in Tateyama mountain range (image by Forest Experimental Station, Forest Bureau, Ministry of Agriculture and Forestry, Japan (1968) with permission from Forestry and Forest Products Research Institute, Japan). **b**. Gray Lowland soil at Yoshioka, Toyama, Toyama Prefecture (image by Hitoshi Nakada). **c**. Non-allophanic Andosols on the Middle Terrace at Hiyoshi, Shinshiro, Aichi Prefecture (image by Hirotatsu Murano). **d**. Sandy Regosols at Hamaoka, Shizuoka Prefecture (image by Obara et al. 2015 with permission from National Institute for Agro-Environmental Sciences, NARO)

Fig. 8.8 Hill of Echizen, Fukui Prefecture, the Hokuriku District. Red-yellow soils are distributed on the Higher Terrace and the Middle Terrace in the Chubu Region (image by Obara et al. 2015 with permission from National Institute for Agro-Environmental Sciences NARO)

and "Middle Terrace" surfaces were built in the mid-Middle Pleistocene, the late Middle Pleistocene, and the early Late Pleistocene, respectively. Brown soil (5YR 4/8—2.5YR 4/6) is generally developed on the Highest Terrace surface, though its thickness and color are variable. Reddish Red-Yellow soil of [5YR 3/6—2.5YR 4/6] and [5YR 4/8—2.5YR 4/6] covers the surface of part of the upper-Higher Terrace and lower-Higher Terrace, respectively. Non-Allophanic Andosols cover the very top of the Middle Terrace everywhere, while Andosols occupy the top of the soil profile on the Middle Terrace in this area.

(3) **Soils on lowlands**

Fluvic soils are generally distributed in the lowlands of the Chubu region. In the lowlands of the Izu Peninsula and the surrounding areas affected by the activity of Quaternary volcanoes, many Alluvial Andosols are redeposited by rivers. Consequently, Wet Andosols are distributed in places with poor drainage and in alluvial plains. This soil is used as paddy fields, while places with relatively good drainage are used as upland fields (Shizuoka Prefecture 2014). In mid-western Shizuoka Prefecture, Gray Fluvic soils are widely distributed in alluvial plains, alluvial fans, and valley-bottom plains and are used as rice paddies; they are also distributed in upland fields and orchard fields. A total of 40% of the soils of the paddy fields of midwestern Shizuoka Prefecture have a gley horizon, the depths of which have increased in recent years as a result of the construction of underdrainage and open-channel drainage to the paddy fields (Shizuoka Prefecture 2014).

The Noubi Plain, an alluvial plain, is distributed among the southern part of Gifu Prefecture, the eastern part of Mie Prefecture, and the western part of Aichi Prefecture. The Kiso River, Ibi River, and Nagara River—the so-called Kiso Three Rivers—flow through this plain. In the downstream part of the Nobi Plain, areas with elevations below sea level are distributed. Many of the branches of the Kiso Three Rivers in these areas are raised-bed rivers. Consequently, Gley Fluvic soils are distributed in the paddy fields of these areas. It has been reported that the gley horizons disappeared after the dredging of riverbed to reduce flooding (Kitamura 2007).

In Hokuriku region, on the Sea of Japan side, Gray Fluvic soils are distributed on its alluvial fan, and Gley Fluvic soils are distributed in lowlands along rivers and coasts. These tow soils account for 70–80% of lowland soil in the Hokuriku region (Sage 1998). In the western and eastern part of Toyama Prefecture, alluvial fans are distributed in the lowlands. Consequently, the soil textures in these areas are coarse, and skeletal Gray Lowland soils are distributed. The plow layer is very thin, and the gravel layer exists at shallow depth, especially in the alluvial fan of the Kurobe river basin in the eastern part of Toyama Prefecture (Fig. 8.7) (Yamada 2007a).

(4) **Sandy soil**

Sandy Regosols are distributed in a sand dune on the coastline from Lake Hamana to Cape Omaezaki (Fig. 8.7)

(Shizuoka Prefecture 2014). This sand dune is 70 km long, 0.5–3 km in width, has a layer thickness that reaches 4–5 m or more and covers an area of 70 ha. The sand of the dune is derived from the Tenryu River and several other rivers in this area. The sand flowing out of the rivers accumulates by coastal current on the lowland along the coast and the alluvial estuary lowland. The dune was formed by sand movement and accumulation caused by strong westerly wind with an average wind speed in winter exceeding 7 m s^{-1}. From the end of the Edo period to the beginning of the Meiji period, fences made of trees and bamboo were installed in the sand dune to prevent sand blowing. Furthermore, the dune was stabilized by the planting of *Pinus thunbergii* (Kuribayashi 1956). In recent years, the supply of sand has decreased due to the revetment of the rivers in this area, and consequently, some areas of the sand dune have become smaller.

Another sand dune is distributed on the coast from the base of the Noto Peninsula to the border of Fukui Prefecture. This sand dune has a length of 90 km, a width of 300–2000 m, and an area of 13,000 ha (Wakabayashi et al., 1976). Sandy Regosols are also distributed on this sand dune (Sengi 1993). In this sand dune, afforestation for erosion control was carried out before the Meiji period, and the reforestation of *P. thunbergii* has been performed since 1911 (Fig. 8.9) (Yamada 2015).

8.2 Typical Agricultural and Forestry Practices and Soil Management

8.2.1 Paddy Fields in the Hokuriku Area

1. **Use and soil management of paddy fields in Fukui Prefecture**

The Hokuriku region is a major producer of paddy rice. Paddy fields account for 90% of the cultivated land in the region, much higher than the average value for the whole of Japan (54.3%; Ministry of Agriculture, Forestry and Fisheries 2017). Large amounts of snowfall prevent crop cultivation in winter, while sufficient meltwater, high temperature, and sufficient sunlight in summer facilitate paddy rice production, which contributes to the high proportion of paddy fields in the Hokuriku region (Ministry of Agriculture, Forestry and Fisheries 2017).

Since 1950, paddy fields have been reformed by creating a wide block from narrow blocks to improve the efficiency of rice production operations. Since 1970, the drainage of paddy fields has been improved by the construction of underdrains and the separation of irrigation and drainage water in order to allow an adjustment of rice production and stable production of upland crops that were implemented to

Fig. 8.9 Sand dunes of the Kaga Coast National Forest distributed on the coast in the Noto peninsula. The coast in 1910 before the reforestation **a**, the coast in 1918 seven years after the reforestation project began **b**, full view of the recent Kaga Coast National Forest **c**. (images of **a** and **b** by Yamada (2015) with permission from Nihon Chisan Chisui Kyoukai and of **c** by Kinki-Chugoku Regional Forest Office 2018)

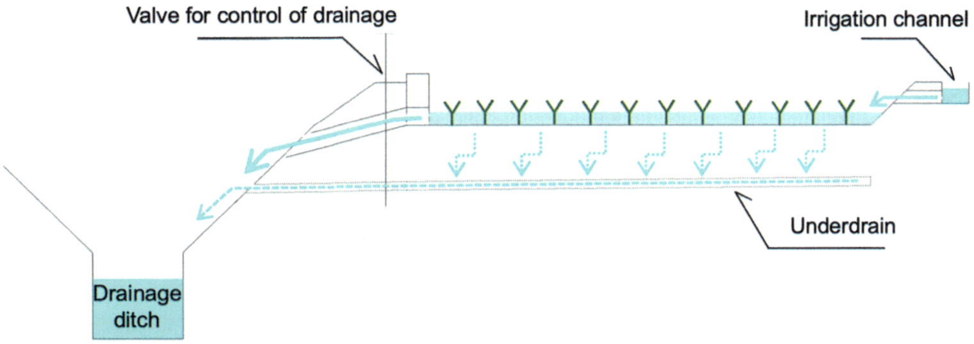

Fig. 8.10 Schematic view of the underdrain and separation of irrigation and drainage water in paddy field. Figure supplied by Hidetaka Sasaki

decrease production excess (Fig. 8.10). As a result, the reformation of fields to a wide block of over 3000 m^2 has been completed in more than 90% of the paddy fields in Fukui Prefecture (compared with 64% for the whole of Japan) (Ministry of Agriculture, Forestry and Fisheries). Additionally, underdrains have been constructed, and irrigation and drainage water flows have been separated in many paddy fields in the prefecture.

(1) Change in soil properties in the paddy fields of Fukui Prefecture

The soil of paddy fields in Fukui Prefecture is mainly alluvial, and most of the soils are wet and clayey. After the implementation of the field improvement program, the drainage of soil was improved. More than 60% of the paddy fields in the prefecture were wet before paddy–upland rotational managements were started in the 1970s, the proportion of wet paddy fields decreased to about 33% in 2000. Furthermore, the depth of the plow layer was reduced by the formation of plow pan that resulted from the use of large machinery in the wide block of paddy fields. Around 1970, 66% of the paddy fields in Fukui Prefecture achieved the target value of plow layer depth (deeper than 150 mm, according to the Soil Fertility Enhancement Act); however, only 36% of paddy fields achieved that target value about 30 years later (Imori et al. 2002).

The soil drying effect, in which soil organic matter decomposition is enhanced by the rewetting of soil after dryness, is exhibited when upland field is reconverted to paddy field. The soil drying effect has become smaller as paddy–upland rotations have been repeated. Additionally, the application rate of nitrogen (N) fertilizer had been decreased to 80% of the standard amount for rice cultivation after barley cultivation. However, the decreased N application rate has not been needed in recent years.

Although phosphate and potassium fertilizers and silicate materials were applied to paddy fields extensively until the 1980s, the application rate has been decreasing since the mid-1990s due to the falling price of rice and the worldwide rise in the prices of fertilizers. Therefore, the contents of available phosphate and exchangeable potassium tend to have decreased in paddy field soil after a peak around 2000 (Fukui Agricultural Experimental Station, 1979–2011; Fig. 8.11).

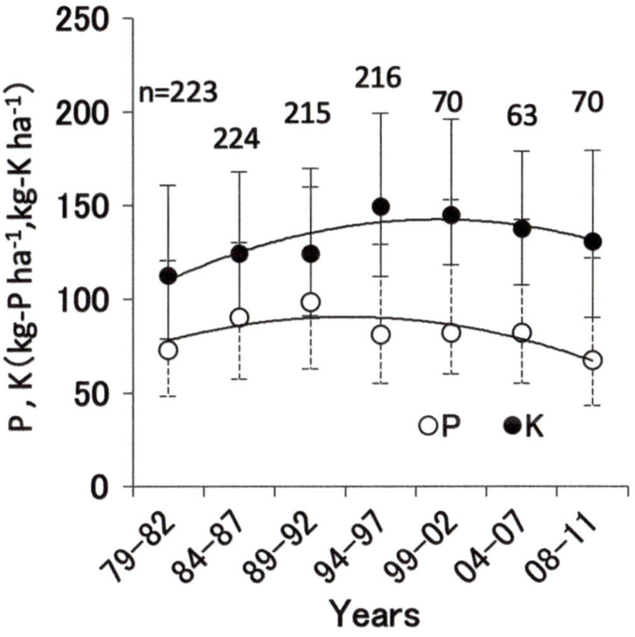

Fig. 8.11 Changes in contents of available phosphate and exchangeable potassium in paddy field soil from 1979 to 2011. Symbols denote the median value, and error bars show the range between the first and third quartiles of data. Values in figure indicate the number of study points. Figure created by Kouichi Hosokawa based on the data in Fukui Agricultural Experiment Station (1979–2011)

(2) Newly developed uses of paddy fields

The deterioration of soil fertility has become a concern due to the repeated implementation of paddy–upland rotation. The rotation of paddy rice and upland crops such as barley and soybean has been conducted in paddy–upland rotational fields where the drainage property is raised by field improvement programs. Recently, vegetables in open ground, such as Japanese leek, cabbage, and broccoli and fruits such as Japanese apricot, Japanese pear, and

Fig. 8.12 Implementation of winter flooding (photographed by K. Inoue in December 2012). (Inoue 2016, with permission from Yokendo)

persimmon have also been cultivated in upland fields converted from paddy fields in order to earn high income.

Japanese apricot is a major fruit crop in Fukui Prefecture, and about 30% of Japanese apricot fields in the prefecture were converted from paddy fields. Although converted fields have been dried by field improvement programs, the groundwater level is still sometimes high, which leads to the presence of standing water on the soil surface. Under these conditions, the roots of the Japanese apricot are difficult to extend due to a high oxygen requirement, and the growth of trees is therefore suppressed (Kanda 2015). In order to improve the draining property in converted fields, drainage ditches have been constructed, and/or the plow pan has been broken with a subsoiler before cultivation.

Winter flooding is a soil management system in which paddy fields are flooded from autumn, after rice harvest, until spring. This system is often implemented to enhance biodiversity, since many waterfowls fly to flooded paddy fields during winter. This management is covered by a direct payment framework from the Japanese government due to its positive effect on environmental conservation (Fig. 8.12).

Although winter flooding goes against the policy to make paddy fields dry to allow efficient operations by machinery and the stable production of upland crops, the area of paddy field under winter flooding has expanded in recent years, as environmentally friendly agriculture is attracting considerable interest. Under conventional rice cultivation, the consecutive period of flooding is generally about 60 days, between just before rice planting and mid-season drainage. Meanwhile, in winter-flooded paddy fields, the flooding period is about 200 days from November to June. It is often difficult to keep paddy fields flooded during autumn and winter as in many cases, the supply of irrigation water from rivers is cut off due to water irrigation rights. The Hokuriku region experiences the highest levels of autumn and winter precipitation of any region in Japan; the average rainfall from November to February in the region is more than 933 mm, compared with 252 mm in Tokyo (Japan Meteorological Agency 2017). The Hokuriku region is a suitable area for farmers to easily implement winter flooding since irrigation is not necessary due to the high rainfall and snowfall during this season.

The growth of rice plants is improved in paddy fields under winter flooding management. The growth of these plants is more vigorous about 1 month after transplanting than it is in non-flooded (drained) fields during winter, and as a result, the yield of rice in winter-flooded paddy fields is equal to or greater than that in non-flooded paddy fields during winter (Inoue 2016). The following factors are possible causes for this increase in rice growth: (1) the root zone is expanded in the plow layer, which is deepened and softened by the long duration of flooding; (2) ammonium-N mineralized from soil organic N is retained in soil without being nitrified and leached under the anoxic soil conditions present in flooded fields; (3) the decomposition process of organic matter is changed (Hosokawa et al. 2012). Additionally, a marked increase in rice growth is seen in paddy fields where the soil organic matter content is high, the soil is clayey, or the plow layer is deep enough. In such fields, rice can be cultivated with a 20% reduction of N fertilization rate

(Inoue 2016). Winter flooding is also expected as a management method for improving soil fertility that was depleted by paddy–upland rotation.

2. Present and future of soil science and plant nutrition in Ishikawa Prefecture

According to "Ishikawa's Visions for Food and Agriculture/Agricultural Area", Ishikawa Prefecture has been working toward devising a system of low-cost rice cultivation technology, the efficient use of rice fields, environmentally friendly agriculture, the multiple use and application of biomass resources, and so on. Challenges specific to the area of soil science and plant nutrition in these visions involve: (1) reducing the fertilization cost by efficient fertilization in response to the soil fertility; (2) solving the problem of poor yield of soybean crops by increasing soil fertility; (3) establishing environmentally friendly farming technology; and (4) promoting the utilization of unused organic materials such as food residue. Researches corresponding to these issues have been ongoing.

(1) Estimation of the amount of phosphate fertilizer in rice cultivation reducible to the soil fertility

After soaring in 2008, the price of fertilizers has been kept higher than pre-2008 levels, which has imposed a constraint for farming. The price rise was particularly remarkable for phosphate fertilizers, and a further rise is predicted due to the declining resources for phosphate fertilizers. It is thus essential to establish versatile and sustainable systems for the utilization and circulation of unused organic materials and to advance technologies for the reuse of phosphate from wastes. Immediate measures are needed, as it will likely take time to develop such new systems and technologies. One possible immediate measure is to determine a minimum amount of fertilizer application based on the soil fertility, which could lead to a reduction in fertilizer expenditure. We are studying the effects of reducing phosphate fertilizer application in paddy rice production according to soil fertility, including its effect on the yield quality and quality of rice.

The average Truog P content of the soils of Ishikawa Prefecture is 156 mg kg^{-1} dry soil; however, approximately 30% of the agricultural land in the prefecture has a Truog P content of less than the target figure of 100 mg kg^{-1} dry soil that is set by the prefecture. An experiment is currently being conducted in a rice field in the author's research center, in which the Truog P content was adjusted to 68 and 118 mg kg^{-1} dry soil by adding calcium dihydrogen phosphate in order to study the effects of four rates of phosphate fertilization on the growth of rice and soil phosphate fertility, namely: 1) conventional (100 kg ha^{-1}); 2) corresponding to the exploited amount (40 kg ha^{-1}); 3) corresponding to the L-type fertilizer (20-10-10) application (20 kg ha^{-1}); and 4) no phosphate fertilizer added.

The experiment will be conducted for five years and is now in the second year. The rice yield in the first year was comparable among the plots and treatments, suggesting that a reduction of the amount of phosphate fertilizer application is possible in soils that have phosphate levels equal to or greater than that of the experimental field in this study, at least for a single year. However, the soil phosphate fertility decreased in the treatment with no phosphate fertilizer application. The sound management of the phosphate fertilizer application should be developed by monitoring the rice yield and soil phosphate content in the following years (Table 8.1) (Kudo and Uno, 2012).

(2) Establishment and extension of environmentally friendly agricultural technologies

In Ishikawa Prefecture, there are several closed water areas with slow water circulation, and the deterioration of water quality is regarded as an environmental problem. Kahokugata, the largest lagoon in the prefecture, is a representative closed water area. The agricultural sector needs to contribute to the improvement of water quality by reducing the load of eutrophic substances. In Kahokugata lagoon, there is reclaimed land where barley, soybean, paddy rice, lotus root, watermelon, cabbage, and so on are planted, and we are working on reducing the load of eutrophic substances in these fields.

So far, for the barley and soybean crops, we have conducted field tests of environmental load-improving cultivation by side-dressing with slow-release fertilizers and have demonstrated that the total amount of nitrogen discharged from the field can be reduced by 34% compared with conventional cultivation (Fig. 8.13). In flooded rice cultivation, we adopted the environmentally-friendly managements of water and fertilizer: the shallow water-puddling and prevention of forced drainage to suppress discharge of turbid water, and side-dressing fertilization with slow-release fertilizer. These managements successfully reduced the amount of discharged total nitrogen and discharged total phosphate by 23–39% and 34–39%, respectively (Fig. 8.14). We are now testing the reducing effect of automatic water-saving irrigation on the load of eutrophic substances in lotus root cultivation. In the future, we plan to introduce machinery for local fertilization in vegetable cultivation, which requires more fertilizers than other crops, and to study the effectiveness of such technology for improving fertilizer utilization and reducing the amount of excess nitrogen loaded into the lagoon (Uno et al. 2010).

Table 8.1 Effect of reduced phosphate fertilizer on rice yield and soil phosphate fertility (2009)

Experimental plot	Treatment	Polished rice yield		Phosphate content in soil (mg kg^{-1} dry soil)			
		t ha^{-1}	Relative to conventional	Before cultivation		After cultivation	
				Truog	Bray II	Truog	Bray II
Low P	Conventional	4.93	100	68	626	70	424
	Exploited	4.85	98			51	287
	L-type	4.86	99			76	262
	No phosphate	4.92	100			35	262
Intermediate P	Conventional	5.05	100	118	723	85	470
	Exploited	4.52	90			73	405
	L-type	4.94	98			76	521
	No phosphate	5.31	105			66	386

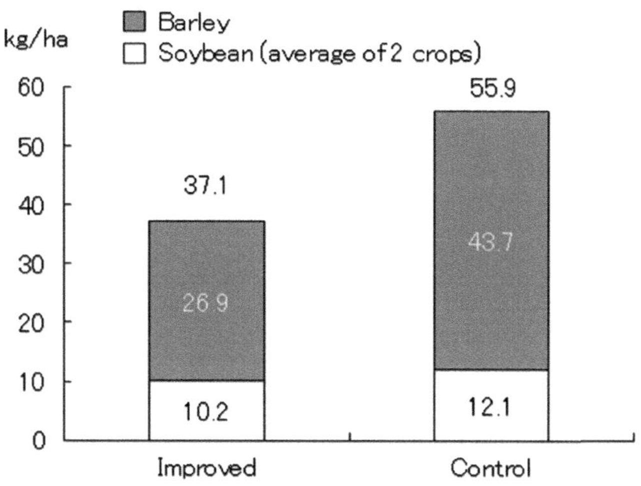

Fig. 8.13 Effect of improved fertilization management on nitrogen discharge from the barely-soy bean crop field (2008–2009). Reproduced by Fumio Uno from Uno et al. (2010)

3. **Countermeasures against cadmium contamination in paddy fields in Toyama Prefecture**

So-called itai-itai disease, a mass cadmium poisoning which occurred in the Jinzu river basin of Toyama Prefecture in early twentieth century, was certified as a pollution disease in 1968.

Cadmium released from the Kamioka Mine, located in Kamioka-cho (now Hida-shi), Gifu Prefecture, flowed into the Jinzu River, and contaminated the paddy fields of the downstream region via agricultural water. The cadmium subsequently accumulated in rice, and its ingestion caused itai-itai disease.

Some parts of the Specifications and Standards of Food and Food Additives, and so on, under the Food Sanitation Act, were amended in 1970, and it was added that the cadmium (Cd) concentration of rice must be less than 1 ppm. In 2010, this reference value was amended to less than or equal to 0.4 ppm.

In 1970, the Act to Prevent Soil Contamination on Agricultural Land was established to take measures against the contamination of agricultural land by specified hazardous substances (initially only cadmium, although copper (Cu) and arsenic (As) were later added).

In response to the establishment of the Act to Prevent Soil Contamination on Agricultural Land, a nationwide investigation of the pollution situation of agricultural fields was conducted. The total area of field with a soil cadmium concentration equal to or greater than the reference value was found to be 7592 ha by the investigation from 1971 to 2015. It has been reported that restoration projects in 7055 ha of cadmium-contaminated field had been completed by the end of the 2016 fiscal year (Ministry of the Environment 2016).

(1) **The restoration of cadmium-contaminated soil using soil dressing methods**

Of the above total area of 7592 ha of cadmium-contaminated field, the Jinzu river basin region contains the largest contaminated area of any region, 1500.6 ha. Here, soil restoration work was conducted by the pollution control special land improvement projects over 33 years from 1979 to 2011.

As countermeasures against contaminated paddy field soil, two types of soil dressing methods were adopted: the restoration method of embedding contaminated soil and

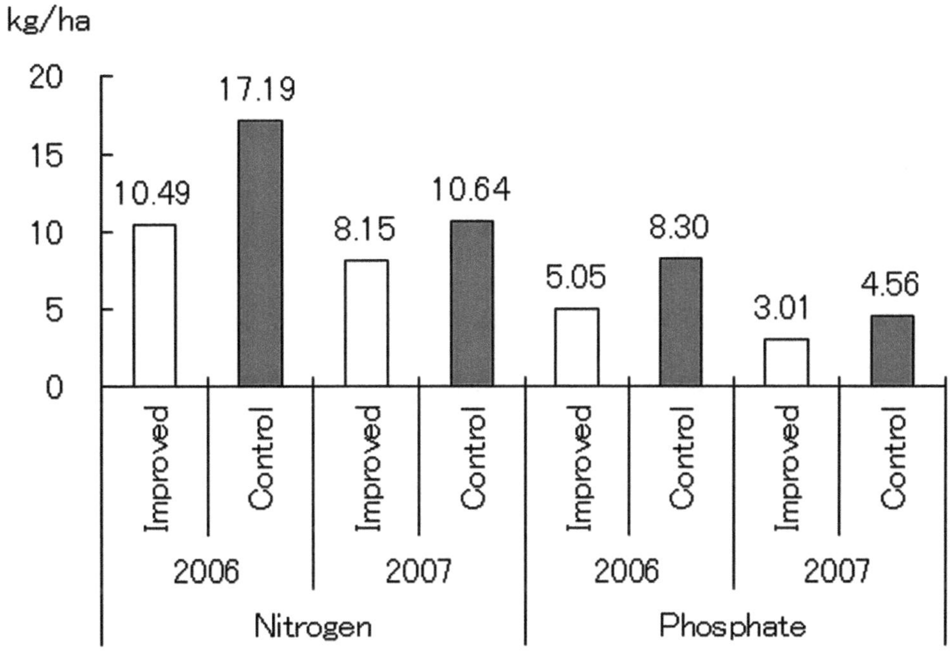

Fig. 8.14 Effect of improved water and fertilization managements on nitrogen and phosphate discharge from the rice field (2006–2007). Reproduced by Fumio Uno from Uno et al. (2010)

topping up soil of another place and the restoration method of topping up with soil of another place (Fig. 8.15). The type of soil dressing method was decided according to several factors, including the availability of carry-out destination for the contaminated soil, the difficulties involved with the installation of roads and waterways for the agricultural field, the groundwater level, and so on. Yamada reported the soil dressing techniques used to restore heavy metal contaminated agricultural soil and their sustainability (Yamada, 2007b).

Additionally, the restoration method in Toyama Prefecture includes installing a plow pan layer immediately under the plow layer; this is because the installed plow pan layer prevents mixing between the contaminated soil in the lower layer and the new soil deposited on top and also prevents soil sinking during machine operation for farm work.

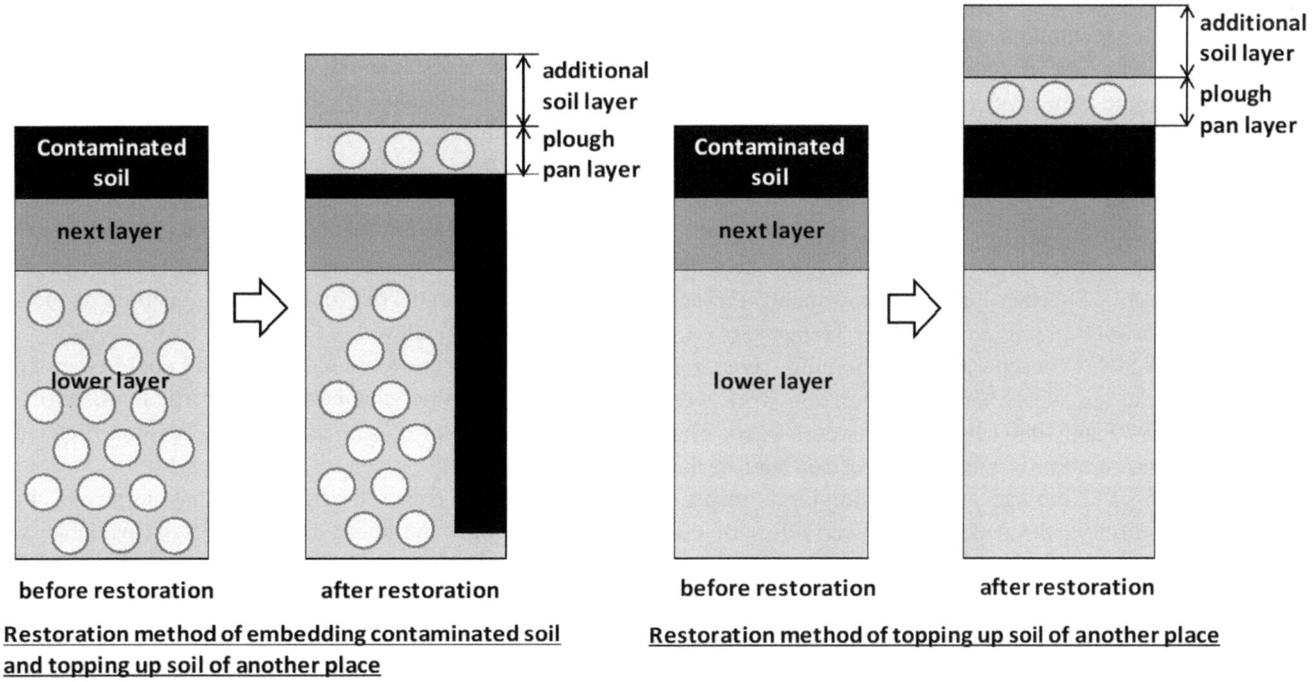

Fig. 8.15 Schematic drawing of the restoration methods. White circles in the lower layer of left panel and in plough pan layers indicate gravel. Created by Jun Koike based on Toyama Prefecture (2012a)

For the added soil, soils from mountains and mountainous areas in the vicinity of the countermeasure area were selected based on the results of analyses of soil physicochemical properties and the amount of soil available. Selection criteria for the replacement soil include: (1) a low content of heavy metals such as Cd, and no other harmful substances; (2) the conventional agricultural production volume should be secured after the restoration of the soil; (3) a large amount of the soil can be collected and transported without difficulty; (4) the area where the additional soil is collected must not be a designated area of buried cultural assets. From the viewpoint of avoiding recontamination and maintaining the agricultural productivity of the arable land, it was desirable that the soil contained as much clay as possible (15% or more) and had a high cation exchange capacity (CEC).

The replacement soil is the lower soil of mountain forest, which is usually low in fertility. Therefore, the soil, as it is, is not suitable for use in the plow layer for cultivating paddy rice. For this reason, it is necessary to improve the soil by applying calcareous materials, phosphoric acid materials, organic materials, and so on, in order to implement stable rice cultivation after the soil dressing. In paddy fields of the Jinzu river basin, 6000 kg ha^{-1} of silicate lime, 6000 kg ha^{-1} of fused phosphate, and 2000 kg ha^{-1} of poultry manure were sprayed after the soil dressing.

(2) **Effects of soil dressing**

Before the restoration work, the average Cd concentration in the plow layer soil of fields in the Jinzu river basin that were subject to the restoration measures was 1.12 mg kg^{-1}, and the average Cd concentration in brown rice of the area was 0.99 mg kg^{-1}. After the restoration work, the average Cd concentration in the plow layer soil and that in brown rice of the area had decreased to 0.13 mg kg^{-1} and 0.08 mg kg^{-1}, respectively. Thus, the effect of the restoration work in reducing cadmium concentrations was clear (Fig. 8.16) (Toyama Prefecture 2012a).

In 1979, we started conducting soil dressing as part of the Pollution Control Special Land Improvement Project ("*Kōgai bōjo tokubetsu tochi kairyō jigyō*" in Japanese). In 2009, we checked Cd concentrations of the plow layer soil and brown rice at six sites where the restoration work was carried out in 1979 and 1980 (three sites for each year). The average Cd concentration in the plow layer soil and in the brown rice was 0.12 mg kg^{-1} and 0.07 mg kg^{-1}, respectively, thus confirming the persistence of the effect of the restoration work.

Please note that the concentration of Cd in the Jinzu River has now declined to the natural environmental level due to the efforts of victims' organizations and companies. It is important to remember that those efforts contributed greatly to the prevention of recontamination.

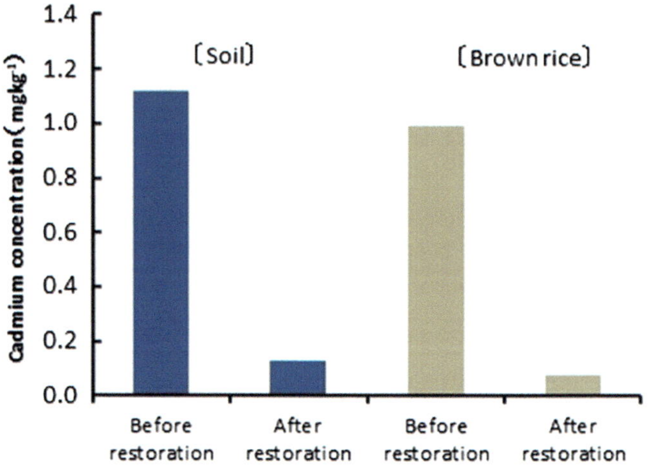

Fig. 8.16 Effect of the restoration work on the decrease of cadmium concentration. The values before restoration show the mean values of the samples of the soil or the brown rice taken from 544 places during the years 1971 and 1976. The values after restoration show the mean values of the soil samples of 1010 places or the brown rice samples of 896 places during the years 1981 and 2010. Created by Jun Koike based on Toyama Prefecture (2012a)

(3) **Crop cultivation after restoration**

The conditions of the soil of the paddy fields reconstructed by soil dressing are quite different from those of the previous paddy fields at the sites. The features of the soil of the new plow layer of the restored paddy fields are as follows: (1) while the capacity to hold fertilizers is large, the content of humic substances is small; (2) water permeability is poor as it is clayey soil; and (3) currently, the nutrient supply is small, including available phosphate, available silicate, and available nitrogen. As the result, the growth of rice plants has the tendencies of fewer stalks, light color, and the varied degree of growth in one field.

Therefore, the Toyama Agriculture and Forestry Promotion Center prepared "rice cultivation guidelines in paddy fields with additional soil." The center focused mainly on teaching so that a conventional yield level could be secured by conducting proper management for irrigation, fertilization, and promotion of soil maturation (Toyama Prefecture 2012a).

According to a report, the more time that has passed since restoration work, the more the content of available soil nitrogen has increased (Toyama Prefecture 2012b). The report showed that the amount of nitrogen fertilizer required to ensure a target yield of 5400 kg ha^{-1} is logarithmically reduced in accordance with the number of elapsed years since the restoration. It also stated that the required amount of nitrogen fertilizer can be reduced by up to 50% 30 years after the restoration compared to the amount of nitrogen required immediately after the restoration.

In Toyama Prefecture, soil dressing was completed in 2011 fiscal year after 33 years of restoration measures. However, as shown by the fact that there are still patients who have itai-itai disease and that the activities of victim groups are still continuing, the problems caused by cadmium pollution have not been solved completely.

The concentration of Cd in the Jinzu River has now returned to the natural environmental level due to the efforts of victim organizations and companies. However, it should not be forgotten that once heavy metal contamination occurs, it causes many losses, and its restoration requires significant amounts of labor and funds. I sincerely hope that the development of the economy and the maintenance of the environment will proceed simultaneously by making use of the lessons learned from the case of Toyama Prefecture.

8.2.2 Forestry in the Central Highland Area

Gifu is an inland prefecture that occupies the extreme western part of the Central Highland area and the center of Honshu Island in the western part of the Hida and Kiso Mountains (Fig. 8.1). From geographical viewpoint, the prefecture is divided into two regions, the northern part (Hida area) and the southern part (Mino area). The elevation ranges from under 0 m a.s.l. in the southern part to over 3000 m a.s.l. in the Hida Mountains. Most areas comprise plateaus and steep mountains, and 82% of the area is forest (Gifu Prefectural Forestry Policy Division, 2017). Therefore, there are various types of soils and forests depending on the elevation.

The forestry industry harvests timbers from plantations of trees suitable for the local climatic and environmental conditions and conducts appropriate thinning and other management. In many areas, according to local conditions, forest management is conducted continuously. Recently, local municipalities have made original regional forest plans, and these municipalities support forestry works in their regions. For example, Gujo City in Gifu Prefecture, is trying to develop a forest zoning plan that takes into account local conditions such as soil type, snow accumulation, and risk of sediment disasters.

Forest management that does not account for local conditions leads to high environmental burden or natural disaster. In the following, some examples are presented in order to review various forests and soils of the Central Highland area, mainly Gifu Prefecture.

Most Podzols are distributed in high mountainous areas, such as the Hida Mountains and Ryohaku Mountains. Under the timberline, subalpine evergreen coniferous forest appears at elevations of 1500 to 2500 m a.s.l. (Namikawa 2005); the main species are fir (*Abies veitchii* var. *veitchii* and *A. mariesii*), Japanese hemlock (*Tsuga diversifolia*) and spruce (*Picea jezoensis* var. *hondoensis*) often accompanied by white pine (*Pinus parviflora*) and birch (*Betula ermanii* var. *ermanii*). Because of the low land productivity and the lack of suitable species and management technology, these high mountainous areas have not been used as afforestation sites. Additionally, because of legal regulations such as the Natural Parks Law, it is thought that forestry work should, for the most part, not be conducted in the future.

After World War II, in an "expansive afforestation stage", broad-leaved forests such as beech (*Fagus crenata*) were cut widely, and coniferous trees such as Japanese cedar (*Cryptomeria japonica*) and Japanese larch (*Larix kaempferi*) were planted at the same time in high-altitude or snowy regions. However, due to the snow cover and its glide pressure, planted trees were damaged, and afforestation was unsuccessful (Aiura, 2014). These places are called "unsuccessful plantations" (Yokoi and Yamaguchi, 1998). Especially in heavy snowfall regions, landslides caused by snow erosion (Fig. 8.17) occurred frequently, and soil pedogenesis has been influenced. These areas faced routine avalanches and the deterioration of water conservation function by soil erosion, accompanied by sediment production derived from this phenomenon (Aiura 2014).

Andosols are mainly distributed in the northeastern and southern parts of the Hida area, while Allophanic Andosols are only developed on the eastern part of Mt. Haku and around Mt. Ontake. Because Andosols are difficult to compact, the risk of subsidence may be increased on forestry work roads (Forestry Agency, 2015). On the other hand, soil subsidence caused by the driving of large forestry machines can lead to the early death of saplings. Figure 8.18 shows the ratio of pores in the soil. The number of coarse pores is remarkably lower in tracks of forestry machine sites than in a control site.

Brown Forest soils are most widespread in mountains or hilly lands in Mino area. This area is steep, and its soils tend to be dry, making it suitable for the afforestation of Japanese cypress (*Chamaecyparis obtusa*). There may be a risk of surface soil erosion in planted cypress forests, depending on management. Surface soil erosion caused by rainfall is decreased by the presence of undergrowth vegetation, as the vegetation and litter cover reduce the impact of raindrops on the soil. Any type of forest, regardless of species, is expected to reduce the impact of raindrops on the soil, but in unmanaged cypress forests, it is difficult to lessen the erosion (Kosugi 2014). The fallen leaves of the cypress are broken into smaller pieces by the throughfall and are transported away from the forest floor by overland flow. There is the risk that large amounts of sediment may erode from the surface due to interference with the invasion or growth of

Fig. 8.17 Landslides by snow erosion in Toyama in 1980s (Photo by courtesy of Hideharu Aiura, Toyama Forestry Research Institute)

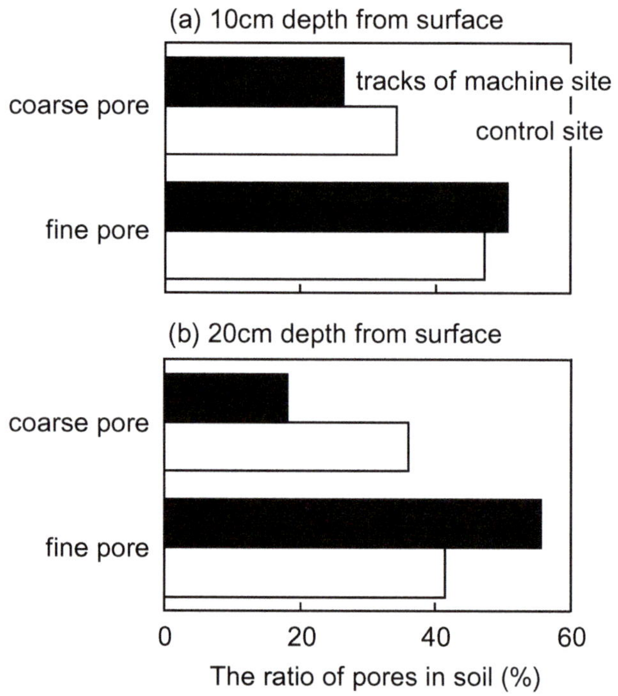

Fig. 8.18 Comparison of the ratio of pores in soil of two sites. Figure supplied by Hitoshi Watanabe

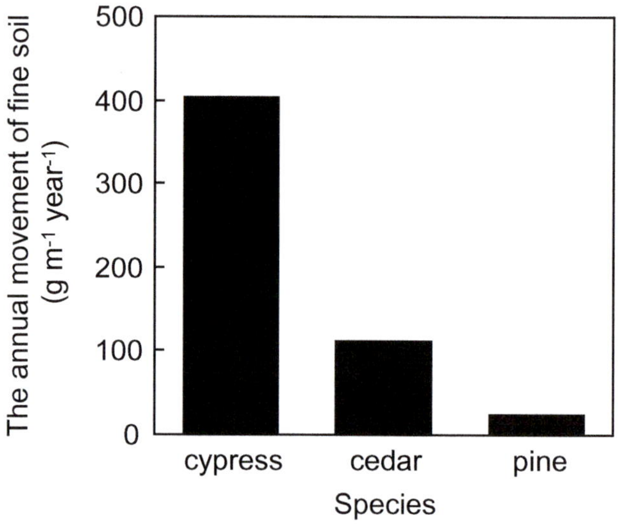

Fig. 8.19 Annual movement of fine soil Modified from Watanabe et al. (2016)

undergrowth, especially in unthinned cypress forests (Kosugi 2014). Figure 8.19 shows influences of planted species on surface soil erosion. On the same steep slope, the movement of fine soil was between 4 and 15 times higher in cypress stands than in stands of cedar or Japanese red pine (*Pinus densiflora*) (Watanabe et al. 2016). In the future, wise management considering the ecosystem is required in cypress forests.

In Sekigahara town, in the far western part of Gifu Prefecture, multi-storied forests have been made, and selection cutting management has been carried out. In the Imasu area (a small watershed area) of the town, traditional forestry management involving the cutting and subsequent planting of trees of suitable diameter has been conducted (Fig. 8.20). This management was mainly an attempt to harvest timbers sustainably from small owned stands. From the perspective of soil management, it was thought that the management had

Fig. 8.20 Multiple-storied forests in Imasu area, Sekigahara, Gifu in 1975. Photo by anonymous, Gifu Prefectural Research Institute for Forests

the effect of preventing the denudation of stands by clear-cutting. Regretfully, traditional management such as this has now been almost completely lost.

8.2.3 Upland Fields in the Tokai Area

1. **Soil management in the Red-Yellow soil region**

The Atsumi Peninsula in Aichi Prefecture has a warm and mild climate. However, the crop productivity in Red-Yellow soil, which is distributed on the diluvial plateau of the peninsula, is low, due to the soil's low humus content, high clay content, and low fertilizer and water holding capacities. Moreover, there are no big rivers within the peninsula to ensure the sufficient supply of agricultural water. In the 1950s, soil was improved by the application of manure compost from animal husbandry that was developed within a short time in combination with new tilling methods such as subsoil disruption (super-deep tilling of more than 1 m in depth was conducted in order to improve drainage when there was a minute clay layer). In addition, irrigation canals from Toyo River, one of the biggest rivers in Aichi Prefecture, were opened in 1968. Under these circumstances, the Atsumi area was transformed to a large vegetable production zone.

However, along with the development of intensive agriculture, the influence of excess fertilization on the environment became difficult to ignore. In the Atsumi Peninsula, a large portion of nutrient salts applied is eluted out into Mikawa Bay, which is largely enclosed. Therefore, it was required to establish vegetable production that can reduce the environmental load, based on proper fertilization, with continuing soil improvement by the application of organic matter.

(1) **Surplus nutrients and environmental loads in vegetable production**

In vegetable production, plant nutrients must be supplied in amounts greater than those that are absorbed by crops, which inevitably generates "surplus nutrients." Figure 8.21 shows the nitrogen (N) balance in the major systems of open-field vegetable cultivation in the Red-Yellow soil region of the Atsumi Peninsula (Makita et al. 2014). Those data indicate the presence of more than 100 kg N ha^{-1} of surplus N in cultivation systems of cabbage (*Brassica oleracea* var. *capitate*) and broccoli (*Brassica oleracea* var. *italica*), where a large amount of crop residues are left in the field, and of spring onion (*Allium cepa*) and maize (*Zea mays* L. var. *saccharata*), where a large amount of N is not used by crops.

Surplus nutrients are dissolved in soil water and eluted into groundwater or rivers. According to the results of our 2-year monitoring of a river in the Red-Yellow soil vegetable field zone, nitrate-N was always supplied via groundwater, while most of the phosphorus (P) was loaded directly from soil in the form of suspended materials during rain events (Kasuya et al. 2010). A positive correlation was observed between the total amount of surplus N in the entire basin, which was estimated from the crop situation, and the nitrate-N concentration in river water. Thus, determining how to manage such surplus nutrients appropriately is important to strike a good balance between productivity and environmental protection.

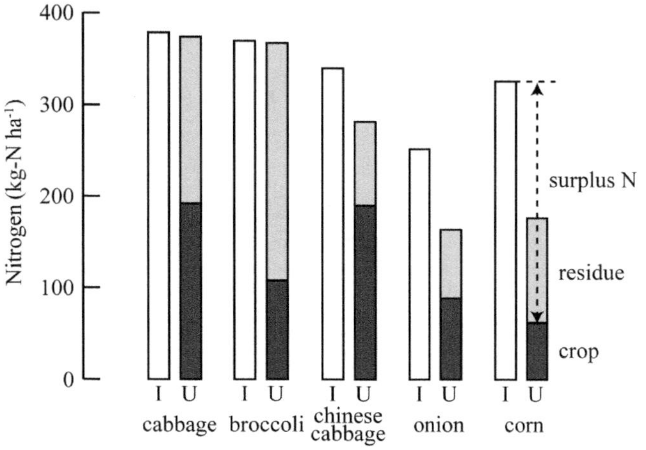

Fig. 8.21 Nitrogen balance of some major vegetables in a Red-Yellow soil upland field in Aichi Prefecture, Japan. I, amount of fertilizer nitrogen applied to soil; U, uptake by plant Modified from Makita et al. (2014). With permission from Japanese Society of Soil Science and Plant Nutrition

(2) Proper fertilization using livestock manure in upland fields

The application of organic matter is an essential agricultural practice in Red-Yellow soils with poor soil physical properties and soil fertility. In the Red-Yellow soil vegetable fields of the Atsumi Peninsula, the application of livestock manure compost originating from animal husbandry is widely conducted. Due to the high contents of beneficial nutrients, the application of livestock manure compost has the potential to reduce the amount of chemical fertilizer, which can lower environmental loads without decreasing productivity.

Aichi-ken Agricultural Research Center investigated the dynamics of nutrients in soil and crop yields in a vegetable upland field where cattle or pig manure compost was applied repeatedly in a field experiment and suggested a method for reducing the amount of chemical fertilizer to a field with continuous application of compost (Table 8.2; Tsuji et al. 2018). The details of this investigation are introduced in the following two paragraphs.

In the field experiment, a cattle manure compost plot, a pig manure compost plot, and a no-compost plot were prepared. Composts were applied in August, and cabbage and maize were cultivated from September to February and from May to July, respectively. The amounts of composts applied were chosen according to the recommended standard of the Aichi Prefectural government (Aichi Prefecture 2016), that is, cattle manure compost was applied at 15 Mg-DW (dry weight) ha^{-1} y^{-1}, while pig manure compost was applied at 10 Mg-DW ha^{-1} y^{-1}. During the first three-year period, the same amount of chemical fertilizer was applied to all the three plots to confirm the increase in yield due to the additional nutrient supply from the composts. Then, assuming that an increase in the available N content in the soil had resulted from the repeated application of compost, the amount of N fertilizer was reduced in a stepwise manner. The rate of application of N fertilizer was determined based on the apparent utilization ratio of compost derived N by crops in the previous year, which was calculated using N balance data (Table 8.2). The N balance was evaluated from crop yield, the amount of N absorbed by crops, soil N content, and the amount of N leached in each crop season. As a result, surplus N was largely reduced without decreasing crop yield relative to the first 3-year records before successive fertilizer reduction was started. Since a portion of surplus N accumulated in the soil, particularly in the cattle manure compost plot, the N elution into groundwater could be largely decreased.

Phosphorus and potassium (K) in manure compost have the following characteristics: (1) the whole amount in

Table 8.2 Average nitrogen balance in a Red-Yellow soil vegetable field with cattle or pig feces compost

Years	Plot*	Input			Uptake		Surplus N kg ha^{-1} y^{-1}	Leaching N kg ha^{-1} y^{-1}
		Fertilizer kg ha^{-1} y^{-1}	Compost kg ha^{-1} y^{-1}	Total kg ha^{-1} y^{-1}	Crop kg ha^{-1} y^{-1}	Residue kg ha^{-1} y^{-1}		
1-3	Cattle	550	287	837	239	194	597	502
	Swine	550	350	900	242	227	658	612
	CF	550	0	550	213	185	337	383
4–6	Cattle	440	278	718	245	174	472	388
	Swine	400	376	776	255	207	521	445
	CF	550	0	550	220	159	330	411
8–10	Cattle	370	267	637	216	183	421	402
	Swine	295	454	749	234	206	515	460
	CF	550	0	550	191	167	359	447

**Cattle*—Cattle feces compost of 15 Mg-DW ha^{-1}, *Swine*—Swine feces compost of 10 Mg-DW ha^{-1}, *CF*—Chemical fertilizer was applied without animal waste compost

compost can be regarded as available to crops, as is those in chemical fertilizer; (2) the K content is higher in cattle manure compost than in pig manure compost, and vice versa for P content; and (3) K is leached from soil, while P is accumulated in soil. Because of these characteristics, from the fourth year onward, cabbage, the first crop grown after compost application, grew well without chemical fertilizer, but K and P fertilizers were needed for maize, the second crop. According to the experimental results, it was concluded that the application of a full amount of P (100 kg P_2O_5 ha^{-1}) with a half amount of K (125 kg K_2O ha^{-1}) relative to the amounts for the no-compost plot is appropriate in the cattle manure compost plot, and that the application of K fertilizer at the same amount as in the no-compost plot (250 kg K_2O ha^{-1}) without P fertilizer application is appropriate to the pig manure compost plot.

As such, it is possible to reduce the cost of fertilization and the load of nutrients on water areas without reducing productivity by estimating appropriate rates of chemical fertilizer application while taking the nutrient supply from livestock manure compost into consideration.

(3) **Recycling of surplus N using green manure**

In the Red-Yellow soil region of the Atsumi Peninsula, green manure is widely used as well as livestock manure compost. Since the autumn/winter production of crops, such as cabbage and broccoli, is popular in this region, many upland fields are left as bare land during the hot and wet summer season. The resulting decrease in soil fertility caused by the active decomposition of soil organic matter and subsequent nitrate N leaching during the summer season is concerning. To prevent nutrients in soil from being leached and to use them as fertilizer for autumn/winter crops, an off-crop season green manure crop cultivation technique was established in a 6-year field experiment. This technique was expected to suppress the decrease in soil fertility and maintain or increase vegetable productivity. Sorghum (*Sorghum bicolor* L. Moench) was grown as the green manure crop and plowed into the soil before cabbage was grown every year. In the first year, the yield of cabbage (Table 8.3) was smaller in the sorghum plot than in the fallow plot. However, the cabbage yield subsequently increased, and the large yield could be maintained even after the rate of N application was reduced by 30 kg N ha^{-1} in the fourth year, equivalent to 10% of the usual application rate (Kasuya and Hiroto, 2010). One of the possible mechanisms behind the larger crop yield in the sorghum plot is the improvement of soil physical properties, as the plowing-in of sorghum may have increased both soil porosity and available moisture, which could encourage the elongation of roots and improve their water absorption.

Table 8.3 Average nitrogen balance in a cabbage field where sorghum was cultivated as summer crop

Years	Cropping in summer	Input (Fertilizer) kg ha^{-1} y^{-1}	Output (Crop) kg ha^{-1} y^{-1}	Surplus N kg ha^{-1} y^{-1}	Cabbage yield Mg ha^{-1}
1–3	Sorghum	300	193	107	67.4
	Fallow	300	185	115	63.7
4–6	Sorghum	270	175	95	63.8
	Fallow	300	170	130	61.5

As the C/N ratio of sorghum is high (28–60), it is not reasonable to expect it to act as a fast-acting N fertilizer. Nevertheless, the yield of cabbage in the sorghum plot was always higher than that in the fallow plot (Table 8.3). In years 4–6, when the rate of N application to the sorghum plot was reduced, the amount of N absorption was similar between the sorghum plot and fallow plot, while the amount of surplus N was smaller in the sorghum plot (Table 8.3). In a lysimeter experiment using ^{15}N-labeled green manure (Kasuya 2007), the increase in the amount of N absorbed by crops was larger than the amount of sorghum N that was plowed in. This observation suggests that the effectiveness of sorghum as N fertilizer is due not only to the N supply from the sorghum itself but also to the acceleration of N cycling in the soil caused by the increase in soil biomass due to the supply of C from the sorghum. The increases in the amount of N absorption by vegetables and the high yields could be interpreted as the results of these sorghum effects.

Although the effect of sorghum on the yield of vegetables was smaller than that of livestock manure compost, the sorghum effect was observed consistently. Thus, the use of sorghum as a green manure crop is recognized as a useful soil management practice in terms of its large reduction of N surplus and N leaching. The use of green manure, mainly sorghum, is conducted in more than 20% of vegetable fields in some active regions and still becoming more widespread. The simultaneous achievement of high productivity and environmental protection will be attained at a higher level through the improvement and widespread use of these soil management techniques.

2. **Cultivation of tea (*Camellia sinensis* (L.) Kuntze) in Shizuoka Prefecture**

The climate of Shizuoka Prefecture is warm with high precipitation. It is generally divided into two types of climate: marine climate in the coastal area and inland climate in the mountainous area (Shizuoka Local Meteorological Office 2018). The climatic differences within the prefecture are large;

the climate is relatively mild in the coastal area, with an annual average temperature of 15–16°C and average annual precipitation of less than 2000 mm yr^{-1}, while in the mountainous area, the annual average temperature is 11–12°C, and the average annual precipitation is 2500–3000 mm yr^{-1}.

The total land area of Shizuoka Prefecture is 777,743 ha. Of this, farmland accounts for 8.6% (67,100 ha). Because the prefecture has a varied climate and topography, various crops are cultivated. The information on soil and agricultural land use in Shizuoka Prefecture has been introduced in detail by Matsumoto (2005).

In Shizuoka Prefecture, tea, citrus, wasabi, and flowers (gerbera, rose, and orchid) have a high market share nationwide. Among these, tea is the representative crop in Shizuoka Prefecture due to its large cultivation area (17,400 ha, 26% of the total farmland area in the prefecture) and the high economic value of its agricultural production (30.6 billion yen, 14% of the total value of agricultural production in Shizuoka Prefecture). Soil management in tea gardens and measures for reducing the environmental load induced by nitrogen fertilizer application are shown below.

(1) **Soil management in tea gardens**

The recommended fertilizer application rates for tea cultivation are set by each prefecture. The recommended level of nitrogen application is about 500 kg N ha^{-1} yr^{-1}, and that of phosphorus (P_2O_5) and potassium (K_2O) is between one third and one half of the nitrogen application rates. For tea crops, the new shoots of the plants are harvested, and since new shoots with high contents of amino acids tend to be high quality, more nitrogen is applied as fertilizer compared with other crops. Nitrogen uptake by tea plants mostly occurs during the period from April to November. Tea plants prefer ammonium–nitrogen (NH_4^+-N) to nitrate–nitrogen (NO_3^--N). The growth and nitrogen content of new shoots increases with the application of NH_4^+-N. The application rates of organic fertilizers in tea gardens are higher than those in other crop lands. In particular, during tea cultivation, organic fertilizers are applied in large amounts in spring and autumn, while fast-acting chemical fertilizers are applied in summer.

Tea plants prefer acidic soils; the optimal soil pH (H_2O) is 4–5. When soil pH exceeds 6, the growth of roots becomes worse, which negatively affects the growth of tea plants. Meanwhile, in significantly lower pH conditions, problems arise including the suppression of root growth, decreases in CEC, leaching of bases, and delays in decomposition of organic matter.

Fertilizer is generally applied in the soil between rows of tea plants, which accounts for only one sixth to one fourth of the total area of the tea garden (Fig. 8.22). The heavy application of nitrogen fertilizer in this narrow area causes the chemical deterioration of soil, including strong soil acidification and the leaching of bases. Additionally, because the soil areas between rows are also used as working roads, the soil becomes compacted by the repeated tread pressure of farmers and work machines, resulting in the degradation of soil physical properties and negatively affecting tea plant growth.

Fig. 8.22 Tea garden and Mt. Fuji (Photo by courtesy of Hiromasa Tanaka, NARO)

The roots of tea plants grow well in autumn, and the late summer and early autumn are therefore the best period for soil improvement. To amend soil acidification, lime is applied at this time. Furthermore, in order to improve the physical properties of the soil, deep plowing (down to about 30 cm in depth) is recommended. However, deep plowing has recently tended to be omitted because of the high workload required. As a result, there is concern about the degradation of soil physical properties. To improve these properties, organic matter such as manure and dried grasses is applied and incorporated into soils. In some tea growing areas, in an effort to obtain grasses for application to tea gardens, semi-natural grassland, which is called "*chagusaba*" in Japanese, has been maintained by farmers around tea gardens. The primary purpose of maintaining the grassland is to improve tea quality, but this has also contributed to the conservation of biodiversity (Inagaki and Kusumoto 2014). Due to the good balance between agricultural production and biodiversity conservation, the "traditional tea–grass integrated system in Shizuoka" was certified as a Globally Important Agricultural Heritage Systems (GIAHS).

(2) **Improvement of nitrogen use efficiency in tea production**

Japanese green tea tends to be traded at high prices for products containing high free amino acid content, one of the most important quality indices for Japanese green tea, and tea plants can store excessive amounts of absorbed nitrogen as amino acids in the plant body. Thus, high rates of nitrogen application can increase the concentrations of amino acids in new shoots. As a result, the amount of N applied to tea gardens has been increased in some tea growing areas.

The nitrogen application rates at tea production sites had often exceeded the recommended amount of nitrogen fertilizer application, that is, the nutrient requirement of tea plants in some tea growing areas. The surplus nitrogen application sometimes increases the yield and quality of tea products but can also cause a decrease in the amount of fine roots of tea plants, causing a vicious cycle of requiring large amounts of nitrogen to maintain the yield and quality of tea. Consequently, environmental problems have arisen. For example, increasing amounts of surplus nitrogen are leached from tea gardens, which increase NO_3^--N concentrations in the surrounding water systems (Nagai 1991). Additionally, the surplus nitrogen application also caused soil acidification and high emission rates of nitrous oxide (N_2O) (Tokuda and Hayatsu 2004), one of the major greenhouse gasses and an ozone-depleting substance.

To address the problems mentioned above, various fertilizer management methods have been developed. Fertilizers with enhanced efficiency have been developed to improve the efficiency of nitrogen use by crops. Among these, coated urea is representative of tea cultivation in Japan. Coated urea gradually releases nitrogen and is used in accordance with the amount required by tea plants. In addition to the use of fertilizers with enhanced efficiency, other new methods of fertilizer application, such as expanding the width of fertilizer application and drip fertigation, which enables the utilization of the root zone under the tea plant canopy, have been proposed in order to increase the efficiency of nitrogen uptake by tea plants.

3. **Pelleting of livestock manure for environmental friendly agriculture in Mie Prefecture.**

Mie Prefecture has 61,000 ha of agricultural land, with paddy fields being the predominant land use (46,000 ha) in this sector. The areas of upland fields and orchards are 8800 ha and 6400 ha, respectively. Fluvic soils occupy 60% of the agricultural land, followed by Andosols (11%) and Red-Yellow soils (10%). Gley Fluvic soils and Gray Fluvic soils are the dominant soil groups in paddy fields, and Non-Allophanic Andosols and Argic Red-Yellow soils are the dominant groups in upland fields and orchards, respectively.

(1) **Soil fertility status**

The Fundamental Guideline for Soil Productivity Improvement, which is based on the Soil Productivity Improvement Law, specifies a standard application rate of manure for paddy fields and upland fields of 10 to 15 Mg ha^{-1} and 15 to 30 Mg ha^{-1}, respectively. However, the application rate of organic manure has been decreasing due to the aging farming population and the increasing farm acreage per farm. In particular, the application rate of manure is very low in paddy fields, because this resource depends on rice straw. According to the National Agricultural Management Statistics, the average application rate of manure in Japanese paddy fields decreased from 5.5 Mg ha^{-1} in 1965 to 1.3 Mg ha^{-1} in 1997. The decrease in the rate of manure application to paddy fields in Mie Prefecture was at the same level as the Japanese average, but a 2-year crop rotation system (rice, wheat, soybean) was widely conducted there in paddy–upland rotation fields. Prolonged non-rice cultivation in paddy fields caused a decrease in soil organic matter and soil productivity. It was thought that soil fertility could be maintained by total rice straw plowing; however, the prolongation of non-rice cultivation in the paddy–upland fields and global warming may increase the decomposition rate of soil organic matter. It is estimated that increased application of manure (15 Mg ha^{-1}) with total rice straw plowing is required in order to maintain soil fertility for the 2-year crop

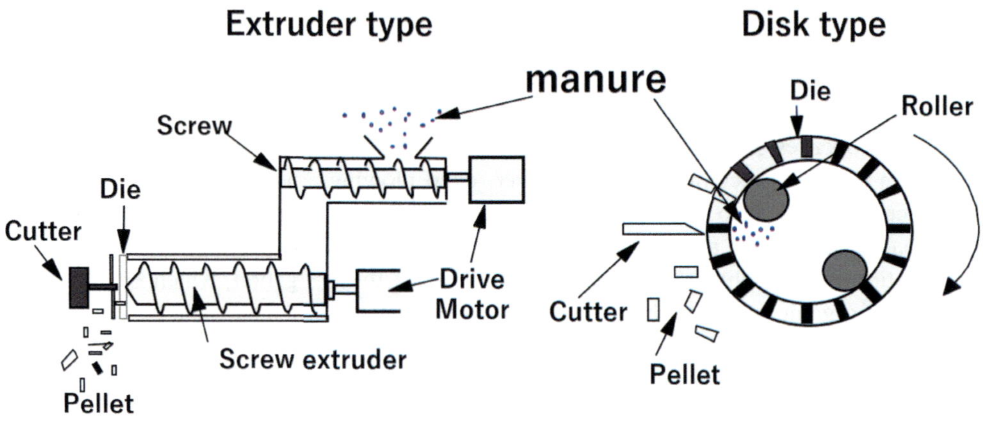

Fig. 8.23 Two types of the extrusion molding Reproduced from Hara (2017). With permission from Japanese Society of Soil Science and Plant Nutrition

rotation system (rice, wheat, soybean) in the paddy–upland rotation fields.

(2) **Trend of livestock manure**

In Japan, about 79 Tg of livestock excreta is generated per year, the nitrogen content of which is almost the same as the total annual nitrogen input from the application of chemical fertilizer. It is estimated that about 160 kg ha^{-1} of the nitrogen generated from livestock excreta is equally input into Japanese agricultural land. However, it is estimated that the input of nitrogen exceeded 300 kg ha^{-1} in some major livestock prefectures. In Mie Prefecture, the nitrogen input from livestock excreta exceeded 300 kg ha^{-1} in some counties. These counties have difficulty in finding alternative fields for avoiding excess nitrogen input from livestock excreta, and there is growing concern about related environmental problems such as nitrate pollution in groundwater. With the price of chemical fertilizer increasing, livestock excreta is again attracting attention as a low-price precursor for the production of fertilizer. However, there are some barriers to the application of livestock manure, for example, the need for specific equipment for field application, the difficulty of manure transportation, and the difficulty of quality control for the livestock manure.

(3) **Pelleting of livestock manure**

The Mie Prefecture Agricultural Research Institute has developed the pelleting of livestock manure using extrusion molding in order to improve the operation of machine spraying and the properties of manure transportation and to control the quality of livestock manure. Currently, two types of extrusion molding (Fig. 8.23) are mainly used for livestock farmering in Japan, with an average diameter and length of manure pellet of 3–5 mm and 6–10 mm, respectively (Hara 2017). The pelleting of livestock manure commonly refers to the extrusion process, where cylindrical pellets are formed by forcing the material through a dye and a cutter. To increase the pellet strength and prevent the deterioration of the pellet quality, the molded pellets are dried to reduce their moisture content to 0.20 kg kg^{-1} or less. The pelleting also decreases the required volume of livestock manure by 10–40%. This weight and volume reduction can lower the transportation cost of livestock manure (Fig. 8.24). The hardness of the pelleted livestock manure is almost same as that of chemical fertilizer, and it is possible to spray pelleted manure using general-purpose equipment (Hara et al. 2004).

Since livestock manure pelleting requires additional production costs, high value added for utilization is sought. Mie Prefecture developed a poultry manure pelleting method, which enables the decomposition rate of uric acid to be controlled by controlling the temperature and moisture

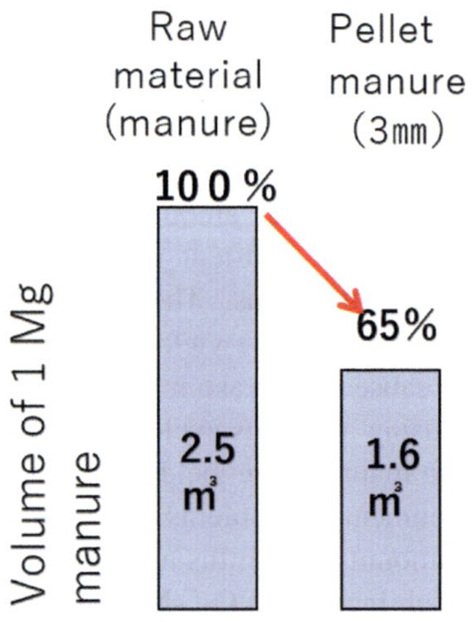

	Water mass	Dry mass
Manure	500 kg	500 kg
Pellet manure	150 kg	800 kg

Fig. 8.24 Improvement effect of the pelleting for transportation Reproduced from Hara (2018)

level of manure fermentation. The pelleting method assures a nitrogen content in the poultry manure pellet of 0.04 kg kg^{-1} or more, and it has high added value.

Acknowledgements The editors of the chapter thank Dr. Hideyuki Umemoto, Mr. Yoshitaka Mukai, Dr. Yusuke Takata, and Prof. Akio Morita for assistance in the preparation of the manuscript.

References

Aichi Prefecture (2016) Recommendation of fertilizers and manure application. http://www.pref.aichi.jp/soshiki/nogyo-keiei/0000085287.html (Jan 2018). (in Japanese)

Aiura H (2014) Landslide by snow erosion. In: Heavy snowy regional forestry technology development council (ed). To create forest of broad-leaved species. J-FIC, Tokyo, pp 188 (in Japanese)

Arai S, Watanabe A et al (1988) Pedochemical studies on Ohnohara peat with special reference to its properties related to Ando soils. Pedologist 32:16–25 (in Japanese with English summary)

Biodiversity Center of Japan, Ministry of the Environment (2018) State of vegetation obtained in the fourth national surveys on the natural environment conducted from 1989–1993. Ministry of the Environment. http://www.biodic.go.jp/vg_map/vg_html/jp/html/vg_map.html. Accessed 5 Mar 2018

Forestry Agency (2015) To create forestry work roads in steep slopes and vulnerable areas. Forestry Agency, Tokyo, p 78 (in Japanese)

Forest Experimental Station, Forest Bureau (1968) Forest soil profile photo collection book, 2nd. Ministry of Agriculture and Forestry, Tokyo, p 84 (in Japanese)

Fukui Agricultural Experiment Station (1979–2011) Report on the soil conservation program (in Japanese)

Gifu Prefectural Forestry Policy Division (2017) Statistical yearbook of forest and forestry in Gifu Prefecture (the 2015 fiscal year). Gifu Prefectural Forestry Policy Division, Gifu, p 148 (in Japanese)

Hara M (2017) Study on pelleting technology of livestock manure compost. Jap J Soil Sci Plant Nutr 88:383–386 (in Japanese)

Hara M (2018) Study on pelleting technology of livestock manure compost. Chubu Dojyou Hiryou Kenkyu 107:2–6 (in Japanese)

Hara M, Furuichi Y, Obata J (2004) Influence of the pelleting condition of livestock manure using a uniaxial extruder to the fertility effect of pelleted manure. Jpn J Soil Sci Plant Nutr 75:223–227 (in Japanese)

Higashino T (2006) Component materials of HM-1 tephra from Shin-Hakusan volcano. Bull Hakusan Nat Conserv Cen. Ishikawa 33:1–6 (in Japanese)

Hosokawa K, Sasaki H, Kogi Y (2012) One example of the rice cultivation method which reduced nitrogenous fertilizer by winter-flooded paddy field and analysis of its factors. Jpn J Soil Sci Plant Nutr 83:700–702 (in Japanese)

Imori H, Makida Y et al (2002) Change of the paddy soil characters and desirable direction of soil and fertility management in Fukui Prefecture. Bull Fukui Agric Exp Stn 39:17–28 (in Japanese with English summary)

Inagaki H, Kusumoto Y (2014) Assessment of GIAHS in Shizuoka: the traditional tea-grass integrated system. J Resour Ecol 5:398–401

Inoue K (2016) Growth, yield and quality of rice in winter-flooded paddy field. Agric Hort 91:160–166 (in Japanese)

Japan Meteorological Agency (2017) Past weather data. Japan meteorological agency. http://www.data.jma.go.jp/gmd/risk/obsdl/#. Accessed 7 Jun 2017

Kamishima T, Takeuchi A (2016) Late Quaternary geomorphology of the Tonami plain and activity of the Tonami-heiya fault zone, Toyama Prefecture, central Japan. Int J Geosci 7:962–976

Kanda M (2015) Improvement of filling materials for drainage in the drained paddy field of the cultivation of Japanese apricot (*Prunus mume* Sieb. et Zucc.). Bull Fukui Agric Exp Stn 52:9–16 (in Japanese with English summary)

Kanda T, Takata Y et al (2016) New soil maps of Hokuriku and Chubu district (1:200,000)-Comparison the comprehensive soil classification system of Japan first approximation with unified soil classification system of Japan (2nd approximation). Pedologist 60:14–31 (In Japanese with English summary)

Kasuya M (2007) Availability of nitrogen from 15 N-labeled green manure in yellow soil. Res Bull Aichi Agric Res Ctr 39:83–87 (in Japanese with English summary)

Kasuya M, Hiroto S (2010) Effect of green manure on nitrogen balance in a cabbage field. Res Bull Aichi Agric Res Ctr 42:141–146 (in Japanese with English summary)

Kasuya M, Banzai K et al (2010) Nitrogen and phosphorus runoffs into a stream from an upland basin with high stocking density in a red-yellow soil area. Jpn J Soil Sci Plant Nutr 81:481–488 (in Japanese with English summary)

Kato T (2002) Geology of central Honshu. In: Editorial board of geology of the Japanese Archipelago (ed) Geology of the Japanese Archipelago, CD-ROM version, Maruzen Publishing, Tokyo (in Japanese)

Kato Y (1975) Age of parent material of soil in the Chubu Region. In: Editorial board of Chubu Branch, Japanese Society of Soil Science and Plant Nutrition (ed) Soil and Agriculture in the Chubu Region, Conference Steering Committee of Japanese Society of Soil Science and Plant Nutrition in 1975, pp 28–29. Mie (in Japanese)

Kimura I (1973a) Pleistocene sediments and geomorphic development in the west coast area of Ise Bay, Japan: Part 2. Bull Aichi Univ Educ (Nat. Sci.) 21:125–150

Kimura I (1973b) Pleistocene sediments and geomorphic development in the west coast area of Ise Bay, Japan: Part 3. Bull Aichi Univ Educ (Nat. Sci.) 22:109–131

Kinki-Chugoku Regional Forest Office (2018) Ishikawa District Forest Office. Ministry of Agriculture, Forestry and Fisheries. http://www.rinya.maff.go.jp/kinki/isikawa/. Accessed 3 May 2018

Kitamura H (2007) Lowland soils and wet-rice farming by using delayed release fertilizer in Noubi plain. In: Editorial board of Japanese Society of Pedology (ed) Love soil and protect soil, Hakubunkan Shinsha Publishers, Tokyo, pp 188–191(in Japanese)

Kondo N (1967) On vertical zonality of mountain soils in the southern part of the Japan South Alps. Pedologist 11:153–169 (in Japanese with English summary)

Kosugi K (2014) Adjustment service of forest, how to adjust the environment? In: The Japanese Forest Society (ed) Forestry as the liberal arts, pp 97–105. Bun'eido publishing, Tokyo (in Japanese)

Kudo T, Uno F (2012) Standards for reducing the amount of phosphorous fertilizer application for paddy fields based on the fertility of the soil. Report of research on agriculture and fishery in Ishikawa Prefecture 14:32 (in Japanese)

Kuribayashi S (1956) Sand dune around Cape Omaezaki. Chiri (Geography) 1:264–271 (in Japanese)

Machida H (1999) The stratigraphy, chronology and distribution of distal marker-tephras in and around Japan. Glob Planet Change 21:71–94

Makita N, Kuno C et al (2014) Fertilizer application rate and nutrient uptake in the major vegetable cultivation area in Aichi Prefecture, Japan. Jpn J Soil Sci Plant Nutr 85:236–240 (in Japanese)

Matsui T, Kato Y (1962) Notes on palaeopedology of red soils in Japan. Daiyonki Kenkyu (Quat. Res.) 2:161–179 (in Japanese with English summary)

Matsumoto M (2005) Agriculture, soil science and plant nutrition in Shizuoka Prefecture. Jp J Soil Sci Plant Nutr 76:673–674 (in Japanese)

Ministry of Agriculture, Forestry and Fisheries (2017) Report on the states of food, agriculture and rural areas in Hokuriku region in 2016. http://www.maff.go.jp/hokuriku/news/print/jyousei/h28.html (in Japanese)
Ministry of Agriculture, Forestry and Fisheries: statistics. http://www.maff.go.jp/j/tokei/ (in Japanese) (Feb 2018)
Ministry of the Environment 2016: Enforcement status of the act to prevent soil contamination on agricultural land in Heisei 28 (the year 2016). http://www.env.go.jp/press/104928.html (in Japanese)
Nagai S (1991) Hydrochemical approach on groundwater pollution. State and issue of inorganic pollution. J Ground Water Hydrol 33:145–154 (in Japanese)
Namikawa K (2005) Vegetation of Subarctic/Subalpine evergreen coniferous forest zone. In: Fukushima T, Iwase T (eds) Vegetation of Japan. Asakura Publishing, Tokyo, pp 90–95(in Japanese)
Naruse Y (1974) Some considerations on the Marine Terraces in the southern Pacific Coast of the Japanese Islands. Bull Osaka Univ Econ 99:89–126 (in Japanese)
Numata M, Miyawaki A, Itow D (1972) Natural and semi-natural vegetation in Japan. Blumea-Biodiversity. Evol Biogeogr Plants 20:435–496
Obara H, Kohyama K et al (2015) Red-Yellow soils. In: Center Natural Resources Inventory (ed) Photographs of soil. Tsukuba, National Institute for Agro-Environmental Sciences, pp 63–69 (in Japanese)
Sage Y (1998) Regional environment and agriculture in the Hokuriku District. In: Japanese Society of Soil Science and Plant Nutrition (ed) Soil and Food, Asakura Publishing, Tokyo, pp 148–151(in Japanese)
Sakai J (1963) Loam deposits in the drainage of Kiso dani Valley (Part 1). Chikyu Kagaku (Earth Sci.) 67:13–20 (in Japanese)
Sakai T (2007) Forest soils of chubu. In: Editorial board of Japanese Society of Pedology (ed) Love soil and protect soil. Hakubunkan Shinsha Publishers, Tokyo, pp 174–177 (in Japanese)
Sengi Y (1993) Distribution of forest soil in Ishikawa Prefecture (1): sandy regosols. Bull Ishikawa-Ken For Exp Stn 24:37–41 (in Japanese)
Shizuoka Local Meteorological Office (2018) Climate characteristics in Shizuoka Prefecture. http://www.jma-net.go.jp/shizuoka/tokusei.html (Feb 2018)
Shizuoka Prefecture (2014) Nature of Shizuoka Prefecture. In: Shizuoka Prefecture (ed) Handbook of soil and plant nutrition. Shizuoka Prefecture, Shizuoka, pp 239–250 (in Japanese)
Tokuda S, Hayatsu M (2004) Nitrous oxide flux from a tea field amended with a large amount of nitrogen fertilizer and soil environmental factors controlling the flux. Soil Sci Plant Nutr 50:365–374
Toya H, Machida H et al (1971) Distribution of alluvial fans in Japan In: Yazawa H, Toya H, Kaizuka S (eds) Alluvial fans-their regional characteristics, Kokon Shoin Publishers, Tokyo, pp 97–120 (in Japanese)
Toyama Prefecture (2012a) Revival of clear stream and fertile land: History of restoration of contaminated agricultural soil of the Jinzu River basin. (in Japanese)
Toyama Prefecture (2012b) Annual report of achievements and diffusion of the agricultural field research of Heisei 24 fiscal year pp 35–36 (in Japanese)
Tsuji M, Yamamoto T et al (2018) Reduction of the amount of chemical fertilizer to cabbage-maize crop system in a yellow soil upland field by continuous application of compost. Jpn J Soil Sci Plant Nutr 89:232–236 (in Japanese)
Uno F, Umemoto H, Kitada K (2010) The effect of improvement of the method of application of the fertilizer for the crop rotation system of soy bean and barley on the nitrogen load. Report of research on agriculture and fishery in Ishikawa Prefecture 12:6 (in Japanese)
Wakabayashi H, Inabe Y, Aoki A (1976) On the chemical and physical properties of the soil and the clay mineral sand dune soil in Ishikawa Prefecture. Sakyu Kenkyu (Sand Dune Res.) 23:21–27 (in Japanese with English summary)
Wakasawa H, Takahashi K (2012) Outline of cultivated soil in Shizuoka Prefecture. In: the Conference of Japan Soil Conservation Measures Project (Dojo Hozen Chosa Jigyo Zenkoku Kyogikai) (ed) Guidebook for cultivated soil in Japan, Japan Soil Association, Tokyo, pp 54–55 (in Japanese)
Watanabe H, Igawahara K et al (2016) Influences of planted species on surface soil movement in stands of Japanese Cypress (*Chamaecyparis obtusa*), Japanese Cedar (*Cryptomeria japonica*), and Japanese Red Pine (*Pinus densiflora*) on a steep slope. J Jpn For Soc 98:193–198 (in Japanese with English summary)
Yamada H (2015) The seashore disaster prevention forest development business of "Kaga Coast National Forest" which preserved an area from a threat of a sandstorm. Suirikagaku (Water Sci.) 59:100–105 (in Japanese)
Yamada N (2007a) Paddy soil in the Toayam Plain. In: Editorial board of Japanese society of pedology. Love soil and protect soil, Hakubunkan Shinsha Publishers, Tokyo, pp 198–199 (in Japanese)
Yamada N (2007b) Leading edge technologies for remedying heavy metal − contaminated agricultural soils. 2. Remediation of heavy metal—contaminated soils by soil dressing, and sustainability of the remediation effects. Jpn J Soil Sci Plant Nutr 78:411–416 (in Japanese)
Yokoi S, Yamaguchi K (1998) Growth of Japanese cedar (*Cryptomeria japonica*) and broad-leaved species in unsuccessful plantations in snowy region. Jpn Soc For Environ 40(2):91–96 (in Japanese)
Yoshida F (1983) The fan and terrace developed around the Yokkaichi city, Mie Prefecture, central Japan—with special reference to the formative age of the Middle Terrace (Mitachi and Sakabe Terrace). Bull Geol Surv Jpn 34:497–511 (in Japanese with English summary)
Yoshikawa T, Kaizuka S, Ota Y (1981) Types and regional characteristics of Japanese plain. In: Yoshikawa T, Kaizuka S, Ota Y (eds) The Landforms of Japan. University of Tokyo Press, Tokyo, pp 139–165

9 Kinki, Chugoku, and Shikoku Regions

Junta Yanai, Mitsuru Toma, Hidetoshi Mochizuki, Naoki Moritsuka, Kentaro Wada, Tomoji Uchiyama, Tsuneyoshi Endo, Yo Toma, Kiyoshi Hiraoka, Naoko Tokuchi, Hiroyuki Hasukawa, Kunihiko Takehisa, Hiroyuki Maki, Minoru Matsuyama, Tetsushi Ohshio, and Masaya Oya

Abstract

The Kinki, Chugoku, and Shikoku regions are located in the west-central part of Japan. The major soil great groups distributed in these regions are Brown Forest soils (47.6%), Red-Yellow soils (16.0%), and Fluvic soils (14.2%) with very limited distribution of Andosols (8.6%), reflecting mountainous geography. Those used for agricultural land are dominated by Fluvic soils (65.5%) followed by Brown Forest soils (14.2%) and Red-Yellow soils (7.8%), reflecting high ratio of paddy (72.8%) in agricultural fields. In these regions, diverse types of agriculture are conducted corresponding to a variety of soil types with various geographical and climatic circumstances, including paddy rice production as well as the production of vegetables (e.g., traditional vegetables), fruits (e.g., orange in Wakayama and Ehime Prefectures and Japanese apricot (*Ume*) in Wakayama Prefecture), and tea in upland fields. Among them "Minabe-Tanabe *Ume* system" and "Nishiawa slope farming system" are designated as globally important agricultural heritage systems (GIAHS) in these regions. Environmentally friendly agriculture has been carried out, which is defined as sustainable agriculture in which crop productivity is well balanced with environmental quality by recycling materials and reducing the input of chemical fertilizer and pesticides. A wide variety of countermea-

J. Yanai (✉)
Graduate School of Life and Environmental Sciences, Kyoto Prefectural University, Kyoto City, Kyoto Prefecture, Japan
e-mail: yanai@kpu.ac.jp

M. Toma
Graduate School of Sciences and Technology for Innovation, Yamaguchi University, Yamaguchi City, Yamaguchi Prefecture, Japan

H. Mochizuki
Division of Agro-Enviroment Research, Western Region Agricultural Research Center, NARO, Fukuyama City, Hiroshima Prefecture, Japan

N. Moritsuka
Graduate School of Integrated Arts and Sciences, Kochi University, Nankoku City, Kochi Prefecture, Japan

K. Wada
Tokushima Agriculture, Forestry and Fisheries Technology Support Centre, Ishii Town, Tokushima Prefecture, Japan

T. Uchiyama
DGCbase, Toyonaka City, Osaka Prefecture, Japan

T. Endo
Faculty of Agriculture, Tottori University, Tottori City, Tottori Prefecture, Japan

Y. Toma
Graduate School of Agriculture, Ehime University, Matsuyama City, Ehime Prefecture, Japan

K. Hiraoka
Western Region Agricultural Research Center, NARO, Zentsuji City, Kagawa Prefecture, Japan

N. Tokuchi
Field Science Education and Research Center, Kyoto University, Kyoto City, Kyoto Prefecture, Japan

H. Hasukawa · K. Takehisa
Shiga Prefecture Agricultural Technology Promotion Center, Ohmihachiman City, Shiga Prefecture, Japan

H. Maki · M. Matsuyama · T. Ohshio
Hyogo Prefectural Technology Center for Agriculture, Forestry and Fisheries, Hyogo Prefectural Govenment, Kasai City, Hyogo Prefecture, Japan

M. Oya
Okayama Prefectural Technology Center for Agriculture, Forestry and Fisheries, Akaiwa City, Okayama Prefecture, Japan

R. Hatano et al. (eds.), *The Soils of Japan*, World Soils Book Series,
https://doi.org/10.1007/978-981-15-8229-5_9

sures has been also accomplished to reduce greenhouse gas emissions and leaching of nitrate, and harmful chemical substances are also discussed.

Keywords

Brown Forest soils • Environmentally friendly agriculture • Fluvic soils • Lake Biwa • Paddy fields

9.1 Outline of the Kinki, Chugoku, and Shikoku Regions

9.1.1 Perspective and Historical Background

The Kinki, Chugoku, and Shikoku regions are located in the west-central part of Honshu Island, Shikoku Island, and surrounding islands, and extend from roughly 32.7–35.8° N latitude and 130.9–136.5° E longitude. The Kinki region is administratively composed of six prefectures—Osaka, Kyoto, Hyogo, Nara, Shiga, and Wakayama—with a total area of 27.3 × 10^3 km^2 (corresponding to 7.2% of the area of Japan) and a total population of 20.68 million people (corresponding to 16.3% of the population of Japan; Ministry of Internal Affairs and Communications, 2016). The Chugoku region is administratively composed of five prefectures—Tottori, Shimane, Okayama, Hiroshima, and Yamaguchi—with a total area of 31.9 × 10^3 km^2 (corresponding to 8.4% of the area of Japan) and a total population of 7.41 million people (corresponding to 5.8% of the population of Japan 2016). The Shikoku region is administratively composed of four prefectures—Tokushima, Kagawa, Ehime, and Kochi—with a total area of 18.3 × 10^3 km^2 (corresponding to 4.8% of the area of Japan) and a total population of 3.82 million people (corresponding to 3.0% of the population of Japan, 2016). The three regions therefore have a combined area of 77.5 × 10^3 km^2 (corresponding to 20.4% of the area of Japan) and a combined population of 31.91 million people (corresponding to 25.1% of the population of Japan 2016).

One of the characteristics of the Kinki, Chugoku, and Shikoku regions is the early acceptance of the culture of the Asian continent due to their proximity. Specifically, rice farming was first introduced from China to Japan (northern Kyushu) at the later stage of the Jomon period (fourth century BC), as evidenced by the Itatsuki ruins in Fukuoka Prefecture and the Nahata ruins in Saga Prefecture. Rice farming was quickly transferred to Eastern Japan and arrived in northern Honshu at the early stage of the Yayoi period (fourth century BC to the middle of the third century AD), as evidenced by the Sunazawa ruins in Aomori Prefecture. Accordingly, it is reasonable to assume that rice farming was introduced in the Kinki, Chugoku, and Shikoku regions at a relatively early stage of the Yayoi period. This would be partly due to their climatic conditions, which are characterized by relatively warm temperature and plentiful precipitation, and to their geography, which consists of many rivers and alluvial plains. Additionally, the pedological conditions of these regions could be another reason for the early introduction of rice farming; specifically, the distribution of volcanic soils (Andosols) with high water permeability was extraordinary sparse and such conditions made it easy to prepare paddy fields, which require the storage of water for rice production. In this sense, these regions must have played a pioneering role during the nascent stage of rice production in Japan.

Another characteristic of the Kinki, Chugoku, and Shikoku regions is their early development as centers of politics, economy, and culture, from (1) the Izumo Dynasty in Shimane Prefecture and (2) the Yamato Dynasty in Nara Prefecture, which named itself "the country of ripening rice" and introduced an ancient land subdivision system, that is, the strip system, and a system of periodic reallocation of paddy fields to farmers, that is, "*Handen-shuju-sei*" (AD 652), for their governance, to (3) the capital of Heijo-kyo in Nara Prefecture (AD 710) and 4) the capital of Heian-kyo in Kyoto Prefecture (AD 794). It is highly probable that the political, economic, and cultural importance of these three regions was also strongly influenced by the specific conditions of these regions, including soil distributions.

The third characteristic of the Kinki, Chugoku, and Shikoku regions is their wide variety of population density, from the second biggest metropolitan and economic center in Japan (around Osaka; the biggest is in the Kanto region, around Tokyo) to largely depopulated areas in some parts of these regions. This variety affects the present land use and soil management, including for agriculture and forestry.

9.1.2 Geomorphology

The landforms of these regions are classified into mountains of the "inner zone" (north of the Median Tectonic Line [MTL]), mountains of the "outer zone" (south of the MTL), and in-between lowlands ranging from the Seto Inland Sea to the Kinki triangle zone (Okada and Chinzei 2004). Distinctive landform divisions in these regions are shown in Fig. 9.1.

Mountains in the inner zone consist of the Tanba mountains (1) in Fig. 9.1) in the north of Kyoto Prefecture and the Chugoku mountains (2), which cover most of the Chugoku region. In this zone, mountains higher than 1000 m above sea level (a.s.l.) are rare and the average elevation is 800–900 m a.s.l. The relief of these mountains is mostly gentle. The Chugoku mountains include backbone ranges and plateaus and extend for more than 300 km from

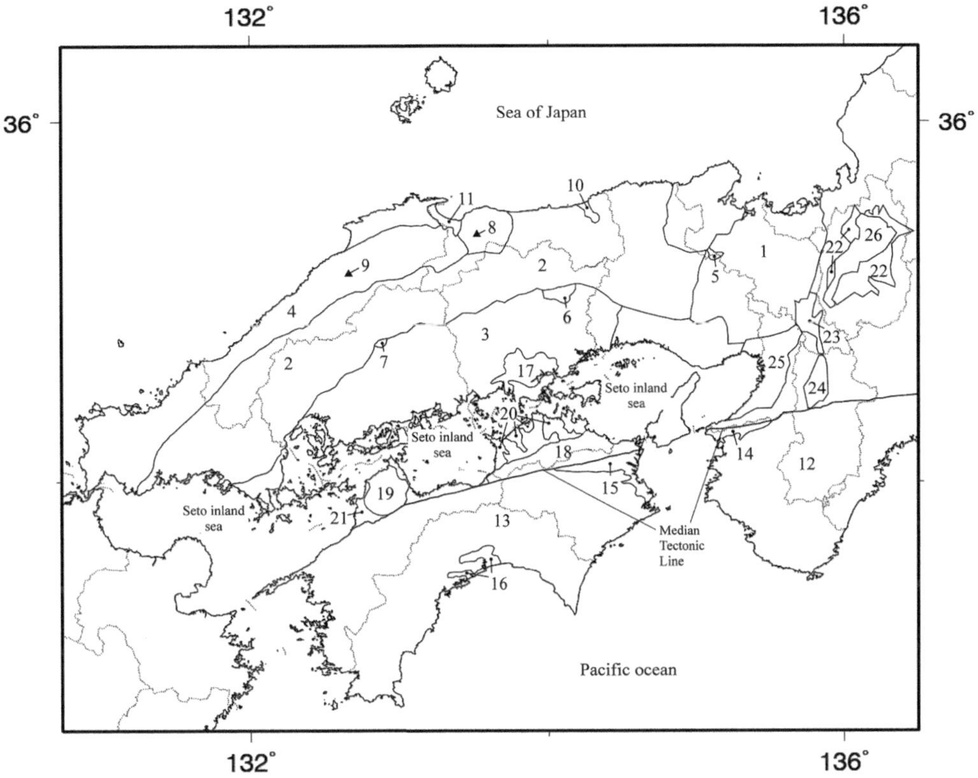

Fig. 9.1 Distinctive landform divisions in Kinki, Chugoku, and Shikoku regions. (1. Tanba mountains, 2. Chugoku mountains, 3. Kibi plateau, 4. Suou-Iwami plateau, 5. Fukuchiyama basin, 6. Tsuyama basin, 7. Miyoshi basin, 8. Mt. Daisen, 9. Mt. Sanbe, 10. Tottori plain, 11. Yonago plain, 12. Kii mountains, 13. Shikoku mountains, 14. Wakayama plain, 15. Tokushima plain, 16. Kochi plain, 17. Okayama plain, 18. Sanuki mountains, 19. Takanawa mountains, 20. Sanuki plain, 21. Matsuyama plain, 22. Oumi basin, 23. Kyoto basin, 24. Nara basin, 25. Osaka plain, and 26. Lake Biwa)

east to west. In the southern backbone range, the Kibi plateau (3) extends from the west of Hyogo Prefecture to Hiroshima Prefecture, with an average elevation of 400–600 m a.s.l. In the north of the backbone range, the Suou-Iwami plateau (4) extends from Shimane Prefecture to Yamaguchi Prefecture, and the average elevation of this plateau is slightly lower than the Kibi plateau. Several relatively wide basins, including the Fukuchiyama (5), Tsuyama (6), and Miyoshi basins (7), along with a number of small basins and lowlands along rivers are located in the Tanba and Chugoku mountains. In this region, a small number of volcanoes were active in the Quaternary Period. Of these, only Mt. Daisen (8) and Mt. Sanbe (9) released abundant tephra (Machida and Arai 1992); the others are small groups of volcanoes located in the Chugoku and Tanba mountains (Machida 2004). Along the coast of the Sea of Japan in the north of the Chugoku mountains, relatively wide alluvial plains, including the Tottori (10) and Yonago (11) plains, are distributed from Tottori Prefecture to the Shimane Peninsula. The area of these plains has been expanded by land reclamation and the sedimentation of deposit soils from the Chugoku mountains, where iron sand was collected from the sixteenth to nineteenth centuries. In the west of the Shimane Peninsula, only a few small alluvial plains are distributed.

Mountains in the outer zone consist of the Kii mountains (12) in the south of the Kii Peninsula and the Shikoku mountains (13) on Shikoku Island. The average elevation of the Kii mountains is 1000–1100 m a.s.l., and that of the Shikoku mountains is 1300–1400 m a.s.l. The central parts of both mountain ranges reach about 2000 m a.s.l. Mountain relief is steep and valleys are deep. In the coastal regions of these mountain ranges, marine terraces are developed and ria coasts are distributed (Ota 2004). Therefore, alluvial plains are limited around these mountains, with the only relatively large plains being the Wakayama (14) and Tokushima (15) plains along the MTL and the Kochi plain (16) in the south of Shikoku Island.

The Setouchi region is a lowland region between the Chugoku mountains and the MTL and is divided into the Chugoku and Shikoku regions by the Seto Inland Sea (*Setonaikai*). In the coast of the Chugoku region, the Okayama plain (17) and other smaller alluvial plains are located from the west of Hyogo Prefecture to the east of Yamaguchi Prefecture. Most of these plains have tidal flats, since tidal currents are gentle and tidal ranges are large. The areas of these plains were expanded by human activities from the sixteenth to the nineteenth centuries, such as the sedimentation of deposits from iron sand collections in the Chugoku mountains and land reclamation for salt production and paddy fields (Naruse 2004). In the Shikoku region side, the Sanuki mountains (18) in Kagawa Prefecture and the Takanawa mountains (19) in Ehime Prefecture are located to the north of the MTL. The highest elevations of these mountains are about 1000 m a.s.l. Small plains such as the

Sanuki (20) and Matsuyama (21) plains are located along the coast of this region. The upper parts of these plains are alluvial fans and the lower parts are deltaic plains (Hirai 2004).

The Kinki triangle zone is a lowland area that includes the Oumi (22), Kyoto (23), and Nara (24) basins and the Osaka plain (25) and is surrounded by small mountains. This lowland area is covered by alluvial sediments or marine sediments of the late Quaternary Period. The Oumi basin lowland includes Lake Biwa (26), the largest freshwater lake in Japan, and alluvial plains are widely located from the east to the south of the lake. The Osaka plain and the Kyoto and Nara basins were submerged under a series of seas and lakes until the middle of the Quaternary Period and were divided by crustal movement in the late Quaternary Period (Sangawa 2004). In the Kyoto basin, alluvial fans are developed in the northeastern part, natural levees in the western part, and back marshes in the southern part. All of the rivers flowing into the Kyoto basin merge in the south of the basin and flow into the Osaka Bay. Due to the narrow exit of the Kyoto basin, a large swamp area was developed upstream of the exit. Here, land reclamation by drainage has been carried out for the development of agricultural land since the fourth century (Tanioka 1964). The central part of the Osaka plain includes an alluvial plain surrounded by hills and the Uemachi plateau. During the Jomon transgression (7–5 kya), the sea area extended to the east of the Uemachi plateau. After the sea level fell, lakes and marshes emerged in this area. Since the fourth century, flood control, land reclamation by drainage, river improvements, and changes to river courses have been conducted to expand agricultural lands. The development of agricultural land was also conducted along the coastal area of the Osaka plain, expanding it to the west.

As a result of the long history of human activity in these regions, artificially changed lands are widely distributed, not only in the Kinki triangle zone but also in the mountains of the inner zone and the Setouchi region.

9.1.3 Soil Distribution

1. **Soils of the whole area**

The soil distribution area and ratio by soil great group in the Kinki, Chugoku, and Shikoku regions, according to the Soil Classification System of Japan (The Fifth Committee for Soil Classification and Nomenclature 2017), are shown in Table 9.1 (calculated based on Obara et al. 2016). The major soil great groups distributed in these regions are Brown Forest soils, Red-Yellow soils, and Fluvic soils, which together occupy about 80% of the land area of the regions (Institute for Agro-Environmental Sciences, NARO 2017).

Brown Forest soils account for 47.6% of the soil area in these three regions, much higher than the national average (33.2%), and are distributed widely in mountainous areas. Red-Yellow soils account for 16.0% of the whole area in the regions, which is also much higher than the national average (7.6%), and are distributed in lower-elevation mountainous areas such as plateaus. A total of 44% of the Red-Yellow soils in Japan is distributed in these regions, and this soil can therefore be regarded as one of the distinctive characteristic soils there. Fluvic soils account for 14.2% of the whole area in these regions, which is close to the national average (13.7%). There are not many large alluvial plains or basins in the regions, but a number of small basins and lowlands along rivers are located throughout mountains in the inner zone. Fluvic soils are distributed in these small basins and lowlands. Andosols, which account for 30.3% of the soil area in Japan, account for only 8.6% of the whole area in the Kinki, Chugoku, and Shikoku regions. This is because only a small number of volcanoes were active in the Quaternary Period in these regions. The small distribution of Andosols is related to the wide distributions of Brown Forest soils and Red-Yellow soils in these regions. Non-allophanic Andosols comprise 84% of the Andosols in these three regions, which is much higher than the national average (23.9%) (Obara et al. 2016). This is related to the fact that Andosols in this region have poor depositions of Holocene tephra. Regosols account for 7.9% of the whole area in these regions, which is similar to the national average (6.9%). Among Regosols, Terrestrial Regosols are dominant, with 36.6% of the Terrestrial Regosols in Japan being distributed in these regions (Obara et al. 2016). This is related to the wide distribution of "*Masa*" soil, which is derived from the residue of weathered granite.

2. **Soils of agricultural land**

The soil distribution area and ratio by soil great group in agricultural lands in the Kinki, Chugoku, and Shikoku regions, according to the Soil Classification System of Japan (The Fifth Committee for Soil Classification and Nomenclature 2017), are shown in Table 9.2 (calculated based on Kanda et al. 2017). Fluvic soils are distributed most widely and occupy 65.5% of agricultural land in these regions, much higher than the national average (47.4%). Fluvic Paddy soils, Gley Fluvic soils, and Gray Fluvic soils account for 44.1%, 23.1%, and 21.3% of Fluvic soils in these regions, respectively (Kanda et al. 2017). These three soil groups are usually used as paddy fields, which is reflected in the high ratio of paddy fields in the total agricultural land area in the three regions (72.8%). In these regions, large

Table 9.1 Distribution area and ratio of soil great groups in Kinki, Chugoku, and Shikoku regions (Calculated based on Obara et al. (2016))

	Kinki, Chugoku, Shikoku		Whole country	
	Area (km^2)	Ratio (%)	Area (km^2)	Ratio (%)
Human-made soils	12.0	0.0	107	0.0
Organic soils	40.0	0.1	4403	1.2
Podzols	31.0	0.0	8297	2.2
Andosols	6723.0	8.9	113,481	30.7
Dark red soils	626.0	0.8	1748	0.5
Lowland soils	11,071.0	14.7	52,085	14.1
Red-yellow soils	18,376.0	24.3	37,294	10.1
Stagnic soils	158.0	0.2	2946	0.8
Brown forest soils	30,789.0	40.7	111,089	30.0
Regosols	6186.0	8.2	26,065	7.0
Rocky land	386.0	0.5	4016	1.1
Others	1161.0	1.5	8645	2.3

plains and basins are only located in the Kinki triangle zone. However, small basins and lowlands along rivers occur throughout mountains in the inner zone, as described previously. There is a long history of human activity in the Kinki triangle, Setouchi region, and these mountain lowlands have long been used as paddy fields. This historical background may have affected the distribution of Fluvic soils. Brown Forest soils are the second most widely distributed soil great group in these three regions and account for 13.3% of agricultural land; this proportion is also higher than the national average (7.8%). This abundance is due to the small distribution of Andosols in these regions compared with other regions, where Andosols are widely used for agricultural land in mountainous areas. Red-yellow soils account for only 8.7% of agricultural land in the Kinki, Chugoku, and Shikoku regions. The proportions of both brown forest soils and Red-Yellow soils in agricultural land are much lower than their proportions in whole area of the region; this is because the proportion of paddy fields is high (72.8%) and these two soils are usually used as upland fields, orchards, and tea gardens. Andosols, which occupy 28.7% of the agricultural land in Japan, occupy only 3.6% in these three regions due to the smaller distribution of this soil.

9.1.4 Land Use

The dominant land use in the Kinki, Chugoku, and Shikoku regions is forestry, with the ratio of forest in regional land area corresponding to 66.3%, 73.3%, and 72.6% in the three regions, respectively (The Forestry Agency 2012); these ratios are close to or greater than the national average of 66.4%, reflecting the mountainous geomorphology of these regions. However, these ratios are variable among prefectures and range from 30.5% in Osaka Prefecture to 84.0% in Kochi Prefecture.

Table 9.2 Distribution area and ratio of soil great groups in agricultural lands in Kinki, Chugoku, and Shikoku regions (calculated based on Kanda et al. (2017))

	Kinki, Chugoku, Shikoku		Whole country	
	Area (1000 ha)	Ratio (%)	Area (1000 ha)	Ratio (%)
Man-made soils	0.0	0.0	0.8	0.0
Organic soils	0.9	0.1	181.5	3.9
Podzols	0.0	0.0	0.6	0.0
Andosols	22.6	3.6	1319.6	28.7
Dark red soils	0.6	0.1	39.9	0.9
Lowland soils	412.6	65.5	2179.7	47.4
Red-yellow soils	54.7	8.7	212.3	4.6
Stagnic soils	41.8	6.6	227.1	4.9
Brown forest soils	83.6	13.3	358.9	7.8
Regosols	13.3	2.1	78.0	1.7

The ratios of agricultural land to regional land area are 8.2, 7.6, and 7.3% in the Kinki, Chugoku, and Shikoku regions, respectively, which are much lower than the national average of 11.3% (Ministry of Agriculture, Forestry and Fisheries 2017a). This is firstly due to the high ratio of forest in these regions and secondly due to the considerable conversion of agricultural land to residential and commercial/industrial areas that has been conducted on the alluvial plains and basins. The ratios of paddy fields to the total agricultural land area are 77.5%, 77.0%, and 64.8% for the three regions, respectively. These ratios are much higher than the national average of 54.5%, suggesting a high dependency on paddy rice production in these regions. The ratios of upland field to the total agricultural land area are 7.9, 15.2, and 12.2% for the three regions, respectively, which are much lower than the national average of 25.7%. The ratios of pastureland to the total agricultural land area are 0.2, 1.3, and 0.4% for the three regions, respectively, which are much lower than national average of 13.5%. The ratios of orchard land to the total agricultural land area are 14.4, 6.4, and 22.5% for the three regions, respectively. These ratios are much higher than the national average of 6.4%, due to the fact that some prefectures have an extraordinarily high ratio of orchard land, such as Wakayama (63.4%) and Ehime (41.1%).

9.2 Agriculture and Forestry

9.2.1 Perspectives

The climate of the Kinki–Chugoku–Shikoku region is mainly divided into three zones: (1) the Sea of Japan side climate, which in winter is characterized by high precipitation, moderately heavy snow, and a small number of sunshine hours; (2) the Seto Inland Sea climate, which is warm and experiences little rain throughout the year, sometimes causing drought; and (3) the Pacific Ocean side climate, which is hot and humid in summer and sunny, warm, and dry in winter. Typhoons sometimes pass through in August and September. Additionally, there are some areas which have a highland climate or basin climate. Based on this variety in climate as well as geography and soil, diverse types of agriculture are conducted in Kinki–Chugoku–Shikoku region.

1. **Paddy fields**

Paddy fields in the Kinki–Chugoku–Shikoku region most commonly contain Gray Fluvic soils. Meanwhile, Andosols are limited to the area around Mt. Daisen, i.e., in Tottori and Shimane Prefectures, and the northern part of Okayama and Hiroshima Prefectures; consequently, the area ratio of these soils is lower than the national average (NARO WARC 2013). Paddy fields occupy flat plains and polders, and 30% are situated in areas with a gradient of less than 1/100 (national average: 44%); in the Chugoku region, that ratio is 26%. Furthermore, the ratio of relatively large fields (>0.3 ha) with improved infrastructure in the Kinki, Chugoku, and Shikoku regions is 55%, 50%, and 24%, respectively, lower than the national average (excluding Hokkaido) of 59% (NARO WARC 2013). Accordingly, fields are often not squared, but rather narrow, which causes disadvantages to production conditions. Due to the various climatic conditions in these regions, farmers cultivate rice in various styles. For example, the southern part of Okayama Prefecture is one of the biggest production regions for dry-seeded rice in Japan, and in Kochi Prefecture, rice is harvested earlier than anywhere else in Japan. There are many terraced paddy fields (“*Tanada*”) in the Kinki–Chugoku–Shikoku region, which are recognized as important cultural landscapes based on the Cultural Properties Protection Law; such terraced fields are present in Asuka Village (Nara Prefecture), Aritagawa Town (Wakayama Prefecture), Okuizumo Town (Shimane Prefecture), Kamikatsu Town (Tokushima Prefecture), and Yusuhara Town (Kochi Prefecture) (Agency for Cultural Affairs 2017). Recently, the feeding of rice plants with whole-crop silage and the feeding of rice cultivation have been increasing through the breeding of special purpose cultivars which have a long culm, high sugar content, and lodging resistance properties, through the development of harvester and dry-seeding machinery.

2. **Upland field**

Although Brown Forest soils are spread most widely in normal upland field in the Kinki–Chugoku–Shikoku region as a whole, the dominant soil type is Gray Fluvic soil in the Kinki region, Red-Yellow soil in the Chugoku region, and Brown Forest soil in the Shikoku region (NARO WARC 2013). Upland fields in the Kinki–Chugoku–Shikoku region include fields converted from paddy fields, and crops such as soybean, wheat and barley, vegetables, flowers, fruit, and tea leaves are cultivated. Because soybean, wheat, and barley are mostly cultivated in fields converted from paddy, the sowing season of these crops is during the summer and autumn rainy season in Japan. Typhoons often pass through these regions, and the upland fields tend to be wet. Therefore, the average yields in this region are lower than the national average. However, the yields of certain crops are close to or higher than the national average due to their cultivation over a wide area, including beer barley in Okayama Prefecture, wheat and naked barley in Kagawa Prefecture, and naked barley in Ehime Prefecture in the Seto

Inland Sea climate zone, and wheat in Shiga Prefecture, where advanced paddy agriculture has been executed (NARO WARC 2013).

In Kyoto, Osaka, Tokushima, Kagawa, and Kochi Prefectures, the production of vegetables accounts for more than 30% of agricultural production (Ministry of Agriculture, Forestry, and Fisheries 2017b) (Table 9.3). Famous traditional vegetables known as "*Kyo-yasai*" are produced in Kyoto Prefecture (Nakajima 2007), and leafy vegetables are cultivated in Osaka Prefecture to meet the demands of large towns. In Tokushima Prefecture, vegetable production for big cities in the Keihanshin metropolitan region, which includes parts of Kyoto, Osaka, and Hyogo Prefectures, is recognized as an important prefectural political tactic (Kuroda 2006), while the production of lotus, carrot, and sweet potato is also traditionally famous. On the other hand, in Kochi Prefecture, greenhouse vegetable cultivation utilizing the warm climate and long daylight time is popular, with the proportion of greenhouse cultivation area to total cultivation area being much higher than the national average (Yamasaki 2005). Many of these greenhouse fields have been converted from paddy field. As Kochi Prefecture is situated far from large towns, vegetables such as eggplant, bell pepper, and cucumber are grown (Yamasaki 2005). Additionally, as a characteristic agriculture in the Kinki–Chugoku–Shikoku region, it is possible to introduce vegetable cultivation in sandy fields, such as sand dunes in Tottori Prefecture, that cultivate crops such as Chinese yam, shallot, and long green onion.

Flower production in the Kinki–Chugoku–Shikoku region as a whole is not particularly popular, but the agricultural production of rose and seedlings for flower gardens in Nara Prefecture and lily in Kochi Prefecture are higher than the national average (Table 9.3).

The ratio of fruit production to total agricultural production in the Kinki–Chugoku–Shikoku region is higher than the national average. This is at least partly due to the fact that the region has few flat areas and a warm climate. The cultivated fruit items are highly various and are very famous in Japan, including citrus fruits (mainly mandarin orange) in Osaka, Wakayama, Hiroshima, Yamaguchi, Tokushima, Kagawa, and Ehime Prefectures; persimmon in Nara, Wakayama, and Ehime Prefectures; Japanese apricot (*Ume*) in Wakayama Prefecture; chestnut in Ehime Prefecture; grape in Osaka, Hyogo, Wakayama, Shimane, Okayama, Hiroshima, Kagawa, and Ehime Prefectures; Japanese pear in Tottori and Tokushima Prefectures; and peach in Wakayama and Okayama Prefectures. Brown Forest soil occurs in orchards in the Shikoku region, in Wakayama Prefecture, and on islands in Hiroshima Prefecture, and Red-Yellow soil is widespread in the Chugoku region (NARO WARC 2013).

Kyoto and Nara Prefectures, which have a long history of the "*Sado*" tea ceremony, produce large quantities of tea leaves. These prefectures account for more than half of the tea production in the Kinki–Chugoku–Shikoku region (NARO WARC 2013).

In the Kinki–Chugoku–Shikoku region, livestock farming accounts for a high proportion of the agricultural production. Pasture production is limited in the highland areas of the Chugoku mountainous area and its area is decreasing (NARO WARC 2013). As mentioned above, the cultivation area of feeding rice and whole-crop silage is constantly increasing, and the domestic feed production is therefore also progressing consistently. Moreover, in flat plains and polders, where the risk of animal injury is low, the cultivation of feeding corn has been started as a test; such cultivation is expected to strengthen the domestic feed supplying ability and to diversify the styles of its production.

3. **Forest**

In the Kinki region, the ratio of forest area to the total land area is the same as the Japanese national average, but the ratios are higher than the national average in the Chugoku and Shikoku regions. According to prefectures, Kochi Prefecture has the highest forest ratio in Japan, 84%, while Osaka Prefecture has the lowest, 31% (Forestry Agency 2017). Despite the high forest ratios in some parts of the Kinki–Chugoku–Shikoku region, forestry production is low. Furthermore, there are no other distinguished forestry production items, except in Tokushima Prefecture, which has the highest shiitake mushroom production in Japan. Forestry production in Osaka and Shiga Prefectures is very low, as is that in the Tokyo metropolitan area and Kanagawa and Okinawa Prefectures (Ministry of Agriculture, Forestry and Fisheries 2017c).

4. **Globally Important Agricultural Heritage Systems**

In the Kinki–Chugoku–Shikoku region, the "*Minabe-Tanabe Ume*" system, which is a sustainable Japanese apricot (*Ume*) production system utilizing hilly slopes covered with low-fertility gravel in Wakayama Prefecture, is recognized as a Globally Important Agricultural Heritage System (GIAHS) by the United Nations Food and Agriculture Organization (FAO). In a brochure published by the Japanese Ministry of Agriculture, Forestry and Fisheries, this system is introduced as follows (Ministry of Agriculture, Forestry and Fisheries 2017d): "The Minabe-Tanabe area is occupied by slopes which have poor-fertility gravel. They produce high-quality *Ume* by reclaiming *Ume* orchards with maintaining coppices of *Quercus phillyraeoides* (ubame oak) on the slope. The coppices function to foster water

Table 9.3 Rank of agricultural production and component ratio (%) in Kinki, Chugoku, and Shikoku regions (Revised from Ministry of Agriculture, Forestry, and Fisheries (2017b))

	Agricultural production (10^8JPY)	Rank									
		1		2		3		4		5	
Prefectures except Hokkaido	76,779	Rice	18.0	Pork	7.6	Beef	7.4	Egg	6.9	Milk	4.9
Kinki	4673	Rice	24.6	Egg	7.0	Orange	6.4	Beef	5.9	Milk	4.5
Chugoku	4380	Rice	23.1	Egg	15.2	Beef	7.3	Milk	7.1	Broiler	4.7
Shikoku	4100	Rice	10.7	Orange	7.1	Egg	6.3	Pork	5.0	Beef	4.5
Shiga	586	Rice	54.4	Beef	10.2	Milk	4.1	Egg	3.6	Soy bean	2.6
Kyoto	719	Rice	22.1	Egg	9.0	Raw tea leaf	6.4	Unprocessed tea leaf	5.1	Milk	4.9
Osaka	341	Rice	22.0	Grape	10.6	Welsh onion	9.1	Egg plant	6.7	Orange	4.7
Hyogo	1608	Rice	26.9	Egg	12.6	Beef	10.9	Onion	7.8	Milk	6.3
Nara	408	Rice	21.8	Persimmon	14.5	Milk	7.1	Strawberry	4.7	Spinach	4.7
Wakayama	1011	Orange	27.4	Plum	10.2	Persimmon	8.5	Rice	7.5	Peach	4.8
Tottori	697	Rice	17.4	Broiler	12.1	Milk	8.3	Japanese pear	8.0	Pork	7.7
Shimane	570	Rice	30.2	Beef	14.0	Milk	12.3	Egg	7.0	Pork	4.9
Okayama	1322	Rice	22.2	Egg	19.4	Grape	10.1	Milk	7.7	Beef	6.5
Hiroshima	1164	Egg	24.6	Rice	19.2	Pork	6.9	Beef	5.8	Milk	5.2
Yamaguchi	627	Rice	32.2	Egg	9.4	Beef	8.1	Broiler	4.8	Milk	3.0
Tokushima	1037	Rice	10.2	Broiler	9.0	Sweet potato	7.9	Beef	7.0	Carrot	6.0
Kagawa	815	Egg	19.3	Rice	12.4	Beef	7.0	Broiler	6.4	Broccoli	5.5
Ehime	1237	Orange	16.4	Rice	10.9	Pork	10.4	Egg	4.9	Iyokan orange	4.9
Kochi	1011	Egg plant	11.5	Rice	9.5	leek	7.5	Ginger	6.9	Cucumber	6.9

resources and prevent landslip, and '*Kishu-Bincho-tan*', wood charcoal, which is widely known for its high quality and hardness, is made from *Quercus phillyraeoides*. Moreover, Japanese honey bees living in the coppices play an important role, carrying the pollen to fruit *Ume*. For the bees, *Ume* trees are valuable because they produce nectar from February when flowers are still small, and a great symbiotic relationship is constructed between them. Seventy percent of workers in the area are connected to the *Ume* industry in some way, and *Ume* support the lives of the citizens as a regional main industry."

9.2.2 Paddy Fields

Rice is generally the most widely grown crop in the Kinki, Chugoku, and Shikoku regions, except for Wakayama and Ehime Prefectures, where fruit cultivation is widespread. The percentage of paddy field area to the total area of agricultural land exceeds 90% in Shiga and Hyogo Prefectures, which are ranked second and third, respectively, among Japan's 47 prefectures for this metric (national average of 54%) (Ministry of Agriculture, Forestry and Fisheries 2017a). This percentage is also higher than 80% in Shimane, Yamaguchi, and Kagawa Prefectures.

The area of individual paddy fields is often small in the Kinki and Shikoku regions, although large prefectural differences exist (Fig. 9.2) (Ministry of Agriculture, Forestry and Fisheries 2017e). For the whole of Japan, an average of 36% of paddy fields has an area of less than 30 a. In contrast, this percentage is higher than 70% in Osaka, Nara, Wakayama, Tokushima, Kagawa, and Ehime Prefectures. Except for these areas, such land fragmentation can be found only in the highly populated Tokyo and Kanagawa Prefectures in the Kanto region.

From a broad perspective, the degree of fragmentation of paddy fields can be explained by the topography and the history of land reclamation. For example, in the Nara basin in Nara Prefecture, rice cultivation has been carried out for

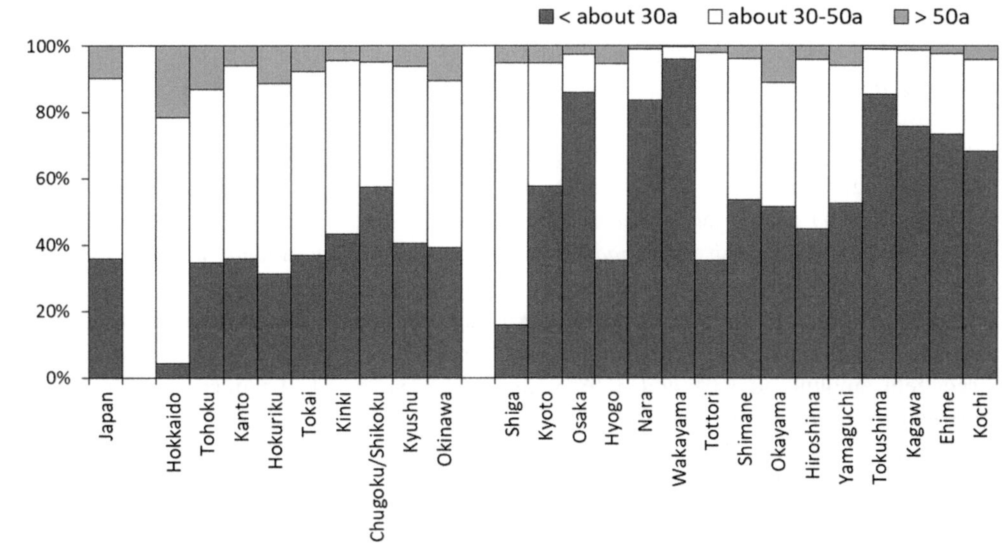

Fig. 9.2 Average size of an individual paddy field in Japan and Kinki, Chugoku, and Shikoku regions. The data is based on the statistical information in 2015 (MAFF 2017e)

more than 1000 years. Most paddy field construction was undertaken during two periods: the eighth to ninth century initiation of the gridded "*jori*" field system and the eighteenth century extension of this system into low marshy areas (Barnes 1986). As shown in Fig. 9.3, the *jori* system can be observed in the regular distribution of rectangular fields, each with an area of about 12 a. In contrast to the small fields in the Nara basin, 11% of the paddy fields in Okayama Prefecture have an area of more than 50 a and 84% of the paddy fields in Shiga Prefecture have an area of more than about 30 a (Fig. 9.2). In the reclaimed land from Lake Dainaka, located to the east of Lake Biwa, Shiga Prefecture, each paddy field has an area of 1.5 ha (Fig. 9.3). In such large fields, rice is produced in a cost-effective way by using big machines, which reduces the labor cost.

While the size of a paddy field is affected by the topography, the cropping system of rice is influenced strongly by the climate. As previously mentioned, the climate in the floodplains of the Kinki, Chugoku, and Shikoku regions can be roughly divided into three areas: the Sea of Japan area in the north, the Seto Inland Sea area in the middle, and the Pacific Ocean area in the south. In Tottori and Shimane Prefectures, located in the north, rice is usually mono-cropped due to snowfall in winter. In Okayama, Yamaguchi, Kagawa, and Ehime Prefectures, located in the middle, rice is often rotated with "*mugi*" (wheat or barley) in a single year. In Osaka, Hyogo, and Tokushima Prefectures, which have easy access to the metropolitan areas of Kyoto, Osaka, and Kobe, collectively called Keihanshin, rice is often rotated with winter vegetables. In the Kochi plain,

Fig. 9.3 Small rectangular paddy fields in the Nara basin in Nara Prefecture (**a**) and large paddy fields in the reclaimed land from Lake Dainaka in Shiga Prefecture (**b**). These aerial photographs, modified by the author from the images of CKK20083-C16-33 (**a**) and CKK20111-C17-3 (**b**) provided under CC BY 4.0 by the Geospatial Information Authority of Japan, were taken in May 15, 2008 (**a**) and September 29, 2011 (**b**)

Kochi Prefecture, located in the south, rice used to be cropped twice a year by making use of the warm climate. However, due to the "*gentan*" rice control policy described in Chap. 1, rice is now mono-cropped by early planting, which enables early harvesting in the beginning of August. In the inland mountainous areas, on the other hand, the climate is usually colder and wetter than in the floodplains. Here, rice is often grown in *Tanada* terraced fields (see Sect. 9.2.1) with a slope of 1:20 or greater. Many of these terraced fields are located in the Chugoku region, and especially in Hiroshima Prefecture, where the area of terraced field amounts to 43% and 26% of the total area of terraced field in Japan, respectively (Ministry of Agriculture, Forestry and Fisheries 2017f).

In the Kinki, Chugoku, and Shikoku regions, the percentage of the area of paddy fields under the *gentan* policy is more than 20% in Tottori and Kochi Prefectures, about 9% in Wakayama Prefecture, and 10–20% in other prefectures (Board of Audit of Japan 2016). In all prefectures, the production of staple rice has been controlled by rotating it with other crops such as vegetables, wheat or barley (*mugi*), and soybean—for example, three croppings of rice, wheat, and soybean in a two-year rotation. In recent years, more farmers have started to grow non-staple rice, such as rice for the production of food and drink items such as Japanese rice wine (*sake*), rice crackers, *miso* paste, and soy sauce, and also rice for feeding domestic animals such as unhulled rice for pigs and chickens and whole-crop silage for cows.

The type and properties of paddy soils in the Kinki, Chugoku, and Shikoku regions are described briefly. The average soil characteristics in each prefecture are summarized in Table 9.4. According to a national survey from 1959 to 1978, a large regional variation existed in the distribution of Gley Fluvic soils in paddy fields. For example, in Kagawa Prefecture, only 3% of the paddy soil was classified as Gley Fluvic soils, whereas that figure was more than 50% in Shimane and Shiga Prefectures. Gley Fluvic soils are formed in wet paddy fields where the water table is high and soil is clayey and impermeable. It is important even now to improve the field permeability by introducing open ditches and underground drainage especially when rice is rotated with upland crops. According to a more recent national survey from 1999 to 2003 (Ministry of Agriculture, Forestry and Fisheries Agricultural Production Bureau 2008), the surface paddy soils in the Kinki, Chugoku, and Shikoku regions showed lower values of total C and cation exchange capacity than those in other regions of Japan. Within these regions, the soil contents of exchangeable K and available P tended to be higher in Kyoto and Kagawa Prefectures.

9.2.3 Upland Fields

In the Kansai region (which includes the Kinki, Chugoku, and Shikoku regions), wheat varieties, beans, and millets are grown mainly in paddy fields utilized for double cropping or paddy–upland rotation, or in dedicated crop fields converted from paddy fields, thus providing production on diverse farmlands adapted to local climatic and soil conditions. In mountainous areas in the Shikoku region, crops are cultivated in small plots located on sloping lands.

1. **Crop cultivation based on the fields originally used as paddy (wheat varieties, beans, and millets)**

To cultivate crops susceptible to moisture damage, such as cereals, beans, and buckwheat (*Fagopyrum esculentum*), it is of utmost importance to improve the drainage of farmland. Depending on the climatic, topographic, and soil conditions, various efforts are being made for enhanced drainage, including the construction of open ditches, underdrains, and mole drains (Fig. 9.4).

(1) **Paddy–upland rotation for wheat and soybean (Shiga Prefecture: Fluvic Paddy soils, etc.)**

In Shiga Prefecture, where paddy fields account for as much as 92% of all farmland, many farmers adopt a three-year four-crop rotation of rice (summer harvest) to rice (summer harvest) to wheat (winter harvest) to soybean (summer harvest) (Kitagawa 2012). For the cultivation of wheat and soybean, Fluvic Paddy soils, Gray Fluvic soils, and Gley Fluvic soils distributed along rivers around Lake Biwa are used, with various measures being taken combining the use of open ditches, underdrains, and mole drains for improving the quantity and quality of production (Shiga Prefecture 2012).

(2) **Soybean production (Hyogo Prefecture, Okayama Prefecture: Gray Fluvic soils, Brown Fluvic soils)**

In Hyogo Prefecture, soybean (*Glycine max*), including large-grain black soybeans, are cultivated, mainly in Gray Fluvic soils. Soybean is susceptible to humidity damage, especially after seeding. Therefore, open ditches, underdrains, and radial-shaped mole drains have been constructed to improve drainage. In Okayama Prefecture, large-grain black soybean is cultivated mainly in Brown Fluvic soils. In recent years, instances of decreased soybean harvest have been reported, indicating a decline in soil fertility. In this region, as in many other regions, a higher frequency of black soybean cultivation has been shown to be correlated with

Table 9.4 Type and properties of paddy soils in Kinki, Chugoku, and Shikoku regions (prefectural averages). The data is cited from several publications on the national soil survey programs in Japanese agricultural land

Region	Prefecture	Gley soil distribution in paddy fields (area %)	Effective sample numbers (min–max)	pH (H_2O)	Total C (g kg^{-1})	Total N (g kg^{-1})	Cation exchange capacity ($cmol_c$ kg^{-1})	Exchangeable Ca ($cmol_c$ kg^{-1})	Exchangeable Mg ($cmol_c$ kg^{-1})	Exchangeable K ($cmol_c$ kg^{-1})	Available phosphate (mgP_2O_5 kg^{-1})	Plow layer thickness (cm)
Kinki	Shiga	53.1	82–159	5.8	20.0	1.88	14.1	8.13	1.83	0.39	258	15.0
	Kyoto	31.8	27–27	6.2	22.3	1.91	14.8	8.58	2.13	0.76	1005	14.3
	Osaka	5.6	41–50	6.0	19.7	1.78	11.0	7.83	1.41	0.51	799	12.6
	Hyogo	12.2	74–74	6.1	22.6	2.13	14.2	7.46	1.42	0.50	559	18.5
	Nara	17.8	19–19	6.2	18.7	2.12	12.2	7.31	1.41	0.42	632	18.2
	Wakayama	7.2	17–18	6.2	18.6	1.96	13.1	5.47	1.57	0.55	781	16.1
Chugoku	Tottori	13.1	22–24	5.8	25.7	2.28	16.3	7.87	1.68	0.34	364	15.3
	Shimane	62.4	39–39	5.7	23.8	1.91	15.3	7.53	1.74	0.65	192	15.1
	Okayama	14.7	63–63	6.0	22.8	2.06	16.0	9.80	1.86	0.53	600	15.4
	Hiroshima	13.5	87–87	5.7	26.2	2.53	16.7	7.13	1.11	0.43	434	16.0
	Yamaguchi	24.8	85–85	5.7	22.1	2.17	14.3	7.79	1.44	0.58	223	15.6
Shikoku	Tokushima	24.7	47–47	6.2	19.1	1.91	12.3	6.55	1.40	0.54	664	16.7
	Kagawa	3.0	52–52	6.2	23.5	2.33	11.8	10.48	1.52	0.64	1063	12.6
	Ehime	15.3	51–51	6.0	25.0	2.10	9.1	5.29	1.11	0.32	350	15.3
	Kochi	19.7	33–33	5.8	21.1	1.98	12.7	5.93	1.04	0.24	309	16.1
Japan		30.8	2664–2824	5.8	26.8	2.39	18.1	8.84	1.68	0.61	354	15.0

Effective sample numbers indicate the minimum and maximum number of soil samples to calculate the prefectural averages except for the area percentage of gley soil
Available phosphate was evaluated by the Truog's extraction method

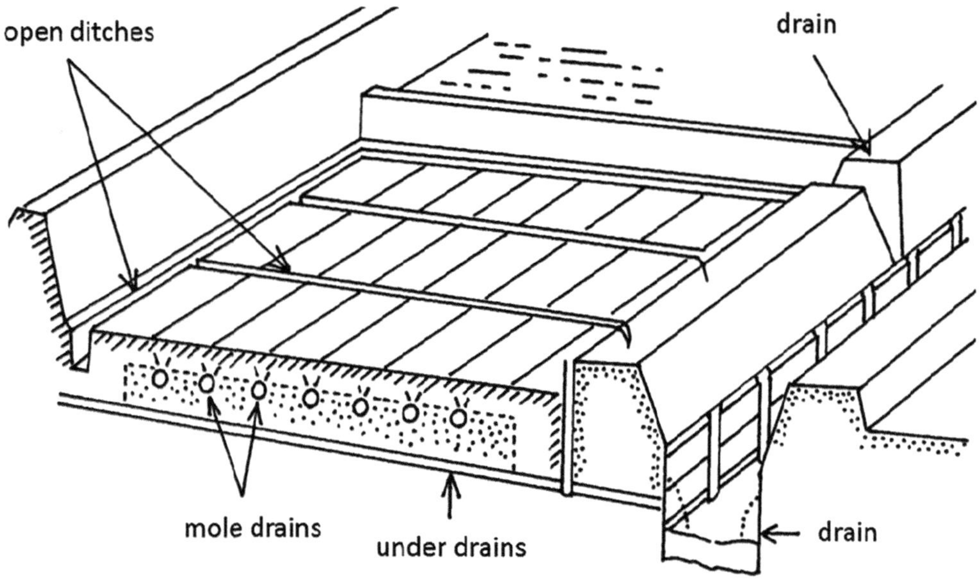

Fig. 9.4 Schematic diagram of farm drainage (Shimane Prefecture 2019)

decreased harvest in paddy–upland rotation fields that have produced black soybean for many years. In order to maintain yield and soil nitrogen fertility, it seems reasonable to set the frequency of black soybean cultivation to 40% or less (Moritsugu and Washio 2016).

(3) **Naked barley production (Ehime Prefecture, Kagawa Prefecture: Gray Fluvic soils)**

Ehime and Kagawa Prefectures, which are located on the coast of the Seto Inland Sea, where the climate is warm and precipitation is relatively low, have been producing naked barley (*Hordeum volgare* var. *nudum*) since ancient times. A total of 52% of the cultivated farmland area for this crop in Japan is located in these two prefectures. With a wide distribution of soil derived from granite and Izumi sandstone, both Ehime and Kagawa Prefectures have an abundance of moderately coarse-grained Gray Fluvic soils with excellent drainage. Being more susceptible to damage by cold temperature, snow, and humidity than wheat, naked barley is suited to the climatic and soil conditions in this region. For this reason, the double cropping of rice and naked barley is widely practiced (Oomori 2012; Tanabe 2012).

2. **Sweet potato production in large-scale soil dressing farmland (Tokushima Prefecture: Sandy Regosols)**

The "*Naruto Kintoki*" is a kind of sweet potato that is a local agricultural specialty in Tokushima Prefecture. This cultivar is characterized by its soft and crumby texture, sweet taste, and vivid violet skin. It is produced on approximately 1100 ha of farmland comprising immature Sandy Regosols (referred to as "*Sunajibata*") stretching across the northern coast of the prefecture (Fig. 9.5). These sandy fields were once paddy fields. However, the Nankai earthquakes in 1946 caused the ground to sink by more than 30 cm, leaving infertile single-crop fields with high salinity. Later, these fields were subjected to soil dressing as sand in the former Yoshino river was pumped in. A total of 90% of these sandy fields were created through soil dressing in which riverbed or seabed sand was accumulated to a thickness of 60–100 cm after the 1960s (Tateishi 2007).

The sand used for soil dressing was derived primarily from crystalline schist containing greenschist (bluestone) distributed in the Sambagawa belt of the Yoshino river basin that cuts through Tokushima Prefecture. The soil management for these sandy fields is characterized by the absence of organic matter application and regular soil dressing with sea sand (referred to as "*Teirezuna*") in order to maintain excellent ventilation and water retention properties.

3. **Traditional crop farming on slopes (Tokushima Prefecture: Brown Forest soils)**

In the west of Tokushima Prefecture, farmers have traditionally produced wheat, upland rice, tubers, maize (*Zea mays*), soba, millet, konjac (*Amorphophallus konjac*), and other crops in the sloping upland fields of mountains. However, due to the recent depopulation of the area, both the variety of produce and the area of farmland have been in decline. In this area, wild grass such as *Miscanthus sinensis* is cut, stored, and applied to the fields to prevent soil loss. Furthermore, furrows are made along contour lines, and runoff soil is pushed back in an effort to continue production

Fig. 9.5 Sweet potato sand field in Tokushima Prefecture. The picture on the bottom right shows the application situation of the maintenance sand. Sea sand is applied every few years to maintain proper sand condition

(Fig. 9.6). These sloping upland fields comprise Brown Forest soils, which contain large amounts of gravel such as crystalline schist, which are prone to flaking. Originating from basic rocks rich in colored minerals, this soil abounds in nutrients (Japanese Society of Pedology 2007). Today, this area is designated as a globally important agricultural heritage system for its "Nishiawa slope farming system."

9.2.4 Vegetable Fields

In the Kansai region (Kinki, Chugoku, and Shikoku regions), the planting areas of the main vegetables (Japanese radish [*Raphanus sativus*], carrot [*Daucus carota*], potato [*Solanum tuberosum*], taro [*Colocasia esculenta*], Chinese cabbage [*Brassica Pekinensis*], cabbage [*Brassica oleracea*],

Fig. 9.6 Soil lifting on steep sloping upland fields of the mountains. The photograph on the upper left shows "Koeguro," which is a pile of the mountain grass

spinach [*Spinacia oleracea*], lettuce [*Lectuca sativa*], onion [*Allium cepa*], green onion [*A. cepa var. Aggregatum*], cucumber [*Cucumis sativus*], eggplant [*Solanum melongena*], tomato [*Solanum lycopersicum*], and green pepper [*Capsicum annuum*]) occupy between 3.5% and 23.0% of the respective agricultural land of Japan (Statistics Department of Minister's Secretariat, Ministry of Agriculture, Forestry and Fisheries 2017). The areas are not larger than those in other regions, as the total area of the Kansai region accounts for approximately 20.6% of the area of Japan. A major cause would be the difficulty of agricultural mechanization due to the lack of flat land and the utilization of most of the limited flat land for paddy farming. However, a wide variety of vegetables, including traditional vegetables, is produced based on appropriate soil management.

1. **Cultivated soils for major vegetables**

Many of the major vegetables produced in the Kansai region are closely related to paddy field agriculture. For instance, in Awaji Island, Hyogo Prefecture, vegetables such as cabbage, onion, and lettuce are combined with the crop rotation system with paddy rice (Fig. 9.7), thereby avoiding problems such as salt accumulation and the occurrence of disease, which are common in continuous vegetable cropping systems. The high-quality crop rotation system, named triple cropping, has become the basis of this production area, which focuses on highly profitable vegetables (e.g., lettuce). In other words, this system is the opposite to that applied in Hokkaido, where spring sowing and autumn harvest areas are represented by the change from paddy rice as the principal crop and wheat as the interim crop (Tano 2017). Additionally, taro is planted after paddy rice under Gray Fluvic soils in Ehime Prefecture (Oomori 2012), and the production of carrot is increased by double cropping with paddy rice under Brown Fluvic soils and Gray Fluvic soils in Tokushima Prefecture. These two prefectures are among the leading production areas in Japan. The weathered granite soil ("*Masa*" soil: Granitic Terrestrial Regosol) distributed over a wide range of the Kansai region is sandy, but is made fine by weathering and consequently often causes drainage failures. Although constant improvement effects are undertaken, such as the construction of underdrains and the continuous application of organic matter, these efforts have not caused permanent improvement, and therefore, this drainage failure can be a future problem in this area.

2. **Vegetable cultivation by overcoming soil problems**

(1) Andosols

Most of the Non-allophanic Andosols spread throughout the mid-western part of Tottori Prefecture are derived from volcanic ash erupted from Mt. Daisen. Non-allophanic Andosols have the disadvantages of strong acidity and high phosphate absorption capacity. In recent years, the chemical properties of these soils have been improved with liming and the addition of phosphate fertilizers. The physical properties of these soils, namely deep effective soil layer and

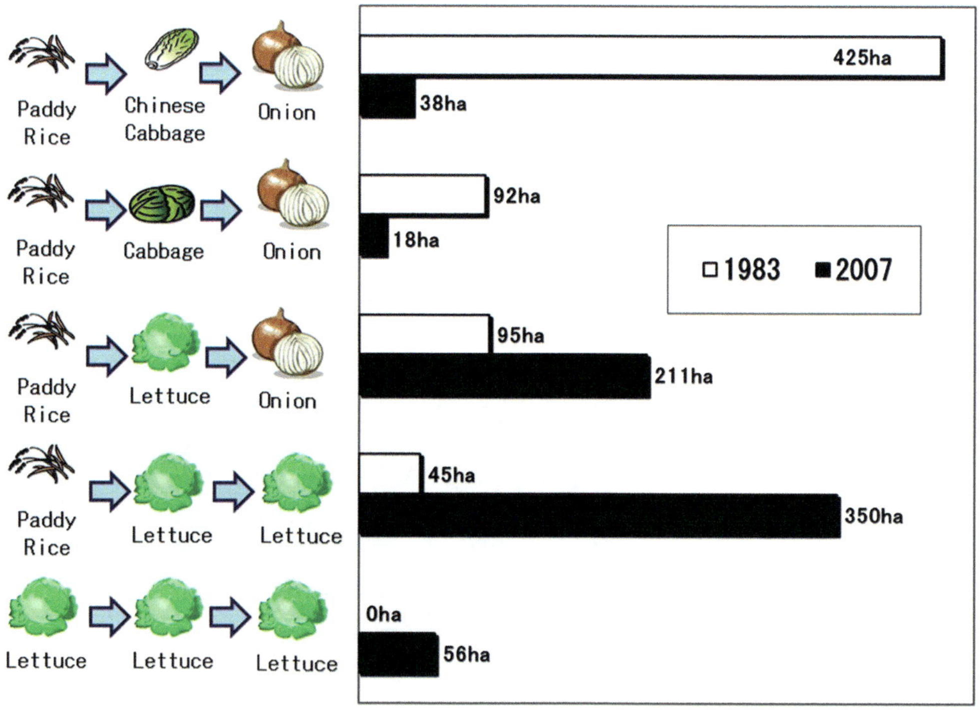

Fig. 9.7 Transition of the triple-cropping system in Awaji area, Hyogo Prefecture. The soil environment is stabilized by incorporating flooded paddy fields while focusing on highly profitable vegetables in order to increase the soil productivity

high soil water holding capacity, are suitable for vegetable cultivation. Furthermore, there are no water shortages, even during drought (which occur once every few years) due to the maintenance of irrigation facilities. This enables stable, high-quality agricultural production. The area has become the largest national production area of watermelon, which is vulnerable to moisture stress, due to the progress of such technological developments.

(2) Sandy Regosols

The soil fertility of Sandy Regosols is extremely low due to their remarkably low clay content and low organic matter accumulation. Although the soil's permeability and drainage are high, its water and nutrient retention properties are low. Since Sandy Regosols are susceptible to drought and nutrient deficiency, they were deemed to be poor soils for farming until the 1940s. However, sand dune agriculture was developed by removing the many difficulties in the 1950s, and nutrient and water management technology was developed by introducing sequential irrigation facilities. This has allowed the development of vegetable production of Sandy Regosols that are easy to be deep-tilled for sufficient root extension, and that possess good drainage with high gas phase distribution with no moisture damage (Yamamoto et al. 2007). For this reason, the sand dune area in the eastern part of Tottori Prefecture has become a main producing area of fresh white shallot in Japan, despite its low soil fertility and lack of moisture and nutrients. The cultivation of shallot has been established on the sand dunes because this crop requires only rain water irrigation for harvesting and is relatively resistant to wind-blown sand. Large-scale farming management is conducted in a sand dune field in the eastern part of Tottori Prefecture. Organic cultivation and chemical fertilizer-saving cultivation have also been attempted in sand dune fields in recent years. Unlike sand dunes in the eastern and central parts of Tottori Prefecture, the sand of the Kyuhin sand dunes in the western part of Tottori Prefecture is fine and rich in organic matter compared with the other sand dune areas. Moreover, the groundwater level is high due to a special groundwater structure, which results in a layer of freshwater above the saltwater. Due to these special soil conditions, white onions can be cultivated without special irrigation facilities such as sprinklers.

(3) Marshy paddy field in polder areas

At polder areas of Nakaumi (a brackish lake) in Shimane Prefecture and Kojima Bay in Okayama Prefecture, drainage measures have been promoted, not only to reduce high groundwater level but also to avoid the influence of seawater. In the Nakaumi polder with acid sulfate soil, production areas have been established for open-field vegetables (e.g., cabbage) by improving of the physical properties of the surface layer by promoting drainage and improving the chemical properties with lime and phosphate fertilizer (Masunaga 2007).

9.2.5 Grassland

Pasture is very sparsely distributed in the Kinki, Chugoku, and Shikoku regions; the pasture in these regions only accounts for around 0.7% of all pasture in Japan (Ministry of Agriculture, Forestry and Fisheries 2017g).

The area of meadows and pastures ranged from 430 to 955 km^2, and the area of seminatural grasslands, which are naturally established and artificially maintained (e.g., through harvesting and burning), is approximately 905 km^2 based on the grassland classification by Matsuura et al. (2012). These areas represent 5–13 and 15% of the total area of meadows and pastures (7220–8480 km^2) and seminatural grasslands (6190 km^2) in Japan, respectively (Matsuura et al. 2012). During the 15 years from 2002 to 2017, the area of meadows in the Kinki, Chugoku, and Shikoku regions decreased by 24.7%, 14.3%, and 24.7%–1.60, 5.37, and 1.69 km^2, respectively (Ministry of Agriculture, Forestry and Fisheries 2017h). These decreases are considerably higher than the national average decrease of the area of all meadows and agricultural land (upland field, orchard, and pasture) over the same period (6.06%); therefore, it is assumed that the area of other grassland, such as pasture and seminatural grassland, also decreased in these three regions.

Brown Forest soils (total area of 564 km^2), Red-Yellow soils (514 km^2), and Andosols (381 km^2) are major soil types in the grassland of the Kinki, Chugoku, and Shikoku regions (Fig. 9.8). However, the contribution of these soil types is different among the three regions (Fig. 9.9). For example, in the Kinki region, soils in grassland are mostly Red-Yellow soils (46%) and Brown Forest soils (35.3%); in the Chugoku region, Andosols and Brown Forest soils cover 29.2% and 25.4% of the grassland, respectively, while in the Shikoku region, Brown Forest soils (38.1%) are the most widespread soil in grasslands.

1. **Kinki region**

The combined area of meadows, pastures, and seminatural grasslands in the Kinki region is 634 km^2, which accounts for 34% of the total grassland area in the Kinki, Chugoku, and Shikoku regions. Brown Forest soils following to Red-Yellow soils are widespread in meadows and pastures, and Brown Forest soils cover 63.4% of the seminatural

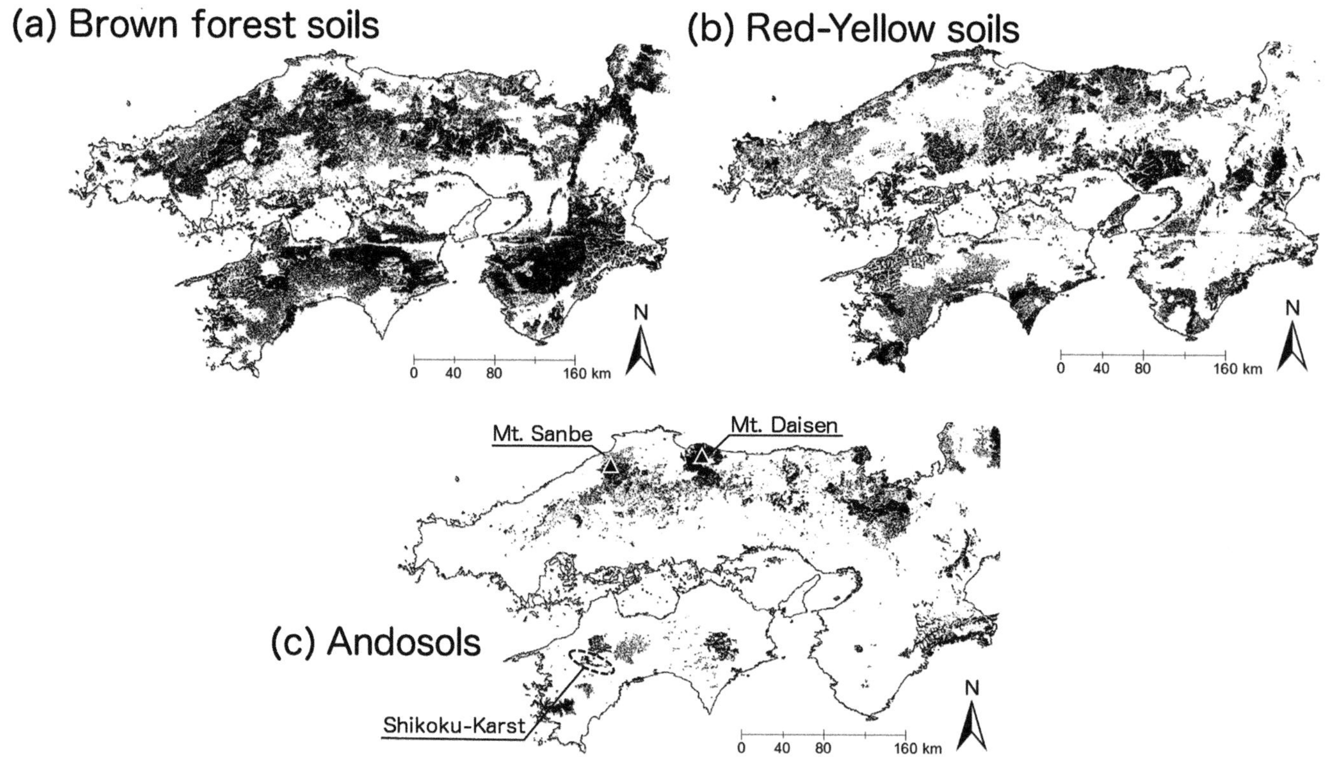

Fig. 9.8 Distribution of Brown Forest soils **a**, Red-Yellow soils **b**, Andosols **c** in Kinki, Chugoku, and Shikoku regions

grasslands (Fig. 9.9b). Brown Forest soils in meadows, pastures, and seminatural grasslands have a low content of humic substances, and soil texture varies from heavy clay to sandy (Fig. 9.10). While Red-Yellow soils have a low content of humic substances, soils with a high clay content are often observed. Andosols are not widely distributed in grassland, accounting for approximately 7%. The content of humic substances in the surface soil of Andosols in seminatural grasslands is more than 10%, higher than that of meadows and pastures, as those of meadows and pastures range from 5 to 10%. The soil texture of Andosols is often heavy clay or clayey in meadows and pastures, while that in seminatural grasslands is clayey or loamy.

2. **Chugoku region**

The Chugoku region has the largest area of meadows, pastures, and seminatural grassland among the Kinki, Chugoku, and Shikoku regions, accounting for 54% (1002 km^2). In meadows and pastures, the prevalence of Andosols varied from 34 to 40%, while Brown Forest soils and Andosols together account for approximately 56% of seminatural grasslands (Fig. 9.9c). Andosols are distributed around Mt. Daisen, Mt. Sanbe, and the Chugoku mountains (Fig. 9.8c). The area of Andosols is the largest in Hiroshima Prefecture, followed by Okayama, Tottori, Shimane, and Yamaguchi Prefectures (Yamamoto 2007). In seminatural grasslands, more than 40% of Andosols contain more than 10% of humic substances, while Andosols in meadows, pastures, and seminatural grasslands have higher concentrations of humic substances (>5%) in surface soil (Fig. 9.10). The soil textures of Andosols in seminatural grasslands are clayey and loamy, while heavy clay and clayey Andosols also account for 24% of meadows and pastures. Brown Forest soils in meadows, pastures, and seminatural grasslands have lower levels of humic substances (<5%) or a buried humic layer and their soil texture is often heavy clay or clayey. The characteristics of the humic substances in Red-Yellow soils is similar to those of Brown Forest soils, and the soil texture is heavy clay, clayey, or loamy in meadows and pastures and clayey or sandy in seminatural grasslands.

3. **Shikoku region**

In the Shikoku region, the area of meadows, pastures, and seminatural grasslands are smaller than those in the Kinki and Chugoku regions, only accounting for 12% (224 km^2) of the combined total of the three regions. One of the reasons for this is that, in this region, large mountainous areas are widely distributed while there are few lowland or flat areas; furthermore, most of the lowland and flat area is used for the cultivation of crops, such as rice, wheat, and vegetables.

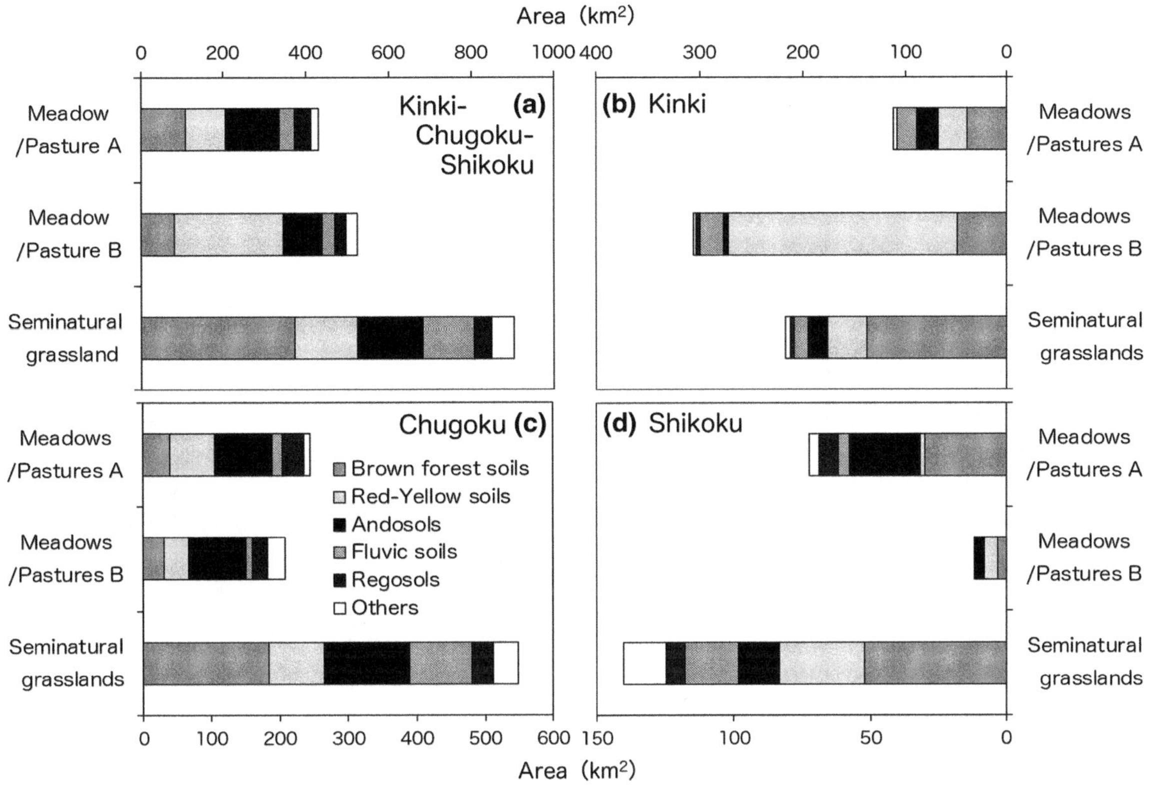

Fig. 9.9 Contribution of soil great group of meadows, pastures, and seminatural grassland in Kinki, Chugoku, and Shikoku region **a** and Kinki **b**, Chugoku **c**, and Shikoku **d** regions. Meadows/pasture A contains only meadows and pastures. Meadows/pastures B contains meadows, pastures, and unidentified artificial grassland (include golf course). Classification of meadows, pastures, and seminatural grassland is followed by the land-use classification of Matsuura et al. (2012). Others of soil great group include Stagnic soils, Eutrosols, Organic soils, human-made soils, and urban and riparian areas

Brown Forest soils and Andosols are mainly distributed in meadows and pastures, while Brown Forest soils and Red-Yellow soils are dominant in seminatural grasslands. In the Shikoku region, upland distributed on hillside slopes, slopes, and plateaus is mainly dominated by Brown Forest soils and Red-Yellow soils (Sakurai 2007), and grassland soils also have a similar soil distribution. Brown Forest soils in meadows and pastures contain lower levels of humic substances and are clayey or loamy in texture in the surface layer (Fig. 9.10). In seminatural grassland, Brown Forest soils contain buried humic layers and higher contents of silt and sand compared to that in meadows and pastures; levels of humic substances are also lower. Red-Yellow soils also often have buried humic layers and lower levels of humic substances. This indicates that most of the grassland in the Shikoku region is distributed on slopes. The soil texture of Red-Yellow soils is mostly clayey, loamy, or sandy in meadows, pastures, and seminatural grasslands. Additionally, in meadows, pastures, and seminatural grasslands, the levels of humic substances in the surface of Andosols are lower compared to those in the Kinki and Chugoku regions, and the soil texture is mostly clayey or loamy. Most of the grasslands in the Shikoku region are distributed in the Shikoku Karst area, located in the highlands of Ehime and Kochi Prefectures (Fig. 9.8c). Therefore, the characteristics of grassland soil properties in Andosols are also limited.

9.2.6 Orchards and Tea Plantations

1. **Cultivation of fruit and tea suitable for the meteorological environment**

The average annual temperature in the Kinki, Chugoku, and Shikoku regions is 13.2–16.8 °C, which is suitable for cultivating fruit trees and tea. The annual rainfall varies from 1115 to 2370 mm and differs greatly from region to region (Japan Meteorological Agency). Fruit trees are cultivated in areas with an annual rainfall of 1100–1500 mm, since the suitable precipitation for main fruit trees is a rainfall of 1300 mm or less from April to October (Ministry of Agriculture, Forestry and Fisheries 2015). The area of fruit tree cultivation in the Kinki, Chugoku, and Shikoku regions accounts for 28% of the main fruit tree cultivation area in

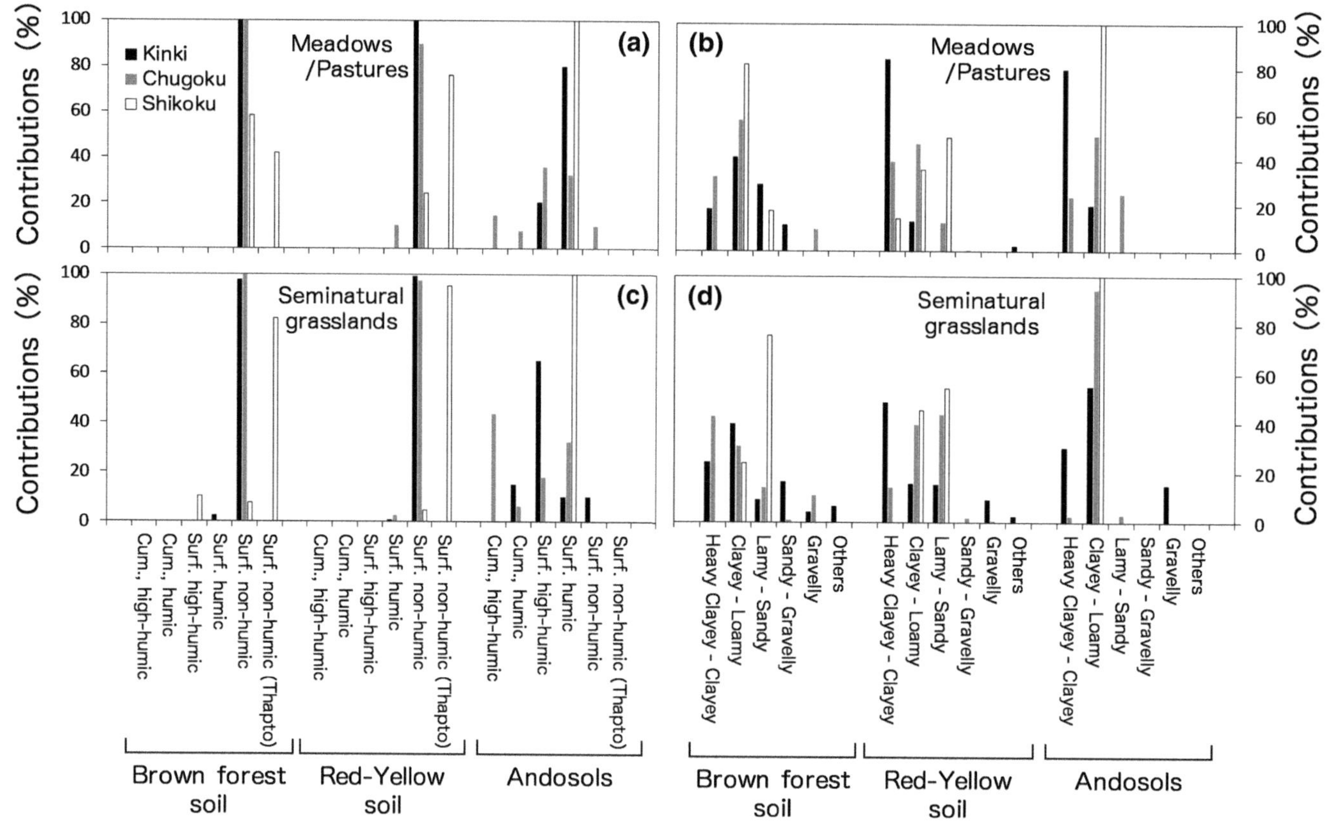

Fig. 9.10 Contribution of humic content and soil texture in meadows/pastures **a**, **b** and Seminatural grasslands **c**, **d**. Meadows/pastures include meadows/pasture A and meadows/pastures B classified by Matsuura et al. (2012). Classification for humus content is cited from the Soil Survey for Maintenance of Farmland Fertility. High humic: humus content >10%; Humic: 5% <humus content <10%; Non-humic: humus content <5%; Surf.: Thickness of surface soil <50 cm. Heavy clayey: HC, LiC, SiC, SC; Clayey: SiCL, CL, SCL; Loamy: SiL, L, FSL, SL; Sandy: LS, S

Japan, 58% of which is a citrus fruit orchard (Table 9.5). The cultivated areas of citrus trees in Ehime and Wakayama Prefectures, the main production areas for citrus in Japan, are 14,030 ha and 9840 ha, respectively, which together accounts for 35% of the citrus cultivation area in Japan (Ministry of Agriculture, Forestry and Fisheries 2017i). Meanwhile, the area of Japanese apricot (*Ume*) cultivation in Wakayama Prefecture, which has cultivated a high yield cultivar, accounts for 34% of the area of Japanese apricot cultivation in Japan, while the area of persimmon cultivation in Wakayama and Nara Prefectures accounts for 22% of its cultivation area in the whole of Japan. Accordingly, the area of orchards in the Kinki, Chugoku, and Shikoku regions as a percentage of the total in Japan is 51% for citrus, 40% for Japanese apricot, 33% for persimmon, 23% for chestnut, 22% for loquat, 16% for peach, 16% for grape, and 14% for Japanese pear. Meanwhile, tea cultivation requires about 1500 mm of annual rainfall, and tea plantations are therefore located in the Yamashiro area in Kyoto Prefecture and Yamato Kogen in Nara Prefecture; the total area of these plantations (2271 ha) accounts for about 5% of the total area of tea plantations in Japan.

The total area of fruit and tea cultivation in the Kinki, Chugoku, and Shikoku regions is 61,548 ha, which accounts for 24% of the total area of fruit and tea cultivation in Japan. However, this percentage is only 13% of the cultivated land area, and 75% of the area in the three regions is paddy fields (Ministry of Agriculture, Forestry and Fisheries 2017j).

2. **Nutrient accumulation in Regosols**

The geology of the Kinki, Chugoku, and Shikoku regions is different to the north and south of the MTL in the east–west direction. To the south of the MTL, the soil originates mainly from sedimentary rocks, and in Wakayama Prefecture, 64% of orchards are occupied by Brown Forest soils, 11% by Red-Yellow soils, and 11% by Fluvic soils (Wakayama Prefecture 2011a). On the other hand, to the north of the MTL, the soil is derived from igneous rocks, and in Shiga Prefecture, 48% of orchards are occupied by Fluvic soils, 19% by Brown Forest soils, 11% by Red-Yellow soils, and 11% by Andosols (Shiga Prefecture 2001). The soils in the hilly and mountainous areas that are used as orchards are immature, and more than 60% of the citrus orchard soils in

Table 9.5 Main fruit tree and tea cultivation area in Kinki, Chugoku, and Shikoku regions

	Citrus unshiu	Other citrus	Apple	Japanese pear	Persimmon	Loquat	Peach	Plum	Japanese apricot	Grape	Chestnut	Tea	
All Japan	42,800	26,000	38,100	12,100	20,300	1270	10,400	3000	15,900	18,000	19,300	42,400	(ha)
Shiga				56						60			
Kyoto				87							444	1570	
Osaka	729									423	146		
Hyogo	168			66		44				275	563		
Nara					1830				316			701	
Wakayama	7580	2260			2570	40	760	292	5440				
Tottori				862	295					71			
Shimane					320					241	104		
Okayama					402		660			1210	337		
Hiroshima	1960	1470	87	140	373				294	286			
Yamaguchi	726	457		190					245		744		
Tokushima	821	990		220					140				
Kagawa	1160	316		40	199	77	206			190	55		
Ehime	6030	8000			651	71	81			156	2130		
Kochi	339	1740				42							

MAFF 2017.7.15 Upper stage: Kinki region, Middle stage: Cyugoku region, Lower stage: Shikoku region

Wakayama Prefecture required soil improvement for low productivity (Wakayama Prefecture 2011b). For this reason, deep plowing has been carried out and organic materials, phosphate materials, and green mulch have been applied. As a result, the content of organic matter in these soils exceeded the target value, 3 g/100 g dry soil, in more than 90% of the orchard and tea plantation soils in Japan (Ministry of Agriculture, Forestry and Fisheries 2008a). Furthermore, the soils in half of the orchards had an excess of calcium and over 30% had an excess of potassium, although 70% were found to be deficient in magnesium. Additionally, the soils of nearly 90% of tea plantations had an excess of potassium, while 60% were deficient in calcium and 70% were deficient in magnesium (Ministry of Agriculture, Forestry and Fisheries 2008b). Levels of soil available phosphate exceed the target value, 30 mg/100 g, in 80% of orchards and 90% of tea plantations, and increased from the 1980s to the 1990s (Ministry of Agriculture, Forestry and Fisheries 2008c). As groundwater pollution due to fertilizer nitrogen was identified, the reduction of fertilizer amount was promoted in consideration of the environment (Umemiya 2004).

9.2.7 Forest Soil and Forestry

The Kinki–Chugoku–Shikoku region is small, but it includes all of the major climate regions in Japan due to large variations in precipitation and distance from the sea. The southern parts of the Kinki and Shikoku regions border the Pacific Ocean, the northern parts of the Kinki and Chugoku regions border the Sea of Japan, and the southern part of the Chugoku region and the northern part of the Shikoku region are close to the Seto Inland Sea. This results in climatic variation. Snow occurs during winter in the region close to the Sea of Japan, which is subject to a coastal climate, and snowfall is very heavy in the northern part of the Kinki region. Along the Seto Inland Sea, precipitation is low all year. The region of heavy snow in the northern part of the Kinki region along the Sea of Japan, and the regions that experience heavy rain, including Mt. Odaigahara on the Kii Peninsula and Kochi Prefecture, receive the highest levels of precipitation in Japan.

These climates are related to topography. For example, in the Seto Inland Sea area, the winter seasonal wind is blocked by the Chugoku mountains and the summer seasonal wind from the Pacific Ocean is blocked by the Shikoku mountains. Consequently, the air is dry all year and precipitation is low.

Mean annual temperatures are relatively high throughout the Kinki–Chugoku–Shikoku region, with temperatures of 16.3, 14.3, and 16.2 °C in the cities of Osaka, Matsue, and Hiroshima, respectively. The soil temperature regime is thermic along the coast and mesic in the mountainous areas (Takata et al. 2011).

1. **Forest and forestry in the Kinki–Chugoku–Shikoku region**

The Kinki–Chugoku–Shikoku region is rich in forest resources as a result of its mild temperatures and precipitation. Forest covers an average of 63% (31–77%) of the Kinki region (close to the national average of 67%), an average of 73% (68–78%) of the Chugoku region, and an average of 70% (47–84%) of the Shikoku region (Forestry Agency 2016). Forest covers only 31% of Osaka Prefecture, the lowest percentage of any Japanese prefecture, but this is the result of human influences rather than natural conditions.

The forests of the Kinki–Chugoku–Shikoku region are mainly evergreen forests consisting of *Castanopsis* and *Quercus* sp. The mountain areas are home to deciduous forest, including *Fagus crenata* and *Quercus cripula* Blume, and intermediate temperate forest; *Abies firma* and *Tsuga sieboldii* grow on the Kii Peninsula and the high-elevation areas of Shikoku Island. Along the Seto Inland Sea, *Pinus densiflora* is the dominant species and is adapted to dry conditions. Forest fires often occur in this dry region, although these events support *Pinus densiflora* regrowth. *P. densiflora*, *Quercus serrata*, and *Quercus acutissima* are popular species in "*Satoyama*" (the Japanese term for areas between mountain foothills and arable flat land) and are influenced by anthropomorphic factors. Additionally, a decrease in biodiversity caused by invading bamboo is a recent major concern.

Plantation forests cover 51, 42, and 53% of the Kinki, Chugoku, and Shikoku regions, respectively, all exceeding the national average of 41%. These plantations are privately owned. Since ancient times, the population and forest usage have been high in the Kansai region. The main plantation species are *Cryptomeria japonica* and *Chamaecyparis obtuse*, most of which were planted between 1950 and 1960. These plantations now need thinning, as in other areas. However, it is difficult to maintain these forests since the price of wood is low and the number of foresters has decreased.

2. **Forest soil in the Kinki–Chugoku–Shikoku region**

The influence of volcanic ash is weaker in the Kinki–Chugoku–Shikoku region than in Eastern Japan and the Kyushu region, except for in Non-allophanic Andosols in the vicinity of Mt. Daisen (Saigusa and Matsuyama 1994; Matsuyama and Saigusa 1994).

As already mentioned, the Kinki–Chugoku–Shikoku region has a long history. Many temples are located in this area and the old capital buildings were built from wood. The Tanakami mountains in Shiga Prefecture once hosted a natural *Chamaecyparis japonica* forest. However, people repeatedly used the wood from this area for building during the Heian period (794–1185 AD), denuding it. In the Tanakami mountains, the bedrock is granite. Once the area was denuded, this granite was readily subjected to erosion, making regrowth difficult; when it rains, the soil is transported into rivers where it accumulates on the riverbed. People built dikes to prevent the river from overflowing its banks, and ultimately the riverbed became higher than the surrounding area. This kind of construction is called "*tenjo-gawa*," which means a river with a bed higher than the surrounding plains. In the denuded mountain area in the Tanakami mountains, afforestation was started by foreign technicians in the Meiji period (1868–1912 AD), and the mountains are now covered by forest. Soil formation processes in relation to the period of afforestation in the Tanakami mountains is reported (Tsukimori et al. 1992).

Generally, almost no nitrogen (N) leaches into streams from forested ecosystems. However, recent reports showed that N concentrations in stream water are increasing (e.g., Vitousek et al. 1997). This is the result of increased fossil fuel emissions, which have increased N deposition in forested ecosystems. The deposited N exceeds the biological demand of the forested ecosystems and ultimately leaches into streams. This phenomenon is called N saturation (Aber et al. 1998).

Makino et al. (2017) found a large difference in N concentrations in streams between two adjacent hills in Toyama Prefecture in the northern part of the Kinki region. The hills are only 5 km apart, a small distance that implies there should be no significant difference in N deposition. The largest N pool is soil, but there was no clear difference in soil type between the two hills. The researchers therefore concluded that N saturation was caused by other factors in addition to N deposition. Further study is needed to clarify the N dynamics in this soil.

9.3 Environmental Conservation

9.3.1 Environmental Conservation in Paddy Fields in Shiga Prefecture

Among the 47 prefectures in Japan, Shiga Prefecture has promoted the production of specially cultivated rice most extensively. This is because Lake Biwa, which is the largest inland lake in Japan and supplies water to about 14 million people in the Kinki region, is located in the center of the prefecture, and also because more than 90% of the agricultural land in the prefecture is classified as paddy field (the second highest percentage in Japan as of 2017); this paddy field is used for the production of rice, wheat, soybean, and vegetables in an open system. The prefecture established regulations for the promotion of environmentally friendly agriculture in 2003. From the following year, the prefecture started to subsidize farmers, mostly rice producers, who fulfilled the requirements for environmentally friendly agriculture, such as limiting the input of chemical fertilizers and pesticides to less than 50% of the conventional dosages. The rice produced by this system is certified locally as environmentally conscious rice ("*kankyo kodawari mai*"). The area of paddy fields under this system has increased to about 46% (2017) of the total paddy fields in Shiga Prefecture.

The major agricultural technical development that has been adopted by such rice producers is classified into two types: (1) the preservation and reuse of irrigation water by plastering ridges, puddling with a shallow water depth, and introducing a circulating irrigation system; and (2) the minimum input and recycling of fertilizer elements by utilizing polyolefin-coated fertilizer and organic fertilizers such as cattle manure, green manure, and rice straw, and by introducing a special rice transplanting machine capable of applying fertilizer concomitantly.

1. **The initial cropping period**

Wise water and fertilizer management is the most fundamental and effective strategy for reducing the outflow of N and P from paddy fields during a cropping period. Management should be conducted with special care during the initial cropping period, since the inflow of water by irrigation and rainfall into a field during a cropping (irrigation) period showed a high positive correlation with the outflow of total N (T-N) and total P (T-P) from the field through surface water drainage and infiltration (Hasegawa et al. 1992). Additionally, the outflow of T-N and T-P occurred more intensively during a period between puddling and 30 days after transplanting than that during the following cropping period (Hasegawa 1992).

One of the techniques for wise water management is puddling and land-leveling with a shallow water depth by using a paddy field harrow. The paddy field harrow enables puddling with a shallow water depth at a faster speed than the conventional rotary harrow, as it has a wider operational width, shorter tillage blades, and a faster speed of blade rotation. Furthermore, puddling with a shallow water depth can decrease the outflow of suspended solids after the puddling, since it causes soil particles to settle faster than in conventional puddling (Tanaka 2001). This technique also helps to prevent forced drainage at the time of transplanting.

Another technique for wise fertilizer management is a localized application of fertilizer to the root zone soil by using a rice transplanting machine capable of applying fertilizer concomitantly. Because fertilizer can be placed a few centimeters below the soil surface with this transplanting machine, the applied chemicals are less susceptible to dissolve in the ponded water and flow out from the field.

When this transplanting technique was combined with polyolefin-coated urea with a controlled availability, the yield of rice obtained after the application of 80 kg N ha^{-1} was comparable to the yield obtained after the split application of quick-release ammonium fertilizer at 100 kg N ha^{-1} (Table 9.6). This is probably due to the enhanced recovery of the applied N (Shibahara et al 1992, 2000).

By using the ^{15}N tracer technique (A value method), Shibahara (2008) further evaluated the fate of N in different types of fertilizers (quick-release N, coated N, and quick-release N plus organic N at 1:1), each of which was applied uniformly or locally in the root zone. The percentage of fertilizer-derived N absorbed by rice plants was highest (49.2%) when coated N was applied locally (Table 9.7). On the other hand, when organic N was applied together with quick-release N, the percentage of fertilizer-derived N that remained in the surface soil after harvest was higher than 40%, and the percentage of the N unrecovered from the plant–soil system (due probably to denitrification and leaching) was lower than 20%. These results suggest that the rate of N application can be reduced most effectively by the localized application of coated N fertilizer at transplanting and that the transfer of the applied N from a paddy field to the surrounding environment can be decreased by the co-application of organic N with quick-release inorganic N.

2. **Rice yield and nutrient flows under the Shiga system**

The environmentally friendly rice production in Shiga Prefecture is regarded as an original cropping system and incorporates a series of the abovementioned technical developments. The rice producers must reduce the amount of pesticides and chemical fertilizers to less than 50% of the conventional dosages, that is, less than seven chemical components for pesticides and less than 40 kg N ha^{-1} for chemical fertilizers. Moreover, farmers must prevent suspended solids from flowing out of their paddy fields by decreasing the depth of ponded water during the period between puddling and transplanting.

We evaluated the effects of this cropping system on rice yield and nutrient flows at the field scale by comparison with the conventional cropping system. The results indicated that the average rice yield over 3 years did not differ significantly between the two cropping systems (about 5500 kg ha^{-1}), while the average outflow loads of T-N, T-P, and suspended solids through surface water drainage could be reduced significantly under the new cropping system (Table 9.8; Shibahara 2010).

Table 9.6 Effect of basal application of polyolefin-coated urea on rice growth, yield, and N uptake (Shibahara et al. 2000)

Fertilizer type	Fertilizer placement	Rate and timing of N application (kg ha^{-1})				Response of rice plants (kg ha^{-1})		
		Basal application	Topdressing between transplanting and panicle formation	Topdressing at panicle formation stage	Total amount	Weight of rice straw	Yield of brown rice	N uptake
Ammonium N	Broadcasting	30	30	40	100	7790	6550	107
Coated urea[b]	Broadcasting	60	–	30	90	7990	6500	108
Coated urea[b]	Banding[c]	50	–	30	80	7370	6480	104

[a]Data obtained at an experimental field in Shiga Prefecture Agricultural Experiment Station. A rice variety Nipponbare was transplanted on May 1, 1996 in a field with a medium to coarse-textured gley lowland soil
[b]Composite NPK fertilizer for basal application containing LP-100-day type coated urea at 80% of N (LP-D80)
[c]Coated urea was band-applied at a 5 cm soil depth and 5 cm aside from a transplanted rice seedling
Based on Shibahara et al. 2000 (Jpn. J. Soil Sci. Plant Nutr., 71 898–902)

Table 9.7 Effect of fertilizer type and application method on the fate of fertilizer N (Shibahara 2008)

Fertilizer type at basal application	Fertilizer placement	Rate of N application (kg ha^{-1})			Recovery rate of fertilizer N (%)		
		Basal application	Topdressing between transplanting and panicle formation	Topdressing at panicle formation stage	Uptake by rice plants	Remaining in the surface soil	Loss from the soil-plant system
50% organic N + 50% uncoated inorganic N	Broadcasting	30	0	30	36.2	44.9	18.9
	Banding	30	0	30	39.7	41.8	18.6
100% uncoated inorganic N	Broadcasting	20	10	20 + 10[b]	37.7	32.5	29.8
	Banding	30	0	20 + 10[b]	40.2	32.1	27.6
45% coated N + 55% uncoated inorganic N[a]	Banding	30	0	20 + 10[b]	49.2	26.4	24.3

[a]Composite NPK fertilizer containing LP-70-day type coated urea at 45% of N
[b]Topdressing at panicle formation stage was carried out twice
Based on Shibahara (2008) (Japanese Research Project for utilizing advanced technologies in agriculture, forestry, and fisheries, 1–121)

3. **For the non-cropping period**

The net efflux load of N during the non-cropping period can be greater than that during the cropping period, as has been reported by many researchers (e.g., Kobayashi et al. 2005). This is because the non-irrigation (non-cropping) period is longer than the irrigation (cropping) period when the field is used for the mono-cropping of rice. Besides this, soil management practices during the fallow period, such as plowing and making ridges by which crop residues on the ground are incorporated into soil, are known to accelerate the rates of mineralization of organic N and nitrification, thereby causing the leaching of nitrate after rainfall events (Tanaka 2001).

It is therefore important to prevent the outflow of N during the non-cropping period in addition to conducting the environmentally friendly practices during the cropping period. We carried out field experiments and found that two countermeasures can be performed effectively during the non-cropping period (Hasukawa et al. 2011). The proposed methods are: (1) the plowing and incorporation of rice straw in December, which is one month later than the normal plowing season in Shiga Prefecture; and (2) capping the underground drainage pipe and closing the water outlet with a cut-off board. A combination of these practices would make it possible to keep the soil conditions moist to wet throughout the non-cropping season and to reduce the

Table 9.8 Rice yield, N uptake and outflow loads of N, P and suspended solids under the conventional production system and the Shiga-type environmentally friendly production system (Shibahara 2010)

Cropping system	Field management			Yield of brown rice[c]		N uptake (kg ha^{-1})	Surface outflow load during a cropping season (kg ha^{-1})[c]		
	Number and method of puddling	Rate of N fertilization (kgN ha^{-1})[a]	Total number of pesticide compounds	Average (S.D.) (kg ha^{-1})	Relative % to control		T-N	T-P	Suspended solids
Conventional system (control)	Twice by conventional method (normal water depth)	60 (60)[b]	13	5670 (320)	(100)	93	3.8	0.43	205
Shiga-type new system	Once with a paddy field harrow (shallow water depth)	60 (30)[b]	4	5530 (300)	98	88	0.8	0.14	13

In both systems, a rice variety Koshihikari was transplanted

[a]Basal N fertilizer was applied uniformly (conventional system) or locally at a 5 cm soil depth and 5 cm aside from a transplanted rice seedling (new system)

[b]Values in the parenthesis indicate the rate of inorganic N applied. Rate of application of inorganic N was halved in the new system

[c]Values for rice yield and surface outflow loads indicate averages over 3 years

Based on Shibahara 2010 (Jpn. J. Water and Waste, 52, 55–62)

outflow of N by lowering the rates of decomposition of rice straw and nitrification.

9.3.2 Environmental Impacts Associated with Agricultural Activities

1. **Generation of carbon dioxide, methane gas, and nitrous oxide**

(1) **Generation of carbon dioxide and methane gas in paddy fields**

The Kansai region, a mild temperate area within Japan which includes the Kinki, Chugoku, and Shikoku regions, has many areas with low precipitation, and in general has conditions that are conducive to the decomposition of organic matter. Especially in the Setouchi region, rainfall is low, and in the four prefectures of Ehime, Kagawa, Okayama, and Hiroshima, the proportion of Gley Fluvic soils (developed in the rainy climate) to the total paddy field area is about 15% (Ishibashi 2004; Tanabe 2007; Tanimoto 2006), which is lower than the average of 23% in the Kansai region. Meanwhile, in the four prefectures affected by the climate of the Sea of Japan, namely, Kyoto, Shiga, Tottori, and Shimane, the proportion of Gley Fluvic soils in paddy fields is as high as 40% (Ito 2005; Miyata 2007; Nakajima 2007). It can be seen that there is a regional difference in the generation of carbon dioxide and methane gas accompanying the decomposition of organic matter in the soil. The generation of these greenhouse gases depends not only on the climate and the type of soil but also on the input quantity of organic materials, water management, and so on. Although no comprehensive investigation has been made into the carbon balance of the agricultural soil, Kanazawa (1998) estimated the potential generation of carbon dioxide during the inundation of paddy fields in the Kansai area to be 7.97–8.17 × 10^{11} g. Soil carbon stocks, which result from soil carbon balance, are continuously being investigated in each prefecture. Looking at the results from the National Agricultural Land Soil Guidebook (Council of Investigation for Soil Conservation 2012), the carbon content of the soil in the Kansai region tends to decline in three prefectures, remain constant or decrease in one prefecture, remain constant in two prefectures, and increase in two prefectures.

(2) **Generation of nitrous oxide**

In addition to spontaneous generation, nitrous oxide is produced from the combustion of substances, by the chemical industry, and so on, and from excess nitrogen in agriculture. The generation of nitrous oxide from farmland occurs when nitrate–nitrogen is denitrified in anaerobic conditions and when ammonia is nitrified in aerobic conditions. In wetland rice cultivation, fertilizer and nitrogen in soil are generally mainly in the ammonia state, and the amount is also small, and it is considered that generation of nitrous oxide there is not high. On the other hand, in farmland, the generation of this gas occurs more frequently when fertilizer is applied in large quantities. In the Kansai region, farmland (including tree gardens) occupies 25% of the total agricultural land, lower than the Japanese average of 46%, and the risk of nitrous oxide generation is therefore relatively low. The reduction of the amount of applied fertilizer is an effective countermeasure to nitrous oxide generation. In the Kansai region, farmland is often over-fertilized, and as a result, in 13 of the 15 prefectures in the region, there are reports of fields holding nutrients that exceed the improvement target of the field (Council of Investigation for Soil Conservation 2012).

2. **Nitrate leaching**

(1) **Concentration of nitrate–nitrogen in groundwater**

Nitrate–nitrogen and nitrite–nitrogen (hereinafter referred to as nitrate) were added to the environmental quality standards for water pollution (items for protecting human health and living environment) in 1999; the standard value was set at 10 mg L^{-1}.

Since the addition of the nitrate standards, the national government and local governments have conducted the monitoring of nitrate, and their findings have shown that the detected concentration is especially high in groundwater. Although the number of wells exceeding the standard values had been increasing until peaking in 2010, it subsequently decreased slightly for five consecutive years (Ministry of the Environment, 2016a). Looking at the results for 2015 in the Kansai region, in a general monitoring survey with 589 measurement points, excess levels of nitrate were detected for 13 points (2.2%). Additionally, in a continuous monitoring survey with 202 measurement points, excess levels were detected for 60 points (29.7%). Nationwide, excess levels of nitrate were detected in 3.5% of measurement points in the general monitoring survey and in 43.8% in the continuous survey. Although it is not possible to simply compare both surveys, the Kansai region seems to have relatively low concentrations of nitrate. As mentioned above, one of the reasons for this is considered to be the fact that the proportion of paddy fields is 75% of total agricultural land and that of upland field is 25%. However, in the coastal area of the Seto Inland Sea, excess levels of nitrate are high. In this area, annual rainfall is as low as 1300 mm, many storage reservoirs are built, and water is repeatedly used. Another reason would be that, as in the coastal area of the Seto Inland Sea, there are many Gray Fluvic soils with good drainage, paddy fields are mainly well drained, and the cultivated area of vegetables and fruit trees is large.

(2) **Techniques for suppressing nitrate leaching**

With regard to the type of fertilizer, effect-controlled fertilization that can control the amount of elution to match the growing season of crops has been developed, and technology to increase nitrogen utilization rate has been introduced.

In the fertilizer application method, side-dressing rice transplanter and simultaneous ridge-forming fertilization machines for the cultivation of vegetables (lettuce (*Lactuca sativa*), cabbage (*Brassica oleracea* var. *capitata*), etc.) have been developed. These are technologies to allow the reduction of fertilizer application by increasing the absorptivity of fertilizer nitrogen by the topical application of fertilizer with controlled fertilization to the root area along with labor saving.

With the fertilization reduction technology based on the cropping system, the use of green manure is also increasing. Green manure plants of the legume family (e.g., hairy vetch [*Vicia villosa*] and Chinese milk vetch [*Astragalus sinicus*]) provide nitrogen fertilization and can thus allow the reduction of fertilization, and additionally improve the physicochemical properties of soil by replenishing organic matter. Nitrate leaching is promoted when the field becomes bare ground in winter, and nitrate leaching can be suppressed by absorbing nitrogen to green manure.

3. **Noxious chemicals**

In the Kansai region, noxious chemicals related to the contamination of agricultural soils include cadmium (Cd), copper (Cu), and arsenic (As) as specified noxious matter, and nickel (Ni) which is abundant in serpentine soils. The administrative specified situations of soil contamination and its area are based on the data of the Ministry of Environment (Ministry of the Environment 2016b).

(1) **Cadmium (Cd)**

In the Kansai region, the main areas of Cd contamination are caused by the inflow of mine wastewater. Such areas are distributed in three prefectures, and to date cover a total of 316 ha. All of these were unspecified as of 2016, with countermeasures such as soil dressing and the implementation of sand settling channels (to prevent repollution by bottom materials) completed.

In the case of paddy rice, the basic method for avoiding Cd uptake by rice plants is constant flooding for three weeks before and after the heading date (six weeks in total). For example, in excessively permeable paddy fields (water requirement in depth $\geq$ 30 mm day^{-1}), in the main Cd-contaminated area of Hyogo Prefecture (Hyogo Prefecture, 1987), careful water management is required so as not to exposure the soil surface. In paddy fields with medium- and coarse-textured gravelly soil, soils can become gradually acidified. Soil pH correction by the application of liming materials is desirable to reduce the risk of Cd contamination in the case of lowering the levels of ponding water. The conventional midseason drainage can cause deep cracks in soil and prevent constant flooding; therefore, in this period, saturated irrigation is recommended.

(2) **Arsenic (As)**

In the Kansai region, the agricultural specified areas of As contamination (soluble As in 1 M HCl $\geq$ 15 mg kg^{-1}) are distributed in two prefectures, over a total of 109 ha. All of these are unspecified as of 2016, with soil dressing and other countermeasures completed. Soil contamination with As is not only related to mining activities; for example, marine clays containing pyrite can also lead to As contamination (Shimada 2009). Nevertheless, the main contaminations of As in paddy fields are caused by mine minerals (e.g., arsenopyrite) and have been spread around basins through river water in the past. About 68% of the abovementioned specified areas are categorized to combined contamination with Cd or copper (Cu), also administratively. In one area near a mine, water restriction was operated in the past after rice transplanting and was supposed to avoid physiological disorders due to excess As (Jpn. Soc. Soil Sci. Plant Nutr. (ed.) 1991)

(3) **Copper (Cu)**

In the Kansai region, the agricultural specified areas of Cu contamination (soluble Cu in 0.1 M HCl $\geq$ 125 mg kg^{-1}) are distributed in four prefectures, covering a total area of 137 ha. The contamination is related to mining. The details are as follows: simple contamination in two prefectures, 62 ha in total; combined contamination with Cd in one prefecture, 67 ha in total. All of these are unspecified as of 2016, with countermeasure constructions completed. From research in two prefectures, the decrease of brown rice yield caused by physiological disorders was estimated to be approximately 10% when the specified level of soil soluble Cu in 0.1 M HCl was 125 mg kg^{-1}. The replacement of surface soil with subsoil and the application of liming materials have been carried out as countermeasures against excess Cu concentrations in soil (Jpn. Soc. Soil Sci. Plant Nutr. (ed.) 1991).

(4) **Nickel (Ni)**

Serpentine soils are distributed in parts of the Chugoku and Shikoku mountains. Weathering soils of serpentinite generally abound in Ni, which is isomorphously substituted with magnesium (Mg) atoms in serpentine. Chromium (Cr) is also accumulated by the same mechanism, but this element is flocculated in heavy minerals, is not highly soluble, and is relatively harmless to the growth of crops (Shishido and Ishida 1999). In areas of serpentine soil, Dark Red soils (Eutrosols), which obtain their color from heavy metal compounds, show relatively high concentrations of Ni or Cr (Morita et al. 1986). In the case of total Ni, the same trend was observed in the serpentine soil areas in Hyogo Prefecture (Tsutaka 1989). In one of the field cases, where soybean was cultivated in soils with excess levels of Ni (exchangeable Ni > 10 mg kg^{-1}), the yellowing or red spotting of leaves was observed. These typical disorders were almost eliminated after soil pH correction by the application of liming minerals (Jpn. Soc. Soil Sci. Plant Nutr. (ed.) 1991)

9.3.3 Soil Management for Environmental Load Reduction

1. **Utilization of controlled-release fertilizer, such as coated urea fertilizer**

The single or multiple application of controlled-release fertilizer can be an option to supply the amount of nutrients that is needed according to the various kind of crops and the growing stage. The utilization of controlled-release fertilizer is a very important tool for the reduction of environmental load. Fertilizer consisting of coated urea, which makes the temperature and relative humidity controlling factors of fertilizer elution, is used extensively in paddy rice cropping (i.e., under flooded conditions). A basal application technique of whole amount which is necessary to crop paddy rice is established at many prefectures. This technique is combined with the technology to apply fertilizer in band shape beside of the rows, which raises the fertilizer utilization rate and also leads to labor saving and the reduction of application rate.

2. **Technology of soil management for environmental load reduction**

(1) **Paddy field**

The technology used to reduce the environmental load from paddy fields consists of two methods. One is shallow water puddling, which involves plowing and irrigating the fields in the shallow water state using the rice paddy harrow in order to prevent compulsion drainage of ponded water. The other one involves the plowing of rice straw in autumn to promote the organization of mineralized nitrogen in soil in the non-irrigated period. A study in Shiga Prefecture clearly showed that the nitrogen flux from a paddy rice field can be reduced by these technologies, and that they can also maintain soil fertility and increase it (Tanaka 2001). A green manure cultivation technique, which utilizes nitrogen in the forage legume grown after paddy rice harvesting as a basal fertilizer for the next crop of rice, has also been developed in three prefectures (Osaka, Hyogo, and Okayama). This technique is used for the distinctive cultivation of paddy rice as well as to allow a decrease in chemical fertilizer application.

(2) Upland fields

In order to achieve high profits in upland fields, particularly for vegetable cropping, fertilization and soil amendment are conducted to a large degree. Therefore, excesses of soil nutrients such as P_2O_5 and CaO, unbalances in soil nutrients, and the leaching of nitrate from upland fields to the environment will be often a problem due to the high application amounts of fertilizer and soil amendments, which leads to a rise in fertilizer cost. Therefore, technical development to decrease the application amounts of fertilizer, such as the utilization of controlled-release fertilizer and the introduction of local fertilization technology, has been advanced in four prefectures (Shiga, Hyogo, Shimane, and Okayama). A method for estimating the amount of fertilizer which has accumulated in soil and to decrease the application amount of fertilizer has also been developed in three prefectures (Osaka, Tokushima, and Kochi). Additionally, an automated pulsating drip irrigation system using a solar pump was developed by the Western Region Agricultural Research Center (Yoshikawa and Nakao 2010). Since the system has a better water and nitrogen use efficiency than the conventional system, yield improvement and a reduction of the amount of nitrogen fertilization can be achieved at the same time for various kinds of vegetables.

(3) Orchards

Clean-cultured soil surface management, in which weeds are not grown, was previously used at orchards. However, due to the availability of passenger-type mowers, the conduction of weed-cultured soil surface management, in which weeds are grown and cut periodically, has been increasing recently. In the Kansai region, because of the presence of many orchards in hilly and semimountainous areas, the introduction of this weed-cultured management contributes to the prevention of soil erosion and nutrient leaching. However, the Kansai region contains mineral-rich soil, and the topsoil layer is shallow. Therefore, there is a fear of water and nutrient competition between fruits and weeds during weed-cultured management in this region. The development of fertilization technology in weed-cultured orchards will be needed from now on. On the other hand, the level of nitrogen fertilizer application is generally high in tea plantations, and accordingly there were fears of nitrate–nitrogen leaching. However, the utilization of controlled-release fertilizer such as coated urea enabled the amounts of fertilization to be reduced substantially (Shiwa et al. 2000).

(4) Evaluation technique for fertilization rate of applied livestock manure

In the Kansai region, particularly the Chugoku and Shikoku regions, dairy and poultry are raised extensively and livestock manure has been widely produced and circulated. Livestock manure was used as a soil amendment in the past, and the fertilization rate was not considered. However, more recently, environmentally friendly agriculture has spread, and the consciousness of fertilizer cost reduction has risen. Therefore, the development of techniques to evaluate the appropriate fertilization rate of the applied livestock manure has been advanced in three prefectures (Okayama Prefecture 2013; Hiroshima; Yamaguchi). As a result, the appropriate fertilization rate of livestock manure has become clear, which is useful for the reduction of fertilizer cost and avoiding the accumulation of soil nutrients.

References

Aber JD, Acdowell W et al (1998) Nitrogen saturation in temperate forest ecosystems—hypothesis revisited. Bioscience 48:921–934

Agency for Cultural Affairs (2017) On important cultural landscapes. (in Japanese)

Barnes GL (1986) Paddy field archaeology in Nara, Japan. J Field Archaeol 13:371–379

Board of Audit of Japan (2016) Present situation of the regulation of rice production (Appended table 10: Percentage of conversion to non-rice crops in every prefecture in Japan). http://report.jbaudit.go.jp/org/h27/ZUIJI7/2015-h27-Z7105-0.htm. (in Japanese)

Council of Investigation for Soil Conservation (2012) The guidebook of agricultural soil in Japan. Japan Soil Association Tokyo 77–99. (in Japanese)

Forest Agency (2016) Annual report on trends in forests and forestry. Tokyo

Forestry Agency (2017) Forest ratio and Planted forest ratio listed by prefectures on March 31, 2012. http://www.rinya.maff.go.jp/j/keikaku/genkyou/h24/1.html. (in Japanese)

Hasegawa K (1992) Studies on the behavior and balance of nitrogen in paddy field near the lake and its influence on environment especially on the water quality. Spec Bull Shiga Agric Exp Stn 17:1–164 (in Japanese with English summary)

Hasukawa H, Shibahara F et al (2011) Runoff load reduction by improved soil management practice in paddy rice non-cropping period. Bull Shiga Pref Agric Tech Promo Cent 50:47–59 (in Japanese with English summary)

Hirai Y (2004) Plains of Shikoku side. In Ota Y, Nause T, Tanaka S, Okada A (eds) Regional geomorphology of the Japanese Islands,

vol 6. Geomorphology of Kinki, Chugoku and Shikoku. University of Tokyo Press. Tokyo, pp 232–239. (in Japanese)
Hyogo Prefecture (1987) The varieties and characteristics of paddy soils in Hyogo. (in Japanese)
Institute for Agro-Environmental Sciences, NARO (2017) Japan Soil Inventory, 1/200000 Soil Map, Tsukuba. http://soil-inventory.dc.affrc.go.jp/download20.php
Ishibashi E (2004) Agriculture and soil management in Okayama prefecture. J Soil Sci Plant Nutr 75:623–624 (in Japanese)
Ito J (2005) Agriculture and soil management in Shimane prefecture. J Soil Sci Plant Nutr 76:931–932 (in Japanese)
Japanese Society of Pedology (ed) (2007) Japanese soils (Dojo o aishi dojo o mamoru-Nihon no dojo-). Hakuyusya, Tokyo, p 256. (in Japanese)
Japanese Society of Soil Science and Plant Nutrition (1991) Toxic metal contamination of soil-situation measure and prospects. Hakuyusha Tokyo. (in Japanese)
Kanazawa K (1998) Estimation of carbon dioxide evolution as the global warming gas evolved from agricultural lands in Japan. Jpn J Farm Work Res 33:155–169 (in Japanese)
Kanda T, Takata Y et al (2017) Development of the 1:50000 digital soil map of cultivated soil in Japan according to the comprehensive soil classification system of Japan-first approximation. Jpn J Soil Sci Plant Nutr 88:29–34 (in Japanese)
Kitagawa T (2012) Actual status of cultivated soil in Shiga prefecture. In: Soil guidebook for arable land in Japan (Zenkoku noukouti dojou gaidobukku). Japan Soil Association, Tokyo, pp 70–71. (in Japanese)
Kobayashi T, Komori N, Tokuda Y (2005) Improved fertilizer application and optimized water management for reducing outflow load of nutrient salts etc. from Gley paddy fields soils. Bull. Shiga Pref. Agric Res Cent Exp Stn 45:13–36 (in Japanese with English summary)
Kuroda Y (2006) Agriculture and soil management in Tokushima prefecture. Jpn J Soil Sci Plant Nutr 77:239–240 (in Japanese)
Machida H, Arai F (1992) Atlas of Tephra in and around Japan. University of Tokyo Press. Tokyo, pp 109–112. (in Japanese)
Machida H (2004) Volcanoes in Chugoku mountains. In: Ota Y, Nause T, Tanaka S, and Okada A (eds) Regional geomorphology of the Japanese islands, vol 6. Geomorphology of Kinki, Chugoku and Shikoku. University of Tokyo Press. Tokyo, pp 153–164. (in Japanese)
Makino S, Tokuchi N et al (2017) Nitrogen dynamics of nitrogen-saturated and unsaturated deciduous forest ecosystems on Toyama Plain. Japan J For Res 99:120–128
Masunaga T (2007) Physico-chemical properties and management of soil at Nakaumi polder. In: Japanese Society of Pedology (ed) Love our soil and save our soil. Hakuyusha, Tokyo, pp 237–242. (in Japanese)
Matsuura S, Sasaki H, Kohyama K (2012) Organic carbon stocks in grassland soils and their spatial distribution in Japan. Grassland Sci 58:79–93. https://doi.org/10.1111/j.1744-697X.2012.00245.x
Matsuyama N, Saigusa M (1994) Distribution of allophanic andosols and non-allophanic andosols in Western Japan. Pedologist 38:2–9 (in Japanese with English abstract)
Ministry of Agriculture, Forestry and Fisheries (2008a) Current situation and problems of agricultural soil. 9. http://www.maff.go.jp/j/study/dozyo_kanri/01/pdf/ref_data1.pdf. (in Japanese)
Ministry of Agriculture, Forestry and Fisheries (2008b) Current situation and problems of agricultural soil. 10. http://www.maff.go.jp/j/study/dozyo_kanri/01/pdf/ref_data1.pdfw. (in Japanese)
Ministry of Agriculture, Forestry and Fisheries (2008c) Current situation and problems of agricultural soil. 11. http://www.maff.go.jp/j/study/dozyo_kanri/01/pdf/ref_data1.pdf. (in Japanese)
Ministry of Agriculture, Forestry and Fisheries (2015) Fruit tree agriculture promotion basic policy. 26. (in Japanese)
Ministry of Agriculture, Forestry and Fisheries (2017a) Area of agricultural land in Japan in 2017 (July 15). http://www.maff.go.jp/j/tokei/kouhyou/sakumotu/menseki/index.html. (in Japanese)
Ministry of Agriculture, Forestry, and Fisheries (2017b) Agricultural gross production and Productional Agricultural Income (National) in 2015. (in Japanese)
Ministry of Agriculture, Forestry, and Fisheries (2017c) Forestry production in 2015. (in Japanese)
Ministry of Agriculture, Forestry, and Fisheries (2017d) Recognized as a globally important agricultural heritage systems. (in Japanese)
Ministry of Agriculture, Forestry and Fisheries (2017e) Present situation of agricultural land improvement in Japan (March, 2015). http://www.maff.go.jp/j/council/seisaku/nousin/bukai/h28_5/attach/pdf/index-13.pdf. (in Japanese)
Ministry of Agriculture, Forestry and Fisheries (2017f) Report on results of 2005 census of agriculture and forestry in Japan. http://www.e-stat.go.jp/SG1/estat/List.do?bid=000001011999&cycode=0. (in Japanese)
Ministry of Agriculture, Forestry and Fisheries (2017g) Area of cultivated land in 2017. www.maff.go.jp/j/tokei/kouhyou/sakumotu/menseki/index.html
Ministry of Agriculture, Forestry and Fisheries (2017h) Statistics of Agriculture, Forestry and Fisheries. (in Japanese)
Ministry of Agriculture, Forestry and Fisheries (2017i) Fruit trees and tea cultivation area in 2017. Statistics of agriculture. 4–8. (in Japanese)
Ministry of Agriculture, Forestry and Fisheries (2017j) The farmland area in 2017. Statistics of agriculture. 4–8. (in Japanese)
Ministry of Agriculture, Forestry, and Fisheries Agricultural Production Bureau (2008) Summary on the national soil survey programs in Japanese agricultural land from 1979 to 2003. Agricultural Production Bureau, Ministry of Agriculture, Forestry and Fisheries. (in Japanese)
Ministry of Internal Affairs and Communications, Department of Statistics (2016) Population estimates. www.stat.go.jp/data/jinsui/2016np/index.htm (October 1, 2016)
Ministry of the Environment (2016a) Investigation result of groundwater quality. http://www.env.go.jp/water/report/h28-03/h28-03_full.pdf. (in Japanese)
Ministry of the Environment (2016b) Enforcement of act to prevent soil contamination on agricultural land in 2015
Miyata I (2007) Agriculture and soil management in Tottori Prefecture. J Soil Sci Plant Nutr 87:103–104 (in Japanese)
Morita Y, Oosumi Y, Natume D (1986) Genesis and classification of dark red soil derived from Serpentinite 1. Bull For For Prod Res Inst 341:27–46. (in Japanese with English summary)
Moritsugu S, Washio T (2016) Studies on the soil fertilization management of black soybean in the upland field converted from paddy field (1) current condition of soil fertility and analysis of yield decrease factor in black soybean fields. Bull Res Inst Agric Okayama Prefectural Technol Center Agric For Fish 7:7–17 (in Japanese)
Nakajima H (2007) Agriculture, soil science and plant nutrition in Kyoto Prefecture. Jpn J Soil Sci. Plant Nutr 78:545–546. (in Japanese)
NARO WARC (2013) Diagrams of agricultures in Kinki-Chugoku-Shikoku Region, 2012, pp 2, 5, 22, 24–25, 31, 38. (in Japanese)
Naruse T (2004) Plains of Sanyo rigion. In: Ota Y, Nause T, Tanaka S, Okada A (eds) Regional geomorphology of the Japanese Islands, vol 6. Geomorphology of Kinki, Chugoku and Shikoku. University of Tokyo Press. Tokyo, pp 220–232. (in Japanese)
Obara H, Takata Y, et al (2016) A new soil map based on comprehensive soil classification system of Japan fiest approximation. Bull Natl Inst Agro-Environ Sci 37:133–148. (in Japanese with English summary)

Okada A, Chinzei K (2004) Major landform of Kinki, Chugoku, and Shikoku. In: Ota Y, Nause T, Tanaka S, Okada A (eds) Regional geomorphology of the Japanese Islands, vol 6. Geomorphology of Kinki, Chugoku and Shikoku. University of Tokyo Press. Tokyo, pp 2–4. (in Japanese)
Okayama Prefecture (2013) Guide of livestock manure proper application. http://www.pref.okayama.jp/site/22/388931.html. (in Japanese)
Oomori T (2012) Status of cultivated soil properties in Ehime Prefecture. In: The council of soil conservation and soil survey in Japan (ed) The guideline for cultivated soils in Japan. Japan Soil Association, Tokyo, pp 96–97. (in Japanese)
Ota Y (2004) Coasts of the outer zone. In: Ota Y, Nause T, Tanaka S, Okada A (eds) Regional geomorphology of the Japanese Islands, vol 6. Geomorphology of Kinki, Chugoku and Shikoku. University of Tokyo Press. Tokyo, pp 296–297. (in Japanese)
Saigusa M, Matsuyama N (1994) Distribution of allophanic and nonallophanic Andosols in Japan. Tohoku J Agri Res 48:75–83
Sakurai K (2007) Overview of upland soils in Shikoku region. Dojyo-wo-aishi, Dojyo-wo-mamoru. Hakuyu-Sya, Tokyo, pp 253–254. (in Japanese)
Sangawa A (2004) Kyoto basin, Nara basin, Osaka Plain, Osaka bay. Outline of region and active faults. In Ota Y, Nause T, Tanaka S, Okada A (eds) Regional geomorphology of the Japanese Islands, vol 6. Geomorphology of Kinki, Chugoku and Shikoku. University of Tokyo Press. Tokyo, p 68. (in Japanese)
Shibahara F (2008) Development of rice cultivation systems for environmentally conscious agriculture and estimation models of the load-reducing effect in the Kinki region. Research and development: research project for utilizing advanced technologies in agriculture, forestry and fisheries, pp 1–121. (in Japanese)
Shibahara F (2010) Development of rice cultivation systems for environmentally conscious agriculture and estimation models of the load-reducing effect. J Water Waste 52:55–62 (in Japanese)
Shibahara F, Takehisa K, Hasegawa K (2000) Development of a simple model for nitrogen uptake and fertilizer application of rice crop. Jpn J Soil Sci Plant Nutr 71:898–902 (in Japanese)
Shibahara F, Tsuji T, Nishimura M (1992) Improvement on nitrogen application efficiency to paddy rice and reduction in the outflow of fertilizer elements by using the coated urea fertilizer—on the nitrogen uptake by paddy rice from 15 N labelled coated urea. Bull Shiga Pref Agric Exp Stn 33:17–29 (in Japanese with English summary)
Shiga Prefecture (2001) Characteristics and distribution of prefectural cultivated soil. 4. (in Japanese)
Shiga Prefecture (Agricultural Administration Fisheries Department Agricultural Management Division) (2012) Cultivation guidelines for wheat varieties and soybean favorable to sell. http://www.maff.go.jp/j/seisan/kankyo/hozen_type/h_sehi_kizyun/sig03.html. (in Japanese)
Shimada N (2009) The essence of problems on groundwater and soil pollutions caused by naturally occurring heavy metals and harmful elements: arsenic. Appl Geol Techn Rep 29:31–59 (in Japanese with English summary)
Shimane Prefecture (2019) Key points of field maintenance for wheat and soybean cultivation. https://www.pref.shimane.lg.jp/industry/norin/gijutsu/nougyo_tech/kenyui/sisin_manyuaru/ine-sisin/ine-sisin.data/P88-89mugidaizuho.pdf (in Japanese)
Shishido N, Ishida T (1999) Crops damage by nickel (Ni) in serpentinite soil. Monthly Rep Civ Eng Res Inst 559:33–37 (in Japanese)
Shiwa M, Yoshizawa K, Ito M (2000) Dissolution of coated urea fertilizer and absorption of tea plants in tea field soil. Tea Res J 88:31–38 (in Japanese with English summary)
Statistics Department of Minister's Secretariat, Ministry of Agriculture, Forestry and Fisheries (2017) Statistics of agriculture, forestry and fisheries. Ministry of Agriculture, Forestry and Fisheries, Tokyo, pp 1–55. (in Japanese)
Takata Y, Kuwagata T et al (2011) Delineation of Japanese soil temperature regime map. Soil Sci Plant Nutr 57:294–302
Tanabe K (2007) Agriculture and soil management in Kagawa Prefecture. Jpn J Soil Sci Plant Nutr 78:331–332 (in Japanese)
Tanabe K (2012) The actual status of cultivated soil in Kagawa prefecture. Soil guidebook for arable land in Japan (Zenkoku noukouti dojou gaidobukku).Japan Soil Association, Tokyo, pp 96–97. (in Japanese)
Tanaka Y (2001) For reduction of pollution loading from paddy fields in Shiga Prefecture. J Agric Sci 56:251–256 (in Japanese)
Tanimoto T (2006) Agriculture and soil management in Hiroshima Prefecture. J Soil Sci Plant Nutr 77:367–368 (in Japanese)
Tanioka T (1964) Developments of plains, Kokin-Shoin. Tokyo, pp 57–70
Tano H (2017) Establishment and sustainable development of agricultural management of triple-cropping in paddy field (I)—Case study of Minami Awaji city, Hyogo Prefecture. Bulletin of the Chiba University of Commerce 54(2):29–48 (in Japanese)
Tateishi H (2007) The story of the formation of agriculture in the Awa area (awagata nougyou monogatari). tokushima nousei club, tokushima, pp 87–88. (in Japanese)
The Fifth Committee for Soil Classification and Nomenclature (2017) Soil classification system of Japan. The Japanese Society of Pedology, pp 1–53
The Forestry Agency (2012) The ratios of forest and planted forest in each prefecture. http://www.rinya.maff.go.jp/j/keikaku/genkyou/h24/1.html
Tsukimori A, Hirai H, Kyuma K (1992) Studies on the changes in soil properties brought about by afforestation in Mt. Tanakami of Shiga Prefecture. Pedologist 36:17–30 (in Japanese with English summary)
Tsutaka T (1989) The agriculture and soil nutrition in Sanyo area (chapter: serpentine soils), pp 90–96. (in Japanese)
Umemiya Y (2004) Present state of fertilizer nitrogen load from orchard. Hort Res 3(2):127–132 (in Japanese)
Vitousek PM, Aber JD et al (1997) Human alteration of the global nitrogen cycles: sources and consequences. Ecol Appl 7:737–750
Wakayama Prefecture (2011a) Guidelines for soil fertilizer application. 51. (in Japanese)
Wakayama Prefecture (2011b) Guidelines for soil fertilizer application. Wakayama Prefecture. 51–54. (in Japanese)
Yamamoto S (2007) Andosols in Chugoku region. In: Dojyo-wo-aishi Dojyo-wo-mamoru (ed) Japanese society of pedology. Tokyo, Hakuyu-Sya, pp 231–237 (in Japanese)
Yamamoto S, Endo T, Honna T (2007) The soil of Tottori sand dunes. In: Dojyo-wo-aishi Dojyo-wo-mamoru (ed) Japanese society of pedology. Tokyo, Hakuyusha, pp 245–251 (in Japanese)
Yamasaki H (2005) Agriculture, Soil Science and Plant Nutrition in Kochi. Jpn J Soil Sci Plant Nutr 76:101–102 (in Japanese)
Yoshikawa (Yamanishi) H, Nakao S (2010) Construction procedure of the automated pulsating drip-Irrigation system using solar pump Miscellaneous Publication of The National Agricultural Research Center for Western Region, 7(Feb), p 21. (in Japanese with English summary)

Kyushu and Okinawa Regions

10

Yusuke Arakawa, Takashi Kusaba, Hideo Kubotera, Ichiro Uezono, Naoko Miyamaru, Yuichi Saeki, Hiroshi Niimi, Keiko Nakano, Nobuhisa Koga, Yoshitaka Hara, Yasunao Yamada, Katsuhiro Inoue, Hirotaka Ihara, Makoto Nagatomo, Koichi Yoshida, Noriko Yamaguchi, Hiroaki Hayashi, Tomohiro Kondo, Tsuyoshi Yamane, Isao Akagi, Kazutoshi Kinjo, Takeo Shima, and Kojiro Mitsugi

Abstract

Kyushu is the most southerly island among Japan's main islands. The landscape of Kyushu has been formed through volcanic activity where volcanic soils (Andosols) are widely distributed in the central and southern part of the island. Okinawa is the southern half of the Nansei Islands, which extends approximately 1200 km from Kyushu to Taiwan. The soils of Okinawa are far different from the soils of Japan's mainland. The three main soils are Red-Yellow soils, Calcareous Eutrosols, and Terrestrial Regosols, all having low soil fertility. Kyushu has a warm and rainy climate, while Okinawa has a subtropical oceanic climate. However, typhoons and torrential rains are frequent in both regions during summer and autumn, leading damage to fields and crops behind them. Kyushu is the second largest food producer on a monetary basis next to Kanto including

Y. Arakawa (✉) · H. Niimi · T. Shima
Division of Upland Farming Research, Kyushu Okinawa Agricultural Research Center, NARO, Miyakonojo City, Miyazaki Prefecture, Japan
e-mail: arakaw@affrc.go.jp

T. Kusaba
Division of Agro-Environment Research, Kyushu Okinawa Agricultural Research Center, NARO (Retired), Koshi City, Kumamoto Prefecture, Japan

H. Kubotera
Institute for Agro-Environmental Sciences, NARO, Tsukuba City, Ibaraki Prefecture, Japan

I. Uezono
Kagoshima Prefectural Institute for Agricultural Development Osumi Branch, Kanoya City, Kagoshima Prefecture, Japan

N. Miyamaru
Sugarcane and Ordinary Crops Section, Okinawa Prefectural Agricultural Research Center, Itoman City, Okinawa Prefecture, Japan

Y. Saeki
Faculty of Agriculture, University of Miyazaki, Miyazaki City, Miyazaki Prefecture, Japan

K. Nakano
Division of Lowland Farming Research, Kyushu Okinawa Agricultural Research Center, NARO, Chikugo City, Fukuoka Prefecture, Japan

N. Koga
Headquarters, NARO, Tsukuba City, Ibaraki Prefecture, Japan

Y. Hara
Kyushu Okinawa Agricultural Research Center, NARO, Koshi City, Kumamoto Prefecture, Japan

Y. Yamada · K. Inoue
Nagasaki Agricultural & Forestry Technical Development Center, Nagasaki Prefectural Government, Isahaya City, Nagasaki Prefecture, Japan

H. Ihara
Division of Agro-Environment Research, Kyushu Okinawa Agricultural Research Center, NARO, Koshi City, Kumamoto Prefecture, Japan

M. Nagatomo
Kagoshima Prefectural Institute for Agricultural Development, Minamisatsuma City, Kagoshima Prefecture, Japan

K. Yoshida
Okinawa Agricultural Technology & Development Co., Ltd., Itoman City, Okinawa Prefecture, Japan

N. Yamaguchi
Division of Soil Science and Plant Nutrition, Central Region Agricultural Research Center, NARO, Tsukuba City, Ibaraki Prefecture, Japan

H. Hayashi
The Kunisaki Peninsula Usa GIAHS Promotion Association, Kunisaki City, Oita Prefecture, Japan

T. Kondo
Faculty of Regional Innovation, University of Miyazaki, Miyazaki City, Miyazaki Prefecture, Japan

Present Address:
T. Kondo
Graduate School of Agriculture, Kyoto University, Kyoto City, Kyoto Prefecture, Japan

T. Yamane
Division of Agro-Environmental Research, Hokkaido Agricultural Research Center, NARO, Sapporo City, Hokkaido, Japan

R. Hatano et al. (eds.), *The Soils of Japan*, World Soils Book Series,
https://doi.org/10.1007/978-981-15-8229-5_10

diverse sectors such as paddy rice, arable crops, vegetables, and fruits. Okinawa, on the other hand, produces sugarcane, pineapple, tropical fruits, and so on by taking advantage of its subtropical climate. Moreover, large-scale livestock production is active through both regions. Like other regions of the globe, environmental issues related to modern agriculture are challenges, including nitrate contamination of groundwater, greenhouse gas emission, soil erosion, and so on. Beside dominating modern agriculture, traditional agriculture with thought-provoking soil management practices still survives in these regions.

Keywords

Kyushu–Okinawa • Andosols • Red-Yellow soils • Large-scale livestock production • Environmental-friendly agriculture

10.1 Outline of Soils in Kyushu and Okinawa Regions

10.1.1 Soil-Forming Factors

The Kyushu–Okinawa region is located in the southwest of the Japanese Islands and consists of Kyushu, islands around Kyushu including Tsushima Island, the Goto Islands, the Amakusa Islands, the Koshikijima Islands, and the Nansei Islands (Figs. 10.1 and 10.2). The region is extensive, ranging from 123 to 132° E (a distance of 850 km) in the east–west direction and from 24 to 35° N (1200 km) in the north–south direction. Although the climate of the Kyushu region is mild as a whole under the influences of the Japan Current and the Tsushima Current, it is further subdivided into several climatic zones due to influences such as those from mountains. Additionally, subtropical regions are distributed in the Nansei Islands. The Kyushu region is located at the junction between the Southwest Japan Arc and the Ryukyu Arc. Volcanoes such as Mt. Aso and Mt. Unzen are located in central Kyushu, and the volcanic line consisting of Mt. Kirishima, Mt. Sakurajima, and Iwo Jima is distributed from the southern part of Kyushu to the Nansei Islands.

I. Akagi
Faculty of Agriculture, Kagoshima University, Kagoshima City, Kagoshima Prefecture, Japan

K. Kinjo
Faculty of Agriculture, University of Ryukyus, Nishihara Town, Okinawa Prefecture, Japan

K. Mitsugi
Kumamoto Prefectural Agricultural Research Center, Kumamoto Prefectural Government, Koshi City, Kumamoto Prefecture, Japan

Generally, climate, vegetation, geology (parent material) and topography greatly influence soil formation. Various soil types are generated depending on the type and intensity of soil-forming factors. For example, one of the soil types that strongly indicates the effects of climate and vegetation is Red-Yellow soil, a weathered soil type with less accumulation of organic matter and low base saturation. In the Kyushu–Okinawa region, Red-Yellow soils are widely distributed in the hilly low mountain area of the Nansei Islands. On the other hand, one type of soil that is strongly affected by geology (parent material) is Andosols. This soil type, derived from volcanic ejecta, is distributed around the volcanoes in the wet subtropical monsoon region from the southern subarctic. In the Kyushu–Okinawa region, this soil is widely distributed in the central and southern part of Kyushu. Holocene fluvial and marine sediments are deposited in the plains around the Ariake Sea. Fluvic soils are distributed in such areas.

1. **Climate and vegetation**

According to the climate classification using the Warmth Index suggested by Kira (1971), the climate of the Kyushu–Okinawa region is classified into the following four climate zones: the cool temperate zone, warm temperate zone, subtropical zone and Setouchi climate zone (Nakamura et al. 1996). The annual average temperature and annual precipitation in each climate zone were cited from a finer climate classification by Ohba (2004). According to the soil temperature regime map by Takata et al. (2011), the cool temperate zone, the warm temperate zone, the subtropical zone, and the Setouchi climate zone almost correspond to the mesic or thermic soil temperature regime, the thermic soil temperature regime, the hyperthermic soil temperature regime, and the thermic soil temperature regime, respectively.

(1) **Cool temperate zone**

The annual average temperature, annual precipitation, and average temperature in January are 15 °C or less, about 2000 mm or more, and 5 °C or less, respectively. This climate zone includes the mountainous area in central Kyushu. Deciduous broad-leaved trees, mainly Japanese beech (*Fagus crenata*) and shiraki (*Neoshirakia japonica*), are distributed. Evergreen coniferous trees, mainly Japanese fir (*Abies firma*) and southern Japanese hemlock (*Tsuga sieboldii*), are distributed in the transition area with the warm temperate zone.

(2) **Warm temperate zone**

The annual average temperature, annual precipitation, and average temperature in January are 15–17 °C, about 1600–

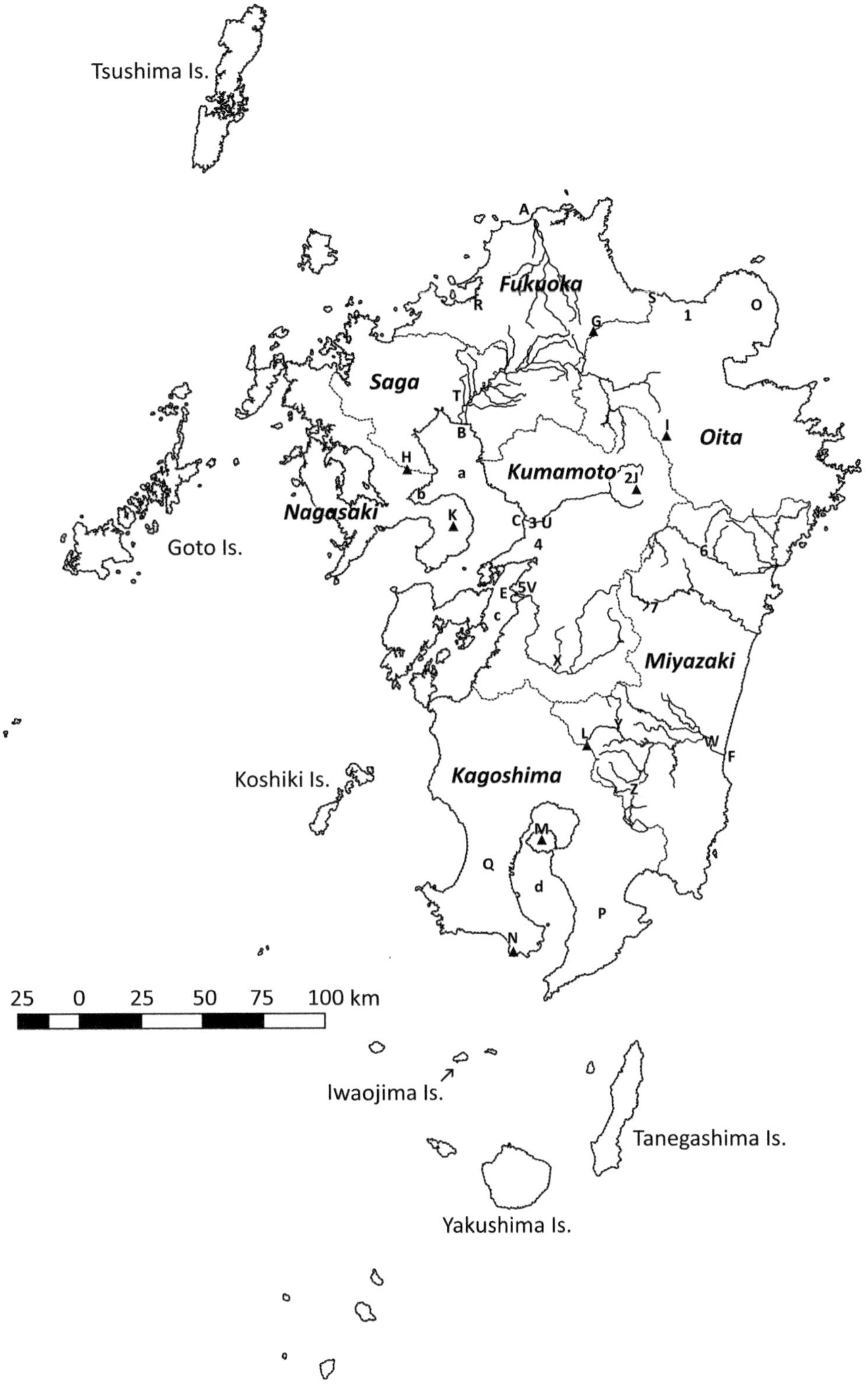

Fig. 10.1 The Kyushu Island and islands around the Kyushu Island. Modified from blank map (2018), Copyright 2018, Geographical Survey Inst. River, (A) Ongagawa, (B) Chikugogawa, (C) Shirakawa, (D) Gokasegawa, (E) Kumagawa, (F) Oyodogawa, Mountain, (G) Hiko, (H) Tara, (I) Kuju, (J) Aso, (K) Unzen, (L) Kirishima, (M) Sakurajima, (N) Kaimondake, Peninsula, (O) Kunisaki, (P) Osumi, (Q) Satsuma, Plain, (R) Fukuoka, (S) Nakatsu, (T) Chikushi, (U) Kumamoto, (V) Yatsushiro, (W) Miyazaki, Basin, (X) itoyoshi, (Y) Kobayashi, (Z) Miyakonojo,Sea, (a) Ariake-kai, (b) Isahaya Bay, (c) Shiranui-kai, (d) Kagoshima Bay, City, town, (1) Usa, (2) Aso, (3) Kumamoto, (4) Uto, (5) Yatsushiro, (6) Takachiho (Toroku), (7) Shiibayama

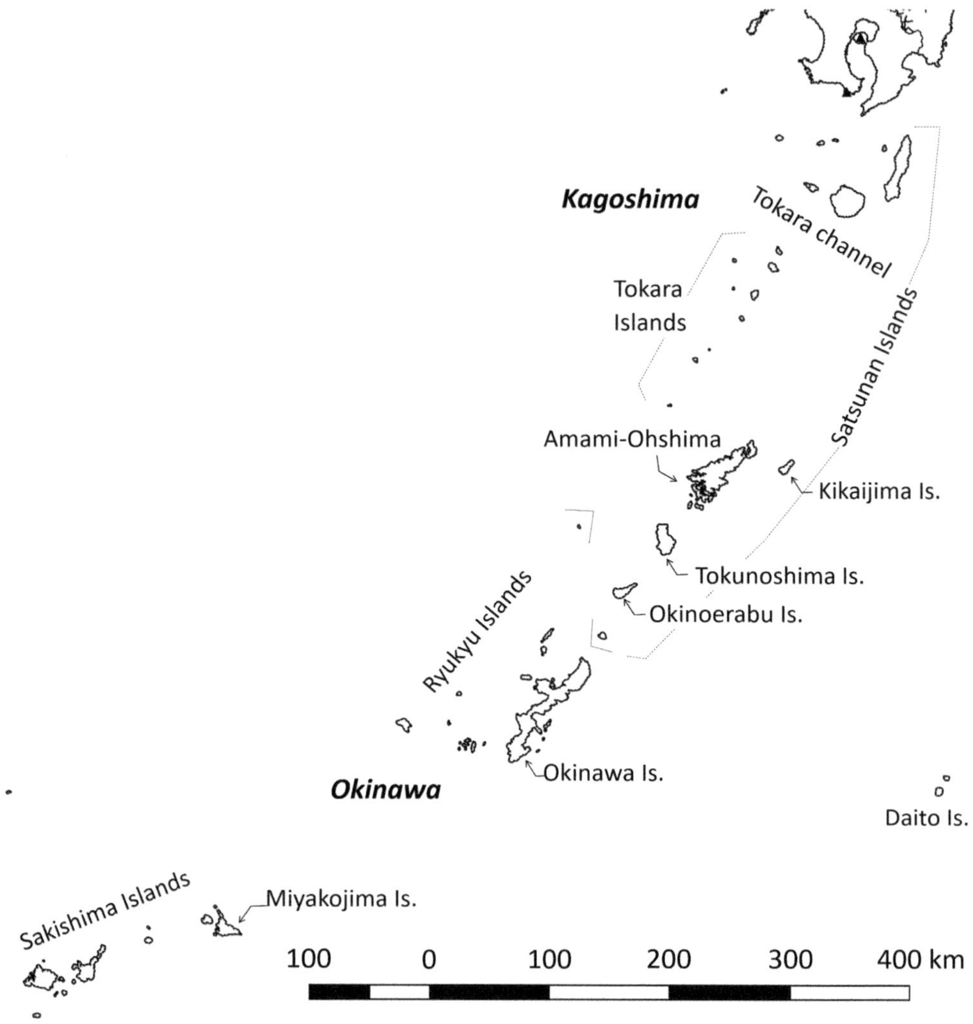

Fig. 10.2 The Nansei Islands. Modified from blank map (2018), Copyright 2018, Geographical Survey Inst.

3000 mm, and 5–8 °C, respectively. This climate zone includes the Nansei Islands to the north of the Tokara channel and the coastal region of Kyushu excluding the Seto Inland Sea coastal region in Oita Prefecture. Deciduous broad-leaved trees typified by chinquapin trees (*Castanopsis cuspidate* spp.) and cinnamon (*Cinnamomum camphora*)are mainly distributed.

(3) **Subtropical zone**

The annual average temperature and annual precipitation are 17 °C or more and about 2000–4000 mm, respectively. This climate zone has monthly average temperatures of 20 °C or more for six to eight months a year and does not experience freezing temperatures. The annual range in temperature and precipitation is small. This climate zone includes the Nansei Islands to the south of the Tokara channel. Among evergreen broad-leaved trees, many tropical plants such as epiphytes and wood ferns grow abundantly.

(4) **Setouchi climate zone**

The annual average temperature, annual precipitation, and average temperature in January are 15–16 °C, 1400–1600 mm, and 5–6 °C, respectively. The Seto Inland Sea coastal region of Oita Prefecture belongs to this climate zone. Evergreen broad-leaved trees are mainly distributed in this zone. This zone experiences less snow in winter and less rainfall in summer.

2. **Geomorphology and geology**

The geomorphology and geology of the Kyushu–Okinawa region is divided into four parts: (1) the central part, characterized by a volcanic graben zone (the Beppu–Shimabara graben) crossing central Kyushu and outstanding volcanoes; (2) the northern part, in the north of central Kyushu; (3) the southern part, in the south of central Kyushu; and (4) the Nansei Islands (Machida et al. 2001). The northern margin of central Kyushu passes from Mt. Hiko through the north of the Chikushi Plain to Mt. Tara. The southern margin of central Kyushu is the Usuki–Yatsushiro Tectonic Line, which is considered as an extension of the Median Tectonic Line.

(1) **Central Kyushu**

Central Kyushu is characterized by numerous volcanoes, normal faults, and seismic activity. This region is crossed by a volcanic area and also contains an old volcanic mountainous area, a small non-volcanic area, and the Kumamoto and Chikushi plains. The volcanic area contains volcanoes such as Mt. Kuju, Mt. Aso, and Mt. Unzen, new and old calderas formed post-Pliocene, a pyroclastic flow plateau, and a lava plateau. The Chikushi Plain–Ariake Sea area is the widest lowland in Kyushu and is considered to be a sedimentary basin formed between the Beppu–Shimabara graben and northern Kyushu.

(2) **Northern Kyushu**

Northern Kyushu consists of the Chikushi Mountains, plains/basins, a western hilly low mountain area, and the islands region. The plains/basins consist of the Fukuoka Plain, the Ongagawa River lowland, and other areas. The western hilly low mountain area consists of hills (Paleogene and Miocene sedimentary rocks) and tableland composed of basalt lava covering the Paleogene and Miocene rocks.

(3) **Southern Kyushu**

Southern Kyushu is subdivided into two parts: one is the area having the geomorphological and geological features of the "outer zone" of southwest Japan, and the other is the area where the volcanoes are distributed. The former consists of the Kyushu Mountains (accretionary complex), the Miyazaki Plain, and the Osumi Peninsula (plutonic rocks). The latter consists of the Kagoshima graben, ranging from the Kobayashi basin to Mt. Kaimon, the Hisatsu/Kitasatu mountainous area, and the Satsuma Peninsula. Numerous calderas and pyroclastic flow plateaus originating from volcanic activity are distributed in the Kagoshima graben.

(4) **The Nansei Islands**

The Nansei Islands are composed of the Satsunan Islands and the Ryukyu Islands, which form the arcuate islands between Kyushu and Taiwan (the Ryukyu Arc), and the Daito Islands, which are located on the eastern side of the Ryukyu Arc and consist of elevated coral reefs (Ota and Kawana 2001). The Ryukyu Arc is divided into the following two chains by topography and geology: (1) an island chain consisting of volcanic islands from Iwo Jima to Iwo Tori Island through the Tokara Islands (also known as the Tokara volcano island chain); and (2) a non-volcanic island chain consisting of the Satsunan Islands and the Ryukyu Islands excluding the Tokara volcano island chain (Kato 2009). This non-volcanic island chain consists of mountains and hills (accretionary complex), hilly land (Neogene strata and Pleistocene calcareous sediment), and Holocene coral reefs.

10.1.2 Soil Distribution

The Kyushu–Okinawa region has an area of 45,000 km^2, equal to 12% of Japan's land area. Of the soil distributed throughout the region, 34% are Andosols, 30% are Brown Forest soils, 14% are Red-Yellow soils, 13% are Fluvic soils, and 2% are Eutrosols (Table 2.4 in Chap. 2). The area of cropland (paddy field, upland field, orchard, and tea field) in the Kyushu–Okinawa region is about 568,000 ha, which represents 15% of Japan's arable land area (Statistics Department, Minister's Secretariat, Ministry of Agriculture, Forestry and Fisheries 2016). Of the agricultural land in the region, 44% is occupied by Fluvic soils, 28% by Andosols, 8% by Red-Yellow soils, and 5% by Eutrosols. Thus, the croplands in the Kyushu–Okinawa region are characterized by a high proportion of Fluvic soils, Red-Yellow soils, and Andosols.

(1) **Fluvic soils**

This soil type is distributed in the lowlands of river basins and is influenced by the geology around the plains and the transportation and sedimentation by rivers. Among the soil physicochemical properties reflected by these characteristics, the mineral composition of the clay fraction and the particle size fraction of the soil have great influences on the water permeability, the nutrient supplying capacity, and other factors. As volcanoes are distributed in Kyushu, the proportion of soil with an amorphous clay fraction is higher there than in other regions (Soil Genesis and Classification Laboratory, National Institute for Agro-Environmental Sciences, 1997). Based on their clay mineral composition, the Fluvic soils in the Kyushu region are further classified into the following four soil groups (Seino 1980). (1) Soils in the reclaimed land around the Ariake Sea such as the Chikushi Plain and the Kumamoto Plain. These soils have a fine texture, are mainly composed of 2:1 clay minerals, and have a high nutrient supplying capacity. Their productivity for paddy rice is high, but drainage measures are required for upland utilization. (2) Soils in the Nakatsu Plain and the Miyazaki Plain. These soils have a medium texture and are mainly composed of 1:1 clay minerals. (3) Soils in the

Fukuoka Plain. These soils have a coarse texture, contain 1:1 clay minerals, and have a weak buffering capacity. (4) Soils in the Hitoyoshi and Miyakonojo Basins. These soils contain 1:1 clay minerals and amorphous clay minerals influenced by volcanic ash and have a high phosphate adsorption coefficient. Fluvic soils are subdivided taxonomically into Fluvic Paddy soils, Gley Fluvic soils, Gray Fluvic soils, Brown Fluvic soils, and so on, based on the effects of irrigation water and groundwater. Fluvic Paddy soils and Brown Fluvic soils are distributed in natural levees, alluvial fans, and in high altitude areas in the Chikugo, Kuma, and Oyodo river basins. Gray Fluvic soils are distributed in the flat areas of the Chikushi, Kumamoto, and Yatsushiro Plains. Gley Fluvic soils are widely distributed in reclaimed areas such as the Ariake Sea and the Shiranui Coast and are also distributed in mountainous areas or valley bottom plains between terraces (Hamazaki 2004). Gley Fluvic soils may be accompanied by a peat layer, a black mud layer, and in some cases, an acid sulfate soil layer in the lower layer.

Soils of the Fluvic soil great group are mainly used as paddy fields. There are many paddy fields reclaimed around the Ariake Sea in northern Kyushu, and the soil productivity is moderate to good. In addition to paddy rice (*Oryza sativa* L.), wheat (*Triticum aestivum* L.), soybean (*Glycine max* L. Merril, strawberry (*Fragaria ananassa* DUCHESNE), and eggplant (*Solanum melongena*) and onion (*Allium cepa* L.) is cultivated in the Chikushi Plain, and cucumber (*Cucumis sativus* L.) and tomato (*Solanum lycopericum* L.) are cultivated in the Miyazaki Plain.

(2) **Andosols**

Soils of this great group are distributed in the areas covered with volcanic ejecta around Mt. Aso, Mt. Kuju, Mt. Unzen, Mt. Sakurajima, Mt. Kirishima, and Mt. Kaimon. Uncultivated Andosols are generally rich in humus, the content of which may exceed 30% (Miyazawa 1986). Soil layers with properties such as high humus content and high phosphate absorption coefficient were generated at the rim of Mt. Aso 10 kya and at the western foot of the rim of Mt. Aso about 24 kya (Yamada and Kubotera 1994). On the other hand, Non-Allophanic Andosols showing strong acidity are scattered around northern Kyushu. Furthermore, it was recently revealed that Non-Allophanic Andosols are also widely distributed in the Kuju Plateau in central Kyushu (Kubotera et al. 2013).

Andosols have low base saturation and a high phosphate absorption coefficient, and are small in volume weight and rich in pores. Therefore, these soils have good water retention and drainage, have low bulk densities, and allow exceptionally easy cultivation. The agricultural land on Andosols is used for upland fields, forage fields, orchards, and tea fields, as this soil type is widely distributed at the feet of volcanic mountains and plateaus.

(3) **Red-Yellow soils**

Red-Yellow soils are distributed on hills and tableland in northern and central Kyushu and on mountainous hills and terraces in the Nansei Islands. Red-Yellow soils are derived from various kinds of rocks and strata such as basalt, andesite, crystalline schist, granite, Paleogene and Neogene sedimentary rocks, Pleistocene strata, and so on. Red-Yellow soils in northern Kyushu are distributed only in high terraces and are considered to be the paleosol formed in the Pleistocene warm period (about 120–130 kya) (Matsui and Kato 1962). On the other hand, Red-Yellow soils in the central and the southern part of the Nansei Islands are considered to have formed under the current climate. Some Red-Yellow soils derived from andesite are difficult to form under the current warm and humid climatic conditions, which causes moderate soil weathering. It has been pointed out that such soils may have been formed by post-volcanic action, such as the intense oxidation or alteration of the soil host rock during or after the lava cooling process (Yoshida et al. 1974). Furthermore, it has been pointed out that the formation of these soils is influenced by eolian dust from the Asian continent based on evidence about the grain size and the chemical properties of the quartz grains contained in the soil (Mizota et al. 1992).

Since Red-Yellow soils are distributed in plateaus and hills, are weakly acidic, and have a low nutrient content and low nutrient-holding capacity; they are mainly used as upland fields, orchards, and tea fields. In the central and the southern part of the Nansei Islands, sugarcane, pineapple (*Ananas comosus* (L.) Merr.), and citrus fruits are cultivated.

10.1.3 Properties and Management of Local Problem Soils

Kyushu has many volcanoes, and volcanic soils are therefore widely distributed. Andosols derived from fine volcanic ash are suitable for upland farming due to their excellent physical properties. However, some volcanic ejecta have unsuitable physical properties. In 1952, a law for the enhancement of disaster prevention and soil improvement for these problematic soils was enforced by the Japanese government. In Kyushu, the soils described below are subject to this law.

Fig. 10.3 A big profile of Shirasu. Plateaus edged with steep cliffs are the representative landform of this pyroclastic deposit. The photo was provided by Dr. M. Nagatomo

(1) **Shirasu** (Fig. 10.3)

"*Shirasu*" (Japanese for "white sand") is pyroclastic flow deposits that are widely distributed in Kagoshima and Miyazaki Prefectures in southern Kyushu. The distribution area of Shirasu in southern Kyushu is 4700 km^2 (Fujimoto et al. 1983). Shirasu consists of a mixture of white sandy volcanic ash and pumice, with various degrees of cementation. Its principal source was the Ito pyroclastic flow from the Aira caldera (equivalent to the northern part of current Kagoshima Bay) at 29–26 kya. Volcanic ash erupted with Shirasu was the Aira TN volcanic ash (AT) and was deposited as far as Hokkaido in Japan, 1400 km away (Machida and Arai 2003). AT is a representative widespread tephra in Japan and is used as a key tephra for the determination of soil horizon age. The thickness of Shirasu reaches up to 150 meters in some areas, and its landforms are tablelands rimmed with steep cliffs known as "*Shirasu-daichi*" (*daichi* is Japanese for "tableland"). From a soil engineering standpoint, Shirasu is an unstable material that is vulnerable to landslides. In cultivated soils, Shirasu is seen as a low-productivity soil with little clay and humus and a small holding capacity for water and nutrients. Shirasu is thickly deposited, unlike other problem soils, and therefore, performing soil improvement by its removal is difficult. The application of organic matter is recommended to improve the productivity of Shirasu.

(2) **Akahoya** (Fig. 10.4)

"*Akahoya*" is a yellow-orange-colored volcanic ash layer that contains abundant bubble-wall-type volcanic glass. Akahoya is called by various local names across its distribution areas, for example, "*Imogo*," "*Onji*," and "*Hoya*." It corresponds to the Kikai-Akahoya tephra (K-Ah) erupted from Kikai Caldera in the East China Sea, 50 km south of Kyushu. Its eruption, with a huge pyroclastic flow known as the Koya pyroclastic flow, occurred at 7.3 kya (Machida and Arai 2003) and caused catastrophic damage to the inhabitants, animals, and vegetation of southern to central Kyushu. K-Ah, which was deposited as far as the Tohoku region, 1300 km away, is a representative widespread key tephra in Japan, similar to the AT. The depth of Akahoya in southern to central Kyushu is mostly between 10 and 40 cm. However, in some areas of the Ōsumi Peninsula, Kagoshima Prefecture, located in the main axis of the deposits, the total depth of Akahoya and related pyroclastic flow deposits and pumice exceeds 1 m. The elongation of plant roots into Akahoya is restricted, and it is therefore evaluated as a low-productivity soil. Additionally, as Akahoya contains a

Fig. 10.4 An Andosol profile with Akahoya in the subsoil. A black buried A horizon called "Kuroniga" lies under Akahoya. The photo was provided by Dr. M. Nagatomo

large amount of allophane and imogolite, the heavy application of phosphate is needed, as in ordinary Andosols, as a countermeasure for phosphate fixation.

(3) **Kora**

"*Kora*" is the name given to indurated volcanic ash and gravel layers that are distributed around Mt. Kaimon volcano, from which it was erupted, in the Satsuma Peninsula, Kagoshima Prefecture. Kora is subdivided into five units, the eruption dates of which are estimated at 4, 3.3, 2.3, 1.5, and 1.0 kya, respectively (Fujino and Kobayashi 1992). The name "Kora" is related to "*koura*" (Japanese for "shell") and refers to the hardness of the deposit. The main problem associated with the presence of Kora in cultivated fields is the inhibition of plant root extension due to its hardness. The removal of Kora by farm machinery has been conducted under national and prefectural projects, and nowadays, its distribution area has become so small that it was designated as a "red data soil" that is considered as endangered (Japanese Society of Pedology 2000). However, soil horizons or tephra layers with the problem of hardness similar to that of Kora are seen in several volcanic areas in Kyushu. Kubotera and Yamada (2000) divided these into two types. One type is the subsurface horizons of Andosols with large amounts of carbon and clay; this type is not hard in a moist state, but when it is exposed at the surface and becomes dry, it shrinks and forms hard soil blocks. This type should be thoroughly pulverized immediately after exposure, and chemical property improvements, for example, those involving the heavy application of phosphate fertilizer, which are commonly conducted in Andosols, are also required. This type includes "*Nigatsuchi*" around Aso volcano in Kumamoto Prefecture and "*Kashinomi*" around Unzen volcano in Nagasaki Prefecture. The existence of a similar type of soil in the Azores, Portugal, was reported by Pinheiro (1997). Another type is indurated tephra layers, such as Kora, which have little humus and clay. This type is hard in both moist and air-dried states, and pulverization and removal by machinery are required for soil improvement. This type includes Kora, "Hanamure" around Mt. Kuju in Oita Prefecture, and "Banban" on the somma of Aso volcano. Indurated volcanic ashes in South America, namely "Tepetate" (Mexico), "Telpetate" (Nicaragua), and "Cangahua" (Ecuador), probably belong to this type.

(4) **Bora**

"*Bora*" is the generic term for the pumice layers with various degrees of weathering that are widely distributed in Kagoshima and Miyazaki Prefectures. A representative Bora unit is the Miike Bora, which was erupted from the Miike crater, Miyazaki Prefecture, during activity of Kirishima volcano 3 kya. Bora was also ejected from Sakurajima volcano, an island in Kagoshima Bay that is still active at present, at several times in the past; three major Bora units were erupted from Sakurajima, in AD 1471, 1779, and 1914 (Kobayashi and Tamaike 2002), and are called "*Bunmei*

Bora," "*An-ei Bora*" and "*Taishou Bora*" in reference to the historical periods in which each was erupted. Bora layers are composed of coarse non-cemented pumice. On slopes, it often slides easily and can cause landslide disasters. In upland fields, Bora inhibits water supply and plant root extension. Therefore, the removal of Bora by machinery is conducted if it appears from shallow depth. On the other hand, weathered Bora is used as a gardening material due to its excellent air and water permeability.

10.2 Methods of Soil Fertility Management

10.2.1 Rapid Evaluation of Available Nitrogen in Upland Soils

Available nitrogen indicates the mineralization potential of soil organic nitrogen that the soil slowly supplies to crops and is an important parameter influencing the crop production potential.

In Japan, the official measurement method of available nitrogen in upland soil has required four weeks of soil incubation under aerobic conditions at 30 °C. However, this incubation method is time consuming and labor intensive, and few institutions provide available nitrogen data for farmers as a soil diagnosis service, which makes it difficult for farmers to determine soil fertility promptly. Although many chemical extraction methods have been proposed, these methods lack versatility and convenience. That is, some methods cannot be applied to certain soil types or require expensive analytical equipment. Thus, these methods have not been adopted universally (Uezono et al. 2010a).

Uezono et al. (2012) developed a method that allows the simple, rapid evaluation of available nitrogen in soil. The method estimates available nitrogen by an incubation method from the extracted nitrogen amount (organic nitrogen + ammonium nitrogen) by heating the soil in hot water (80 °C) for 16 h extraction. This heating extracts easily decomposable organic matter. At the same time, some of the extracted organic matter is decomposed to ammonium nitrogen during heating, and the amount of extracted ammonium nitrogen increases over time. The source of the ammonium nitrogen is protein-like nitrogen compounds that are decomposed by the heat treatment. Therefore, in order to estimate the amount of available nitrogen with high accuracy, it is necessary to calculate the total amount of heating-extracted nitrogen using both the organic nitrogen and ammonium nitrogen. The available nitrogen can be estimated from the amount of nitrogen or organic carbon extracted from the soil. The reason for this is that the hot water extraction method provides a constant value of C/N ratio of around 9 regardless of the soil type and production and management history (Curtin et al. 2006; Moriizumi et al.

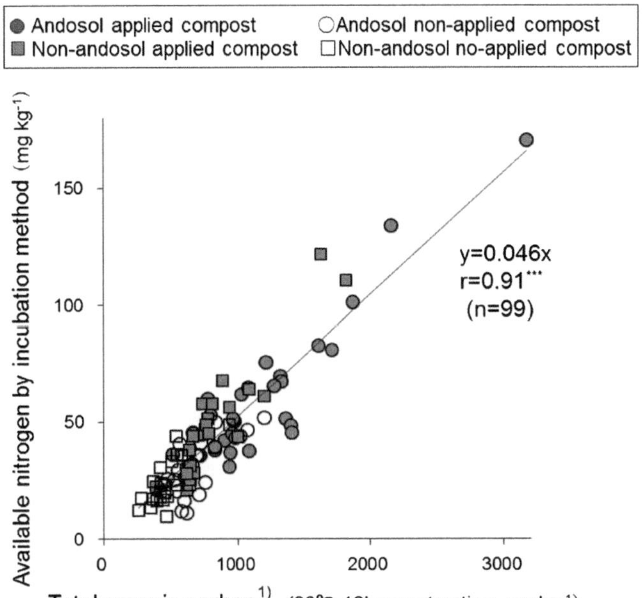

Fig. 10.5 Relationship between total organic carbon at 80 °C-16 h hot water extraction and available nitrogen by incubation method. Reprinted with translation from Uezono et al. (2010) Fig. 2 with permission from Japanese Society of Soil Science and Plant Nutrition. [1)]measured by using a total organic carbon analyzer by Shimadzu, and numerical value per dry soil

2015). From these results, the hot water extraction is likely to extract easily decomposable organic matter that particularly undergoes nitrogen mineralization.

For this reason, the organic carbon extracted by the abovementioned hot water extraction (80 °C for 16 h) can be used as an assessment index, since it has a strong positive correlation with available nitrogen measured by the incubation method (Fig. 10.5).

However, for the analysis of extracted organic nitrogen and organic carbon, pretreatment, such as acid decomposition, or expensive analytical equipment is necessary.

Therefore, ways to simplify the current analytical methods were examined. It was found that the chemical oxygen demand (COD) of an extract solution has a very strong positive correlation with the level of organic carbon.

More simply, the amount of available nitrogen can be estimated from the COD value of the extract (Uezono et al. 2010b). Since the COD value can be measured using a commercial simplified measuring kit, the measurement can be made by farmers themselves (Fig. 10.6). The measurement method is known as oxidation with potassium permanganate in alkalinity and visual colorimetric method. Referring to Fig. 10.6. First day: (1) Soil sampling in upland field. (2) Weighing of the soil sample. (3) Pouring 50 mL of hot water (80 °C) onto the soil. (4) Heat treatment at 80 °C for 16 h. Second day: (5) The soil is taken out by a pot and cooled. (6) The extract is diluted five times. (7) The COD test

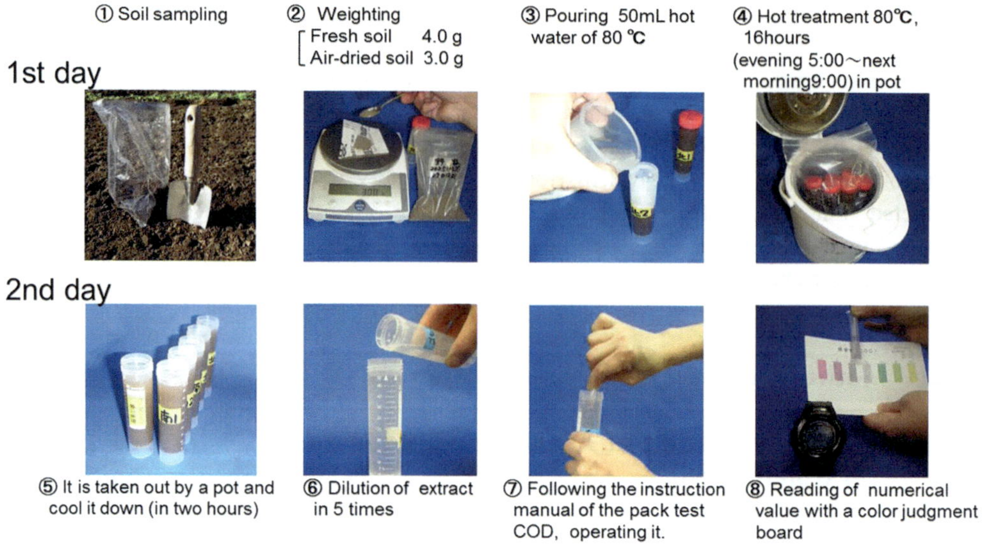

Fig. 10.6 Manipulation procedure of the 80 °C-16 h water extracting method in the technical guidance spot. Available nitrogen is calculated by the following equation. Available nitrogen (mg kg^{-1} of the air-dried soil) = Measurements × dilution magnification × (quantity of the 1000/soil weight) × (water weight/1000 of the water) × 0.034. The photo was provided by Ichiro Uezono

pack is operated following the instruction manual. (8) Reading of numerical value using a color judgment board. This simplified measuring method of for available nitrogen has the following characteristics: (1) it does not require demanding chemical analysis; (2) the results can be obtained as early as the next day after soil sampling at upland fields; (3) it does not require expensive chemical analyzers and measurement can be undertaken using easily obtainable instruments; (4) it does not require the use of any poisonous substances; and (5) it can be applied to various types of soil, including soil to which compost has been applied.

10.2.2 Utilization of Green Manure

(1) **Importance of green manure in the Nansei Islands**

The Nansei Islands are located south of Kyushu. The main soils of these islands are Red-Yellow soils, Calcareous Eutrosols, and Terrestrial Regosols, all of which have low soil fertility. This area has a subtropical climate, and consequently, the decomposition of organic matter occurs faster than in other areas in Japan. As such, the application of organic matter is important to improve soil fertility.

Green manure refers to a plant that is cultivated and then incorporated into soil to improve the soil fertility. In the Nansei Islands, leguminous plants have been cultivated as green manure for a long time, since the subtropical climate is suitable for plant growth. In a book about a farming technique in the kingdom of Ryukyu in the eighteenth century, it was recommended to incorporate mung bean (*Vigna radiata* (L.) Wilczek) or adzuki bean (*Vigna angularis* (Wild.) Ohwi et Ohashi) as green manure for the improvement the soil fertility.

(2) **Major green manures in the Nansei Islands**

In the Nansei Islands, the fallow period is generally from spring to summer in order to avoid typhoon damage, and green manures are cultivated during this period. Currently, sunn hemp (*Crotalaria juncea* L.) and sorghum (*Sorghum bicolor* (L.) Moench) are cultivated widely in this area, and sunflower (*Helianthus annus* L.) is also cultivated in some parts of this area to improve not only soil fertility but also the rural landscape.

Sunn hemp is often cultivated before the summer planting of sugarcane (*Shaccharum officinarum* L.), which is the main crop in the Nansei Islands. Sunn hemp is an annual leguminous plant native to India. The plant height is approximately 2 m (Fig. 10.7), and the dry weight and nitrogen content of the aerial part are 7 Mg ha^{-1} and 20 g kg^{-1}, respectively, at 70 days after sowing (Miyamaru et al. 2008a). Unlike sunn hemp, sorghum is often cultivated before horticultural crops such as chrysanthemum (*Chrysanthemum morifolium* Ram.) and pumpkin (*Cucurbita moschata*). It shows a luxuriant development and the dry yield is often more than 10 Mg ha^{-1} (Goto and Nagata 2000; Miyamaru et al. 2008a). Although the dry yield of sunflower is lower than that of sunn hemp or sorghum, the cultivation of sunflower increases the yield of succeeding plants by promoting arbuscular mycorrhizal formation on roots (Arihara and Karasawa 2000).

Fig. 10.7 Sunn hemp (*Crotalaria juncea* L.). Reprinted from Miyamaru (2003) with permission from the author

(3) **Effects of green manure application**

A field study was carried out for nine years in order to examine the effects of the long-term application of sunn hemp and sorghum on the soil properties of Calcareous Eutrosols on Tokunoshima, in the Nansei Islands (Goto and Nagata 2000). The average yield of succeeding potato (*Solanum tuberosum* L.) increased by 11–18% with green manure application. Soil concentrations of total carbon, total nitrogen, and exchangeable potassium increased with the successive application of green manure. The increase in the level of total carbon was higher with sorghum application due to a larger carbon input than for sunn hemp. Additionally, the successive application of sorghum improved the soil physical properties, the bulk density decreased, and the porosity of the soil increased.

On Okinawa Island, a field study was carried out for ten years on Marlitic Terrestrial Regosols to examine the effects of the long-term application of sesbania (*Sesbania cannabina* L. Ritz), leguminous green manure, and cattle manure compost (Miyamaru et al. 2012). The average yield of succeeding sweetcorn was 1% and 7% higher following the application of sesbania and compost, respectively, compared to a control with no application of organic matter. Although no remarkable differences were observed in the soil physical properties, the carbon and nitrogen contents in the soil microbial biomass (Fig. 10.8) and the microbial activity were significantly higher in the soil with sesbania or compost application than in the control, while the level of mineralizable nitrogen also increased. The total carbon and nitrogen contents of the soil increased with the successive application of organic matter, but the increase was small. There were no remarkable differences between the long-term application of sesbania and compost on soil properties. Therefore, it is considered that the application of green manure and compost is effective for the improvement of soil fertility in the Nansei Islands.

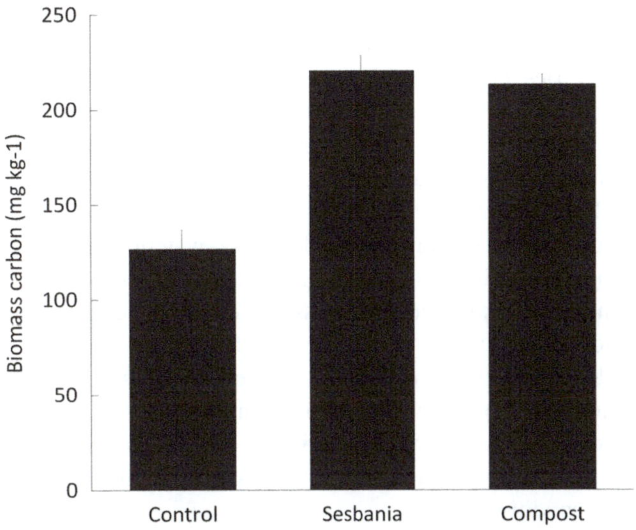

Fig. 10.8 Comparison of soil microbiol biomass by organic matter application. Modified from Miyamaru et al. (2012) Fig. 7 with permission from Japanese Society of Soil Science and Plant Nutrition

As mentioned in Sect. 10.4.5, soil erosion from farmland is one of the most serious problems in the Nansei Islands, especially in the area where Red-Yellow soils are widely distributed. The cultivation of green manure as a cover crop reduced soil erosion from farmland (Nagumo et al. 2006). Thus, green manure is effective not only for the improvement of soil fertility but also for soil conservation.

10.2.3 Soybean-Rhizobium Management

Soybean (*Glycine max* L. Merril) is one of the most important crops for Japanese traditional foods, being used in products such as soybean curd, soy paste, soy sauce, and so on. Currently, most soybean production in Japan is conducted in rotational upland fields or fields converted from paddy rice field (upland paddy field). Upland paddy fields account for more than 80% of soybean fields in Saga and Fukuoka Prefectures. In Japan, paddy rice cultivation in alluvial soil is a major production system, and wheat and soybean are recommended as the rotational crops. Rice production is conducted mostly under waterlogged conditions during the cultivation period. These anaerobic conditions create soil reducing dissolved oxygen content and oxidative–reductive potential and cause soil bacteria to undergo anaerobic respiration. The effectiveness of energy acquisition by anaerobic respiration depends on bacteria species and strains. Soil bacteria that are highly effective in terms of energy acquisition will be dominant under anaerobic conditions.

The genus *Bradyrhizobium* is a group of soil bacteria that are active in symbiotic nitrogen fixation with soybean. The nitrogen-fixing ability of soybean Bradyrhizobia depends on the strains present and the environmental conditions involved in the symbiosis. The major soybean-nodulating rhizobia are *Bradyrhizobium japonicum, Bradyrhizobium diazoefficiens, Bradyrhizobium elkanii*, and *Ensifer* (*Sinorhizobium*) *fredii*. The major endosymbionts in Japan are considered to be *B. japonicum*, *B. diazoefficiens*, and *B. elkanii* in acidic–neutral soils and *E. fredii* in alkaline soils (Saeki et al. 2006; Suzuki et al. 2008; Saeki et al. 2013). Currently, the nature of the nitrogen cycle around the soybean rhizosphere is considered to be that shown in Fig. 10.9. It is considered that rhizobia will be involved in most nitrogen cycles except for nitrification. Many bacteria, including Bradyrhizobia, undergo anaerobic respiration, such as nitrate respiration (commonly known as denitrification), which reduces nitrate to nitrogen gas. Denitrification activity varies depending on the strain of Bradyrhizobia (Sameshima-Saito et al. 2006). Some strains display complete denitrification activity and are able to release dinitrogen gas, while others show incomplete denitrification and release nitrous acid (NO_2^-), nitric oxide (NO), or nitrous oxide (N_2O). The *B. diazoefficiens* strain USDA110^T shows a high symbiotic nitrogen fixation and is expected as a useful inoculant, and is also known as a soybean-bradyrhizobium indicating complete denitrification activity.

Recently, the cultivation of soybean in upland paddy fields has indicated a protective effect against the effects of global warming by the mitigation of N_2O release from soybean fields (Itakura et al. 2013). In the case of Bradyrhizobia, *B. diazoefficiens* such as strain USDA110^T that perform complete denitrification will acquire approximately 30% more energy than bacteria that perform incomplete denitrification until N_2O (Thauer et al. 1977). In Japan, the dominant soybean-nodulating Bradyrhizobia in upland paddy fields belong to a cluster of *B. diazoefficiens* USDA110^T which perform complete denitrification (Saeki et al. 2013; Shiina et al. 2014). Waterlogging management will enable the Bradyrhizobia with complete denitrification ability to be dominant in the soil (Saeki et al. 2017). Additionally, waterlogging management can cause the

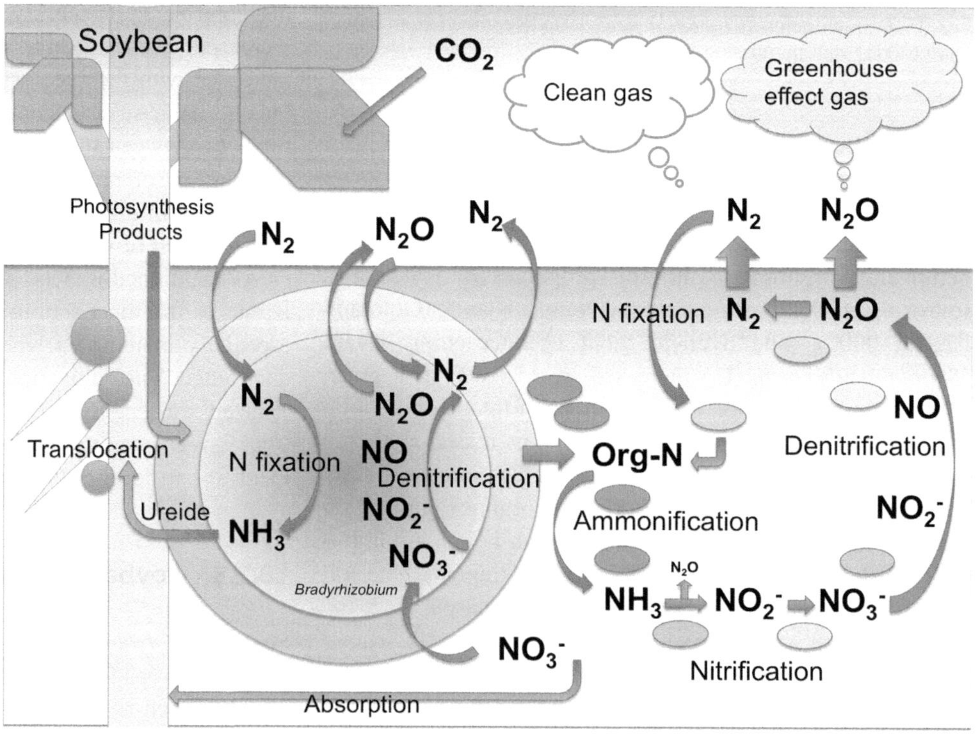

Fig. 10.9 Nitrogen cycle around soybean rhizosphere. Reprinted with translation from Saeki (2018) with permission from the author

neutralization of soil acidity under reductive conditions. An almost-neutral pH is also suitable for activities of Bradyrhizobia infecting host soybean. This phenomenon is one of the merits of soybean cultivation in upland paddy fields.

Most soybean production in Saga and Fukuoka Prefectures is conducted on upland paddy fields. This agricultural system involves the management of soybean Bradyrhizobia, both to increase the soybean yield and to mitigate the emission of N_2O gas from soybean fields. If soybean germination damage caused by excess soil water can be avoided by the construction of ridges in the field, the rotational crop production of paddy rice and soybean will allow good agricultural performance for Asian monsoon regions.

10.2.4 Manure Application and Organic Agriculture

In Kyushu, organic amendments made from crop residue (rice straw, chaff, wheat straw, etc.) and animal manure and food residues discharged from food processing factories are frequently used. On the other hand, improper manure management can cause harm to the environment or lead to a significant risk of air and water pollution, as manure contains substantial quantities of N, much of which is in inorganic forms. For example, Niimi (2002) reported that the extensive application of dairy cattle slurry barnyard manure to upland fields resulted in significant losses of *N* by ammonia volatilization, nitrate leaching, and nitrous oxide emission. Accordingly, it is necessary to develop circular agriculture, in which organic resources are used effectively for both sustainable production and environmental protection such as the prevention of groundwater contamination. Additionally, organic farming is gradually expanding in Kyushu. Organic farming involves the use of organic fertilizers; however, the nitrogen supply from organic fertilizers is often not sufficiently absorbed by crops, even though nitrogenous organic constituents are fully contained, since the nitrogen in organic constituents is absorbed by crops just after mineralization by soil microorganisms. This is the major reason that organic farming is less productive than conventional farming (Seufert et al. 2012).

One way to improve productivity in organic farming is to co-apply different types of organic fertilizers with different mineralization characteristics. Gunjikake and Kubo (1996) investigated the nitrogenous mineralization characteristics of several organic fertilizers with kinetics analysis (Table 10.1). They demonstrated that the co-application of composted cattle manure and rapeseed oil cake to adapt to

Table 10.1 The kinetic parameters of *N* mineralization of organic amendments, obtained from the first-order kinetic model $[N_m = N_0(1 - e^{-kt}) + C]$

Class	Total *N* of the organic amendment $g{\cdot}kg^{-1}$	Labile nitrogen[a] N_0, mg/100 g	Reaction rate coefficient at 25 °C k, day^{-1}	Apparent activation energy *Ea*, cal/mol	Intercept of Y axis[b] C, mg/100 g
Soybean meal	75.2	97.8	0.179	14,570	−1.2
Rapeseed meal	59.4	84.1	0.158	16,490	0.1
Meat and bone meal	76.1	82.2	0.127	19,020	22.5
Steamed bone meal	43.3	70.1	0.203	21,600	6.7
Fish meal	43.2	65.9	0.075	14,720	0.3
Dried blood meal	136	79.6	0.071	27,500	1.6
Dried cell fertilizer	77.2	37.9	0.076	20,140	65.4
Cattle manure compost	14.3	6.6	0.007	22,710	1.2
Swine manure compost	20.4	8.6	0.020	17,430	19.2
Wild grass compost	26.3	15.3	0.004	22,770	5.7

Translated from Gunjikake and Kubo (1996) with permission from the authors

[a]Labile nitrogen (mg/100 g dried soil), corresponding to N-mineralization potential under organic amendment (150 mg N basis) is incubated with an Andosol (100 g dried)

[b]Intercept of *Y* axis (mg/100 g dried soil), corresponding to initial level of mineral nitrogen obtained from calculation

the nitrogen absorbing pattern in crops ensured the growth of melon (*Cucumis melo* L.) and tomato (*Solanum lycopersicum* L.) without applying any chemical fertilizers. Similar results were obtained for the co-application of composted cattle manure, dried chicken manure, and rapeseed oil cake to sweet pepper (*Capsicum annuum* L. var. "grossum") (Ono et al. 1998). Furthermore, Inoue (2015) showed that the yield of onion (*Allium cepa* L.) with the co-application of composted cattle manure, composted chicken manure, and rapeseed oil cake doubled compared with the yield with the application of composted cattle manure only and was equal to or greater than the yield with the co-application of chemical fertilizer and composted cattle manure.

Organic fertilizer with low nitrogen mineralization characteristics (e.g., composted animal manure) can provide inorganic nitrogen over one cropping season. Therefore, in a cabbage–sweet potato double cropping system, composted animal manure for which the amount of mineralized nitrogen had been predicted for the cabbage cropping season was applied once to cabbage in winter, and a small amount of fertilizer was applied to the subsequent sweet potato. Throughout the three-year experiment, yields in both crops under composted animal manure were similar to those under chemical fertilizer, and a high fertilizer nitrogen recovery rate was obtained (76% and more than 100% for composted cattle and chicken manure, respectively) (Matsumoto 2007). During a four-year experiment, the single application of concentrated stillage from *shochu*, a Japanese liquor, under a daikon radish (*Raphanus sativus* var. longipinnatus)–sweet potato double cropping system (one application to both crops in a single year) resulted in yields of both radish and sweet potato that were as high as those obtained under the application of chemical fertilizer to every crop, while the fertilizer nitrogen recovery rate was 75% throughout the experiment (Niimi et al. 2016). Furthermore, phosphorus and potassium were not accumulated in the soil.

Thus, in organic farming, it is possible to obtain high nitrogen use efficiency and stable production by constructing crop rotation systems introducing crops such as sweet potato which can grow with limited nitrogen availability in soil as a subsequent crop.

10.2.5 Soil Physical Properties

Although flooding is a prerequisite for paddy fields, the increase in the use of tractors in the early 1960s accelerated efforts toward the reformation of well-drained paddy fields to ensure machine operability. Furthermore, when the production of rice began to be regulated in 1969, soybean began to be used as a rotational crop. As a result, the number of paddy–upland rotation fields increased. In order to allow paddy fields to be alternately flooded and drained, the consolidation of farmland including regional drainage accommodations has been implemented (Haraguchi and Wakasugi 2014). The promotion of upland crop cultivation in rice fields involves a reduction in the length of the flooding period, and in recent years, this has led to changes in the properties of paddy fields that had maintained their productivity despite continuous cropping. From the Soil-Environmental Monitoring Project, which was implemented nationwide from 1979, and the long-term testing of cultivated land, there have been reports of a decline in levels of soil available nitrogen, which has led to an awareness of declining soil fertility in paddy fields, as mentioned by Sumida et al. (2005). On the other hand, no change trends were identified on a national level in terms of physical condition. For example, within Kyushu and Okinawa, reports from Nagasaki Prefecture showed a high soil bulk density in surveys from 1994–1998, as well as an increase in the soil coefficient of permeability in surveys from 2009–2013 (Inoue et al. 1999). Furthermore, reports from Miyazaki Prefecture showed declines in the solid ratio, bulk density, and compactness as well as a rise in the air ratio during 1984–1989 (Akagi et al. 2000). In the Kagoshima Prefecture, no changes in bulk density were observed during the 30-year period before 2007 (Nishi et al. 2013). Although these reports did not exclusively find unfavorable phenomena, in terms of productivity, there are concerns that the worsening physical conditions could be causing a decline in soil fertility.

Overwetting is one of the major problems for the yield for upland crops cultivated in paddy fields. Although topography and hydrographical conditions are difficult to control in a field, double cropping (rice and wheat) has been implemented in northern Kyushu because of the warm climate. Yoshida and Adachi (1985) carried out a detailed soil survey in a deltaic plain between a natural levee and a hill. They confirmed a mutual dependence among the topography, hydrographical condition, soil distribution, and the growth and yield of wheat. Even though the utiliazation rate of paddy fields (total planted area of crops in paddy fields/paddy field area × 100) has been declining in recent years, this figure in Kyushu area was nevertheless 113% in 2016, which is clearly different from other areas that have ratios under 100% (Kyushu Regional Agricultural Administration Office 2018). Odahara et al. (2012) reported that in areas where either wetland rice or soybean was planted in the summer and wheat varieties were planted in the winter, soybean yield decreased as the number of soybean plantings increased (increasing upland usage ratio), which had a strong positive correlation with the surface soil macropore ratio. Although it was previously considered that the physical condition of surface soil changes dramatically due to tilling, those fundamental changes were difficult to observe on the farmer's fields; this report uncovered only one physical aspect of soil fertility decline. As planting is also carried out in the winter in this area, this could cause dryness to be more severe in the cultivated land than in other areas. These changes occurring in the paddy fields of northern Kyushu give us a glimpse of the possible future conditions of paddy fields north of Kyushu, where cultivation is expected to continue. While Japan's paddy fields have been assessed to require both drainage and flooding functionality, this raises the question of how to maintain the unique permanence of paddy fields while managing functionality to increase the production of paddy–upland rotation fields, as well as how to revitalize cultivated land that is in danger of losing its permanence.

10.3 Typical Agricultural Practices and Soil Management

10.3.1 Agricultural Production

As in other regions in Japan, agriculture in the Kyushu–Okinawa region, southwest Japan, includes diverse sectors such as the production of paddy rice, arable crops, vegetables, and fruits, as well as livestock farming (dairy farming and meat production). In 2015, the total agricultural output on a monetary basis in Kyushu (excluding Okinawa Prefecture) accounted for 19.8% of total Japanese domestic agricultural output (Kyushu Regional Agriculture Administration Office 2017a). Of the prefectures in Kyushu and Okinawa, the total agricultural output is highest in Kagoshima (which ranks third among all 47 prefectures in Japan), Miyazaki (fifth), and Kumamoto (sixth) Prefectures. The farmland area per household in Kyushu is increasing year by year and reached 2.11 ha in 2016. Agriculture in the Kyushu–Okinawa region, being influenced by a warm climate and ample rainfall, can be roughly divided into a paddy rice (*Oryza sativa* L.) region in northern Kyushu, a livestock farming area in southern Kyushu, and a sugarcane (*Saccharum officinarum* L.)-based subtropical region in Okinawa Prefecture and part of Kagoshima Prefecture.

Paddy rice is cultivated in the whole of the Kyushu–Okinawa region. In particular, the contribution of agricultural output from paddy rice production to the total agricultural output is high in Fukuoka and Saga Prefectures in northern Kyushu and Kumamoto Prefecture in central Kyushu (Kyushu Regional Agriculture Administration Office 2017a). Due to long-term decreasing trends in per capita rice consumption in Japan, soybean (*Glycine max* L. Merril) is widely grown instead of rice in paddies. Additionally, in the main rice-producing areas, which include Fukuoka and Saga Prefectures, wheat (*Triticum aestivum* L.) and barley (*Hordeum vulgare* L.) are grown as a winter crop after rice or soybean is harvested. As a result, the production of soybean, wheat, and barley is high in the rice-producing areas in northern Kyushu. The warm climate and ample rainwater allow farms to produce early delivering rice in southern Kyushu and to perform the double cropping of rice in Okinawa Prefecture. However, since around 1990, extremely high temperatures in the rice-growing season have often decreased the quality of rice grains (e.g., whitish immature grains) (Morita 2008). As a countermeasure to these high temperatures, the cultivation of new rice varieties resistant to high temperatures, such as "*Nikomaru*," is increasing in northern Kyushu. As Kyushu is geographically close to the Asian continent, rice producers there often suffer from serious virus-mediated disease spread by migratory planthoppers flying from the continent (Otsuka et al. 2010).

As mentioned earlier, the production of soybean, wheat, and barley is high in rice-producing areas in northern Japan. Since soybean is cultivated in paddy fields, the high moisture content in the soil caused by a long spell of rainy weather limits the soybean grain yield (Tasaka and Sasaki 2005). Wheat grain produced in northern Kyushu is mainly used for noodles. Many efforts are being made to increase domestic wheat production, such as the use of new wheat varieties for bread making (Tasaka and Sasaki 2005). Sweet potato (*Ipomoea batatas* L.) is also an important crop produced in arable land in southern Kyushu, where volcanic ash-derived soils are widely distributed and typhoons occur frequently. Sweet potatoes produced in southern Kyushu have three main uses: (1) table use (grocery store stock); (2) material for starch production; and (3) material for *shochu* (a Japanese distilled alcoholic beverage) production. Due to a recent rise in the consumption of shochu in Japan, the production of sweet potatoes used for this purpose has increased. In subtropical regions such as Okinawa Prefecture and the Amami Islands (part of Kagoshima Prefecture), sugarcane is a staple crop. Approximately, 70% of farm households in Okinawa Prefecture cultivate sugarcane (Okinawa General Bureau 2017).

Vegetables are widely produced across Kyushu and Okinawa. In particular, agricultural output from the vegetable sector is high in Fukuoka, Kumamoto, and Miyazaki Prefectures (Kyushu Regional Agriculture Administration Office 2017a). Vegetables with a high national production share in Kyushu include green pepper (*Capsicum annuum* L. var. *grossum*), tomato (*Solanum lycopericum* L.), taro (*Colocasia esculenta* (L.) Schott.), Japanese white radish (*Raphanus sativus* L. var. *longipinnatus*), and cucumber (*Cucumis sativus* L.). For fruits, a wide variety of citrus fruits such as mandarin oranges (*Citrus unshiu* (Swingle) Marcow.) are produced in Kyushu, which accounts for approximately 30% of citrus production in Japan (production year of 2016). The production of tea (*Camellia sinensis* (L.) Kuntze) is also at its most active across Kyushu, particularly in Kagoshima Prefecture. The Kyushu region accounts for about 40% of domestic tea production (production year of 2016).

Livestock farming (dairy farming and meat production) is the largest agricultural sector in Kyushu (Kyushu Regional Agriculture Administration Office 2017a). In Kagoshima and Miyazaki Prefectures, in southern Kyushu, the agricultural output from the livestock farming sector exceeds 60% of the total prefectural agricultural output. The total number of beef cattle, pork pigs, and broiler chickens is higher in Kyushu than in other regions in Japan.

10.3.2 Paddy Farming System in Northern Kyushu

The ratio of paddy field to arable land is 58% in the whole of Kyushu, but about 80% in Saga and Fukuoka Prefectures, in northern Kyushu (Table 10.2). Paddy fields are mainly located in lowlands, as mentioned in Sect. 10.1.2, where Fluvic soils are mainly distributed, accounting for 75% of paddy fields in Kyushu and 90% in Saga and Fukuoka Prefectures (National Agriculture and Food Research Organization 2017). The main crops grown in paddy fields are rice, soybean, wheat, and barley. In northern Kyushu, double cropping with block rotation of these crops is popular. The area of rice cultivation excluding for feed production is 165,700 ha in the whole of Kyushu. Paddy fields that rice is cultivated account for 53% of the total cultivated paddy fields in the whole of Kyushu. Lately, the use of rice for whole crop silage and as a feed grain has been encouraged by national policy, the production areas of those have increased to 23,000 ha and 6700 ha, respectively. The cultivation of soybean in paddy fields rather than rice is also politically encouraged. The area of soybean cultivation is 22,200 ha. The ratio of soybean-cultivated area to paddy field area is 7% in Kyushu, but is 20% in Saga Prefecture and 13% in Fukuoka Prefecture. Soybean is cultivated mostly in northern Kyushu, generally once every two or three years. The average yield of soybean for recent five years between 2009 and 2015 excluding the lowest and highest year was between 1.9 and 2.1 t ha^{-1} in Saga and Fukuoka Prefectures and about 1.5 t ha^{-1} for the whole of Kyushu. Wheat and barley are cultivated in paddy fields in winter. The total area of wheat and barley cultivation (excluding for feed production) is 55,400 ha. The ratio of wheat and barley cultivated area to paddy field area is 18% in Kyushu, 49% in Saga Prefecture, and 33% in Fukuoka Prefecture. Wheat and barley are cultivated mostly in northern Kyushu, in proportions of around 60% and 40%, respectively. The paddy field cultivation utilization rate (percentage of the average number of cultivated paddy fields in a year) is 102% in Kyushu, 144% in Saga Prefecture (the highest in Japan), and 121% in Fukuoka Prefecture (the second highest). Fukuoka and Saga Prefectures are ranked second and third in Japan, respectively, in terms of wheat output, and third and fourth in Japan for soybean output. The productivity of paddy fields of northern Kyushu is high.

The quality of rice grains is often more important than rice yield. Excessive nitrogen application often deteriorates rice palatability, and consequently, farmers attempt to practice minimal fertilization. On the other hand, rice grain ripening and development is often deteriorated in years with high temperatures in low-fertility soils. To avoid such deterioration, soil fertility management, suitable depth tillage, and the improvement of soil physical properties are important (Morita et al. 2016). A decline in soil fertility caused by an increase in soybean cultivation has been a concern (Odahara et al. 2012; Inoue et al. 2014). The

Table 10.2 Area of paddy field and crops cultivation by prefecture in Kyushu in fiscal year 2016

	Fukuoka	Saga	Nagasaki	Kumamoto	Oita	Miyazaki	Kagoshima	Total
Area of cultivated field (ha)	83,900	52,600	48,000	1,12,000	56,100	67,600	1,20,400	5,40,600
Area of paddy field (ha)	66,500	42,800	22,000	67,700	40,000	36,700	38,600	3,14,400
Ratio of paddy field (%)	79	81	46	60	71	54	32	58
Area of rice cultivation (ha)	36,000	24,800	12,000	33,800	21,300	16,800	21,000	1,65,700
Ratio of rice cultivation (%)	54	58	55	50	53	46	54	53
Area of soybean cultivation (ha)	8430	8370	438	2680	1720	261	355	22,200
Ratio of soybean cultivation (%)	13	20	2	4	4	1	1	7
Area of wheat and barley cultivation (ha)	21,700	20,800	1890	6950	4900	180	210	56,600
Ratio of wheat and barley cultivation (%)	33	49	9	10	12	0	1	18

Data Source Ministry of Agriculture, Forestry, and Fisheries (2016b, 2017a, b, c)

adequate application of compost is effective for maintaining soil fertility (Kitagawa et al. 2009; Fujitomi et al. 2014) and also enables the reduction of fertilizer application (Haruguchi et al. 2009; Fukuda et al. 2001; Ishii et al. 2016). As livestock manure is not sufficiently applied in paddy fields in northern Kyushu, it is encouraged to incorporate rice straw and wheat straw into paddy soil without burning to maintain soil fertility (Kitagawa et al. 2009). In Saga Prefecture in 2016, half of rice straw was removed from paddy fields, for various purposes, and half was plowed into the soil. The percentage of rice-cultivated fields in which rice straw was burned was just 4% in the prefecture. Compared with rice straw, the usage of wheat and barley straws out of fields is not common. These straws are plowed in 80% of their cultivated field. The percentage of cultivated fields in which straw was burned was about 10%. Plowing straw suppresses the initial growth of crops, but also suppresses weed incidence and increases rice yield by improving the ripening (Hideshima et al. 2016).

To allow two crops a year to be produced in northern Kyushu, it is important to keep the soil fertility high. In order to achieve this without increasing cultivation costs, various locally unutilized organic matter should be applied. Therefore, it is necessary to estimate the (highly variable) fertility of paddy fields to which various organic matters are applied, and to establish a fertilization method to give high-quality and high-yield products in individual paddy fields.

10.3.3 Vegetable Production on Isahaya Bay Reclaimed Land

(1) **Agriculture on reclaimed land**

Isahaya Bay is located on the northwest coast of the Ariake Sea in northwest Kyushu. The surrounding low-lying area is geographically and topographically susceptible to typhoons and torrential rain. This area has been affected by natural disasters many times, such as the Great Flood of Isahaya, which took place on 25 July 1957. Therefore, in order to enhance the effectiveness of regional disaster prevention and to develop large amounts of flat farmland with higher productivity, the National Project of Isahaya Bay Land Reclamation, which was conducted between 1989 and 2008, implemented a double dike system of land reclamation in which both ocean and land are blocked. This project developed 672 ha of farmland, 12 ha of housing land, 2600 ha of flood regulation pond, 126 ha of bank land, and 132 ha of road and canal land. Upland irrigation was introduced using water in regulated pondage in all fields. This allows for large-scale vegetable farming and horticultural enterprise. The farmlands are leased to immigrants, who were selected from various applicants.

(2) **Features Isahaya Bay reclaimed soil**

The reclaimed land of Isahaya Bay is located on a sea coast, and its soil parent materials are marine deposits. Most soils are classified as Gley Fluvic soils immediately after land drying, being rich in smectite-a 2:1 type clay minerals and classified as clayey in soil texture. The soil has a high stickiness and low permeability, and consequently, drainage systems have been installed at 10-m intervals in fields. The soil has a high nutrient-holding capacity, high levels of exchangeable bases, and is rich in available phosphate, however has a low organic matter content. In particular, its high concentration of chloride ions and alkaline soil pH has caused concerns about crop cultivation.

(3) **Soil management up to the time when farming starts**

Terai et al. (2006) examined the desalinization and drainage measure of the heavy reclaimed soil and found that the desalinization to a level that made upland farming possible by green manure cultivation for two years or during four crop seasons in conjunction with field drainage using supplementary drainage (Table 10.3).

This achievement was possible due to the following reasons: (1) An annual rainfall of approximately 2000 mm; (2) the increase of evapotranspiration caused by the green manure crop; and (3) the occurrence of large-scale soil cracking caused by the shrinkage of clay during drying.

(4) **Soil management after the initiation of farming**

After desalinization and the improvement of permeability on all farmland, farming was started in the Isahaya Bay reclaimed land in April 2008. In 2016, the total growing area was 1117 ha, with onions, lettuce, wheat (*Triticum aestivum* L.), barley (*Hordeum vulgare* L.), carrots, cabbages, broccoli (*Brassica oleracea* var. *italica*), and perilla (*Perilla frutescens* var. *crispa*) being cultivated. The local government provides a guidance of large-scale environmental-friendly agricultural practices in the Isahaya Bay reclaimed land (Nagasaki Prefecture 2013). Farmers are encouraged to utilize organic fertilizers such as oil cake, cattle manure compost pellets, and poultry litter to halve nitrogen application.

Table 10.3 Soil chemical properties after green manure four season cropping. (survey in 2002) Translated in part from Yamada et al. (2007) with permission from the Nagasaki Agricultural and Forestry Technical Development Center

Sampling date	Cropping system	Soil depth	Air-dried soil		w-Cl[a]	ESP[b]	Total carbon
		(cm)	pH (H_2O)	EC (1:5)			
			(1:2.5)	(dS m^{-1})	(mg kg^{-1})	(%)	(%)
Jul. 2000	Before seeding	0–15	7.5	1.80	463	22.9	1.52
		15–35	7.6	3.60	4336	66.7	1.53
		35–50	8.0	5.60	8802	91.2	1.53
May 2002	Sorghum–barley	0–16	6.8	0.37	100	6.9	1.67
		16–38	6.4	1.24	727	13.0	1.58
	Sorghum-Italian ryegrass	0–12	6.5	0.32	68	3.3	2.14
		12–22	6.6	0.55	90	10.5	1.72
		22–43	6.6	1.32	768	24.1	1.55
Soil diagnosis standard			6.0–6.5	≦0.3	≦100	≦10	1.72≦

[a]Water-soluble chloride ion concentration
[b]Exchangeable sodium percentage

After ten years of farming, the groundwater level in the soils was lowered, and the soil became Gray Fluvic soils. The soil content of available phosphate and exchangeable calcium was observed to be reduced.

10.3.4 Horticulture and Soil Management in Northern Kyushu

The greenhouse cultivation of vegetables such as strawberry and asparagus is successfully conducted in northern Kyushu.

(1) **Strawberry**

The total area of strawberry cultivation in Kyushu is 1477 ha, accounting for 27.1% of the agricultural land area in Kyushu and 33% of strawberry production in Japan. This cultivation obtains a high productivity of 3.57 t ha^{-1}. Although soil culture is predominant in northern Kyushu, table top culture was developed ahead of Japan due to the requirements of saving labor (Fig. 10.10).

Strawberry can be cultivated in any soil, but prefers that with suitable air permeability and water-holding capacity. Strawberry is shallow-rooted, and its roots have a large oxygen demand. Since continuous rotary tilling tends to make the topsoil layer shallow, the effective soil layer must be expanded by forming high ridges or conducting deep tillage.

The appropriate range of the soil chemical properties for strawberry cultivation is 5.5–6.5 for pH, 45–75% for base saturation percentage, and 200–600 mg P_2O_5 kg^{-1} for available phosphate concentration. The appropriate range of electrical conductivity (EC) is 0.3–0.5 dS m^{-1} (the upper limit is 1.0) for volcanic ash soils and 0.2–0.4 dS m^{-1} (the upper limit is 0.8) for non-volcanic ash soils. Strawberry is one of the most sensitive crops to salt injury, and soil solutions with low salt concentration are therefore suitable for strawberry roots. In the case of soil culture, as the strawberry variety changed from "Toyonoka" through "Sachinoka" to "Yumenoka," the rate of nitrogen fertilizer application for the basal dressing has been reduced, which has resulted in the emphasis of side dressing.

(2) **Asparagus**

The area of agricultural land under asparagus cultivation in Kyushu is 461 ha, producing an annual yield of 9178 t, accounting for 8.4 and 31.5%, respectively, of the total in Japan. Asparagus productivity is extremely high, at 19.9 t ha^{-1} (Inoue 2008). Asparagus used to be cropped only in early spring because the microthermal climate in this period is suitable; however, in about 1990, mother stem standing cultivation was developed in northern Kyushu, which allowed for cropping up to October, thus increasing the number of crops per unit area fivefold or more (Inoue 2001). This is the greatest technical revolution that has been achieved for asparagus cultivation, and the technology is becoming widespread both domestically and internationally.

Green asparagus semi-forcing cultivation is a type of mother stem cultivation in which a few stems per stock are elongated at the end of the crop of buds in spring and buds that sprout on the base of the standing stems in summer and autumn are cropped (Fig. 10.11; Inoue 1996). It is important to nurture stumps and encourage the accumulation of

Fig. 10.10 Table top cultivation system for strawberry. The photo was provided by Kastuhiro Inoue

Fig. 10.11 The mother furn cultivation of Asparagus. Reprinted from Inoue (2019), with permission from the author

nutrients into storage roots after cropping is completed in order to increase yield the next spring. Therefore, it is necessary to establish appropriate annual cultivation management technology, which allows for high-yield stable production for ten years or more.

Basic properties of asparagus include high-salinity tolerance, optimum growing pH of 6.0–6.5, and low acid resistance. Farmers apply a large amount of nitrogen, 50 g m^{-2} or more as chemical fertilizer, in fields where 10 kg m^{-2} of fully matured compost of is applied, and the split application of fertilizer that uses base fertilizer before crop of buds in spring and continuously uses additional fertilizer during the crop of buds in spring is widely used (Inoue 2005). As a result, the chemical properties of the surface soil include an EC of 0.93 dS m^{-1}, a humus content of 6.5%, an available phosphate content of 2460 mg P_2O_5 kg^{-1}, a base saturation percentage of 130%, and an exchangeable potassium content of 3.8 $cmol_c$ kg^{-1}, all of which are high; this indicates significant accumulation of nutrients.

10.3.5 Horticulture in Southern Kyushu

In this section, the three prefectures in southern Kyushu (Kumamoto, Miyazaki, and Kagoshima) and Okinawa Prefecture are collectively referred to as southern Kyushu. In 2008, the growing area of greenhouse crops in southern Kyushu was 84.5 km^2 (vegetables: 61.4 km^2; fruits:

12.9 km^2; flowers: 10.2 km^2) (Ministry of Agriculture, Forestry, and Fisheries 2015). The major greenhouse crops in this region are tomato (*Solanum lycopersicum*), melon (*Cucumis melo L*), watermelon (*Citrullus lanatus*), cucumber (*Cucumis sativus* L.), strawberry (*Fragaria* L.), sweet pepper (*Capsicum annuum* L.), chrysanthemum (*Chrysanthemum* L.), and citrus fruits inherent to this area such as "Banpeiyu" pummelo (*Citrus grandis, Citrus maxima*) and "Shiranui" mandarin ((*Citrus unshiu* × *C. sinensis*) × *C. reticulata*). Fruit trees of tropical origin such as mango (*Mangifera indica*) are also cultivated in greenhouses.

The physicochemical properties of Japanese soils were investigated nationwide in the Soil-Environmental Monitoring Project from 1979 to 1998 (Obara and Nakai 2004). This survey identified the excessive accumulation of exchangeable bases and phosphorus in the soils of greenhouse crops. Due to the extreme increase in the prices of phosphate ores for fertilizer in 2008, the amount of phosphate fertilization was subsequently reduced. Regarding fertilizer components other than phosphorus, efforts to make application rates more appropriate have been addressed by local governments. The proper fertilization values for crops including tomato, watermelon, eggplant (*Solanum*), sweet pepper, okra (*Abelmoschus esculentus*), chingen sai (*Brassica rapa var. chinensis*), lettuce (Lactuca sativa), orange (*Citrus unshiu*), mango, and chrysanthemum are published on the website of each prefecture (e.g., Okinawa Agricultural Research Center 2015; Mimaki 2016).

In the cultivation of tomato, an important crop in southern Kyushu, active fostering and suppressive cultivation modes are conducted from autumn to spring. Tomato yellow leaf curl disease and other soil diseases and pests are likely to result from repeated and long-term cultivation (Nagatomo 2004). In tomato and other crops, "soil solarization," in which phytopathogens in the soil are killed by raising the soil temperature by covering the soil surface with transparent film in the summer, has been used practically (Shiraki 2007). In Miyazaki Prefecture, a method to stabilize the disinfection effect of soil solarization and to reduce the use of chemical fertilizer by applying some condensed grain distillers soluble of "*shochu*" spirit prior to soil solarization was documented (Miyazaki Agricultural Research Institute 2016b).

In order to increase the productivity of greenhouse crops, in recent years, facilities equipped with CO_2 gas application systems, drip irrigation and liquid fertilization systems, and so on, have been established. Large-scale facilities are in operation in Kumamoto City, and fertilization methods adapted to new cultivation environments are also studied (Kumamoto Prefectural Agricultural Research Center, 2017).

10.3.6 Upland Crop Production in Southern Kyushu

Southern Kyushu, which extends more than 600 km from south to north, contains Kagoshima and Miyazaki Prefectures, which have average yearly temperatures of 19.2 °C and average annual rainfall of approximately 2000–3000 mm. Agricultural production in both prefectures mainly includes sweet potato (*Ipomoea batatas* L.), vegetables, livestock farming, flowering plants, and tea (*Camellia sinensis* (L.) Kuntze.) by taking advantage of the mild weather and vast area of agricultural fields. This region is distant from mass consuming regions, and crops such as potatoes (*Solanum tuberosum* L.), taro (*Colocasia esculenta* (L.) Schott), and fava beans (*Vicia faba* L.) are shipped in a relay system.

Upland fields in Kagoshima Prefecture consist of Andosols, which account for 67% of all upland fields, Red-Yellow soils (14%), and Eutrosols (8%). Upland fields in Miyazaki Prefecture are located on hilly areas generated in the Paleogene period, Pleistocene terraces, and the Shirasu Plateau in the southwest of the prefecture, and Andosols with a base material of volcanic ash account for 90% of these fields. Additionally, a very small amount of dune immature soil and Brown Fluvic soils are distributed along river banks and in river basins, respectively.

According to national and prefectural soil monitoring projects conducted during 1979–2008, the content of exchangeable calcium, as measured during six survey cycles, had a decreasing trend in vegetable fields and was almost constant in staple crop fields and forage crop fields. The soil content of exchangeable magnesium also showed a decreasing trend. The content of exchangeable potassium was constant in staple crop fields and vegetable fields and tended to decrease in other fields. The content of available phosphate showed a leveling off trend in staple crop fields and a rising trend in vegetable, forage crop, and tea fields (Fig. 10.12; Nishi et al. 2013).

The content of soil available phosphate has been increasing in southern Kyushu, and it is excessively accumulated. This has been causing concern about a decline in productivity caused by crop physiological disorders, and so on. Implementing proper fertilization while taking into consideration, the nutrients that have accumulated in soils allow the maintenance of soils and the reduction of fertilization cost.

Nagatomo et al. (2017) considered the proper amount of fertilizer for phosphate in Chinese cabbage (*Brassica rapa* var. *pekinensis*) in Andosols. They set up two plots with different levels of soil phosphate fertility—100 and

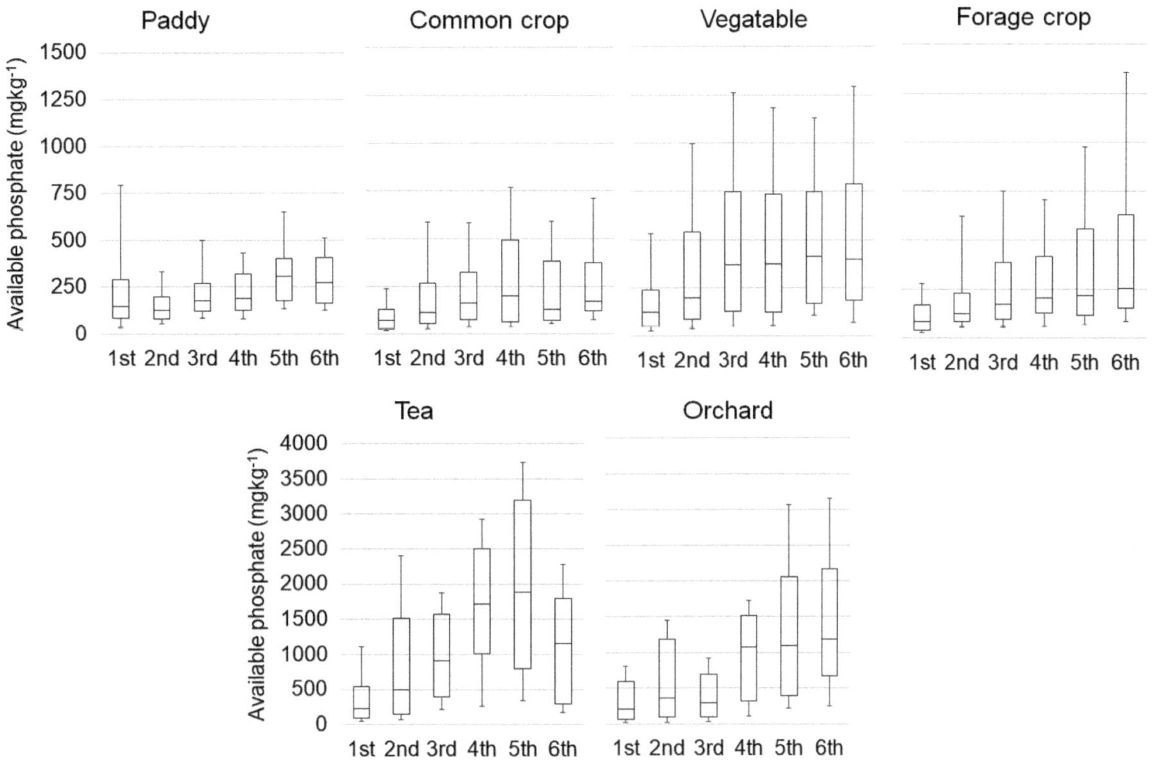

Fig. 10.12 Changes in available phosphate contents in soil for each field category. Modified from Nishi et al. (2013), with permission from Kagoshima Prefectural Institute for Agricultural Development

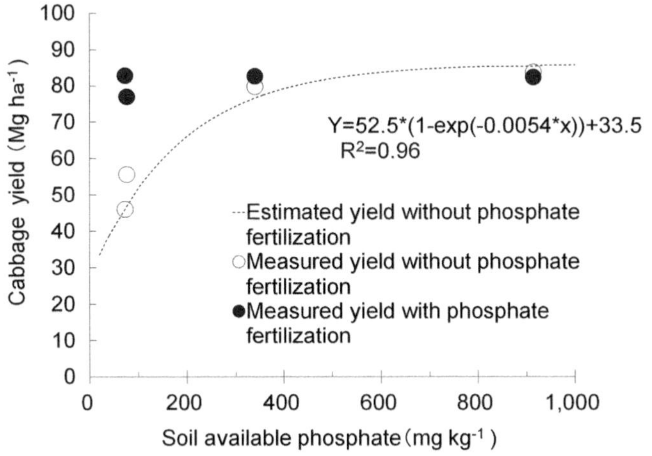

Fig. 10.13 Chinese cabbage yield as a function of soil available phosphate. Reprinted with translation from Nagatomo et al. (2017) Fig. 1 with permission from Japanese Society of Soil Science and Plant Nutrition

1000 mg kg^{-1}, then two phosphate fertilization treatments, 0 and 200 kg ha^{-1}, were conducted in each plot for ten years. Figure 10.13 shows the relationship between phosphorous application rate and yield of Chinese cabbage for the different phosphorous fertility levels. As Fig. 10.13 shows, the yield increase is gradually decreased with the phosphate fertility increasing. The result shows that the soil content of available phosphate of approximately 500 mg kg^{-1} or more showed the small effect of fertilization and allowed the reduction of phosphate fertilization.

10.3.7 Soil Management in the Nansei Islands

The main crops that are cultivated in the Nansei Islands are sugarcane (*Shaccharum officinarum* L.), potato (*Solanum tuberosum* L.), and sweet potato (*Ipomoea batatas* L.), the harvested or planted area of which is 23,000 ha, 1800 ha, and 1000 ha, respectively. The other main cultivated crops are leaf tobacco (*Nicotiana tabacum* L.; 950 ha), chrysanthemum (*Chrysanthemum morifolium* Ram.; 830 ha), and rice (*Oryza sativa* L.; 800 ha).

The main soils of the Nansei Islands are Argic Red-Yellow soils (local name: "*Kunigami mahji*"), Calcarious Eutrosols (local name: "*Shimajiri mahji*"), and Marlitic Terrestrial Regosols (local name: "*Jahgaru*"), while the main soils of the islands of Tanegashima and Yakushima, located in the north part of the Southwest Islands, are Andosols and Brown Forest soils, respectively (Obara et al. 2011, 2016). The Argic Red-Yellow soils distributed in the Nansei Islands are acidic, and their soil texture is clay loam to light clay. The Calcarious Eutrosols are acidic to alkaline and their soil texture is mainly heavy clay. The Marlitic

Terrestrial Regosols contain free calcium carbonate, resulting in a soil pH of about 8, and their soil texture is mainly light clay to silty clay. The common characteristics of these soils are their low organic matter contents.

The conventional fertilizer application levels for sugarcane are 200–300 kg ha^{-1} of nitrogen, 60–120 kg ha^{-1} of phosphate, and 60–100 kg ha^{-1} of potassium, although there are some differences depending on region and cropping type (Okinawa Prefecture 2014). According to the Soil-Environmental Monitoring Project conducted between 1979 and 1999, the available phosphate contents in sugarcane fields increased from 200 to 600 mg kg^{-1} over these 20 years (Kuniyoshi and Gima 2004). Accordingly, the reduction of phosphate fertilizer has been recommended.

The main soil chemical and physical properties affecting sugarcane yield in the Nansei Islands are soil available nitrogen level and plow layer depth (Yoshida et al. 2016). The soil available nitrogen content is significantly lower in the Nansei Islands than in other parts of Japan; the average soil available nitrogen level in Japanese upland soils is 53.0–59.0 mg kg^{-1}, while that in the sugarcane fields in Kitadaito Island (Nansei Islands) is 13.6 mg kg^{-1} (Yoshida et al. 2016).

Generally, the clay contents of soils in the Nansei Islands are high, and the subsoils tend to be easily compacted (Tokashiki 1993). Consequently, hardpan forms under the weight of large machinery (such as cane harvesters), which has harmful effects on sugarcane growth (Kameya 2006). It is important to increase the plow layer depth by applying organic matter such as compost or green manure or by performing subsoiling or deep tillage for soil management in the Nansei Islands.

Studies on the long-term application of organic matter in a sugarcane field showed that the total soil carbon content increased and the sugarcane yield increased by 10–22% after the application of 50 Mg ha^{-1} or more (Goto and Nagata 2008). Moreover, Miyamaru et al. (2012) reported that the carbon and nitrogen contents of soil microbial biomass significantly increased after the application of organic matter, which increased the level of soil available nitrogen in upland soils. Furthermore, the application of cane molasses, a byproduct of sugarcane production, also increased the levels of soil available nitrogen and improved the sugarcane yield by 20–30% (Yoshida et al. 2017). It was suggested that easily decomposable organic matter such as sucrose in molasses increased the amount of microbial biomass and as a result increased the level of soil available nitrogen.

The destruction of the hardpan layer is generally performed by subsoiling. Additionally, it is reported that subsoiling increases sugarcane yield by 12% (Oshiro and Hamakawa 1980). It has also been noted that subsoiling improves the soil water-holding capacity and drainage conditions as well as softening the soil (Onaga and Gibo 1984; Shinzato et al. 2013). Sugarcane is a deep-rooted crop and the effect of irrigation on the yield is therefore high, although in some cases negative effects on sugarcane growth are seen due to poor drainage. Therefore, subsoiling is suitable as a soil management method for sugarcane fields.

10.3.8 Globally Important Agricultural Heritage Systems

The modern agricultural system has succeeded in supplying food for a growing population but has also resulted in environmental problems such deforestation, water pollution, climate change, and so on throughout the globe, causing the extinction of unique local cultures and the destruction of landscapes and biodiversity. Nevertheless, it is evident that in many parts of the world, unique systems of agriculture have been maintained. Traditional agriculture, including knowledge systems, adapted technologies, landscapes, and biodiversity, continues to be protected from the adverse effects of modernization.

The Food and Agriculture Organization of the United Nations (FAO) is designating certain areas as globally important agricultural heritage systems (GIAHS) in order to maintain and preserve such agricultural systems in an integrated manner for the next generation. GIAHS are defined as "remarkable land use systems and landscapes which are rich in globally significant biological diversity evolving from the co-adaptation of a community with its environment and needs and aspirations for sustainable development" (FAO 2002).

As of 2018, 50 sites in 20 countries in the world, 11 of which are in Japan, had been designated as GIAHS. Three sites have been designated in Kyushu: the Aso Grasslands, the Kunisaki Peninsula Usa area, and the Takachihogo-Shiibayama area. The following subsections introduce the soil management in these sustainable agricultural systems.

(1) The Aso Grasslands

The Aso Grasslands in Kumamoto prefecture, located in central Kyushu, is the largest grassland area in Japan, covering about 220 km^2 (Miyabuchi and Sugiyama 2008). The Aso Grasslands are a secondary (or semi-natural) grassland composed mostly of *Miscanthus* (Japanese pampas grass), *Pleioblastus* (warm temperate dwarf bamboo), and *Zoysia* (Miyabuchi and Sugiyama 2016). The Aso Grasslands continue under the balance of receiving artificial disturbance and natural reformation. Since the Aso Grasslands are connected with both livestock rearing and local farming through the use of the biomass of the grasslands as green manure and compost, the region has been jointly managed by every village and sustainably maintained for a long time by mowing, grazing, or burning. The burning-off of fields is customarily carried out in early spring (Aso regional GIAHS Executive Committee 2014).

A description of the Aso Grasslands is given in the "*Nihon Shoki*," which was edited around the eighth century, and the "*Engishiki*," edited around the tenth century, which shows that the grasslands have been used as a pasture since around these times (Otaki 1997). From these ancient documents, we can interpret that the Aso Grasslands have existed for over one thousand years. However, Ogura et al. (2002) suggested that the Aso Grasslands have been intentionally burned for over 10,000 years based on the analysis of microscopic charcoal in soil samples from the northern caldera rim of Aso volcano. Similarly, Miyabuchi and Sugiyama (2008) also suggested that sustainable maintenance of the Aso Grasslands by burning has continued over 10,000 years based on a phytolith analysis of a tephra sequence.

The soil type of the Aso region is Andosol. Andosols are formed from tephras or pyroclastic materials (Shoji et al. 1993). Matsuyama and Saigusa (1994) showed that Allophanic Andosols that have a clay fraction dominated by allophane and imogolite are mainly distributed, while Non-Allophanic Andosols are scattered, in this area. Additionally, Kubotera et al. (2015) analyzed the top 15 cm of soil around the Aso region and showed that Non-Allophanic Andosols were densely distributed on the northwestern Aso somma. In Non-Allophanic Andosols, the clay fractions are mostly dominated by 1:2 clay minerals and the active Al consists largely or wholly of Al complexed with humus (Shoji et al. 1985). In Allophanic Andosols, aluminum toxicity does not occur frequently, but the large amount of exchangeable Al in Non-Allophanic Andosols may lead to acid injury to plants due to aluminum toxicity (Saigusa et al. 1980; Matsuyama et al. 2005).

Andosols in the Aso region have high carbon contents of about 200 Mg C ha^{-1}. Toma et al. (2010) indicated that ash and charcoal formed by burning appeared to contribute to carbon accumulation in soil. Furthermore, Toma et al. (2013) suggested that the semi-natural, C4 (*Miscanthus,* etc.)-dominated Aso Grassland system serves as an important carbon sink and is worthy of future conservation.

The Aso Grasslands have adapted to this low-fertility soil and have contributed to local agriculture and the prevention of global warming.

(3) **Kunisaki Peninsula Usa**

The Kunisaki Peninsula Usa area in Oita Prefecture is located in the northeast of the island of Kyushu, and its distinct geographical features, ecosystem, and agricultural culture have been preserved and were consequently designated as a GIAHS by the FAO in 2013 (The Kunisaki Peninsula Usa GIAHS Promotion Association 2017). In particular, this area is characterized by low rainfall in winter, steep rivers that extend in every direction from the mountains, and coverage by soil that is easily penetrated by rainfall. These conditions pose a challenge for farmers, who are constantly struggling with water shortages when trying to irrigate rivers.

One of the important products in this GIAHS is dried log wood-cultivated shiitake mushroom (*Lentinula edodes*) that uses sawtooth oak (*Quercus acutissima*) as a log wood. This traditional cultivation system depends on sustainable forestry by coppicing to produce enough log wood. In broad-leaved forests in the mountainous part of the Kunisaki Peninsula Usa area, even when cut down in autumn, fully developed sawtooth oak trees have the characteristic of shooting up from the stump in the next spring and re-growing in 15–20 years with appropriate management (Fig. 10.14a and b). Log woods that hosted shiitake mushroom were covered by twigs so as to avoid sunlight and stacked on the mountainside for up to two years where the sprouting of sawtooth oak would take place again. During this period, the shiitake mycelium proliferates within log wood, which is finally taken from the mountainside to a scrap wood site where mushroom harvesting takes place (Fig. 10.14b).

As the stumps of sawtooth oak can survive for more than 100 years and their roots grow further and faster in the soils and absorb more nutrients year after year, the surface soil continues to be full of vigorous growth. Due to the abundance of fallen leaves that accumulate on the soil surface every year, dark Brown Forest soil occurs beneath the abundant sawtooth oak broad-leaved forests in the Kunisaki Peninsula Usa area which are rich in organic matter plus inorganic nutrients and have a high water-holding capacity that increases each year (Fig. 10.14c). By storing the rainwater that this sawtooth oak forest accumulates in roughly 1200 small-scale reservoirs, farmers can continue to farm paddy fields and carry out forestry and fishery, and the unique ecosystem can be permanently preserved (Hayashi 2014).

Fig. 10.14 Typical sawtooth oak forests and its soil profile in the Kunisaki Peninsula Usa Area. **a** Roughly twenty-year old sawtooth oak forest just before logging (Photographed 13 November 2016). This sawtooth oak forest has been repeatedly cut for log wood for Shiitake production for at least 60 years. **b** Reforestation of the same sawtooth oak forest after coppicing (Photographed 23 November 2017). Several sprouts vigorously grow from the cut stump. Log woods where Shiitake fungus was inoculated were stacked between stumps on the mountainside under the shading conditions indicated by the white arrow. **c** Soil profile under a stump that shows the dark brown A horizon (about 20 cm) and brown B horizon (about 40 cm). An abundance of leaf litter from sawtooth oak trees covers the soil surface. The white arrows show the roots of the sawtooth oak tree. The photo was provided by Hiroaki Hayashi

(4) **Takachihogo-Shiibayama**

The Takachihogo-Shiibayama GIAHS site is situated in the northwest of Miyazaki Prefecture and is composed of Takachiho Town, Hinokage Town, Gokase Town, Morotsuka Village, and Shiiba Village. The site is well known for many locations related to Japanese mythology. The site lies in the northern part of the Kyushu-Sanchi Mountains, whose peaks reach 1000–1700 m a.s.l. The annual average air temperature and precipitation are 14 °C and 2200 mm, respectively, although these vary depending on the elevation. Brown Forest soils and Andosols dominate the Takachihogo-Shiibayama site.

Due to the presence of the mountains, the extent of agricultural flat land is limited; 92% of the land is forest, while just 3% is used for agriculture (Fujikake 2017). A total of 58% of forest land is planted forest; 72% of this planted forest is Japanese cedar (*Cryptomeria japonica*), 12% is hinoki cypress (*Chamaecyparis obtusa*), and 11% is sawtooth oak (*Quercus acutissima*), the latter of which is used for shiitake mushroom (*Lentinula edodes*) production. Together, this planted forest and natural forest form a "mosaic-patterned" forest, since the area of each planted forest is small; this forest pattern is one of the characteristics of the Takachihogo-Shiibayama site. Another characteristic of the site is its rice terraces. As the rivers run through deep valleys, long irrigation canals in mountainous areas are necessary for rice terraces in the site (Takeshita 2017). Many major canals in the mountains were constructed from the middle of the nineteenth century to the beginning of the twentieth century, after which the area of rice terraces increased.

The complex agricultural and forestry system in the Takachihogo-Shiibayama site is relatively new. Before the establishment of the current system, the most important agriculture was shifting cultivation.

In the 1950s, the area of shifting cultivation in the Kyushu-Sanchi area was 2500 ha, but rapidly decreased

Table 10.4 Soil pH and mineral contents under shifting cultivation at Shiiba Village

Duration after burning (years)	pH (H_2O)	Soil mineral contents							
		NO_3–N	NH_4–N	Total-C	Total-N	Ex-Ca	Ex-Mg	Ex-K	Truog-P
		(mg/100 g soil)	(mg/100 g soil)	(%)	(%)	($cmol_c$/kg)	($cmol_c$/kg)	($cmol_c$/kg)	(P_2O_5 mg/100 g soil)
0.3	4.99	2.97	5.17	17.3	1.21	5.92	1.64	1.59	3.20
1.3	5.07	1.17	1.68	11.1	0.82	3.43	0.66	1.15	1.40
3.3	4.90	1.88	1.81	14.5	1.02	2.26	0.58	1.13	0.16
16	4.62	0.78	2.07	13.3	0.97	1.05	0.38	0.81	0.24
>40	4.56	1.73	4.91	20.5	1.38	2.35	0.55	1.29	0.20

after that time (Sasaki 1970); now, shifting cultivation remains in only a few places, including Shiiba Village. One of the reasons for the rapid decline in shifting cultivation in the site is thought to be cedar planting after shifting cultivation, which has been conducted since the 1940s (Shiiba and Uthumi 2010). According to the traditional method of shifting cultivation in Shiiba Village, the field is burned on a sunny day in early August, buckwheat (*Fagopyrum esculentum*) is cultivated in the first year, Japanese millet (*Echinochloa esculenta*) and foxtail millet (*Setaria italica*) are cultivated in the second year, and adzuki bean (*Vigna angularis*) and soybean (*Glycine max*) are cultivated in the third and fourth year, and the fallow period is about 30 years (Fujiwara 1998).

In this shifting cultivation, as in general shifting cultivation, soil pH and the contents of soil minerals such as P, Ca, Mg, and K are increased as a result of the ash produced by burning (Table 10.4) (Kondo and Saeki 2017). Due to these effects, the yield of crops can be maintained at a relatively high level. For example, the yield of Japanese millet, which was the most important crop in the site, was reported to be 1980 kg ha^{-1} (Fujiwara 1998).

Traditional rotation cropping might be a reasonable method judging by soil chemical analysis. In the first and second year, soil N and P contents were high, and so Japanese millet and foxtail millet, which were staple foods, as well as buckwheat, could be cultivated. In the third and fourth years, soil P content was relatively high although N content was low, and so adzuki bean and soybean, which can fix N, could be cultivated. On the other hand, in the fifth year, soil P content was low, and so the field had to be returned to fallow due to the difficulty of crop cultivation. Although the rotation cropping was conducted empirically, the method was reasonable and people were able to make a stable living from mountain land.

10.4 Agriculture and Environmental Issues

10.4.1 Manure Management and Composting

The Kyushu region holds the largest populations of beef cattle, swine, and broiler chicken in Japan; the national share of these livestock was 35.6, 31.5, and 48.1%, respectively, in 2016 (Kyushu Regional Agriculture Administration Office 2017b). Massive amounts of excrement are produced as by-products from livestock farms, which are densely distributed in Miyazaki and Kagoshima Prefectures, southern Kyushu (Table 10.5); however, most farms lack cropland area for the proper utilization of manure nutrients. Therefore, sustainable manure management for the prevention of environmental pollution is necessary, which involves proper

Table 10.5 Estimated Livestock excrements amount in Kyushu

Prefecture	Survey year	(*Unit* 1000 ton)	Destined to composting %
Fukuoka	2015	725	90.8
Saga	2015	829	89.8
Nagasaki	2014	1576	92.4
Kumamoto	2015	2821	N/A
Oita	2014	1012	70.0
Miyazaki	2014	4160	68.5
Kagoshima	2014	5685	68.1

Data source Fukuoka Prefecture (2017), Kagoshima Prefecture (2016), Kumamoto Prefecture (2016), Miyazaki Prefecture (2017), Nagasaki Prefecture (2017), Oita Prefecture (2016), Saga Prefecture (2016)

treatment followed by the distribution of manure as organic fertilizer.

Manure can be treated by techniques such as drying, anaerobic digestion, composting, and so on. Of these methods, composting is the most popular in Japan (Table 10.5), which makes the application of manure more manageable. Composting is generally defined as the biological oxidative decomposition of organic constituents in wastes under controlled conditions. A high temperature (greater than 60 °C) generated during manure composting contributes to killing weed seeds and pathogens, the evaporation of water, and the resulting production of sanitary compost.

(1) **Pelletized compost**

As mentioned in 8.2.3, compost granulation contributes to the correction of regional manure surpluses in concentrated livestock areas if the usable product is transported and applied to the soil where organic amendment is lacking for maintaining soil fertility.

A further advantage of compost granulation is that it results in a greater phosphorus (P) recovery rate (Arakawa 2012a, b). In a study, P availability between single superphosphate and two forms of compost, pelletized compost, a 1:1 mixture of cattle manure compost and poultry litter, and the corresponding ungranulated compost, was examined using komatsuna (*Brassica Campestris* L.) and buckwheat (*Fagopyrum esculentum* Moench) as indicator crops. Total P uptake increased in a curvilinear manner with increasing application rates for all three fertilizer materials, while the average P recovery rate was much higher with pelletized compost than with ungranulated compost for both crops, and was comparable to that of superphosphate. Meanwhile, the researchers observed the plant roots were winding outside of the compost granules. The factors affecting improved P efficiency were investigated by Takahashi et al. (2016). The fate of P was investigated in an incubation study using pelletized and ground pelletized poultry manure compost as amendments to an Andosol under five levels of moisture content. Compost amendment increased levels of Truog P and labile inorganic P in the soil, as earlier reported. A greater increase of these P fractions, which was mainly attributable to a decreased soil level of phytate-like organic P, was observed in the soil amended with pelletized compost compared with the soil amended with ground compost. The researchers also observed that P mineralization was enhanced under higher moisture content. The improvement in the availability of compost P in pelletized compost can be explained as follows: (1) Organic amendments prevent the rapid fixation of available P in soil by blocking P adsorption sites (Haynes and Mokolobate 2001); (2) phytate-like organic P in the compost may be satisfactorily mineralized under high moisture content of the pellets; (3) inorganic P inside of the pellets may be fixation-avoidable since it does not contact with soil colloids; then (4) crop roots may efficiently absorb P from the compost granules.

(2) **Compost-mixed compound fertilizer**

In accordance with the Fertilizer Regulation Act (Act No. 127 of 1950), it had been prohibited to fortify compost with normal fertilizers, such as nitrogenous mineral fertilizers. However, since 2014, a new official fertilizer specification, "compost-mixed compound fertilizer," was enacted, and now, it is possible to register and manufacture fertilizers using composted manure and/or composted food waste for fertilizer materials in the course of business. For this granulated fertilizer, the amount of nutrients are balanced, and the minimum amount of each nutrient is guaranteed and clearly indicated, which makes it usable in the same way as other compound fertilizers (Katoh 2017). Encouraging the increased use of this type of fertilizer in Japan is expected to reduce the dependence on imported fertilizer resources and reduce the cost of fertilizer in agriculture systems.

10.4.2 Nitrate Contamination of Groundwater

On Kyushu, welded tuff layers with a very low water permeability are mostly found on the edge of pyroclastic deposit areas, and rainwater is ample (more than 2000 mm of annual precipitation), and these geographical and meteorological conditions allow the storage of ample groundwater in large parts of the island. Groundwater is extremely important for domestic water supply in the southern part of Kyushu. In particular, the Kumamoto region and the Miyakonojo basin region depend on groundwater for almost 100% of their domestic water supply, which supports populations of about 1 million and 240,000, respectively.

Some of the Nansei Islands, such as Miyako, Okinoerabu, and Kikai Islands, are composed of limestones derived from coral. Since there is a layer with a very low water permeability under the limestone layer, groundwater is retained in the below ground part of those islands. Agriculture and daily life in those islands are largely dependent on groundwater.

Despite the importance of groundwater in communities in Kyushu and some of the Nansei Islands, nitrate (NO_3^-) contamination of groundwater has long been a local environmental issue. For example, in Kumamoto Prefecture, 21% of 370 wells surveyed in 2012 contained NO_3^- concentrations higher than 10 mg N L^{-1} (the water quality standard in Japan) (Kumamoto Prefecture 2013). High NO_3^-

concentrations in well water are particularly observed in the northern part of Kumamoto Prefecture, which contains arable land, horticulture, and livestock farming areas. Furthermore, in the Miyakonojo basin region in Miyazaki Prefecture, where cropping and livestock farming are the primary industries, NO_3^- concentrations in well water also exceeded 10 mg N L^{-1} in 16% of 782 wells surveyed in 1998 (Miyazaki Prefecture 2004).

Many studies have been conducted in wide areas of Kyushu to identify the mechanism of NO_3^- contamination in groundwater and to establish efficient countermeasures in agricultural areas. Lysimeter experiments conducted in Miyakonojo City showed that the amount of NO_3^- leaching in different soils was due to cumulative water permeability following fertilizer application (Kobayashi et al. 1994). ^{15}N-labeling experiments revealed that NO_3^- contamination in water in a well (18.5 m depth) could be projected by cumulative precipitation (4158 mm) in a farmland with long-term application of slurry from dairy cattle (Niimi 2002). By analyzing isotopes of nitrogen, oxygen, carbon, sulfur, and hydrogen in groundwater and by measuring dissolved oxygen and dissolved organic carbon contents in groundwater, the dynamics of NO_3^- in groundwater were studied in the Kumamoto region and the Miyakonojo basin region. As a result, the regional-scale transport and disappearance of NO_3^- in groundwater was elucidated (Mori et al. 2016). The information obtained was provided to administrators of local governments to establish viable countermeasures for NO_3^- contamination in groundwater.

In Kumamoto Prefecture, an item on NO_3^- was added to the "Kumamoto Prefectural Ordinance on Groundwater Conservation" in 2012 to promote the appropriate application of fertilizers and management of animal wastes to reduce the NO_3^- contamination of groundwater (Kumamoto Prefecture 2000). Additionally, Kumamoto Prefecture established a framework which awards certification to agricultural products produced under farming practices which contribute to groundwater conservation, thus allowing consumers to purchase such agricultural products selectively to support activities for NO_3^- reduction (Kumamoto Prefecture 2015).

In the Miyakonojo basin region, the "Council on Groundwater Conservation in the Miyakonojo Basin" was first organized by local municipalities and Miyazaki University in 1995 and continued to monitor NO_3^- concentration in groundwater (Miyazaki Prefecture 2004). In 2003, the Miyazaki prefectural government established a council to reduce NO_3^- contamination in groundwater in the Miyakonojo basin region and in 2004 announced concrete plans to reduce NO_3^- concentrations in all well waters in the region to below 10 mg N L^{-1} (Miyazaki Prefecture 2004). The council was organized by a wide range of personnel and formulated comprehensive countermeasures regarding activities such as animal waste management, fertilizer application, and household waste-water management. As a result of these activities, the percentage of wells whose water had NO_3^- contents exceeding 10 mg N L^{-1} in 2017 was only 4.8%, one-third what it was in 1998 (Miyakonojo City 2017).

10.4.3 Greenhouse Gas Emission

The appropriate management of livestock manure and its effective utilization are important issues. In this section, studies concerning the emissions of greenhouse gases, nitrous oxide (N_2O) in particular, from soils applied with livestock manure are introduced.

Niimi (2002) investigated the nitrogen dynamics in the Andosol field of the Kyushu–Okinawa Agricultural Research Center, National Agriculture and Food Research Organization (Miyakonojo, Miyazaki Prefecture), where forage corn (*Zea mays* L.) and Italian rye grass (*Lolium multiflorum* Lam.) were cropped year-round for five years from 1992 to 1997. In the experiment, cattle manure slurry fertilizer was applied to the crops at three different rates for each crop cultivation, 60 Mg ha^{-1} (standard), 150, and 300 Mg ha^{-1}, and N_2O flux was also measured. The mean cumulative N_2O–N emission rates for five years accounted for 0.4% of the total nitrogen amount applied in the 60 and 150 Mg ha^{-1} plots, close to the average value in Japanese fields (0.62%; Akiyama et al. 2006), but accounted for 4.6% in the 300 Mg ha^{-1} plots, which is comparable with the value observed in acidic tea fields, where N_2O is emitted at the highest rate in Japan (e.g., Tokunaga et al. 1996).

Additionally, Uezono et al. (2013) studied the relationship between cumulative N_2O emission rates from soils to which hog and poultry manure composts were applied and the amount of acid detergent (AD) soluble nitrogen in the manure composts, by incubation experiments. There was a positive correlation ($r = 0.88$, $p < 0.01$) between the emission rates and the amount of AD soluble nitrogen, which was thus shown to be a possible indicator for estimating N_2O emission rates from soils treated with livestock manure composts.

As for the N_2O emissions after the application of Manure compost pellets (MCPs) to soils, it was reported that N_2O emissions were 3–9 times higher after the application of MCPs than after the application of ordinary manure compost or chemical fertilizer during the cultivation period (Inoue and Shibukawa 2008; Yamane and Yamada 2009). In these studies, N_2O emission peaked soon after the application of MCPs (within a week) and the peak emission was considered to derive from the denitrification in pellets. However, Yamane et al. (2011) reported that cumulative N_2O emission rates during the cultivation period decreased by 30–91% when pellets of nitrogen-enriched manure composts (nitrogen-enriched MCPs) were applied to soils, compared with emission rates with the application of ordinary MCPs (non-nitrogen-enriched). The nitrogen-enriched manure composts were made by blowing the ammonia emitted during the composting process into matured composts (Tanaka 2009). It is important to clarify the mechanism of the decreased N_2O emissions in such MCPs in order to develop techniques for decreasing N_2O emissions from ordinary MCPs, which is a future challenge.

10.4.4 Heavy Metal Contamination

Mining and refining activities can cause serious heavy metal contamination to soil in localized areas. Examples include the arsenic (As) pollution that occurred in Toroku, a small mountain village in Miyazaki Prefecture, Kyushu. At the Toroku mine, As-bearing ores had been mined and smelted until 1962. The smelting plant was very poor and primitive, lacking a dust-collecting system. As a result, effluent gases containing arsenic trioxide leaked from the smelter and diffused throughout the entire Toroku region. Additionally, slag containing As was thrown into the Toroku River, which flows through the Toroku region, and polluted the river water. The As-contaminated river water was then used as domestic and irrigation water. Consequently, the environment of Toroku and its surrounding areas were completely polluted with As. Many residents living these areas developed chronic As poisoning, and a large number of them died. The agricultural soils along the Toroku River were also severely contaminated with As. The growth of crops, including paddy rice, was severely suppressed due to As toxicity, and the yields decreased remarkably. The As concentrations in the paddy (seven sites) and upland (one site) soils of the Toroku region reached an average of 274 mg kg^{-1} (range: 44.2–424 mg kg^{-1}) and 1302 mg kg^{-1}, respectively, according to a soil survey in 1970–1971 (Miyazaki Prefecture 1972). In the paddy soils approximately 50 km downstream from the Toroku mine, the As concentrations were 228 and 635 mg kg^{-1} (Saito 1964). In 1973, the Japanese government officially recognized the area as a "polluted area by As" and undertook soil restoration of the contaminated agricultural fields.

Soil capping (i.e., the capping of contaminated soils with uncontaminated soils brought from another place) has been generally used as a method for the restoration of As-contaminated paddy soils in Japan. Upon planning the soil restoration project of the Toroku region, examinations of capping materials and their thickness were conducted to achieve an appropriate remediation effect. The examinations were carried out on lysimeter fields (area of 1 m $\times$ 0.95 m) filled with As-contaminated soil for three years. The study found that capping with over 7.5 cm of Andosol from neighboring areas could prevent the As toxicity from inhibiting the growth of paddy rice and reduce the As concentration of soil at 0–15 cm depth to less than 15 mg kg^{-1} As extracted 1.0 mol L^{-1} HCl (Miyazaki Agricultural Research Institute 1977). The soil restoration project was completed in 1983, and monitoring and assessment has been continuously performed there since.

In Kyushu, the greenhouse cultivation of fruit vegetables has flourished since the 1970s as a result of farmers taking advantage of the warm climate in winter. Recently, it has become clear that the concentration of heavy metals, e.g., cadmium (Cd), chrome (Cr), copper (Cu), and zinc (Zn), in the soils of these greenhouse fields has been slowly increasing (Akagi and Chishaki 2015). The concentrations of total (hot HNO_3–$HClO_4$ extractable) and water-extractable Cd, Cr, Cu, and Zn in the soils collected from 23 greenhouse field sites and from 11 sites of adjacent non-agricultural field are shown in Fig. 10.15. The concentrations of total and water-extractable Cd in the greenhouse field soils ranged from 82.4 to 438 μg kg^{-1} (median: 293 μg kg^{-1}) and from 0.520 to 2.97 μg kg^{-1} (median: 1.75 μg kg^{-1}), respectively. These Cd concentrations were approximately three times higher than those in the adjacent non-agricultural field soils. Similarly, the concentrations of Cr, Cu, and Zn in the greenhouse field soils were significantly higher than those in the adjacent non-agricultural field soils. These heavy metal concentrations were strongly related to the concentration of total phosphate in the soils (Fig. 10.16), which indicates that there was a link between phosphate fertilizer use and an increase of Cd, Cr, Cu, and Zn concentrations. Additionally, the livestock manure used in Japan also contains small amounts of Cu and Zn. Therefore, if a large amount of livestock manure is used continuously, it may also contribute a source of Cu and Zn to the soil. The increase in the concentration of toxic heavy metals, including Cd, in agricultural soils is undesirable for sustainably producing safe food. In order to avoid the ineffectual application of phosphate fertilizer, which can cause an increase in heavy metal concentration in soils, soil researchers in Kyushu are working to make a new improved fertilizer application guideline considering the phosphate fertility levels of soils today.

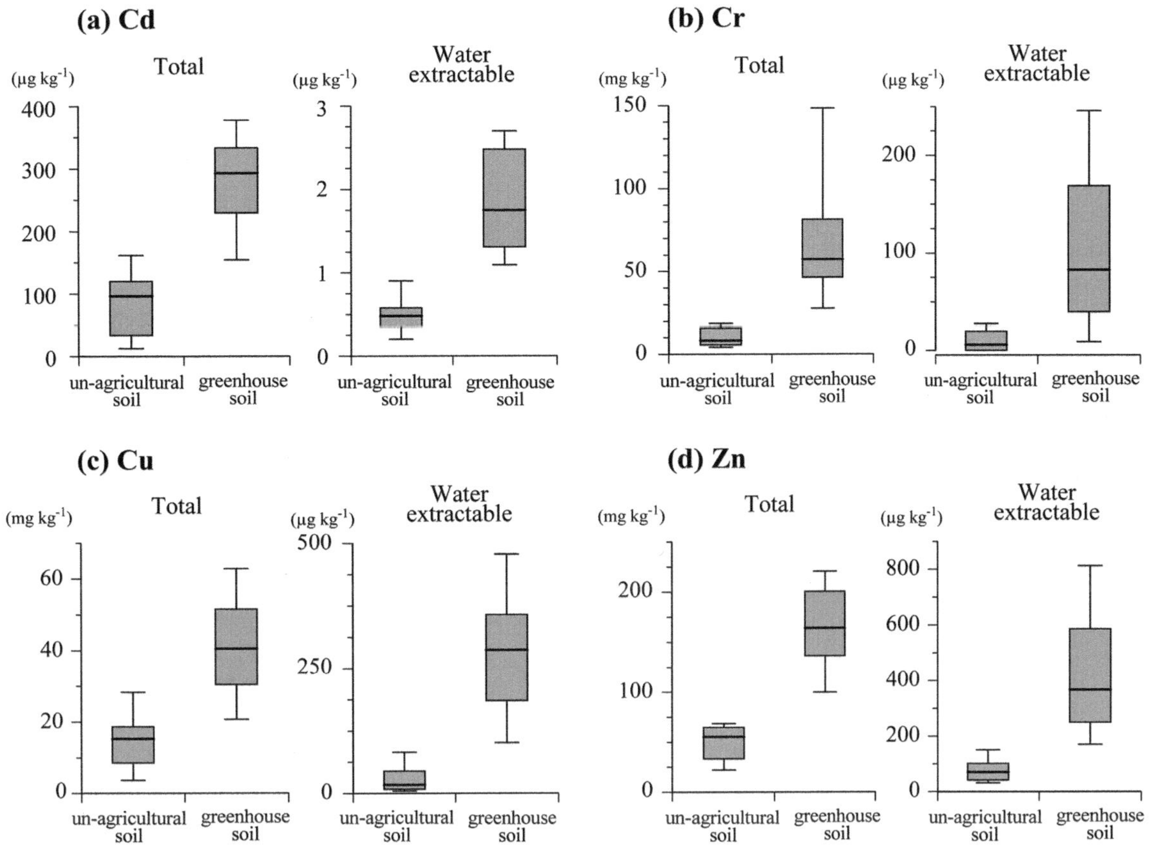

Fig. 10.15 Concentration of total (hot HNO_3–$HClO_4$ extractable) and water-extractable **a** cadmium, **b** chrome, **c** copper and **d** zinc in un-agricultural and greenhouse soils. Created by the author based on data from Akagi and Chishaki (2015). The median is shown as solid line, while the box shows the 25–75 percentile range. Whiskers extend to 10–90 percentile

10.4.5 Soil Erosion and Its Control

(1) **Properties of soil erosion in the Ryukyu Islands**

Argic Red-Yellow soils in the Ryukyu Islands are classified as *Typic Hapludults*. This soil covers approximately 50% of the area of the Ryukyu Islands. The soil is acidic and has a base saturation ratio of less than 50%, as the contents of carbon and cation exchange capacity are about 5 g kg^{-1} and about 10 $cmol_c$ kg^{-1}, respectively. The major clay minerals in this soil are kaolin minerals, mica, intergraded chlorite-vermiculite, and vermiculite (Tokashiki et al. 1982). The soil aggregates in this soil can be easily disintegrated by rainwater, as the levels of carbon and exchangeable cations in this soil are low. This indicates that the content of water-stable aggregates is very low in this soil. Therefore, soil erosion is caused by rainwater as a result of poorly developed soil aggregates.

There are two causes of soil erosion: natural and human causes. The former is based on landform, geology, precipitation, and rainfall intensity, whereas the latter is based on farming activities and civil engineering activities. The amount of soil erosion on the Ryukyu Islands was very low before 1950, because soil erosion was only due to natural factors. However, since 1950, the amount of soil erosion has increased due to increasing human activities (Onaga et al. 1999). Mountains covered in Red-Yellow soil were over-deforested in order to develop farmland, the US military base, and tourism facilities. Over-deforestation in these areas led to soil erosion, and the soil that was eroded from these areas caused extensive marine pollution (Onaga et al. 1999). A recent Okinawa prefectural government report stated that soil erosion mainly originated from farmland. It suggested that the soil erosion is related to the climate and farming activity in the Ryukyu Islands. The Ryukyu Islands are hit by typhoons during the rainy season in April to June, which leads to heavy rains. A clay crust is subsequently formed in the surface soil of the farmland that is exposed to rain and wind during the rainy season; rainwater cannot percolate through the subsoil owing to the formation of a clay crust. As a result, the surface soil of farmland was eroded by water, which flowed into an estuary (Onaga et al. 1994). Increased amounts of daily rainfall and surface runoff resulted in increasing amounts of soil erosion (Yoshinaga

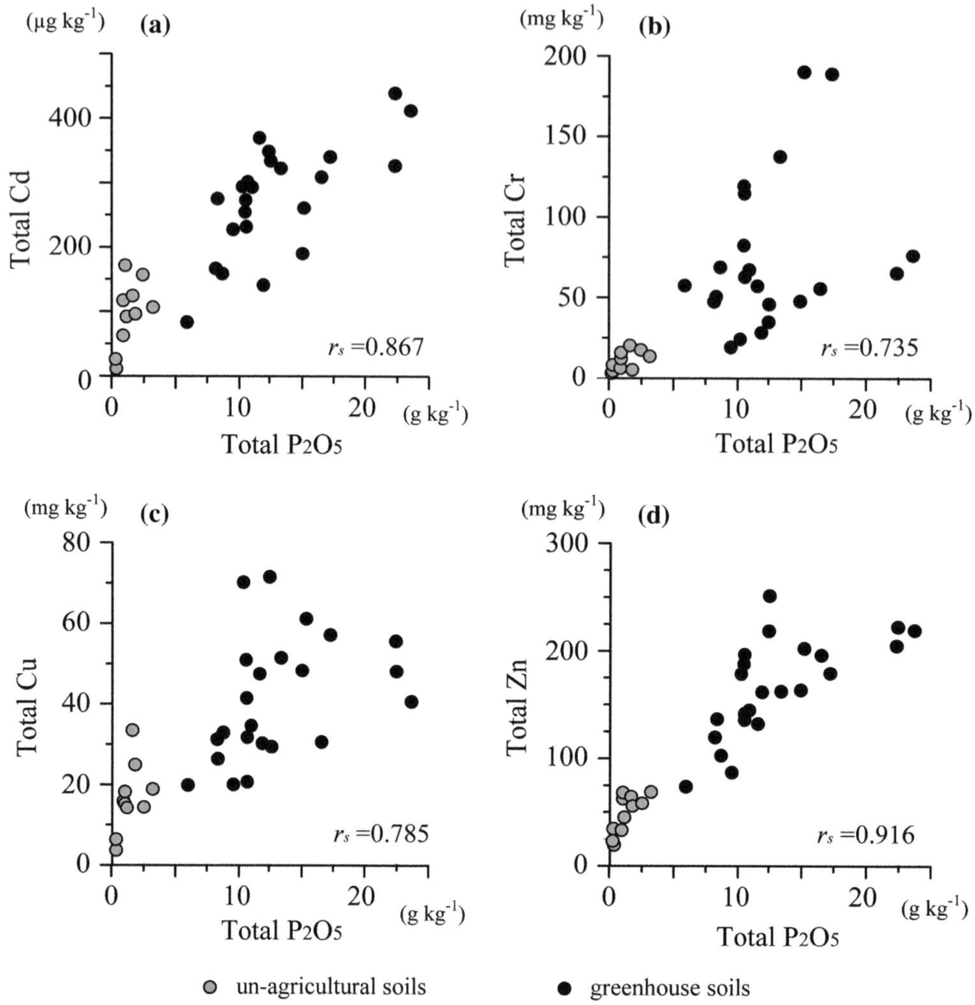

Fig. 10.16 Relationship between total phosphate concentration and **a** total cadmium concentration, **b** total chrome concentration, **c** total copper concentration, **d** total zinc concentration in the soils collected in un-agricultural and greenhouse field. Created by the author based on data from Akagi and Chishaki (2015) versus Spearman rank-correlation coefficient

and Onaga 1993). This suggests that the amount of soil erosion during the rainy season was higher than that in other seasons. Additionally, other factors affecting soil erosion on the Ryukyu Islands were also investigated (Yoshinaga et al. 2004; Kinjo and Tokashiki 2012; Nishimura 1996a, b). For example, potassium chloride, calcium carbonate, and calcium sulfate, which are contained in chemical fertilizers, are possible causes of an increase in soil dispersion on farmland. This increasing soil dispersion led to increased soil erosion.

(2) **Marine pollution due to soil erosion and reduction efforts**

There are three national parks on the Ryukyu Islands. The marine and terrestrial life in these parks is protected and managed by the Ministry of Environment of the Japanese government. However, as mentioned above, soil erosion from farmland led to estuary marine pollution. Omija et al. (1995) reported that the eroded soil that flowed into the estuary killed coral and sea urchin larvae, suppressed the activity of plankton, and altered the behavior of fish. This indicates that marine pollution due to soil erosion impaired biological diversity.

The Okinawa prefectural regulation on soil erosion was enacted in 1994. This regulation requires supervisors of construction work (civil engineering work) that is being conducted in areas with Red-Yellow soil to take measures to reduce soil erosion. The amount of soil erosion as calculated using the universal soil loss equation before and after enacting the regulation was 521,000 and 298,000 t year^{-1}, respectively (Higa et al. 1995; Nakasone et al. 1998). These reports indicate that the amount of soil erosion gradually decreased (Table 10.6). In particular, the amount of soil erosion from areas under construction decreased drastically

Table 10.6 Soil erosion before and after enacting the regulation

	Before enacting the regulation[a]	After enacting the regulation[b]
	Soil erosion -t/year- (%)	Soil erosion -t/year- (%)
Total	521,000 (100%)	298,000 (100%)
Farmland	321,000 (61%)	255,000 (85%)
US. Military work	25,000 (4.9%)	11,000 (3.6%)
Civil engineering work	167,000 (32%)	25,000 (8.3%)
Other	7000 (1.3%)	7000 (2.3%)

Created by the author based on data from [a]Higa et al. (1995) and [b]Nakasone et al. (1998)

owing to the measures taken by civil engineering contractors. However, the amount of soil eroded from farmlands did not decrease after enacting the regulation (Table 10.6). In fact, the amount of soil erosion from farmland was the highest after the regulation was enacted. It is very important to decrease the soil erosion from farmland. Specific objectives have been proposed for reducing soil erosion further in Okinawa Prefecture. For instance, the amount of soil erosion in 76 estuaries is being monitored in order to protect sea creatures; this will reduce soil erosion by 93,000 tons by 2021 in these estuaries. There are two measures that will be used for achieving this objective: protecting farmland and treating muddy water generated from farmland (Gima 2016). The former will be done via the introduction of mulch and the flattening of sloping farmland, whereas the latter will be done by laying an underground drain and establishing a sedimentation tank for gathering eroded soil.

Several studies have been conducted concerning measures for reducing soil erosion on farmland, especially on materials for protecting surface soil from rain and wind (Tokashiki et al. 1994; Yoshinaga and Onaga 1993). The materials investigated were green manure (Nagumo et al. 2006; Miyamaru et al. 2008b), compost (Yang et al. 1998), and dead Japanese pampas grass (*Miscanthus sinensis*), and were used as mulch (Yoshinaga et al. 2006). Polyvinylalcohol (PVA) was also investigated for enhancing the amount of soil aggregates in surface soil on farmland (Shinjo et al. 1993).

10.4.6 Groundwater Recharge via Agricultural Practices

In Japan, nearly 60 billion m^3 of water is consumed by the agricultural sector annually, with much of it being drawn from rivers. This agricultural water is used to irrigate local areas through a capillary-like network of small and large waterways, which exerts a great influence on the hydrological cycle in these areas. The agricultural water sourced from the upper reaches of rivers is used for irrigation, and it ultimately recharges the groundwater, flows back into rivers, and is reused in the lower reaches for agricultural, industrial, and/or domestic uses.

The Kumamoto region, located in the center of Kyushu, consists of Kumamoto City and ten municipalities. With a population of about one million and an area of 1041 km^2, the region is almost entirely dependent on groundwater for its domestic water supply; currently, this region has no alternative source of water. Aside from the groundwater being naturally recharged by rainfall, much of the groundwater in this area can be traced back to irrigation water that seeps into the ground.

As the area of paddy rice-planted land has been reduced due to acreage reduction policies, and as the area of farmland has decreased due to the abandonment of cultivation and urbanization in recent years, the amount of groundwater recharge has become smaller. To cope with this situation, an artificial groundwater recharge project has been executed since 2004 through the cooperation with the Council for Sustainable Water Use in Agriculture. The project provides subsidies to farmers who flood their converted paddy fields with water from the Shirakawa River every day for one to three months during fallow periods, that is, after planting forage rice, before planting soybean, and/or before planting vegetables such as carrots for fall/winter cropping. Currently, approximately 400 ha of land is exposed to flooding. Accordingly, the decline in the groundwater level has slowed, resulting in positive outcomes (Oshima 2010).

The artificial flooding increased the exchangeable base contents in the plow layer (Arakawa and Yamamoto 2012). The concentrations of potassium, calcium, and magnesium

Table 10.7 Available nitrogen, phosphate, and exchangeable base concentration in plowed soil

	pH	Inorg.N[a]	Av.N[b]	Bray2-P	Ex-Ca	EX-Mg	Ex-K
		mg^{-1}			$cmol_c kg^{-1}$		
Control fieild (unflooded)							
28 April	6.11	13.6	39.2	422	12.4	3.33	1.09
11 August	6.35	10.4	35.8	386	12.2	3.24	0.92
Test fieild (flooded)							
28 April	6.26	16.5	49.2	391	14.3	5.08	0.93
11 August	6.46	23.4	45.2	432	16	6.18	0.98

Reprinted from Arakawa and Yamamoto (2012) with permission from Taylor & Francis (4575190077617)
Notes [a]NH_4-N and NO_3-N(Inorganic nitorogen).[b]Inorgaric nitorogen increase after four weeks of incubation

are high in the water of the Shirakawa River (Kobayashi 1960). Thus, the farmlands exposed to artificial flooding with the said water benefited as the exchangeable base contents in the plow layer increased compared with the state before flooding (Table 10.7). Additionally, compared with the pre-flooding state, the levels of nitrate–nitrogen, available nitrogen assessed by an incubation method, and available phosphate (Table 10.7) either remained unchanged or increased. This could be attributed to factors such as algal growth on flooded farmlands and the suppressed decomposition of soil organic matter under anaerobic conditions. This practice has facilitated the saving of fertilizer for summer-sown carrots after drainage.

10.5 Natural Disasters and Soil

10.5.1 The 2016 Kumamoto Earthquake

In April 2016, several large earthquakes hit central Kyushu, mainly Kumamoto Prefecture. In a series of the earthquakes, named "the 2016 Kumamoto Earthquake" by the Japan Meteorological Agency (JMA), earthquakes with a JMA seismic intensity of 7 (scales from 0 to 7) occurred twice (magnitude of 6.5 on 14 April and magnitude of 7.3 on 16 April). The 2016 Kumamoto Earthquake was the first in the history of earthquake observation in Japan to contain two huge earthquakes with a JMA seismic intensity of 7 in a series of earthquakes.

The 2016 Kumamoto Earthquake seriously damaged crops, agricultural land, and agricultural facilities (Kanamori 2017). The damage to agricultural land caused by the earthquake included ground cracks, uneven ground surface, and liquefaction in paddies, arable lands, and greenhouses. Large ground cracks and uneven ground surface, which required large-scale reconstruction, occurred mainly in areas along the active faults that caused the earthquake. Meanwhile, liquefaction occurred not only in areas along the active faults but also across large areas of Kumamoto Prefecture, including in coastal polders.

In several paddy fields where uneven ground surface occurred as a result of the earthquake, paddy rice (*Oryza sativa* L.) or soybean (*Glycine max* L. Merril) was compared in terms of growth and grain yield at convex and concave parts within the field. In a paddy, rice growth was found to be greater in concave parts than in convex parts (Wakiyama et al. 2019). In a different paddy field, however, a large number of plants were missing, possibly as a result of excess water depth and feeding damage by *Pomacea canaliculata* (Lamarck). In the same paddy, rice plants showed a deep leaf color, and grain protein content was high in the convex part of the paddy, while weeds were flourishing in the concave part of the paddy. In soybean, there was no significant difference in plant growth parameters, such as length of main stems, between convex and concave parts within a field until August (Nomiyama et al. 2018). However, at the mature stage after September, leaf color turned lighter, leaf temperature increased, and some soybean plants failed to ripen at the concave part of the field. As a result of the excess moisture injury to soybean, final grain yield decreased by 50% at the concave part of the field.

As the coastal polder areas in Kumamoto Prefecture are Japan's leading greenhouse-based producing areas for tomato (*Solanum lycopericum* L.), the farmers in liquefaction-affected coastal polder areas were planning to resume agricultural production that year, despite the serious damage to tomatoes caused by the liquefaction at the moment of the earthquake. To allow the prompt resumption of agriculture, emergency soil surveys were conducted in summer 2016 (Inoue et al. 2019; Koga et al. 2019). The impacts of liquefaction on the chemical and physical characteristics of soils were investigated in four greenhouses. Although sand content increased and water permeability slightly decreased in the soils where liquefaction had occurred in some greenhouses, the changes in soil chemical and physical properties were often minor and were not expected to seriously decrease plant growth (Koga et al. 2019). Sugiyama et al. (2019) reported that there was no clear difference in plant growth parameters affecting tomato yields, such as the number of flowers, between liquefaction-affected and unaffected plots in the greenhouse. The possible reason for the minor impact on soils and crops in greenhouses in this earthquake was that sand released as a result of liquefaction was visually detectable on the ground surface, but the total amount of sand released was substantially limited.

10.5.2 Coastal Flood Caused by Typhoon No. 18

The Kyushu and Okinawa regions are frequently hit by typhoons and often suffer from storm surge disasters. Typhoon No. 18, in September 1999, produced coastal flooding through a combination of a stormy wind exceeding 50 m/s and high tide, which caused seawater, including water-bottom sediment, flowing back from rivers to flow onto agricultural lands near the coast in Uto-gun (Uto-shi) and Yatsushiro-gun in Kumamoto Prefecture within Shiranui Bay. No yield was obtained from the paddy fields, greenhouses, and orchards just before harvest. Furthermore, a high level of salt remained in the soil, and it was necessary to promptly remove deposits and salt and restart farming.

To remove salt from high-salinity soils, soils must be washed with a sufficient amount of fresh water. However, the salt removal effect decreased in fields such as those with poor drainage, clayey soil, and with hardpan in the lower layer of the surface; therefore, the drainage performance of the field should be ensured. Yamamoto et al. (1996) reported that permeability was improved and the salt removal effect was increased by allowing excessive water on the ground to be moved to a collecting head of ordinary underdrain via a combination of ordinary underdrainage and sub-underdrainage or mole underdrainage construction. It was also reported that in addition to the construction of ordinary underdrain and mole underdrain, the combination of plowing by stubble cultivators can increase permeability with the pan cracked (Fig. 10.17), which allowed the concentration of Cl^- to drop below the safety standard of paddy rice, 28 mmol kg^{-1} (100 mg/100 g) and which indicated improvement in the salt removal effect (Kumamoto Prefectural Agricultural Research Center, 2001).

Furthermore, for fields where Na-type clay has been formed by seawater and permeability has been degraded, it is necessary to implement measures for the removal of Na^+. Although Cl^- can be easily removed by washing with freshwater, Na^+ must be replaced by calcium ions (Ca^{2+}) in order to be removed completely, since it is absorbed by soil particles (Nishimura and Toride 1999). Ootuka et al. (1994) reported that salt removal was accelerated by mixing calcium sulfate ($CaSO_4$) into paddy soils into which seawater had flowed, which controlled the reduction in permeability. Moreover, the salt removal effect depending on the type of

Fig. 10.17 Stubble cultivator (the photo was kindly provided by Sugano farm machinery MFG Co. Ltd. https://www.sugano-net.co.jp/products/stubble/)

calcareous materials was considered, and it was reported that calcium sulfate with a high solubility was effective for soils with a pH of 6 or more, while calcium carbonate with a low solubility showed less of an effect (Kumamoto Prefectural Agricultural Research Center 2001). Note that the application of calcium carbonate for acidity correction is often used if the soil pH is 6 or less.

References

Akagi I, Chishaki N (2015) Heavy metal concentrations and effect of phosphate fertilizer application in greenhouse soil samples. J Environ Chem 25:199–206 (in Japanese with English summary)

Akagi Y, Kawano M, Nonaka S et al (2000) The status and change of the properties of cultivated soil in Miyazaki Prefecture as basic survey for 20 years. Bull Miyazaki Agric Exp Stn 35:3–25 (in Japanese)

Akiyama H, Yan X, Yagi K (2006) Estimations of emission factors for fertilizer-induced direct N_2O emissions from agricultural soils in Japan: summary of available data. Soil Sci Plant Nutr 52:774–787

Arakawa Y (2012a) Pelletization of composted manure improves the phytoavailability of contained phosphorus. Jpn J Soil Sci Plant Nutr 83:249–255 (in Japanese with English summary)

Arakawa Y (2012b) Compost pelletization increases the phyto availability of phosphorus in compost. In: Trémier et al (eds) ORBIT2012-Global assessment of organic resources and waste management 8th international conference proceedings. Weimar, Germany. ISBN: 3-935974-35-3, pp 48–54

Arakawa Y, Yamamoto K (2012) Improvement of soil fertility by fallow flooding for groundwater conservation on the middle reaches of the Shirakawa River in Kumamoto, Japan. Arch Agron Soil Sci 58:S213–S218

Arihara J, Karasawa T (2000) Effect of previous crops on arbuscular mycorrhizal formation and growth of succeeding maize. Soil Sci Plant Nutr 46:43–51

Aso regional GIAHS Executive Committee (2014) Value as world agriculture inheritance. http://www.giahs-aso.jp.e.mn.hp.transer.com/value/ (May, 2019 in Japanese)

Curtin D, Wright CE et al (2006) Hot water-extractable nitrogen as an indicator of soil nitrogen availability. Soil Sci Soc Am J 70:1512–1521

FAO (2002) Report of the FAO workshop of potential stakeholders and steering committee on globally-important ingenious agricultural heritage systems. Rome, Italy, 5–7 August 2002. FAO Land and Water Division, FAO, Rome, Italy

Fujikake I (2017) Extraction of agriculture and forestry complex dominant areas in Japan and characteristics of the GIAHS. In: GIAHS site (ed) Reports of studies on Takachihogo-Shiibayama GIAHS research group. University of Miyazaki, Miyazaki, pp 1–10. (in Japanese)

Fujimoto H, Haruyama M, Murata S (1983) Shirasu. In: Japanese Geotechnical Society Kyushu Branch (ed) Local Soils in Kyushu Okinawa. Kyushu University Press, Fukuoka, pp 145–169. (in Japanese)

Fujino N, Kobahashi T (1992) Mode of sedimentation of the Kora beds from Kaimondake volcano. Rep Fac Sci, Kagoshima Univ, Earth Sci Biol 25:69–83 (in Japanese with English summary)

Fujitomi S, Kuroyanagi N et al (2014) Effect of organic matter application on soil carbon accumulation in double-cropped paddy fields in Fukuoka. Bull Fukuoka Agric Res Cent 33:1–6 (in Japanese with English summary)

Fujiwara H (1998) Visit to the village practicing shifting cultivation. In: Research for origin of rice cultivation. Iwanami Shoten, Tokyo, pp 84–112. (In Japanese)

Fukuda K, Kunieda E et al (2001) Decomposition of organic matter and absorption of crop in paddy field. Kyushu Agricultural Research 63:59 (in Japanese)

Fukuoka Prefecture (2017) Kachikuhaisetubutu no riyosokushin wo hakarutameno keikaku (The prefectural plan for promoting animal manure usage) http://www.pref.fukuoka.lg.jp/uploaded/life/267660_52642243_misc.pdf (May, 2019)

Gima Y (2016) The properties of agricultural land in Okinawa prefecture. Pedologist 60:71–75 (in Japanese)

Goto S, Nagata S (2000) Effect of crotararia, sorghum and pampas grass incorporated as green manure on the yield of succeeding crops and soil physical and chemical properties. Jpn. J Soil Sci Plant Nutr 71:337–344 (in Japanese with English summary)

Goto S, Nagata S (2008) Effect of long-term application of compost on the sugarcane yeild and soil physical and chemical properties in dark red soil. Jpn. J Soil Sci Plant Nutr J Soil Sci Plant Nutr 79:9–16 (in Japanese with English summary)

Gunjikake N, Kubo K (1996) Research nitrogen mineralization characteristics and application method of organic matters for melon and tomato vegetables. Res Bull Kumamoto Prefectural Agric Res Center 5:46–55 (in Japanese)

Hamazaki T (2004) Soil distribution and properties. In: Japanese Society of Soil Science and Plant Nutrition Kyushu Branch (ed) Agriculture and soil science and plant nutrition in Kyushu and Okinawa. Japanese Society of Soil Science and Plant Nitrition Kyushu Branch, Fukuoka, pp 13–16. (in Japanese)

Haraguchi N, Wakasugi K (2014) A new framework for study of irrigated paddy rice and upland corps rotation farming and its relation to soil and plant nutrition science. 5. Water management system required for upland crops rotation farming. Jpn J Soil Sci Plant Nutr 85:264–267 (in Japanese)

Haruguchi S, Sakanashi J et al (2009) Available application of cattle feces manure to successive cultivation of rice and wheat in paddy field of lowland in Kumamoto prefecture. Res Bull Kumamoto Prefectural Agric Res Center 16:11–19 (in Japanese with English summary)

Hayashi H (2014) Understandings of relationships between agriculture and biodiversity in Kunisaki GIAHS. J Resour Ecol 5:395–397

Haynes RJ, Mokolobate MS (2001) Amelioration of Al toxicity and P deficiency in acid soils by addition of organic residues: a critical review of the phenomenon and mechanisms involved. Nutr Cycling Agroecosyst 59:47–63

Hideshima Y, Arima S et al (2016) Effects of the method of processing wheat or barley straw in the double cropping system on weed incidence and growth of paddy rice. Jpn J Crop Sci 85:122–129 (in Japanese with English summary)

Higa E, Omija T et al (1995) Quantity of soil run-off on Okinawa Island. Ann Rep Okinawa Prefectural Inst Health Environ 29:83–88 (in Japanese)

Inoue K (1996) Differential localization of main nutrients in semi-forcing asparagus (*Asparagus officinalis* L.) grown in year-round-harvesting-cultures. Bull Nagasaki Agric Forest Exp Station Sect. of Agric 23:31–45

Inoue K (2001) Asuparagasu-Seinan danchi. (Asparagus- warm southwest region of Japan), Nogyo gijyutsu taikei (Soil and fertilizer), 6–1. Rural Culture Association Japan, Tokyo, pp. 258-2-11. (in Japanese)

Inoue K (2005) Soil behavior, the adequate amount of fertilizer and the simple analyzing method of nitrate nitrogen for green asparagus production by whole season cropping system. Bull Nagasaki Agric Forest Exp Station Sect. of Agric 31:1–14 (in Japanese with English summary)

Inoue K (2008) Improving the green asparagus production system in Japan. Acta Hortic. https://doi.org/10.17660/actahortic.2008.776.9
Inoue K (2019) Application of humic acid liquid fertilizer increases asparagus yield. Zen-NOH Green Report, 597. (in Japanese)
Inoue K, Fujiyama M et al (2014) The changes of several physico · chemical properties of cultivated soils in Nagasaki prefecture. Part 3: changes in the 30 years of paddy soil, upland soil and orchard soil. Bull Nagasaki Agric Forest Tech Develop Center 5:85–92 (in Japanese with English summary)
Inoue K, Nagao Y et al (1999) The changes of several physico-chemical properties of paddy soils in Nagasaki prefecture 2. Bull Nagasaki Agric Forest Exp Station Sect Agric 25:69–80 (in Japanese)
Inoue K, Shibukawa H (2008) Nitrous oxide emission from upland soils applied pelletized poultry manure compost. Jpn J Soil Sci Plant Nutr 79:181–182 (in Japanese)
Inoue K, Wakasugi K et al (2019) Geophysical investigation of farmlands damaged by the 2016 Kumamoto Earthquake. Mem Kyushu Okinawa Agric Res Cent 95:18–29 (In Japanese with English summary)
Inoue K (2015) Method of fertilizer application for onion based on nitrogen fertilizer efficiency of rapeseed meal and new estimation method of composted poultry. Bull Nagasaki Agric Forest Techn Develop Center 6:103–111 (in Japanese)
Ishii K, Fujiyama M et al (2016) Influence of long-term application of chaff cow manure compost on the yield and inorganic components of rice plant by coarse-textured gray lowland soils. Bull Nagasaki Agric Forest Techn Develop Center 7:81–97 (in Japanese with English summary)
Itakura M, Uchida Y et al (2013) Mitigation of nitrous oxide emissions from soils by *Bradyrhizobium japonicum* inoculation. Nat Clim Change 3:208–212
Japanese Society of Pedology (2000) Toward the conservation of vanishing soils in Japan—conservation of red data soils, p 21. Japanese Society of Pedology, Tsukuba. (in Japanese)
Kagoshima Prefecture (2016) Kagoshimaken ni okeru kachikuhaise-tubutu no riyosokushin wo hakarutameno keikaku (The prefectural plan for promoting animal manure usage in Kagoshima) http://www.pref.kagoshima.jp/ag07/sangyo-rodo/nogyo/tikusan/kankyo/documents/24620_20160321013309-1.pdf (May, 2019)
Kameya S (2006) Agriculture, Soil Science and plant nutrition in Okinawa prefecture. Jpn J Soil Sci Plant Nutr 77:695–697 (in Japanese)
Kanamori N (2017) Damages in agriculture caused by the 2016 Kumamoto Earthquake and actions taken by extension departemnts. Rep Kyushu Br Crop Sci Soc Japan 83:57–60 (in Japanese)
Kato Y (2009) Sea around the Ryukyu Arc and the Daito Ridge. In: The geological society of Japan (ed) Regional geology of Japan 8, Kyushu and Okinawa Region. Asakura Publishing Co., Tokyo, pp 407–411. (in Japanese)
Katoh M, Arakawa Y et al (2017) Development of compost-mixed compound fertilizer and perspectives. Jpn. J Soil Sci Plant Nutr 88:272–276 (in Japanese)
Kira T (1971) Nature considered from the viewpoint of ecology. Kawade Shobo Shinsya, Tokyo, pp 105–141. (in Japanese)
Kinjo K, Tokashiki Y (2012) Effects of the $CaCO_3$ application on the water dispersible colloids of red yellow soils from Okinawa Island. Jpn J Soil Sci Plant Nutr 83:594–598 (in Japanese)
Kitagawa H, Tsuchiya K et al (2009) Effect of long-term organic matter application to rice yield components of warm region in Japan in hot year. In: The 227th meeting of the crop science society of Japan, pp 156–157. (in Japanese)
Kobayashi J (1960) A chemical study of the average quality and characteristics of river water of Japan. Ber Ohara Inst Landwirtschafft Biol 11:313–358 (in Japanese)
Kobayashi T, Tameike T (2002) History of eruptions and volcanic damage from Sakurajima volcano, Southern Kyushu, Japan. Quatern Res 41:269–278 (in Japanese with English summary)
Kobayashi Y, Niimi H et al (1994) Fate of fertilizer nitrogen applied in an area with high precipitation in Southern Kyushu. JARQ 28 (1):12–19
Koga N, Mitsugi K et al (2019) Liquefaction impacts on soil properties in greenhouse farming areas affected by the 2016 Kumamoto Earthquake. Mem Kyushu Okinawa Agric Res Cent 95:42–48 (in Japanese with English summary)
Kondo T, Saeki Y (2017) Evaluation of shifting cultivation by soil chemical analysis. In: GIAHS research group (ed) Reports of studies on Takachihogo-Shiibayama GIAHS site. University of Miyazaki, Miyazaki, pp 21–30. (in Japanese)
Kubotera H, Yamada I (2000) Characteristics of the induration of tephra-derived soils in Kyushu, Japan (3) Properties and typology of five indurated soils. Soil Sci Plant Nutr 46:365–379
Kubotera H, Kusaba T et al (2015) Properties of the surface horizon of nonallophanic Andosols on the Kuju Plateau and the northern Aso somma, Kyushu. Jpn J Soil Sci Plant Nutr 86:17–23 (in Japanese with English summary)
Kubotera H, Kusaba T, Shishibe I (2013) Distribution and horizons sequence of non-allophanic Soil in Kuju Plateau, Kyushu, Japan. Pedologist 57:72–80 (in Japanese with English summary)
Kumamoto Prefectural Agricultural Research Center (2017) Effect of carbon dioxide application rate on fruit enlargement and fruit yield on promoted tomato. https://www.pref.kumamoto.jp/common/UploadFileOutput.ashx?c_id=3&id=6666&sub_id=6&flid=109082 (December, 2017; in Japanese)
Kumamoto Prefecture (2000) (2012 amendment): Kumamoto prefectural ordinance on groundwater conservation (in Japanese)
Kumamoto Prefecture (2013) Report on water quality in 2012. http://www.pref.kumamoto.jp/kiji_585.html (February, 2018; in Japanese)
Kumamoto Prefecture (2015) Kumamoto prefectural ordinance on cultivating soil and groundwater. http://www1.g-reiki.net/kumamoto/act/frame/frame110010506.htm (February, 2018; in Japanese)
Kumamoto Prefecture (2016) Kumamotoken ni okeru kachikuhaise-tubutu no riyosokushin wo hakarutameno keikaku (The prefectural plan for promoting animal manure usage in Kumamoto) https://www.pref.kumamoto.jp/common/UploadFileOutput.ashx?c_id=3&id=18235&sub_id=1&flid=92318 (May, 2019)
Kuniyoshi K, Gima Y (2004) Okinawa prefecture In: Japanese Society of Soil Science and Plant Nutrition Kyushu Branch (ed) Agriculture and soil science and plant nutrition in Kyushu and Okinawa. Japanese Society of Soil Science and Plant Nitrition Kyushu Branch, Fukuoka, pp 55–58. (in Japanese)
Kyushu Regional Agriculture Administration Office (2017a) Annual report on food, agriculture and rural community in Kyushu in fiscal year 2016. http://www.maff.go.jp/kyusyu/kikaku/kikaku.html (February, 2018; in Japanese)
Kyushu Regional Agriculture Administration Office (2017b) Annual report on food, agriculture and rural community in Kyushu in fiscal year 2016. Trends. http://www.maff.go.jp/kyusyu/kikaku/attach/pdf/h28jyouseihoukoku-36.pdf (January, 2018; in Japanese)
Kyushu Regional Agricultural Administration Office (2018) Aggregate of planted area of crops and utilization rate of cultivated land. In: The 64th Statistical yearbook of Agriculture, Forestry and Fisheries in Kyushu, 45. http://www.maff.go.jp/kyusyu/toukei/hensyu/attach/pdf/64nenpou-53.pdf (May, 2019)
Machida H, Arai F (2003) Atlas of Tephra in and around Japan. University of Tokyo Press, Tokyo, pp 58–70. (in Japanese)
Machida H, Moriwaki H et al (2001) Outline of large terrain and division of topography. In: Machida H, Ota Y, Kawana T et al

(eds) Regional Geomorphology of the Japanese Islands, vol 7 Geomorphology of Kyushu and the Ryukyus. University of Tokyo Press, Tokyo, pp 2–6. (in Japanese)

Matsu T, Kato Y (1962) Note on palaeopedology of red soils in Japan. Quat Res 2:161–179 (in Japanese with English summary)

Matsumoto J (2007) A study on clarification of nitrogen load and countermeasures in the upland distributed in the Shirasu plateau of Southern Kyushu. Bull Kagoshima Prefectural Inst Agric Develop Agric Res 1:99–180 (in Japanese)

Matsuyama N, Saigusa M (1994) Distribution of allophanic andosols and nonallophanic andosols in western Japan. Pedologist 38:2–9 (in Japanese with English summary)

Matsuyama N, Saigusa M et al (2005) Acidification and soil productivity of allophanic Andosols affected by heavy application of fertilizers. Soil Sci Plant Nutri 51:117–123

Mimaki N (2016) Reduction of fertilizer application rate by means of evaluating phosphorus and potassium contents in cattle manure compost by extracting those elements with citric acid solution. https://www.pref.kumamoto.jp/common/UploadFileOutput.ashx?c_id=3&id=6666&sub_id=4&flid=71834 (December, 2017; in Japanese)

Ministry of Agriculture, Forestry and Fisheries (2015) Report about farming area of greenhouse horticulture 2014. http://www.maff.go.jp/j/seisan/ryutu/engei/sisetsu/haipura/setti.html (December, 2017) (in Japanese)

Ministry of Agriculture, Forestry and Fisheries, Rural Development Bureau (2016a) Overview of countermeasures for special soil areas. http://www.mlit.go.jp/common/001151253.pdf (May 2019; in Japanese)

Ministry of Agriculture, Forestry and Fisheries (2016b) Circumstances surrounding rice in Kyushu area. (in Japanese). http://www.maff.go.jp/kyusyu/seisan/sinko/pdf/inemeguzih28.pdf (November, 2017)

Ministry of Agriculture, Forestry and Fisheries (2017a) Kyushu food, agriculture, and rural situation report in 2016. (in Japanese). http://www.maff.go.jp/kyusyu/kikaku/h28jyouseihoukoku.html (November, 2017)

Ministry of Agriculture, Forestry and Fisheries (2017b) Circumstances surrounding soybeans in the Kyushu region. (in Japanese). http://www.maff.go.jp/kyusyu/seiryuu/daizu/attach/pdf/kyushudaizu-1.pdf (November, 2017)

Ministry of Agriculture, Forestry and Fisheries (2017c) Circumstances surrounding wheat and barley in the Kyushu region. (in Japanese). http://www.maff.go.jp/kyusyu/seiryuu/bakusaku/attach/pdf/kyushu_bakusaku2-3.pdf (November, 2017)

Miyabuchi Y, Sugiyama S (2008) A 30,000-year phytolith record of a tephra sequence at the southeastern foot of Aso volcano, Japan. J Geogr 117:704–717 (in Japanese with English summary)

Miyabuchi Y, Sugiyama S (2016) 90,000-year phytolith records from caldera rim to western foot of Aso Volcano, Japan: implications for vegetation history since catastrophic eruption. Quatern Int 397:392–403

Miyakonojo City (2017) Current status on groundwater quality in the Miyakonojo Basin region. http://cms.city.miyakonojo.miyazaki.jp/display.php?cont=120827141658 (February, 2018; in Japanese)

Miyamaru N (2003) Okinawa no satoukibi to ryokuhi (2). Satouruijoho 79:1–5 (in Japanese)

Miyamaru N, Gima Y et al (2008a) Growth and decomposition property of several green manures in Okinawa. J Okinawa Agric 42:21–27 (in Japanese with English summary)

Miyamaru N, Gima Y et al (2008b) Growth property and prevention effect of green manure for soil erosion. Jpn J Soil Sci Plant Nutr 79:496–499 (in Japanese)

Miyamaru N, Iha S et al (2012) Effects of long-term application of green manure and compost on the physical, chemical and biological properties of Jahgaru soil in Okinawa. Jpn J Soil Sci Plant Nutr 83:280–287 (in Japanese with English summary)

Miyazaki agricultural research institute (1977) Examination for reducing As toxic damage to rice plant. Rep Result Examination Rem Arsenic Soil Contam 1973–1975:79–113 (in Japanese)

Miyazaki agricultural research institute (2016a) Effect of condensed grain distillers soluble of "shochu" spirit as fertilizer. https://www.pref.miyazaki.lg.jp/contents/org/nosei/mae-station/result/index.html (December, 2017; in Japanese)

Miyazaki agricultural research institute (2016b) Miyazaki method soil solarization plus technology which performs solar heat soil disinfection after applying condensed grain distillers soluble of "shochu" spirit. https://www.naro.affrc.go.jp/karc/prefectural_results/files/28_5_01.pdf (May 2019; in Japanese)

Miyazaki prefecture (1972) Summary of social medicine studies on mining pollution around Toroku district. Environ Health Rep 15:13–60 (in Japanese)

Miyazaki Prefecture (2004) Council to reduce NO_3^- contamination in groundwater in the Miyakonojo basin region. (in Japanese)

Miyazaki Prefecture (2017) Miyazakiken ni okeru kachikuhaisetubutu no riyosokushin wo hakarutameno keikaku (The prefectural plan for promoting animal manure usage in Miyazaki)

Miyazawa K. (1986) Productivity of andosols and its improvement. Kagaku to Seibustu. https://doi.org/10.1271/kagakutoseibutsu1962.25.97 (in Japanese)

Mizota C, Shimoyama S et al (1992) Sources and ages of fine-grained soils developed on gentle slopes in northern Kyushu. Quat Res 31:101–111 (in Japanese with English Abstract)

Mori K, Tawara Y et al (2016) Numerical modeling for simulating fate and reactive transport processes of nitrogen in watershed and discussion on applicability to actual fields. J Groundwater Hydrol 58:63–86 (in Japanese)

Moriizumi M, Uezono I et al (2015) Decomposition characteristics of hot-water-extractable soil organic nitrogen during aerobics soil incubation. Jpn J Soil Sci Plant Nutr 86:8–16 (in Japanese with English abstract)

Morita S (2008) Prospect for developing measures to prevent high-temperature damage to rice grain ripening. Jpn J Crop Sci 77:1–12 (in Japanese with English summary)

Morita S, Wada H, Matsue Y (2016) Countermeasures for heat damage in rice grain quality under climate change. Plant Prod Sci 19:1–11

Nagasaki Prefecture (2013) The guidance of large scale environmental-friendly agricultural practices in the Isahaya bay reclaimed land (supplemental edition) https://www.pref.nagasaki.jp/e-nourin/nougi/section/03reclaimed_land//tebiki2013.pdf (in Japanese; May, 2019)

Nagasaki Prefecture (2017) Nagasakiken ni okeru kachikuhaisetubutu no riyosokushin wo hakarutameno keikaku (The prefectural plan for promoting animal manure usage in Nagasaki) https://www.pref.nagasaki.jp/shared/uploads/2017/08/1502087521.pdf (in Japanese; May,2019)

Nagatomo M, Uezono I et al (2017) Phosphorus fertilizer recommendation based on soil available phosphate in andosol. Jpn J Soil Sci Plant Nutr 88:45–52 (in Japanese)

Nagatomo M (2004) Occurrence and control of leaf yellowing on oriental hybrid lily. Kyushu Agric Res 66:60 (in Japanese)

Nagumo F, Issaka RN, Hoshikawa A (2006) Effects of tillage practices combined with mucuna fallow on soil erosion and water dynamics on Ishigaki Island, Japan. Soil Sci Plant Nutr 52:676–685

Nakamura K, Kimura R, Uchijima Z (1996) Climate of Japan. Iwanami Shoten, Tokyo, pp 55–61. (in Japanese)

Nakasone K, Higa E et al (1998) Estimation of soil loss in Okinawa prefecture (II). Ann Rep Okinawa Prefectural Inst Health Environ 32:67–72 (in Japanese)

NARO (National Agriculture and Food Research Organization) (2017) Japanese soil inventory. http://soil-inventory.dc.affrc.go.jp/ (November, 2017; in Japanese)

Niimi H (2002) Mechanism and measures of environmental pollution caused by livestock waste nitrogen in the upland farming belt in Southern Kyushu (1) Dynamics of nitrogen in upland field applied with a large amount of slurry barnyard manure. J Jpn Agric Syst Soc 18:161–173 (in Japanese with English summary)

Niimi H, Suzuki T et al (2016) Yield and sustainability of the Japanese radish-sweet potato double cropping system in combination with continuous ridge use and organic liquid fertilizer application. Jpn J Org Agric Sci 8:46–58 (in Japanese)

Nishi H, Morita S et al (2013) Status and changes for 30 years of the physical and chemical properties of cultivated soils distributed over Kagoshima Prefecture. Bull Kagoshima Prefectural Inst Agric Develop 7:47–61 (in Japanese with English summary)

Nishimura T (1996a) Changes in infiltration and ion exchange subsequent to Gypsum application to a Japanese acid soil. Trans JSIDRE 184:167–173

Nishimura T (1996b) Effects of application on erodibility of the Kunigami Mahji soil. Trans JSIDRE 185:99–105

Nishimura T, Toride N (1999) Fundamentals and applications of soil colloid science (12) hydravlic properties and soil erosion. J Agric Eng Soc Jpn 67:277–284 (in Japanese)

Nomiyama R, Matsuo N et al (2018) Effects of the roughness of paddy field after the Kumamoto Earthquake on growth and yield of soybean. Jpn J Crop Sci 87:176–182 (In Japanese with English summary)

Obara H, Ohkura T et al (2011) Comprehensive soil classification system of Japan first approximation. Bull Natl Inst Agro-Environ Sci 29:1–73 (in Japanese with English summary)

Obara H, Takata Y et al (2016) A new soil map of Japan based on comprehensive soil classification system of Japan first approximation. Bull Natl Inst Agro-Environ Sci 37:133–148 (in Japanese with English summary)

Obara H, Nakai M (2004) Available phosphate of Arable Lands in Japan. Jpn J Soil Sci Plant Nutr 75:59–67 (in Japanese with English summary)

Odahara K, Fukushima Y et al (2012) The soil fertility status and soybean productivity in paddy-upland rotation fields in Japan's Chikugo River basin. Jpn J Soil Sci Plant Nutr 83:405–411 (in Japanese with English summary)

Ogura J, Yamamoto S, Ikeda A (2002) The origin of the grassland of Aso region, Kyushu Japan, by microscopic charcoal analysis. Summ Res Using AMS at Nagoya Univ. 13:236–240 (in Japanese with English summary)

Ohba K (2004) Climate features. In: Japanese Society of Soil Science and Plant Nitrition Kyushu Branch (ed) Agriculture and soil and plant nutrition in Kyushu and Okinawa. Japanese Society of Soil Science and Plant Nitrition Kyushu Branch, Fukuoka, pp 3–6. (in Japanese)

Oita Prefecture (2016) Kachikuhaisetubutu no riyosokushin wo hakarutameno keikaku (The prefectural plan for promoting animal manure usage)

Okinawa Agricultural Research Center (2015) Reduction of phosphorus fertilizer for balsam pear on the Jahgaru soil. http://www.pref.okinawa.jp/arc/_userdata/fukyu/2015/shido27_02.pdf (December, 2017; in Japanese)

Okinawa General Bureau (2017) Annual report on agriculture, forestry and fisheries in Okinawa in fiscal year 2016. http://www.ogb.go.jp/-/media/Files/OGB/Nousui/area-gaiyo/201708/h28_honban_all.pdf (May 2019; in Japanese)

Okinawa Prefectre (2014) Sugarcane cultivation guidelines, p 12 (in Japanese)

Omija T, Higa E et al (1995) Soil erosion and conservation of coral community on the Okinawa Islands. J Jpn Soc Irrig Drain Reclam Eng 63:267–272 (in Japanese)

Onaga K, Gibo S (1984) Special soil of Japan—special soil of Okinawa (Mahji, Jahgaru). J Jpn Soc Irrig Drain Reclam Eng 52:517–524 (in Japanese)

Onaga K, Komesu R, Arakaki A (1999) The outline of red soil low and erosion control measure in Okinawa. Sci Bull Fac Agric Univ Ryukyus 46:71–82 (in Japanese)

Onaga K, Yoshinaga A, Shimo M (1994) Infiltration characteristics of Kunigami maaji—in relation to soil erosion. Sci Bull Fac Agric Univ Ryukyus 41:269–277 (in Japanese)

Ono T et al (1998) Effects of organic materials-application by blending on sweet Pepper's growing in summer and autumn. Kyushu Agric Res 60:48 (in Japanese)

Ootuka N, Fukuda K et al (1994) Osmotic desalting effect of fresh water artificial flooding and application of lime (gypsum) to paddy in which seawater had flowed. Kyushu Nogyo Kenkyu Seika joho 9:485–486 (in Japanese)

Oshima I (2010) Administration for groundwater management in the Kumamoto area. J Groundwater Hydorol. https://doi.org/10.5917/jagh.52.49 (in Japanese with English summary)

Oshiro K, Hamakawa K (1980) Soils and improvements in Okinawa. Shinpousyuppan, Naha, pp 89–118 (in Japanese)

Ota Y, Kawana N (2001) Outline of the topography and geology of the Nansei Islands. In: Machida H, Ota Y, Kawana T, Moriwaki H, Nagaoka S (eds) Regional geomorphology of the Japanese Islands, vol 7 Geomorphology of Kyushu and the Ryukyus. University of Tokyo Press, Tokyo, pp 26–31. (in Japanese)

Otaki N (1997) Grassland and human activity. Ichinomiya Choshi (History of Ichinomiya town, Kumamoto prefecture, Japan) 10. Ichinomiya Cho, Kumamoto, pp 42–46. (in Japanese)

Otsuka A, Matsumura M et al (2010) The 2008 overseas mass migration of the small brown planthopper, Laodelphax striatellus, and subsequent outbreak of rice stripe disease in western Japan. Appl Entomol Zool 45:259–266

Pinheiro J (1997) Characterization of three andosols of the Azores archipelago (Portugal) showing different types of indurated horizons. Suelos Volcanicos Endurecidos, III Simposio Internacional, pp 116–120

Saeki Y, Aimi N et al (2006) Diversity and geographical distribution of indigenous soybean-nodulating bradyrhizobia in Japan. Soil Sci Plant Nutr 52:418–426

Saeki Y, Nakamura M et al (2017) Effect of flooding and the nosZ gene in bradyrhizobia on bradyrhizobial community structure in the soil. Microbes Environ 32:154–163

Saeki Y, Shiro S et al (2013) Mathematical ecology analysis of geographical distribution of soybean-nodulating bradyrhizobia in Japan. Microbes Environ 28:470–478

Saeki Y (2018) Yuyo biseibutsu 1- Chisso kotei saikin In: Toyota K (ed) Dojyo-biseibutsugaku (Soil microbiology). Asakura-shoten, Tokyo, pp 51–74

Saga Prefecture (2016) Sagaken ni okeru kachikuhaisetubutu no riyosokushin wo hakarutameno keikaku (The prefectural plan for promoting animal manure usage in Saga) https://www.pref.saga.lg.jp/kiji00351259/3_51259_23991_up_x50o6jle.pdf (May, 2019)

Saigusa M, Shoji S, Takahashi T (1980) Plant root growth in acid Andosols from northeastern Japan: 2. Exchange acidity y1 as a realistic measure of aluminum toxicity potential. Soil Sci 130:242–250

Saito B (1964) Arsenic concentration of paddy soils. Abs Ann Meet Jpn Soc Soil Sci Plant Nutr 10:A30 (in Japanese)

Sameshima-Saito R, Chiba K, Minamisawa K (2006) Correlation of denitrifying capability with the existence of nap, nir, nor and nos

genes in diverse strains of soybean bradyrhizobia. Microbes Environ 21:174–184
Sasaki K (1970) Comparative geographical studies on shifting cultivation. Doctor thesis of Kyoto Univ. (in Japanese)
Seino K (1980) Rice and wheat cultivation in Kyushu. In: Kyushu National Agricultural Experiment Station (ed) Shashin de miru Kyushu no dojo to nogyo (Soil and agriculture in Kyushu seen in photos). Kyushu National Agricultural Experiment Station, Chikugo, pp 37–38. (in Japanese)
Seufert V, Ramankutty N, Foley JA (2012) Comparing the yields of organic and conventional agriculture. Nature 485:229–232
Shiiba Y, Uthumi Y (2010) Swidden agriculture at Okawachi Settlement in Shiiba Village. Bull Kyushu Univ For 91:34–39 (in Japanese with English summary)
Shiina Y, Itakura M et al (2014) Relationship between soil type and N2O reductase genotype (nosZ) of indigenous soybean bradyrhizobia: nosZ-minus populations are dominant in andosols. Microbes Environ 29:420–426
Shinjo T, Komiya Y et al (1993) Prevention of reddish soil loss in the Kunigami maaji area by membrane of chemical stabilization. Trans JSIDRE 166:97–104 (in Japanese)
Shinzato Y, Fukami K et al (2013) Effects of drainage improvement for sugarcane fields by subsurface drainage and subsoiling. J Jpn Soc Agric Mach 75:426–433 (in Japanese with English summary)
Shiraki M (2007) The effectiveness and prevalence of the solarization method constructed in Miyazaki. Plant Prot 61:14–17 (in Japanese)
Shoji S, Ito T et al (1985) Properties of nonallophanic Andosols from Japan. Soil Sci 140:264–277
Shoji S, Nanzyo M, Dahlgren R (1993) Volcanic ash soils: genesis, properties and utilization. Elsevier, Amsterdam; Tokyo, pp 1–5, 44–45, 54
Soil genesis and classification laboratory, National Institute for Agro-Environmental Sciences (1997) Typology and regional characteristics of clay mineral composition in lowland soil in Japan. http://www.naro.affrc.go.jp/archive/niaes/sinfo/result/result13/result13_02.html (in Japanese; May 2019)
Statistics Department, Minister's Secretariat, Ministry of Agriculture, Forestry and Fisheries (2016) Area by type of cultivated land. In: The 90th statistical yearbook. Ministry of Agriculture, Forestry and Fisheries, Tokyo, pp 348–349. (in Japanese)
Sugiyama M, Koba T et al (2019) Effects of liquefaction on the growth of the next planting tomatoes in the 2016 Kumamoto Earthquake. Mem Kyushu Okinawa Agric Res Cent 95:86–92 (in Japanese with English summary)
Sumida H, Kato N, Nishida M (2005) Depletion of soil fertility and crop productivity in succession of paddy rice-soybean rotation. Bull Natl Agric Res Cent Tohoku Reg 103:39–52 (in Japanese with English summary)
Suzuki K, Oguro HT et al (2008) Diversity and distribution of indigenous soybean-nodulating rhizobia in the Okinawa Islands, Japan. Soil Sci Plant Nutr 54:237–246
Takahashi S, Ihara H, Karasawa T (2016) Compost in pellet form and compost moisture content affect phosphorus fractions of soil and compost. Soil Sci Plant Nutr 62:399–404
Takata Y, Kuwagata T et al (2011) Delineation of Japanese soil temperature regime map. Soil Sci Plant Nutr 57:294–302
Takeshita S (2017) Studies on history and topological characteristics of rice terrace and irrigation canal in the mountains at the GIAHS. In: Reports of studies on Takachihogo-Shiibayama GIAHS site (ed) GIAHS research group, University of Miyazaki, pp 11–20. (in Japanese)
Tanaka A (2009) Odor emission control during composting cattle manure by using compost deodorization and utilization of nitrogen enriched compost. J Jpn Assoc Odor Environ 40:229–234 (in Japanese with English summary)
Tasaka K, Sasaki Y (2005) Recent trend of agriculture and farm mechanization in Kyushu Okinawa Region. J Jpn Soc Agr Math 67:23–27 (in Japanese)
Terai T, Yamada Y et al (2006) Improvement of soil physical and chemical properties by green manure cultivation during early period of farm managements in Isahaya Bay polder. Jpn J Soil Sci Plant NutrJ Soil Sci Plant Nutr 77:321–328 (in Japanese)
Thauer RK, Jungermann K, Decker K (1977) Energy conservation in chemotrophic anaerobic bacteria. Bacteriol Rev 41:100–180
The Kunisaki Peninsula Usa GIAHS promotion Association (2017) What are GIAHS/What is Kunisaki Peninsula Usa integrated forestry, agriculture and fisheries system|GIAHS—Sawtooth Oak Forests and Irrigation Ponds Connecting the Kunisaki Peninsula Usa Integrated Forestry Agriculture and Fisheries System. http://www.kunisaki-usa-giahs.com/en/about_giahs/ (February 2018)
Tokashiki Y (1993) The characteristic properties of the shimajiri mahji and jahgaru soil in okinawa prefectre. Pedologist 37:99–112 (in Japanese)
Tokashiki Y, Shimo M, Oya K (1982) Clay minerals and some physical properties of soil sediments eroded from reclamation land. Sci Bull Fac Agric Univ Ryukyus 29:7–18 (in Japanese)
Tokashiki Y, Yamada T et al (1994) Physical properties of the surface and the runoff soils of made land in Okinawa Island. Jpn J Soil Sci Plant Nutr 65:115–125 (in Japanese)
Tokunaga T, Tanizaki T et al (1996) Emission of nitrous oxide from tea field soil. Bull Yamaguchi Agric Expt Stn 47:59–66 (in Japanese with English summary)
Toma Y, Clifton-Brown J et al (2013) Soil carbon stocks and carbon sequestration rates in seminatural grassland in Aso region, Kumamoto, Southern Japan. Glob Change Biol 19:1676–1687
Toma Y, Fernandez F et al (2010) Aboveground plant biomass, carbon, and nitrogen dynamics before and after burning in a seminatural grassland of *Miscanthus sinensis* in Kumamoto, Japan. GCB Bioenergy 2:52–62
Uezono I, Nishina K et al (2013) Estimate of N_2O emission of hog and poultry manure compost from amount of acid detergent soluble nitrogen. Jpn J Soil Sci Plant Nutr 84:104–107 (in Japanese)
Uezono I, Kato N, Moriizumi M (2010a) Applicability of rapid analysis by 80 °C-16 h hot water extraction for estimating available nitrogen in upland soil in Japan. Jpn J Soil Sci Plant Nutr 81:39–43 (in Japanese)
Uezono I, Kato N, Moriizumi M (2010b) Rapid analysis of available nitrogen in upland soil by simple measuring of chemical oxygen demand of 80 °C-16 h hot water extract. Jpn J Soil Sci Plant Nutr 81:252–255 (in Japanese)
Uezono I, Kato N, Moriizumi M (2012) Accuracy of various chemical extraction methods for estimating available nitrogen in upland soils and a comparison of the characteristics of extracted organic matter. Jpn J Soil Sci Plant Nutr 81:555–563 (in Japanese with English summary)
Wakiyama Y, Sakae S et al (2019) Property of rice growth cultivated on paddy fields damaged by the Kumamoto Earthquake in 2016. Mem Kyushu Okinawa Agric Res Cent 95:49–66 (in Japanese with English summary)
Yamada I, Kubotera H (1994) Tephras and Soils in Aso Region, Central Kyushu, Japan. Report of Kyushu National Agricultural Experiment Station, vol 83. Kyushu National Agricultural Experiment Station, Chikugo, pp 1–35. (in Japanese)
Yamamoto S, Yanai M et al (1996) Studies on the method of soil improvement in the Kasaoka Bay polder (2) Effect of the underdrain for the drainage and the salts exclusion. Bull Okayama Prefectural Agric Exp Station 14:71–83 (in Japanese)
Yamane T, Wakiyama Y, Kubotera H (2011) Nitrous oxide emissions from nitrogen-enriched cattle manure compost pellets applied to an Andosol. Jpn J Soil Sci Plant Nutr 82:234–238 (in Japanese)

Yamane T, Yamada I (2009) Nitrous oxide emission from cattle manure compost pellets applied to Andosols. Jpn J Soil Sci Plant Nutr 80:403–407 (in Japanese)

Yang J, Gibo S, Tokashiki Y (1998) Erosion control of Kunigami-maaji soil by applying of different organic matters. Trans JSIDRE 198:161–168

Yoshida K, Miyamaru N, Kawanaka T et al (2016) Low nitrogen availability and shallow plow layer decrease sugarcane (*Saccharum officinarum* L.) productivity in Kitadaito Island, Japan. Soil Sci Plant Nutr 62:504–510

Yoshida K, Miyamaru N, Kawanaka T et al (2017) Effects of application of molasses on the sugarcane growth, yield and chemical properties in a red-yellow soil on Kitadaito Island. Jpn J Soil Sci Plant Nutr 88:509–518 (in Japanese with English summary)

Yoshida M, Adachi T (1985) The effects of the hydrologic environment on soil properties in a lowland soil area. Bull Kyushu Agric Exp Station 24:33–52 (in Japanese)

Yoshida M, Yamaguchi M (1974) Studies on a red-colored soil derived from basaltic andesite of Shibayama V. An Introduction of techniques of rock magnetism. Soil Sci Plant Nutr 20, 117–125

Yoshinaga A, Onaga K (1993) Suspended soil from reclaimed farmland of Kunigami maji area in Okinawa. Trans JSIDRE 168:105–110 (in Japanese)

Yoshinaga A, Sakai K, Tokashiki Y (2004) The research on influence of chemical fertilizer for dispersion, flocculation and sedimentation of the Kunigami maaji soil. Trans JSIDRE 229:63–69 (in Japanese)

Yoshinaga A, Sakai K, Nakandakari T et al (2006) Effectiveness of mulch using Japanese pampas grass for preventing sediment runoff. Trans JSIDRE 249:83–89 (in Japanese)

Correction to: Soil Classification and Distribution

Hitoshi Shinjo and Yusuke Takata

Correction to: Chapter 3 in: R. Hatano et al. (eds.), *The Soils of Japan*, World Soils Book Series, https://doi.org/10.1007/978-981-15-8229-5_3

In the original version of the book, text that was inadvertently published in Table 3.1 (Chapter 3) has been deleted. The correction chapter and the book have been updated.

Table 3.1 Corrleation of soil classification system of Japan (2017) with WRB (2015) and Soil Taxonomy (2014)

Soil classification system of Japan (2017)			World reference base for soil resources (2015)	UUSDA Soil Taxonomy (2014)
Great group				
	Group			
		Subgroup		
A. 【Human-made soils】			Technosols, Regosols	(Entisols)
	Artifactural soils		Technosols	(Udorthents)
		Organic	Gabric Technosols	
		Ekrantic	Linic Technosols Ekranic Technosols	
		Mineral	Urbic Technosols Spolic Technosols	
	Reformed soils		Regosols (Transportic)	(Udorthents)
		Upland	Regosols (Transportic)	
		Lowland	Regosols (Transportic)	
B. 【Organic soils】			Histosols	Histosols
	Peat soils		Histosols	Histosols
		Sapric	Saplic Histosols	Haplosaprists
		High-moor	Fibric Histosols Hemic Histosols	Shagnofibrists Haplofibrists Haplohemists
		Transitional-moor	Fibric Histosols Hemic Histosols	Haplofibrists Haplohemists

(continued)

The updated version of this chapter can be found at https://doi.org/10.1007/978-981-15-8229-5_3

R. Hatano et al. (eds.), *The Soils of Japan*, World Soils Book Series,
https://doi.org/10.1007/978-981-15-8229-5_11

Table 3.1 (continued)

Soil classification system of Japan (2017)			World reference base for soil resources (2015)	UUSDA Soil Taxonomy (2014)
Great group				
	Group			
		Subgroup		
		Low-moor	Fibric Histosols Hemic Histosols	Haplofibrists Haplohemists
C. 【Andosols】			Andosols, Podzols	Andisols, Spodosols
	Podzolic Andosols		Andic Podzols	Aquods, Humods, Orthods, Aquands, Vitrands, Udands
		Epi-peaty	Andic Histic Podzols	Histic Endoaquods, Histic Endoaquands
		Epi-pseudogleyic	Andic Stagnic Podzols	Lithic Epiauods, Andic Epiaquods, Typic Epiaquods, Vitraquands, Melanaqunds, Epiaquands
		Aquic	Andic Gleyic Podzols	Lithic Epiauods, Andic Epiaquods, Typic Epiaquods, Aquic Haplorthods, Vitraquands, Melanaqunds, Epiaquands, Aquic subgroups in Andiosols
		Haplic	Andic Podzols	Andic Haplohumods, Andic Haplorthods, Andisols except Aquands and Aquic subgroups
	Regosolic Andosols		Vitric Andosols	Vitrands, Aquands
		Aquic	Gleyic Vitric Andosols	Vitraquands
		Humic	Melanic Vitric Andosols Umbric Vitric Andosols	Humic Udivitrands
		Thapto-humic	Vitric Andosols	Thaptic Udivitrands
		Haplic	Vitric Andosols	Typic Udivitrands
	Gleyed Andosols		Gleysols (Andic)	Aquands
		Peaty	Histic Gleysols (Andic)	Endoaquands
		Cumulic	Dystric Gleysols (Andic)	Melanaquands Endoaquands
		Haplic	Dystric Gleysols (Andic)	Typic Endoaquands
	Wet Andosols		Gleyic Silandic Andosols Gleyic Aluandic Andosols	Aquands
		Peaty	Histic Gleyic Silandic Andosols Histic Gleyic Aluandic Andosols Thapto-histic Gleyic Silandic Andosols Thapto-histic Gleyic Aluandic Andosols	Histic Endoaquands
		Thapto-upland	Gleyic Silandic Andosols Gleyic Aluandic Andosols	Typic Endoaquands
		Endofluvic	Gleyic Silandic Andosols Gleyic Aluandic Andosols	Typic Endoaquands
		Cumulic	Histic Gleyic Silandic Andosols Histic Gleyic Histic Aluandic Andosols Gleyic Silandic Andosols (Melanic) Gleyic Aluandic Andosols (Melanic)	Melanaquands Endoaquands

(continued)

Table 3.1 (continued)

Soil classification system of Japan (2017)			World reference base for soil resources (2015)	UUSDA Soil Taxonomy (2014)
Great group				
	Group			
		Subgroup		
		Haplic	Gleyic Silandic Andosols Gleyic Aluandic Andosols	Typic Endoaquands
	Non-allophanic Andosols		Aluandic Andosols	Alic Hapludands Melanudands
		Anthraquic	Aluandic Andosols	Anthraquic Melanudands Anthraquic Hapludands
		Cumulic	Aluandic Andosols (Melanic) Aluandic Andosols	Pachic Melanudands Alic Hapludands
		Thapto-humic	Aluandic Andosols	Alic Hapludands
		Humic	Aluandic Andosols	Alic Hapludands
		Brown-humic	Aluandic Andosols (Fulvic)	Pachic Fulvudands Thaptic Fulvudands Typic Fulvudands Hydric Fulvudands
		Aquic	Gleyic Aluandic Andosols	Aquic Hapludands Oxyaquic Hapludands
		Haplic	Aluandic Andosols	Alic Hapludands
	Allophanic Andosols		Silandic Andosols	Udands
		Anthraquic	Silandic Andosols	Anthraquic Melanudands Anthraquic Hapludands
		Thapto-upland	Silandic Andosols	Typic Hapludands Ultic Hapludands
		Endofluvic	Silandic Andosols	Typic Hapludands
		Cumulic	Siluandic Andosols (Melanic) Siluandic Andosols	Pachic Melanudands Typic Hapludands
		Cumulic	Silandic Andosols	Thaptic Hapluands
		Humic	Silandic Andosols	Typic Hapludands
		Brown-humic	Silandic Andosols (Fulvic)	Pachic Fulvudands Thaptic Fulvudands Typic Fulvudands Hydric Fulvudands Typic Hapludands Hydric Hapludands
		Aquic	Gleyic Silandic Andosols	Aquic Hapludands Oxyaquic Hapludands
		Haplic	Silandic Andosols	Typic Hapludands Hydric Hapludands
D. 【Podzols】			Podzols	Spodosols
	Podzols		Podzols	Spodosols
		Epi-peaty	Histic Podzols	Histic Epiaquods Histic Endoaquods
		Aquic	Gleyic Podzols	Endoaquods Aquic Haplorthods
		Epi-pseudogleyic	Stagnic Podzols	Epiaquods
		Pseudogleyic	Stagnic Podzols	Aquic Haplorthods

(continued)

Table 3.1 (continued)

Soil classification system of Japan (2017)			World reference base for soil resources (2015)	UUSDA Soil Taxonomy (2014)
Great group				
	Group			
		Subgroup		
		Haplic	Haplic Podzols	Haplorthods Haplocryods
E. 【Fluvic soils】			Fluvisols, Anthrosols	Inceptisols, Entisols
	Fluvic Paddy soils		Hydragric Anthrosols (Fluvic)	Anthraquic Eutrudepts Aeric Epiaquepts
		Albic	Hydragric Anthrosols (Fluvic)	Anthraquic Eutrudepts Aeric Epiaquepts
		Epi-gleyed	Hydragric Anthrosols (Fluvic)	Typic Epiaquepts
		Endoaeric	Hydragric Anthrosols (Fluvic)	Anthraquic Eutrudepts Aeric Epiaquepts
		Aquic	Hydragric Anthrosols (Fluvic)	Typic Epiaquepts
		Haplic	Hydragric Anthrosols (Fluvic)	Anthraquic Eutrudepts Aeric Epiaquepts
	Gley Fluvic soils		Fluvic Gleysols	Aquents, Aquepts, Aquepts
		Thionic	Fluvic Thionic Gleysols	Sulfic Endoaquepts Sulfic Hydraquents Sulfic Fluvaquents
		Peaty	Fluvic Histic Gleysols	Thapto-Histic Hydraquents Thapto-Histic Fluvaquents
		Humic	Fluvic Umbric Gleysols	Typic Hydraquents Mollic Fluvaquents
		Epi-gray	Fluvic Gleysols	Typic Fluvaquents
		Strong	Fluvic Reductigleyic Gleysols	Typic Hydraquents
		Mottled	Fluvic Oxygleyic Gleysols	Typic Hydraquents Typic Fluvaquents
	Gray Fluvic soils		Gleyic Fluvisols	Aquepts, Aquents
		Thionic	Gleyic Fluvisols	Sulfaqepts Sulfic Endoaquepts Sulfic Fluvaquents
		Peaty	Gleyic Histic Fluvisols	Fluvaquentic Endoaquepts Thapto-Histic Fluvaquents
		Humic	Gleyic Umbrisols	Humaquepts Mollic Fluvaquents
		Epi-gleyed	Gleyic Fluvisols	Fluvaquentic Endoaquepts Typic Endoaquepts Typic Fluvaquents
		Gleyic	Gleyic Fluvisols	Typic Endoaquepts Typic Fluvaquents
		Thapto-andic	Gleyic Fluvisols	Aquandic Endoaquepts Aquandic Fluvaquents
		Haplic	Gleyic Fluvisols	Fluvaquentic Endoaquepts Typic Fluvaquents Typic Psammaquents
	Brown Fluvic soils		Fluvisols	Udifluvents, Psamments
		Aquic	Fluvisols (Oxyaquic)	Oxyaquic Udifluvents Aquic Udipsamments

(continued)

Table 3.1 (continued)

Soil classification system of Japan (2017)			World reference base for soil resources (2015)	UUSDA Soil Taxonomy (2014)
Great group				
	Group			
		Subgroup		
		Humic	Gleyic Umbrisols	Mollic Udifluvents
		Protoanthraquic	Fluvisols (Oxyaquic)	Oxyaquic Udifluvents Aquic Udipsamments
		Haplic	Fluvisols	Typic Udifluvents Typic Udipsamments
	Regosolic Fluvic soils		Fluvisols	Udifluvents, Psamments
		Aquic	Fluvisols (Oxaquic)	Oxyaquic Udifluvents Oxyaquic Udipsamments
		Haplic	Fluvisols	Typic Udifluvents Typic Udipsamments
F. 【Red-Yellow soils】			Alisols, Acrisols, Cambisols	Udults, Udepts
	Argic Red-Yellow soils		Alisols, Acrisols, Lixisols	Udults
		Anthraquic	Alic Stagnosols Acric Stagnosols	Anthraquic Paleudults Aquic Hapludults
		Albic	Stagnic Albic Alisols Albic Acrisols	Typic Paleudults Aquic Paleudults Typic Hapludults Aquic Hapludults
		Pseudogleyic	Stagnic Alisols Stagnic Acrisols	Aquic Paleudults Aquic Hapludults
		Aquic	Gleyic Alisols Gleyic Acrisols	Aquic Paleudults Aquic Hapludults
		Humic	Alic Umbrisols Acric Umbrisols	Typic Haplohumults Humic Hapludults
		Reddish	Chromic Alisols Chromic Acrisols	Typic Paleudults Typic Hapludults
		Haplic	Haplic Alisols Haplic Acrisols	Typic Paleudults Typic Hapludults
	Cambic Red-Yellow soils		Cambisols	Udepts
		Anthraquic	Stagnosols	Anthraquic Eutrudepts
		Albic	Gleyic Cambisols	Aquic Dystrudepts Typic Dystrudepts
		Pseudogleyic	Stagnic Cambisols	Aquic Dystrudepts Oxyaquic Dystrudepts
		Aquic	Gleyic Cambisols	Aquic Dystrudepts Oxyaquic Dystrudepts
		Humic	Cambic Umbrisols	Humic Dystrudepts
		Reddish	Chromic Cambisols	Typic Dystrudepts
		Andic	Dystric Cambisols	Andic Dystrudepts
		Haplic	Dystric Cambisols	Oxyaquic Dystrudepts Typic Dystrudepts
G. 【Stagnic soils】			Gleysols, Stagnosols, Anthrosols	Aquepts, Aquults, Aquents
	Stagnogley soils		Gleysols, Anthrosols	Epiaquepts, Endoaquepts, Endoaquents

(continued)

Table 3.1 (continued)

Soil classification system of Japan (2017)			World reference base for soil resources (2015)	UUSDA Soil Taxonomy (2014)
Great group				
	Group			
		Subgroup		
		Anthraquic	Anthraquic Gleysols Hydroargic Anthrosols	Typic Epiaquepts
		Epi-peaty	Histic Gleysols	Typic Epiaquepts Typic Endoaquents
		Humic	Umbric Gleysols	Typic Humaquepts Humaquptic Endoaquents
		Haplic	Haplic Gleysols	Typic Epiaquepts Typic Endoaquents
	Pseudogley soils		Stagnosols, Gleysols	Aquepts, Aquults, Aquents
		Anthraquic	Anthraquic Stagnosols	Typic Epiaquepts Typic Epiaquults
		Groundwater-aquic	Endogleyic Stagnosols Anthraquic Gleysols Haplic Gleysols	Typic Endoaquepts Typic Endoaquents
		Humic	Umbric Stagnosols	Typic Humaquepts Typic Umbraquults
		Aeric	Haplic Stagnosols	Aeric Epiaquepts Aeric Epiaquults
		Haplic	Haplic Stagnosols	Typic Epiaquepts Typic Epiaquults
H. 【Eutrosols】			Luvisols, Cambisols	Udalfs, Udepts
	Magnesian Eutrosols		Luvisols, Cambisols	Udalfs, Udepts
		Argic	Leptic Luvisols (Chromic) Leptic Luvisols (Rhodic) Haplic Luvisols (Chromic) Haplic Luvisols (Rhodic)	Typic Paleudalfs Typic Rhodualfs Typic Hapludalfs
		Haplic	Leptic Cambisols (Eutric) Haplic Cambisols (Eutric)	Lithic Eutrudepts Typic Eutrudepts
	Calcareous Eutrosols		Luvisols, Cambisols	Udalfs, Udepts
		Argic	Leptic Luvisols Haplic Luvisols	Typic Paleudalfs Typic Rhodualfs Typic Hapludalfs
		Haplic	Leptic Cambisols (Eutric) Haplic Cambisols (Eutric)	Lithic Eutrudepts Typic Eutrudepts
I. 【Brown Forest soils】			Camibisols, Stagnosols	Udepts
	Brown Forest soils		Camibisols, Stagnosols	Udepts
		Anthraquic	Anthraquic Stagnosols	Anthraquic Eutrudepts Aquic Dystrudepts
		Andic	Dystric Cambisols	Andic Dystrudepts Andic Eutrudepts Lithic Dystrudepts
		Podzolic	Dystric Cambisols	Typic Dystrudepts Lithic Dystrudepts
		Humic	Cambic Umbrisols	Humic Dystrudepts

(continued)

Table 3.1 (continued)

Soil classification system of Japan (2017)			World reference base for soil resources (2015)	UUSDA Soil Taxonomy (2014)
Great group				
	Group			
		Subgroup		
			Dystric Cambisols (Humic)	
		Thapto-red-yellow	Dystric Cambisols Haplic Alisols	Typic Dystrudepts Typic Paleudults Inceptic Hapludults Typic Hapludults
		Aquic	Gleyic Cambisols (Oxaquic)	Aquic Dystrudepts Oxyaquic Dystrudepts Lithic Dystrudepts
		Epi-gleyed	Gleyic Cambisols	Aquic Dystrudepts Oxyaquic Dystrudepts Lithic Dystrudepts
		Eutric	Eutric Cambisols	Lithic Eutrudepts Typic Eutrudepts
		Haplic	Dystric Cambisols	Typic Dystrudepts Lithic Dystrudepts
J. 【Regosols】			Regosols, Arenosols, Leptosols, Phaeozems	Entisols, Mollisols
	Volcanogeous Regosols		Tephric Regosols	Orthents
		Aquic	Gleyic Tephric Regosols	Aquic Udorthents
		Haplic	Tephric Regosols	Vitrandic Udorthents
	Sandy Regosols		Arenosols	Udipsamments
		Calcaric	Calcaric Arenosols	Typic Udipsamments
		Aquic	Gleyic Arenosols	Aquic Udipsamments Oxyaquic Udipsamments
		Haplic	Arenosols	Typic Udipsamments
	Lithosols		Leptosols	Udorthents, Rendolls
		Calcaric	Calcaric Rendzic Leptosols Calcaric Leptosols	Lithic Haprendolls Lithic Udorthents
		Aquic	Leptosols (Gleyic)	Lithic Udorthents
		Haplic	Leptosols	Lithic Udorthents
	Terrestrial Regosols		Regosols	Udorthents
		Marlitic	Calcaric Regosols Calcaric Leptosols	Typic Udorthents Lithic Udorthents
		Calcaric	Calcaric Regosols Calcaric Phaeozems	Typic Haprendolls Typic Udorthents
		Para-lithic	Dystric Regosols Dystric Leptosols	Typic Udorthents Lithic Udorthents
		Granitic	Skeletic Regosols Skeletic Leptosols	Typic Udorthents Lithic Udorthents
		Haplic	Dystric Regosols	Typic Udorthents

Please note that it shows only representative classification names corresponding to each classification and does not cover all cases

Index

R. Hatano et al. (eds.), *The Soils of Japan*, World Soils Book Series,
https://doi.org/10.1007/978-981-15-8229-5

GPSR Compliance

The European Union's (EU) General Product Safety Regulation (GPSR) is a set of rules that requires consumer products to be safe and our obligations to ensure this.

If you have any concerns about our products, you can contact us on ProductSafety@springernature.com

In case Publisher is established outside the EU, the EU authorized representative is:

Springer Nature Customer Service Center GmbH
Europaplatz 3
69115 Heidelberg, Germany

Batch number: 10731667

Printed by Printforce, the Netherlands